Intermediate Mathematics: Book I
For College Students

by Farhad Ghassemi Tari, Ph.D.

DORRANCE
PUBLISHING CO
EST. 1920
PITTSBURGH, PENNSYLVANIA 15238

Dorrance Publishing Co
585 Alpha Drive
Suite 103
Pittsburgh, PA 15238
Visit our website at *www.dorrancebookstore.com*

ISBN: 979-8-8868-3258-7
eISBN: 979-8-8868-3654-7

PREFACE

Intermediate Mathematics I Intermediate Mathematics I and its companion volume, Intermediate Mathematics II, systematically elucidate essential concepts and tools crucial for college students aspiring to build a solid foundation for advanced mathematical topics such as algebra, calculus, and differential equations.

The study of mathematics becomes enjoyable when approached systematically, progressing step by step through learning its concepts and tools. These two books aim to furnish readers with a comprehensive understanding of mathematics, its practical applications in computation, and its significance in real-life scenarios. The objective is to equip young mathematicians with the requisite knowledge to effectively communicate with peers across all branches of advanced mathematical topics.

Distinguished by their unique approach, these textbooks commence each chapter with a quote from a romantic poet, infusing an artistic flair into the initiation of the mathematical journey. Research employing tools like functional magnetic resonance imaging has demonstrated that pondering poetry activates specific areas of the brain, thereby enhancing our capacity to interpret reality.

Within these books, each topic is accompanied by numerous examples, urging readers to actively engage with the material by solving problems rather than passively reading about them. As a result, these texts emphasize not only the content of mathematics but also the process of learning and applying mathematical concepts. Through this active learning approach, readers develop essential mathematical skills that will serve them well in future studies.

Moreover, the books illustrate the real-world applications of each major topic. After presenting concepts and tools, each chapter delves into various real-life scenarios where these mathematical principles find practical use. Additionally, every chapter concludes with 60 multiple-choice questions, strategically designed to foster deeper comprehension and interaction with the material.

This organization structure renders the contents of these books ideal for individuals seeking to cultivate a robust understanding of mathematics through active learning and practical application. These books are particularly suitable for those who:

- Are in their freshman year at a college or university.
- Aspire to prepare for GMAT and GRE quantitative subjects.
- Seek for preparation of SAT and ACT quantitative subjects.

This is the first volume of Intermediate Mathematics comprises five chapters, each focusing on fundamental mathematical concepts. volume of intermediate mathematics containing five chapters.

Chapter one is dedicated to numbers and arithmetic of numbers. It introduces and presents the definition of the natural numbers, whole numbers, rational numbers, etc. Then, it covers arithmetic operations on numbers involving fractions, radicals, exponents and so on. It also presents number's divisibility rules, arithmetic of series, measurement units, and real-life applications of numbers. At the end of this chapter sixty multiple-choice questions and their answer keys are provided.

Chapter two introduces the principle of analytical geometry. It starts with Cartesian coordinate system and how to plot points and lines in this system. Then, in section two it presents the concepts and tools for determining distances between two points, a point, and a line, two parallel lines etc. Following the description of lines section three covers plot of some common graphs, circles, ellipses, hyperbolas, and parabolas. Section four of this chapter investigates many selected real-life applications of analytical geometry. Finally, section five presents sixty multiple-choice questions and their answer keys.

Chapter three deals with fundamental topics of the algebraic subject, equalities, and inequalities in one variable. The first section looks at algebraic expressions and explains operations on algebraic expressions involving exponents, radicals, logarithms, etc. Different methods of interpolation and extrapolation are the contents of section two. The linear equation and linear inequalities on one variable and their solution methods are explained in section three. Following linear equations and inequalities some other types of equalities and inequalities such as absolute value, polynomial, rational, radical, and logarithmic equations, and inequalities are investigated in section four. As the pattern of the previous two chapters, real-life applications of equalities and inequalities and sixty multiple-choice questions along with their answer keys are the subjects of the next section.

Chapter four delves into a detailed exploration of polynomial types, polynomial arithmetic, and methods for solving polynomial equations and inequalities. Section one introduces polynomial classifications with a focus on binomial expressions. Section two primarily covers methods of factoring polynomials, both with leading coefficients of one or more than one. Quadratic equations and inequalities are the subjects of the third section, where solving methods are explained. Section four tackles polynomial equations and inequalities with degrees greater than two, as well as those involving rational expressions, exponential expressions, and radicals. Real-world applications of polynomials are discussed in section five, followed by sixty multiple-choice questions and their answers in section six.

Chapter five centers on the study of mathematical functions, which are widely utilized in science, engineering, and various mathematical fields. Section one begins by distinguishing between relations and functions and proceeds to explain function arithmetic, along with different types of function asymptotes. Section two explores function composition, the inverse of a function, and rules for determining concavity. Section three delves into the definition of derivatives, elementary rules of differentiation for various function types, second derivatives, and methods for finding local and global extrema. In section four, integration rules, common formulas, and properties of definite integrals are presented. Real-life applications of different function types are provided in section five, while section six concludes with sixty multiple-choice questions and their answer keys.

I hoped that these two complementary books would introduce readers to more advanced mathematical topics, making abstract ideas encountered later more meaningful.

Oxnard, California. 2024
Farhad G. Tari, Ph.D.
Emails: ghasemi@sharif.edu
ghassemi8@gmail.com

Table of Contents

The minute I heard my first love story,
I started looking for you,
not knowing how blind that was.
Lovers don't finally meet somewhere.
They're in each other all along.
Rumi

Chapter 1

Arithmetic of Numbers and Preliminary

Contents

Introduction

The foundational principle of number theory, as articulated by Carl Friedrich Gauss in 1801, asserts that any integer greater than 1 can be expressed uniquely as the product of prime numbers. **Numbers**, mathematical entities utilized for counting, measuring, and labeling, form the basis of arithmetic, often interchangeably referred to as number theory. Arithmetic involves operations such as addition, subtraction, multiplication, division, and exponentiation, which are used for calculations involving numbers. The study or application of these operations is termed arithmetic. Additionally, the term "arithmetic" may also denote number theory, which investigates the properties of numbers.

The term "arithmetic" originates from the Greek word "arithmos," meaning "number." It represents the oldest and most fundamental branch of mathematics, serving various purposes from basic counting to intricate scientific and business calculations.

The objective of this chapter is to revisit several topics, including sets of numbers and the properties of operations used in manipulating numbers. Many of these topics are crucial for understanding more advanced algebraic subjects. It is assumed that readers have encountered these topics in previous courses, thus this chapter primarily serves as a review. However, given the significance of these topics in comprehending other algebraic concepts, we will ensure a thorough understanding by providing a quick review within this chapter. The following is a brief overview of the material covered in this chapter.

1. Numbers

Numbers are fundamental arithmetic values utilized to represent quantities in our daily lives. Often referred to as numerals, they are essential for tasks such as counting objects, indicating dates, telling time, managing finances, and more. Numbers serve various purposes, from measurement to labeling, and their properties enable them to undergo arithmetic operations.

In this chapter, we introduce numbers and sets of numbers, covering arithmetic operations, their applications in arithmetic and geometric series, and other related topics.

1.1. Sets of Numbers as Subsets

A set of numbers encompasses elements that are all numerical. Beginning with the natural numbers, we have expanded each set to encompass a broader range, implying a subset relationship between the sets of numbers. To visualize these relationships, refer to the following diagram:

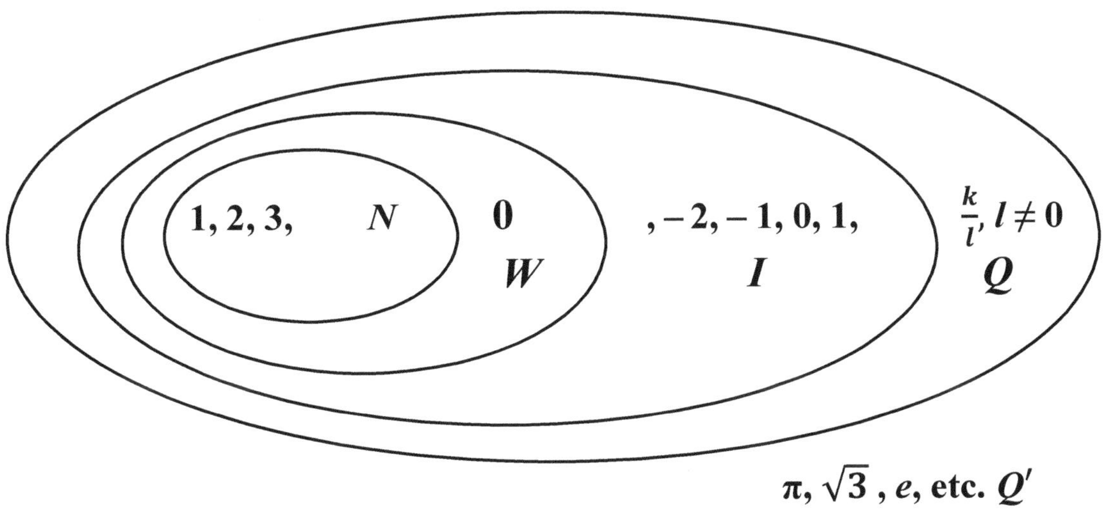

N: the set of natural numbers.

W: the set of whole numbers.

I: the set of integers.

Q: the set of rational numbers.

Q′: the set of irrational numbers.

Natural Numbers: Natural number which is also named "Counting Numbers" is a set of numbers such as: **1, 2, 3, 4, 5, ⋯**. Note, the use of three dots at the end of the list is a notation to indicate that the list keeps going forever.

Whole Numbers: Natural numbers together with "zero" that is: 0, 1, 2, 3, 4, 5, ⋯.

Integer Numbers: Whole numbers plus negatives. In fact, including negative versions of counting in whole number we get integer numbers, that is: – 4, – 3, – 2, – 1, 0, 1, 2, 3, 4, ⋯.

The next generalization that we can make is to include rational numbers. While it is unlikely that a farmer owns a fractional number of cows, many other things in real life are measured in rational numbers, like a half-cup of sugar. If we add rational numbers to the set of integers, we get the set of rational numbers.

Rational Numbers: All numbers of the form $\frac{b}{a}$, where a and b are integers ($b \neq 0$). In fact, including fractions to the set of integers, we get the set of rational numbers. Notice that the word "rational" contains the word "ratio". The bottom number of the rational number is called the denominator, and the top number is called the numerator. Although one may think that the set of rational numbers would cover every possible instance, that is not so. There are numbers, called irrational numbers, that cannot be expressed as a rational number.

Irrational Numbers: The **irrational numbers** are all the real numbers which are not rational numbers. Among irrational numbers are all square roots of natural numbers, other than of perfect squares, the ratio π of a circle's circumference to its diameter. Notice that $\pi = 3.14159$.

- Cannot be expressed as a ratio of integers.
- Decimals never repeat or terminate (while rational numbers always do one or the other).

Real Numbers: Real numbers are just the combination of rational and irrational numbers, in the number system. In general, all the arithmetic operations can be performed on these numbers, and they can be represented in the number line, also.

- Rational Numbers + Irrationals Numbers.
- All points on the number line.
- Or all possible distances on the number line.

Including irrational numbers to the rational numbers, we attain the complete set of real numbers. In measuring an amount of something, such as a weight, a volume, or the distance between two points, the measurement numbers will always be represented by the real number.

Rewriting Numbers in a Set Notation

Definition: A set is any well-defined list, collection, or class of objects. The objects in sets can be anything such as: a collection of some numbers 1, 2, 3, 5, 7, the students in a class, the outcome of rolling a dice, solutions of an equation, etc.

Notation:

- {} enclose a set.
- Sets will usually be denoted by capital letters; $A, B, U, X, Y, \cdots$.
- A set is not defined by order of elements; that is $A = \{1, 3, 5, 7, 9\} = \{3, 1, 9, 7, 5\}$.
- A set which does not contain any element is called the **null set or the empty set**. $\emptyset$ is a symbol representing an empty set.
- A symbol that expresses the phrase "to be a member of" is called the belongs to symbol. $x \in C$ denotes x is an element of set C.

Set-Roster Notation and Set Builder Notation

There are essentially two ways to specify a particular set.

1. To list all its elements if it is possible, use **set-roster notation**.
2. To stat the property which characterizes the elements in the set, use **set builder notation**.

Consider the set denoted by A, then it may appear as, set-roster notation: $A = \{2, 4, 6, 8, 10\}$. Note that the elements are separated by commas and enclosed in a bracket$\{\ \}$. Set A can be represented in another form, set builder notation; $A = \{x/\ x$ takes all the even natural number less than or equal to $10\}$. As another example let us consider set B, which is shown as $B = \{1, 3, 5, 7, 9\}$. Then set B can be represented in another form; $B = \{x/\ x$ takes all the odd natural number less than $10\}$.

Example
Write set $C = \{6, 8, 10, 12, 14, 16, 18\}$in form of set builder notation.

Solution
Set C can be written as:

$C = \{x/\ x$ takes all the even natural numbers greater than or equal to 6 and less than or equal to$18\}$.

Example
Write set $D = \{$a, e, i, o, u$\}$ in form of set builder notation.

Solution
Set D can be written as:

$D = \{x/\ x$ is the vowels of the alphabet$\}$.

Example
Rewrite set $E = \{x/\ x^2 = 16\}$ using the same form of set builder notation.

Solution
We can rewrite E as

$E = \{x/\ x = \pm 4\}$.

Example
Rewrite set $F = \{x/\ (x - 1)^2 = 25\}$ using the same form of set builder notation.

Solution
We can rewrite F as

$F = \{x/\ (x - 1)^2 = 25\}$

$F = \{x/\ (x - 1) = \pm 5\}$, or

$F = \{x/\ x = \pm 5 + 1\}$, or

$F = \{x/\ x = 6$ and $x = -4\}$.

There are some reasons that we must combine sets. Two basic methods for combining sets are **intersection** and **union**.

Definition Considering two sets A and B;

- The intersection of A and B is denoted by $A \cap B$ and is defined as:
  - $A \cap B = \{x|\ x \in A$ and $x \in B\}$.
- The union of A and B is denoted by $A \cup B$ and is defined as:
  - $A \cup B = \{x|\ x \in A$ or $x \in B$ (or both)$\}$.

Presenting Numbers by Set

1. The Empty Set: $\emptyset = \{\} = \{\, x \,/\, x \neq x\}$. This is the set with no elements. Like the number `0,' it plays a vital role in mathematics.
2. The Natural Numbers: N={1, 2, 3, ⋯}. The periods of ellipsis here indicate that the natural numbers contain 1, 2, 3, `and so forth'.
3. The Whole Numbers: W = {0, 1,2,3, ⋯}.
4. The Integers: $I = \{\cdots, -3, -2, -1, 0, 1, 2, 3, \cdots\}$.
5. The Rational Numbers: $Q = \{\frac{a}{b} \,/\, a \in \mathrm{I}, b \in \mathrm{I}, \text{and } b \neq 0\}$. Rational numbers are the ratios of integers (provided the denominator is not zero!) It turns out that another way to describe the rational numbers is $Q = \{\, x \,/\, x$ possesses a repeating or terminating decimal representation$\}$.
6. The Real Numbers: $R = \{\, x \,/\, x$ possesses a decimal representation$\}$.
7. The Irrational Numbers: $P = \{\, x \,/\, x$ is a non-rational real number.$\}$ Said another way, an irrational number is a decimal which neither repeats nor terminates.
8. The Complex Numbers: $C = \{a + bi$ where $a, b \in \mathrm{R}$ and $i = \sqrt{-1}\}$. For the complex number $a + bi$, a is called the real part, and b is called the *imaginary part*. Despite their importance, the complex numbers play only a minor role in the text.

An Ordered Set

Real numbers have the property that they are ordered, which implies that given any two different numbers we can always find that one is greater or less than the other. In other word, for any two real numbers a and b, one and only one of the following three statements hold:

- a is less than b, (is represented as $a < b$).
- a is equal to b, (is represented as $a = b$).
- a is greater than b, (is represented as $a > b$).

1.2. The Number Line

Arranging the ordered nature of the real numbers along a line (the line contains an infinite number of points which are packed so closely together that they form a solid line).

On **a number line, numbers** to the right of 0 are positive. Numbers to the left of 0 are negative, as shown in the following figure. The arrows on the ends indicate that the line keeps going forever in both directions.

Given any two numbers on a number line, the one on the right is always larger, regardless of its sign (positive or negative).

- Every real number corresponds to a distance on the number line, starting at zero.
- Negative numbers correspond to distances on the left of zero, and positive numbers are distances to the right of zero.

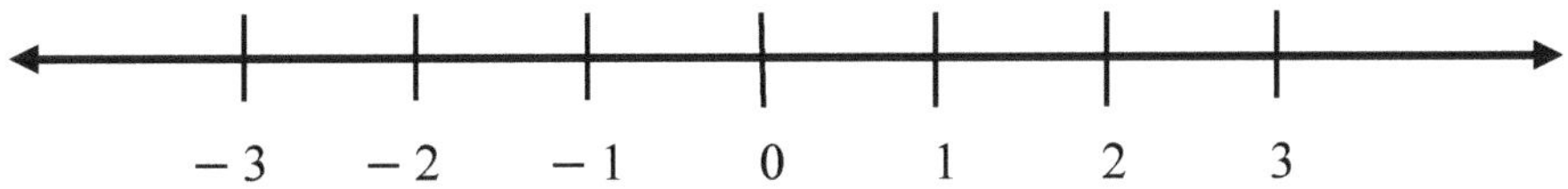

Interval Notation

Let a and b be real numbers with $a < b$.

Set of Real Numbers	Interval Notation	Region on the Real Number Line
$\{x / a < x < b\}$	(a, b)	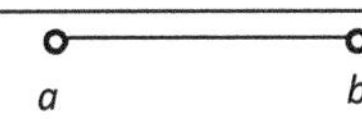
$\{x / a \le x < b\}$	$[a, b)$	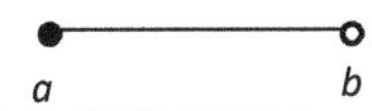
$\{x / a < x \le b\}$	$(a, b]$	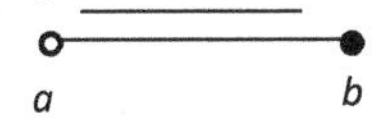
$\{x / a \le x \le b\}$	$[a, b]$	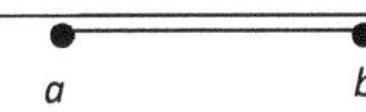
$\{x / x < b\}$	$(-\infty, b)$	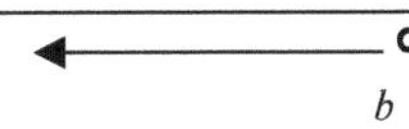
$\{x / x \le b\}$	$(-\infty, b]$	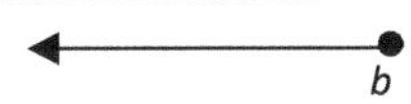
$\{x / x > a\}$	(a, ∞)	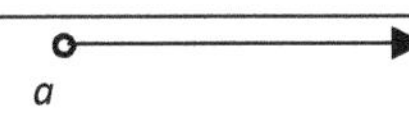
$\{x / x \ge a\}$	$[a, \infty)$	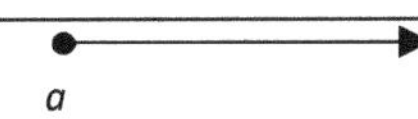
$\mathbb{R}$	$(-\infty, \infty)$	

Example

Express the following sets of numbers using interval notation and graph them as a set of numbers on the number line.

a) $\{x| \, x \le -1 \text{ or } x \ge 1\}$.

b) $\{x| \, x \ne 7\}$.

c) $\{x| \, x \ne \pm 4\}$.

d) $\{x| -2 < x \le 3 \text{ or } x = 7\}$.

Solution

a) The inequality $x \le -1$ corresponds to the interval $(-\infty, -1]$. Similarly, the inequality $x \ge 1$ corresponds to the interval $[1, \infty)$. Since we are asked to describe the real numbers x in one of these or the other, we have $\{x| \, x \le -1 \text{ or } x \ge 1\} = (-\infty, -1] \cup [1, \infty)$. To graph this set of numbers on the number line, we have:

$(-\infty, -1] \cup [1, \infty)$

b) Let us first graph the set $\{x| \, x \ne 7\}$. To achieve this, we can shade the entire real number line except $x = 7$, in which the excepted $x = 7$ is designated by an open circle (see the line below). Looking at the line, we see the real numbers are divided into two intervals, $(-\infty, 7)$ and $(7, \infty)$. Since the values of x could be in either one of these two intervals, we conclude.

$\{x|\ x \neq 7\} = (-\infty, 7) \cup (7, \infty)$.

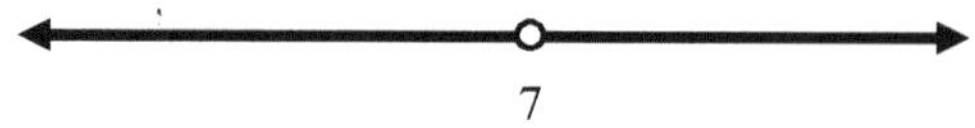

$(-\infty, 7) \cup (7, \infty)$

c) Like the above description, we first graph the set $\{x|\ x \neq \pm 4\}$. we can now shade the entire real number line except $x = \pm 4$, in which the excepted $x = \pm 4$ is designated by two open circle around 4 and − 4 (see the line below). Looking at the line, we see the real numbers are divided into three intervals, $(-\infty, -4)$, $(-4, 4)$ and $(4, \infty)$. Since the values of x could be in either one of these three intervals, we conclude that $\{x|\ x \neq \pm 4\} = (-\infty, -4) \cup (-4, 4) \cup (4, \infty)$.

$(-\infty, -4) \cup (-4, 4) \cup (4, \infty)$

d) Graphing the set $\{x|\ -2 < x \leq 3 \text{ or } x = 7\}$, we obtain one interval, $(-2, 3]$ along with a single number, $\{7\}$, in which $x = 7$ is designated by a solid filled circle (see the line below). Therefore, the answer can be written as: $\{x|\ -2 < x \leq 3 \text{ or } x = 7\} = (-2, 3] \cup \{7\}$.

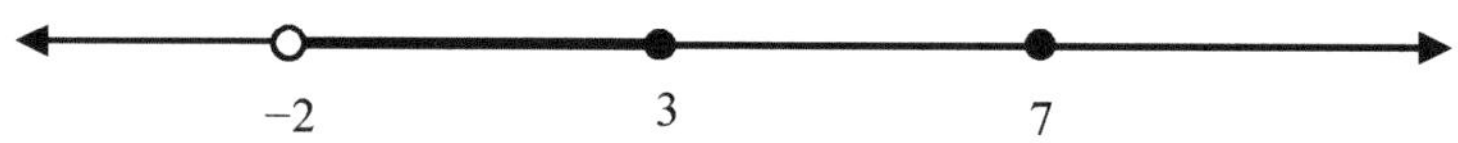

$(-2,3] \cup \{7\}$

Positive and Negative Numbers Rules

Adding, multiplying, and dividing numbers can produce positive or negative values depending on their original signs. Following are the positive and negative rules:

$(+) + (-)$ = the sign of the larger number

$(-) + (-)$ = negative number.

$(+) + (+)$ = positive number.

$(-) \times (+)$ = negative number.

$(-) \times (-)$ = positive number.

$(-) \div (-)$ = positive number.

$(-) \div (+)$ = negative number.

Example
Find the sum of − 37 and 62.

Solution

We learned $(+) + (-)$ = the sign of the larger number.

$(-37) + (+62) = 25$.

Example
Find the sum of − 45 and 31.

Solution

We know (+) + (−) = the sign of the larger number.

(− 45) + (+31) = − 14.

Example

Find the sum of − 12 and − 28.

Solution

We know (−) + (−) = negative number

(− 12) + (− 28) = − 40.

Example

Find the sum of 16 and 32.

Solution

We know (+) + (+) = positive number

(+16) + (+32) = 48.

Example

Find the product of − 4 and 8.

Solution

We know (−) × (+) = negative number

(− 4) × (+8) = − 32.

Example

Find the product of − 3 and − 11.

Solution

We know (−) × (−) = positive number

(− 3) × (− 11) = 33.

Example

Find the value of − 24 divide by − 6.

Solution

We know (−) ÷ (−) = positive number

(− 24) ÷ (− 6) = 4.

Example

Find the value of − 36 divide by 12.

Solution

We know (−) ÷ (+) = negative number

(− 36) ÷ (+12) = 3.

1.3. Order of Operations

When we face an expression such as $2 + 11 \times 9 - 44 \div 4 + 11 \times 12 - 7$, it makes pretty a difference how we choose the sequence of operations to perform. We require a set of rules that would guide us to one unique value for this type of expression. Operations in mathematical expressions are evaluated in a systematic order, which can be simplified using the abbreviation **PEMDAS**:

- Parentheses.
- Exponents.
- Multiplication.
- Division.
- Addition.
- Subtraction.

Parentheses- Parentheses have a great many specialized meanings in mathematics. It is a symbol used to denote grouping anything inside symbols such as ( ), [ ], { }, | | and $\sqrt{\ }$.

Exponents- it is also other special functions such as *log*, *ln*, *sin*, *cos*, etc.

Multiplication and Division- they are performed from left to right. The left-to-right order does not matter where only multiplication is involved, but it matters for division.

Addition and Subtraction- they are also performed from left to right. Again, the left-to-right order does not matter if only addition is involved, but it matters for subtraction.

Example
Simplify $5 + 4\,(11-7)^2$.

Solution

$5 + 4\,(11 - 7)^2$	Parenthesis first
$5 + 4\,(4)^2$	Exponents
$5 + 4\,(16)$	Multiply
$5 + 64$	Add
69	Our Solution.

It is especially important to remember to multiply and divide from left to right.

Example
Simplify $27 \div 9 \times 4$.

Solution

$27 \div 9 \times 4$	Divide first (left to right!)

3×4 — Multiply

12 — Our Solution.

Example

Simplify $3\{6^2 - 7[32 - 4(3^2 + 1)](-1)\}$.

Solution

$3\{6^2 - 7[32 - 4(3^2 + 1)](-1)\}$	Innermost parenthesis, exponents first
$= 3\{6^2 - 7[32 - 4(9 + 1)](-1)\}$	Add inside those parentheses
$= 3\{6^2 - 7[32 - 4(10)](-1)\}$	Multiply inside innermost parenthesis
$= 3\{6^2 - 7[32 - 40](-1)\}$	Subtract inside those parentheses
$= 3\{6^2 - 7[-8](-1)\}$	Exponents next
$= 3\{36 - 7[-8](-1)\}$	Multiply left to right, sign with the number
$= 3\{36 + 56(-1)\}$	Finish multiplying
$= 3\{36 - 56\}$	Subtract inside parenthesis
$= 3\{-20\}$	Multiply
$= -60$	Our Solution.

Example

Simplify $\dfrac{2^5-(-8)\times 2}{\frac{15}{5}-1}$.

Solution

$\dfrac{2^5-(-8)\times 2}{\frac{15}{5}-1}$	Exponent in the numerator, divide in denominator
$= \dfrac{32-(-8)\times 2}{3-1}$	Multiply in the numerator, subtract in denominator
$= \dfrac{32-(-16)}{2}$	Add the opposite to simplify numerator, denominator is done
$= \dfrac{48}{2}$	Reduce, divide
$= 24$	Our Solution.

Example

Simplify $4 + 3(10 - 2) + (-5 + 4)^2$.

Solution

$4 + 3(10 - 2) + (-5 + 4)^2$	
$= 4 + 3(8) + (-1)^2$	we evaluated parenthesis first
$= 4 + 3(8) + (+1)$	we evaluated exponent
$= 4 + 24 + 1$	we evaluated multiplication
$= 4 + 25$	we evaluated summation
$= 29$	our solution.

Example
Evaluate $x - zx(5 - x)\frac{z}{3}$ when $x = -1$ and $z = -3$.

Solution
Replacing all x's with 6 and z's with 2, we have.

$x - zx(5 - x)\frac{z}{3}$

$= x - zx(5 - x)\frac{z}{3} = (-1) + (-3)(-1)(5 - (-1))\frac{(-3)}{3}$

$= x - zx(5 - x)\frac{z}{3} = -1 + (-3)(-1)(6)(-1)$ we multiplied left to right

$= x - zx(5 - x)\frac{z}{3} = -1 + 3(6)(-1)$ we multiplied left to right

$= x - zx(5 - x)\frac{z}{3} = -1 + 18(-1)$

$= x - zx(5 - x)\frac{z}{3} = -1 - 18$

$= x - zx(5 - x)\frac{z}{3} = -19.$

Therefore, the solution is -19.

1.4. Rational Numbers and Fractions

A **Fraction** which is called a **rational number**, represents a part of a whole or, more generally, any number of equal parts. Mathematically, a rational number is illustrated as; $\frac{a}{b}$, where a and b are whole numbers ($b \neq 0$) and a is the numerator and b is the denominator. In other words, if a number can be expressed as a fraction where both the numerator and the denominator are integers, the number is a rational number. The denominator b is the number of equal parts that the whole has been divided into, and the numerator a specifies how many parts are included.

For example, if I have eaten 1 apple out of a basket containing 3 apples, then I have eaten one third of the basket apples. The denomination is 3 and I had one of them: $\frac{1}{3}$.

Comparison of Rational Numbers

Number comparison problems may be presented in different formats, and we are asked which is larger or smaller or whether are equivalent. As essential method to solve such problems is to convert the numbers to the same format.

Note that for negative numbers, higher numbers are basically smaller.

Example
Is 5.28 greater than, equal to, or less than $\frac{28}{5}$.
Solution
First convert $\frac{28}{5}$ to a decimal format.

$28 \div 5 = 5.60.$

Therefore 5.6 is greater than 5.28.

Example
Compare $3\frac{2}{5}$ and $\frac{19}{5}$.

First convert $3\frac{2}{5}$ to $\frac{17}{5}$.

Therefore $\frac{17}{5}$ is less than $\frac{19}{5}$.

Example
Compare $5\frac{3}{5}$ and 5.11.

First convert $5\frac{3}{5}$ to $\frac{28}{5}$. Then convert $\frac{28}{5}$ to 5.40.

Therefore $5\frac{3}{5}$ is greater than 5.11.

Example
Designate which is rational and which is irrational.

a) $\frac{1}{2}$.
b) $\frac{1}{3}$.
c) $\frac{5}{11}$.
d) $\frac{5}{7}$.
e) $\sqrt{2}$.
f) π.

Solution

a) $\frac{1}{2} = 0.5$ Rational (terminates)

b) $\frac{1}{3} = 0.333333\overline{3}$ Rational (repeats)

c) $\frac{5}{11} = 0.4545454\overline{545}$ Rational (repeats)

d) $\frac{5}{7} = 0.714285\overline{714285}$ Rational (repeats)

e) $\sqrt{2} = 1.41421356...$ Irrational (neither repeats nor terminates)

f) $\pi = 3.14159265...$ Irrational (never repeats or terminates).

Mixed Number Notation

A way of representing an improper rational number like, $\frac{7}{5}$ is as a mixed number like $1\frac{2}{5}$, which is read as "one and two-fifths". We may use a more proper representation by converting improper rational numbers to regular rational numbers.

Conversion of Mixed Fractions to Improper Fraction

The following two steps are the method for converting mixed fractions to improper fractions:

1. Multiply the integer part with the denominator of the fraction.
2. Add the result to the nominator of the fraction.

The general formula is:

$$a\frac{b}{c} = \frac{ac+b}{c}.$$

Example
Convert $3\frac{4}{7}$ to an improper fraction.

Solution
Using the above formula, we have:

$$3\frac{4}{7}$$

$$3\frac{4}{7} = \frac{(3 \times 7) + 4}{7}$$ we multiplied 3 to 7 and added to 4

$$3\frac{4}{7} = \frac{21 + 4}{7}$$

$$3\frac{4}{7} = \frac{25}{7}.$$

Example
Convert $9\frac{3}{5}$ to an improper fraction.

Solution
Using the above formula, we have:

$$9\frac{3}{5}$$

$$9\frac{3}{5} = \frac{(9 \times 5) + 3}{5}$$ we multiplied 9 to 5 and added to 3

$$9\frac{3}{5} = \frac{45 + 3}{7}$$

$$9\frac{3}{5} = \frac{48}{5}.$$

Conversion of Improper Fractions to Mixed Fractions

1. Divide the nominator by denominator get the integer part.
2. Put the remainder of division over the old denominator to get the fractional part.

Example
Convert $\frac{35}{11}$ to a mixed number.

Solution
Using the latter steps, dividing 35 by 11, we get integer 3, and reminder of 2: then we have:

$$\frac{35}{11} = 3\frac{2}{11}.$$

Example
Convert $\frac{78}{15}$ to a mixed number.

Solution
Using the latter steps, dividing 78 by 15, we get integer 5, and reminder of 3: then we have:

$$\frac{78}{15} = 5\frac{3}{15}.$$

Equivalent Rational Numbers Property

Equivalent rational numbers are rational numbers that have the same value. That is; if a, b, and c are numbers were,

$$b \neq 0, c \neq 0, \text{ then } \frac{a}{b} = \frac{a}{b} \times \frac{c}{c} = \frac{a}{b}.$$

For example, $\frac{1}{3} = \frac{1}{3} \times \frac{2}{2} = \frac{2}{6}$. The rational number $\frac{2}{6}$ has the same value as $\frac{1}{3}$. Also, $\frac{1}{3} = \frac{1}{3} \times \frac{3}{3} = \frac{3}{9}$. As another example, $\frac{5}{8} = \frac{5}{8} \times \frac{2}{2} = \frac{10}{16}$, , $\frac{5}{8} = \frac{5}{8} \times \frac{3}{3} = \frac{15}{24}$, $\frac{5}{8} = \frac{5}{8} \times \frac{8}{8} = \frac{40}{64}$, $\frac{5}{8} = \frac{5}{8} \times \frac{10}{10} = \frac{50}{80}$, all the $\frac{10}{16}$, $\frac{40}{64}$ and $\frac{50}{80}$ each have the same value as $\frac{5}{8}$.

A rational number is considered as a simplified rational number, if there are no common factors, other than 1, in its numerator and denominator.

For example, $\frac{1}{2}$, $\frac{5}{7}$, and $\frac{11}{13}$ are each a simplified rational number, while, $\frac{6}{9}$, $\frac{15}{21}$, and $\frac{110}{143}$ are not a simplified rational numbers, because, $\frac{6}{9} = \frac{2 \times 3}{3 \times 3}$, $\frac{15}{21} = \frac{5 \times 3}{7 \times 3}$, and $\frac{110}{143} = \frac{11 \times 10}{11 \times 13}$.

Before presenting some examples, we need to know what prime numbers are and what composite numbers are.

Prime number: A **prime number is** a numeral that is greater than 1 and cannot be divided evenly by any other number except 1 and itself. Numbers such as 2, 3, 5, 7, 11, 13, ··, are examples of a prime number.

If a number can be divided evenly by any other number not counting itself and 1, it is not prime number and is referred to as a composite number. For example, 4, 9, and 18 are not a prime number because the first one can be divided evenly by 2, the second one can be divided evenly by 3, and the third one can be divided evenly by 2, 3, 6 or 9.

Composite number: When a number can be divided up exactly it is a **Composite Number.** Numbers such as 4, 6, 8, 9, 10, 12, 14, 15, 16, 18, 20, 21, ··· are examples of a composite number.

Example,
Simplify each:

a) $\frac{3}{5}$.
b) $\frac{24}{9}$.
c) $\frac{120}{480}$.

Solution

a) $\frac{3}{5}$, it is a simplified rational number since there are no common factors of 3 and 5.

b) $\frac{24}{9}$.

$$\frac{24}{9} = \frac{2 \times 2 \times 2 \times 3}{3 \times 3}$$

we rewrote 24 and 9 as the products of prime numbers

$$\frac{24}{9} = \frac{2 \times 2 \times 2}{3}$$

we deleted the common 3

$$\frac{24}{9} = \frac{8}{3}.$$

c) $\frac{120}{480} = \frac{2 \times 2 \times 2 \times 3 \times 5}{2 \times 2 \times 2 \times 2 \times 2 \times 3 \times 5}$

we rewrote 120 and 480 as the products of prime numbers

$$\frac{120}{480} = \frac{1}{2 \times 2}$$

we deleted the common 2, 3, and 5

$$\frac{120}{480} = \frac{1}{4}.$$

1.5. Multiplying and Dividing Ration Numbers

Multiplying Rational Numbers -To multiply rational numbers, we simply multiply the numerators and multiply the denominators.

> **Rational Numbers Multiplication**
> Let a, b, c, and d, be four numbers, where $b \neq 0$, and $d \neq 0$, then
> $$\frac{a}{b} \times \frac{c}{d} = \frac{a \times c}{b \times d}.$$

Note that, when multiplying rational numbers, the properties of positive and negative numbers still apply.

Example

Multiply each:

a) $\frac{2}{3} \times \frac{7}{5}$.

b) $\frac{2}{3} \times \frac{-11}{9}$.

c) $\frac{2}{3} \times \frac{x^2}{5}$.

Solution

a) $\frac{2}{3} \times \frac{7}{5}$.

$$\frac{2}{3} \times \frac{7}{5} = \frac{2 \times 7}{3 \times 5}$$

we multiplied the numerators by 7 and multiplied the denominators by 5

$$\frac{2}{3} \times \frac{7}{5} = \frac{14}{15}.$$

b) $\frac{2}{3} \times \frac{-11}{9}$.

$$\frac{2}{3} \times \frac{-11}{9} = \frac{2 \times (-11)}{3 \times 9}$$

we multiplied the numerators by (-11) and multiplied the denominators by 9

$$\frac{2}{3} \times \frac{-11}{9} = -\frac{22}{27}.$$

c) $\frac{2}{3} \times \frac{x^2}{5}$.

$\frac{2}{3} \times \frac{x^2}{5} = \frac{2 \times x^2}{3 \times 5}$ we multiplied the numerators by x^2 and multiplied the denominators by 5

$\frac{2}{3} \times \frac{x^2}{5} = \frac{2\,x^2}{15}.$

A useful concept for dividing rational numbers is the concept of reciprocal.

Reciprocal- The **reciprocal** of a rational number is obtained by inverting the rational number. That is, placing the numerator in the denominator and the denominator in the numerator. For example, the reciprocal of $\frac{4}{5}$ is $\frac{5}{4}$, the reciprocal of $\frac{135}{717}$ is $\frac{717}{135}$, and the reciprocal of $\frac{a}{b+2}$ is $\frac{b+2}{a}$.

Dividing Rational Numbers -To divide rational numbers, we multiply the first rational number by the reciprocal of the second.

> **Dividing Rational Numbers**
> Let a, b, c, and d be four numbers, where $b \neq 0$, $c \neq 0$, and $d \neq 0$, then to divide rational numbers, we multiply the first rational number by the reciprocal of the second.
>
> $$\frac{a}{b} \div \frac{c}{d} = \frac{a}{b} \times \frac{d}{c} = \frac{ad}{bc}.$$

Note that, when multiplying rational numbers, the properties of positive and negative numbers still apply.

Example

Divide each of the following:

a) $\frac{4}{5} \div \frac{9}{5}.$

b) $\frac{12}{13} \div \frac{-17}{26}.$

c) $\frac{7}{8} \div \frac{3a}{11b}.$

Solution

a) $\frac{4}{5} \div \frac{9}{5}.$

$\frac{4}{5} \div \frac{9}{5} = \frac{4}{5} \times \frac{5}{9}$ we multiplied the first rational number to the reciprocal of the second

$\frac{4}{5} \div \frac{9}{5} = \frac{4 \times 5}{5 \times 9}$ we multiplied the numerators by 5 and multiplied the denominators by 9

$\frac{4}{5} \div \frac{9}{5} = \frac{4}{9}.$ we deleted the common 5.

b) $\frac{12}{13} \div \frac{-1}{26}.$

$\frac{12}{13} \div \frac{-17}{26} = \frac{12}{13} \times \frac{26}{-17}$ we multiplied the first rational number to the reciprocal of the second

$\frac{12}{13} \div \frac{-17}{26} = \frac{12 \times 26}{13 \times (-17)}.$ we multiplied the numerators by 26 and multiplied the denominators by (- 17)

$\frac{12}{13} \div \frac{-17}{26} = \frac{12 \times 2 \times 13}{13 \times (-17)}$ we wrote 26 as 2 multiplied to 13

$\frac{12}{13} \div \frac{-17}{26} = \frac{12 \times 2}{(-17)}$ we deleted the common 13

$\frac{12}{13} \div \frac{-17}{26} = -\frac{24}{17}.$

c) $\frac{7}{8} \div \frac{3a}{11b}.$

$\frac{7}{8} \div \frac{3a}{11b} = \frac{7}{8} \times \frac{11b}{3a}$ we multiplied the first rational number to the reciprocal of the second

$\frac{7}{8} \div \frac{3a}{11b} = \frac{7 \times 11b}{8 \times 3a}$ we multiplied the numerators by $11b$ and multiplied the denominators by $3a$

$\frac{7}{8} \div \frac{3a}{11b} = \frac{77b}{24a}.$

1.6. Add and Subtract Rational Numbers

To add or subtract rational numbers, we add or subtract numerators when denominators are the same. In general, to add or subtract rational numbers they must have a common denominator.

Add and Subtract Rational Numbers

- Let *a*, *b*, and *c* be four numbers, where $c \neq 0$, then $\frac{a}{c} + \frac{b}{c} = \frac{a+b}{c}$
- Let *a*, *b*, and *c* be four numbers, where $c \neq 0$, then $\frac{a}{c} - \frac{b}{c} = \frac{a-b}{c}.$

In cases in which rational numbers do not have the same denominators, we need to find the **least common denominator (LCD)** to make them somehow with the same denominator. For finding the **LCD**, let us first present some definitions and some concepts.

1.6.1. Finding the Common Denominator

To find the common denominator, one simple answer is to multiply the current denominators together. For example, given two rational numbers $\frac{5}{7}$ and $\frac{13}{28}$, we will have two equivalent rational numbers $\frac{5}{7} \times \frac{28}{28} = \frac{140}{196}$ and $\frac{13}{28} \times \frac{7}{7} = \frac{91}{196}$ with a common denominator.

Similarly, given three rational numbers $\frac{2}{3}$, $\frac{5}{8}$, and $\frac{7}{9}$, we will have three equivalent rational numbers $\frac{2}{3} \times \frac{8}{8} \times \frac{9}{9} = \frac{144}{216}$, $\frac{5}{8} \times \frac{3}{3} \times \frac{9}{9} = \frac{135}{216}$, and $\frac{7}{9} \times \frac{3}{3} \times \frac{8}{8} = \frac{168}{216}$ with a common denominators. That is all fine, but can we make a common denominator with smaller number? Let us consider two rational numbers $\frac{1}{3}$ and $\frac{1}{6}$. We have found a common denominator by multiplying the current denominators together, that is $\frac{1}{3} \times \frac{6}{6} = \frac{6}{18}$ and $\frac{1}{6} \times \frac{3}{3} = \frac{3}{18}$ in this case. However, we can have $\frac{1}{3} \times \frac{2}{2} = \frac{2}{6}$ and $\frac{1}{6}$ as an equivalent rational number for $\frac{1}{3}$ and $\frac{1}{6}$, which are rational numbers with smaller common denominators.

Note that any common denominator will work, but usually the lowest common denominator is preferable because it makes the rest of the calculation as simple as possible. We will describe the method of finding the lowest common denominator, after presenting the **least common multiple (LCM)** concept.

1.6.2. Least Common Multiple (LCM)

The **least common multiple**, lowest common multiple, or smallest common multiple of two integers a and b, usually denoted by lcm(a, b), is the smallest positive integer that is divisible by both a and b.

How to Find Least Common Multiple (LCM)

There are multiple ways to find the least common multiple. The most popular methods are:

- Factoring prime factors of each number.
- Listing multiples.
- Greatest common factor (GCF) or greatest common divisor (GCD).

Finding LCM by Factoring Prime Factors

The steps below show how to find the LCM by factoring prime factors.

1. Find all the prime factors of each given number.
2. List all the prime numbers found, as many times as they occur most often for any one given number.
3. Multiply the list of prime factors together to find the LCM.

Example
Find the LCM of 24 and 300 by factoring the prime factors of each number.

Solution

- Prime factorization of 24 = $2 \times 2 \times 2 \times 3$.
- Prime factorization of 300 = $2 \times 2 \times 3 \times 5 \times 5$.
- Using each prime number as frequently as it appears most frequently, we consider $2 \times 2 \times 2 \times 3 \times 5 \times 5 = 600$.

Therefore, the LCM of 24 and 300 is 600.

Example
Find the LCM of 3, 9, and 24 by factoring the prime factors of each number.

Solution

- Prime factorization of 3 = 3.
- Prime factorization of 9 = 3×3.
- Prime factorization of 24 = $2 \times 2 \times 2 \times 3$.
- Using all prime numbers found as often as each occurs most often, we take $2 \times 2 \times 2 \times 3 \times 3 = 72$.
- Therefore, the LCM of 3, 9, and 24 is 72.

Finding the LCM by Listing Multiples

The steps below show how to find the LCM by **listing multiples**.

1. List the multiples of each number until at least one of the multiples appears on all lists.

2. Find the smallest number that is on all the lists.
3. This number is the LCM.

Example
Find the least common multiple of 4 and 6.

Solution
We first need to find multiples of 4 and multiples of 6.
Here are multiples of 4: 4, 8, 12, 16, 20, 24, 28, 32, 36, 40, 44, 48, ⋯.

Here are multiples of 6: 6, 12, 18, 24, 30, 36, 42, 48, 54, ⋯.
Common multiples of 4 and 6 are the numbers that are in both lists:

12, 24, 36, 48, ⋯.

Among this list, the smallest number is 12, hence the least common multiple is 12.

Example
Find the least common multiple of 6, 7 and 21.

Solution
We first need to find multiples of 6, 7 and 21.

Here are multiples of 6: 6, 12, 18, 24, 30, 36, 42, 48, 54, 60, 66, 72, ⋯.
Here are multiples of 7: 7, 14, 21, 28, 35, 42, 49, 56, 63, 70, 77, ⋯.
Here are multiples of 21: 21, 42, 63, 84, 105, 126, 147, ⋯.

Common multiples of 6, 7, and 21 are the numbers that are in all lists:

42, 84, 126, ⋯.

Among these lists, the smallest number that is on the lists is 42. Therefore, the LCM of 6, 7 and 21, is 42.

Finding the LCM by the Greatest Common Factor (GCF)

The third viable method for finding the LCM of some given integers is using the greatest common divisor. This is also frequently referred to as the greatest common factor (GCF). Given LCM (a, b), the procedure for finding the LCM using GCF is to divide the product of the numbers a and b by their GCF, i.e. $(a \times b)/\text{GCF}(a, b)$. When trying to determine the LCM of more than two numbers, a, b, and c, find the LCM of a and b where the result will be q. Then find the LCM of $\boldsymbol{c}$ and q. The result will be the LCM of all three numbers. However, we need first to demonstrate how to find the GCF. The greatest common factor (GCF) of a set of numbers is the largest number that can evenly divided into all elements of that set. To find the GCF of two numbers:

- List the prime factors of each number (including 1).
- Multiply those factors both numbers have in common. If there are no common prime factors, the GCF is 1.

1.6.3. Greatest Common Factor

The **greatest common factor (GCF)** of a set of numbers is thc largest number that can evenly divided into all elements of that set.

To find the GCF of two numbers:

1. List the prime factors of each number (including 1).
2. Multiply those factors both numbers have in common. If there are no common prime factors, the GCF is 1.

Examples
Find the GCF of 15 and 25.

Solution

Find factors of 15.

15: 1, 3, 5, 15.

Find factors of 25.

25: 1, 5, 25.

The factors that are common to the two numbers are 1, and 5.

The greatest common factor is 5.

Examples
Find the GCF of 18 and 24.

Solution

Find factors of 18.

18: 1, 2, 3, 6, 9, 18.

Find factors of 24.

24: 1, 2, 3, 4, 6, 8, 12, 24.

The factors that are common to the two numbers are 1, 2, 3, and 6.

The greatest common factor is 6.

Examples
Find the GCF of 121 and 44.

Solution

Find factors of 121.

121:1, 11, 121.

Find factors of 44.

44: 1, 2, 4, 11, 22, 44.

The factors that are common to the two numbers are 1, and 11.

The greatest common factor is 11.

Examples
Find the GCF of 15, 30, and 105.

Solution

Find factors of 15.

15:1, 3, 5, 15.

Find factors of 30.

30: 1, 2, 3, 5, 6, 10, 15, 30.

Find factors of 105.

105: 1, 3, 5, 7, 15, 21, 35, 105.

The factors that are common to all three numbers are 1, 3, 5, and 15.

The greatest common factor is 15.

Now let us look at the LCM. See the examples below.

Example
Find the LCM of 308 and 1176 by the greatest common factor.

Solution
We first need to find the GCF of 308 and 1176. So, we list the prime factors of the values.

Factors of 308: 2^2, 7, 11

Factors of 1176: 2^3, 3, 7^2.

The greatest common factor is $2^2 \times 7$.

Then using the formula for finding the LCM by the GCF we have:

LCM (308, 1176) = $\frac{308 \times 1176}{2^2 \times 7}$

we multiplied two numbers, 308×1176, and divided it by the GCF = $2^2 \times 7$

LCM (308, 1176) = $\frac{362{,}208}{4 \times 7}$

LCM (308, 1176) = $\frac{362{,}208}{28}$

LCM (308, 1176) = 12,936.

Example
Find the LCM of 21, 14, and 38 by the greatest common factor.

Solution
First, we find GCF (14, 38) which is 2. Then using the formula for finding the LCM by the GCF we have:

LCM (14, 38) = $\frac{38 \times 14}{2}$ = 266.

Similarly, we find GCF (21, 266) which is 7. Then using the formula for finding the LCM by the GCF we have:

$$\text{LCM}(21, 266) = \frac{21\times266}{2} = 798.$$

Then,

$$\text{LCM}(21, 14, 38) = 798.$$

Therefore, the LCM of 21, 14, and 38 is 798.

Up to this point, we have explored the process of finding the LCM of two or more numbers. We now can look at the initial discussion regarding adding and subtracting rational numbers. To add or subtract rational numbers, we add or subtract numerators when denominators are the same. Given two or more rational numbers, the LCM of the denominators is the LCD of the rational numbers.

Now, we can perform the addition and subtraction of rational numbers using the LCD for finding the common denominators of rational numbers. To see how, let us present some examples.

Example

Find the LCD for these two rational numbers, $\frac{5}{18}$ and $\frac{7}{30}$.

Solution

Using the factorization method, we can find LCM and hence the LCD of rational numbers. Factoring the denominators, we have:

$$18 = 2 \times 3 \times 3$$

$$30 = 2 \times 3 \times 5.$$

The LCM is $2 \times 3\times 3 \times 5 = 90$, and hence LCD is 90. We now can assemble LCD,

$$2\times3\times3\times5= 90$$

$$\text{LCD} = 90.$$

After we determine the LCD of two rational numbers, we need to convert the rational numbers to equivalent rational numbers with the LCD. To find the equivalent rational numbers with the LCD, we multiply each rational number (top and bottom) by whatever is needed to build up the denominator to the LCD: Putting these steps together allows us to add and subtract rational numbers because they have the same denominators. So,

$$\frac{5}{18} = \frac{5}{18}\times\frac{5}{5} = \frac{25}{90}$$

we multiplied the nominator and the denominator by 5, which is the LCD divided by the original denominator, $\frac{\text{LCD}}{18} = \frac{90}{18} = 5$

$$\frac{7}{30} = \frac{7}{30}\times\frac{3}{3} = \frac{21}{90}$$

we multiplied the nominator and the denominator by 3, which is the LCD divided by the original denominator, $\frac{\text{LCD}}{30} = \frac{90}{30} = 3$.

Therefore, the two rational numbers, $\frac{5}{18}$ and $\frac{7}{30}$ are equivalent to $\frac{25}{90}$ and $\frac{21}{90}$ with a common denominator.

Example

Find $\frac{5}{18} + \frac{7}{30}$.

Solution
As we found in the previous example, the LCD of two rational numbers is 90. We must multiply each rational number (top and bottom) by whatever is needed to build up the denominator to the LCD.

$\frac{5}{18}+\frac{7}{30}=\frac{5}{18}\left(\frac{5}{5}\right)+\frac{7}{30}\left(\frac{3}{3}\right)$ we found $\frac{5}{5}$, and $\frac{3}{3}$ to make denominators of the first rational number and the second rational numbers to build up the denominators to the LCD, respectively

$\frac{5}{18}+\frac{7}{30}=\frac{25}{90}+\frac{21}{90}$

$\frac{5}{18}+\frac{7}{30}=\frac{25+21}{90}$ since the denominators are the same, we just added the nominations

$\frac{5}{18}+\frac{7}{30}=\frac{46}{90}$

$\frac{5}{18}+\frac{7}{30}=\frac{23\times 2}{45\times 2}$

$\frac{5}{18}+\frac{7}{30}=\frac{23}{45}.$ we deleted the common 2

Therefore, summation of $\frac{5}{18}$ and $\frac{7}{30}$ is $\frac{23}{45}$.

Example
Find $\frac{53}{20}-\frac{47}{18}$.

Solution
We first find the LCM of the denominators; (20, 18).

- Multiples of 20 = 20, 40, 60, 80, 100, 120, 140, 160, 180, 200, 220, 240,⋯.
- Multiples of 18 = 18, 36, 54, 72, 90, 108, 126, 144, 162, 180, 198, 216, 234,⋯.
- We find the smallest number, bolded number, that is on all the above lists.

Therefore, the LCM of 20 and 18 is 180.

As we found the LCM the LCD of 20 and 18 is 180. We must multiply each rational number (top and bottom) by whatever is needed to build up the denominator to the LCD

$\frac{53}{20}-\frac{47}{18}=\frac{53}{20}\left(\frac{9}{9}\right)-\frac{47}{18}\left(\frac{10}{10}\right)$ we found $\frac{9}{9}$ and $\frac{10}{10}$ to make denominators of the first, and second rational number respectively to build up the denominators to the LCD

$\frac{53}{20}-\frac{47}{18}=\frac{477}{180}-\frac{470}{180}$

$\frac{53}{20}-\frac{47}{18}=\frac{477-470}{180}$ the denominators are the same, so we just subtracted the nominations

$\frac{53}{20}-\frac{47}{18}=\frac{7}{180}.$

Example
Find $\frac{17}{21}+\frac{6}{7}-\frac{5}{6}$.

Solution
We first find the LCM (21, 6, 7).

- Multiples of 6 = 6, 12, 18, 24, 30, 36, 42, 48, 54, 60, 66,⋯.
- Multiples of 7 = 7, 21, 28, 35, 42, 49, 56, 63, 70, 77,⋯.
- Multiples of 21 = 21, 42, 63, 84, 105, 126,⋯.
- We find the smallest number, bolded number, that is on all the above lists.

Therefore, the LCM of 6, 7, and 21 is 42.

As we found the LCM the LCD of 6, 7, and 21 is 42. We must multiply each rational number (top and bottom) by whatever is needed to build up the denominator to the LCD

$$\frac{17}{21}+\frac{6}{7}-\frac{5}{6}=\frac{17}{21}\left(\frac{2}{2}\right)+\frac{6}{7}\left(\frac{6}{6}\right)-\frac{5}{6}\left(\frac{7}{7}\right)$$

we found $\frac{2}{2}$, $\frac{6}{6}$, and $\frac{7}{7}$ to make denominators of the first, second and third rational numbers respectively to build up the denominators to the LCD

$$\frac{17}{21}+\frac{6}{7}-\frac{5}{6}=\frac{34}{42}+\frac{36}{42}-\frac{35}{42}$$

$$\frac{17}{21}+\frac{6}{7}-\frac{5}{6}=\frac{34+36-35}{42}$$

since the denominators are the same, we just subtracted the nominations

$$\frac{17}{21}+\frac{6}{7}-\frac{5}{6}=\frac{35}{42}.$$

1.7. Properties of Real Numbers

Commutative Properties

By the commutative property numbers may be added in any order without affecting the sum.

$a+b=b+a.$

Similarly, by the commutative property of multiplication numbers may be multiplied in any order without affecting the product.

$a\times b=b\times a.$

Distributive Property

By the distributive property, the product of a factor times a sum is the sum of the factor times each term in the sum.

$a\,(b+c)=(a\times b)+(a\times c)=ab+ac.$

Identity Properties

By the identity property of addition, there is a unique number, named the additive identity “0” that, when it is added to a number, results in the original number.

$a+0=a.$

Similarly, by the identity property of multiplication, there is a unique number, called the multiplicative identity “1” that, when multiplied by a number, results in the original number.

$a\times 1=a.$

Inverse Properties Addition

By the inverse property of addition, for every real number a, there is a unique number, called the additive inverse, denoted $-a$, that, when added to the original number, results in the additive identity, 0.

$a + (-a) = 0.$

Inverse Properties Multiplication

By the inverse property of multiplication, every nonzero real number a has a multiplicative inverse, or reciprocal, denoted $\frac{1}{a}$, such that, when multiplied by a number, results in, 1.

$a \times (\frac{1}{a}) = 1.$

	Addition	Multiplication
Commutative Property	$a + b = b + a$	$a \times b = b \times a$
Associative Property	$a + (b + c) = (a + b) + c$	$a \times (bc) = (ab) \times c$
Distributive Property	$a \times (b + c) = (a \times b) + (a \times c) = ab + ac$	
Identity Property	$a + 0 = a$	$a \times 1 = a$
Inverse Property	$a + (-a) = 0$	$a \times (1/a) = 1$

Example

Simplify the following expression.

$6 + (12 + x).$

Solution

Using the associative property, we can simplify it.

$6 + (12 + x).$

$6 + (12 + x) = (6 + 12) + x.$

$6 + (12 + x) = (18) + x.$

$6 + (12 + x) = 18 + x.$

Example

Simplify the following expression,

$18a\,(\frac{7}{9}).$

Solution

First, let us use the cumulative property.

$18a\,(\frac{7}{9}) = \frac{7}{9}\,(18a).$

Next, we will use the associative property.

$18a\,(\frac{7}{9}) = [\frac{7}{9} \times (18)] \times a$

$18a\left(\frac{7}{9}\right) = (7 \times 2) \times a$

$18a\left(\frac{7}{9}\right) = 14a.$

Example
Simplify the following expression,

$$\frac{5}{6}\left(\frac{12}{15}a + \frac{18}{25}b - \frac{6}{5}c + \frac{24}{25}\right).$$

Solution
Using the distributive property, we rewrite the following expression:

$$\frac{5}{6}\left(\frac{12}{15}a + \frac{18}{25}b - \frac{6}{5}c + \frac{24}{25}\right) = \frac{5}{6}\left(\frac{12}{15}a\right) + \frac{5}{6}\left(\frac{18}{25}b\right) - \frac{5}{6}\left(\frac{6}{5}c\right) + \frac{5}{6}\left(\frac{24}{25}\right).$$

Now, using the identity property, we rewrite the following expression:

$$\frac{5}{6}\left(\frac{12}{15}a + \frac{18}{25}b - \frac{6}{5}c + \frac{24}{25}\right) = \frac{2}{3}a + \frac{3}{5}b - c + \frac{4}{5}.$$

Example
Use the distributive property to factor the following expression:

$81a - 27b + 18.$

Solution

$81a - 27b + 18 = 9(9a - 3b + 2).$

Example. Use the properties of real numbers to rewrite and simplify each expression. State which properties apply.

a) $5 \times 8 + 5 \times 7.$

b) $12 + (11 + 8).$

c) $\frac{7}{9} \times \left(\frac{9}{7} \times \frac{2}{3}\right).$

d) $100 \times [(0.85 + (-5.47)].$

Solution

a) $5 \times 8 + 5 \times 7.$

$= 5 \times (8 + 7)$ we used the distributive property

$= 5 \times 15$

$= 75.$

Therefore, our solution is 75.

b) $12 + (11 + 8).$

$12 + (11 + 8) = (12 + 11) + 8$ we used the associative property

$12 + (11 + 8) = 23 + 8$

$12 + (11 + 8) = 31.$

Therefore, our solution is 31.

c) $\frac{7}{9}\times\left(\frac{9}{7}\times\frac{2}{3}\right)$.

$\frac{7}{9}\times\left(\frac{9}{7}\times\frac{2}{3}\right) = (\frac{7}{9}\times\frac{9}{7})\times\frac{2}{3}$ we used associative property of multiplication

$\frac{7}{9}\times\left(\frac{9}{7}\times\frac{2}{3}\right) = (1)\times\frac{2}{3}$

$\frac{7}{9}\times\left(\frac{9}{7}\times\frac{2}{3}\right) = \frac{2}{3}$ we used identity property of multiplication.

Therefore, our solution is $\frac{2}{3}$.

d) $100 \times [(0.85 + (-5.47)]$.

$100 \times [(0.85 + (-5.47)] = (100\times0.85) + (100\times5.47)$ we used the distributive property

$100 \times [(0.85 + (-5.47)] = (85) + (547)$

$100 \times [(0.85 + (-5.47)] = 632$.

Therefore, our solution is 632.

Example
Simplify $9x - 4y - 6(x - 2y)$.

Solution

$9x - 4y - 6(x - 2y)$

$9x - 4y - 6(x - 2y) = 9x - 4y - 6x + 12y$ we used the distributive property.

$9x - 4y - 6(x - 2y) = 3x + 8y$ we combined like terms, $9x - 6x$ and $-4y + 12y$.

Therefore, the solution is $3x + 8y$.

Example
Simplify $6x^2 - 5x + 9xy + 10 - x^2 + 3x - 2xy - 11$.

Solution

$6x^2 - 5x + 9\text{xy} + 10 - x^2 + 3x - 2\text{xy} - 11$

$6x^2 - 5x + 9\text{xy} + 10 - x^2 + 3x - 2\text{xy} - 11 = 5x^2 - 2x + 7xy - 1$ we combined like terms $6x^2 - x^2$ $-5x + 3x$, $9xy - 2xy$ and $10 - 11$.

Therefore, our solution is $5x^2 - 2x + 7xy - 1$.

Example
Simplify $-(7x - 3y + 8)$.

Solution

$-(7x - 3y + 8)$

$-(7x - 3y + 8) = -1\ (7x - 3y + 8)$ negative can be thought of as -1

$-(7x - 3y + 8) = -7x + 3y - 8$ we multiplied each term by -1.

Therefore, our solution is $-7x + 3y - 8$.

Example
Simplify $7(3x - 2) - 6$.

Solution

$7(3x - 2) - 6$

$7(3x - 2) - 6 = 21x - 14 - 6$ we distributed 7, multiplied each term

$7(3x - 2) - 6 = 21x - 20$ we combined like term, $-14 - 6$.

Therefore, our solution is $21x - 20$.

1.9. Exercises

1. Fill in the chart below:

Set of Real Numbers	Interval Notation	Region on the Real Number Line
$\{x \mid -1 \le x < 5\}$		
	$[0, 3)$	

$\{x \mid -5 < x \le 0\}$		
	$(-3, 3)$	

$\{x \mid x \le 3\}$		
	$(-\infty, 9)$	

$\{x \mid x \ge -3\}$		
	$[2, \infty)$	

2. Find the indicated intersection or union and express your answers in the set interval notation.
   a) $(-1, 5] \cap [0, 8)$.

   b) $(-1, 1) \cup [0, 6]$.

   c) $(-\infty, 4] \cap (0, \infty)$.

   d) $(-\infty, 0) \cap [1, 5]$.

   e) $(-\infty, 0) \cup [1, 5]$.

   f) $(-\infty, 5] \cap [5, 8)$.

3. Write the set using the set interval notation.
   a) $\{x \mid x \neq 8\}$.

   b) $\{x \mid x \neq -5\}$.

   c) $\{x \mid x \neq 9\}$.

   d) $\{x \mid x = \pm 3\}$.

   e) $\{x \mid x \le -2 \text{ or} \ge 3\}$.

f) $\{x| -2 < x \leq 4\}$.

g) $\{x|x \leq 2 \text{ or } x = 7\}$.

h) $\{x|-6 < x \leq 2 \text{ or } 8\}$.

4. Use the properties of real numbers to rewrite each expression and state which properties apply.

a) $\left(-\frac{23}{5}\right) \times \left[2 \times \left(-\frac{5}{23}\right)\right]$.

b) $5 \times (6.2 + 0.4)$.

c) $18 - [45 \div (-15)]$.

d) $\frac{17}{18} + \left[\frac{4}{9} + \left(-\frac{17}{6}\right)\right]$.

e) $6 \times (-3) + 6.3 \times (-5)$.

f) $2 \times \left(\frac{4}{7} + 6\right) - \left(\frac{8}{7} + 13\right)$.

g) $\frac{11}{15} + \left[\frac{6}{7} + 3 \times \left(-\frac{11}{15}\right)\right]$.

h) $\left[\frac{9}{19} + \left(-\frac{13}{9}\right)\right] + \frac{13}{3}$.

i) $3.6 + (3.4 - 2)\left(\frac{10}{7}\right)$.

j) $\frac{5}{6} \times \left(\frac{7}{9} \times \frac{6}{5}\right)$.

k) $100 \times [0.75 + (-2.38)]$.

l) $80 \times [0.25 + (-1.27) + 0.66]$.

5. Express the sets of numbers using interval notation and graph on the number line.

a) $\{x|\ x \leq -3 \text{ or } x \geq 2\}$.

b) $\{x|\ x = 5\}$.

c) $\{x|\ x \neq \pm 6\}$.

d) $\{x| -4 < x \leq 5 \text{ or } x = 9$.

6. What kind of number are in the following?

a) - 6.

b) $-\frac{24}{10}$.

c) $-\frac{36}{9}\sqrt{4}$.

d) $2\frac{3}{4}$

e) 5.

7. Compare the following numbers.

a) -9 and -12.

b) $-\frac{24}{10}$ and $-2\frac{7}{12}$.

c) $5\frac{3}{5}$ and 5.12.

d) $\frac{3}{9}\sqrt{4}$ and 0.78.

8. Find the GCF of the following numbers:
    a) 100 and 25.
    b) 124 and 64.
    c) 18 and 138.
    d) 27 and 36.
    e) 160 and 84.
    f) 124 and 24.
    g) 18 and 138.
    h) 27 and 36.

9. Find the LCM of 48, and 240 by factoring the prime factors of each number.

10. Find the LCM of 36, 64 and 320 by factoring the prime factors of each number.

11. Find the LCM by 45, 60, and 90 by listing multiples.

12. Find the LCM by 120, 80, and 48 by listing multiples.

13. Find the LCM by 40, 120, and 60 by the greatest common factor (GCF).

14. Find the LCM by 16, 48, and 72 by the greatest common factor (GCF).

15. Simplify the following expressions:
    a) $\frac{-36(2)}{-3(6)}$.
    b) $\frac{-16(3)}{-9(4)} - \frac{24}{6}$.
    c) $\frac{-3}{-9(2)} - \frac{28}{3} + \frac{-15(2)}{-6(3)}$.
    d) $\frac{8}{-9} - \frac{11(4)}{6} + \frac{-12(-4)}{-6(2)}$.
    e) $\left(\frac{4}{5} \times \frac{15}{4}\right) - \left(\frac{5}{7} + \frac{3}{14}\right)$.
    f) $12 + (38 \div 10)^2$.
    g) $11 + 2(9-5)^2 - (4-1)(13-17)$
    h) $9 \times 2 + (9-7)(8-5)^2$
    i) $16 \div 4 + 2(9-6)^2 - (4-2)^3$.
    j) $2[(4+3)^2(6-1)^2 + 6] - (5-2)^2$.
    k) $8[(5-3)^2 + 6](14-11) - (3+2)^2(5-1)^2$.

l) $3[(4-1)^3+\frac{8-7}{2(5-3)}]-(4-2)(13-7)^2$.

m) $4x + x(17-13)$ where $x = 2$.

n) $2y - (3)2y - 17$ where $y = 1$.

o) $2a(34) - 18a \div 6$ where $a = -1$.

p) $8b - 4b(3) + (a-1)3^2$ where $b = -3$ and $a = 4$.

q) $(93 \div 31)(6a - 4b) + a(b-2)$ where $a = 1$ and $b = -2$.

r) $3t(2-4t) \div 2t + t^2 - (2t + 8) \div 2$ where $t = 3$.

s) $\frac{18-2^2}{(2^3-9)-(7-4)^3} + (8-5)^3(18-14)^2$.

t) $\frac{(12-2^2)(2^3-6)}{(22-4^2)-(7-3)^2} + \frac{(8-5)^3(27-23)^2}{2^4-6}$.

16. Is 11.78 greater than, equal to, or less than $\frac{95}{8}$?

17. Convert $12\frac{7}{9}$ to a rational number.

18. Convert $\frac{23}{9}$ to a mixed number.

19. Find each.

a) $\frac{13}{14} + \frac{5}{7} - \frac{12}{21}$.

b) $\frac{7}{9} + \frac{9}{12} - \frac{11}{15} + \frac{5}{6}$.

c) $\frac{5}{8} + \frac{7}{12} - \frac{11}{16} - \frac{17}{24}$.

d) $\frac{11}{15} + \frac{16}{25} - \frac{7}{10}$.

20. Multiply each.

a) $\frac{9}{17} \times \frac{14}{15}$.

b) $\frac{6}{11} \times \frac{-14}{9} \times \frac{7}{8}$.

c) $\frac{4}{7} \times \frac{1}{2x} \times \frac{5-x}{3x^2}$ where $x = 3$.

d) $\frac{a-2b}{3} \times \frac{4}{(a+1)} \times \frac{b}{2a-4b}$ where $a = 4$ and $b = 1$.

e) $\frac{(5k-2h)^2}{16} \times \frac{64}{(10k-4h)} \times \frac{kh}{8(15-6h)}$ where $k = 1$ and $h = 3$.

21. Divide each.

a) $\frac{7}{11} \div \frac{14}{9}$.

b) $(\frac{9}{25} \div \frac{18}{15})(-\frac{3}{2})$.

c) $\frac{2(14)}{5(7)} \div \frac{6}{(-5)^2} \times \frac{3}{10}$.

d) $\frac{2}{3-7} \div \frac{1}{6.7-3.3+2-1.4}$.

22. Use the properties of real numbers to rewrite and simplify each expression.

a) $12 \times 9 + 12 \times 5 - 6 - 24 \times 7$.

b) $\frac{11}{19} \times \left(\frac{19}{11} \times \frac{7}{8}\right)$.

c) $\frac{17}{23} \div \left(\frac{8}{23} \times \frac{17}{4}\right)$.

d) $10 \times [0.6 + \frac{3}{10}(-3.7) + \frac{5}{100}(10.2)]$.

2. Exponents and Radical Expressions

In this section, we delve into the concepts of exponent expressions and radical expressions.

2.1 Exponent Expressions

Exponential notation is composed of a base and an **exponent**. It is a "shorthand" way of writing repeated multiplication and indicates that the base is a factor, and the exponent is the number of times the factor is used in the multiplication. For example, a^n, involving two numbers, the base a and the exponent or power n. When n is a positive integer, exponentiation corresponds to repeated multiplication of the base: that is, a^n is the product of multiplying n bases:

$$a^n = \underbrace{a \times a \times a \times \cdots \times a}_{n \text{ times}}$$

The Product Rule of Exponents

For any real number a, and natural numbers m and n, the product rule of exponents states that:

$$a^m \times a^n = a^{m+n}.$$

The Quotient Rule of Exponents

For any real number a and natural numbers m and n, such that $m > n$, the quotient rule of exponents states that:

$$\frac{a^m}{a^n} = a^{m-n}.$$

The Power Rule of Exponents

Given any real number a, and positive integers m and n, the power rule of exponents states that:

$$(a^m)^n = a^{m.n}.$$

The Zero Exponent Rule of Exponents

Given any nonzero real number, a, the zero-exponent rule of exponents states that:

$$a^0 = 1.$$

The Negative Rule of Exponents

Given any nonzero real number, a, and natural number n, the negative rule of exponents states that:

$$a^{-n} = \frac{1}{a^n}.$$

The Power of a Product Rule of Exponents

Given any real numbers a and b and any integer n, the power of a product rule of exponents states that:

$$(ab)^n = a^n b^n.$$

The Power of a Quotient Rule of Exponents

Given any real numbers a and b and any integer n, the power of a quotient rule of exponents states that:

$$\left(\frac{a}{b}\right)^n = \frac{a^n}{b^n}.$$

Example

Simplify each expression:

a) $a^5 \cdot a^6$.

b) $3x \times 6x^4$.

c) $5(ab)^2 \times 4(ab)^5$.

d) $3d^2 \times 2d^3 \times 5d^7$.

Solution

a) Using the property, $a^m . a^n = a^{m+n}$,

$$a^5 a^6 = a^{11}.$$

b) Using the cumulative property and using the property of exponents, $a^m . a^n = a^{m+n}$,

$$3x \times 6x^4 = (3 \times 6)(x \times x^4)$$

$$3x \times 6x^4 = 18x^5.$$

c) Using the cumulative property and using the property of exponents, $a^m . a^n = a^{m+n}$,

$$5(ab)^2 \times 4(ab)^5 = (5 \times 4)[(ab)^2 \times (ab)^5]$$

$$5(ab)^2 \times 4(ab)^5 = 20(ab)^7$$

$$5(ab)^2 \times 4(ab)^5 = 20a^7b^7.$$

d) Using the cumulative property and using the property of exponents, $a^m . a^n = a^{m+n}$,

$$3d^2 \times 2d^3 \times 5d^7 = (3 \times 2 \times 5)(d^2 \times d^3 \times d^7)$$

$$3d^2 \times 2d^3 \times 5d^7 = 30d^{12}.$$

Example

Simplify each expression:

a) $\frac{x^5}{x^3}$.

b) $\frac{15^8}{15^6}$.

c) $\frac{(ab)^9}{(ab)^6}$.

d) $\frac{7t^4}{14\ ^6}$.

e) $\frac{4c^5}{2c^5}$.

Solution

a) Using the negative rule of exponent, $a^{-n} = \frac{1}{a^n}$, we will have.

$\frac{x^5}{x^3}$

$\frac{x^5}{x^3} = x^5 \times \frac{1}{x^3}$

$\frac{x^5}{x^3} = x^5 \times x^{-3}$ we applied the negative rule of exponents

$\frac{x^5}{x^3} = x^2$ we applied the product rule of exponents.

b) Using the negative rule of exponent, $a^{-n} = \frac{1}{a^n}$, we will have.

$\frac{15^8}{15^6}$

$\frac{15^8}{15^6} = 15^8 \times \frac{1}{15^6}$

$\frac{15^8}{15^6} = 15^8 \times 15^{-6}$

$\frac{15^8}{15^6} = 15^2 = 225.$

c) Using the negative rule of exponent, $a^{-n} = \frac{1}{a^n}$, we will have.

$\frac{(ab)^9}{(ab)^6} =$

$\frac{(ab)^9}{(ab)^6} = (ab)^9 \times \frac{1}{(ab)^6}.$

$\frac{(ab)^9}{(ab)^6} = (ab)^9 \times (ab)^{-6}$ we applied the negative rule of exponents

$\frac{(ab)^9}{(ab)^6} = (ab)^3$ we applied the product rule of exponents

$\frac{(ab)^9}{(ab)^6} = a^3b^3.$

d) We first perform some basic simplification operations. So:

$\frac{7t^4}{14t^6}$

$\frac{7t^4}{14t^6} = \frac{7}{14} \times \frac{t^4}{t^6}$

$\frac{7t^4}{14t^6} = \frac{7}{7 \times 2} \times \frac{t^4}{t^6}$

$\frac{7t^4}{14t^6} = \frac{1}{2} \times \frac{t^4}{t^6}$

$\frac{7t^4}{14\ ^6} = \frac{1}{2} \times t^4 \times \frac{1}{t^6}.$

Using the negetive property of exponents, $a^{-n} = \frac{1}{a^n}$, we will have:

$\frac{7t^4}{14\ ^6} = \frac{1}{2} \times t^4 \times \frac{1}{t^6}$

$\frac{7t^4}{14t^6} = \frac{1}{2} \times t^4 \times t^{-6}$ we applied the negative rule of exponents

$\frac{7t^4}{14\ \ 6} = \frac{1}{2} \times t^{-2}$ we applied the product rule of exponents

$\frac{7t^4}{14\ \ 6} = 0.5t^{-2}.$

e) Using the quotient property of exponents, $\left(\frac{a}{b}\right)^n = \frac{a^n}{b^n}$, we will have:

$\frac{4c^5}{2c^5}$

$\frac{4c^5}{2c^5} = \frac{4}{2} \times \frac{c^5}{c^5}$

$\frac{4c^5}{2c^5} = 2 \times c^5 \times \frac{1}{c^5}$

$\frac{4c^5}{2c^5} = 2 \times c^5 \times c^{-5}$ we applied the negative rule of exponents

$\frac{4c^5}{2c^5} = 2 \times c^{5-5}$ we applied the product rule of exponents

$\frac{4c^5}{2c^5} = 2 \times c^0.$

Using zero exponent property of exponents, $a^0 = 1$, we will have:

$\frac{4c^5}{2c^5} = 2 \times 1 = 2.$

Example

Simplify each expression:

a) $(-3ab)^4$.

b) $(5xy^3)^0$.

c) $(3k^{-4})^3$.

d) $(5\frac{m^5}{m^2})^3$.

Solution

a) Using power product rule of exponents, $(ab)^n = a^n b^n$, we will have.

$(-3ab)^4$

$(-3ab)^4 = (-3)^4 a^4 b^4$ we applied the power product of exponents

$(-3ab)^6 = 81a^4b^4$.

b) Using power of a product property of exponents $(ab)^n = a^n b^n$, we will have.

$(5xy^3)^0$

$(5xy^3)^0 = (5)^0 x^0 (y^3)^0$ we applied the power product rule of exponents

$(5xy^3)^0 = 1 \times 1 \times 1$ we applied the zero-exponent rule of exponents

$(5xy^3)^0 = 1$.

c) Using power of a product property of exponents $(ab)^n = a^n b^n$, we will have.

$(3k^{-4})^3$

$(3k^{-4})^3 = (3)^3 \times (k^{-4})^3$ we applied the power product of exponents

$(3k^{-4})^3 = 27 \times (\frac{1}{k^4})^3$ we applied the negative rule of exponents

$(3k^{-4})^3 = 27 \times (\frac{1}{k^{4\times 3}})$ we applied the power rule of exponents

$(3k^{-4})^3 = 27 \times (\frac{1}{k^{12}})$ we applied the power product rule of exponents

$(3k^{-4})^3 = \frac{27}{k^{12}}.$

d) Using the quotient rule of exponents, we will have.

$(5\frac{m^5}{m^2})^3$

$(5\frac{m^5}{m^2})^3 = (5)^3 \times (\frac{m^5}{m^2})^3$ we applied the power product of exponents

$(5\frac{m^5}{n^2})^3 = (5)^3 \times (\frac{m^{5\times 3}}{m^{2\times 3}})$ we applied the power rule of exponents

$(5\frac{m^5}{n^2})^3 = 125 \times (\frac{m^{15}}{m^6})$ we applied the power product rule of exponents

$(5\frac{m^5}{n^2})^3 = 125 \times (m^{15-6})$ we applied the quotient rule of exponents

$(5\frac{m^5}{n^2})^3 = 125 \times m^9.$

2.2. Scientific Notation

Scientific notation is the way that scientists easily handle very large numbers or very small numbers. For example, instead of writing 0.0000000056, we write 5.6×10^{-9}. So, how does this work? We can think of 5.6×10^{-9} as the product of two numbers: 5.6 (the digit term) and 10^{-9} (the exponential term).

A number is written in scientific notation if it is written in the form $a \times 10^n$, where $1 \leq |a| < 10$ and n is an integer number.

Here we present some examples of scientific notation.

- $1 = 1 \times 10^0/$
- $10 = 1 \times 10^1.$
- $100 = 1 \times 10^2.$
- $1000 = 1 \times 10^3.$
- $320{,}000 = 3.2 \times 10^5.$
- $48{,}000{,}000 = 4.8 \times 10^7.$
- $0.1 = 1 \times 10^{-1}.$
- $0.01 = 1 \times 10^{-2}.$
- $0.001 = 1 \times 10^{-3}.$
- $0.0001 = 1 \times 10^{-4}.$

- $0.000065 = 6.5 \times 10^{-5}$.
- $0.00000027 = 2.7 \times 10^{-7}$.
- $0.00000000465 = 4.65 \times 10^{-9}$.

Example
Multiply (5×10^3) and (4.67×10^8).

Solution
We need to perform the multiplication by the following steps:

Step 1: $5 \times 4.67 = 23.35$.

Step 2: Next, we add exponents, $3 + 6 = 11$.

Step 3: The result of the above two steps, is, 23.35×10^{11}.

Step 4: Answer is 2.335×10^{12}.

Example
Divide (12×10^7) and (5×10^3).

Solution

Step 1: $12 \div 5 = 2.4$.

Step 2: Next, we subtract exponents, $7 - 3 = 4$.

Step 3: The answer is 2.4×10^4.

Addition and subtraction require that the exponents be the same, so in many cases some manipulation will be needed.

Example
Add (4.7×10^6) and (9.042×10^9).

Solution

Step 1: We need to change 4.7×10^6 to 0.0047×10^9.

Step 2: Next, we need to add 0.0047 to 9.042, that is $0.0047 + 9.042 = 9.0467$.

Step 3: The answer is 9.0467×10^9.

Example
Subtract (1.3×10^6) from (9.1×10^7).

Solution

Step 1: Again, we need to change 1.3×10^6 to 0.13×10^7.

Step 2: Next, we need to subtract 0.13 from 9.1, that is $9.1 - 0.13 = 8.97$.

Step 3: The answer is 8.97×10^7.

2.3. Radical Expressions

The principal square root of a is written as $\sqrt{a}$. The symbol is called a radical, the term under the symbol is called the radicand, and the entire expression is called a radical expression.

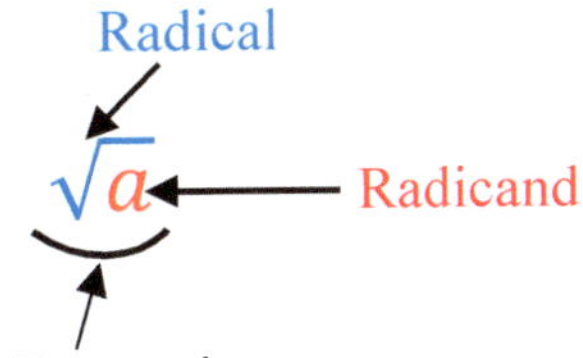

Principal Square Root

The **principal square root** of a is the nonnegative number that, when multiplied by itself, equals a. It is written as a **radical expression**, with a symbol called a radical over the term called the radicand: $\sqrt{a}$.

The Product Rule of Square Roots

If a and b are nonnegative, the square root of the product ab is equal to the product of the square roots of a and b.

$$\sqrt{ab} = \sqrt{a}.\sqrt{b}.$$

The Quotient Rule of Square Roots

The square root of the quotient $\frac{a}{b}$ is equal to the quotient of the square roots of a and b, where $b \neq 0$.

$$\sqrt{\frac{a}{b}} = \frac{\sqrt{a}}{\sqrt{b}}.$$

Note that while we can "break up" products and quotients under a radical we cannot do the same thing for sums or differences. In other words,

$$\sqrt{a+b} \neq \sqrt{a} + \sqrt{b}, \text{ and}$$
$$\sqrt{a-b} \neq \sqrt{a} - \sqrt{b}.$$

Example
Simplify each expression.

a) $\sqrt{64a}$.

b) $\sqrt{81a^3}$.

c) $\sqrt{36a^5}$.

d) $\sqrt{32a^4}$.

Solution

a) $\sqrt{64a}$.

$\sqrt{64a} = \sqrt{8^2 a}$

$\sqrt{64a} = 8\sqrt{a}$ we applied the principle of square roots.

b) $\sqrt{81a^3}$.

$\sqrt{81a^3} = \sqrt{9^2 a \times a^2}$ we applied the product rule of square roots

$\sqrt{81a^3} = 9a\sqrt{a}$ we applied the principle of square roots.

c) $\sqrt{36a^5}$.

$\sqrt{36a^5} = \sqrt{6^2 \times a \times a^2 \times a^2}$ we applied the product rule of square roots

$\sqrt{36a^5} = 6a^2\sqrt{a}$ we applied the principle of square roots.

d) $\sqrt{32a^4}$.

$\sqrt{32a^4} = \sqrt{2 \times 16 \times a^2 \times a^2}$ we applied the product rule of square roots

$\sqrt{32a^4} = \sqrt{2 \times 4^2 \times a^2 \times a^2}$

$\sqrt{32a^4} = 4 \times a \times a \times \sqrt{2}$ we applied the principle of square roots

$\sqrt{32a^4} = 4a^2\sqrt{2}$ We applied the product rule of exponents.

Example

Simplify $\sqrt{9a^3} \times \sqrt{8a^2}$.

Solution

We can use the product rule of square roots to combine the two radicals into one radical and then we will see if there is any simplification that needs to be done.

$$\sqrt{9a^3} \times \sqrt{8a^2} = \sqrt{(9a^2) \times (8a^3)}$$

$$\sqrt{9a^3} \times \sqrt{8a^2} = \sqrt{(9 \times 8)(a^2 \times a^3)}$$

$$\sqrt{9a^3} \times \sqrt{8a^2} = \sqrt{72a^5}.$$

Now that it is in this form, we can do some simplification.

$$\sqrt{9a^3}\,\sqrt{8a^2} = \sqrt{9 \times 4 \times 2 \times a^5}$$

$$\sqrt{9a^3}\,\sqrt{8a^2} = \sqrt{3^2 \times 2^2 \times 2 \times a^2 \times a^2 \times a}$$

$\sqrt{9a^3}\,\sqrt{8a^2} = 3 \times 2 \times a \times a\sqrt{2 \times a}$ we applied the principal rule of square roots

$\sqrt{9a^3}\,\sqrt{8a^2} = 6a^2\sqrt{2a}$ we applied the product rule of exponents.

Example

Simplify $\frac{\sqrt{27a^5}}{\sqrt{3a^4}}$.

Solution

We can use the quotient rule of square roots to combine the two radicals into one radical and then we will see if there is any simplification that needs to be done.

$$\frac{\sqrt{27a^5}}{\sqrt{3a^4}} = \sqrt{\frac{27a^5}{3a^4}}$$

$$\frac{\sqrt{27\;^5}}{\sqrt{3a^4}} = \sqrt{\frac{27}{3} \times \frac{a^5}{a^4}}$$

$$\frac{\sqrt{27a^5}}{\sqrt{3a^4}} = \sqrt{\frac{3 \times 9}{3} \times \frac{a \times a^4}{a^4}}$$ we applied the power product rule of exponents

$$\frac{\sqrt{27a^5}}{\sqrt{3a^4}} = \sqrt{9 \times a}.$$

Now that it is in this form, we can do some simplification.

$$\frac{\sqrt{27\;^5}}{\sqrt{3a^4}} = \sqrt{9 \times a}$$

$$\frac{\sqrt{27a^5}}{\sqrt{3a^4}} = \sqrt{3^2 \times a}$$

$$\frac{\sqrt{27a^5}}{\sqrt{3a^4}} = 3\sqrt{a}$$ we applied the principal rule of square roots.

Example

Simplify $\frac{5^2-4}{7} - \sqrt{11-2}$.

Solution

$$\frac{5^2-4}{7} - \sqrt{11-2} = \frac{5^2-4}{7} - \sqrt{9}$$

$$\frac{5^2-4}{7} - \sqrt{11-2} = \frac{5^2-4}{7} - 3$$ we simplified the radical

$$\frac{5^2-4}{7} - \sqrt{11-2} = \frac{25-4}{7} - 3$$ we simplified the exponent

$$\frac{5^2-4}{7} - \sqrt{11-2} = \frac{21}{7} - 3$$ we simplified subtraction in the numerator

$$\frac{5^2-4}{7} - \sqrt{11-2} = 3 - 3$$ we simplified the division

$$\frac{5^2-4}{7} - \sqrt{11-2} = 0$$ this is our solution.

Example

Write each of the following products with a single base. Do not simplify further.

a) $\frac{\left(-2\sqrt{5}\right)^{14}}{\left(-2\sqrt{5}\right)^{9}}$.

b) $\frac{(\sqrt{t})^{23}}{(\sqrt{t})^{15}}$.

c) $\frac{\left(z\sqrt{2}\right)^{5}}{z\sqrt{2}}$.

Solution

Using the quotient rule of exponents, we can simplify each expression.

a) $\frac{(-2\sqrt{5})^{14}}{(-2\sqrt{5})^{9}}$.

$\frac{(-2\sqrt{5})^{14}}{(-2\sqrt{5})^{9}} = (-2)^{(14-9)}$ we applied the quotient rule of exponents

$\frac{(-2\sqrt{5})^{14}}{(-2\sqrt{5})^{9}} = (-2\sqrt{5})^{5}$

$\frac{(-2\sqrt{5})^{14}}{(-2\sqrt{5})^{9}} = -32\sqrt{5^5}$.

b) $\frac{(\sqrt{t})^{23}}{(\sqrt{t})^{15}}$.

$\frac{(\sqrt{t})^{23}}{(\sqrt{t})^{15}} = (\sqrt{t})^{(23-15)}$ we applied the quotient rule of exponents

$\frac{(\sqrt{t})^{23}}{(\sqrt{t})^{15}} = (\sqrt{t})^{8}$.

c) $\frac{(z\sqrt{2})^{5}}{z\sqrt{2}}$.

$\frac{(z\sqrt{2})^{5}}{z\sqrt{2}} = (z\sqrt{2})^{(5-1)}$ we applied the quotient rule of exponents

$\frac{(z\sqrt{2})^{5}}{z\sqrt{2}} = (z\sqrt{2})^{4}$

$\frac{(z\sqrt{2})^{5}}{z\sqrt{2}} = z^4(\sqrt{2})^{4}$.

2.4. Principal n^{th} Root

If a is a real number with at least one n^{th} root, then the **principal n^{th} root** of a, written as $\sqrt[n]{a}$, is the number with the same sign as a that, when raised to the nth power, equals a. For example, -3 is the 5th root of -243 because $(-3)^5 = -243$. The index of the radical is n. For square roots we typically drop the index.

A General Note: Principal n^{th} Root
If a is a real number with at least one n^{th} root, then the principal n^{th} root of a, written as $\sqrt[n]{a}$, is the number with the same sign as a that, when raised to the nth power, equals a. The index of the radical is n.

Example
Simplify each of the following:

a) $\sqrt[5]{128}$.

b) $\sqrt[4]{3}\,\sqrt[4]{27}$.

c) $\sqrt[3]{\frac{27x^9}{64}}$.

d) $5\sqrt[4]{3} - \sqrt[4]{48}$.

Solution

a) Since $(-2)^7 = -128$, therefore, our solution is -2.

b) First express the product as a single radical expression. $\sqrt[4]{3 \times 27} = \sqrt[4]{81}$. Since $3^4 = 81$, therefore, our solution is 3. Another way of getting the solution is
$\sqrt[4]{3 \times 27} = \sqrt[4]{3 \times 3^3} = \sqrt[4]{3^4} = 3$.

c) $\sqrt[3]{\frac{27x^9}{64}}$.

$$\sqrt[3]{\frac{27\ ^9}{64}} = \frac{\sqrt[3]{27x^9}}{\sqrt[3]{64}}$$ we wrote the quotient as two radical expressions

$$\sqrt[3]{\frac{27\ ^9}{64}} = \frac{\sqrt[3]{3^3x^9}}{\sqrt[3]{4^3}}$$

$$\sqrt[3]{\frac{27\ ^9}{64}} = \frac{\sqrt[3]{3^3x^3x^3x^3}}{\sqrt[3]{4^3}}$$

$$\sqrt[3]{\frac{27x^9}{64}} = \frac{3(x)(x)(x)}{4}$$

$$\sqrt[3]{\frac{27\ ^9}{64}} = \frac{3x^3}{4} = \frac{3}{4}x^3 = 0.75x^3.$$

d) $5\sqrt[4]{3} - \sqrt[4]{48}$.

e) $5\sqrt[4]{3} - \sqrt[4]{48} = 5\sqrt[4]{3} - \sqrt[4]{3 \times 16}$.

$5\sqrt[4]{3} - \sqrt[4]{48} = 5\sqrt[4]{3} - \sqrt[4]{3 \times 2^4}$

$5\sqrt[4]{3} - \sqrt[4]{48} = 5\sqrt[4]{3} - 2\sqrt[4]{3}$

$5\sqrt[4]{3} - \sqrt[4]{48} = (5-2)\sqrt[4]{3}$

$5\sqrt[4]{3} - \sqrt[4]{48} = 3\sqrt[4]{3}$.

Additionally, we can also write the general rational exponent in terms of radicals as follows.

$$a^{\frac{m}{n}} = \left(a^{\frac{1}{n}}\right)^m = \left(\sqrt[n]{a}\right)^m, \text{ or}$$

$$a^{\frac{m}{n}} = (a^m)^{\frac{1}{n}} = \sqrt[n]{a^m}.$$

Example

Simplify $\sqrt[9]{a^6}$.

Solution

Using the above equations, we have.

$$\sqrt[9]{a^6} = (x^6)^{\frac{1}{9}}$$

$$\sqrt[9]{a^6} = a^{\frac{6}{9}}$$

$$\sqrt[9]{a^6} = a^{\frac{2}{3}}$$

$$\sqrt[9]{a^6} = \sqrt[3]{a^2}.$$

Example

Simplify $\sqrt[4]{32a^9b^5c^{12}}$.

Solution

$$\sqrt[4]{32a^9b^5c^{12}}$$

$$\sqrt[4]{32a^9b^5c^{12}} = \sqrt[4]{2^4 \times 2 \times a^4 \times a^4 \times a \times b^4 \times b \times c^4 \times c^4 \times c^4}$$

$$\sqrt[4]{32a^9b^5c^{12}} = 2 \times a \times a \times b \times c \times c \times c\sqrt[4]{2 \times a \times b}$$

$$\sqrt[4]{32a^9b^5c^{12}} = 2a^2bc^3\sqrt[4]{2ab}.$$

Example

Simplify $\sqrt[5]{75a^7}\,\sqrt[5]{5a^4}\,\sqrt[5]{125a^2}$.

Solution

We can use the product rule of principle n^{th} radical to combine the three radicals into one principle n^{th} radical and then we will see if there is any simplification that needs to be done.

$$\sqrt[5]{75a^7}\,\sqrt[5]{5a^4}\,\sqrt[5]{125a^2}$$

$$\sqrt[5]{75a^7}\,\sqrt[5]{5a^4}\,\sqrt[5]{125a^2} = \sqrt[5]{(75a^7)(5a^4)(125a^2)}$$

$$\sqrt[5]{75a^7}\,\sqrt[5]{5a^4}\,\sqrt[5]{125a^2} = \sqrt[5]{(75 \times 5 \times 125) \times (a^7 \times a^4 \times a^2)}$$

$$\sqrt[5]{75a^7}\,\sqrt[5]{5a^4}\,\sqrt[5]{125a^2} = \sqrt[5]{(46{,}875) \times (a^{13})}.$$

Now that it is in this form, we can do some simplification.

$$\sqrt[5]{75a^7}\,\sqrt[5]{5a^4}\,\sqrt[5]{125a^2} = \sqrt[5]{(15 \times 3{,}125) \times (a^5 \times a^5 \times a^3)}$$

$$\sqrt[5]{75a^7}\,\sqrt[5]{5a^4}\,\sqrt[5]{125a^2} = \sqrt[5]{(15 \times 5^5) \times (a^5 \times a^5 \times a^3)}$$

$$\sqrt[5]{75a^7}\,\sqrt[5]{5a^4}\,\sqrt[5]{125a^2} = 5 \times a \times a\sqrt{15 \times a^3}$$

$$\sqrt[5]{75a^7}\,\sqrt[5]{5a^4}\,\sqrt[5]{125a^2} = 5a^2\sqrt{15a^3}.$$

Example

Simplify $\frac{\sqrt[4]{64\quad^9}}{\sqrt[4]{4b^3}}$.

Solution

We can use the quotient rule of principle n^{th} radical to combine the two radicals into one principle n^{th} radical and then we will see if there is any simplification that needs to be done.

$$\frac{\sqrt[4]{64\ \ {}^{9}}}{\sqrt[4]{4b^3}}$$

$$\frac{\sqrt[4]{64b^9}}{\sqrt[4]{4b^3}} = \sqrt[4]{\frac{64b^9}{4b^3}}$$

$$\frac{\sqrt[4]{64b^9}}{\sqrt[4]{4b^3}} = \sqrt[4]{16b^6}.$$

Now that it is in this form, we can do some simplification.

$$\frac{\sqrt[4]{64b^9}}{\sqrt[4]{4b^3}} = \sqrt[4]{16b^6}$$

$$\frac{\sqrt[4]{64b^9}}{\sqrt[4]{4b^3}} = \sqrt[4]{2^4 \times b^4 \times b^2}$$

$$\frac{\sqrt[4]{64\ \ {}^{9}}}{\sqrt[4]{4b^3}} = 2b\sqrt[4]{b^2}.$$

2.5. Exercises

1. Write each of the following products with a single base. Do not simplify further.
   a) $5t^5\, t^2 \frac{1}{t^5} \times t^7$.
   b) $(-4)^2(16)(-8)^5(\frac{1}{4^3})$.
   c) $(4-2x)^2 \times (2-x)^{-3} \times (2-x)^4$.
   d) $(16-4k)^3 \times \frac{1}{(4-k)^5} \times 12\,(4-k)^7$.

2. Write each of the following expressions with a single base. Do not simplify further.
   a) $\frac{(2)^{12}}{(-2)^9}$.
   b) $\frac{(t-4)^{23}}{(2t-8)^{15}}$.
   c) $\frac{(z\sqrt{4})^8}{(2z\sqrt{4})^3}$.

3. Evaluate the expression $256^{-\frac{5}{2}}$.

4. Write each of the following products with a single base. Do not simplify further.
   a) $(-12)^4 \times (-24) \times (-12)^6$.
   b) $(4-\frac{4}{x}) \times (1-\frac{1}{x})^6 \times (1-\frac{1}{x})^{-3} \times (1-\frac{1}{x})^4$.
   c) $\frac{3(5t)^2y^4x^5(2t)^3}{2x^6y^3(2\,t)^2t^4}$.

5. Simplify each radical expression.
   a) $\sqrt{16a}$.
   b) $\sqrt{27a^5}$.
   c) $\sqrt{72(ab^2)^3}$.
   d) $\sqrt{18ab^3c^5}$.

6. What kind of number is $-\frac{24}{8}\sqrt{4}$.

7. Simplify $\frac{6^2-3}{9} - \sqrt{18-2}$.

8. Simplify each of the following:
   a) $\sqrt[6]{4{,}096}$.
   b) $\sqrt[7]{63}\,\sqrt[7]{243}$.
   c) $\sqrt[3]{\frac{108x^6}{1{,}331}}$.
   d) $8\sqrt[3]{625} - 5\sqrt[3]{5}$.

9. Simplify the following expression:

a) $\frac{-36(2)}{-3(6)}$.

b) $\frac{-16(3)^2}{-9(4)} - \frac{24}{6}$.

c) $\frac{-3}{-9(2)} - \frac{28}{3} + \frac{-15(2)}{-6(3)}$.

d) $\frac{8}{-9} - \frac{11(4)}{6} + \frac{-12(-4)}{-6(2)}$.

e) $\left(\frac{4}{5} \times \frac{10}{2}\right) - \left(\frac{6}{7} + \frac{5}{14}\right)$.

f) $12 + (33 \div 10)^2$.

g) $12 + 3(9 - 4)^2 - (4 - 1)(13 - 10)$.

h) $9 \times 2 + (6 - 4)(7 - 4)^2$.

i) $12 \div 4 + 2(9 - 5)^2 - (4 - 1)^3$.

j) $2[(4 + 3)^2(9 - 4)^2 + 6] - (2 + 3)(5 - 1)^2$.

k) $8[(5 - 3)^2(18 - 14)^2 + 6](14 - 11) - (4 + 2)^2(5 - 1)^2$.

l) $3[(4 - 2)^3(9 - 4)^2 + \frac{8 - 5}{(5 - 2)(2 + 1)}] - (2 + 3)(42 - 36)^2$.

m) $15 \div (2 + 3) - 6(16 - 6)^3(11 - 6)(13 - 10)$.

n) $[6 \div 3 + 4 - \left(\frac{2}{3} + \frac{5}{4} - \frac{6}{15}\right)](12 - 8)^3$.

o) $\frac{(-2^3)(2^3 - 5)}{3^3 - 88 + (7 - 4)^2(9 - 7)^3} + \frac{(7 - 4)^3(23 - 21)^2}{2^4 - 5}$.

10. Evaluate $p^2(q^3 + 16)$ where $p = 2$ and $q = 4$.

11. Given $\frac{x}{z} + zx(5 - zx)\frac{3x}{6}$ where $x = -8$ and $z = -2$. Replace all x's with -8 and z's with -2.

12. Simplify the following expression where $x = 1$ and $y = -1$.

a) $5x^2 - 4x + 3y - x^2 - 6x + 5y$.

b) $2x^2 + 3x + 4 - y^2 + 3y - x^2 - 7x + 3y^2 + 4y - 3$.

c) $(\frac{x^2 - 2}{6} - \sqrt{17 - 8} + \frac{y^2 - 1}{3} - \sqrt{9}$.

d) $12 - (3x - 2y + 4) + 3(2x + y)(x + 5)$.

13. For the following exercises, simplify the given expression. Write answers with positive exponents.

a) $\frac{a^3a^2}{a}$, where $a = 17$.

b) $\frac{m^3n^2}{m^{-2}}$, where $m = 5$ and $n = 10$.

c) $\left(\frac{x^{-3}}{y}\right)^{-5}$, where $x = 3$ and $y = 27$.

d) $(w^0x^5)^{-1}$, where $x = 6$.

e) $\frac{m^4}{n^0}$, where $m = 4$.

f) $y^{-4}(y^2)^3$, where $y = 2$.

g) $\frac{m^3n^4}{m^{-2}n^{-5}}$, where $m = 2$ and $n = 3$.

h) $\left(\frac{x^{-3}}{y}\right)^{-5} \times \left(\frac{x^3}{y^4}\right)$, where $x = 5$ and $y = 2$.

14. Simplify the exponential expression. Assume that variables represent nonzero real numbers.

a) $\left(\frac{9x^{-4}y^{-3}z^5}{5x^2y^{-3}z^{-2}}\right)^{-2}$.

b) $\left(\frac{5uv^4}{7w}\right) \div \left(\frac{25u^2v^3}{14\ ^{2}}\right)$.

c) $(3a^2b^3)\left(\frac{4a^2b^2c^4}{9a^2c^5}\right)^{-1}$.

d) $\left(\frac{36\ ^{2}}{17hk^2t^5}\right) \div \left(\frac{2k^2h^2t^4}{5h^5}\right)(34k^2t^3)\left(\frac{2}{5t^2}\right)^2$.

15. Evaluate each expression when $x = -1$ and $y = 2$.

a) $(x - 4y) \times (2x + 5y)^{-2} \times 12\,(1 - 2x)$.

b) $\frac{4(7x + 2y^2)}{(y + 3)(x^3 - 1 + 3y)} + \frac{(23 + 17x)^2}{y^2 + 2x} - 3x$.

c) $\frac{x^3 - 3y}{x^5} - \sqrt{y + x} + \frac{x^2 - 4y}{3\sqrt[3]{4y + 9x}} - \frac{1}{12}\sqrt{9y + 2x}$.

d) $-\frac{\sqrt[4]{3x + 42y}}{45x} + \frac{\sqrt{2x + 9y}}{25} - \frac{\sqrt[3]{3x - 12y}}{5x - 5y}$.

e) $\left(\frac{8x^2y}{12x + 5y^2}\right)(8x^3)^2 \div \left(\frac{36}{9y^2}\right)\left(\frac{2x}{5y}\right)^{-2}$.

3. Divisibility Rules

Divisibility rules help us to determine if a number is divisible by another without going through the actual division process such as the long division method. In this section, we look at the divisibility rules for 2, to 10. Note that every number is divisible by 1. Hence, divisibility rule for 1 does not have any condition. Any number divided by 1 will give the number itself, irrespective of how large the number is. For example, 3 is divisible by 1 and 3000 is also divisible by 1 completely. Additionally, if the numbers in question are numerically small enough, we may not need to use the rules to test for divisibility. However, for numbers whose values are large enough, we want to have some rules to serve as "shortcuts" to help us figure out if they are indeed divisible by each other.

Let us first present the definition of divisible. A number a is divisible by the number b if $a \div b$ has a remainder of zero or "divisible by" means "when you divide one number by another the result is a whole number". For example, 14 divided by 7 is exactly 2 which implies that its remainder is zero (the result is a whole number), while 15 is not divisible by 7, because $15 \div 7 = 2.14$ (the result is not a whole number). A divisibility rule is a shorthand way of determining whether a given integer is divisible by a fixed divisor without performing the division, usually by examining its digits.

3.1. Divisibility Rule for 2

If a number is even or a number whose last digit is an even number, i.e.: 0, 2, 4, 6, 8, it is always completely divisible by 2. For example, 3,206 is an even number and is divisible by 2 but 3,205 is not an even number, hence it is not divisible by 2. The procedure to check whether 3,206 is divisible by 2 or not is as follows:

- Take the number 3,206.
- Take the last digit 6 and divide it by 2.
- The last digit, 6, is divisible by 2.
- Therefore, the number 3,206 is divisible by 2.

Example
Check if the numbers 534, 2,348, 5,789, 912, and 75 are divisible by 2.

Solution
Consider all three numbers, 534, 2,348, and 912. The last digit of each of three numbers, is an even number. Hence, 534, 2,348, and 912 are divisible by 2.

Now consider 5,789, and 75. The last digit of each of two numbers (9 qnd 5), is NOT an even number. Hence, 5,789, and 75. are NOT divisible by 2.

3.2. Divisibility Rule for 3

A number is divisible by 3 if the sum of the digits of the number is divisible by 3. For example, 38,214 is completely divisible by 3 because the sum of its digits, $3 + 8 + 2 + 1 + 4 = 18$ and the sum is divisible by 3, $\frac{18}{3} = 6$. But, 38,213 is NOT divisible by 3, since the sum of its digits, $3 + 8 + 2 + 1 + 3 = 17$, it is NOT divisible by 3, $\frac{17}{3} = 5.67$.

Example
Check if the numbers 6,437, 357, 21,789, 972, and 63 are divisible by 3.

Solution
Consider all four numbers, 357, 21,789, 972, and 63. The sum of digits for each of four numbers is divisible by 3 and all the four numbers are divisible by 3, as seen below.

For 357: $3 + 5 + 7 = 15$, where $\frac{15}{3} = 5$.

For 21,789: $2 + 1 + 7 + 8 + 9 = 27$, where $\frac{27}{3} = 9$.

For 972: $2 + 7 + 9 = 18$, where $\frac{18}{3} = 6$.

For 63: $6 + 3 = 9$, where $\frac{9}{3} = 3$.

Now consider 6,437. The sum of its digits, $6 + 4 + 3 + 7 = 20$ is NOT divisible by 3, $\frac{20}{3} = 6.67$. Hence, 6,437 is NOT divisible by 3.

3.3. Divisibility Rule for 4

A number is divisible by 4 if the last two digits of the number are divisible by 4 or all two last digits are zeros. For example, 9,724 is completely divisible by 4 because the last two-digits, 24 is divisible by 4, $\frac{24}{4} = 6$. But 9,725 is NOT divisible by 4, since the last two-digit, 23 is NOT divisible by 4, $\frac{23}{4} = 5.75$.

Example
Which of the six numbers 102,756, 61,432, 8,565, 3,878, 4,182 and 146 is divisible by 4?

Solution
Consider all three numbers, 102,756, 61,432, and 148. The last two digits for 102,756, 61,432, and 148 are 56, 32, and 48, respectively. Each of these numbers is divisible by 4, because each of their last two digits are divisible by 4, as shown below.

For 102,756: the last two digits are 56, where $\frac{56}{4} = 14$.

For 61,432: the last two digits are 32, where $\frac{32}{4} = 8$.

For 148: the last two digits are 48, where $\frac{48}{4} = 12$.

Now consider 8,565, 3,878, and 4,182. The last two digits for 8,565, 3,878, and 4,182 are 65, 78, and 82, respectively. Each of these numbers is NOT divisible by 4, because each of their last two digits are NOT divisible by 4, as shown below.

For 8,565: the last two digits are 65, where $\frac{65}{4} = 16.25$.

For 3,878: the last two digits are 78, where $\frac{78}{4} = 19.5$.

For 4,182: the last two digits are 82, where $\frac{82}{4} = 20.5$.

3.4. Divisibility Rule for 5

A number is divisible by 5 if the last digit of the number is 0 or 5. For example, 7,165 is completely divisible by 5 because its last digit is 5. Also, 11,430 is divisible by 5 because its last digit is 0. But neither 7,164 nor 11,429 are NOT divisible by 5 since the last digit of 7,164 is neither 5 nor 0, and similarly, the last digit of 11,429 is neither 5 nor 0.

Example
Which of the four numbers 73,530, 9,852, 1,135, and 580 is divisible by 5?

Solution
Consider all three numbers, 73,530, 1,135, and 580. The last digit for each of 73,530, and 580, is 0 and hence they both are divisible by 5. Also, the last digit of 1,135 is 5 and hence this number is divisible by 5 as well.

Considering 9,852, we see that its last digit is neither 5 nor 0. So, 9,852 is NOT divisible by 5.

3.5. Divisibility Rule for 6

A number is divisible by 6 if the number is divisible by both 2 and 3. For example, 738 is completely divisible by 6 because it is divisible by 2 since its last digit is 8, an even number, and it is also divisible by 3 since the sum of its digits is 18, which $\frac{18}{3} = 6$ is divisible by 3. Considering 736 we realize that it is NOT divisible by 6, because it is divisible by 2, since its last digit is 6, an even number, but it is not divisible by 3 since the sum of its digits $7 + 3 + 6 = 16$ is not divisible by 3, $\frac{16}{3} = 5.33$.

Similarly, consider 737. It is NOT divisible by 6, because it is not divisible by 2, since its last digit is 7, which is not an even number, additionally, it is not divisible by 3 since the sum of its digits $7 + 3 + 7 = 17$ is not divisible by 3, $\frac{17}{3} = 5.67$ though the latter divisibility was irrelevant in this case.

Example
Which of the four numbers 5,274, 2,352, 1,236, and 892 is divisible by 6?

Solution
Consider the three numbers, 5,274, 2,352, and 1,236. The last digit for each of 5,274, 2,352, and 1,236, is even and hence each is divisible by 2. The sum of digits for each of 5,274, 2,352, and 1,236, is $5 + 2 + 7 + 4 = 18$, $2 + 3 + 5 + 2 = 12$, and $1 + 2 + 3 + 6 = 12$, respectively which each is divisible by 3 since for 5,274, we have $\frac{18}{3} = 6$, for 2,352, we have $\frac{12}{3} = 4$, and for 1,236 we have $\frac{12}{3} = 4$. Since every of these three numbers is divisible both by 2 and 3, they are divisible by 6.

Now considering 892, we see that although it is divisible by 2, the last digit, 2 is even. but it is not divisible by 3 since the sum of its digits is, $8 + 9 + 2 = 19$, which is not divisible by 3, $\frac{19}{3} = 5.33$. Hence 892 is NOT divisible by 6 because it is NOT divisible by both 2 and 3.

3.6. Divisibility Rules for 7

The rule for divisibility by 7 is a little more complicated. The following steps describe how we can find a number is divisible by 7 or not.

Step 1 Take the last digit of the number and drop it from the original number.

Step 2 Double the dropped digit.

Step 3 Subtract it from the new number formed by removing the last digit of the original number.

Step 4 Repeat the above steps until the number is reduced to two digits.

Step 5 If the two-digit number is divisible by 7, then the original number is divisible by 7. Otherwise, it is not.

Example
Which of the three numbers 3,024, 2,352, and 1,563 is divisible by 7?

Solution
Consider 3,024. We need to follow the steps to find whether this number is divisible by 7 or not. So,

3,024	
302	we dropped the last digit
$4 \times 2 = 8$	we double the dropped digit
$302 - 8 = 294$	we subtracted the doubled digit from the new number formed by removing the last digit
294	the remaining number is not a two-digit number. Hence, we must repeat the process
29	we dropped the last digit
$4 \times 2 = 8$	we double the dropped digit
$29 - 8 = 21$	we subtracted the doubled digit from the new number formed by removing the last digit
21	we have reached a two-digit number
$21 = 3 \times 7$	we see that the two-digit number is 3 times 7.

Therefore, 3,024 is divisible by 7.

Now consider 2,352. Again, we need to follow the steps to find whether this number is divisible by 7 or not. So,

2,352	
235	we dropped the last digit
$2 \times 2 = 4$	we double the dropped digit
$235 - 4 = 231$	we subtracted the doubled digit from the new number formed by removing the last digit
231	the remaining number is not a two-digit number. Hence, we must repeat the process
23	we dropped the last digit

$1 \times 2 = 2$ — we double the dropped digit

$23 - 2 = 21$ — we subtracted the doubled digit from the new number formed by removing the last digit

21 — we have reached a two-digit number

$21 = 3 \times 7$ — we see that the two-digit number is 3 times 7.

Therefore, 2,352 is divisible by 7.

We are left to check 1,563. Like the above, we need to follow the steps to find whether this number is divisible by 7 or not. So,

1,563

156 — we dropped the last digit

$3 \times 2 = 6$ — we double the dropped digit

$156 - 6 = 150$ — we subtracted the doubled digit from the new number formed by removing the last digit

150 — the remaining number is not a two-digit number. Hence, we must repeat the process

15 — we dropped the last digit

$0 \times 2 = 0$ — we double the dropped digit

$15 - 0 = 15$ — we subtracted the doubled digit from the new number formed by removing the last digit

15 — we have reached a two-digit number

$\frac{15}{7} = 2.14$ — we see that the two-digit number is not divisible by 7.

Therefore, 2,563 is NOT divisible by 7.

3.7. Divisibility Rules for 8

A number is divisible by 8 If the last three digits of a number are zeros or if the last three digits is divisible by 8. For example, 4,704 is completely divisible by 8 because the last three-digit, 704 is divisible by 8, $\frac{704}{8} = 88$. But 4,705 is NOT divisible by 8, since the last three-digit, 705 is NOT divisible by 8, $\frac{705}{8} = 88.125$.

Example
Which of the four numbers 23,256, 52,328, 480, and 18,574 is divisible by 8?

Solution
Consider the first number 23,256. The last three-digit is 256, which is divisible by 8, $\frac{256}{8} = 32$.

Consider the second number 52,328. The last three-digit is 328, which is divisible by 8, $\frac{328}{8} = 41$.

Consider number 480. It has three digits. Since $\frac{480}{8} = 60$, it is divisible by 8.

Finally, consider 18,574. The last three-digit is 574, which is not divisible by 8, since $\frac{574}{8} = 71{,}75$.

3.8. Divisibility Rule for 9

A number is divisible by 9 if the sum of the digits is divisible by 9. The rule for divisibility by 9 is like divisibility rule for 3. For example, 2,295 is divisible by 9 because the sum of its digits, $2 + 2 + 9 + 5 = 18$, is divisible by 9, $\frac{18}{9} = 2$. But 2,296 is NOT divisible by 9 since the sum of its digits, $2 + 2 + 9 + 6 = 19$, which is NOT divisible by 9, $\frac{19}{9} = 2.11$.

Example
Which of the four numbers 73,152, 928,143, 7,839, and 6,107 is divisible by 9?

Solution
Let us consider the first number, 73,152.

Sum of its digits: $7 + 3 + 1 + 2 + 5 = 18$

18 is divisible by 9, $\frac{18}{9} = 2$.

Therefore, 73,152 is divisible by 9.

Considering the second number, 928,143.

Sum of its digits: $9 + 2 + 8 + 1 + 4 + 3 = 27$

27 is divisible by 9, $\frac{27}{9} = 3$.

Therefore, 928,143 is divisible by 9.

Considering the third number, 7,839.

Sum of its digits: $7 + 8 + 3 + 9 = 27$

27 is divisible by 9, $\frac{27}{9} = 3$.

Therefore, 7,839 is divisible by 9.

Finally, considering the third number, 6,107.

Sum of its digits: $6 + 1 + 0 + 7 = 14$

14 is not divisible by 9, $\frac{14}{9} = 1.56$.

Therefore, 6,107 is not divisible by 9.

3.9. Divisibility rule for 10

A number is divisible by 10 if its last digit is 0. For example, number 2,630 is completely divisible by 10 because its last digit is 0. But the number 2,631 is NOT divisible by 10 since its last digit, 1, is NOT 0.

Example
Which of the three numbers 340, 48,685, 654,348, and 1,735,320 is divisible by 10?

Solution

- For 340, its last digit is 0. So, it is divisible by 10.
- For 48,685, its last digit is 5. So, it is NOT divisible by 10.
- For 654,348, its last digit is 5. So, it is NOT divisible by 10.
- For 1,735,320, its last digit is 0. So, it is divisible by 10.

The table below summarizes the rules of divisibility.

A number is divisible by	Definition	Example
2	The last digit is an even number.	2,458 8 is divisible by 2
3	The sum of the digits is divisible by 3.	123 $1 + 2 + 3 = 6$ 6 is divisible by 3
4	The last two digit form a number that is divisible by 4.	4,524 24 is divisible by 4
5	The last digit is either a 5 or a 0 (zero).	12,390 or 3,475 both 0 and 5 are divisible by 5
6	The number is divisible by BOTH 2 and 3.	24 24 is divisible by BOTH 2 and 3
7	You double the last digit and subtract the sum from the rest of the number and set an answer that is divisible by 7.	672 $2 \times 2 = 4$ and $67 - 4 = 63$. So 63 is divisible by 7
8	The last three digits from the number that is divisible by 8.	1,816 816 is divisible by 8
9	The sum of all the digits is divisible by 9.	153 $1 + 5 + 3 = 9$ 9 is divisible by 9
10	The number ends in a 0 (zero).	257,890 0 (zero) is divisible by 10

Now let us look at some examples for the divisibility of numbers in which more than one digit is missing.

Example
For what values of a and b the number $\overline{6{,}3ab}$ is divisible by both 4 and 5?

Solution
The divisibility rule of 5 tells us that the last digit is either 0 or 5. Hence, $b = 0$ or $b = 5$. Also, the divisibility rule of 4 tells us that the last two-digit is divisible by 4. Since a two-digit number that is divisible by 4 must be an even number, b cannot be 5 and hence b =0. The two-digit numbers divisible by 4 are, 00, 20, 40, 60, and 80. Therefore, a can take 0, 2, 4, 6, and 8. Hence below shows the five distinct solutions:

$\overline{6,3ab} = 6,300.$

$\overline{6,3ab} = 6,320.$

$\overline{6,3ab} = 6,340.$

$\overline{6,3ab} = 6,360.$

$\overline{6,3ab} = 6,380.$

Example

For what values of a and b the number $\overline{72,5ab}$ is divisible by both 4 and 10?

Solution

The divisibility rule of 10 tells us that the last digit is 0. Hence the last digit, b, must be 0, $b = 0$. Also, the divisibility rule of 4 tells us that the last two-digit is divisible by 4. The two-digit numbers divisible by 4 are 20, 40, 60, and 80.

Therefore, a can takes 2, 4, 6, and 8. Hence below shows the four distinct solutions:

$\overline{72,5ab} = 72,520.$

$\overline{72,5ab} = 72,540.$

$\overline{72,5ab} = 72,560.$

$\overline{72,5ab} = 72,580.$

Example

For what values of a and b the number $\overline{6,a8b}$ is divisible by both 3 and 5?

Solution

The divisibility rule of 5 tells us that the last digit is either 0 or 5. Hence, $b = 0$ or $b = 5$. To find a, we need to consider each of the values of b separately. So,

For $b = 0$, we will have:

we learned a number is divisible by 3 if the sum of the digits of the number is divisible by 3. Hence, considering $\overline{6,a80}$ the sum of its digits is $6 + a + 8 + 0 = 14 + a$, must be divisible by 3. Since $0 \leq a \leq 9$, for $14 + a$ to be divisible by 3 is:

$14 + a = 15$

$14 + a = 18$

$14 + a = 21.$

For $14 + a = 15$, we will have:

$14 + a = 15$

$14 + a - 14 = 15 - 14$ to isolate a, we add -14 to both sides

$a = 15 - 14$

$a = 1.$

For $14 + a = 18$, we will have:

$14 + a = 18$

$14 + a - 14 = 18 - 14$ to isolate a, we add -14 to both sides

$a = 18 - 14$

$a = 4.$

For $14 + a = 21$, we will have:

$14 + a = 21$

$14 + a - 14 = 21 - 14$ to isolate a, we add -14 to both sides

$a = 21 - 14$

$a = 7.$

Therefore, for $\overline{6,a80}$, the values that a must take is 1, 4, and 7. Therefore, the number $\overline{6,a80}$ can be either 6,180, 6,480, or 6,780.

For $b = 5$, we will have:
Again, we know a number is divisible by 3 if the sum of the digits of the number is divisible by 3. Hence, considering $\overline{6,a85}$ the sum of its digits is $6 + a + 8 + 5 = 19 + a$, must be divisible by 3. Since $0 \leq a \leq 9$, for $19 + a$ to be divisible by 3 is:

$19 + a = 21$

$19 + a = 24$

$19 + a = 27.$

For $19 + a = 21$, we will have:

$19 + a = 21$

$19 + a - 19 = 21 - 19$ to isolate a, we add -19 to both sides

$a = 21 - 19$

$a = 2.$

For $19 + a = 24$, we will have:

$19 + a = 24$

$19 + a - 19 = 24 - 19$ to isolate a, we add -19 to both sides

$a = 24 - 19$

$a = 5.$

For $19 + a = 27$, we will have:

$19 + a = 27$

$19 + a - 19 = 27 - 19$ to isolate a, we add -19 to both sides

$a = 27 - 19$

$a = 8.$

Therefore, for $\overline{6,a85}$, the values that a must take is 2, 5, and 8. Hence the number $\overline{6,a85}$ can be 6,285, 6,585, or 6,885.

Divisibility Rules

For 2	The last digit is an even number.
For 3	Sum of the digits is divisible by 3.
For 4	The last two digit is divisible by 4 or it has zeros in two last digits.
For 5	The last digit is either a 0 or a 5.
For 6	The number is divisible by both 2 and 3.
For 7	The difference between twice the unit digit of the number and its remaining part of the number should be multiple of 7 or to be 0.
For 8	The last three digits is divisible by 8 or be three zeros.
For 9	Sum of all the digits is divisible by 9.
For 10	The number ends in a 0 (zero).

3.10. Exercises

1. Which of the following eight numbers is divisible by 2?
   a) 37.
   b) 245.
   c) 1,234.
   d) $\overline{3a,bc8}$.
   e) $\overline{54,ab5}$.
   f) $\overline{aa,bc9,140}$.
   g) $\overline{1,24a,bc4}$.
   h) $\overline{3,abc,dea,be2}$.

2. Which of the eight numbers is divisible by 3?
   a) 255.
   b) 975.
   c) 1,234.
   d) 32,643.
   e) 234,179.
   f) 2,861,777.
   g) 5,481,369.
   h) 1,431,834,327.

3. Which of the eight numbers is divisible by 4?
   a) 354.
   b) 692.
   c) $\overline{a,b86}$.
   d) $\overline{ab,c48}$.
   e) $\overline{abc,d56}$.
   f) $\overline{3,aab,c88}$.
   g) $\overline{451,abb,b08}$.
   h) $\overline{7,abc,dab,b76}$.

4. Which of the six numbers is divisible by 5?
   a) 165.
   b) 518.
   c) $\overline{a,b70}$.
   d) $\overline{ab,b85}$.

e) $\overline{abc,c50}$.

f) $\overline{abc,dac,dd5}$.

5. Which of the eight numbers is divisible by 6?

a) 552.

b) 426.

c) 8,324.

d) 45,693.

e) *abc*,176.

f) 2,861,436.

g) 36,937,362.

h) 27,567,834,342.

6. Which of the eight numbers is divisible by 7?

a) 294.

b) 1,018.

c) 2,016.

d) 7,560.

e) 6,762.

f) 4,563.

g) 7,918.

h) 28,392.

7. Which of the eight numbers is divisible by 8?

a) 6,104.

b) 53,000.

c) $\overline{abc,016}$.

d) $\overline{a,add,568}$.

e) $\overline{a,bcd,874}$.

f) $\overline{ab,bcc,867}$.

g) $\overline{aab,cdd,008}$.

h) $\overline{a,bbb,ccd,064}$.

8. Which of the eight numbers is divisible by 9?

a) 7,479.

b) 60,345.

c) 445,419.

d) 4,023,885.

e) 8,241,653.

f) 43,518,327.

g) 639,819,099.

h) 2,326,563,063.

9. Which of the six numbers is divisible by 10?

a) 2,145,870.

b) 75,842,860.

c) $\overline{aa,bcc,ad5}$.

d) $\overline{ab,bcd,ab0}$.

e) $\overline{baa,acc,a30}$.

f) $\overline{a,bab,dcc,aa0}$.

10. For what values of a and b the number $\overline{6{,}7ab}$ is divisible by 6?

11. For what values of a and b the number $\overline{5,a2b}$ is divisible by 9?

12. For what values of a and b the number $\overline{14{,}8ab}$ is divisible by both 4 and 5?

13. For what values of a and b the number $\overline{a,27b}$ is divisible by both 3 and 4?

14. Which of the three numbers $\overline{73,abc,336}$, $\overline{ab,216}$, and $\overline{aa,354}$ is divisible by 8?

15. For what values of a and b the number $\overline{23,abb}$ is divisible by both 8 and 5?

4. Arithmetic and Geometric Series

A sequence is a set of numbers in a specific order. The sum of any sequence of numbers is called a series. The two simplest sequences to work with are arithmetic and geometric sequences. Arithmetic and geometric sequences have many interesting mathematical properties. In this section we look at their formulas and their mathematical properties.

4.1. Arithmetic Sequences

A sequence in which the difference between consecutive terms is constant is an arithmetic sequence. The difference between consecutive terms in an arithmetic sequence, $a_k - a_{k-1} = d$, is the common difference, where $k \geq 2$.

Examples of an arithmetic sequence are presented below:

a) 5, 9, 13, 17, 21, 25, ⋯   9−5   13−9   17−13   21−17 ⋯ $a_k - a_{k-1}$

4 4 4 4 4   4   4   4   4 $d = 4$

b) 6, 13, 20, 27, 34, 41, ⋯   13−6   20−13   27−20   34−27 ⋯ $a_k - a_{k-1}$

7 7 7 7 7   7   7   7   7 $d = 7$

c) 11, 8, 5, 2, -1, -4, ⋯   11 − 8   8 − 5   5 − 2   − 1− 2 ⋯ $a_k - a_{k-1}$

−3−3−3−3−3   − 3   − 3   − 3   − 3 $d = -3$.

Example
Designate each sequence which is an arithmetic sequence, and if so, indicate its common difference.

a) 2, 8, 14, 20, 26, 32, ⋯

b) 5, 9, 12, 15, 19, 23, ⋯

c) 13, 8, 3, − 2, − 7, − 12, ⋯.

Solution
Determine the difference of the consecutive terms of each sequence.

a) 8 − 2,   14 − 8   20 − 14, 26−20, 32 − 26
6   6   6   6   6 ⟹ an arithmetic sequence with $q = 6$.

b) 9 − 5,   12 − 9,   15 - 12   19 - 15   23 − 19
4   3   3   4   4 ⟹ it is not an arithmetic sequence.

c) 8 −13   3 − 8   − 2 − 3   −7− (−2)   −12− (−7)
− 5   − 5   − 5   − 5   − 5 ⟹ an arithmetic sequence with $d = -5$.

Example
Write the first four terms for each sequence, with the given the first term and the given d.

a) $a_1 = 5$, and $d = 12$.

b) $a_1 = 9$, and $d = -2$.

c) $a_1 = 17$, and $d = -4$.

d) $a_1 = -9$, and $d = 5$.

Solution

To find the answer of each sequence, we must start with the first term and add *d*. Then we add *d* to that result to get the next term, and so on.

a)

a_1	a_2	a_3	a_4
5	$5 + 12 = 17$	$17 + 12 = 29$	$29 + 12 = 41$
5	17	29	41.

b)

a_1	a_2	a_3	a_4
9	$9 + (-2) = 7$	$7 + (-2) = 5$	$5 + (-2) = 3$
9	7	5	3.

c)

a_1	a_2	a_3	a_4
17	$17 + (-4) = 13$	$13 + (-4) = 9$	$9 + (-4) = 5$
17	13	9	5.

d)

a_1	a_2	a_3	a_4
-9	$-9 + 5 = -4$	$-4 + 5 = 1$	$1+5 = 6$
-9	-4	1	6.

4.1.1.The nth Term of an Arithmetic Sequence

To find a formula for the general term of an arithmetic sequence, let us write the first few terms, assuming the first term is a_1 and the common difference is *d*. Through this we will find the pattern and can find the formula.

1	2	3	4	5
a_1	$a_1 + d$	$a_1 + d + d$	$a_1 + d + d + d$	$a_1 + d + d + d + d$
a_1	$a_1 + d$	$a_1 + 2d$	$a_1 + 3d$	$a_1 + 4d.$

The above pattern reveals that $0d$ is added to the first term, $1d$ is added to the second term, $2d$ is added to the third term, $3d$ is added to the fourth term, $4d$ is added to the fifth term. Hence, we can conclude that the number of *d*s that were added to k^{th} is one less than the number of the term. This leads us to the following formula. $a_n = a_1 + (n - 1)\, d.$

$$a_n = a_1 + (n - 1)\, d.$$

For example, if $a_1 = 7$ and $d = 3$, the 11th term can be attained by:

$a_n = a_1 + (n - 1)\, d$

$a_{11} = 7 + (9) \times 3$

$a_{11} = 7 + 27$

$a_{11} = 34.$

In some cases, we may ask to find the common difference where two distinct terms which are given. For example, if we are asked to find the common difference where a_7 and a_{10} given, it is obvious that we must add d three times (the difference of the 10th term and 7th term) to a_7 for reaching to a_{10}. In other word,

$a_8 = a_7 + d$

$a_9 = a_8 + d$

$a_9 = a_7 + d + d$ — a_8 is substituted by $a_8 = a_7 + d$

$a_{10} = a_9 + d$

$a_{10} = a_7 + d + d + d$ — a_9 is substituted by $a_9 = a_7 + d + d$

$a_{10} = a_7 + 3d.$

From the above observation, a more general formula can be deducted. That is, $a_h = a_k + (h - k)d$, given $h > k$. therefore, $d = (a_h - a_k)/(h - k)$.

$$d = \frac{a_h - a_k}{h - k}.$$

Example

Find the requested case using other given information.

a) $a_4 = 10$ and $d = 3$, find a_8.
b) $a_3 = 2$ and $a_6 = 17$, find d.
c) $a_3 = 5$ and $a_8 = 30$, find a_1.

Solution

a) $a_n = a_1 + (n - 1)\,d$

$10 = a_1 + (4 - 1) \times 3$ — we plugged $a_4 = 10$ $n = 4$, and $d = 3$ into the equation

$10 = a_1 + (3) \times 3$

$10 = a_1 + 9$

$10 - 9 = a_1 + 9 - 9$ — we added -9 to both sides

$10 - 9 = a_1$

$a_1 = 1.$

To find a_8 we need to use the formula, $a_n = a_1 + (n - 1)\,d$. So,

$a_n = a_1 + (n - 1)\,d$

$a_8 = 1 + (8 - 1) \times 3$ — we plugged $a_1 = 1$, $n = 8$, and $d = 3$ into the equation

$a_8 = 1 + (7) \times 3$

$a_8 = 1 + 21$

$a_8 = 22.$

b) To find d, we need to use the formula $d = \frac{a_h - a_k}{h - k}$. So,

$d = \frac{a_h - a_k}{h - k}$

$d = \frac{17 - 2}{h - k}$ we plugged $a_h = a_6 = 17$, and $a_k = a_3 = 2$ into the equation.

Since $a_h = a_6$, and $a_k = a_3$ we will have $h = 6$, and $k = 3$. So,

$d = \frac{17 - 2}{6 - 3}$ we plugged $h = 6$, and $k = 3$ into the equation

$d = \frac{15}{3}$

$d = 3$.

c) To find a_1, we first need to determine d.

$d = \frac{a_h - a_k}{h - k}$

$d = \frac{30 - 5}{h - k}$ we plugged $a_h = a_8 = 30$, and $a_k = a_3 = 5$ into the equation.

Since $a_h = a_8$, and $a_k = a_3$ we will have $h = 8$, and $k = 3$. So,

$d = \frac{30 - 5}{8 - 3}$ we plugged $h = 8$, and $k = 3$ into the equation

$d = \frac{25}{5}$

$d = 5$.

Now we can use either a_3 or a_8 to obtain a_1. So,

$a_3 = a_1 + (n - 1) \times d$

$5 = a_1 + (n - 1) \times 5$ we plugged $a_3 = 5$, and $d = 5$, into the equation.

Since the index of a_3 is 3 and the index of a_1 is 1, their difference determines the value of n. Hence $n = 3 - 1 = 2$. So,

$5 = a_1 + (n - 1) \times 5$

$5 = a_1 + (2 - 1) \times 5$ we plugged $n = 2$, into the equation

$5 = a_1 + (1) \times 5$

$5 = a_1 + 5$

$5 - 5 = a_1 + 5 - 5$ we added -5 to both sides

$0 = a_1$.

Therefore, $a_1 = 0$.

4.1.2. Sum of the First *n* Terms of an Arithmetic Sequence

It is every so often useful to find a formula for finding the sum of an arithmetic sequence. Here we show how this formula including only the first and the last term can be developed. The sum of the first n terms of any arithmetic sequence is written as:

$S_n = a_1 + a_2 + a_3 + \cdots + a_n.$

Using the common difference, d, we can rewrite the sum as:

$$S_n = a_1 + (a_1 + d) + (a_1 + 2d) + \cdots + a_n \cdots. \qquad (1)$$

Now we may realize that if the last term of an arithmetic sequence is a_n, we simply can have its direct preceding term as $a_{n-1} = a_n - d$. Based on this the above formula can be written as:

$$S_n = a_1 + \cdots + (a_n - 2d) + (a_n - d) + a_n,$$

or reshuffling the terms and rewriting it from the last term to the first term, we have:

$$S_n = a_n + (a_n - d) + (a_n - 2d) + \cdots + a_1 \cdots. \qquad (2)$$

Now, if we add Equation (1) to Equation (2) we will have:

$$S_n = a_1 + (a_1 + d) + (a_1 + 2d) + \cdots + a_n$$

$$+S_n = a_n + (a_n - d) + (a_n - 2d) + \cdots + a_1$$

$$2S_n = (a_1 + a_n) + (a_1 + a_n) + (a_1 + a_n) + \cdots + (a_1 + a_n).$$

In the above equation, there are n sums of $(a_1 + a_n)$ on the right side of the equation, so we rewrite the right side as $n\,(a_1 + a_n)$.

$$2S_n = n\,(a_1 + a_n)$$

$$S_n = \frac{n}{2}(a_1 + a_n).$$

Therefore, the formula for finding sum of an arithmetic sequence in terms of the first term and the last term is:

$$\boxed{S_n = \frac{n}{2}(a_1 + a_n).}$$

Substituting the formula for obtaining the nth term of the arithmetic sequence, $a_n = a_1 + (n - 1)d$, into the above equation, we find the formula below for finding sum of n terms of an arithmetic sequence

$$\boxed{S_n = \frac{n}{2}[2a_1 + (n - 1)d].}$$

Example

Find the sum of the first 36 terms of the arithmetic sequence:

a) 11, 15, 19, 23, 27, $\cdots$.

b) 22, 28, 34, 40, 46, $\cdots$.

c) 47, 44, 41, 38, 35, $\cdots$.

Solution

To obtain the sum of 36 terms we use the formula $S_n = \frac{n}{2}(a_1 + a_n)$. However, since we do not have the 36th term we must first use the formula $a_n = a_1 + (n-1)\,q$ to obtain the last term.

a) Realizing q is equal to 4, $a_{36} = 11 + (36 - 1) \times 4 = 11 + (35 \times 4) = 11 + 140 = 151$.

$$S_{36} = \frac{36}{2}(11 + 151)$$

$$S_{36} = 18\,(162)$$

$$S_{36} = 2916.$$

b) Realizing q is equal to 6, $a_{36} = 22 + (36 - 1) \times 6 = 22 + (35 \times 6) = 22 + 210 = 232$.

$$S_{36} = \frac{36}{2}(22 + 232)$$

$$S_{36} = 18(254)$$

$$S_{36} = 4572.$$

c) Realizing q is equal to -3, $a_{36} = 47 + (36 - 1) \times (-3) = 47 + (35) \times (-3) = 47 + (-105) = -58$.

$$S_{36} = \frac{36}{2}(47 - 58)$$

$$S_{36} = 18\,(-11)$$

$$S_{36} = -198.$$

Example

Find the sum of the first 16 terms of the arithmetic sequence, 3, 8, 13, 18, 23, 28, 33, ⋯.

Solution

To obtain the sum of n terms of the arithmetic sequence, $S_n = \frac{n}{2}[2a_1 + (n - 1)d]$, we first need to find the common difference, d. We have learned the common difference of an arithmetic sequence can be find by a term minus its preceding term. Taking the second term, 8 its preceding term is 3. Hence, $d = 8 - 3 = 5$, in this case.

Now we can use formula $S_n = \frac{n}{2}[2a_1 + (n - 1)d]$ to find the sum of the first 16 terms of the arithmetic sequence with $a_1 = 3$, $n = 16$, and $d = 5$. So,

$$S_n = \frac{n}{2}[2a_1 + (n - 1)d]$$

$$S_{16} = \frac{16}{2}[2(3) + (16 - 1)(5)]$$ we plugged $a_1 = 3$, $n = 16$, and $d = 5$ into the equation

$$S_{16} = 8[6 + (15)(5)]$$

$$S_{16} = 8(6 + 75)$$

$$S_{16} = 8(81)$$

$$S_{16} = 648.$$

Therefore, the sum of the first 16 terms of the arithmetic sequence is $S_{16} = 648$.

4.2. Geometric Sequences

A sequence in which the ratio between consecutive terms is constant is a geometric sequence. The ratio between consecutive terms in a geometric sequence, $a_k / a_{k-1} = r$, is the common ratio, for $k \geq 2$.

Examples of a geometric sequence are presented below:

a) 5, 10, 20, 40, 80, 160, ⋯

×2 ×2 ×2 ×2 ×2

$\frac{10}{5}$ $\frac{20}{10}$ $\frac{40}{20}$ $\frac{80}{40}$ ⋯ $\frac{a_k}{a_{k-1}}$

2 2 2 2 $r = 2.$

b) 6, -12, 24, -48, 96, 192,

(-2) ×(-2) ×(-2) ×(-2) ×(-2)

$-\frac{12}{6}$ $-\frac{24}{12}$ $-\frac{48}{24}$ $-\frac{96}{48}$ ⋯ $\frac{a_k}{a_{k-1}}$

-2 -2 -2 -2 $r = -2.$

Example

Designate each sequence which is a geometric sequence, and if so, indicate its common ratio.

a) 2, 8, 32, 128, 516, 2042, ⋯.

b) 1, 4, 16, 64, 256, 1024, ⋯.

c) 99, 33, 11, $\frac{11}{3}$, $\frac{11}{9}$, $\frac{11}{27}$, ⋯.

Solution

Determine the difference of the consecutive terms of each sequence.

a) $\frac{8}{2}$, $\frac{32}{8}$ $\frac{128}{32}$, $\frac{516}{128}$, $\frac{2042}{516}$.

4 4 4 4.03 ⟹ it is not a geometric sequence.

b) $\frac{4}{1}$ $\frac{16}{4}$ $\frac{64}{16}$ $\frac{256}{64}$ $\frac{1024}{256}$.

4 4 4 4 4 ⟹ it is a geometric sequence with $r = 4$.

c) $\frac{99}{33}$ $\frac{33}{11}$ $\frac{\frac{11}{3}}{11}$ $\frac{\frac{11}{9}}{\frac{11}{3}}$ $\frac{\frac{11}{27}}{\frac{11}{9}}$.

$\frac{1}{3}$ $\frac{1}{3}$ $\frac{1}{3}$ $\frac{1}{3}$ $\frac{1}{3}$ ⟹ it is a geometric sequence with $r = \frac{1}{3}$.

4.2.1. General Term (n^{th} Term) of a Geometric Sequence

As we showed how to obtain a formula for the general term of an arithmetic sequence, we can as well find a formula for the general term of a geometric sequence. Let us write the first few terms of this sequence along with its common ratio to observe its pattern.

a_1 a_2 a_3 a_4 a_5

a_1 ra_1 ra_2 ra_3 ra_4

a_1 ra_1 r^2a_1 r^3a_1 r^4a_1.

Observing the last row of the above pattern, we see that the first terms is simply denoted a_1, the second term is multiplied by r, the third term is multiplied by r^2, the third term is multiplied by r^3, and so on. Based on this pattern, a_1 is multiplied by $r^{k-1}$to obtain the kth term. This conducts us to the following formula to obtain the nth term of a geometric sequence:

$$a_n = r^{n-1} a_1.$$

Example

a) Find the ninth term of a geometric sequence starting with 3 and having the common ratio 2.

b) Find the tenth term of a geometric sequence starting with 64 and having the common ratio $\frac{1}{2}$.

c) Find the eighth term of a geometric sequence starting with 36 and having the common ratio -3.

Solution

Using the formula $a_n = r^{n-1} a_1$, we can fulfill all above three requests.

a) $a_n = r^{n-1} a_1$.

$a_9 = 2^{(9-1)} \times 3$ — we plugged $n = 9$, $r = 2$, and $a_1 = 3$ into the equation

$a_9 = 2^8 \times 3$

$a_9 = 256 \times 3$

$a_9 = 768$.

b) $a_n = r^{n-1} a_1$.

$a_{10} = \left(\frac{1}{2}\right)^{(10-1)} \times 64$ — we plugged $n = 10$, $r = \frac{1}{2}$, and $a_1 = 64$ into the equation

$a_{10} = \frac{1}{512} \times 64$

$a_{10} = \frac{1}{8}$.

c) $a_n = r^{n-1} a_1$.

$a_8 = (-3)^{(8-1)} \times 36$ — we plugged $n = 8$, $r = -3$, and $a_1 = 36$ into the equation

$a_8 = (-3)^7 \times 36$

$a_8 = (-2187) \times 36$

$a_8 = 78{,}732$.

Example

Given the fourth term of a geometric sequence equal to 320, where its first term is 5, find the common ratio of this geometric sequence.

Solution

We have learned that the formula for finding the nth term of a geometric sequence is $a_n = r^{n-1} a_1$, in which a_n is the nth term, a_1 is the first term and r is the common ratio of the geometric sequence. Using this formula, we will have.

$a_n = r^{n-1} a_1$

$a_4 = r^{4-1} a_1$

$320 = r^{4-1} \times 5$ — we plugged $n = 4$, $a_4 = 320$, and $a_1 = 5$ into the equation

$\frac{1}{5} \times 320 = \frac{1}{5}(r^3 \times 5)$ we multiplied both sides by the reciprocal of the variable

$\frac{320}{5} = r^3$

$\frac{5 \times 64}{5} = r^3$

$64 = r^3$

$\sqrt[3]{64} = r$

$\sqrt[3]{4^3} = r$

$r = 4$ we took the radicand out.

Therefore, the common ratio of this geometric sequence is $r = 4$

4.2.2. Sum of the First *n* Terms of a Geometric Sequence

Using the similar procedure that we found the sum of the arithmetic sequence; we can develop a formula for finding the sum of the geometric sequence. The sum of the first n terms of a geometric sequence can be written as $S_n = a_1 + a_2 + a_3 + \cdots + a_n$. Now using the common ration to rewrite this sum we have:

$$S_n = a_1 + ra_1 + ra_2 + \cdots + ra_{n-1} \quad \text{or}$$

$$S_n = a_1 + ra_1 + r^2a_1 + \cdots + r^{(n-1)}a_1. \qquad (1)$$

Multiplying both sides of the above equation by $-r$ we have:

$$-rS_n = -ra_1 - r^2a_1 - r^3a_1 - \cdots - r^n a_1. \qquad (2)$$

Next, we add up the last two equations, Equation (2) and Equation (1).

$$\begin{array}{l} S_n = a_1 + \cancel{ra_1} + \cancel{r^2a_1} + \cdots + \cancel{r^{(n-1)}a_1} \\ -rS_n = -\cancel{ra_1} - \cancel{r^2a_1} - \cdots - \cancel{r^{(n-1)}a_1} - r^n a_1 \\ \hline S_n - rS_n = a_1 - r^n a_1 \\ S_n(1 - r) = a_1(1 - r^n). \end{array}$$

To obtain the formula for S_n, we simply divide both side by $(1 - r)$.

$$\boxed{S_n = \frac{a_1(1-r^n)}{1-r}.}$$

Example

What is the sum of the first 10 terms of the geometric sequence 2, 6, 18, 54, 162, 486, ⋯.

Solution

Knowing it is a geometric sequence, we can easily find r by selecting a term and dividing it by the preceding term, that is, $r = \frac{6}{2} = 3$. Now using the above formula, $S_n = \frac{a_1(1 - r^n)}{1 - r}$, we have:

$$S_n = \frac{a_1(1 - r^n)}{1 - r}$$

$S_{10} = \frac{2(1-3^{10})}{1-3}$ we plugged $a_1 = 2, r = 3$, and $n = 10$ into the equation

$S_{10} = \frac{2(1-59{,}049)}{1-3}$

$S_{10} = \frac{2(-59{,}048)}{-2}$

$S_{10} = \frac{-118{,}098}{-2}$

$S_{10} = \frac{-118{,}098}{-2} \times \frac{-1}{-1}$ we multiplied the right-hand side by $\frac{-1}{-1}$ to eliminate the negative signs

$S_{10} = 59{,}049.$

Therefore, the sum of the first 10 terms of the geometric sequence is $S_{10} = 59{,}049$.

Example
What is the sum of the first 5 terms of the geometric sequence 1, 4, 16, 64, 256, 1024, ⋯.

Solution
Knowing it is a geometric sequence, we can easily find r by selecting a term and dividing it by the preceding term, that is, $r = \frac{16}{4} = 4$. Now using the above formula, $S_n = \frac{a_1(1-r^n)}{1-r}$, we have:

$S_n = \frac{a_1(1-r^n)}{1-r}$

$S_5 = \frac{1(1-4^5)}{1-4}$ we plugged $a_1 = 1, r = 4$, and $n = 5$ into the equation

$S_5 = \frac{1(1-1024)}{-3}$

$S_5 = \frac{-102}{-3}$

$S_5 = \frac{-1023}{-3} \times \frac{-1}{-1}$ we multiplied the right-hand side by $\frac{-1}{-1}$ to eliminate the negative signs

$S_5 = \frac{1023}{3}$

$S_5 = 341.$

Therefore, the sum of the first 5 terms of the geometric sequence is $S_5 = 341$.

4.2.3. Sum of an Infinite Geometric Series

The development of a formula for finding the sum of an infinite geometric series is beyond of this manuscript subjects. Therefore, we give the formula without explaining how we can obtain it.
The sum of an infinite geometric series, S, in which the first term is a_1, and its common ratio is r, is:

$S = \frac{a_1}{1-r}$	if $\lvert r\rvert < 1$
S is not defined	if $\lvert r\rvert \geq 1$.

Example
Find the sum of an infinite geometric series, presented as:

2, 10, 50, 250, 1250, 6250, $\cdots$.

Solution
Since $r = \frac{10}{2} = 5$, since $r > 1$, the sum is not defined.

Example
Find the sum of an infinite geometric series, presented as:

80, 40, 20, 10, 5, $\frac{5}{2}$, $\cdots$.

Solution
Since $r = \frac{40}{80} = \frac{1}{2} < 1$, using the formula for finding the sum of an infinite geometric series is $S = \frac{a_1}{1-r}$, where, $a_1 = 80$, and $r = \frac{1}{2}$. So,

$$S = \frac{a_1}{1-r}$$

$$S = \frac{80}{1-\frac{1}{2}}$$ we plugged $a_1 = 80$, and $r = \frac{1}{2}$, into the equation

$$S = \frac{80}{\frac{1}{2}} = 80 \div \frac{1}{2}$$

$$S = 80 \times 2$$ we used the dividing rational numbers rule

$$S = 160.$$

Therefore, the sum of the infinite geometric series is $S = 160$.

4.3. Exercises

1. Determine which sequence is an arithmetic series, and if so, determine the common difference.
   a) 65, 50, 35, 20, 5, -10, $-25, \cdots$.
   b) 10, 11, 12, 13, 14, 15, $16, \cdots$.
   c) $\frac{2}{3}, 1, \frac{4}{3}, \frac{5}{3}, 2, \frac{7}{3}, \frac{8}{3}, 3, \cdots$.
   d) 1, 6, 11, 16, 20, 25, $32, \cdots$.
   e) 22, 16, 10, 4, -2, -7, $-12, \cdots$.
   f) 17, 21, 25, 29, 33, $37, \cdots$.

2. Determine which sequence is a geometric series, and if so, determine the common ratio.
   a) 3, 12, 48, 192, 768, $3072, \cdots$.
   b) 96, 48, 24, 12, 6, 3, $\frac{3}{2}, \cdots$.
   c) 81, 27, 9, 3, 1, $\frac{2}{3}, \frac{5}{9}, \cdots$.
   d) 1, 5, 10, 50, 250, $1250, \cdots$.
   e) 84, -42, 21, $-\frac{21}{2}, \frac{21}{4}, \frac{21}{8}, \cdots$.
   f) 54, 18, 6, 2, 1, $\frac{1}{3}, \frac{4}{9}, \cdots$.

3. If $a_1 = 3$ and $d = -2$, find the 11th term of the arithmetic series.

4. If $a_1 = \frac{1}{3}$ and $d = 5$, find the 12th term of the arithmetic series.

5. If $a_1 = 2$ and $d = 4$, find the 10th term of the arithmetic series.

6. If $a_1 = 22$ and $d = \frac{1}{2}$, find the 9th term of the arithmetic series.

7. If $a_1 = 88$ and $r = \frac{1}{2}$, find the 7th term of the geometric series.

8. If $a_1 = 36$ and $r = -2$, find the 10th term of the geometric series.

9. If $a_1 = 99$ and $r = 3$, find the 8th term of the geometric series.

10. Given an arithmetic series with the known information, find the requested.
    a) $a_4 = 18$ and $d = 3$, find a_9.
    b) $a_5 = 12$ and $a_9 = 32$, find d.
    c) $a_6 = 36$ and $a_8 = 12$, find a_1.

11. Given a geometric series with the known information, find the requested.
    a) $a_1 = 10$, $r = 2$, find a_6.

b) $a_1 = 3$, $r = \frac{1}{2}$, find a_5.

c) $a_6 = 128$, $r = 2$, find a_1.

d) $a_1 = 2$, $a_5 = 32$, find r.

e) $a_1 = 8{,}192$, $r = -\frac{1}{4}$, find a_4.

12. Find the sum of the first 16 terms of the arithmetic sequence:

a) 4, 12, 20, 28, 36, ⋯.

b) 11, 14, 17, 20, 23, ⋯.

c) 37, 34, 31, 28, 25,⋯.

d) $\frac{1}{2}, 1, 1\frac{1}{2}, 2, 2\frac{1}{2}, \cdots$.

13. Find the sum of the first 12 terms of the geometric sequence:

a) 4, 8, 16, 32, 64,⋯.

b) 64, 32, 16, 8, 4,⋯.

c) $36, -12, 4, -\frac{4}{3}, \frac{4}{9}, \cdots$.

d) $\frac{1}{2}, \frac{1}{10}, \frac{1}{50}, \frac{1}{250}, \frac{1}{1{,}250}, \cdots$.

e) $10{,}000, 1000, 100, 10, 1, \frac{1}{10}, \cdots$.

14. Find the sum of an infinite geometric series.

a) $99, 33, 11, \frac{11}{3}, \frac{11}{9}, \cdots$.

b) 512, 256, 128, 64, 32,⋯.

c) $2{,}916, -972, 324, -108, \cdots$.

d) $\frac{1}{2}, \frac{1}{6}, \frac{1}{18}, \frac{1}{54}, \frac{1}{162}, \cdots$.

e) $1{,}024, 256, 64, 16, 4, 1, \frac{1}{4}, \cdots$.

5. Measurement Units

Measurement is to find a number that shows magnitude of a quantity. A **unit of measurement** is a definite magnitude of a quantity, defined and adopted by convention or by law, that is used as a standard for measurement of the same kind of quantity. Any other quantity of that kind can be expressed as a multiple of the unit of measurement. For example, length is a physical quantity. The meter is a unit of length that represents a definite predetermined length.

The Need for a Common Language

Every field of science involves taking measurements, understanding them, and communicating them to others. In other words, we all must speak the same basic language. Whether we are a chemist, a physicist, a biologist, an engineer, or even a medical doctor, we need a consistent way of communicating size, mass, shape, temperature, time, amount, energy, power, and speed.

In human history, various unit systems were developed and used in different regions and cultures. Currently, the global standard of measurement is the **International System of Units (SI)**, which is a modern form of the metric system. Although SI is intended for global use, it has not been fully adopted, and some other systems of measurement are still used in parts of the world.

5.1. Conversion of units

Conversion of units is the conversion between different units of measurement for the same quantity, typically through multiplicative **conversion factors**.

In mathematics, it is required to convert the units while solving many problems. To carry out the required calculations, mathematical conversions are needed. For example, to find the area of a triangle in the square centimeter, it is necessary to convert the height and the base in centimeters. If a base is given in centimeters, and the height is given in meter, you are asked the area in square centimeters, then you need to convert the height in centimeter. Therefore, it is necessary to learn the conversion of units, to convert a unit into the required unit.

A conversion factor is a number used to change one set of units to another, by multiplying or dividing. When a conversion is necessary, the appropriate conversion factor to an equal value must be used. For example, to convert inches to feet, the appropriate conversion value is 12 inches equal 1 foot. To convert minutes to hours, the appropriate conversion value is 60 minutes equal 1 hour. Some more examples of conversion factors are as follows:

For volume: 1 gallon = 3.78541 litters.

1 cubic centimeter = 1000 cubic millimeters.

For mass: 1 kilogram = 1000 grams

1 pound = 453.592 grams.

For area: 1 square yard = 0.83612736 square meters.

1 acre = 4046.8564224 square meters.

For power: 1 BTU = 0.2520 kilo calories.

1 BTU/hr. = 1.0551 kilo Joules.

Table 5.1. illustrate some of the important units and their associated conversion factors.

Table 5.1. Units and Conversion Factors

Dimension	**American**	**SI**
length	inch/foot/yard/mile	meter
mass	ounce/pound/ton	gram
volume	cup/pint/quart/gallon	liter
force	pound-force	newton
pressure	pound-force per square inch	pascal
work and energy	cal/British thermal unit	joule
temperature	Fahrenheit	kelvin
charge	Faraday	coulomb

The abbreviated factors are defined in Table 5.2.

Table 5.2. Abbreviations and Names of SI and English Units

Abbreviation	Unit name	Abbreviation	Unit name
atm	atmosphere	*g*	gram
bbl	barrel	*kg*	kilogram
btu	British thermal unit	*ton*	metric ton
bu	bushel	*ml*	milliliter
cwt	hundredweight	*L*	liter
cal	calorie	m^3	cubic meter
°F	degree Fahrenheit	*in*	inch
ft	foot	*lb*	pound
gal	gallon	*lbf*	pound force
hp	horsepower	*ln tn*	long ton
°C	degree Centigrade	*pdl*	poundal
g	gram	*sh tn*	short ton
kg	kilograms	*yd*	yard
mm	millimeter	*in/s*	inch per second
cm	centimeter	in^2	square inch
m	meter	in^3	cubic inch
km	kilometer	*J*	joules
cm^2	square centimeter	*N*	newton
m^2	square meter	*W*	watt

Table 5.3. presents the conversion multiplication factors to convert from English units to SI units (and vice versa) for basic measurements of length.

Table 5.3. Length conversion of common units

	SI System				US Standard Unit (English Unit)			
	millimeter (*mm*)	centimeter (*cm*)	meter (*m*)	kilometer (*km*)	inch (*in*)	foot / feet (*ft*)	yard (*yd*)	mile (*mi*)
1 millimeter (*mm*)	1	0.1	0.001	0.000001	0.0394	0.0033	0.001	6.214×10^{-7}
1 centimeter (*cm*)	10	1	0.01	0.00001	0.3937	0.0323	0.0109	6.214×10^{-6}
1 meter (*m*)	1000	100	1	0.001	39.3701	3.281	1.0936	6.214×10^{-4}
1 kilometer (*km*)	1000000	100000	1000	1	39370.08	3280.84	1093.61	0.6214
1 inch (*in*)	25.4	2.54	0.0254	0.000025	1	0.0833	0.0278	1.578×10^{-5}
1 foot / feet (*ft*)	304.8	30.48	0.3048	0.000305	12	1	0.3333	$1.899\ 10^{-4}$
1 yard (*yd*)	914.4	91.44	0.9144	0.000914	36	3	1	5.68×10^{-4}
1 mile (*mi*)	1609344	160934.4	1609.344	1.609344	63360	5280	1760	1

Table 5.4. presents the conversion multiplication factors to convert from English units to SI units (and vice versa) for basic measurements of area.

Table 5.4. Area conversion of common units

	SI System			US Standard Unit (English Unit)				
	square centimeter (cm^2)	square meter (m^2)	hectare (ha)	square inch (in^2)	square foot (ft^2)	square yard (yd^2)	square mile (mi^2)	acre (ac)
1 square centimeter (cm^2)	1	0.0001	1.0×10^{-8}	1.0×10^{-12}	0.1550	0.0011	0.00012	3.86×10^{-11}
1 square meter (m^2)	10000	1	0.0001	1.0×10^{-6}	1550.003	10.7639	1.19599	3.86×10^{-7}
1 hectare (ha)	1.00×10^{8}	10000	1	0.01	15500031.0	107639.1	11959.9	3.86×10^{-3}
1 square kilometer (km^2)	1.00×10^{10}	1000000	100	1	1.55×10^{9}	10763910.4	1195990.1	0.386
1 square inch (in^2)	6.4516	0.00065	6.45×10^{-8}	6.45×10^{-10}	1	0.0069	0.000772	2.49×10^{-10}
1 square foot (ft^2)	929.0304	0.0929	9.29×10^{-6}	9.29×10^{-8}	144	1	0.1111	3.59×10^{-8}
1 square yard (yd^2)	8361.2736	0.8361	8.4×10^{-5}	8.36×10^{-7}	1296	9	1	3.23×10^{-7}
1 square mile (mi^2)	258998811	2589991	258.99	2.58999	4.014×10^{9}	27878400.0	3097600.0	1
1 acre (ac)	40468564.2	4046.86	0.4047	0.004047	6272640	43560	4840	0.00156

Table 5.5. presents the conversion multiplication factors to convert from English units to SI units (and vice versa) for basic measurements of weight.

Table 5.5. Weight conversion of common units.

	SI System			US Standard Unit (English Unit)					
	gram (*g*)	kilogram (*kg*)	ton metric (*ton*)	pound (*lb*)	ounce (*oz*)	carat	short *ton* (US)	long *ton* (UK)	grain (*gr*)
1 gram (*g*)	1	0.001	0.000001	0.0022	0.035	5	0.0000011	9.8×10^{-7}	18.43
1 kilogram (*kg*)	1000	1	0.001	2.2046	35.27	5000	0.0011	9.8×10^{-4}	15432.36
1 metric ton (*ton*)	1000000	1000	1	2204.62	35273.96	5000000	1.1023	0.9842	7000
1 pound (*lb*)	453.59	0.4536	0.00045	1	16	2267.96	0.0005	4.5×10^{-4}	437.5
1 ounce (*oz*)	28.35	0.0283	0.000028	0.0625	1	141.75	0.00003	2.8×10^{-5}	3.09
1 carat	0.2	0.0002	0.0000002	0.0004	0.0071	1	2.2×10^{-7}	1.8×10^{-7}	1.4×10^{7}
1 short *ton* (US)	907184.74	907.18474	0.90718	2000	32000	4535923.7	1	0.8929	1.568×10^{7}
1 long *ton* (UK)	1016046.9	1016.0469	1.0160	2240	35840	5080234.5	1.12	1	15432358.4
grain (*gr*)	0.06	0.000064	0.00014	0.0023	0.32	7.14×10^{-8}	6.4×10^{-8}	6.5×10^{-8}	1

Table 5.6. presents the conversion multiplication factors to convert from English units to SI units (and vice versa) for basic measurements of volume.

Table 5.6. Volume conversion of common units

	SI System			US Standard Unit (English Unit)					
	milliliter (*ml*)	liter (*L*)	cubic meter (m^3)	cubic inch ((in^3)	cubic foot/feet (ft^3)	pint (*pt*)	quart (*qt*)	gallon (*gal*)	barrel (*bbl*)
1 milliliter (*ml*)	1	0.001	0.000001	0.0610	0.000035	0.0021	0.001	0.00000098	0.0000084
1 liter (*L*)	1000	1	0.001	61.02	0.035	2.113	1.057	0.00098	0.0084
1 cubic meter (m^3)	1000000	1000	1	61023.74	35.31	2113.38	1056.69	0.00045	8.386
1 cubic inch (in^3)	16.39	0.016	0.000016	1	0.00057	0.035	0.017	0.00003	0.00014
1 cubic foot/feet (ft^3)	28316.85	28.32	0.028	1728	1	59.844	29.92	0.0000002	0.237
1 pint (*pt*)	473.18	0.473	0.00047	28.87	0.016	1	0.5	0.893	0.0040
1 quart (*qt*)	946.35	0.946	0.00095	57.75	0.033	2	1	1	0.0079
1 gallon (*gal*)	3785.41	3.785	0.0038	231	0.134	8	4	0.984	0.0317
1 barrel (*bbl*)	119240.47	119.24	0.119	7276.5	4.211	252	126	31.5	1

Table 5.7. presents converting factors for some of the most important units.

Table 5.7. conversion of some of the most important units

Dimension	American	SI
length	$1ft = \frac{1}{3}yd = 12in$	$1ft = 0{,}3048m$; $1mile = 1609.34m$ $1km = 1000$m $1mile = 1.60934km$
area	$1ft^2 = 144in^2$	$1ft^2 = 0{,}092903m^2$
volume	$1ft^3 = 1728in^3 = 6.22882gal$ (UK) $= 7.48052gal$ (US)	$1ft^3 = 0.0283169m^3$ $1bu = 3.523907\times 10^{-2}m^3$ lag (US) $= 3.785412 \times 10^{-3}m^3$ $lgal$ (US) $= 3.785412ltr$
speed	$1ft/s$ $1knot = 1.150785mile/h = 1.6877ft/s$	$1ft/s = 0.3048m/s$
acceleration	$1ft/s^2$	$1ft/s^2 = 0.3048m/s^2$
mass	$1lb = cwt/112$; $1\ sh\ tn = 2000lb$ $1slug = 32.174lb$; $1in\ tn = 2240lb$	$1lb = 0.453592kg$ $1slug = 14.5939kg$
force	$1lbf$ $1pdl = 0.031061lbf$	$1lbf = 4{,}44822N$ $1pdl = 0.138255N$
work	$1ft\ lb = 0.323832cal$ $1btu = 252.16cal = 778.17ft\ lb$	$1ft\ lb = 1.35582J$ $1btu = 1.05506kJ$
pressure	$1lb/ft^2 = 6.9444 \times 10^{-3}lb/in^2$ $1lb/in^2 = 0.068046atm$ $1atm = 29.92\ in\ Hg = 33.90ft\ water$	$1lb/ft^2 = 47.88N/m^2$ $1lb/in^2 = 6894.76N/m^2$ $1atm = 1.01325bar$
density	$1lb/ft^3 = 5.78704 \times 10^{-4}lb/in^3$ $1lb/gal = 6.22882lb/ft^3$	$1lb/ft^3 = 16.0185kg/m^3$ $1lb/gal = 99.7633kg/m^3$
temperature	$32°F = 0°C$; $212°F = 100°C$ $°K = °F + 255.9$	$1°F = 0.5556°C$ $°K = °C + 373.15$
power	$1ft\ lb/s = 1.8182 \times 10^{-3}\ hp$ $1ft\ lb/s = 1.28505 \times 10^{-3}btu/s$	$1ft\ lb/s = 1.35582W$
electric charge	$1Faraday$	$96485.3coulomb$

Now that we are given the units converting factors, let us describe how to use them. Consider a case of converting *km*/*h* (kilometers per hour) to *m*/*s* (meters per second). We need to do this: A kilometer has 1000 meters, and an hour has 3600 seconds, so a kilometer per hour is:

$$\frac{1000}{3600} = 0.2777m/s.$$

How did I know to make it $\frac{1000}{3600}$ and not 3600 × 1000? The trick is to do the conversions as rational numbers. The steps for using the rational number are as below.

Step 1 Write the conversion as a rational number that equals 1.

Step 2 Multiply the rational number to the unit that needs to be converted (leaving all units in).

Step 3 Cancel any units that are both top and bottom.

Example
Convert 5 kilometer (*km*) to meter (*m*).

Solution
From the converting factors table, we have $1km = 1000m$. Hence, we need to write the rational number of 1. So,

$$\frac{1000m}{1km} = 1.$$

We now multiply the $5km$ by rational number 1. So,

$5km$

$$5km = 5km \times \frac{1000m}{1km}$$

$$5km = \frac{5km \times 1000m}{1km}$$

$$5km = \frac{5\times 1000m}{1}$$

we canceled *km* from the nominator and the denominator

$$5km = 5\times \frac{1000m}{1} = 5{,}000m.$$

Now let us look at an example with a little more complicated conversion, the case that we wanted to convert *km/h* to *m/s.*

Example
Convert 54*km/h* to *m/s*.

Solution
This case requires two conversions: converting *km/h* to *m/h*, and converting *m/h* to *m/s*. Using the rational number 1 we need one rational number $\frac{1000m}{1km} = 1$, and another rational number $\frac{1h}{3{,}600s} = 1$. For the first conversion we will have:

$54km/h$

$$54km/h = 54km/h \times \frac{1000m}{1km}$$

$$54km/h = 54km/h \times \frac{1000m}{1km}$$

$$54km/h = \frac{54km/h \times 1000m}{1km}$$

$$54km/h = \frac{54/h\times 1000}{1}$$

we canceled *km* from the nominator and the denominator

$$54km/h = 54/h \times \frac{1000m}{1} = 54{,}000m/h.$$

For the second conversion we will have.

$54{,}000m/h$

$54{,}000m/h = 54{,}000m/h \times \frac{1h}{3{,}600}$

$54{,}000m/h = \frac{54{,}000m/h \times 1h}{3{,}600s}$

$54{,}000m/h = \frac{54{,}00m}{3{,}600s}$ we canceled *km* from the nominator and the denominator

$54{,}000m/h = \frac{54{,}00m}{3{,}600s} = 15m/s.$

Therefore, $54km/h$ is $15m/s$.

Example

According to Mazda company advertisement the 2021 Mazda3 has a combined city and highway fuel consumption rate 30 *mile/gal*. Compute the equivalent value of this fuel consumption rate in units of *km/ltr*. Round the answer to one decimal place.

Solution

Obviously, this conversion will involve two parts: *mile* to *km* and US *gal* to *ltr*. Our converting factors table provides numbers for both conversions. Let us first convert $30mile/gal$ to km/gal. So,

$30mile/gal$

$30mile/gal = 30mile/gal \times \frac{1.60934km}{1mile}$

$30mile/gal = 30mile/gal \times \frac{1.60934km}{1mi}$

$30mile/gal = \frac{30mile/gal \times 0.1.60934km}{1mile}$ we canceled *km* from the nominator and the denominator

$30mile/gal = 30/gal \times \frac{1.60934km}{1} = 48.2802km/gal.$

Now let us convert $48.2802km/gal$ to km/ltr. So,

$48.2802km/gal$

$48.2802km/gal = 48.2802km/gal \times \frac{1gal}{3.785412ltr}$

$48.2802km/gal = \frac{48.2802km/\cancel{gal} \times 1\cancel{gal}}{3.785412ltr}$

$48.2802km/gal = \frac{48.2802km}{3.785412ltr}$ we canceled *km* from the nominator and the denominator

$48.2802km/gal = 12.754km/ltr.$

Therefore, the fuel consumption rate of Mazda3 is $12.8km/gal$.

Example:

If 200 people threw away an average of 36 pounds of trash every week, find how many metric tons or of trash did they threw every week. Also convert the total trash to its equivalent kilograms.

Solution

Let us first determine the total trash for the week expressed in pounds. If 200 people each throw away 36 pounds, we can find the total by multiplying these two numbers. So,

total trash $(lb.) = 200 \times 36 = 7{,}200 lb$.

Then we need to convert 7,200 pounds to metric tons. From table 5.5 we get 1 metric ton is equal to 2204.62 pounds. In this case we find the conversion factor as follow:

$$\frac{7{,}200 lb}{1} \times \frac{1 ton}{2{,}204.62 lb}$$

$$= \frac{7{,}200}{2{,}204.62} = 3.266.$$

Therefore, the total trash people throw every week is $3.226 ton$.

To find the total trash for the week expressed in kilograms, again we refer to Table 5.5 and in which we get 1 metric ton is equal to 1000 kilograms. Hence, we need to multiply the resulted tons by 1000. So,

$$\frac{3.226\ ton}{1} \times \frac{1000\ kg}{1\ ton}$$

$$= 3.226 \times 1000 = 3{,}226.$$

Therefore, the total trash people throw every week is $3{,}226 kg$.

Example:

One bottle holds 3,288 milliliters while another one holds 3.18 litters. Find the difference in capacity between two bottles.

Solution

As can be easily realized the two measurements are in different units. Hence, we can convert both units to liters and then compare them. From table 5.6 we get 1 liter equal to 1000 milliliters. In this case we find the conversion factor as follow:

$$\frac{3{,}288 ml}{1} \times \frac{1 L}{1000 ml}$$

$$= \frac{3{,}288}{1000} = 3.288.$$

Therefore, the first bottle holds 3.288 liters. Since the second bottle holds 3.18 litters, to get the difference in capacity between two bottles, we need to subtract the latter number from the first one. So,

$3.288 - 3.18 = 0.108.$

Therefore, the difference in capacity between two bottles is $0.108L$ or $0.108L \times 1000 \frac{ml}{L} = 108 ml$.

Example:

Normal body temperature has been commonly accepted as $37.0°C$ (although it varies depending on time of day and method of measurement, as well as among individuals). What is this temperature on the kelvin scale and on the Fahrenheit scale?

Solution

From Table 5.7., the formula for converting centigrade to its equivalent kelvin scale is, $°K = °C + 273.15$. In this case, $C = 37.0°$. So,

$°K = °C + 273.15$

$°K = 37.0 + 273.15$ we plugged $C = 37.0°$ into the equation

$°K = 310.15.$

Therefore, we obtained a body temperature on the kelvin scale equal to $310.15°K$.

To find the temperature on the Fahrenheit scale, again we refer to Table 5.7. in which we found the conversion factor of $1°F = 0.5556°C$. This can also be obtained using a formula, $\frac{F-32}{180} = \frac{C}{100}$ or $°F = \frac{9}{5}°C + 32$. In this case, $C = 37.0°$. So,

$$°F = \frac{9}{5}°C + 32$$

$$°F = \frac{9}{5}(37.0) + 32$$ we plugged $C = 37.0°$ into the equation

$$°F = \frac{9 \times 37}{5} + 32$$

$$°F = \frac{333}{5} + 32$$

$$°F = 66.6 + 32 = 98.6.$$

Therefore, we obtained body temperature on the Fahrenheit scale equal to $98.6°F$.

5.2. Exercises

1. A yard measures $81f$ long (L) and $45f$ wide (W). How many meters long and width is the yard?

2. The classroom temperature, t, is $23^{\circ}c$. what is the room temperature in Fahrenheit.

3. Convert the speed of $v = 16m/s$ to speed in term of km/h.

4. Roxana has 130 ounces of marbles in a box. She adds 120 ounces of more marble into it. How much marble does she have in total in pounds?

5. Antony traveled 46 miles to visit his mother.
   a) How many yards did he travel?
   b) How many kilometers did he travel?

6. Mania wants to buy 2 gallons of alcohol solution but can only find quart containers for sale. How many quarts does she need?

7. Peter's house backyard has a rectangular shape with 62 feet long by 28 feet wide. He plans to use grass mats to carpet the entire backyard. The grass mat costs $36 per square meter. What will be the total cost of the carpet?

   Hint: Area of rectangle = length × width

8. The mass of a basketball ball is 625 grams. Convert this mass to ounces.

9. How many milliliters of a soft drink are contained in a $12.0oz$ can?

10. Write conversion factors for the number of:
    a) Gallons in 1 liter.
    b) Liters in 1 cubic foot.
    c) Grams in 1 pound.
    d) Grams in 1 ounce.
    e) Meters in feet.

11. Convert $4670dm$ to centimeters and meters.

12. Convert $340yd$ to meters and decimeters.

13. Convert $65^{\circ}C$ to Fahrenheit.

14. Convert $50lb$ to grams and kilograms.

15. Convert $76oz$ to milliliters and liters.

6. Basic Concept of Percentage

Basic concepts of **percentage** are essential for everyone Percentages are used widely and in many different areas. For example, discounts in shops, bank interest rates, rates of inflation and many statistics in the media are expressed as percentages. Percentages are important for understanding the financial aspects of everyday life.

A percentage is a portion of a whole expressed as a number between 0 and 100 rather than as a rational number. All of something is 100 percent, half of it is fifty percent, none of something is zero percent. To calculate the percentage of a value we need simply divide the value by the total value and then multiply the result by 100. For example, if in a class of 100 students, 90 students passed in a subject then the percentage of students who passed in the exam is 90%. Had the total number of students been 200, the percentage would have reduced to 45%. Mathematically, it can be calculated as $(\frac{90}{200}) \times 100 = 45\%$.

Percentage Formula

Formula to calculate percentage is:

$$\text{Percentage} = \frac{\text{Value}}{\text{Total Value}} \times 100.$$

Every percentage problem has three possible unknowns or variables:

- Percentage.
- Part.
- Base.

To solve any percentage problem, you must be able to identify these variables. For example, if 60% of 80 is 48, then 60 is the percentage, 80 is the base, and 48 is the part. Similarly, if 22% Of 300 is 66, then 22 is the percentage, 300 is the base, and 66 is the part.

Variables of Percentage

- **Percentage.**
- **Base.**
- **Part.**

Example: If 20% Of 45 is 9, then:

- 20 is the percentage.
- 45 is the base.
- 9 is the part.

Percentage × Base = Part.

Percentage can be expressed in decimal, in rational number or as a whole number. For example, 65% can be expressed as in decimal 0.65, or as in rational number $\frac{65}{100} = \frac{13}{20}$. Similarly, 20% can be expressed as in decimal 0.20, or as in rational number $\frac{20}{100} = \frac{1}{5}$.

Percentage is a dimensionless number used to tell how many parts per hundred are being considered. It is often denoted using the percent sign, "%". The reference point for calculating percentage is taken as 100.

> **Percentage**
>
> Percentage can be defined as "a part out of each hundred" or in other words "for every hundred".

To understand percentages, we need only remember that percent means "out of one hundred". A square that is cut into 100 smaller squares can be used to model percentages, as illustrated in the example below.

Example
Below given are four grids of chocolate bars. What percent of each "White Chocolate Bar" has "Dark Chocolate Bar"?

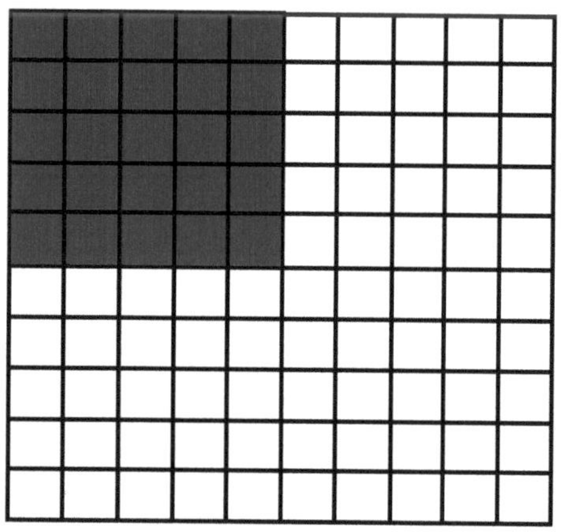

Chocolate Bar (I)

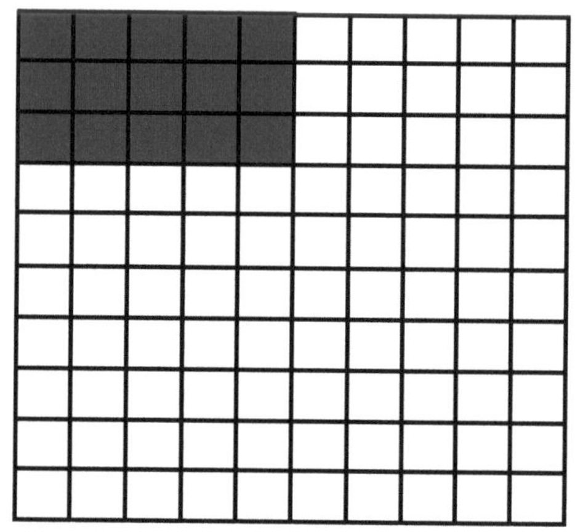

Chocolate Bar (II)

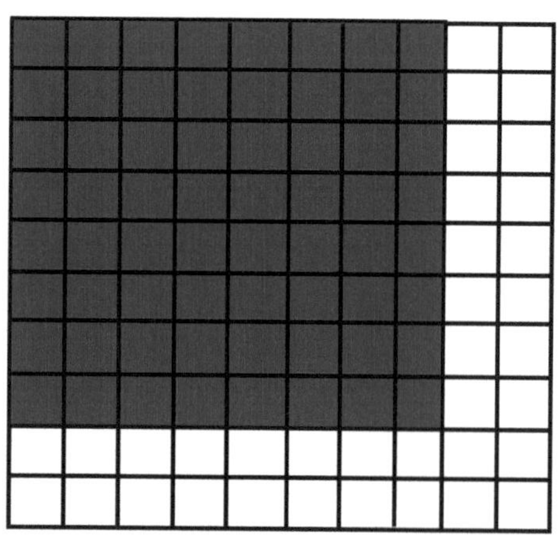

Chocolate Bar (III)

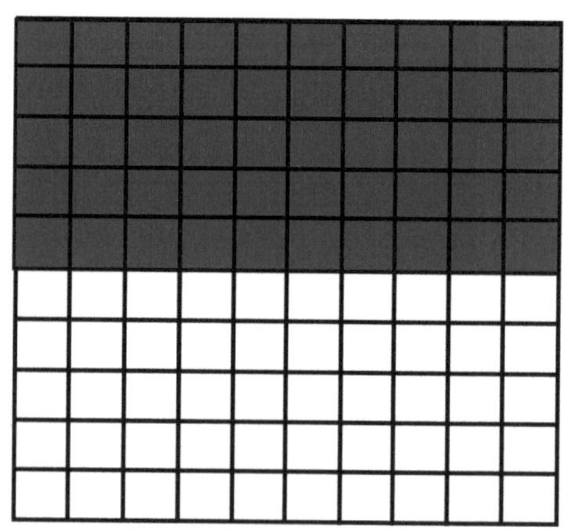

Chocolate Bar (IV)

Solution
Each grid above has 100 blocks. For each white chocolate bar, the ratio of the number of dark chocolate boxes to the total number of white chocolate bars can be represented as a rational number. Or the percentage of white chocolate bars as well as dark chocolate bars can be expressed as whole number, as in decimal, or as in rational number. So,

Chocolate Bar (I): 25 dark and 75 white.

25 darks per 100 or 25%, 0.25, or $\frac{25}{100} = \frac{1}{4}$

75 white per 100 or 75%, 0.75, or $\frac{75}{100} = \frac{3}{4}$.

Chocolate Bar (II): 15 dark and 85 white.

15 darks per 100 or 15%, 0.15, or $\frac{15}{100} = \frac{3}{20}$

85 white per 100 or 85%, 0.85, or $\frac{85}{100} = \frac{17}{20}$.

Chocolate Bar (III): 64 dark and 36 white.

64 darks per 100 or 64%, 0.64, or $\frac{64}{100} = \frac{16}{25}$

36 white per 100 or 36%, 0.36, or $\frac{36}{100} = \frac{9}{25}$.

Chocolate Bar (IV): 50 dark and 50 white.

50 dark per 100 or 50%, 0.50, or $\frac{50}{100} = \frac{1}{2}$

50 white per 100 or 50%, 0.50, or $\frac{50}{100} = \frac{1}{2}$.

Note that, in each case we have two types of chocolates, dark and white. Since each is a portion of a whole, their total gives us whole. That is, the total of their percentage, as whole number is 100%, and as in decimal, or rational number is equal to 1. So,

For Chocolate Bar (I):
The total percentage of dark bars and white bars is; as a whole number 25% + 75% = 100%, as in decimal 0.25 + 0.75 = 1, as in rational number, $\frac{1}{4} + \frac{3}{4} = \frac{4}{4} = 1$.

For Chocolate Bar (II):
The total percentage of dark bars and white bars is; as a whole number 15% + 85% = 100%, as in decimal 0.15 + 0.85 = 1, as in rational number, $\frac{3}{20} + \frac{17}{20} = \frac{20}{20} = 1$.

For Chocolate Bar (III):
The total percentage of dark bars and white bars is; as a whole number 64% + 36% = 100%, as in decimal 0.64 + 0.36 = 1, as in rational number, $\frac{16}{25} + \frac{9}{25} = \frac{25}{25} = 1$.

For Chocolate Bar (IV):
The total percentage of dark bars and white bars is; as a whole number 50% + 50% = 100%, as in decimal 0.50 + 0.50 = 1, as in rational number, $\frac{1}{2} + \frac{1}{2} = \frac{2}{2} = 1$.

Example
In a class of 50 students, the final grade of 20% of students are A, 30% are B, 40% are C, and 10% are D. What are the number of students who are graded A, graded B, graded C and graded D?

Solution
Since it is given that the final grade of 20% out of 50 students are A, we will have:

$$\text{Number of students with grade of A} = \frac{20}{100} \times 50 = \frac{20 \times 50}{100} = \frac{1{,}000}{100} = 10.$$

Similarly, given that the final grade of 30% out of 50 students are B, 40% are C, and 10% are D, we will have:

$$\text{Number of students with grade of B} = \frac{30}{100} \times 50 = \frac{30 \times 50}{100} = \frac{1{,}500}{100} = 15.$$

$$\text{Number of students with grade of C} = \frac{40}{100} \times 50 = \frac{40 \times 50}{100} = \frac{2{,}000}{100} = 20.$$

$$\text{Number of students with grade of D} = \frac{10}{100} \times 50 = \frac{10 \times 50}{100} = \frac{500}{100} = 5.$$

6.1. Important Formulas for Percentage Calculation

Here we have provided a set of basic concepts, tips, and shortcuts on how to solve percentage problems easily and quickly. The following is a list of important formulas for percentage:

1. Percent implies "for every hundred". The symbol "%" is read as percentage and *a*% is read as "*a* per cent".

2. To calculate p % of a

   $$\left(\frac{p}{100}\right) \times a = \frac{p \times a}{100}.$$

3. p% of a = a% of p.

4. To find what percentage of x is y:

   $$\frac{y}{x} \times 100.$$

5. To convert percent to decimal, move the decimal point two places to the left. For example, 65% = 0.65.

6. To convert percent to rational number, place the percent number in the numerator and 100 in the denominator. For example, $65\% = \frac{65}{100} = \frac{13}{20}$.

7. To convert decimal to percent, move decimal point two places to the right, For example, 1.65 = 165%.

8. To convert rational number to percent, first convert rational number to decimal, then follow the directions to convert from decimal to percent. For example, $\frac{13}{20} = 0.65$ and 0.65 = 65 %.

9. To calculate percentage change in value:

   $$\text{Percentage change} = \frac{\text{change}}{\text{initial value}} \times 100.$$

10. Increase N by $p\% = N(1 + \frac{p}{100})$.

11. If the value of an item goes up by $p\%$, the percentage reduction to be now made to bring it back to the original point is $\frac{100p}{100+p}\%$.

12. If the value of an item goes down by $p\%$, the percentage increment to be now made to bring it back to the original point is $\frac{100p}{100-p}\%$.

13. If A is $p\%$ more than B, then B is $\frac{100p}{100+p}\%$ less than A.

14. If A is $p\%$ less than B, then B is $\frac{100p}{100-p}\%$ more than A.

15. If A is $p\%$ of C and B is $q\%$ of C, then $A = \frac{p}{q} \times B$.

16. If two numbers are respectively $p\%$ and $q\%$ less than a third number, then the first number is $(\frac{100-p}{100-q} \times 100)\%$ of the second and the second number is $(\frac{100-q}{100-p} \times 100)\%$ of the first number.

17. If two numbers are respectively $p\%$ and $q\%$ more than a third number, then the first number is $(\frac{100+p}{100+q} \times 100)\%$ of the second and the second number is $(\frac{100+q}{100+p} \times 100)\%$ of the first number.

18. If a number is first increased by $p\%$ and then decreased by $q\%$ the percentage change will be,

$$\text{Percentage change} = \left(p - q - \frac{pq}{100}\right)\%.$$

19. If a number is first decreased by $p\%$ and then increased by $q\%$ the percentage change will be,

$$\text{Percentage change} = \left(-p + q - \frac{pq}{100}\right)\%.$$

20. If a number is first decreased by $p\%$ and then decreased by $q\%$ the percentage change will be,

$$\text{Percentage change} = \left(-p - q + \frac{pq}{100}\right)\%.$$

21. If a number is first increased by $p\%$ and then increased by $q\%$ the percentage change will be,

$$\text{Percentage change} = \left(p + q + \frac{pq}{100}\right)\%.$$

22. If two numbers A and B are multiplied to get a product and if A is changed by $p\%$ and B is changed by $q\%$, then the net % change in the product $(A \times B)$ is given $(p + q + \frac{pq}{100})\%$.

23. If a value A increases by $a\%$, $b\%$, $c\%$ and so on up to $n\%$, the total rise in % is given by the formula:

$$\text{Final output} = A(1+\frac{a}{100})(1+\frac{b}{100})(1+\frac{c}{100}) \cdots (1+\frac{n}{100}).$$

24. If a value A decreases by $a\%$, $b\%$, $c\%$ and so on up to $n\%$, the total decline is given by the formula:

$$\text{Final output} = A\left(1-\frac{a}{100}\right)\left(1-\frac{b}{100}\right)\left(1-\frac{c}{100}\right) \cdots \left(1-\frac{n}{100}\right).$$

25. If the price of an item decreases by $p\%$, then the increase in consumption so that the expenditure remains same, which is: $(\frac{100p}{100-p})\%$.

26. If the price of an item increases by $p\%$, then the reduction in consumption so that the expenditure remains same, which is: $(\frac{100p}{100+p})\%$.

27. Formulas for percentage of population growth.

If the current population is P and it increases at a rate of $r\%$ per annum, then:

Population after n years = $P\left(1+\frac{r}{100}\right)^n$

Population n years ago = $\frac{P}{\left(1+\frac{r}{100}\right)^n}$.

28. Formulas for Depreciation

If the current value of an asset is P and it depreciates at a rate of $r\%$ per annum, then:

Value of asset after n years = $P\left(1-\frac{r}{100}\right)^n$

Value of asset n years ago = $\frac{P}{\left(1-\frac{r}{100}\right)^n}$.

29. Percentage–Ratio equivalence:

Ratio	Percentage Equivalence	Ratio	Percentage Equivalence	Ratio	Percentage Equivalence
$\frac{1}{2}$	$\frac{1}{2}\times 100 = 50\%$	$\frac{2}{5}$	$\frac{2}{5}\times 100 = 40\%$	$\frac{5}{7}$	$\frac{5}{7}\times 100 = 71.43\%$
$\frac{1}{3}$	$\frac{1}{3}\times 100 = 33.33\%$	$\frac{2}{7}$	$\frac{2}{7}\times 100 = 28.57\%$	$\frac{5}{9}$	$\frac{5}{9}\times 100 = 55.56\%$
$\frac{1}{4}$	$\frac{1}{4}\times 100 = 25\%$	$\frac{2}{9}$	$\frac{2}{9}\times 100 = 22.22\%$	$\frac{6}{7}$	$\frac{6}{7}\times 100 = 85.71\%$
$\frac{1}{5}$	$\frac{1}{5}\times 100 = 20\%$	$\frac{3}{4}$	$\frac{3}{4}\times 100 = 75\%$	$\frac{7}{9}$	$\frac{7}{9}\times 100 = 77.78\%$
$\frac{1}{6}$	$\frac{1}{6}\times 100 = 16.67\%$	$\frac{3}{5}$	$\frac{3}{5}\times 100 = 60\%$	$\frac{8}{9}$	$\frac{8}{9}\times 100 = 88.89\%$
$\frac{1}{7}$	$\frac{1}{7}\times 100 = 14.29\%$	$\frac{3}{7}$	$\frac{3}{7}\times 100 = 42.86\%$	$\frac{9}{10}$	$\frac{9}{10}\times 100 = 90\%$
$\frac{1}{8}$	$\frac{1}{8}\times 100 = 12.50\%$	$\frac{3}{8}$	$\frac{3}{8}\times 100 = 37.5\%$	$\frac{10}{11}$	$\frac{10}{11}\times 100 = 90.91\%$
$\frac{1}{9}$	$\frac{1}{9}\times 100 = 11.11\%$	$\frac{4}{5}$	$\frac{4}{5}\times 100 = 80\%$	$\frac{10}{13}$	$\frac{10}{13}\times 100 = 76.92\%$
$\frac{1}{10}$	$\frac{1}{10}\times 100 = 10\%$	$\frac{4}{7}$	$\frac{4}{7}\times 100 = 57.14\%$	$\frac{10}{17}$	$\frac{10}{17}\times 100 = 58.82\%$
$\frac{2}{3}$	$\frac{2}{3}\times 100 = 66.67\%$	$\frac{4}{9}$	$\frac{4}{9}\times 100 = 44.44\%$	$\frac{10}{19}$	$\frac{10}{19}\times 100 = 52.63\%$

Based on the above 24-formula for different cases of percentage calculation, following, we will present at least one example for each case.

Example
What is 84% of 500?

Solution
The statement 84% of 500, can be written as:

$$\frac{84}{100} \times 500$$

$$= \frac{84 \times 500}{100}$$

$$= \frac{42{,}000}{100} = 420.$$

Example
Convert each of percent, 38%, 6%, 95%, and 50% into rational number.

Solution
For 38%, we will have:

$$38\% = \frac{38}{100} = \frac{19}{50}.$$

For 6%, we will have:

$$6\% = \frac{6}{100} = \frac{3}{50}.$$

For 95%, we will have:

$$95\% = \frac{95}{100} = \frac{19}{20}.$$

For 50%, we will have:

$$50\% = \frac{50}{100} = \frac{1}{2}.$$

Example
Convert each of percent, 78%, 1.39%, and 34%, into decimal.

Solution
For 78%, we will have:

$$78\% = \frac{78}{100} = 0.78.$$

For 1.39%, we will have:

$$1.39\% = \frac{139}{100} = 1.39.$$

For 34%, we will have:

$$34\% = \frac{34}{100} = 0.34.$$

Example
Find 15% of 54 plus 10% of 49.

Solution
To find 15% of 54 we will have:

$$\frac{15}{100} \times 54$$

$$= \frac{15 \times 54}{100}$$

$$= \frac{810}{100} = 8.10.$$

Similarly, to find 10% of 49 we will have:

$$\frac{10}{100} \times 49$$

$$= \frac{10 \times 49}{100}$$

$$= \frac{49}{100} = 4.90.$$

The sum of 15% of 54 and 10% of 49 is:

$$8.10 + 4.90 = 13.$$

Example
Find what percentage of 92 is 23.

Solution
We have learned that, for finding what percentage of x is y, the formula is: $\frac{y}{x} \times 100$. So,

$$\frac{y}{x} \times 100$$

$$\frac{23}{92} \times 100$$ we plugged $y = 23$ and $x = 92$ into the equation

$$\frac{2{,}300}{92} = 25.$$

Therefore, 25 percent of 92 is 23. Or 23 is 25% of 92.

Example
12 is what percentage of 64?

Solution
The required percentage is obtained by:

$$\frac{12}{64} \times 100$$

$$= \frac{1{,}200}{64} = 18.75\%.$$

Example
Brazilian coffee production is forecasted for Marketing Year 2020/21 (July-June) at a record of 67.9 million 60*kg* bags, an increase of 15 percent compared to last year's output. Find the last year Brazilian coffee production in term of million 60*kg* bags.

Solution
We have learned that, If the value of an item goes up by $p\%$, the percentage reduction to be now made to bring it back to the original point is $\frac{100p}{100+p}\%$. So,

$\frac{100p}{100+p}\%$

$\frac{100 \times 15}{100+15}\%$ we plugged p = 15% into the equation

$= \frac{1{,}500}{115}\%$

$= \frac{1{,}500}{115}\%$

= 13.04%.

We also learned that if N decrease by p% the new value is $N(1 - \frac{p}{100})$. In this case, N = 67.9 million 60kg bags and we obtained p = 13.04%. So,

last year Brazilian coffee production = $N(1 - \frac{p}{100})$

last year Brazilian coffee production = $67.9(1 - \frac{13.04}{100})$ we plugged N = 67.9 and p = 13.04 into the equation

last year Brazilian coffee production = 67.9(1 – 0.1304)

last year Brazilian coffee production = 67.9(0.8696)

last year Brazilian coffee production = 59.05.

Therefore, the last year Brazilian coffee production is 59.05 million 60kg bags.

To check our answer, we know the last production increased by 15%. Hence, given the forecast for Marketing Year 2020/21 (July-June) at a record of 67.9 million 60kg bags, if we multiply our answer by 1 + 0.15, we must get 67.9. So,

$59.05(1 + 0.15) \overset{?}{\Leftrightarrow} 67.9$

$59.05(1.15) \overset{?}{\Leftrightarrow} 67.9$

$67.9075 \overset{?}{\Leftrightarrow} 67.9$

$67.9 \approx 67.9$.

Example

Sheila makes \$180 a week from her job. She earns a raise and now makes \$220 a week. What is the percentage increase?

Solution

First, we find the amount of increase in her earnings. So,

amount of increase = \$220 – \$180 = \$40.

Now if we divide the amount of increase by her initial earning and multiply the result by 100 we will get the percent increase. So,

percent increase = $\frac{40}{180} \times 100\%$

percent increase = $\frac{4000}{180}$= 22.22%.

Therefore, the percent increase is 22.22%.

Example

The population of the town is 2.865 million. If it increases at the rate of 4% per annum:

a) What will be its population after 2 years?

b) What was its population 2 years ago?

Solution

We have learned if the current population is P and it increases at a rate of r% per annum, then we will have:

$$\text{Population after } n \text{ years} = P\left(1 + \frac{r}{100}\right)^n$$

$$\text{Population } n \text{ years ago} = \frac{P}{\left(1 + \frac{r}{100}\right)^n}.$$

Using the above formulas, we can answer the requested questions. So,

a) To find population after 2 years, we have:

$$\text{Population after } n \text{ years} = P\left(1 + \frac{r}{100}\right)^n$$

$$\text{Population after } n \text{ years} = 2.865\left(1 + \frac{4}{100}\right)^2$$ we plugged $P = 2.865$, $r = 4$ and $n = 2$ into the equation

$$\text{Population after } n \text{ years} = 2.865(1 + 0.04)^2$$

$$\text{Population after } n \text{ years} = 2.865(1.04)^2$$

$$\text{Population after } n \text{ years} = 2.865(1.0816)$$

$$\text{Population after } n \text{ years} = 3.098784.$$

Therefore, the town population after 2 years is 3.098784 million, or 3,098,784.

b) To find population 2 years ago, we will have:

$$\text{Population } n \text{ years ago} = \frac{P}{\left(1 + \frac{r}{100}\right)^n}$$

$$\text{Population } n \text{ years ago} = \frac{2.865}{\left(1 + \frac{4}{100}\right)^2}$$ we plugged $P = 2.865$, $r = 4$ and $n = 2$ into the equation

$$\text{Population } n \text{ years ago} = \frac{2.865}{(1 + 0.04)^2}$$

$$\text{Population } n \text{ years ago} = \frac{2.865}{(1.04)^2}$$

$$\text{Population } n \text{ years ago} = \frac{2.865}{1.0816}$$

$$\text{Population } n \text{ years ago} = 2.648854.$$

Therefore, the town population 2 years ago was 2.648854million, or 2,648,854.

Example

The price of a commodity increased by 30%. By what percentage should it be reduced to bring it to actual level?

Solution
We learned if the value of an item goes up by $p\%$, the percentage reduction to be now made to bring it back to the original point is $\frac{100p}{100 + p}\%$. Using this formula, we will have:

$$\frac{100p}{100 + p}\%.$$

$$\frac{100 \times 30}{100 + 30}\%.$$

$$\frac{3000}{130}\% = 23.08.$$

Therefore, the percentage should it be reduced to bring it to actual level is 23.08%.

Example
If Calvin's salary is first increased by 14% and thereafter decreased by 11%, what is the net change in his salary?

Solution
We are given a formula which implies, if a number is changed (increases or decreases) successively by $p\%$ and $q\%$, then net % change is given by, $(p + q + \frac{pq}{100})\%$ for successive increase and $(p - q - \frac{pq}{100})\%$ for successive decrease. In this case considering the 11% decrease as $-$ 11% increase, we can use $(\frac{p + q + pq}{100})\%$ to obtain the answer. So,

The net change in Calvin salary = $(p + q + \frac{pq}{100})\%$

The net change in Calvin salary = $(14 - 11 + \frac{14(-11)}{100})\%$ we plugged $p = 14$, and $q = -11$ into the equation

The net change in Calvin salary = $(4 - 11 + \frac{-154}{100})\%$

The net change in Calvin salary = $(3 - 1.54)\%$

The net change in Calvin salary = 1.46%.

Therefore, the net change in Calvin salary is 1.46 %. Note, the answer we get is a positive value, so Calvin salary increases by 1.46%.

Example
The steward salary is 25% more than Billy's salary. Billy's salary is what percentage less then Steward's salary?

Solution
There are two ways to solve this problem. The first way is using basic percent definitions and percent concepts. The second way is using the formula presented in list of "important formulas for percentage, item 17". So,

Approach I

If we let Billy's salary be 100, then Steward's salary is 25% more and hence is $1.25 \times 100 = 125$. Billy's salary is 25 less than Steward, hence we need to divide the difference by the Steward's salary and then multiply the result by 100 to get the difference in percentage. So,
The difference in percentage= $\frac{25}{125} \times 100\% = \frac{2500}{125}\% = 20\%$.

Therefore, Billy's salary is 20% less than Steward's salary.

Approach *II*

Using the given formula which implies, if A is $p\%$ more than B, then B is $\frac{100}{100+p}\%$ less than A. So,

Percent B is less than $A = \frac{100p}{100+p}\%$

Percent Billy's salary is less then Steward's salary$= \frac{100\times25}{100+25}\%$ we plugged $p = 25$ into the equation

Percent Billy's salary is less then Steward's salary$= \frac{2500}{125}\%$

Percent Billy's salary is less then Steward's salary$=$ 20%.

Therefore, Billy's salary is 20% less than Steward's salary.

Example

Carin's height is 11% shorter than her brother's height. Her brother's height is what percentage taller than Carin's height?

Solution

We learned, if A is $p\%$ less than B, then B is $\frac{100}{100-p}\%$ more than A. So,

Percent B is more than $A = \frac{100p}{100-p}\%$

Percent Carin's brother height taller than Carin height $= \frac{100\times11}{100-11}\%$ we plugged $p = 11$ into the equation

Percent Carin's brother height taller than Carin height $= \frac{1100}{89}\%$

Percent Carin's brother height taller than Carin height $=$12.36%.

Therefore, percent Carin's brother's height taller than Carin's height is 12.36%.

Example

Marvin's weight is 195*lb*. Gordon's weight is 96% of Steven's weight and Marvin's weight is 104% of Steven's weight. What is Gordon's weight?

Solution

We learned if A is $p\%$ of C and B is $q\%$ of C, then $A = \frac{p}{q}\times B$. Using this formula, if we correspond Gordon's weight to A, Steven's weight to C, and Marvin's weight to B, we can find Gordon's weight as:

$A = \frac{p}{q}\times B$

Gordon weight $= \frac{96}{104}\times 195$ we plugged $p = 96$, $q = 104$, and $B = 195$ into the equation

Gordon weight $= \frac{96\times 195}{104}$

Gordon weight $= \frac{18{,}720}{104}$

Gordon weight = 180.

Therefore, Gordon's weight is $180lb$.

Example
If a side of a rectangle is increased by 15% and the other side is increased by 18% its area is increased by AI, find the value of AI.

Hint: The area of a rectangle is calculated by multiplying the width (or side) by the length (the other side) of a rectangle.

Solution
We learned if two numbers A and B are multiplied to get a product and if A is changed by $p\%$ and B is changed by $q\%$, then the net % change in the product ($A \times B$) is given $(p + q + \frac{pq}{100})\%$.

Using this formula, if we correspond one side to A, and the other side to B, we can find AI:

Percent change in the product $(A \times B) = (p + q + \frac{pq}{100})\%$.

$AI = (15 + 18 + \frac{15 \times 18}{100})\%$ — we plugged $p = 15$, and $q = 18$ into the equation

$AI = (33 + \frac{270}{100})\%$

$AI = (33 \times \frac{100}{100} + \frac{270}{100})\%$ — we multiplied the first term in the right side by $\frac{100}{100}$, to make rational numbers with a common denominator

$AI = (\frac{3{,}300}{100} + \frac{270}{100})\%$

$AI = (\frac{3{,}570}{100})\%$

$AI = 35.7\%$.

Therefore, the area of the square has increased by 35.7%.

Example
After the COVID-19 epidemy the price of a specific face mask increased by 40%. By what percentage should it be reduced to bring it to the initial price?

Solution
Let assume that the initial price is 100. So,

New price = $100 + 100 \times 40\%$

New price = $100 + 100 \times 0.40$

New price = $100 + 40 = 140$.

The difference between the new price and the initial price is $140 - 100 = 40$. Hence, the price needs to be reduced by 40 but from the new price, 140. So,

The percentage reduction $= \frac{40}{140} \times 100\% = \frac{4{,}000}{125}\% = 32\%$.

Therefore, the percentage should it be reduced to bring the new price it to the initial price is 32%.

Example
The population of the city was 1.5 million in 2017. It increased at the rate of 3% in 2018, 4% in 019, and 5% in 2020. What was the population of this city be in 2020?

Solution
We know that if a value *A* increases by *a*%, *b*%, *c*% and so on up to *n*%, the total rise in % is given by the formula equal to $A\left(1+\frac{a}{100}\right)\left(1+\frac{b}{100}\right)\left(1+\frac{c}{100}\right)\cdots\left(1+\frac{n}{100}\right)$.

Using this formula we will have:

$$\text{Final value of } A = A\left(1+\frac{a}{100}\right)\left(1+\frac{b}{100}\right)\left(1+\frac{c}{100}\right)\cdots\left(1+\frac{n}{100}\right)$$

$$\text{Population in 2020} = 1.5\left(1+\frac{3}{100}\right)\left(1+\frac{4}{100}\right)\left(1+\frac{5}{100}\right)$$

we plugged A =1.5, a = 3, b= 4, and c = 5 into the equation

$$\text{Population in 2020} = 1.5(1+0.03)(1+0.04)(1+0.05)$$

$$\text{Population in 2020} = 1.5(1.03)(1.04)(1.05)$$

$$\text{Population in 2020} = 1.5(1.12476)$$

$$\text{Population in 2020} = 1.68714.$$

Therefore, the city population in 2020 was 1.68714 million, or 1,687140.

Example
Sabrina paid $35,000 to purchase a new car in 2016. Based on the Blue Book, the price of her car decreases 9% in the first year, 8% in second year, 6% in third year, and 4% in fourth year. What is the price of a car at the end of 2020?

Solution
We know that if a value *A* decreases by *a*%, *b*%, *c*% and so on up to *n*%, the total decline in % is given by the formula equal to $A\left(1-\frac{a}{100}\right)\left(1-\frac{b}{100}\right)\left(1-\frac{c}{100}\right)\cdots\left(1-\frac{n}{100}\right)$. Using this formula we will have:

$$\text{Final value of } A \text{ after 4 decreases} = A\left(1-\frac{a}{100}\right)\left(1-\frac{b}{100}\right)\left(1-\frac{c}{100}\right)\left(1-\frac{d}{100}\right)$$

$$\text{The car price in 2020} = 35{,}000\left(1-\frac{9}{100}\right)\left(1-\frac{8}{100}\right)\left(1-\frac{6}{100}\right)\left(1-\frac{4}{100}\right)$$

we plugged A = 35,000, a = 9, b= 8, c = 6 and d = 4 into the equation

$$\text{The car price in 2020} = 35{,}000(1-0.09)\,(1-0.08)\,(1-0.06)\,(1-0.04)$$

$$\text{The car price in 2020} = 35{,}000\,(0.91)\,(0.92)\,(0.94)\,(0.96)$$

$$\text{The car price in 2020} = 35{,}000\,(0.75549)$$

$$\text{The car price in 2020} = 26{,}442.15.$$

Therefore, Sabrina's car price in 2020 was $26,442.15.

Example
If the price of basmati rice goes up by 12%, then what would be the percentage decrease in the quantity consumed so that the total expenditure on basmati rice remains the same?

Solution
Let P be the price of basmati rice, and q_1 be the quantity consumed and q_2 be the quantity consumption after reduction. The new price of basmati rice due to 12% increase is:

Initial price = P

Amount increased = $P \times 12\%$

Amount increased = $P \times \frac{12}{100}$

Amount increased = $0.12P$

New price = Initial price + Amount increased

New price = $P + 0.12P = 1.12P$.

The expenditure before price increase is:

$P \times q_1$.

The expenditure after price increase is:

$1.12P \times q_2$.

Since the total expenditure must be the same, we will have:

$P \times q_1 = 1.12P \times q_2$

$\frac{1}{P} \times P \times q_1 = \frac{1}{P} \times 1.12P \times q_2$ — to eliminate P, we multiplied both sides by $\frac{1}{P}$

$\frac{P}{P} \times q_1 = \frac{P}{P} \times 1.12q_2$

$q_1 = 1.12q_2$

$q_1 - q_2 = 1.12q_2 - q_2$ — to find the quantity reduction, we added $-q_2$ to both sides

$q_1 - q_2 = 0.12q_2$.

To find the percentage decrease in the quantity we need to divide their difference, $q_1 - q_2$, by the initial consumption, q_1 and then multiply the result by 100. percent reduction. Note that from the above relations we found $q_1 - q_2 = 0.12q_2$, and $q_1 = 1.12q_2$. So,

percentage decrease in the quantity = $\frac{q_1 - q_2}{q_1} \times 100\%$

percentage decrease in the quantity = $\frac{0.12q_2}{1.12q_2} \times 100\%$ — we plugged $q_1 - q_2 = 0.12q_2$ and $q_1 = 1.12q_2$ into the equation

percentage decrease in the quantity = $\frac{0.12 \times 100}{1.12}\%$

percentage decrease in the quantity = $\frac{12}{1.12}\%$

percentage decrease in the quantity = 10.71%.

Therefore, the percentage decrease in the quantity consumed is 10.71%.

Example
If the price of extra virgin olive oil decreases by 15%, then what would be the percentage increase in the quantity consumed so that the total expenditure on extra virgin olive oil remains the same?

Solution

Let P be the price of extra virgin olive oil, and q_1 be the quantity consumed and q_2 be the quantity consumption after raise. The new price of extra virgin olive oil due to 15% reduction is:

Let the initial price = P

Amount decreased = $P \times 15\%$

Amount decreased = $P \times \frac{15}{100}$

Amount decreased = $0.15P$

New price = Initial price − Amount decreased

New price = $P - 0.15P = 0.85P$.

The expenditure before price decrease is:

$P \times q_1$.

The expenditure after price decrease is:

$0.85P \times q_2$.

Since the total expenditure must be the same, we will have:

$P \times q_1 = 0.85P \times q_2$

$\frac{1}{P} \times P \times q_1 = \frac{1}{P} \times 0.85P \times q_2$ — to eliminate P, we multiplied both sides by $\frac{1}{P}$

$\frac{P}{P} \times q_1 = \frac{P}{P} \times 0.85q_2$

$q_1 = 0.85q_2$

$q_1 - q_2 = 0.85q_2 - q_2$ — to find the quantity raise, we added $-q_2$ to both sides

$q_1 - q_2 = -0.15q_2$

$(-1)(q_1 - q_2) = (-1)(-0.15q_2)$ — to obtain $q_2 - q_1$, we multiplied (-1) to both sides

$q_2 - q_1 = 0.15q_2$.

To find the percentage increase in the quantity we need to divide the new consumption quantity minus the initial consumption quantity, $q_2 - q_1$, by the initial consumption quantity, q_1 and then multiply the result by 100. percent reduction. Note that from the above relations we found $q_2 - q_1 = 0.15q_2$, and $q_1 = 0.85q_2$. So,

percentage raise in the consumption = $\frac{q_2 - q_1}{q_1} \times 100\%$

percentage raise in the consumption = $\frac{0.15q_2}{0.85\ _2} \times 100\%$ — we plugged $q_2 - q_1 = 0.15q_2$ and $q_1 = 0.85q_2$ into the equation

percentage raise in the consumption = $\frac{0.15 \times 100}{0.85}\%$

percentage raise in the consumption = $\frac{15}{0.85}\%$

percentage raise in the consumption = 17.65%.

Therefore, the percentage increase in the quantity consumed is 17.65%.

Example

Flight Safe Inc. purchased a jumbo jet for \$180 million. Suppose this company depreciates its airplane at the rate of 10% per annum. What is the value of the jumbo jet after 2 years?

Solution

We are given if the current value of an asset is P and it depreciates at a rate of r% per annum, then, the formula to find the value of the asset after n years is:

$$\text{Value of asset after } n \text{ years} = P\left(1 - \frac{r}{100}\right)^n.$$

Using this formula, we can find the value of the jumbo jet after 2 years. So,

$$\text{Value of asset after } n \text{ years} = P\left(1 - \frac{r}{100}\right)^n$$

$$\text{Value of jumbo jet after 2 years} = 180\left(1 - \frac{10}{100}\right)^2$$

we plugged $P = 180$, $r = 10$ and $n = 2$ into the equation

$$\text{Value of jumbo jet after 2 years} = 180(1 - 0.10)^2$$

$$\text{Value of jumbo jet after 2 years} = 180(0.90)^2$$

$$\text{Value of jumbo jet after 2 years} = 180(0.81)$$

$$\text{Value of jumbo jet after 2 years} = 145.8.$$

Therefore, the value of jumbo jet after 2 years is \$145.8 million. Or \$145,800.000.

Example

The price of an office building in the depreciation accounting records of BLD Company after 2 years is \$840,000. Suppose this company depreciates its buildings at the rate of 5% per annum. It is Find the initial value of the office building.

Solution

We are given if the current value of an asset is P and it depreciates at a rate of r% per annum, then, the formula to find the value of the asset n years ago is:

$$\text{Value of asset } n \text{ years ago} = \frac{P}{\left(1 - \frac{r}{100}\right)^n}.$$

We are given the office building value, depreciated after two years. Using this formula considering $n = 2$, we can find the initial value of the office building. So,

$$\text{Value of asset } n \text{ years ago} = \frac{P}{\left(1 - \frac{r}{100}\right)^n}$$

$$\text{Value of the office building 2 years ago} = \frac{840{,}000}{\left(1 - \frac{5}{100}\right)^2}$$

we plugged $P = 840{,}000$, $r = 5$ and $n = 2$ into the equation

$$\text{Value of the office building 2 years ago} = \frac{840{,}000}{(1 - 0.05)^2}$$

$$\text{Value of the office building 2 years ago} = \frac{840{,}000}{(0.95)^2}$$

$$\text{Value of the office building 2 years ago} = \frac{840{,}000}{0.9025}$$

$$\text{Value of the office building 2 years ago} = 930{,}748.$$

Therefore, the initial value of the office building was $930,748.

Example

Erin makes $980 per week. Her earnings are 15% more than her brother's earnings per week. Find what percent less her brother's spending is than Erin's earning.

Solution

We learned if A is $p\%$ more than B, then B is $\frac{100p}{100+p}\%$ less than A. Using this formula, we can find the percent Erin brother's earning is less than her earning. So,

Percent Erin brother earning less than her earning = $\frac{100p}{100+p}\%$

Percent Erin brother earning less than her earning = $\frac{100(15)}{100+15}\%$ — we plugged $p = 15$ into the equation

Percent Erin brother earning less than her earning = $\frac{1500}{115}\% = 13.04\%$.

Therefore, Erin brother earns 13.04% less than Erin's earning.

Example

The two numbers are respectively 20% and 50% more than the third number.

a) What % is the first number of the second number?
b) What % is the second number of the first number?

Solution

a) We learned if two numbers are respectively $p\%$ and $q\%$ more than a third number, then the first number is $\left(\frac{100+p}{100+q} \times 100\right)\%$ of the second and the second number is $\left(\frac{100+q}{100+p} \times 100\right)\%$ of the first number. So, using $\left(\frac{100+p}{100+q} \times 100\right)\%$, we can find the percent of the first number is the second number. Hence, we will have:

$\left(\frac{100+p}{100+q} \times 100\right)\%$

$\left(\frac{100+20}{100+50} \times 100\right)\%$ — we plugged $p = 20$, and $q = 50$ into the equation

$\left(\frac{120}{150} \times 100\right)\% = 80\%$.

Therefore, 80% of the first number is the second number.

b) Using $\left(\frac{100+q}{100+p} \times 100\right)\%$, we can find the percent of the second number is the first number. Hence, we will have:

$\left(\frac{100+q}{100+p} \times 100\right)\%$

$\left(\frac{100+50}{100+20} \times 100\right)\%$ — we plugged $q = 50$, and $p = 20$ into the equation

$\left(\frac{150}{120} \times 100\right)$%= 125%.

Therefore, 125% of the second number is the first number.

Example
The two numbers are respectively 15% and 10% less than the third number.

a) What % is the first number of the second number?
b) What % is the second number of the first number?

Solution

a) We learned if two numbers are respectively p% and q% less than a third number, then the first number is $\left(\frac{100-p}{100-q} \times 100\right)$% of the second and the second number is $\left(\frac{100-q}{100-p} \times 100\right)$% of the first number. So, using $\left(\frac{100+p}{100+q} \times 100\right)$%, we can find the percent of the first number is the second number. Hence, we will have:

$\left(\frac{100-p}{100-q} \times 100\right)$%

$\left(\frac{100+15}{100+100} \times 100\right)$% we plugged $p = 15$, and $q = 10$ into the equation

$\left(\frac{115}{110} \times 100\right)$%= 104.5%.

Therefore, 104.5 % of the first number is the second number.

b) Using $\left(\frac{100-q}{100-p} \times 100\right)$%, we can find the percent of the second number is the first number. Hence, we will have:

$\left(\frac{100-q}{100-p} \times 100\right)$%

$\left(\frac{100+10}{100+15} \times 100\right)$% we plugged $q = 10$, and $p = 15$ into the equation

$\left(\frac{110}{115} \times 100\right)$%= 95.65%

Therefore, 95.65% of the second number is the first number.

6.2. Exercises

1. What is 72% of 489?

2. Convert each of percent, 45%, 126%, 5%, and 87% into rational number and convert each of decimal, 0.28, 1.15, 0.09, and 0.83, into percent.

3. Find 32% of 124 plus 17% of 78.

4. Find out what percentage of 318 is 82.

5. The number 42 is what percentage of 111?

6. In a class of 30 students, the final grade of 20% students is *A*, 30% is *B*, 40% is *C*, and 10% is *D*. What are the number of students who are graded *A*, graded *B*, graded *C* and graded *D*?

7. Fatima used to make \$264 a week from her new job. After 4 months she got a raise and now makes \$316 a week. What is the percentage increase?

8. The price of a commodity increased by 16%. By what percentage should it be reduced to bring it to the current level?

9. If my salary went up by 24% and then decreased by 22%, what is the net percent change?

10. If the price of beef goes up by 12%, then what should be the percentage decrease in the quantity consumed so that the total expenditure on sugar remains the same?

11. In an electronic shop Smart Full HD-LED TV is marked at \$890. It was discounted 18% for Christmas Eve and a further 15% discounted for New Year's Eve. What is the current price of a Smart Full HD-LED TV?

12. If the current value of a car is \$24,000 and it depreciates at a rate of 12% per annum, then find.
    a) The value of car after 3 years.
    b) The value of a car was 2 years ago.

13. In a boutique the dress trousers are priced \$88. Finally, they went on sale for \$64. What is the percent decrease?

14. The two numbers are respectively 12% and 16% less than the third number.
    a) What % is the first number of the third number?
    b) What % is the second number of the first number?

15. The two numbers are respectively 22% and 14% more than the third number.
    a) What % is the first number of the second number?
    b) What % is the second number of the first number?

16. Sonya makes \$4,480 per month. Her earnings are 13% more than her friend's earnings. Find out what percent her friend's earnings are less than Sonya's earnings.

17. The population of a small town is 340,000. If it increases at the rate of 6% per annum,
    a) What will be its population after 4 years?
    b) What was its population 2 years ago?

18. If Ted's salary is first increased by 18% and thereafter decreased by 8%, what is the net change in his salary?

19. Lili salary is 16% more than Betty's salary. Betty's salary is what percentage less than Lili's salary?

20. A family had dinner in a restaurant and paid \$60 for food. They also had to pay 9.5% sales tax and 10% for the tip. How much did they pay for the dinner?

21. In a small town the population is 65,000. If the male population increases by 4% and the female population increase by 5% then the total population at the end of the year become 67,940. Find the total number of females in the town.

22. If the side of a square is increased by 16% and its area is increased by P%, find the value of P.

Hint: The area of a square is calculated by the product of the length of each side with itself.

23. The price of a car in 2017 was \$28,000. The price of this car decreases 12% in 2018, 10% in 2019, and 7% in 2020. What is the price of a car at the end of 2020?

24. Two numbers are respectively 18% and 30% more than the third number,
    a) What % is the first number of the second number?
    b) What % is the second number of the first number?

25. If the price of brown sugar decreases by 11%, then what should be the percentage increase in the quantity consumed so that the total expenditure on brown sugar remains the same?

7. Real World Applications of Numbers

Numbers permeate our daily existence beyond mere abstract symbols or classroom concepts. They wield profound importance in the real world, propelling technological advancements and underpinning scientific breakthroughs. This article explores the ubiquitous application of numbers across various domains, including finance, engineering, and medicine. By comprehending the pervasive influence of numbers, we gain insight into their omnipresence and pivotal role in shaping modern society.

7.1. Example: Simple Average

US Warehouse Inc. maintains inventory across eight distinct categories, outlined in the table below:

No.	Item number	Stock item name	Cost price ($)
1	CH10001	Chrome plated items	986
2	BR30005	Bronze medals	1,143
3	BR30009	Bronze objects in art	1,826
4	PL50011	Plastic tables & chairs	324
5	PL50014	Plastic toys	328
6	AL80002	Aluminum pots	785
7	CO20006	Copper plates	687
8	GO90021	Gold objects in art	2,472

They wish to know the average price of their stock.

Solution

The simple average is the summation of the values of items divided by the number of items. So,

$$\text{Simple average price} = \frac{\text{sum of the items' value}}{\text{number of items}}$$

$$\text{Simple average price} = \frac{986 + 1{,}143 + 1{,}826 + 324 + 328 + 785 + 687 + 2{,}742}{8}$$

$$\text{Simple average price} = \frac{8{,}551}{8}$$

$$\text{Simple average price} = 1{,}068.875.$$

Therefore, the average price of their stock is $1,068.88 (rounded by hundredth decimals).

7.2. Example: Weighted Average

A teacher has a grading system in which 10 percent of a student's grade will be based on assignments, 25 percent on the first midterm exam, 25 percent on the second midterm exam, and 40 percent on the final exam. What is the grade for a student that has scored 82 percent on assignments, 73 percent on the first midterm exam, 78 percent on the second midterm exam, and 86 percent on the final exam?

Note: To find that student's grade means computing the weighted average.

Solution

For finding the weighted average we must multiply each number by its weight and sum them up and then dividing the resulted summation by the sum of all weights. Mathematically, we have: $WA = \frac{\sum_{i=1}^{n} w_i a_i}{\sum_{i=1}^{n} w_i}$, where *WA* is the weighted average, *n* is the number of terms to be averaged, a_i is the data values to be averaged, and w_i is the weight applied to term *i*. Using this formula, we can fine the student's grade. So,

$$WA = \frac{\sum_{i=1}^{n} w_i a_i}{\sum_{i=1}^{n} w_i}$$

$$WA = \frac{0.10(82) + 0.25(73) + 0.25(78) + 0.40(86)}{0.10 + 0.25 + 0.25 + 0.40}$$

we plugged precents of the grading system and the student's grades into the equation

$$WA = \frac{8.20 + 18.25 + 19.50 + 34.40}{0.10 + 0.25 + 0.25 + 0.40}$$

$$WA = \frac{80.35}{1} = 80.35.$$

Therefore, the student's grade is 80.35 which is rounded as 80.

7.3. Example: Simple Moving Average

Gordon, a stock trader, wants to calculate the simple moving average for Stock BBB. He has looked at the closing prices of the stock for the last eight days. The closing prices for Stock BBB for the last eight days are presented in the table below.

Day – 8	Day – 7	Day – 6	Day – 5	Day – 4	Day – 3	Day – 2	Day – 1
$52.92	$53.32	$53.46	$52.94	$53.58	$54.28	$55.52	$53.86

a) What would be the simple moving average of the stock BBB prices if Gordon decides to consider a period of the last four days?

b) What would be the simple moving average of the stock BBB prices if Gordon decides to consider a period of the last six days?

Solution

The moving average, *MA*, is calculated by adding up all the data points, a_i, during a specific time period and dividing the sum by the number of time periods, *n*. Mathematically, we have: $MA = \frac{\sum_{i=1}^{n} a_i}{n}$, where *MA* is the moving average, *n* is the number of terms in the time period, and a_i is the data values to be averaged. Using this formula, we can answer the requested questions.

a) By the moving average formula, we need to add the prices for day – 4 to day – 1 and divide the result by 4 (number of days in a four-day period). So,

$$MA = \frac{\sum_{i=1}^{n} a_i}{n}$$

$$MA = \frac{53.58 + 54.28 + 55.52 + 53.86}{4}$$

we plugged the stock prices for the last four days and $n = 4$ into the equation

$$MA = \frac{217.24}{4} = 54.31.$$

Therefore, the moving average of the stock prices for the period of the last four days is 54.31.

b) Again, by the moving average formula, we need to add the prices for day – 6 to day – 1 and divide the result by 6. So,

$$MA = \frac{\sum_{i=1}^{n} a_i}{n}$$

$$MA = \frac{53.46 + 52.94 + 53.58 + 54.28 + 55.52 + 53.86}{6}$$ we plugged the stock prices for the last four days and $n = 6$ into the equation

$$MA = \frac{323.64}{6} = 53.94.$$

Therefore, the moving average of the stock prices for a period of the last four days is 54.94.

7.4. Example: Units Conversion

Highway regulations in a certain jurisdiction state that a truck with two axles may have a maximum weight of 16000 pounds on the front axle and 20000 pounds on the rear axle. Convert these weights to their equivalents in kilograms.

Solution

From our table of unit conversion factors, we have

1 pound = 1 lb = 0.454 kg.

The template for converting the maximum front axle weight is

16000 lb. = 16000 $lb \left(\frac{kg}{lb}\right)$.

Filling in the numbers and doing the arithmetic gives

16000 lb = 16000 $lb\ (\frac{0.454\ kg}{1\ lb})$

$= (16000\ lb)(0.454)\ kg$ — we deleted "lb" from the nominator and denominator

= 7264kg.

The calculation for the maximum rear axle weight goes the same way:

20000 lb = 20000 $lb\ (\frac{0.454\ kg}{1\ lb})$

$= (20000\ lb)(0.454)\ kg$ — we deleted "lb" from the nominator and denominator

= 9080 kg.

Hence, these regulations correspond to a maximum weight of 7,264 kg on the front axle and 9,080 kg on the rear axle.

7.5. Example: Landscaping

Joe has a project to build a planter box with a length of 9 feet, the width of 3 feet and the height of 1.1 feet. He also plans to purchase some bagged soil mix to fill up the box. Each bag can fill a volume of 0.30 cubic feet, $V_b = 0.3$, weighs 32lbs, and costs $2.99.

a) Find the number of soil mix bags he may need to fill the planter box.

b) Find his cost for this project.

Solution

a) First, we need to calculate the volume of the dirt needed to fill the planter box. Considering 0.1 residual height for watering planter box, he wants to have of an empty space 0.1 feet depth at the top of planter box. Hence, the height of soil to be poured in the box is 0.1 less than the height of box. Or,

Soil height in the box – the box height minus the residual height = 1.1 – 1 = 1 foot

We now need to find the volume of the soil, V_s. So,

V_s = length × width × height

$V_s = 9ft \times 3ft \times 1ft$ we plugged, length = $9ft$, width = $3ft$ and the height = $1ft$ into the equation

$V_s = 27ft^3$.

To find the number of soil bags, we need simply to divide the volume of the required soil, V_s by the volume of the soil in the bag, V_b. So,

The number of soil bags $= \frac{V_s}{V_b} = \frac{27}{0.3} = 90$.

Therefore, the number of soil mix bags Joe needs to fill the planter box is 90 bags.

b) To find the cost of the planter box project, we need simply to multiply the price of each bag by the number of bags he requires to purchase. So,

Joe's project cost = \$2.99 × 90 = \$260.10.

Therefore, Joe must pay \$260.10 for purchasing 90 soil mix bags.

7.6. Example: Filling a Pool up

Suppose a pool holds 50,000 liters and we have a hose to fill it up. If it takes 35 hours and 24 minutes to fill the pool, what would be the flowrate of the hose in terms of liters per minutes?

Solution

To compute the flow rate of the hose, we need first to convert the filling time to minutes. Knowing ever hour is 60 minutes, we will have.

$35 \text{ hours} \times \frac{60 \text{ minutes}}{1 \text{ hour}} = 2{,}100 \text{ minutes}$

35 hours + 24 minutes = 2,100 minutes + 24 minutes = 2,124 minutes

We can now calculate the flowrate by dividing the pool volume by the time it takes to be filled. So,

$\text{Hose flowrate} = \frac{\text{Pool volume}}{\text{time to fill the pool}}$

$\text{Hose flowrate} = \frac{50{,}000 \text{ liters}}{2{,}124 \text{ minutes}} = 23.54 \text{ liters/minutes}$

Therefore, the hose flowrate is 23.54 liters per minute.

7.7. Example: Machine Production Rate

Suppose a manufacturing workshop has received an order 89,832 units of a specific plastic chair. There are 4 plastic injection molding machines in this workshop, 3 of which are newer and can produce 20 chairs per minute (*cpm*). The fourth machine is older and can manage about 16 chairs a minute. Can these four machines be set up to handle the order in less than 20 hours?

Solution

To obtain the solution we need to add up the machine rates for each of the injection molding machines to get the total possible injection output per minute:

Output per minute, the three newer machines = $20cpm \times 3$ machines = $60\ cpm$

Output per minute, the older machine = $16cpm \times 1$ machine = $16cpm$

Output per minute, all the machines = $60cpm + 16cpm = 0cpm.$

To find the number of minutes for producing the order, we need to divide the number of chairs in the order by the output rate of the four machines. So,

$$\text{Minutes to handle the order} = \frac{\text{Total number of chairs in order}}{\text{Output per minute of all mchines}}$$

$$\text{Minutes to handle the order} = \frac{\text{89,832 chiars}}{\text{76 cpm}} = 1{,}182 \text{ minutes.}$$

Since it is asked the time of handling the order in hour, we need to convert the production time of handling the order in terms of hours. So,

$$\text{Hours to handle the order} = \text{Minutes to handle the order} \times \frac{\text{1 hour}}{\text{60 minutes}}$$

$$\text{Hours to handle the order} = 1{,}182 \times \frac{\text{1 hour}}{\text{60 minutes}} = 19.7 \text{ hours.}$$

Therefore, it takes 19.7 hours to produce 89,832 chairs, or 19 hours 42 minutes (the decimal portion of time is 0.7, which is $0.7 \times \frac{\text{60 minutes}}{\text{1 hour}} = 42$ minutes). Hence, these four machines can be setup to handle the order in less than 20 hours.

7.8. Example: Shopping Discounted Items

Catrina is shopping in a supermarket. She usually purchases discounted items. There are four items having considerable discount tags. Items 1 and 2 are taxable items and are tagged as $8.68 with a 25% discount and $6.50 with a 20% discount, respectively. Items 3 and 4 are nontaxable items and are tagged as $6.70 with a 30% discount and $8.25 with a 40% discount, respectively. The sales tax is 8.0%. If she has $23.75 in her purse, can she purchase all four items? Round the answers to the nearest cents.

Solution

We need to calculate the sale prices of items 1 and 2. Item 1 is tagged $25\% = \frac{25}{100} = 0.25$, and item 2 is tagged $20\% = \frac{20}{100} = 0.20$. So,

Discount of item 1 = $8.68 × 0.25 = $2.17

Sale price of item 1 before sales tax = $8.68 – $2.17 = $6.51

Sales tax of item 1 = $6.51× 0.08 = $0.52

Sale price of item 1 after sales tax = $6.51 + 0.52 = $7.03

Discount of item 2 = $6.50 × 0.20 = $1.30

Sale price of item 2 before sales tax = $6.50 – $1.30 = $5.20

Sales tax of item 1 = $5.20 × 0.08 = $0.42

Sale price of item 1 after sales tax = \$5.20 + 0.42 = \$5.62.

Similarly, we calculate the sale prices of items 3 and 4.

Discount of item 3 = \$6.70 × 0.30 = \$2.01

Sale price of item 3 = \$6.70 – \$2.01 = \$4.69

Discount of item 4 = \$8.25 × 0.40 = \$3.30

Sale price of item 4 = \$8.25 – 3.30 = \$4.95.

To find the sale price for all four items we simply need to add up the sale prices of the four items. So,

Total price of the four items = Sale price of item 1 + Sale price of item 2 + Sale price of item 3 + Sale price of item 4

Total price of the four items = \$7.03 + \$5.62 + \$4.69 + 4.95 — we plugged the sale prices of the four items into the equation

Total price of the four items = \$22.29.

She has \$23.75 in her purse which is greater than \$22.29. So, she can afford to purchase all four items and has \$1,46 left in her purse.

7.9. Example: Population Growth

According to the United Nations, Department of Economic and Social Affairs, global human population growth amounts to around 85 million annually, or 1.10% per year. Global human population in 2020 is 7.8 billion and it is expected to keep growing. What will be the global population after 5 years?

Solution

The population can be written as a geometric sequence in which we have the value of a_1 and we are looking for a_5. We have learned the formula to obtain the nth term of a geometric sequence is $a_n = r^{n-1} a_1$. Hence, using this formula to find a_5, we first need to determine the value of the common ratio, r. we also learned in a geometric sequence, the common ration is determined by dividing a term by its preceding term, $r = \frac{a_{k+1}}{a_k}$. Since the population growth rate 1.10% it means that given a_k and a_{k+1} as the two consecutive terms of the geometric sequence, $a_{k+1} = a_k\,(1 + 1.10\%)$ or $a_{k+1} = a_k\,(1 + 0.011)$. Hence, the common ration is determined by:

$$r = \frac{a_{k+1}}{a_k}$$

$$r = \frac{a_k\,(1+0.011)}{a_k}$$ we plugged $a_{k+1} = a_k\,(1 + 0.011)$ into the equation

$$r = \frac{a_k\,(1+0.011)}{a_k}$$ we eliminated a_k from top and bottom

$$r = (1 + 0.011)$$

$$r = 1.011.$$

To find a_5, we now can use the formula $a_n = r^{n-1} a_1$. So,

$$a_n = r^{n-1} a_1$$

$a_5 = (1.011)^{5-1} \times 7.8$ we plugged $a_1 = 7.8$, $n = 5$, and $r = 0.011$ into the equation

$a_5 = (1.011)^4 \times 7.8$

$a_5 = (1.0796) \times 7.8$

$a_5 = 8.42.$

Therefore, the global population after 5 years is 8.42 billion (rounded by two decimals).

7.10. Example: Exercising Duration

After Clark knee surgery, his physio tarpits has asked he returns to his jogging program slowly. The physio tarpits suggested jogging for 10 minutes each day for the first week. After each week thereafter, she suggested he might increase that time by 5 minutes per day. Find out how many minutes per day Clark will be jogging after 8 weeks.

Solution
It is realized that the suggested program is an arithmetic sequence with $a_1 = 10$, $n = 8$, and $d = 5$, where we are looking to find a_8. The formula relating a_n and a_1 is $a_n = a_1 + (n - 1)d$. Using this formula, we will have.

$a_n = a_1 + (n - 1)d$

$a_8 = 10 + (8 - 1) \times 5$ we plugged $a_1 = 10$, $n = 8$, and $d = 5$, into the equation

$a_8 = 10 + (7) \times 5$

$a_8 = 10 + 35$

$a_8 = 45.$

Therefore, Clark will be jogging 45 minutes per day in 8 weeks.

7.11. Example: Construction Project

Steven plans to construct a trapezoidal shaped structure in his garden. He starts with the longer side of the trapezoid with a row of 86 bricks. Each row must be decreased by 2 bricks on each end and the construction should stop at the 21st row. Find the number of bricks Steven needs to buy.

Solution
Given the longer side of trapezoid is started with 86 bricks and each row is decreased by 4 bricks (2 bricks on each), we will come up with a sequence below.

86, 82, 78, 74, 70, 66, 62, ⋯.

The above sequence is an arithmetic series with the common difference, $d = 86 - 82 = -4$. Since trapezoidal shaped structure is completed at the Steven stops at the 21st row, hence the above sequence arithmetic has 21 terms, or $n = 21$. To find the number of bricks Steven needs to buy, we need to use the formula for finding the sum of 'n' terms of an arithmetic series, $S_n = \frac{n}{2}[2a_1 + (n - 1)d]$ with $a_1 .= 86$, $n = 21$ and $d = -4$. So,

$S_n = \frac{n}{2}[2a_1 + (n-1)d]$

$S_n = \frac{21}{2}[2(86) + (21-1)(-4)]$ we plugged $a_1 = 86$, $n = 21$ and $d = -4$ into the equation

$S_n = \frac{21}{2}[172 + (20)(-4)]$

$S_n = \frac{21}{2}(172 - 80)$

$S_n = \frac{21}{2}(92)$

$S_n = \frac{21(92)}{2}$

$S_n = \frac{1{,}932}{2} = 996.$

Therefore, the number of bricks Steven needs to buy is 996.

7.12. Example: Sales Commission

Jack, a sophomore student at American International College, has three summer job options to choose from: working at a used car dealership, an electronic store, or the school's restaurant.

At the used car dealership, Jack would work 15 hours per week and earn $4 per hour plus a 1% commission on his car sales. The average sale per week at the dealership is $14,000.

Alternatively, Jack could work at the electronic store, also for 15 hours per week, earning $7 per hour plus a 4% commission on sales. The electronic store's projected weekly sales amount to $2,350.

As another option, Jack could earn a fixed rate of $195 per week by working 15 hours washing dishes at the school's restaurant. Jack's objective is to choose the job that pays him the highest income.

Determine which of the three job options offers the highest income for Jack.

Solution

Let us find his earnings from the jobs with the commission payments.

For the car dealership job, letting S_1 to be the amount of car sales for Jack's earnings.

His hourly earnings = 15 ×$4 = $60

His commission = 1% × S_1 = 0.01 × S_1 = 0.01S_1

Jack earning = hourly earning + commission = $60 + 0.01$S_1$.

For the electronic store job, letting S_2 to be the amount of electronic items sales, Jack earning is

His hourly earnings = 15 ×$7 = $105

His commission = 4% × S_2 = 0.04 × S_2 = 0.04S_2

Jack earning = hourly earnings + commission = $105 + 0.04$S_2$.

Jack earns $195 for the school restaurant job.

To compare Jack's earnings, we need to compare his possible two-job earnings at a time. Comparing Jack earnings from the car dealership with his earnings from the resultant job, we will have:

$$\$60 + 0.01S_1 \overset{?}{\Leftrightarrow} \$195$$

$$\$60 + 0.01(\$14{,}000) \overset{?}{\Leftrightarrow} \$195$$ we plugged the last summer sales estimate into the inequality

$$\$60 + \$140 \overset{?}{\Leftrightarrow} \$195$$

$$\$200 \geq \$195.$$

As it is seen, Jack working with car dealership earns \$200 while his earning from the school restaurant is \$195. Hence working with a car dealership is preferable.

We now need to compare the two jobs working for a car dealership and working for the electronic store. So,

$$\$105 + 0.04S_2 \overset{?}{\Leftrightarrow} \$200$$

$$\$105 + 0.04(\$2{,}350) \overset{?}{\Leftrightarrow} \$200$$ we plugged the estimated sales of the electronic store into the inequality

$$\$105 + \$94 \overset{?}{\Leftrightarrow} \$200$$

$$\$199 \leq \$200.$$

As it is seen, Jack working with the electronic store earns \$199 while his earnings from the used car dealership are \$200. Hence working with a car dealership is preferable as well.

Therefore, working with car dealership is the best choice for jack among the three job options.

7.13 . Example: Economy Growth

Suppose a governor of a state is willing to find the effect of a tax rebate on the state economy growth. He asked his economy advisors to conduct a computer digital simulation to find out the multiple effects of a \$1,200 tax rebate. The government statistics say that each household will spend 85% of the rebate on goods and services. Find the multiplier effect of their experiment on the state economy.

Solution

Every time money goes into the economy, 85% of it is spent and is then in the economy to be spent. Again, 85% of this money is then spent on the economy. This situation continues and so leads us to an infinite geometric series with terms shown below.

$$1200,\ 1200(0.85),\ 1200(0.85)(0.85),\ 1200(0.85)\ (0.85)\ (0.85),\ \cdots,\ \text{or}$$

$$1200,\ 1200(0.85),\ 1200(0.85)^2,\ 1200(0.85)^3,\ \cdots.$$

To find the multiple effect of their experiment on the state economy, we need to use the formula for the sum on an infinite geometric sequence, $S = \frac{a_1}{1-r}$, with $a_1 = 1200$, and the common ratio of $r = 0.85$. So,

$$S = \frac{a_1}{1-r}$$

$S = \frac{1200}{1-0.85}$ we plugged $a_1 = 1200$ and $r = 0.85$ into the equation

$S = \frac{1200}{0.15}$

$S = \frac{1200}{0.15} = 8{,}000.$

Therefore, the total effect of the $1,200 received by each household will be a $8,000 growth in the state economy.

7.14. Example: Bank Withdrawal and Deposit

Suppose you have a bank account and have $100 in your bank account on November 1. You deposit $200 on November 2, and later the same day you spend $250 at Center City Mall on your debit card. Illustrate the account numbers activity on the number line.

Solution

Below, we represent the deposit with line numbers.

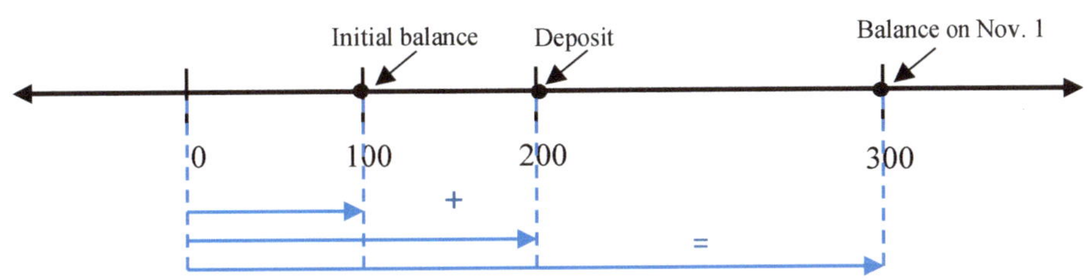

The withdrawal line number is.

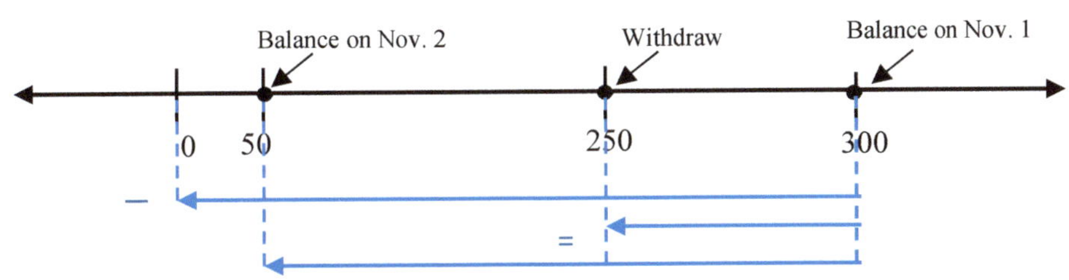

7.15. Example: The LCM Application

Edward visits his physiotherapist every 30 days and Pitter visits the same physiotherapist every 45 days. Edward and Pitter meet each other in the physiotherapist office today. After how many days will they meet each other in the physiotherapist's office again?

Solution

The problem can be solved using the least common multiple of the time span each visits the physiotherapist, time that that will occur the same time (common). Hence, we need to list the prime factors of the values to find their LCM.

Factors of 30: $2 \times 3 \times 5$

Factors of 45: $3 \times 3 \times 5$.

The LCM of 30 and 45 is: $2 \times 3 \times 3 \times 5 = 90$.

Therefore, they will be in the physiotherapist's office together after 90 days.

7.16. Exercises

1. A $25km$ long, $8m$ wide, two-lane highway is to be paved with a $5cm$ thick layer of asphalt. A fleet of three dump trucks is to be employed, each with an empty mass of $20tons$ and a carrying capacity of $20m^3$. Asphalt with a density of $0.72g/cm^3$ will be used.
   a) Find the total volume of asphalt needed.

   b) Find the number of dump truck loads required.

   c) Find the gross mass of each truck.

   Hint: the volume of a right rectangular prism is calculated by length × width × height.

2. An insect population is growing in such a way that each new generation is 1.4 times as large as the previous generation. Suppose there are 8,000 insects in the first generation.
   a) How many will there be in the third generation?

   b) What will be the total number of insects in the fourth generations?

3. There are 85 green balls, 55 black balls, and 130 red balls out of the total of 270 balls in a bag. Find the percentage of:
   a) Green balls in a bag.

   b) Black balls in a bag.

   c) Red balls in a bag.

4. In a US International College, 40% of all students are male, and 20% of all students are majoring in electrical engineering. If 10% of male students are electrical engineering majors, what percentage of electrical engineering majors are male?

5. A small machine shop has 18 employees. Three of those employees are paid $28 per hour, six of them are paid $32 per hour, four are paid $36, and the rest of employees are paid $24 per hour. What is the average hourly wage of these 18 employees?

6. Charli has built a tower with blocks that use 21 for the bottom row. Each row has 3 fewer blocks than the previous row. Suppose that there are 5 rows in the tower.

   How many blocks are used for the top row?

7. A man bought a car for $38,600 and paid a down payment of $6,000. He decided to pay for the car in 60 monthly installments given by an arithmetic sequence. After paying the 36 installments, the man died while owing the car dealer $11,500 of the total debt. What is the amount that he paid for each installment.

8. Stock ABS was trading at 120, 134, 146, 141, 132, 154, 153 in US dollars for the previous seven trading days. Based on the given numbers, find.
   a) The 4-daily moving averages for the given data.

   b) The 6-daily moving averages for the given data.

9. Suppose Tom took five courses in past Fall semester at American International College, and made the grades shown in the table below.

Course	Credit	Grade
Physic	4	B
Algebra	3	A
English literature	2	A
Sociology	4	C
Computer programming	3	D

The college uses the scale: $A = 4$, $B = 3$, $C = 2$, $D = 1$, $F = 0$, to compute the students' GPA. Find Tom's GPA for the past Fall semester.

10. Samantha has two pieces of cloth. One piece is 66 inches wide, and the other piece is 78 inches wide. She wants to cut both pieces into strips of equal width that are as wide as possible. How wide should she cut the strips?

11. A governor in a small African country wants to find the effect of tax rebate on the country economy growth. The government statistics showed that each household will spend 75% of the rebate in goods and services. Find the multiple effect of 8,000 African pounds' tax rebate on the country economy growth.

12. Suppose Mat Milling Machin Shop has received an order 360,000 units of a specific Brans bolts. There are 3 thread milling machines in this workshop. One of these machines is a high-performance machine that can produce 120 bolts per minute. The two other thread milling machines are lower-performance and can produce 98 bolts per minute. Find the number of hours these three machines are working simultaneously to produce the requested order.

13. Suppose an above ground liquified natural gas (LNG) storage tank has the capacity of $40{,}000m^3$ and we have a pump to fill it up. If it takes 64 hours and 48 minutes to fill the tank, what would be the flowrate of pump in terms of liters per minute?

14. You are driving along and see that your tank is only $\frac{3}{8}$ full. The tank holds 16 gallons of gas. Your car goes $21\frac{1}{4}$ miles for every gallon of gas. How far can you drive before you run out of gas?

15. National Natural Park in one of the African countries covers $\overline{1a4b}$ acres of land. If the area of the park is divisible by the numbers 5, 8 and 9, find a and b.

16. Sheela has 38 pennies 35 nickels, 32 dimes and 46 quarters in her ceramic piggy bank saving moneybox. She wants to buy a Disney Princess Royal Shimmer doll worth \$21.99. Can she afford to purchase the doll?

17. Margret has \$6 in her purse, and she needs to pay a friend \$3.00 and she receives from another friend \$2.00.
    a) Illustrate the money numbers activity on the number line.
    b) What would be the answer if she had \$7 in her purse?

18. Rose is preparing herself for a jogging marathon race. She covered $5.8km$ on the first day and she increased the distance by $0.6km$ every day. Find the total distance she covered in 16 days.

19. A gasoline tank had 3,825 liters of gasolinc. The volume of gasoline decreased by 15, 30, 60 liters over the next three days respectively. Following the same pattern, how long it will take for the tank to become empty.

8. Multiple Choice Questions

1. A doctor prescribed 24 pills for his patient to be taken in the first week. Given that the patient should decrease the dosage by 6 pills every week, find the week in which he will stop taking the medicine completely.

   A. In the third week
   B. In the fourth week
   C. In the fifth week
   D. In the sixth week

2. Sam works in a grocery store. He stacks tomato paste cans in rows, where there are 36 cans in the first row, 34 in the second, 32 in the third, and so on. Find the row which has exactly 26 cans.

   A. In the third week
   B. In the fourth week
   C. In the fifth week
   D. In the sixth week

3. An eight-year housing development project consists of constructing new town houses in Oxnard suburbs. 180 townhouses were built in the first year and 340 in the fourth year. If the number of town houses built each year is represented as an arithmetic sequence, how many town houses will be built in total over the 8 years?

   A. 500 town houses
   B. 490 town houses
   C. 470 town houses
   D. 460 town houses

4. Find the average of the series: 118, 302, 176, 211 and 330?

   A. 229
   B. 231
   C. 233
   D. 235

5. In Class I there are 10 students of average age 16 years and in Class II there are 18 students of average age 15 years. What is the average age of both the classes (the answer is rounded by 1 decimal)?

   A. 15.5
   B. 15.4
   C. 15.3
   D. 15.2

6. The average age of the three girls is 15 years and their ages are in proportion 3:5:7. What is the age in years of the youngest girl?

A. 9
B. 12
C. 15
D. 14

7. The average of 10 numbers is 36. If each number is increased by 3, what will the new average be?

A. 39
B. 36
C. 33
D. 30

8. Table below illustrate the weather temperature for the last week. What is the average temperature of the last four days?

Weekday	Sunday	Monday	Tuesday	Wednesday	Thursday	Friday	Saturday
Temperature °F	68	72	66	64	67	69	74

A. 67 °F
B. 67.5 °F
C. 68 °F
D. 68.5 °F

9. Evaluate the expression $81^{-\frac{3}{2}}$.

A. $-\frac{1}{729}$
B. -729
C. $\frac{1}{729}$
D. 729

10. Evaluate the expression $\sqrt[4]{625}$.

A. 125
B. 25
C. 5
D. $\sqrt[3]{5}$

11. Simplify the exponential expression $\frac{(2 \times 9^3)^3}{9^{15}}$.

A. $\frac{2}{9^6}$
B. $\frac{2}{9^9}$
C. $\frac{8}{9^9}$

D. $\frac{8}{9^6}$

12. Simplify the expression $\frac{(5)^3}{25} + \frac{(5)^3}{5} - \frac{26(5)^3}{125}$.

A. 4
B. -4
C. $\frac{4}{125}$
D. $-\frac{4}{125}$

13. Oranges are priced at \$1.40 per pound (*lbs.*). How much one pays for 2500*g.* of orange? (assume1000*g.* = 2.2*lbs.*)

A. \$5.50
B. \$7,70
C. \$1.59
D. \$11.55

14. Jack, Bob, and Tony start to jog around a circular stadium. They complete their rounds in 36 seconds, 48 seconds, and 56 seconds, respectively. After how many seconds will they be together at the starting point?

A. 336*sec.*
B. 504*sec.*
C. 1008*sec.*
D. 2016*sec.*

15. Mrs. Olson has three classes. Each class has 18, 12 and 24 students, respectively. Mrs. Olson wants to divide each class into groups so that every group in every class has the same number of students and there are no students left over. What is the maximum number of students she can put into each group?

A. 4 students
B. 8 students
C. 2 students
D. 6 students

16. Simplify $15 - 2(4a - 3b + 7) + 3(a + 2b)$.

A. $1 - 5a + 6b$
B. $1 - 7a + 12b$
C. $1 - 7a + 6b$
D $1 - 5a + 12b$

17. Simplify $\frac{5}{9} - (x - \frac{3}{18}y + \frac{2}{9}) + 3(x + \frac{5}{18}y)$.

A. $\frac{1}{3} + 2x + y$

B. $\frac{1}{3} + 2x + 2y$

C. $\frac{1}{3} + 2x + \frac{2}{3}y$

D $\frac{1}{3} + 4x + \frac{2}{3}y$

18. Simplify $\frac{7}{3} \times k - \frac{1}{14} \times 21h + \frac{3}{21} \times 14k + \frac{1}{3}$.

A. $\frac{1}{3} + \frac{3}{2}h + \frac{4}{3}k$

B. $\frac{1}{3} + \frac{2}{3}h + \frac{13}{3}k$

C. $\frac{1}{3} + \frac{3}{2}h + \frac{13}{3}k$

D. $\frac{1}{3} + \frac{2}{3}h + \frac{4}{3}k$

19. Simplify $-5\sqrt{17-15} + 3 - \sqrt{8} + \sqrt{9-7} - 5$.

A. $-5\sqrt{2} - 2$

B. $6\sqrt{2} - 5$

C. $-6\sqrt{2} - 2$

D. $5\sqrt{2} - 5$

20. Simplify $\sqrt[3]{81} + 4\sqrt[3]{13-10} - 3\sqrt[3]{27} + 2$.

A. $5\sqrt[3]{3} - 7$

B. $5\sqrt[3]{3} - 1$

C. $7\sqrt[3]{3} - 1$

D. $7\sqrt[3]{3} - 7$

21. A theater has 36 rows of seats. The first row has 25 seats, and the number of seats in each subsequent row is increased by 3 seats. What is the total number of seats in the theater?

A. 2,970 seats

B. 2,890 seats

C. 2,870 seats

D. 2,790 seats

22. Which of the following represents $16^{\frac{3}{2}}$ as an integer?

A. 4

B. 16

C. 48

D. 64

23. Which of the following represents $1 + 27^{\frac{2}{3}}$ as an integer?

A. 8
B. 9
C. 10
D. 12

24. Which of the following represents $\sqrt[3]{81}$ as an integer?

A. $\sqrt[3]{3}$
B. $3\sqrt[3]{3}$
C. $3\sqrt[3]{9}$
D. $\sqrt[3]{9}$

25. What is the equivalent of $\left(\frac{5}{18} \div \frac{10}{27}\right) - \frac{1}{16}$?

A. $\frac{11}{16}$
B. $\frac{13}{16}$
C. $\frac{11}{4}$
D. $\frac{13}{4}$

26. What is the equivalent of $\left(\frac{5}{6} + \frac{1}{12} - \frac{2}{3}\right) \div \frac{7}{4}$?

A. $\frac{1}{7}$
B. $\frac{8}{21}$
C. $\frac{4}{7}$
D. $\frac{4}{21}$

27. What is the equivalent of $\frac{2^4}{5} + \frac{14}{25} - \frac{4^2}{5} - \frac{3^2}{25}$?

A. $-\frac{1}{5}$
B. $\frac{1}{5}$
C. $\frac{1}{25}$
D. $-\frac{1}{25}$

28. Charli works on a project in his physic course to make an AC electric generator using copper wire, and cardboard. He needs to wrap the wire around and around a rectangular prism with the height of $7in.$, the width of $2.5in.$, and the length of 3.5 $in.$ made from cardboard, as shown in the figure below. Charli uses 20.5 feet of wire from which he needs to leave 3 inches on each end for connecting to any electric device. How many times does he need to wrap the wire around the cardboard prism?

Hint: The perimeter of prism base is $2w + 2l$.

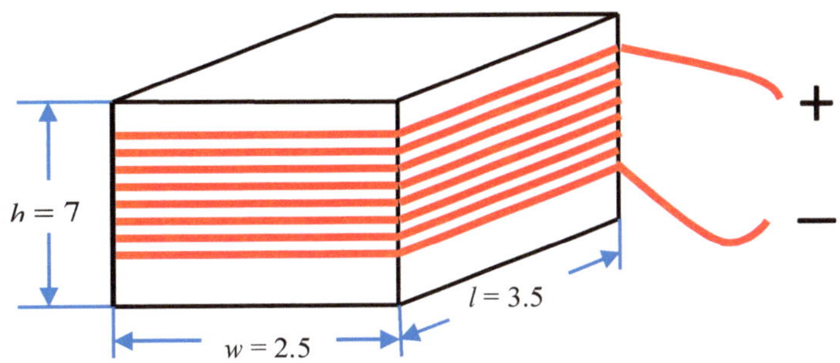

A. 20
B. 22
C. 24
D. 30

29. Write set $A = \{15, 17, 19, 21, 23, 25, \cdots\}$ in the form of set builder notation?

A. $A = \{x / x$ takes all the even integers greater than or equal to $15\}$
B. $A = \{x / x$ takes all the odd integers greater than or equal to $15\}$
C. $A = \{x / x$ takes all the odd integers greater than $15\}$
D $A = \{x / x$ takes all the even integers greater than or equal to $25\}$

30. Express set $\{x | x < -3 \text{or } x \geq 5\}$ using the set interval notation.

A. $(-\infty, -3] \cup (5, \infty)$
B. $(-\infty, -3] \cup [5, \infty)$
C. $(-\infty, -3) \cup [5, \infty)$
D $(-\infty, -3) \cup [5, \infty]$

31. Compare $7\frac{9}{8}$ and 8.11.

A. $7\frac{9}{8}$ is greater than 8.11
B. $7\frac{9}{8}$ is less than 8.11
C. $7\frac{9}{8}$ is equal to 8.11
D They are comparable

32. Compare $\frac{27}{8}$ and $3\frac{3}{8}$.

A. $\frac{27}{8}$ is greater than $3\frac{3}{8}$
B. $\frac{27}{8}$ is less than $3\frac{3}{8}$
C. $\frac{27}{8}$ is equal to $3\frac{3}{8}$
D They are comparable

33. which of the following is not a correct statement?

A. $a + b = b + a$ is the inverse property of addition

B. $a + 0 = a$ is the identity property of addition

C. $a \times (\frac{1}{a}) = 1$ is the inverse property of multiplication

D $a \times 1 = a$ is the identity property of multiplication

34. Find the indicated union $(-7, 1) \cup [1, 9] \cup (-\infty, 7]$ and express your answers in interval notation.

A. $(-7, 9]$
B. $(-\infty, 7]$
C. $[-7, 7]$
D $(-\infty, 9]$

35. Find the indicated intersection $(-\infty, 2] \cap (0, 12)$ and express your answers in interval notation.

A. $(-\infty, 12]$
B. $(2, 12)$
C. $[2, 12)$
D $(-\infty, 12)$

36. Which of the following expressions has the same value as $\sqrt{0.36} \times \sqrt{3}$?

A. $\frac{6}{5}\sqrt{3}$
B. $\frac{3}{10}\sqrt{3}$
C. $\frac{6}{8}\sqrt{3}$
D $\frac{3}{5}\sqrt{3}$

37. Apples are priced at \$3.60 per pound. How much one pays for four bags of 1,500g. of apple? (assume $1\,kg. = 2.2lbs.$)

A. \$10.49
B. \$12.66
C. \$9.99
D \$11.88

38. A rectangular solid as shown below has dimensions 5, a, and b, where a and b are integers. Which of the following CANNOT be the areas of three different faces of this solid?

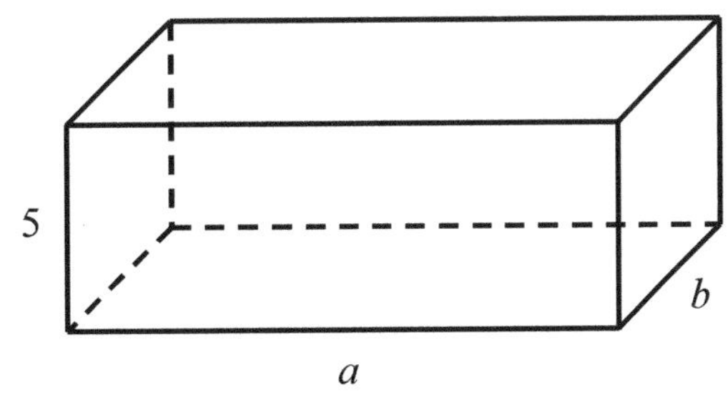

A. 25, 20, 45
B. 15, 10, 30
C. 20, 25, 36
D 35, 40, 30

39. A house appliance store had some toasters. 20% of toasters are sold and still there are 180 toasters left in the store. Originally the number of toasters were in the store is.

A. 144 toasters
B. 200 toasters
C. 900 toasters
D 320 toasters

40. Three candidates, Alice, Fay, and Kelly contested an election and received 15,500, 12,400 and votes 18,600, respectively. What percent of the total votes did Kelly get?

A. 80%
B. 60%
C. 50%
D 40%

41. Which of the following expression is equivalent to $a^{\frac{2}{8}}$?

A. $\sqrt{a^4}$
B. $\sqrt[4]{a}$
C. $\frac{a}{4}$
D $\frac{2}{a^8}$

42. Which of the following expressions has the same value as $\sqrt[3]{\frac{2}{27}} \times \sqrt[3]{4}$?

A. $\frac{2}{3}$
B. $\frac{1}{3}\sqrt[3]{6}$
C. $\frac{1}{9}\sqrt[3]{4}$
D $\frac{4}{\sqrt[3]{27}}$

43. Which of the following is divisible by three numbers5, 8 and 9?

A. 856,440
B. 336,680
C. 754,840
D 136,774

44. Which of the following is divisible by all the three numbers 3, 4 and 5?

A. 74,250
B. 192,630
C. 92,760
D 523,365

45. Edward is 6 feet and 2 inches tall. If 1 foot is equal to approximately 0.3 meters, and 1 inch is 2.54 centimeters then which of the following is closest to Edward's height in meters?

A. 2.15 meters
B. 1.85 meters
C. 2.05 meters
D 1.79 meters

46. A group of patients is recruited for a clinical trial. Their heights, recorded in centimeters, are illustrated in the table below. Two more patients are recruited to study. After these patients join, the mean height is $170cm$. Which of the following could be the heights of the two new patients?

179	169	175	164	172	165	154	168	173	171

A. $148cm$. and $178cm$.
B. $168cm$. and $175cm$.
C. $169cm$. and $181cm$.
D $154cm$. and $180cm$.

47. Which of the following is equivalent to $3^2 \times \frac{1}{243} \times 9^4 \times \frac{1}{27}$?

A. $\frac{1}{9}$
B. 27
C. $\frac{1}{3}$
D 9

48. The monthly salary of Linda is \$3,800. If her salary is raised by 15%, what will be her new salary?

A. \$4,370
B. \$3,815
C. \$3,915
D \$4,471

49. A group of 68 students was polled as to whether they enjoy geometry class, trigonometry class, both, or neither. The results are shown in the table below.

	Geometry	Trigonometry
Enjoy	38	19
Do not enjoy	30	49

Based on the above data, which of the following conclusions is correct?

A. The ratio of those who do not enjoy geometry class to those who do not enjoy trigonometry class is 5:7
B. The ratio of those who enjoy geometry class to those who enjoy trigonometry class is 2:1
C. The ratio of those who do not enjoy geometry class to those who enjoy trigonometry class is 9:7
D The ratio of those who enjoy geometry class to those who do not enjoy trigonometry class is 7:9

50. Which number is divisible by both 2 and 7?

A. 18,046
B. 11,748
C. 18,864
D 22,192

51. Robyn took an exam consisting of 80 multiple-choice questions. In this exam students are asked to select only correct answers from the choices of four listed options. For each correct answer, they earn 1 point and for each incorrect answer they penalize by $\frac{1}{3}$ of a point ($-\frac{1}{3}$ points for each incorrect answer). For each question left unanswered they get 0 points. If Robyn answers 72 questions and gets 57 of them correct, how many points does she earn?

A. $51\frac{1}{3}$
B. 52
C. $53\frac{2}{3}$
D 67

52. If the average of 7, 12, and k is 18, what is the value of k?

A. 18
B. 35
C. 16
D 54

53. Find the sum of the sequence $1, \frac{1}{3}, \frac{1}{3^2}, \frac{1}{3^3}, \cdots$.

A. 2

B. $\frac{3}{2}$

C. $\frac{4}{3}$

D $\frac{10}{9}$

54. Suppose we have a jar which is filled with water. Another similar jar is filled with a solution of 50% orange syrup and 50% water. When both the jars are emptied into a bucket, what will be ratio of water to orange syrup in the bucket?

A. 1:1
B. 2:1
C. 3:1
D 4:1

55. Suppose we have a jar containing water and strawberry syrup in the ratio 8:5. Another similar jar contains water and strawberry syrup in the ratio 9:4. When both the jars are emptied into a bucket, what will be the percentage of strawberry syrup in the bucket?

A. 47.06%
B. 52.94%
C. 48.06%
D 51.94%

56. Kris spends on a fancy hotel that cost $550 per night. If he stayed in that particular hotel for four nights, but his bank only lets him withdraw $400 at a time, how many visits to the ATM must Kris have made in order to cover the cost of his hotel stay?

A. 3
B. 4
C. 5
D 6

57. In a high school student consumes 74.5 pounds of sugar per year. As part of a new nutrition plan, each member of a track team plans to lower the sugar he or she consumes by at least 18% for the coming year. Assuming each track member had consumed sugar at the level of a typical high school student and will adhere to this plan for the coming year, what is the maximum number of pounds of sugar to be consumed by each track team member in the coming year?

A. $55lb.$
B. $61lb.$
C. $63lb.$
D $66lb.$

58. Among 331 million citizens in the United States 42% have valid passports. Of these passport-holding individuals, 45% take international outbound trips each year. Of those who travel

internationally every year, approximately 18% visit European countries. How many U.S. citizens visit European countries every year?

A. 11,260,620
B. 1,126,062
C. 112,606
D 11,261

59. The sum of the three numbers is 180. If the greatest of these numbers is twice of the sum of the other two, what is the value of the greatest of these numbers?

A. 120
B. 116
C. 108
D 100

60. What is the least common multiple (LCM) of 10, 12, 15?

A. 24
B. 30
C. 60
D 300

8.1. Key to the Multiple-Choice Questions

Question No.	Answer
1	C
2	D
3	D
4	B
5	B
6	A
7	A
8	D
9	C
10	C
11	D
12	A
13	B
14	C
15	D
16	D
17	A
18	C
19	C
20	D

Question No.	Answer
21	D
22	D
23	C
24	B
25	A
26	A
27	B
28	A
29	B
30	C
31	A
32	C
33	A
34	D
35	C
36	D
37	D
38	C
39	B
40	D

Question No.	Answer
41	B
42	A
43	A
44	C
45	B
46	C
47	D
48	A
49	B
50	A
51	B
52	B
53	B
54	C
55	B
56	D
57	B
58	A
59	A
60	C

When I am with you, we stay up all night.
When you're not here, I can't get to sleep.
Praise God for these two insomnias!
And the difference between them.
Rumi

Chapter 2

Principles of Analytic Geometry

Contents

Introduction

Analytic geometry, also known as coordinate geometry or Cartesian geometry, investigates geometric concepts using coordinate systems as a framework. Its significance lies in establishing a correlation between geometric shapes and algebraic equations. Key topics covered include determining midpoints and distances, identifying parallel and perpendicular lines on a coordinate plane, segment division, calculating distances between lines and points, among others. The Cartesian coordinate system is commonly employed to manipulate equations representing planes, straight lines, and figures, primarily in two and occasionally three dimensions.

This branch of mathematics finds extensive application in various fields, both directly and indirectly. Analytic geometry plays a pivotal role in physics, engineering, aviation, rocketry, space science, and spaceflight. It serves as the groundwork for numerous modern geometric disciplines, encompassing algebraic, differential, discrete, and computational geometry.

The integration of algebra and geometry became feasible with the advent of a coordinate system, facilitating the expression of geometric concepts such as points and lines in algebraic terms like real numbers and equations. The central concept of analytic geometry revolves around using a coordinate system to associate geometric points with real numbers. By assigning a unique set of real numbers to each point, geometric figures like lines, circles, and conics can be represented through algebraic equations.

At the core of analytic geometry lies the Cartesian coordinate system, which serves as its foundation. This chapter commences by introducing the Cartesian coordinate system, followed by discussions on graphing on the rectangular coordinate, understanding line slopes, applying the midpoint and distance formulas identifying parallel and perpendicular lines on the coordinate plane, and presenting significant graphical representations. Furthermore, real-world applications of analytic geometry will also be explored.

1. Cartesian Coordinates

The introduction of Cartesian coordinates in the 17th century by René Descartes (Latinized as Cartesius) marked a significant breakthrough in mathematics, as it established the first systematic connection between Euclidean geometry and algebra. Through the Cartesian coordinate system, geometric figures can be represented by Cartesian equations, which are algebraic expressions involving the coordinates of the points lying on the figure.

Cartesian coordinates serve as the cornerstone of analytic geometry, offering insightful geometric interpretations for various other mathematical branches, including linear algebra, differential geometry, multivariate calculus, and group theory, among others. One familiar application is the concept of the graph of a function. Moreover, Cartesian coordinates are indispensable tools in numerous applied disciplines dealing with geometry, such as astronomy, physics, engineering, and more.

In a Cartesian coordinate system, each point in a plane is uniquely specified by a pair of numerical coordinates, representing the signed distances from the point to two fixed, perpendicular, and oriented lines. These lines are referred to as coordinate axes, and their intersection point defines the origin of the system at the ordered pair (0, 0). Alternatively, the coordinates can be viewed as the positions of the perpendicular projections of the point onto the two axes, expressed as signed distances from the origin.

A typical Cartesian Cartesian coordinate system consists of x and y axes, each having a unit of length (e.g., feet, meters, miles). Any point within the coordinate system is described by its distance relative to both the x and y axes, denoted as (x, y). The axes intersect at the origin, where both x and y values are zero, labeled as (0, 0).

Imagine two real number lines crossing at a right angle crossing each other at 0 (see the below graph).

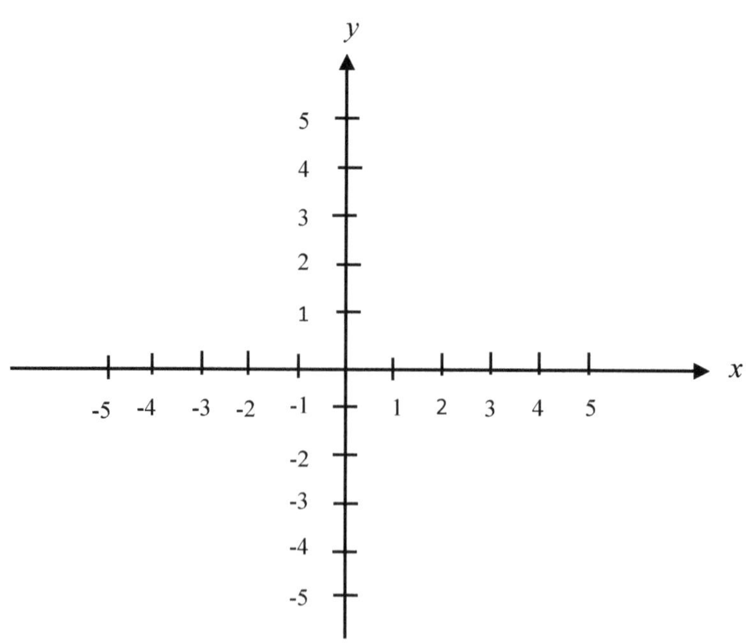

The horizontal number line is generally called the x-axis and the vertical number line is usually called the y-axis. As the number line, these axes extend off indefinitely in both directions. These two number lines allow to locate the positions of points off of the number lines as well as points on the lines themselves. Now, consider point P on the next coordinate axes.

To use the numbers on the axes to locate this point, we must drop a vertical line from the x-axis to P and extend a horizontal line from the y-axis to P. This process is sometimes called 'projecting' the point P to the x- axis and respectively the y- axis.

For example, consider a point, P, using the ordered pair $(3, -5)$. The first number in the ordered pair is usually called the abscissa or x-coordinate and the second is usually called the ordinate or y-coordinate. As the result, the ordered pair $(3, -5)$ comprise the Cartesian coordinates of the point P.

Plotting an Ordered Pair (x, y) in the Coordinate Plane

- Determine the x-coordinate. Beginning at the origin, move horizontally, the direction of the x-axis, the distance given by the x-coordinate.
    - If the x-coordinate is positive, move to the right.
    - If the x-coordinate is negative, move to the left.

- Determine the y-coordinate. Beginning at the x-coordinate, move vertically, the direction of the y-axis, the distance given by the y-coordinate.
    - If the y-coordinate is positive, move up.
    - If the y-coordinate is negative, move down.

- Draw a point at the ending location. Label the point with the ordered pair.

In practice, the distinction between a point and its coordinates is unclear; for example, we often speak of 'the point $(-3, 5)$.' We can think of $(-3, 5)$ as instructions on how to project P from the origin $(0, 0)$ by moving 3 units to the left (because it has negetive sign) and 5 units downwards.

Note that the order in the ordered pair is important. That is if we wish to plot the point $(3, -5)$, we will move to the right 3 units (because it has positive sign) from the origin and then move upwards 5 units, as below.

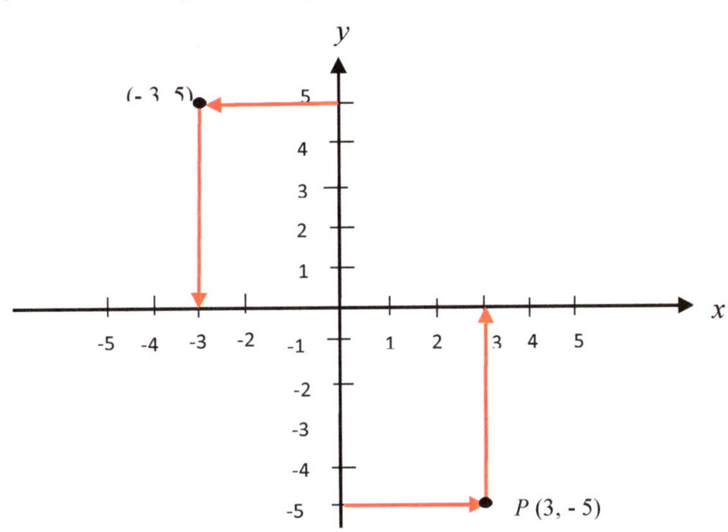

When we contemplate a Cartesian Coordinate Plane, we refer to a set of all possible ordered pairs, $(x; y)$, as x and y take values from the real numbers. A summary of facts about Cartesian coordinates is as described as below:

Facts about the Cartesian Coordinate Plane

A two-dimensional plane where the:

- x-axis is the horizontal axis.
- y-axis is the vertical axis.
- $(a; b)$ and $(c; d)$ represent the same point in the plane if and only if $a = c$ and $b = d$.
- $(x; y)$ lies on the x-axis if and only if $y = 0$.
- $(x; y)$ lies on the y-axis if and only if $x = 0$.
- The origin is the point $(0; 0)$. It is the only point common to both axes.

Example

Plot the following points:

- Point $A(2, 3)$.
- Point $B(4, 5)$
- Point $C\left(-\frac{7}{2}, \frac{5}{2}\right)$.
- Point $D\left(-4, -\frac{9}{2}\right)$.
- Point $E(0, -3)$.

Solution

To plot these points, we start at the origin and move to the right if the x-coordinate is positive; and to the left if it is negative. Similarly, we move up if the y-coordinate is positive or down if it is negative. Also, if the x-coordinate is 0, we start at the origin and move along the y-axis only, and if the y-coordinate is 0, we move along the x-axis only.

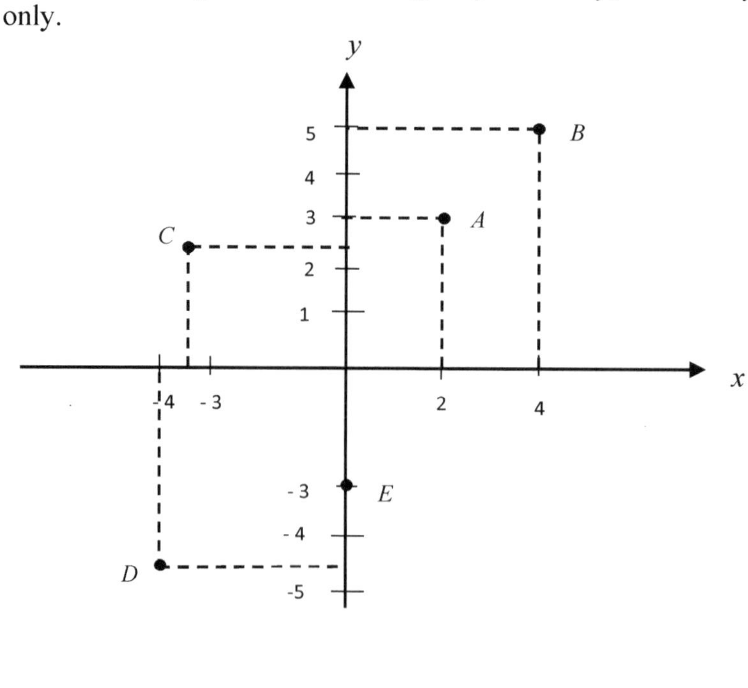

The axes divide the plane into four distinct regions, called quadrants as below:

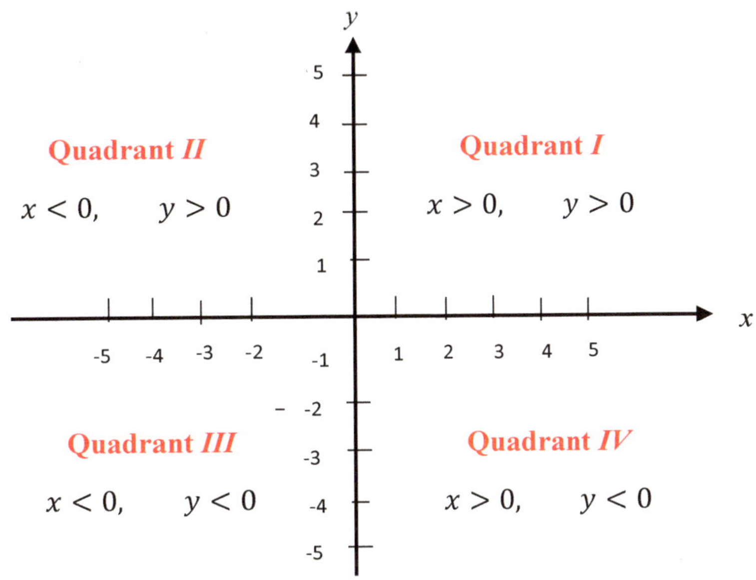

For example, (3, 1) lies in **Quadrant *I***, (– 2, 3) in **Quadrant *II***, (– 4, – 5) in **Quadrant *III*** and (2, – 1) in **Quadrant *IV***. If a point other than the origin happens to lie on the axes, we naturally refer to that point as lying on the positive or negative x-axis (if $y = 0$) or on the positive or negative y-axis (if $x = 0$). For example, (0; 6) lies on the positive y-axis whereas (– 14, 0) lies on the negative x-axis. Notice that such points do not belong to any of the four quadrants.

1.1. Symmetry on Cartesian Coordinates

Two points (a, b) and (c, d) in the plane are said to be:

- Symmetric about the x-axis if $a = c$ and $b = -d$.
- Symmetric about the y-axis if $a = -c$ and $b = d$.
- Symmetric about the origin if $a = -c$ and $b = -d$.

Graphically,

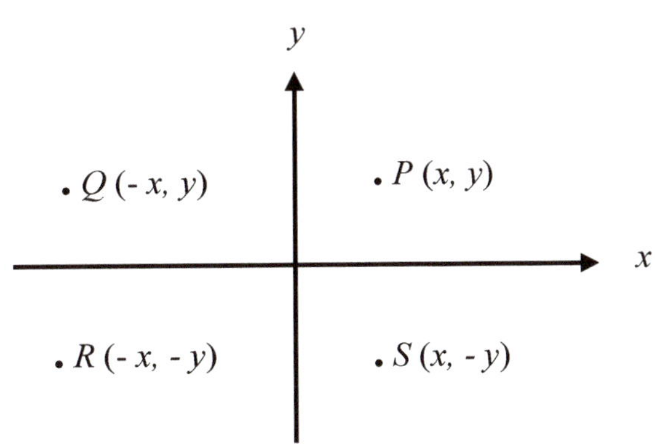

In the above figure, P and S, are symmetric about the x-axis, as are Q and R; P and Q, are symmetric about the y-axis, as are R and S; and P and R, are symmetric about the origin, as are Q and S.

Example:

Let P be the point (– 1, 4). Find the points which are symmetric to P about the:

a) x-axis.

b) y-axis.

c) origin.

Solution.

a) To find the point symmetric about the x-axis, we replace the y-coordinate with its opposite sign to get $Q\,(-1, -4)$.

b) To find the point symmetric about the y-axis, we replace the x-coordinate with its opposite sign to get $R\,(1, 4)$.

c) To find the point symmetric about the origin, we replace the x- and y-coordinates with their opposite signs to get $S\,(1, -4)$.

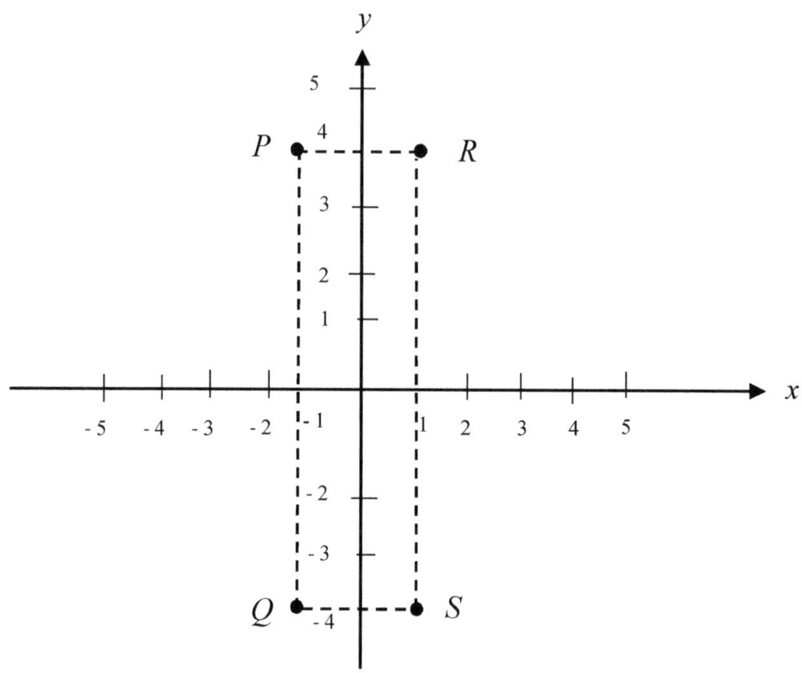

Example

Let P be the point $(-7, 8)$. Find the points which are symmetric to P about the:

a) x-axis.

b) y-axis.

c) Origin.

Solution.

a) To find the point symmetric about the x-axis, we replace the y-coordinate with its opposite to get $(-7, -8)$.

b) To find the point symmetric about the y-axis, we replace the x-coordinate with its opposite to get $(7, 8)$.

c) To find the point symmetric about the origin, we replace the x- and y-coordinates with their opposites to get $(7, -8)$.

From the previous example, we can explain the concept of reflection. Starting with the point $(-7, 8)$ and imagining the x-axis is a mirror, then the reflection of $(-7, 8)$ across the x-axis will be located at $(-7, -8)$. Also, pretending the y-axis is a mirror, the reflection of $(-7, 8)$ across the y-axis will be spotted at $(7, 8)$. Now, reflecting across the x-axis and then the y-axis, we would first get $(-7, 8)$ to $(-7, -8)$ then to $(7, -8)$, which this point, $(7, -8)$, is a symmetric point about the origin.

When you reflect a point across the x-axis, the x-coordinate remains the same, but the y-coordinate is transformed into its opposite (its sign is changed). When you reflect a point across the y-axis, the y-coordinate remains the same, but the x-coordinate is transformed into its opposite. When you reflect a point in the origin, both the x-coordinate and the y-coordinate are negated (their signs are changed). In other words, in a point reflection in the origin, the image of the point (x, y) is the point $(-x, -y)$. We summarize and generalize this process below.

Transformations: Reflections

- The reflection of the point (x, y) across the y-axis is the point $(-x, y)$.
- The reflection of the point (x, y) of the point (x, y) across the x-axis is the point $(x, -y)$.
- The reflection in the origin, the image of the point (x, y) is the point $(-x, -y)$.
- The reflection of the point (x, y) across the line $y = x$ is the point (y, x).
- The reflection of the point (x, y) across the line $y = -x$ is the point $(-y, -x)$.

1.2. Distance Between Two Points on a Line

The **distance between two points** is defined as the length of the line segment connecting those points. When considering points on the Cartesian plane, their distance can be determined by measuring the length of the straight line connecting their coordinates. In mathematical terms, the distance between two points on a line can be calculated as the absolute value of the difference between their respective coordinates. Since it involves taking the absolute value, the distance is always positive. For points $P(a)$ and $Q(b)$ on a line, the distance between them is represented as $d = |b - a|$, and we denote it as $d(P, Q)$.

Example

Given points $P(-1)$, $Q(5)$, $R(-4)$, find:

a). The distance between P and Q.

b) The distance between P and R.

c). The distance between Q and R.

Solution

a). The distance between P and Q is:

$$d(P, Q) = |5 - (-1)| = |5 + 1| = 6.$$

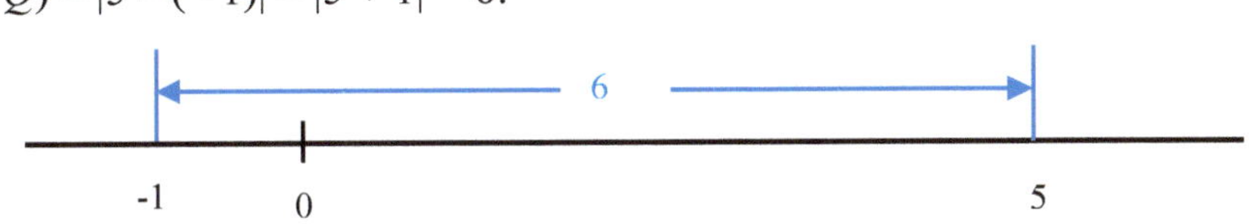

b) The distance between P and R is:

$$d(P, R) = |-4 - (-1)| = |-4 + 1| = |-3| = 3.$$

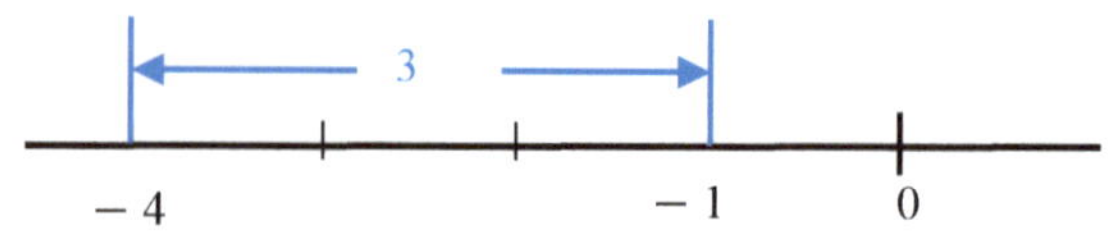

c). The distance between Q and R is:

$d(Q, R) = |-4, -5)| = |-9| = 9.$

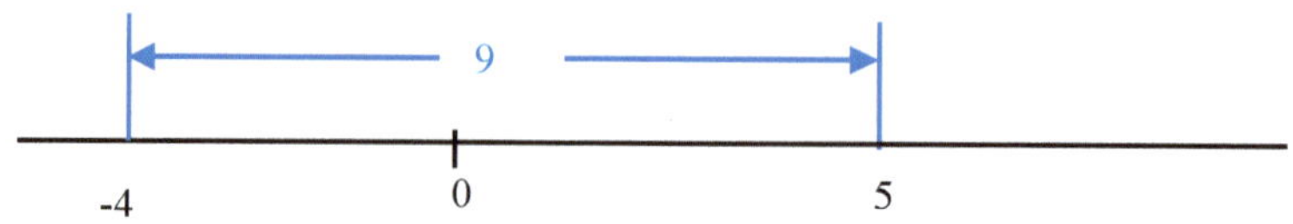

1.3. Midpoint Between Two Points on the x Axis

The **Midpoint between two points on the x Axis** is a point that is equidistant from both endpoints. Also, it is the centroid of both segment and of the endpoints. It bisects the segment. For points $P(a)$, $Q(b)$ on the x axis, the midpoint of a and b is given by $M = \frac{a+b}{2}$.

Example

Given points $P(-1)$ and $Q(7)$, find the midpoint of the line segment connecting P to Q.

Solution

The midpoint of the line segment connecting P to Q is:

$$M = \frac{7 + (-1)}{2} = \frac{6}{2} = 3.$$

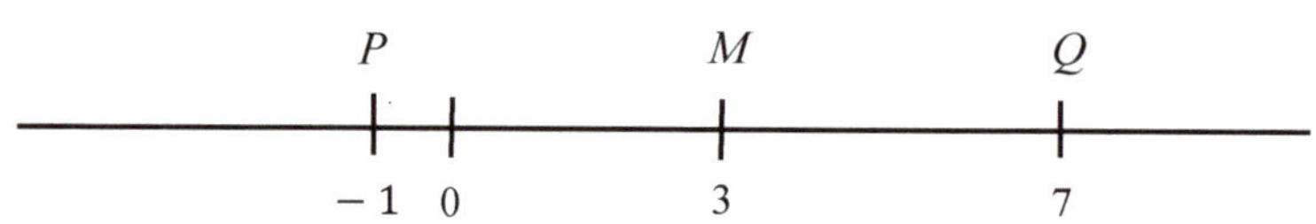

1.4. Distance Between Two Points

Suppose we have two points, $P\ (x_0, y_0)$ and $Q\ (x_1, y_1)$; in the plane. By the distance d between P and Q, we refer the length of the line segment joining P with Q. Now, to create an algebraic formula to compute the distance between these two points, consider the common situation below on the left.

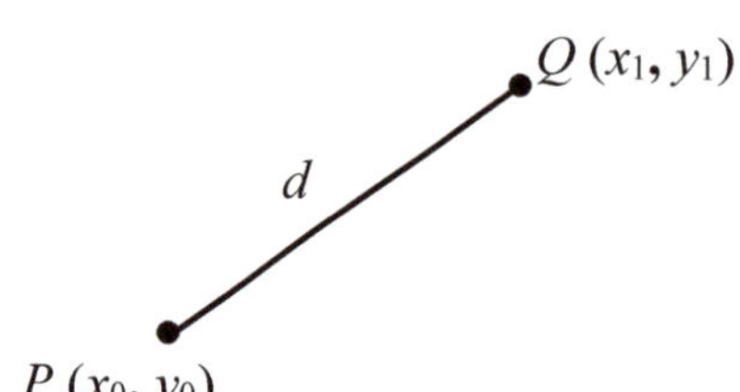

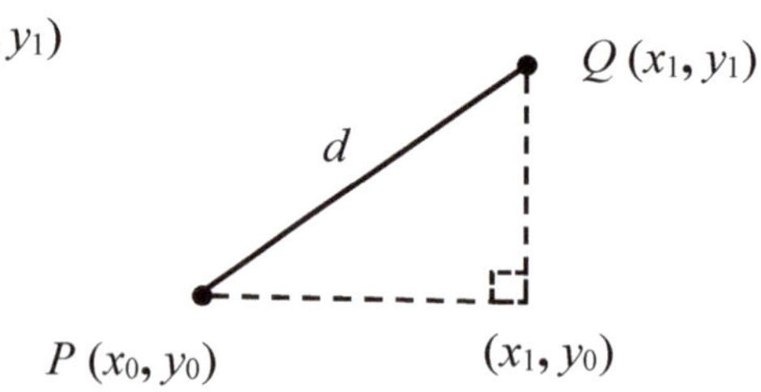

We can visualize a right triangle whose hypotenuse has length d as drawn above on the right. From the latter figure, we see that the lengths of the legs of the triangle are $|x_1 - x_0|$ and $|y_1 - y_0|$ so the Pythagorean Theorem gives us: $|x_1 - x_0|^2 + |y_1 - y_0|^2 = d^2$, or $(x_1 - x_0)^2 + (y_1 - y_0)^2 = d^2$. Thus the distance d, is given by the formula below::

$$d = \sqrt{(x_1 - x_0)^2 + (y_1 - y_0)^2}.$$

Example
Find and simplify the distance between $P(-4, 1)$ and $Q(-1, 5)$.

Solution.

$$d = \sqrt{(x_1 - x_0)^2 + (y_1 - y_0)^2}$$

$$d = \sqrt{(-4-(-1))^2 + (1-5)^2}$$

we plugged $x_1 = -4, x_0 = -1, y_1 = 1$, and $y_0 = -5$, into the equation

$$d = \sqrt{(-3)^2 + (-4)^2}$$

$$d = \sqrt{9+16}$$

$$d = \sqrt{25}$$

$$d = \pm 5.$$

Since distances cannot be negetive, therefore $d = 5$.

1.5. Midpoint of a Line Segment

Given two points, $P(x_1, y_1)$ and $Q(x_2, y_2)$, the midpoint M of P and Q is defined to be the point on the line segment of PQ whose distance from P is equal to its distance from Q.

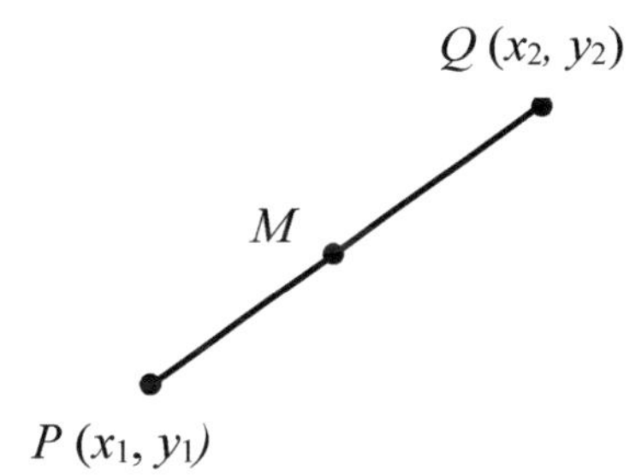

If we think of reaching M by going `halfway over' and `halfway up' we get the following formula.

$$M = (\frac{x_1+x_2}{2}, \frac{y_1+y_2}{2}).$$

Example
Find the midpoint of the line segment connecting $P(-4, 1)$ and $Q(2, -3)$.

Solution.
Using the midpoint formula, we have:

$$M = (\frac{x_1+x_2}{2}, \frac{y_1+y_2}{2})$$

$$M = (\frac{-4+2}{2}, \frac{1+(-3)}{2})$$

we plugged $x_1 = -4$, $x_2 = 2$, $y_1 = 1$, and $y_2 = -3$ into the equation

$$M = (\frac{-2}{2}, \frac{-2}{2}) = (-1, -1)$$

$$M = (-1, -1).$$

Therefore, the midpoint of the line segment connecting P and Q is $M = (-1, -1)$.

Example
If $a \neq b$, prove that the line $y = x$ equally divides the line segment with endpoints (a, b) and (b, a).

Solution
To prove the claim, we use midpoint formula. So,

$$M = \left(\frac{x_1+x_2}{2}, \frac{y_1+y_2}{2}\right)$$

$$M = \left(\frac{a+b}{2}, \frac{b+a}{2}\right)$$

$$M = \left(\frac{a+b}{2}, \frac{a+b}{2}\right).$$

Since the x and y coordinates of this point, M, are the same, we can conclude that the midpoint, M, lies on the line $y = x$, as required.

1.6. Rates and Slope

Slope is the ratio of the vertical and horizontal changes between two points on a surface or a line. If coordinates of any two points of a line are given, then the rate of change is the ratio of the change in the y-coordinates to the change in the x-coordinates.

Example
Suppose that the temperature in the morning is $45^\circ F$, then in the afternoon the temperature measures $60^\circ F$ (F stands for Fahrenheit temperature). Find the change in temperature.

Solution
The change in temperature is found by taking a difference.

Change in temperature = (Afternoon temperature) – (Morning temperature) = $60^\circ F - 45^\circ F = 15^\circ F$

Therefore, there was a ten degree increase in temperature from morning to afternoon.

Calculating the Change in a Quantity.

To calculate the change in a quantity, subtract the earlier measurement from the later measurement.

Let T represent the temperature. We may use the symbolism ΔT to represent the change in temperature. For the change in temperature from morning to afternoon, we would write $\Delta T = 15^\circ F$.

In the above example, the temperature depends on the time. This makes the temperature the dependent variable and time the independent variable. It is traditional to place the independent variable on the horizontal axis and the dependent variable on the vertical axis.

Slope - Rate

Slope is used to describe the measurement of steepness of a straight line. In different situations, slope may be referred to as steepness, incline, pitch, or grade (gradient). Slope is also described as a rate of change. A line with a large slope, such as 30, is very steep. A line with a small slope, such as 0.05 is very flat. We will also use slope to describe the direction of the line. A line that goes up from left to right will have a positive slope and a line that goes down from left to right will have a negative slope.

As a measure steepness, it referred as how fast the line rises compared to how far the line runs, or it describes slope as the fraction rise run. Rise would be a vertical change, or a change in the y-values. Run would be a horizontal change, or a change in the x-values. So, another way to describe slope would be the fraction change in y change in x. It turns out that if we have a graph, we can draw vertical and horizontal lines from one point to another to make what is called a slope triangle. The sides of the slope triangle give us our slope. That is slope = $\frac{\text{Raise}}{\text{Run}}$. The slope is positive, if the rise and the run are both positive or both are negative.

Run is a horizontal change, or a change in the x-values and raise is a vertical change or a change in the y-values. Direction is important when it comes to determining slope. It is important to pay attention to whether you are moving up, down, left, or right; that is, if you are moving in a positive or negative direction. If you go up to get to your second point, the rise is positive. If you go down to get to your second point, the rise is negative. If you go right to get to your second point, the run is positive. If you go left to get to your second point, the run is negative.

The Run and the Raise Signs

- **The positive run**
  If we move horizontally toward the positive direction of x-axis, or moving to the right, then we get a positive run.
- **The negative run**
  If we move horizontally toward the negative direction of x-axis, or moving to the left, then we get a negative run.
- **The positive raise**
  If we move vertically toward the positive direction of y-axis, or moving up, then we get a positive raise.
- **The negative raise**
  If we move vertically toward the negative direction of y-axis, or moving down, then we get a negative raise.

Slope can be expressed in various ratio forms: Slope is traditionally designated by the letter "m".

Slope

The slope of a line is the rate at which the dependent variable is changing with respect to the independent variable. If the dependent variable is y and the independent variable is x, then the slope of the line is:

$$m = \text{Slope} = \frac{\Delta y}{\Delta x}.$$

Rise would be a vertical change, or a change in the y-values.
Run would be a horizontal change, or a change in the x-values.

$$m = \text{Slope} = \frac{\Delta y}{\Delta x} = \frac{\text{Raise}}{\text{Run}}.$$

The following examples show how we find the slope of using these concepts.

Example
Compute the slope of the line passing through the points $P(5, -3)$ and $Q(-1, 7)$.

Solution
To calculate the slope of the line through the points $P(5, -3)$ and $Q(-1, 7)$, we must calculate the change in both the independent and dependent variables. We will do the calculation in two different ways.

1. We subtract the coordinates of point $P(5, -3)$ from the coordinates of point $Q(-1, 7)$.
$$\text{Slope} = \frac{7 - (-3)}{-1 - 5} = \frac{7 + 3}{-6} = \frac{10}{-6} = \frac{5}{-3} = -\frac{5}{3}.$$

2. We subtract the coordinates of point $Q(-1, 7)$ from the coordinates of point $P(5, -3)$.
$$\text{Slope} = \frac{-3 - 7}{5 - (-1)} = \frac{-10}{5 + 1} = \frac{-10}{6} = \frac{-5}{3} = -\frac{5}{3}.$$

As it is seen, regardless of the direction of subtraction, the slope is $-\frac{5}{3}$.

Therefore, our solution is slope $= -\frac{5}{3}$.

Example
Given the graph of a line find its slope.

Solution
To find the slope, we consider the y change, rise, and the x change, run. Drawing these lines in makes a slope triangle that we can use to count from one point to the next the graph goes down 2, right 3.

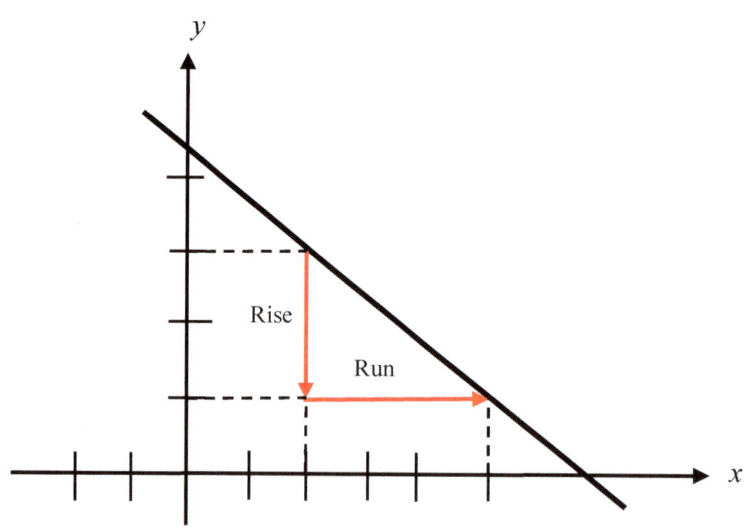

Since we moved 2 units, vertically toward the negative direction of y-axis (moved down), this is rise -2. Similarly, we moved 3 units, horizontally toward the positive direction of x-axis (moved to the right), this is run 3. As a fraction it would be, $\frac{-2}{3} = -\frac{2}{3}$.

Therefore, our solution is slope $= -\frac{2}{3}$.

Example
Given the graph of a line find its slope.

Solution
Here considering the *y* change as rise, and the *x* change as run, we draw *x*-*y* axis to make the slope triangle. Now we can count from one point to the next, the graph goes left 4 and down 1.

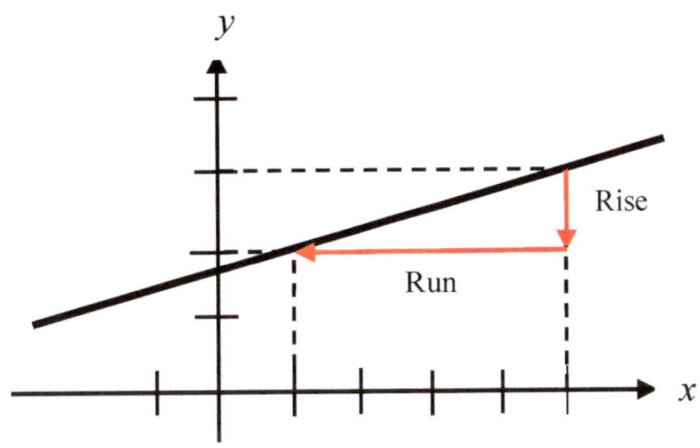

This is rise – 1, run – 4. As a fraction it would be, $\frac{-1}{-4} = \frac{1}{4}$. Our solution is slope $= \frac{1}{4}$.

Notice that finding the slope is independent of the selected points. It does not matter which two points you pick on the line to calculate its slope. See the following example.

Example
A line that crosses the point $P(2, 1)$ and $Q(6, 3)$. Find the slope of the line.

Solution
Let us first sketch the line passing through two points $P(2, 1)$ and $Q(6, 3)$. Here is the figure illustrating the graph of the line.

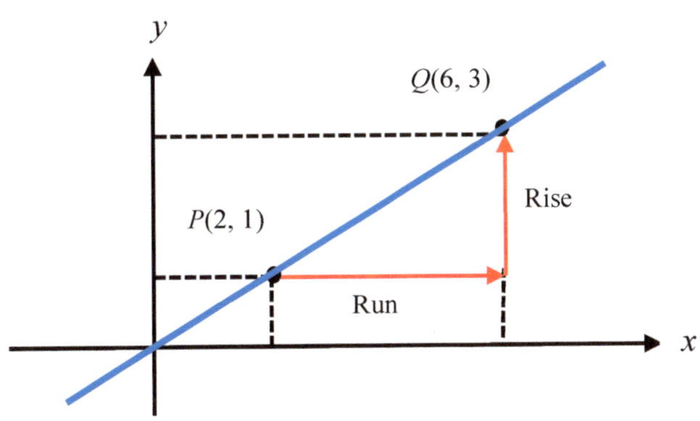

We start from point $P(2, 1)$ and move horizontally until in line with the point $Q(6, 3)$. The run is,

$$\text{Run} = |6 - 2| = 4.$$

Now, we continue moving but from here we vertically until we reach to the point $Q(6, 3)$. The raise is,

$$\text{Raise} = |3 - 1| = 2.$$

Using the slope formula, we find the slope of the line. So,

$m = \text{Slope} = \frac{\textbf{Raise}}{\textbf{Run}}.$

$m = \text{Slope} = \frac{\textbf{2}}{\textbf{4}} = \frac{\textbf{1}}{\textbf{2}}$ We plugged Raise = 3, and Run = 6 into the equation.

Therefore, the slope of the line that crosses the point $P(2, 1)$ and $Q(6, 3)$ is $m = \frac{\textbf{1}}{\textbf{2}}$.

As mentioned earlier, finding the slope is independent of successive moving directions (as it is seen in the two latter examples), and the selected points (as we will show in the example below).

Example
Using the graph of the line from the previous example, find the slope of the line by choosing two other arbitrary points.

Solution
Let us consider the graph of the line. Below we present the graph of the line from the previous example.

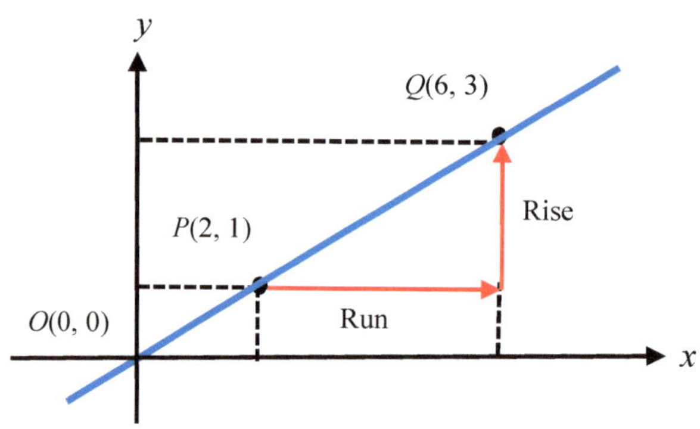

Now, let us consider points $O(0, 0)$ and $Q(6, 3)$. We follow the same procedure, but we start from $Q(6, 3)$. Starting from this point, we move vertically down until in line with the point $O(0, 0)$. The raise is,

Raise = |3 – 0| = 3.

Now, we continue moving horizontally until we reach point $O(0, 0)$. The run is,

Raise = |6 – 0| = 6.

Using the slope formula, we find the slope of the line. So,

$m = \text{Slope} = \frac{\textbf{Raise}}{\textbf{Run}}$

$m = \text{Slope} = \frac{\textbf{3}}{\textbf{6}} = \frac{\textbf{1}}{\textbf{2}}$ We plugged Raise = 3, and Run = 6 into the equation.

The above calculations can be better illustrated, using the graph of the line again. Below, we present the graph of the line.

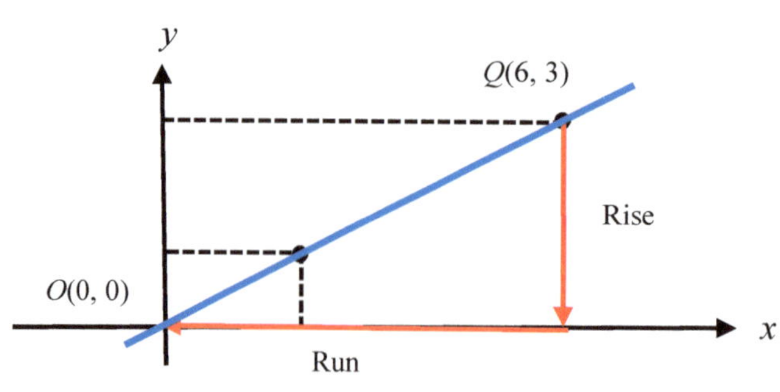

Also, we can find the slope using the concept of raise-run. Hence, this is rise – 3, run – 6. As a fraction it would be, $\frac{-3}{-6} = \frac{3}{6}$. Our solution is slope $= \frac{1}{2}$.

Therefore, by using different points and different method we found the slope $= \frac{1}{2}$.

1.7. Interpreting the Steepness of a Line

The slope of a line is a number that tells us how steep the line is. If slope is a number that measures the steepness of a line, then one would expect that a steeper line would have a larger slope. Also, lines go uphill have positive slopes and lines that go downhill have negetive slopes. Additionally, if the magnitude (absolute value) of the slope of a line (with a negetive slope) is greater it moves downhill more quickly, and if the magnitude of the slope of a line (with a positive slope) is greater it moves uphill more quickly.

The direction of a line is either increasing, decreasing, horizontal or vertical. A line increases when its slope is positive. A line decreases when its slope is negative. A line is horizontal when its slope is zero. When the slope is undefined the line is vertical.

Line Direction

- A line is increasing if it goes up from left to right. The slope is positive, .
- A line is decreasing if it goes down from left to right. The slope is negative, .
- If a line is horizontal the slope is zero.
- If a line is vertical the slope is undefined.

Let us present some examples for illustrating lines directions.

Example

Suppose we are given two lines L_1, and L_2. The line, L_1, passing through the point $P(3, 2)$ and $Q(-3, -1)$, and the line L_2, passing through the point $R(3, 4)$ and $S(1, -3)$. Calculate the slope of each line and interpret the results.

Solution

The graphs of the two lines L_1 and L_2 are shown below. Note that L_1 is less steep than L_2. Remembering that the slope of the line is the rate at which the dependent variable is changing with respect to the independent variable. Both in L_1 and L_2, the dependent variable is y and the independent variable is x.

Again, remembering that m = Slope $= \frac{\Delta y}{\Delta x}$, for $P(3, 2)$ and $Q(-3, -1)$ we have:

$$\text{The slope of } L_1 = \frac{\Delta y}{\Delta x} = \frac{-1-2}{-3-2} = \frac{-3}{-5} = \frac{3}{5} = 0.60.$$

For $R(3, 4)$ and $S(1, -3)$ we have:

$$\text{The slope of } L_2 = \frac{\Delta y}{\Delta x} = \frac{-3-4}{1-3} = \frac{-7}{-2} = \frac{7}{2} = 3.50.$$

Note that both lines go uphill, and both have positive slopes. Also, note that the slope of L_2 is greater than the slope of L_1. This is consistent with the fact that L_2 is steeper than the L_1.

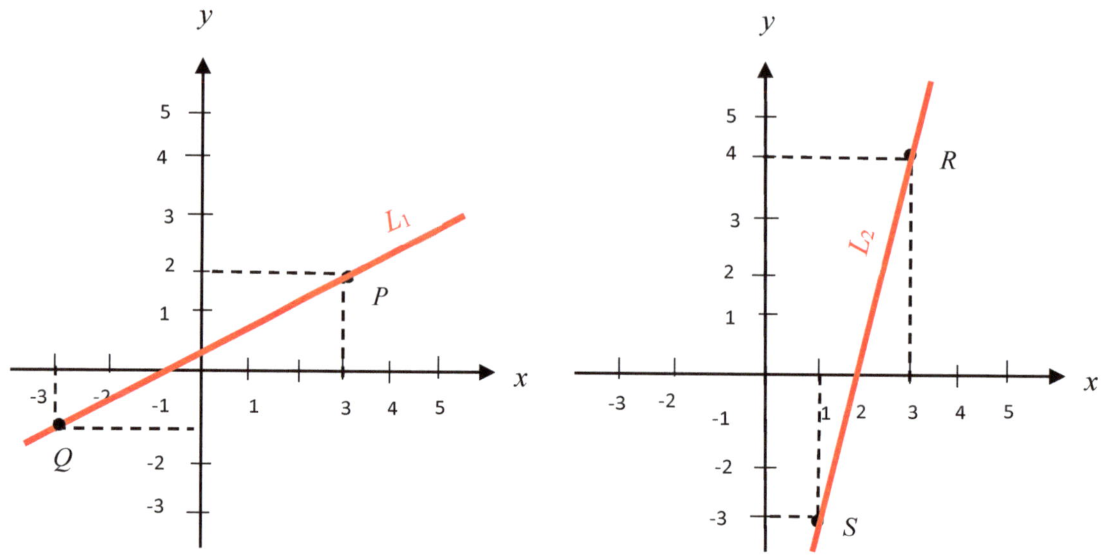

Example
Suppose we are given two lines L_1, and L_2. The line, L_1, passing through the point $P(4, 1)$ and $Q(-2, 3)$, and the line L_2, passing through the point $R(-2, 4)$ and $S(1, -3)$. Calculate the slope of each line and interpret the results.

Solution
The graphs of the two lines L_1 and L_2 are shown below. Note that L_1 is less steep than L_2. Note that the L_1 goes downhill less quickly than L_2.

Remembering that the slope of the line is the rate at which the dependent variable is changing with respect to the independent variable. Both in L_1 and L_2, the dependent variable is y and the independent variable is x.

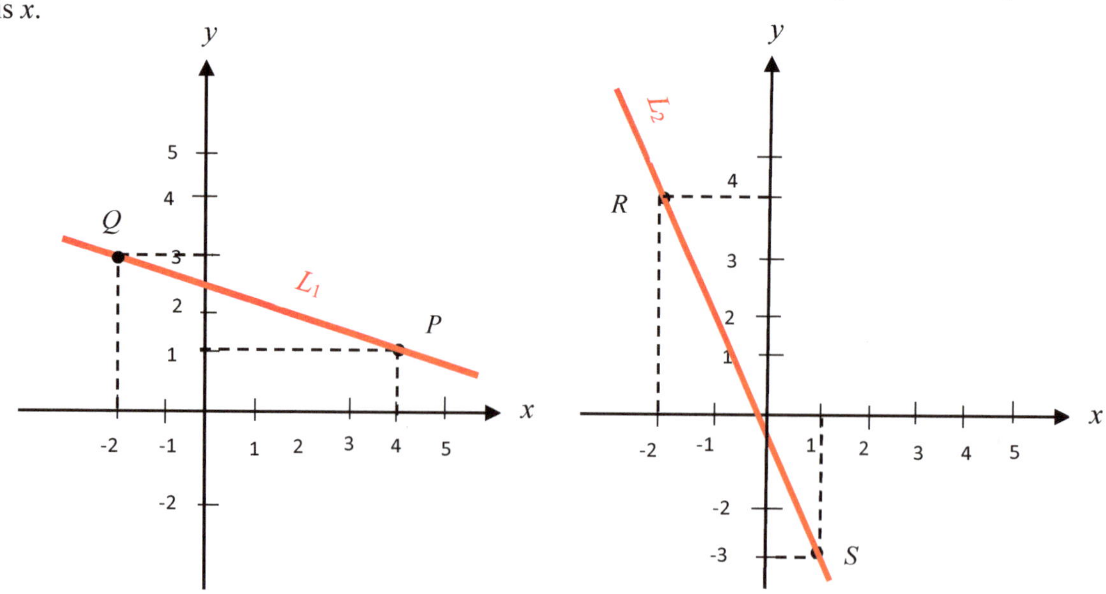

Again, remembering that m = Slope $= \frac{\Delta y}{\Delta x}$, for $P(4, 1)$ and $Q(-2, 3)$ we have:

The slope of $L_1 = \frac{\Delta y}{\Delta x} = \frac{3-1}{-2-4} = \frac{2}{-6} = -\frac{1}{3}$.

For $R(-2, 4)$ and $S(1, -3)$we have:

The slope of $L_2 = \frac{\Delta y}{\Delta x} = \frac{-3-4}{1-(-2)} = \frac{-7}{1+2} = -\frac{7}{3}$.

Note that both lines go downhill, and both have negetive slopes. Also, note that the magnitude (absolute value) of the slope of L_2 is greater than the magnitude of the slope of L_1. This is consistent with the fact that L_2 moves downhill more quickly than L_1.

1.8. Horizonal and Vertical Lines

A **horizontal line** is a line which has zero slope value. All points on a horizontal line have the same y value. Horizontal lines have the general form $y = a$.

Vertical lines go up and down and are in the form of $x = b$ where b represents the shared x coordinate of all points.

The characteristics of the horizontal and the vertical lines are as follows:

- Horizontal lines have the general form $y = a$.
- Vertical lines have the general form $x = b$.
- A vertical line is perpendicular to a horizontal line. Additionally, a horizontal line is perpendicular to a vertical line.
- A horizontal line is a line which has zero slope value.
- The equations of horizontal line and vertical line only have one variable.
- Horizontal lines do not cross each other.
- Vertical lines do not cross each other.

The following figure illustrates graphs of a horizonal line, $y = a$, and a vertical line $x = b$.

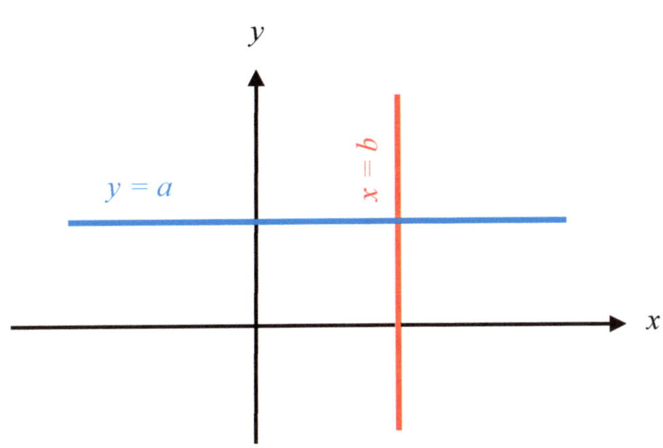

Slope of a Line, Summary of Facts

- The slope of a line is the rate at which the dependent variable is changing with respect to the independent variable. If y is the dependent variable and x is the independent variable, then the slope is m = Slope $= \frac{\Delta y}{\Delta x}$, where Δy is the change in y and Δx is the change in x.

- If a line has positive slope, then the line slants uphill. If two lines have positive slope, then the line with the larger slope rises more quickly.

- If a line has negative slope, then the line slants downhill. If two lines have negative slope, then the line having the slope with the larger magnitude falls more quickly.

- Horizontal lines have slope zero.

- Vertical lines have undefined slope.

1.9. Intercepts

In analytical geometry, **intercepts** are points where a curve intersects an axis. Two fundamental intercepts are the x-intercept and the y-intercept. The x-intercept denotes where a line crosses the x-axis, while the y-intercept represents the point where a line intersects the y-axis. Mastery of these concepts' aids in graphing lines and equations effectively. By understanding x and y intercepts, one gains insights into problem-solving techniques involving graphical representations. Let's delve into the significance of x and y intercepts, their utility, and how they facilitate problem resolution.

The y-intercept: The point where the graph crosses the y-axis ($x = 0$).

The x-intercept: The points where the graph crosses the x-axis ($y = 0$).

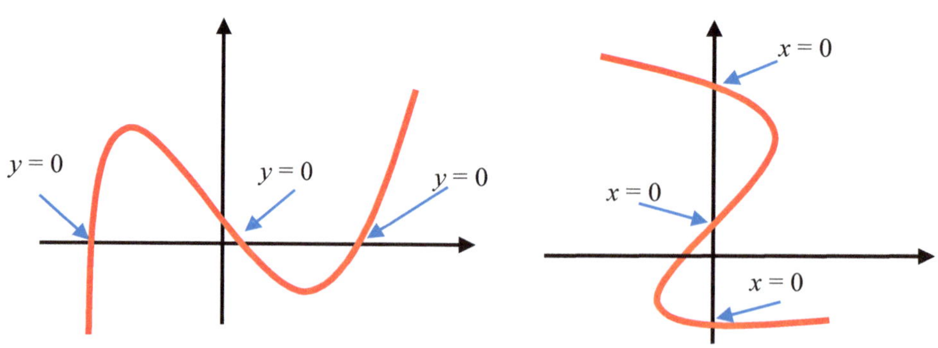

1.9.1. Slope-Intercept Form of a Line

Let us start with the definition of the y-intercept of a line. The point $(0, b)$ where the graph of a line crosses the y-axis is called the y-intercept of the line.

Based on this definition we can now assemble the slope-intercept formula for a line with y-intercept $(0, b)$ and with slope = m. Consider the graph below, in which (x, y) is an arbitrary point on the line.

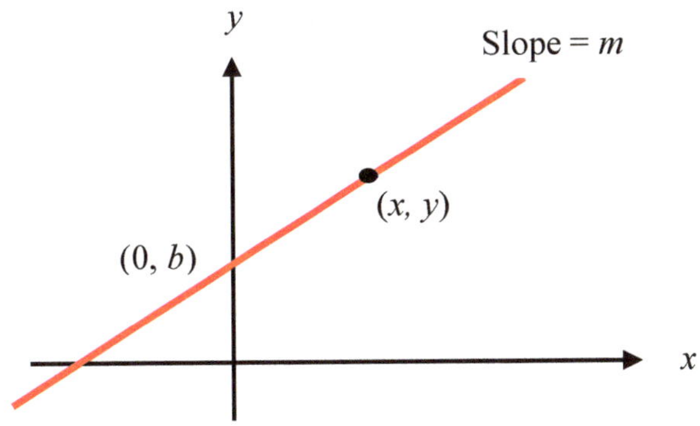

As we have learned, the slope of the line is the rate at which the dependent variable, y, is changing with respect to the independent variable, x. So,

$$\text{Slope} = m = \frac{\Delta y}{\Delta x}.$$

To calculate both the change in y and the change in x, we must subtract the coordinates of the point $(0, b)$ from the point (x, y). So,

$$m = \frac{\Delta y}{\Delta x}$$

$$m = \frac{y - b}{x - 0}$$ we substituted $\Delta y = y - b$ and $\Delta x = x - 0$ into the equation

$$m = \frac{y - b}{x}$$

$$x \times m = x \times \frac{y - b}{x}$$ we multiplied both side of the equation by x

$$mx = y - b$$

$$y = mx + b.$$

Therefore, the equation of the line is $y = mx + b$.

> **The Slope-Intercept Form of a Line.**
> The equation of the line with y intercept $(0, b)$ and slope m is:
>
> $$y = mx + b.$$

Example

Plot the line with y-intercept $(0, 3)$ and slope = $m = -\frac{4}{3}$ and write the slope-intercept form of its equation.

Solution

First, we plot the y-intercept $(0, 3)$. Next, starting at $(0, 3)$ and using slope $m = -\frac{4}{3}$, we move down four units, and move right three units reaching at point $(3, -1)$. Now having two points of the line we can draw the line passing through these two points (see the figure below).

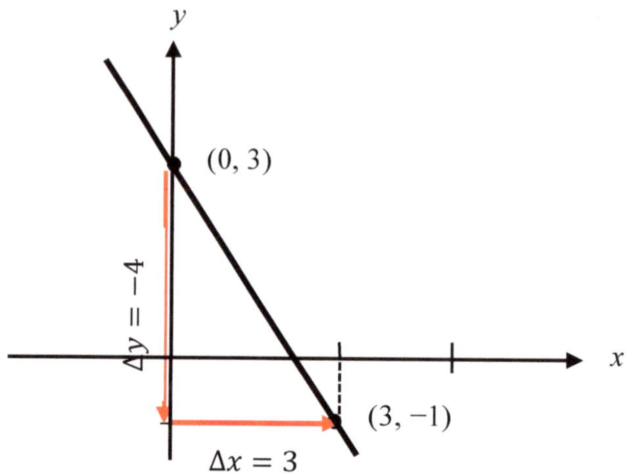

To write the slope-intercept form of its equation, since the y-intercept is (0, 3), we conclude that $b = 3$. Further, the slope is $-\frac{4}{3}$. Substituting them into the slope-intercept form of the line, $y = mx + b$, we will have:

$$y = -\frac{4}{3}x + 3.$$

Therefore, the slope-intercept form of the line is $y = -\frac{4}{3}x + 3$.

1.10. Point Slope Form

Up until now, we have been using the slope-intercept form of a linear equation to describe linear functions. Here, we will learn another way to write a linear equation. It is two write the formula based on the **point-slope form**.

$$y - y_1 = m(x - x_1).$$

The point-slope form is derived from the slope formula. So,

$$m = \frac{y - y_1}{x - x_1}$$

$$(x - x_1)m = (x - x_1)\frac{y - y_1}{x - x_1}$$ we multiplied both sides by $(x - x_1)$.

$$(x - x_1)m = (y - y_1)\left(\frac{x - x_1}{x - x_1}\right)$$

$$(x - x_1)m = (y - y_1)(1)$$

$$y - y_1 = m(x - x_1)$$ we just rearranged terms.

Point-Slope Form of a Linear Equation

The point-slope form of a linear equation takes the form,

$$y - y_1 = m(x - x_1),$$

where m is the slope, x_1 and y_1 are the x and y coordinates of a specific point through which the line passes.

Example
Write the point-slope form of a line with a slope of 5 which passes through the y-intercept $P(0, 10)$.

Solution
We learned the equation of the line with y intercept $(0, b)$ and slope m is $y = mx + b$. The slope is given as $m = 5$, and $b = 10$. Using this equation, we will get,

$y = mx + b$

$y = 5x + 10$ we plugged $m = 5$, and $b = 10$ into the equation.

Also, we learned the point-slope form of a linear equation is $y - y_1 = m(x - x_1)$. To convert the above linear equation with to a point-slope form, we realize in this case, $y_1 = 10$, $x_1 = 0$, and $m = 5$. So,

$y - y_1 = m(x - x_1)$

$y - 10 = 5\,(x - 0)$ we plugged $y_1 = 10$, $x_1 = 0$, and $m = 5$ into the equation

$y - 10 = 5x$

$y = 5x - 10.$

Therefore, the point - slope form of the line with a slope of 5, which pass through the y-intercept $P(0, 10)$ is, $y - 10 = 5x$.

Example
Find the point-slope form of the line from its graph below.

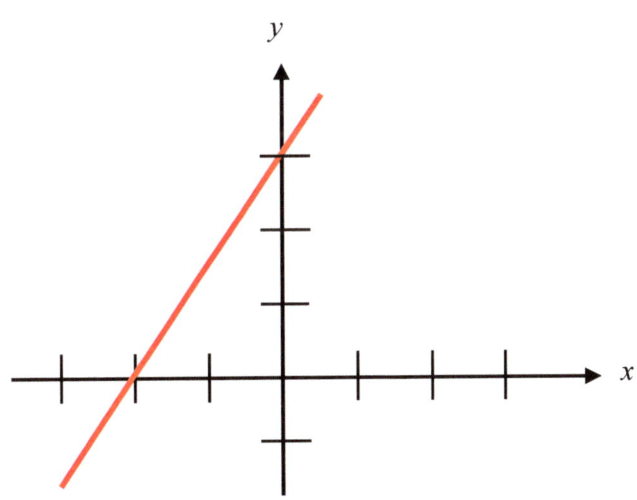

Solution
Let us identify an arbitrary point on the line. One of the convenient points is the x-intercept of the line. Referring to the graph of the line, shown in the figure (a) below, this point is $P(-2, 0)$.

To find the point-slope form of the line, we need to find its slope. Finding the slope graphically, using the concept of the "raise-over-run", that is, $m = \frac{\text{raise}}{\text{run}}$. This gives us $m = \frac{\text{raise}}{\text{run}} = \frac{3}{2}$, as shown in the figure (b) below.

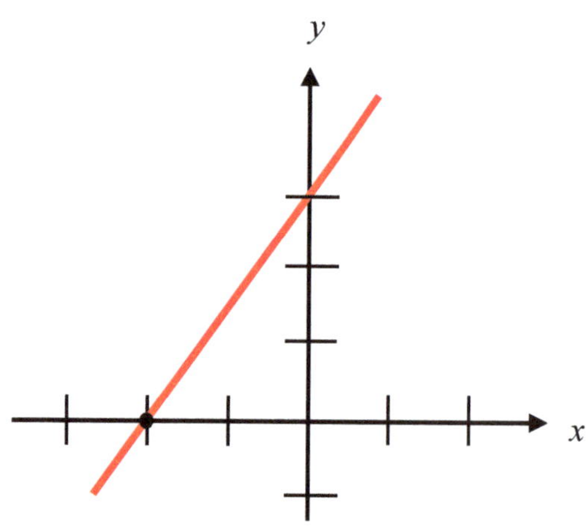

Figure (a)

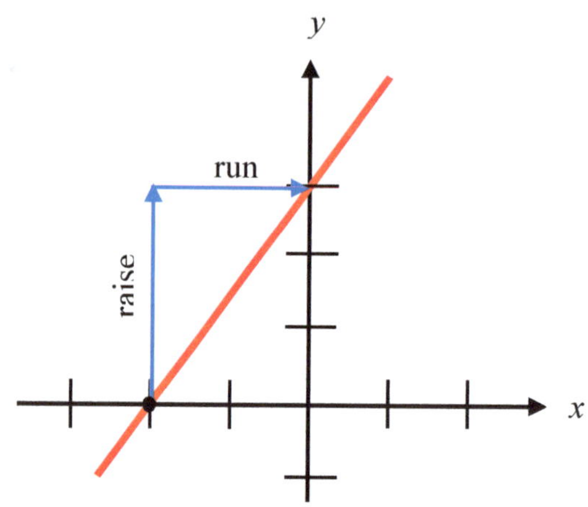

Figure (b)

Now, we have a point $P(-2, 0)$ that has coordinates of $x_1 = -2$ and $y_1 = 0$. Also, we have a slope $m = \frac{3}{2}$, which enable us to write the point-slope form of the line. So,

$y - y_1 = m\,x - x_1)$

$y - 0 = \frac{3}{2}[x - (-2)]$ we plugged $x_1 = -2$, $m = \frac{3}{2}$, and $y_1 = 0$ into the equation

$y = \frac{3}{2}x + 2.$

Therefore, the point-slope form of the line is $y = \frac{3}{2}y = \frac{3}{2}x + 2.$

Example

Write the equation of a line with a slope of 4 which passes through the point (3, 5).

Solution

The slope is given as $m = 4$, and the point (3, 5) has coordinates of $x_1 = 3$ and $y_1 = 5$. Using point-slope formula, we will get,

$y - y_1 = m\,x - x_1)$

$y - 5 = 4(x - 3)$ we plugged $x_1 = 3$, $m = 4$, and $y_1 = 5$ into the equation

$y - 5 = 4x - 12.$

Therefore, the point-slop form of the line with a slope of 4 which pass through the point $P(3, 5)$ is,

$y - 5 = 4x - 12.$

To find the standard form of the line, we will have,

$y - 2 = 7x$

$y - 5 + 5 = 4x - 12 + 5$ to isolate y, we added 5 to both sides

$y = 4x - 7.$

Therefore, the equation of a line with a slope of 4 which passes through the point (3, 5) is $y = 4x - 7$.

1.11. Line Equation from 2 Points

The equation of the line passing through two points (x_A, y_A) and (x_B, y_B) is given by,

$$y - y_A = \frac{y_B - y_A}{x_B - x_A}(x - x_A).$$

Or equivalently,

$$y - y_B = \frac{y_B - y_A}{x_B - x_A}(x - x_B).$$

Example

Find the equation of the line passing through points $A(-1, 4)$ and $B(2, -5)$.

Solution

Using the formula $y - y_A = \frac{y_B - y_A}{x_B - x_A}(x - x_A)$, we will get,

$$y - y_A = \frac{y_B - y_A}{x_B - x_A}(x - x_A)$$

$$y - 4 = \frac{-5-4}{2-(-1)}[x - (-1)]$$ we plugged $Y_A = 4$, $Y_B = -5$, $X_A = -1$, and $X_B = 2$ into the equation

$$y - 4 = \frac{-9}{2+1}(x + 1)$$

$$y - 4 = (-3)(x + 1)$$

$$y - 4 = -3x - 3$$

$$y - 4 + 4 = -3x - 3 + 4$$ to isolate y, we added 4 to both sides

$$y = -3x + 1.$$

Therefore, the equation of the line passing through the points $A(-1, 4)$ and $B(2, -5)$ is $y = -3x + 1$.

To illustrate that the second formula will give the same result, let us use the other formula. So,

$$y - y_B = \frac{y_B - y_A}{x_B - x_A}(x - x_B)$$

$$y - (-5) = \frac{(-5)-4}{2-(-1)}(x - 2)$$ we plugged $Y_B = -5$, $X_A = -1$, and $X_B = 2$ into the equation

$$y + 5 = \frac{-9}{2+1}(x - 2)$$ $Y_A = 4$,

$$y + 5 = (-3)(x - 2)$$

$$y + 5 = -3x + 6$$

$$y + 5 - 5 = -3x + 6 - 5$$ to isolate y, we added -5 to both sides.

$$y = -3x + 1$$

Therefore, the equation of the line passing through the points $A(-1, 4)$ and $B(2, -5)$ is $y = -3x + 1$.
As we expected, we obtained the same result.

1.12. Graphing Lines

Graphing a line in the xy-axis is one of the most basic skills in analytic geometry. In this section, we are going to discuss three basic methods that are commonly used for graphing a straight line. The first method is "constructing a table of values". This is the most straightforward method in graphing a line.

To plot points on a graph, we compile a table of x-values and their corresponding y-values. This process allows us to generate coordinates, which represent points on the graph. Subsequently, we plot these points using the Cartesian coordinate system. By organizing x and y values in a table, we can efficiently visualize the relationship between variables and depict it graphically we need to do is to prepare a table of x-values and y-values to obtain points, and to plot these points.

To load the table, we select some values of x and then evaluate those values in the given equation to get the corresponding values of y. Finally, we plot those points in the xy-axis, and connect them with a straight line.

The second method is "using the slope and y-intercept", where $y = mx + b$. We start by plotting the y-intercept, $(0, b)$, and we find another point using the slope (m) with the y-intercept at the reference point. We than connect the two points with a straight line.

The third method is "connecting the x and y-intercepts". In this method, to get the x-intercept, we set y equal to zero and solve for x, and to determine the y-intercept, we set x equal to zero and solve for y. We than connect the two points with a straight line.

In the following, we present the three methods in more detail along with some examples.

Example
Graph the linear equation $y = 3x - 4$ using a table of values.

Solution

Method 1: Using the Table of Values

In method 1, we will employ the construction of the table values. The table of values is a graphic chart that helps us determine two or more points that can be used to create our graph.

Here is a table of values for the equation, $y = 3x - 4$.

x	$y = 3x - 4$	y	Order pair
0	$3(0) - 4$	-4	$(0, -4)$
1	$3(1) - 4$	-1	$(1, -1)$
2	$3(2) - 4$	2	$(2, 2)$

We now plot the three points in the xy-axis.

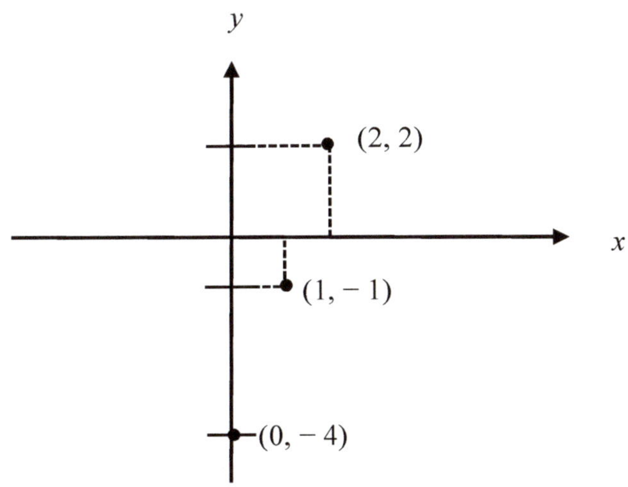

Connecting the three points we get the graph of the line. Here is the graph of the line:

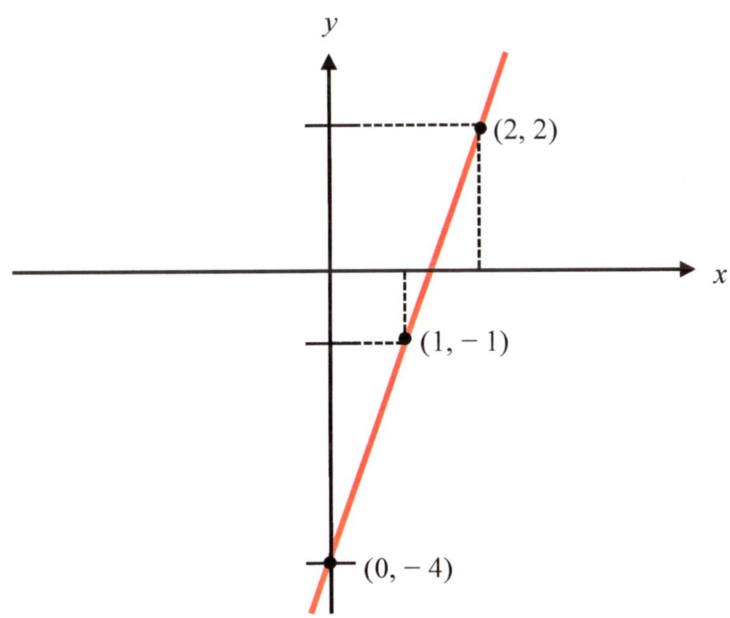

Method 2: Using the Slope and y-Intercept.

To graph a line using its slope and y-intercept, we need to make sure that the equation of the line is in the slope-intercept form, $y = mx + b$. From this format, we can easily read off both the values of the slope and y-intercept. The slope is just the coefficient of x which is m, and the y-intercept is the constant term b (see the figure below).

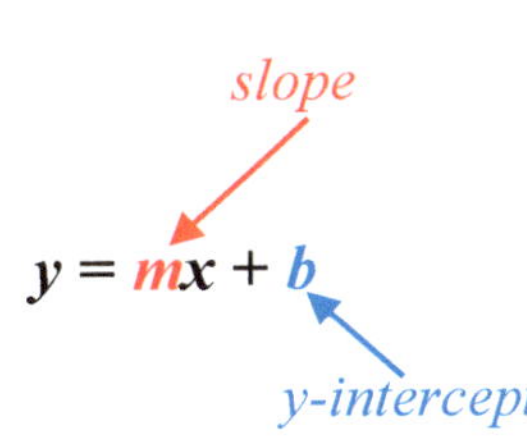

To graph a line using the slope and y-intercept, we need to perform the steps below.

Step 1. Plot the y-intercept, $(0, b)$, on the xy-axis. Recall that, this point always lies on the y-axis.

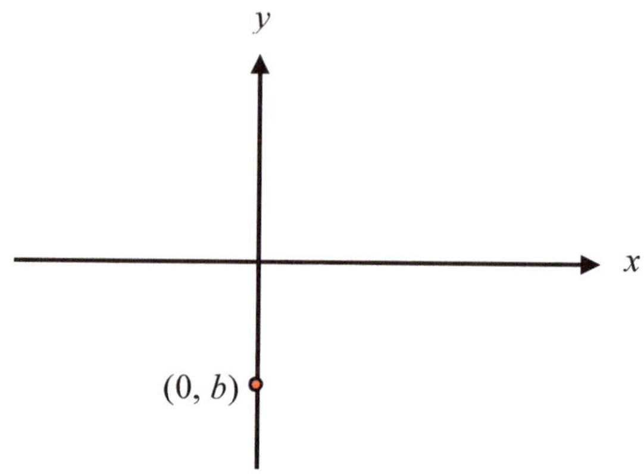

Step 2. Starting from the y-intercept, find another point using the concept of raise over run of slopes.

$$\text{Slop } = m = \frac{\Delta y}{\Delta x} \Longrightarrow \frac{\text{raise}}{\text{run}}.$$

Recall that: direction shows how we go from one point to another. The numerator, raise, tells us how many units to go up/down, while the denominator, run, tells us how many units to move left/right.

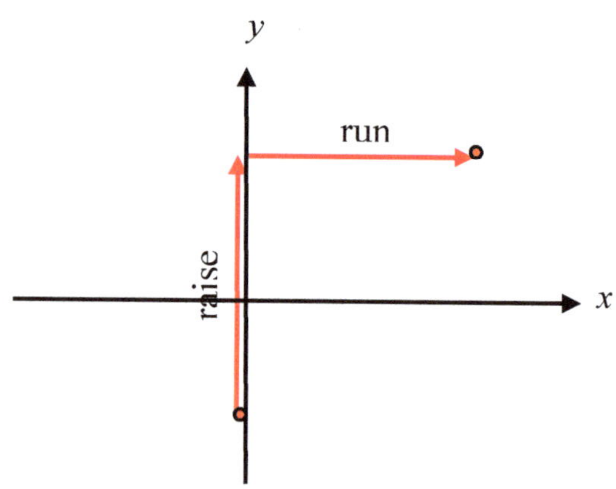

Step 3. Connect the two points generated by the y-intercept and the slope by a straight line. This will give us the graph of the line.

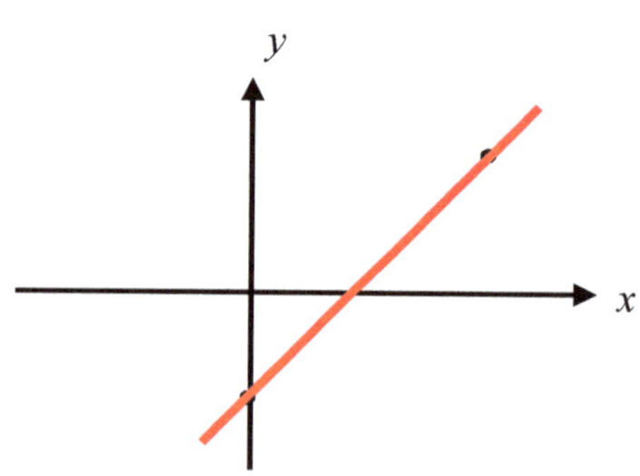

Step 4. Check your answer by selecting a test point. Plot the test point. If it lies on the line you drew, you can be pretty sure your work is correct.

Connect the two points generated by the y-intercept and the slope by a straight line. This will give us the graph of the line.

> **Checking the Result**
>
> To guarantee you did not make a mistake, select the test point $T(x, y)$, by selecting an easy value for x (or for y) and find the y-value (x-value). Plot the test point $T(x, y)$. If the test point lies on the line you drew, you can be quite sure your work is correct.

Example
Graph $y = 2x + 4$ using slope and y-intercept.

Solution
We outline the general steps for graphing a line using slope and y-intercept, given $m = 2$ and the y-intercept $(0, b)$ is $(0, 4)$. By Step 1, we need to Plot the y- intercept $(0, 4)$, in the xy-axis. By Step 2, we start from the y-intercept to find another point using the concept of raise over run of slopes.

$$\text{Slop } = m = 2 = \frac{\Delta y}{\Delta x} = \frac{2}{1} \Longrightarrow \frac{\text{raise}}{\text{run}} = \frac{2}{1} \qquad (1)$$

Or

$$\text{Slop } = m = 2 = \frac{\Delta y}{\Delta x} = \frac{-2}{-1} \Longrightarrow \frac{\text{raise}}{\text{run}} = \frac{-2}{-1} \qquad (2)$$

Figure (1) illustrates the slope formula (1), where $\frac{\text{raise}}{\text{run}} = \frac{2}{1}$, and Figure (2) shows the slope formula (2). As we will later see, using either one will come up with the same graph for the line. So,

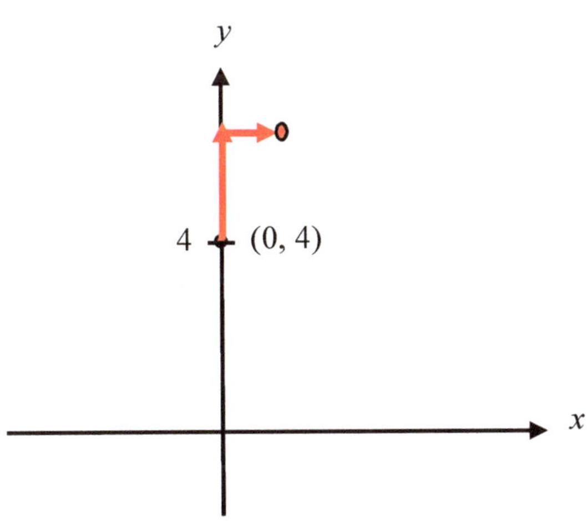

Figure (1)

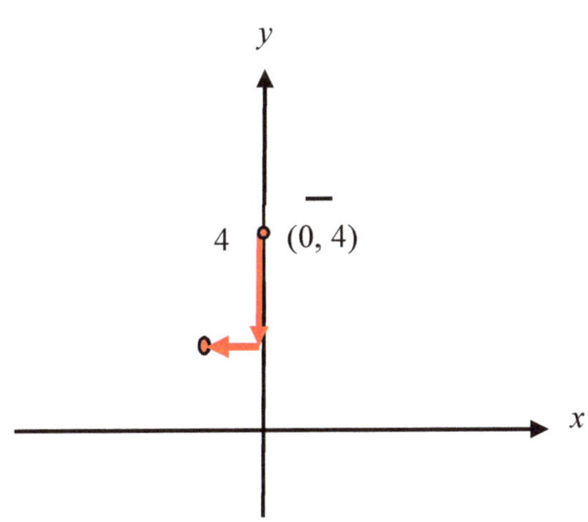

Figure (2)

Now, by Step3, we need to connect the two points generated by the y-intercept and the slope by a straight line. Using the graph in Figure (1), we will get the graph of the line shown in Figure (3) and using the graph in Figure (2), we will have the graph of the line as illustrated in Figure (4).
\

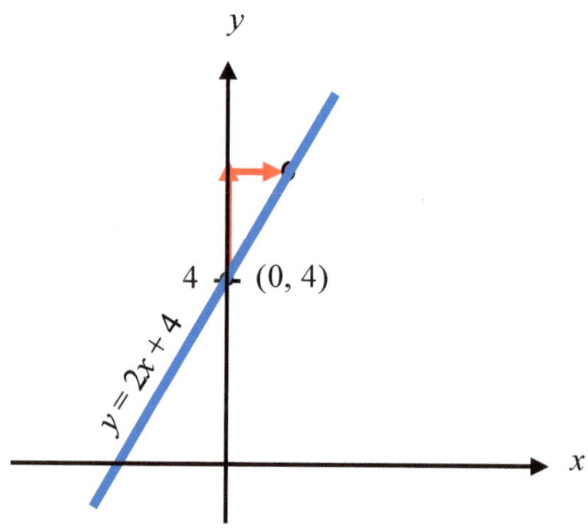

Figure (3)

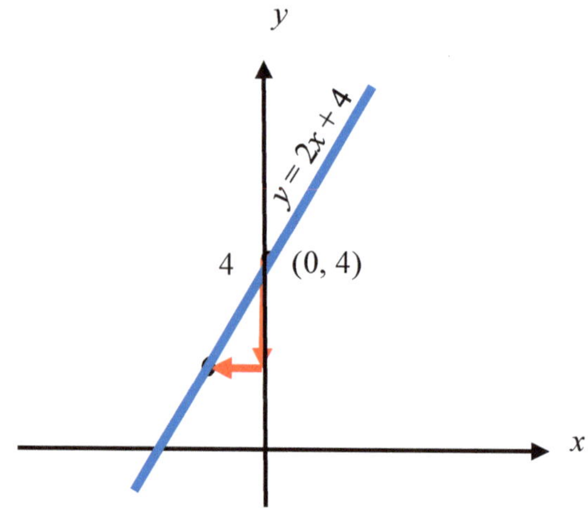

Figure (4)

As we mentioned earlier, we will see that using either the graph in Figure (1) or the graph in Figure (2), we came up with the same graph for the line as shown in Figure (3) and Figure (4). To check this, let us find the x-intercept of the lines in Figure (3) as well as in Figure (4). By letting $y = 0$, we will have.

$$y = 2x + 4$$

$$0 = 2x + 4$$ we plugged $y = 0$ into the equation

$$0 - 4 = 2x + 4 - 4$$ to isolate the variable we added – 4 to both sides

$$-4 = 2x$$

$$\frac{1}{2}(-4) = \frac{1}{2}(2x)$$ we multiplied both sides by the reciprocal of the variable

$$\frac{-4}{2} = \frac{2}{2}(x)$$

$$-2 = x$$

Therefore, the x-intercept is $x = -2$. Now if we plot this point on the graph in Figure (3) we realize that it is placed on the line $y = 2x + 4$.

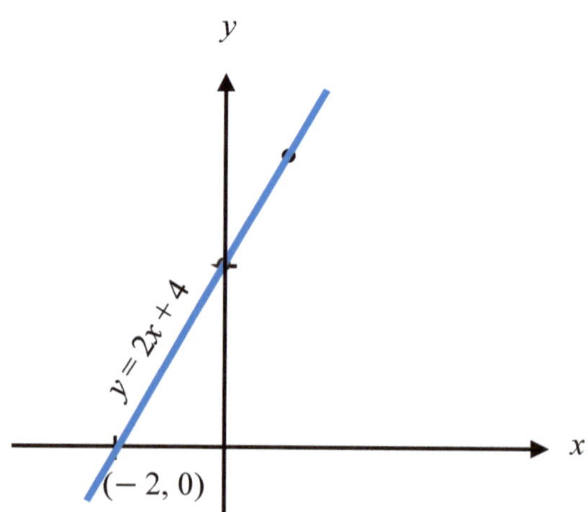

Figure (3)

Now, given the x-intercept is $x = -2$, if we plot this point on the graph in Figure (4) we realize that it is also placed on the line $y = 2x + 4$.

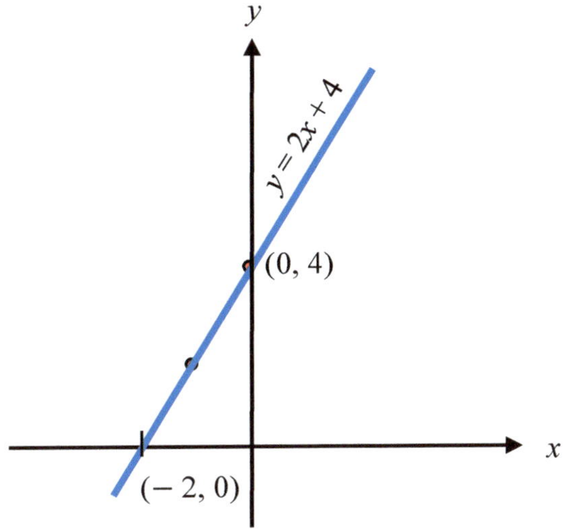

Figure (4)

We can conclude that, the line in Figure (3) and Figure (4) both are the graph of the line $y = 2x + 4$.

Another Method: Connecting the x and y-Intercepts

Another excellent method to graph a line in the xy-axis to use its intercepts. Intercepts are points of the line that are found on the x and y axes. There are two kinds of intercepts.

- The first one is called the x-intercept because it is the point of the line located on the horizontal axis (x-axis).
- The second is the y-intercept which is the point of the line located on the vertical axis (y-axis).

Since the x-intercept is a point where the line crosses the x-axis, it is a point with a y-value of zero.

y value is always zero

***x-intercept*: (x, 0)**

Similarly, since the y-intercept is a point where the line crosses the y-axis, it must be a point with an x-value of zero.

x value is always zero

***y-intercept*: (0, y)**

Recall that, we only need two points to graph a line. The x-intercept $(x, 0)$ and y-intercept $(0, y)$ are the two convenient points which can easily be solved algebraically from the equation of the line. That is,

- to solve for the x-intercept, let $y = 0$ then solve for x,
- to solve for the y-intercept, let $x = 0$ then solve for y.

having the two points, we just connect them with a straight line. Here is a quick diagram that gives you the idea.

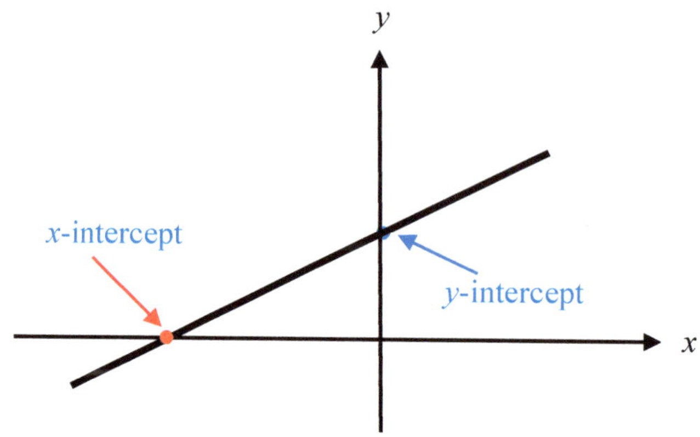

Example
Graph $3x + 4y - 9 = 0$ using the x and y-intercepts.

Solution
Recall that, we only need two points such as P and Q to graph a line. Using the x and y-intercept we need to find two points; one-point P, for x-intercept and the other point Q, for y-intercept. We learned to find the x-intercept, and the y-intercept of the line. So, we need:

- to solve for the x-intercept, by letting $y = 0$ then solve for x,
- to solve for the y-intercept, by letting $x = 0$ then solve for y.

Hence, to find the x-intercept, we let $y = 0$ in the equation, then solve for x. So,

$3x + 4y - 9 = 0$

$3x + 4(0) - 9 = 0$ we plugged $y = 0$ into the equation

$3x - 9 = 0$

$3x - 9 + 9 = 0 + 9$ to isolate the variable we added 9 to both sides

$3x = 9$

$\frac{1}{3} \times 3x = \frac{1}{3} \times 9$ we multiplied both sides by the reciprocal of the variable

$\frac{3}{3}x = \frac{9}{3}$

$x = \frac{3 \times 3}{3}$

$x = 3.$

Therefore, one of the points is $P(3, 0)$.

To find the y-intercept, we let $x = 0$ in the equation, then solve for y. So,

$3x + 4y - 9 = 0$

$3(0) + 4y - 9 = 0$ — we plugged $x = 0$ into the equation

$4y - 9 = 0$

$4y - 9 + 9 = 0 + 9$ — to isolate the variable we added 9 to both sides

$4y = 9$

$\frac{1}{4} \times 4y = \frac{1}{4} \times 9$ — we multiplied both sides by the reciprocal of the variable

$\frac{4}{4}y = \frac{9}{4}$

$y = \frac{9}{4}$

$y = 2.5.$

Therefore, the other point is $Q(0, 2.5)$.

Now, be having the two points $P(3, 0)$ and $Q(0, 2.5)$, we need:

- To plot them on the xy-axis
- To connect them with a straight line.

By connecting the two points $P(3, 0)$ and $Q(0, 2.5)$, we will get the graph of the line $3x + 4y - 9 = 0$. Here is the graph of the line $3x + 4y - 9 = 0$.

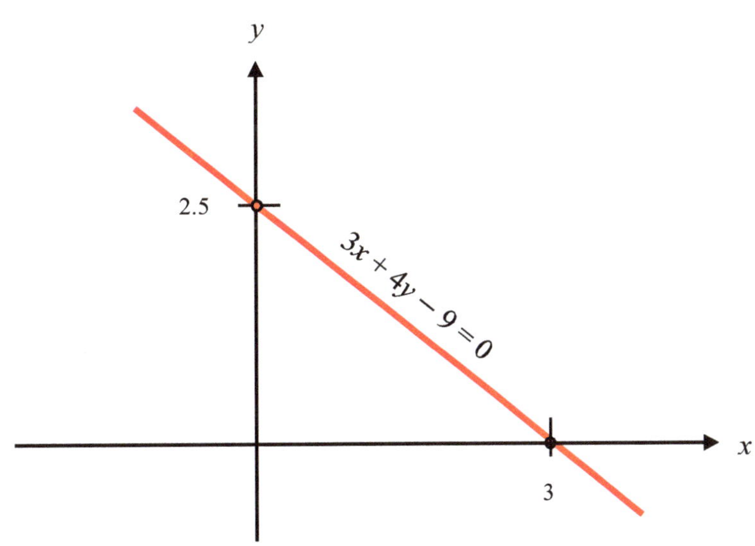

1.13. Exercises

1 Graph the points $A(1, 5)$, $B(-4, 6)$, $C(7, -2)$, $D(8, 0)$, $E(0, -5)$, $F(-9, 0)$ on a Cartesian coordinate.

2. Let P be the point $(3, -7)$. Find the points which are symmetric to P about the:
   a) x-axis.
   b) y-axis.
   c) Origin.

3. Given points $P(3)$, $Q(-6)$, $R(-9)$, $S(5)$ on a line, find:
   d) The distance between P and Q.
   e) The distance between P and R.
   f) The distance between Q and R.
   g) The distance between P and S.
   h) The distance between R and S.
   i) The distance between Q and S.

4. Find the radius of a circle, given that the center is at $O(-2, -3)$ and the point $P(4, -5)$ lies on the circle.

5. Find the length of a chord on a circle, with the end points of $A(1, 4)$ and $B(-2, 1)$.

6. Given $RS = 13$, $R(3, 9)$ and $S(8, y)$. Find y.

7. Given point $A(x, -5)$, find the two points below which have the same distance of 12 units from point $A(x, -5)$.
   a) $B(3, 2)$.
   b) $C(-2, 1)$.

8. Given points $P(-7)$, $Q(1)$ on a line, find the midpoint of the line segment PQ.

9. Find and simplify the distance between
   a) $N(-3, 7)$ and $O(6, -8)$.
   b) $R(-4, 0)$ and $S(2, -9)$.
   c) $T(-1, 5)$ and $U(8, 0)$.
   d) $P(0, -7)$ and $Q(-2, -4)$.

10. Find the missing value of h in the points $(5, 7)$ and $(1, h)$ if its midpoint is at $(3, -2)$.

11. Find the missing value of a in the points $(a, 2)$ and $(6, 9)$ if its midpoint is at $(3, 5.5)$.

12. Compute the slope of the line passing through the points $P(-2, 8)$ and $Q(6, -3)$.

13. Find the missing value of a, if the slope of the line is $m = -3$, and the line passing through two points $P(3, 1)$ and $Q(a, 4)$.

14. Given two lines. One line, L_1, passing through the point $P(7, -2)$ and $Q(-1, -8)$, and the other line L_2, passing through the point $R(2, 5)$ and $S(0, -1)$. Calculate the slope of each line and interpret the results (which one goes downhill or uphill more quickly).

15. Plot the line with y-intercept (0, 6) and slope $= m = -\frac{7}{5}$ and write the slope-intercept form of its equation.

16. Plot the line with x-intercept (4, 0) and slope $= m = \frac{1}{3}$.

17. Write the equation of the line passing through (2, 0) and (– 1, 5), and then plot the graph of the line.

18. Write the point-slope form of a line with a slope of 3 which passes through the y-intercept $P(0, 5)$.

19. Write the point-slope form of a line with a slope of 5 which passes through the point (– 3, 8).

20. Write the point-slope form of a line with a slope of $-\frac{3}{4}$ which passes through the point (4, – 11).

21. Write the point-slope form of a line with a slope of $\frac{1}{3}$ which passes through the point (– 12, – 5).

22. Find the point-slope form of the line from its graph below.

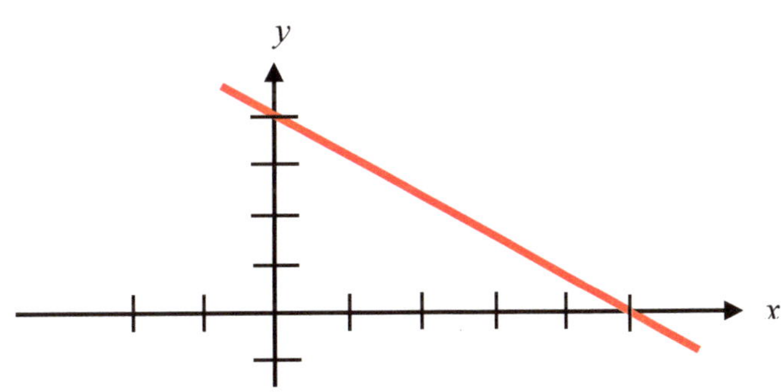

23. Find the point-slope form of the line from its graph below.

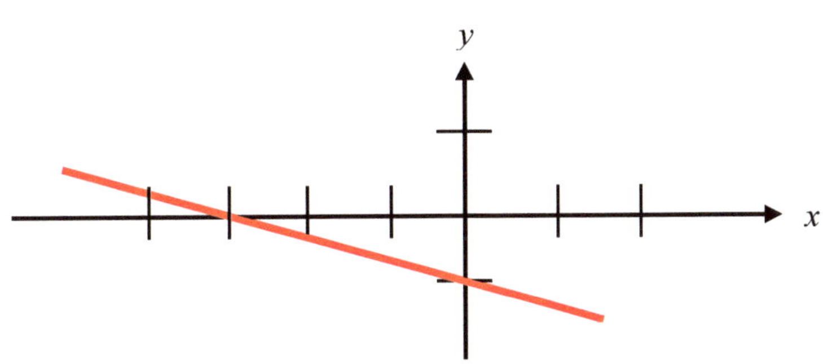

24. Find the equation of the line passing through the points,
   a) $A(-3, 8)$ and $B(0, 6)$.
   b) $C(-2, 0)$ and $D(5, -3)$.

c) $E(-6, -3)$ and $F(10, -8)$.

d) $P(-7, 9)$ and $Q(3, -2)$.

25. Graph $5x + 3y - 12 = 0$ by constructing a table of values.

26. Graph each line equation below using the slope and y-intercept.

a) $2y = 7x - 8$

b) $x - 3y - 6 = 0$

c) $\frac{3}{5}x + \frac{1}{3}y = 15$.

27. Graph each line equation below by connecting the x and y-intercepts.

a) $7x + 14y - 18 = 0$

b) $5x + \frac{3}{10}y = 12$

c) $\frac{2}{3}x + \frac{1}{3}y - 8 = 0$.

2. Lines

A line is a geometric object that extends infinitely in opposite directions, possessing only length and no width. It is comprised of an unending set of points arranged in a straight path.

2.1. Straight Lines

A straight line is a line which is not curved or bent. In analytic geometry, there are three different types of straight lines: parallel lines, intersecting lines and coincident lines. Two lines that lie exactly on top of each other or we can say that one line is exactly lying on top of another line are called coincident lines. Parallel lines are lines in a plane that are always the same distance apart. Parallel lines never intersect. Lines that are non-parallel intersect at a unique point. When two or more lines cross each other in a plane, they are called intersecting lines. The intersecting lines share a common point, which exists on all the intersecting lines, and this is called the point of intersection. Intersecting lines can be perpendicular lines. Perpendicular lines are lines that intersect at a right (90 degrees) angle. The lines which coincide or lie on top of each other are called coincident lines.

In this section we study the basic methods for finding the equation and graphing a pair straight line in a coordinate system. We start with an intersecting line. Then we will present the other two types: parallel lines, and perpendicular lines.

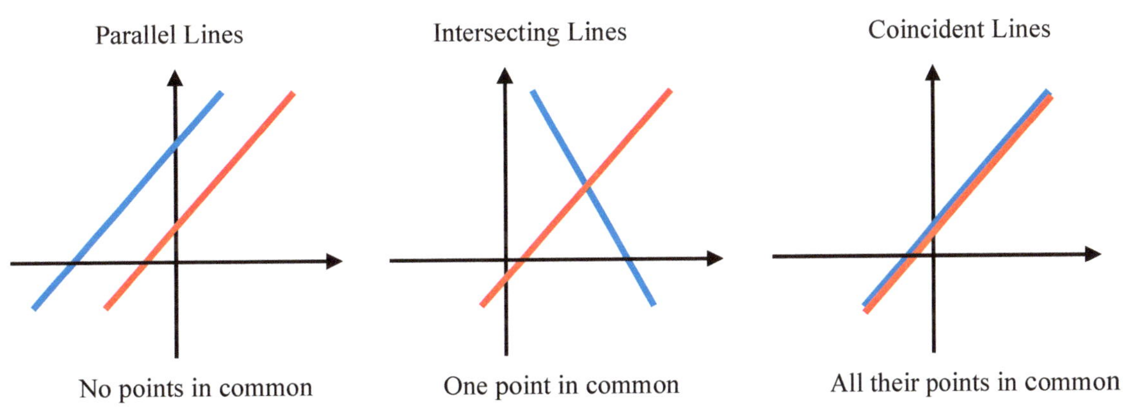

2.1.1. Intersecting Lines

Lines that are non-parallel intersect at a unique point. Two lines are said to be **intersect lines** if they cut each other at a point. Two lines can have at most 1 point of intersection. In the figure below lines L_1 and L_2 intersect each other at point P. Three or more lines that meet at a single point are said to be concurrent and the point of intersection is the point of concurrency.

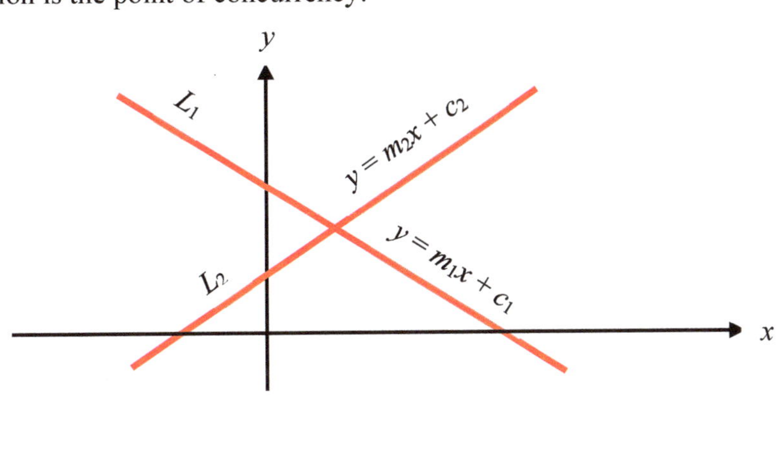

To find the intersection of two lines, called point of intersection, you first need the equation for each line. At the intersection, x and y have the same value for each equation. This means that the equations are equal. We can therefore solve for x. Substitute the value of x in one of the equations and solve for y.

To find the point of intersection, consider the two lines in the figure below. These two lines are represented by the equations $a_1x_1 + b_1y_1 + c_1 = 0$ and $a_2x_2 + b_2y_2 + c_2 = 0$, respectively.

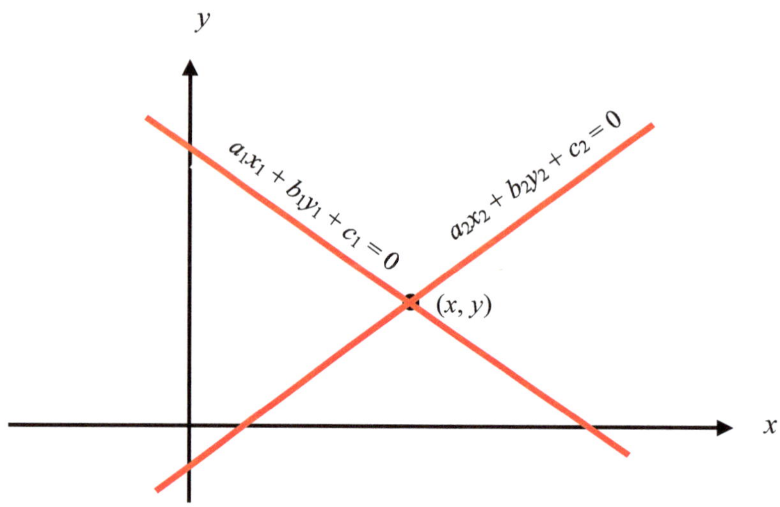

By solving the two equations the xy-coordinates for the point of intersection of two lines is.

$$(x, y) = \left(\frac{b_1c_2 - b_2c_1}{a_1b_2 - a_2b_1}, \frac{a_2c_1 - a_1c_2}{a_1b_2 - a_2b_1}\right).$$

Example

Given the two intersecting lines with the following equations, find the xy-coordinates for the point of intersection of two lines.

L_1: $3x - 2y - 6 = 0.$

L_2: $3x + 4y - 15 = 0.$

Solution

Intersection points of L_1 and L_2 can be calculated using the formula for the point of intersection of two lines, with $a_1 = 3$, $b_1 = -2$, $c_1 = -6$, $a_2 = 3$, $b_2 = 4$, and $c_2 = -15$. So,

$$(x, y) = \left(\frac{b_1c_2 - b_2c_1}{a_1b_2 - a_2b_1}, \frac{a_2c_1 - a_1c_2}{a_1b_2 - a_2b_1}\right)$$

$$(x, y) = \left(\frac{(-2)(-15) - (4)(-6)}{3(4) - (3)(-2)}, \frac{(3)(-6) - 3(-15)}{3(4) - (3)(-2)}\right)$$ we plugged $a_1 = 3$, $b_1 = -2$, $c_1 = -6$, $a_2 = 3$, $b_2 = 4$, and $c_2 = -15$ into the equation

$$(x, y) = \left(\frac{30 - (-24)}{12 - (-6)}, \frac{-1 \;\; - (-45)}{12 - (-6)}\right)$$

$$(x, y) = \left(\frac{30 + 24}{12 + 6}, \frac{-18 + 45}{12 + 6}\right)$$

$$(x, y) = \left(\frac{54}{18}, \frac{27}{18}\right)$$

$$(x, y) = \left(\frac{3 \times 18}{18}, \frac{3 \times 9}{2 \times 9}\right)$$

$(x, y) = \left(3, \frac{3}{2}\right)$.

Therefore, the *xy*-coordinates for the point of intersection of two lines is $(x, y) = \left(3, \frac{3}{2}\right)$.

2.1.2. Parallel Lines

Parallel lines are lines in a plane that are always the same distance apart. Parallel lines never intersect. The slopes of parallel lines are the same. That is, two lines $L_1 = m_1 x + b_1$ and $L_2 = m_2 x + b_2$ are parallel if and only if $m_1 = m_2$.

Parallel Lines

- Two lines are parallel lines if they do not intersect.
- Two lines $L_1 = m_1 x + b_1$ and $L_2 = m_2 x + b_2$ are parallel if and only if $m_1 = m_2$.

Graphs of parallel lines are illustrated below.

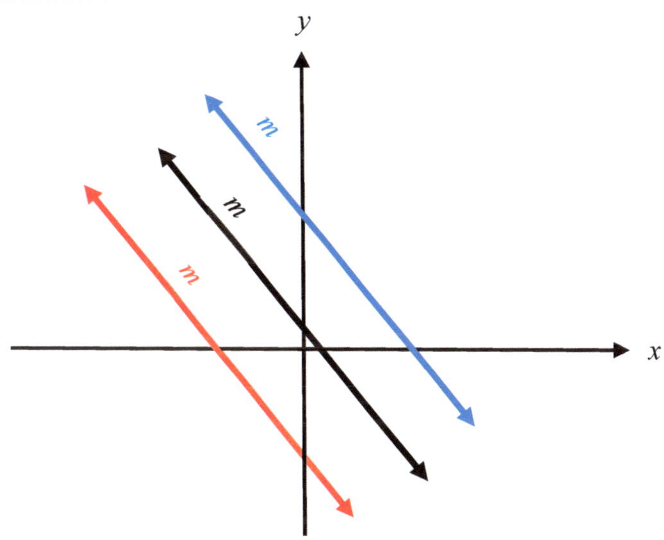

Example
Suppose we know that the slope of line L_1 is 5. We also know that its *y*-intercept is (0, 5). Write the general equation of lines parallel to the given line.

Solution
Any other line with a slope of 5 will be parallel to L_1. The lines L_2's, which are formed by all values *b*, will be parallel to L_1.

Lines L_2's: $y = 5x + b$, for $-\infty < b < \infty$.

Example
Find the equation of the line that is:

- Parallel to $y = 3x + 6$.
- And passes though the point (2, 9).

Solution
The slope of the line $y = 3x + 6$ is 3. The parallel line needs to have the same slope of 3. Now, given the point (2, 9) and the slope of 3, we can find the equation using the point-slope equation of a line. So,

$y - y_1 = m(x - x_1)$

$y - 9 = 3(x - 2)$ we plugged $y_1 = 9$, $m = 3$, and $x_1 = 2$ into the equation

$y - 9 = 3x - 6$

$y - 9 + 9 = 3x - 6 + 9$ to isolate y, we added 9 to both sides

$y = 3x + 3.$

Therefore, the equation of the line parallel to $y = 3x + 6$, and passes though the point (2, 9) is $y = 3x + 3$.

Example
Find the equation of the line passing through point $P(-2, -1)$ and is parallel to the line from its graph below, and then sketch its graph.

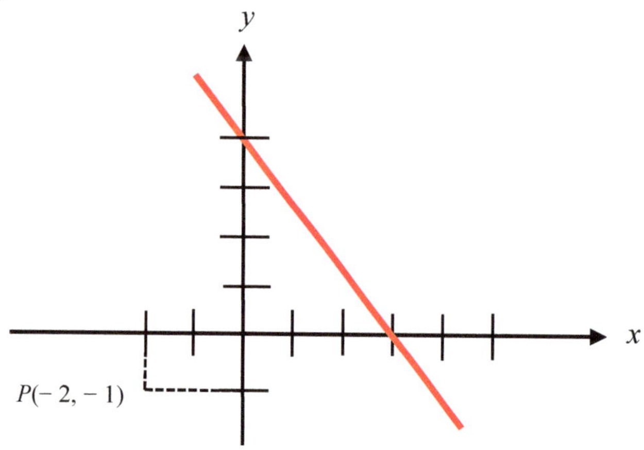

Solution
We have learned to write the equation of a line using point-slope formula, and we need to have its slope and a point the line passing through it. We also learned if two lines are parallel, they must have the same slope. Hence, we need to find the slope of the line graphed above. Using the raise over run concept we can find the required slope as it is seen below.

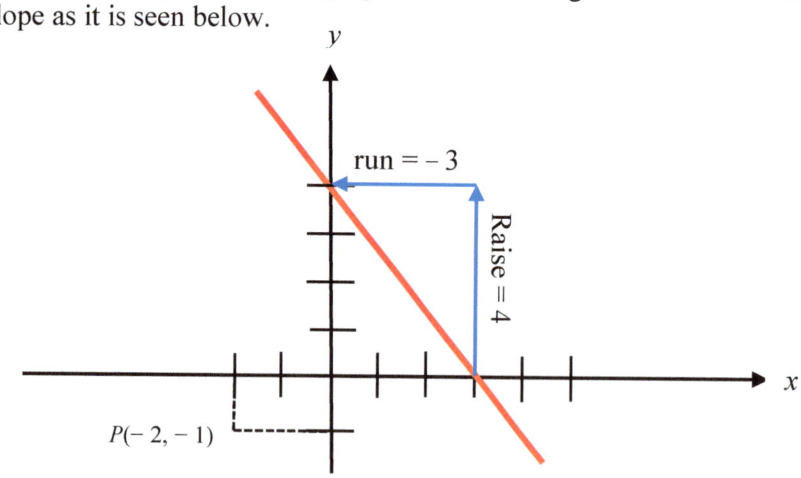

As it is seen we obtained the raise = 4 and the run = – 3 (since we moved to the left). Hence, the slope is:
$\text{slope} = m = \frac{\text{raise}}{\text{run}} = \frac{4}{-3} = -\frac{4}{3}.$

Now we use the point-slope formula to find the equation of the requested line. Given the point $P(-2, -1)$ and the slope $m = -\frac{4}{3}$, we will have.

$y - y_1 = m(x - x_1)$

$y - (-1) = (-\frac{4}{3})[x - (-2)]$ — we plugged $y_1 = -1$, $m = -\frac{4}{3}$, and $x_1 = -2$ into the equation

$y + 1 = (-\frac{4}{3})(x + 2)$

$y + 1 = -\frac{4}{3}x - \frac{4 \times 2}{3}$

$y + 1 = -\frac{4}{3}x - \frac{8}{3}$

$y + 1 - 1 = -\frac{4}{3}x - \frac{8}{3} - 1$ — to isolate y, we added -1 to both sides

$y = -\frac{4}{3}x - \frac{8}{3} - 1$

$y = -\frac{4}{3}x - \frac{8}{3} - 1 \times \frac{3}{3}$ — to make a common denominator, we multiplied the last term in the left-side by $\frac{3}{3}$

$y = -\frac{4}{3}x - \frac{8}{3} - \frac{3}{3}$

$y = -\frac{4}{3}x + \frac{-8-3}{3}$

$y = -\frac{4}{3}x + \frac{-11}{3}.$

$y = -\frac{4}{3}x - \frac{11}{3}.$

Therefore, the equation of the requested line is $y = -\frac{4}{3}x - \frac{11}{3}$.

To convert the equation with integer coefficient and integer constant term, we need to perform some extra arithmetic operations. So,

$y = -\frac{4}{3}x - \frac{11}{3}$

$3 \times y = 3 \times (-\frac{4}{3}x - \frac{11}{3})$ — we multiplied both sides by 3

$3y = -\frac{4 \times 3}{3}x - \frac{11 \times 3}{3}$

$3y = -4x - 11.$

We now need to sketch the graph of the resulting line. We already have a point on this line, $P(-2, -1)$. We simply need to select another point of this line. Letting 0 be the x-coordinate of this point, the y-coordinate is:

$3y = -4x - 11$

$3y = (-0)\,x - 11$ — we plugged the x-coordinate = 0 into the equation

$3y = -11$

$\frac{1}{3} \times 3y = \frac{1}{3} \times (-11)$ — we multiplied both sides by the reciprocal of the variable

$y = -\frac{11}{3} = -3.67.$

To sketch the graph of the line, having the two points (− 2, − 1), and (0, − 3.67), we need to plot them on the *xy*-coordinates and then connect them. So,

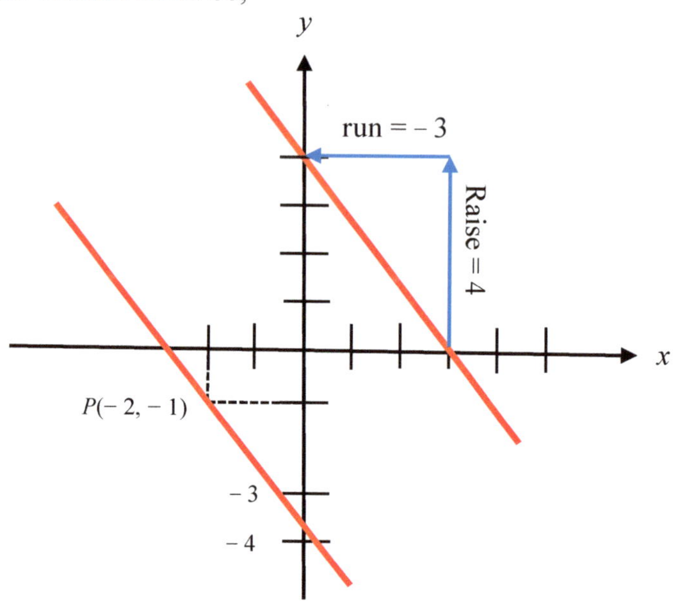

2.1.3. Perpendicular Lines

Perpendicular lines are lines that intersect at a right (90 degrees) angle. That is, two lines $L_1 = m_1 x + b_1$ and $L_2 = m_2 x + b_2$ are perpendicular if and only if $m_1 \times m_2 = -1$, or $m_2 = -\frac{1}{m_1}$.
Graphs of perpendicular lines are illustrated below.

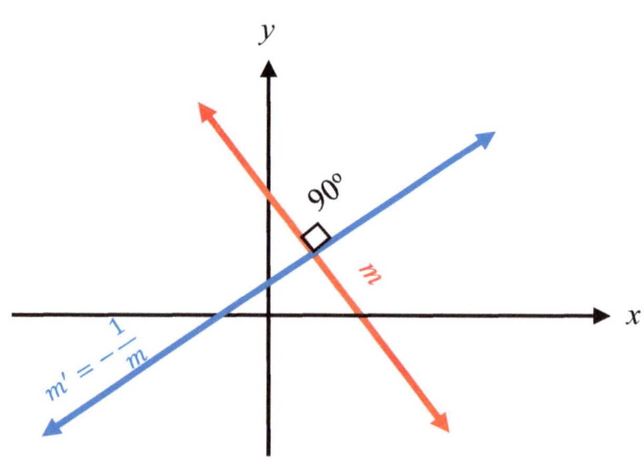

Perpendicularity can be shown to be symmetric, meaning if a first line is perpendicular to a second line, then the second line is also perpendicular to the first. For this reason, we may speak of two lines as being perpendicular (to each other) without specifying an order.

2.1.4. Relationship to Parallel Lines

If two lines, L_1 and L_2, are both perpendicular to a third line L_3, the angle of the third line are right angle. Therefore, in analytic geometry, any two lines that are both perpendicular to a third line are parallel to each other. Conversely, if one line is perpendicular to a second line, it is also perpendicular to any line parallel to that second line (see the figure below).

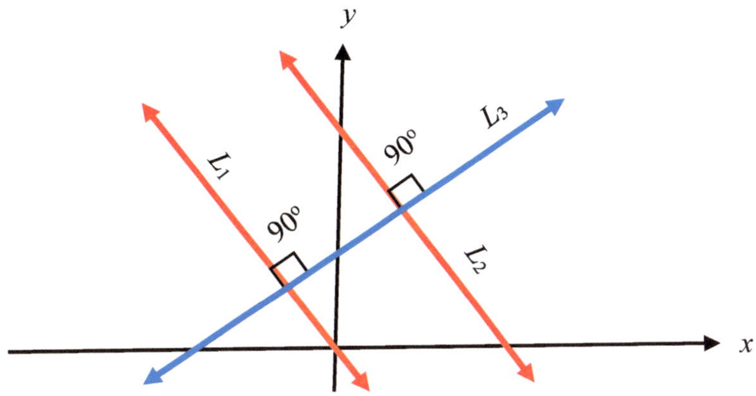

Example
Given the slope of line L_1 is 9, write the general equations of lines perpendicular to line L_1.

Solution
Lines with a slope of $-\frac{1}{9}$ are perpendicular to L_1. All lines with the constant c's are perpendicular to L_1.

Lines L_2's: $y = -\frac{1}{9}x + c,$ for $-\infty < c < \infty.$

Example
Find that the two lines of with the equations $3x + 2y = 7$ and $-3x + 2y = 13$ are parallel, perpendicular.

Solution
Let us convert each equation in the slope-intercept form, $y = mx + b,$ in which their y-intercept is $(0, b)$ and their slope is m. So,

for $3x + 2y = 7$

$3x + 2y - 3x = 7 - 3x$ to isolate y, we added $-3x$ to both sides

$2y = 7 - 3x$

$\frac{1}{2} \times 2y = \frac{1}{2} \times (7 - 3x)$ we multiplied both sides by the reciprocal of y

$y = \frac{1}{2} \times (7 - 3x)$

$y = \frac{7}{2} - \frac{3}{2}x.$

The slope of equations $3x + 2y = 7$ is $-\frac{3}{2}.$

For $-3x + 2y = 13$

$-3x + 2y + 3x = 13 + 3x$ to isolate y, we added $3x$ to both sides

$2y = 13 + 3x$

$\frac{1}{2} \times 2y = \frac{1}{2} \times (13 + 3x)$ we multiplied both sides by the reciprocal of y

$y = \frac{1}{2} \times (13 + 3x)$

$y = \frac{13}{2} + \frac{3}{2}x.$

The slope of equations $-3x + 2y = 13$ is $\frac{3}{2}$. We now can conclude:

- Two lines are parallel if and only if their slopes are the same. In this case, $-\frac{3}{2}$ and $\frac{3}{2}$ are not the same. Hence the given two lines are not parallel.
- Two lines are perpendicular if the product of their slopes is -1. In this case, $(-\frac{3}{2}) \times \frac{3}{2} = -\frac{9}{4}$. Thus, the given two lines are not perpendicular.

Example
Write the equation of a line that contains the point (2, 5) and is perpendicular to the line $3y - 6x + 12 = 0$.

Solution
Let us first covert the equation in the slope-intercept form, $y = mx + b$. So,

$3y - 6x + 12 = 0$

$3y - 6x + 12 + 6x - 12 = 0 + 6x - 12$ to isolate y, we added $6x - 12$ to both sides

$3y = 6x - 12$

$\frac{1}{3} \times 3y = \frac{1}{3} \times (6x - 12)$ we multiplied both sides by the reciprocal of y

$y = \frac{1}{3} \times (6x - 12)$

$y = \frac{6}{3}x - \frac{12}{3}$

$y = 2x - 4.$

The slope of equations $3y - 6x + 12 = 0$ is 2.

We have learned that two lines with slopes of m_1 and m_2, are perpendicular if $m_2 = -\frac{1}{m_1}$. In this case, we found the slope of the line $3y - 6x + 12 = 0$, is $m_1 = 2$. Hence the slope of the line perpendicular to this line is $m_2 = -\frac{1}{m_1} = -\frac{1}{2}$.

Now, we are given a point (2, 5) and we obtained the slope of $-\frac{1}{2}$, so now we simply need to use the point-slope form of a linear equation $y - y_1 = m(x - x_1)$, to find the requested line equation. So,

$y - y_1 = m(x - x_1)$

$y - 5 = (-\frac{1}{2})(x - 2)$ we plugged $y_1 = 5$, $x_1 = 2$, and $m = -\frac{1}{2}$ into the equation

$y - 5 = -\frac{1}{2}x + \frac{2}{2}$

$y - 5 = -\frac{1}{2}x + 1$

$y - 5 + 5 = -\frac{1}{2}x + 1 + 5$ to isolate y, we added 5 to both sides

$y = -\frac{1}{2}x + 6.$

Therefore, the equation of the line that contains the point (2, 5) and is perpendicular to the line $3y - 6x + 12 = 0$ is $y = -\frac{1}{2}x + 6$.

Example
Find the equation of the line passing through point $P(3, 1)$ and is perpendicular to the line from its graph below.

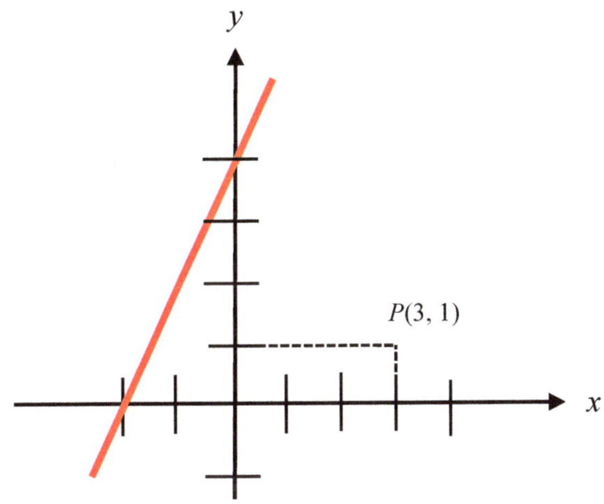

Solution
We have learned that, to write the equation of a line using point-slope formula, we need to have its slope and a point on the line. In this case, the given line graphed above, its slope, m_1, can be found using the raise over run concept, as illustrated in the graph below.

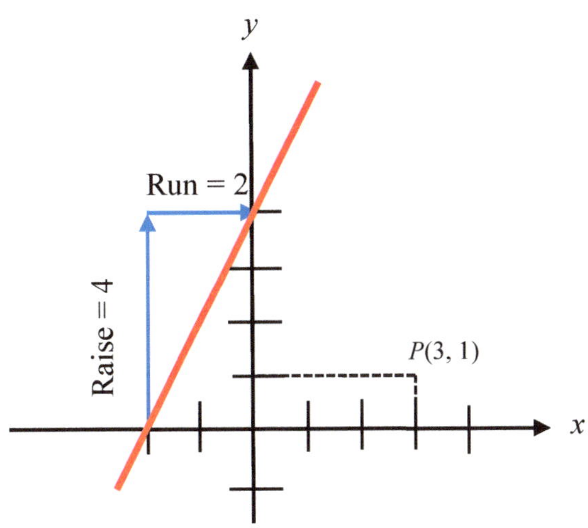

As it is seen, the raise is 4 and the run is 2. Hence, the slope of this line is $m_1 = \frac{4}{2} = 2$.

To find the slope of the requested line, m_2, we learned that if two lines are perpendicular, the product of their slopes is equal to -1, $m_1 \times m_2 = -1$, or $m_2 = -\frac{1}{m_1}$. In this case, $m_1 = 2$ and hence $m_2 = -\frac{1}{m_1} = -\frac{1}{2}$.

Given the point $P(3, 1)$, and the obtained slope of $m_2 = -\frac{1}{m_1} = -\frac{1}{2}$, we simply use the point-slope formula, $y - y_1 = m(x - x_1)$, to find the equation of the requested line. So,

$y - y_1 = m(x - x_1)$

$y - 1 = (-\frac{1}{2})(x - 3)$ — we plugged $y_1 = 1$, $x_1 = 3$, and $m = -\frac{1}{2}$ into the equation

$y - 1 = -\frac{1}{2}x + \frac{3}{2}$

$y - 1 + 1 = -\frac{1}{2}x + \frac{3}{2} + 1$ — to isolate y, we added 1 to both sides

$y = -\frac{1}{2}x + \frac{3}{2} + 1$

$y = -\frac{1}{2}x + \frac{3}{2} + 1 \times \frac{2}{2}$ — to make a common denominator, we multiplied the last term in the left-side by $\frac{2}{2}$

$y = -\frac{1}{2}x + \frac{3}{2} + \frac{2}{2}$

$y = -\frac{1}{2}x + \frac{3+2}{2}$

$y = -\frac{1}{2}x + \frac{5}{2}$.

Therefore, the equation of the requested line is $y = -\frac{1}{2}x + \frac{5}{2}$.

To convert the equation with integer coefficient and integer constant term we need to perform some extra arithmetic operations. So,

$y = -\frac{1}{2}x + \frac{5}{2}$

$2 \times y = 2 \times (-\frac{1}{2}x + \frac{5}{2})$ — we multiplied both sides by 2

$2y = -\frac{2 \times 1}{2}x + \frac{2 \times 5}{2}$

$2y = -x + 5$.

2.2. Distance of Point from a Line

In Euclidean geometry, the distance from a point to a line is defined as the shortest distance from that point to any point on an infinite straight line. This distance is calculated as the length of the line segment joining the given point to the nearest point on the line, which represents the perpendicular distance from the point to the line. The formula for computing this distance can be derived and expressed in various ways. Understanding the distance from a point to a line is valuable in practical scenarios, such as determining the separation between two objects like two buildings in real-life applications.

The formula for calculating it can be derived and expressed in several ways. Since the derivation of the formula is beyond the scope of this book, we present only the result.

Consider a line by the equation $ax + by + c = 0$, where a, b and c are real constants with $a \neq 0$ and $b \neq 0$, the distance, d. from the line to a point $P(x_0, y_0)$ is, $d = \frac{|ax_0 + by_0 + c|}{\sqrt{a^2 + b^2}}$.

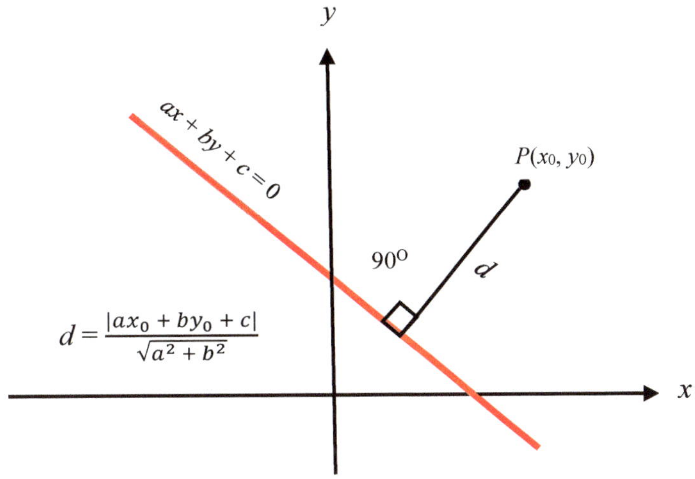

Example

Find the distance from point $P(1, -\frac{1}{2})$ to the line $y = 3$.

Solution

Approach 1- For distance from a point (x_0, y_0) to a line by the equation $ax + by + c = 0$, we have $d = \frac{|ax_0 + by_0 + c|}{\sqrt{a^2 + b^2}}$. In this case,

We are given the point $P(1, -\frac{1}{2})$ and the line $y = 3$ that can be written as $y - 3 = 0$. Hence, we have, $a = 0$, $b = 1$, $c = -3$, $-\frac{1}{2}$. So,

$$d = \frac{|ax_0 + by_0 + c|}{\sqrt{a^2 + b^2}}$$

$$d = \frac{|(0)(1) + (1)\left(-\frac{1}{2}\right) + (-3)|}{\sqrt{0^2 + 1^2}}$$ we plugged $a = 0$, $b = 1$, $c = -3$ and $(x_0, y_0) = (1, -\frac{1}{2})$ into the equation

$$d = \frac{\left|0 - \frac{1}{2} - 3\right|}{1}$$

$$d = \left|-\frac{1}{2} - 3\right|$$

$$d = \left|-\frac{1}{2} - 3 \times \frac{2}{2}\right|$$

$$d = \left|-\frac{1}{2} - \frac{6}{2}\right|$$

$$d = \left|\frac{-1 - 6}{2}\right|$$

$$d = \left|\frac{-7}{2}\right| = 3.5.$$

Therefore, the distance from point $P(1, -\frac{1}{2})$ to the line $y = 3$ is 3.5.

Approach 2- We draw a line that connects the point $P(1, -\frac{1}{2})$ and intersects the line at a perpendicular angle at point $H(1, 3)$, as shown in the figure below.

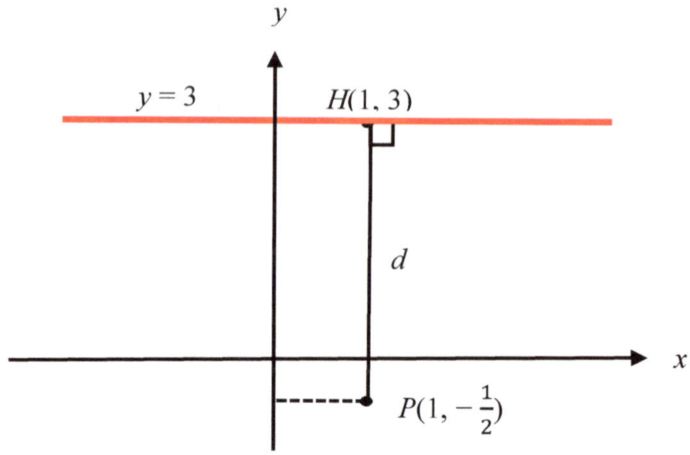

Now, the vertical distance from the point $P(1, -\frac{1}{2})$ to the line $y = 3$ will be the vertical distance from the point $P\,(1, -\frac{1}{2})$ to the point $H\,(1, 3)$. Since the x-coordinate of these two points are the same, 1, the difference of these two points is,

$$d = y_2 - y_1$$

$$d = 3 - (-\frac{1}{2})$$ we plugged $y_2 = 3$ and $y_1 = -\frac{1}{2}$ into the equation

$$d = 3 \times \frac{2}{2} + \frac{1}{2}$$ $\frac{2}{2}$

$$d = \frac{6}{2} + \frac{1}{2} =$$

$$d = \frac{7}{2} = 3.5.$$

Therefore, the distance from point $P\,(1, -\frac{1}{2})$ to the line $y = 3$, is we expected to be the same as the distance obtained by the first approach), that is 3.5.

Example

Find the distance between point $P\,(\frac{1}{2}, \frac{1}{2})$ to the line $x = -\frac{1}{2}$.

Solution

Approach 1- We learned that the distance from a point (x_0, y_0) to a line given by the equation $ax + by + c = 0$, we have $d = \frac{|ax_0 + by_0 + c|}{\sqrt{a^2 + b^2}}$. In this case, we are given the point $P\,(\frac{1}{2}, \frac{1}{2})$ and the line $x = -\frac{1}{2}$ that can be written as $x + \frac{1}{2} = 0$. Hence, we have, $a = 1$, $b = 0$, $c = \frac{1}{2}, \frac{1}{2}\frac{1}{2}$. So,

$$d = \frac{|ax_0 + by_0 + c|}{\sqrt{a^2 + b^2}}$$

$$d = \frac{|(1)(\frac{1}{2}) + (0)(\frac{1}{2}) + (\frac{1}{2})|}{\sqrt{1^2 + 0^2}}$$ we plugged $a = 1$, $b = 0$, $c = \frac{1}{2}$ and $(x_0, y_0) = (\frac{1}{2}, \frac{1}{2})$ into the equation

$$d = \frac{\left|\frac{1}{2}+0+\frac{1}{2}\right|}{1}$$

$$d = \left|\frac{1}{2}+\frac{1}{2}\right|$$

$$d = \left|\frac{1+1}{2}\right|$$

$$d = = \left|\frac{2}{2}\right| = 1.$$

Therefore, the distance from point $P\left(\frac{1}{2}, \frac{1}{2}\right)$ to the line $x = -\frac{1}{2}$ is 1.

Approach 2- We draw a line that connects the point and intersects the line at a perpendicular angle at point $H\left(-\frac{1}{2}, \frac{1}{2}\right)$, as shown in the figure below.

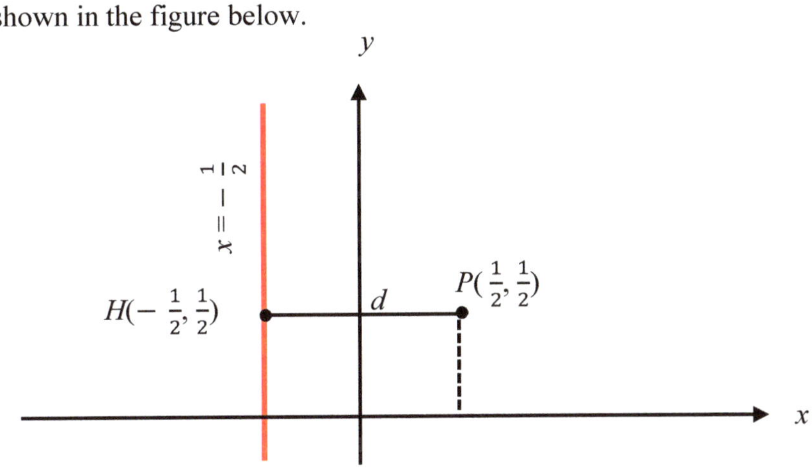

Now, the horizontal distance from the point $P\left(\frac{1}{2}, \frac{1}{2}\right)$ to the line $x = -\frac{1}{2}$ will be the horizonal distance from the point $P\left(\frac{1}{2}, \frac{1}{2}\right)$ to the point $H\left(-\frac{1}{2}, \frac{1}{2}\right)$. Since the y-coordinate of these two points are the same, $\frac{1}{2}$, the difference of these two points is,

$$d = x_2 - x_1$$

$$d = \frac{1}{2} - \left(-\frac{1}{2}\right)$$ we plugged $x_2 = \frac{1}{2}$ and $x_1 = -\frac{1}{2}$ into the equation

$$d = \frac{1}{2} + \frac{1}{2}$$

$$d = \frac{2}{2} = 1.$$

Therefore, the distance from point $P\left(\frac{1}{2}, \frac{1}{2}\right)$ to the line $x = -\frac{1}{2}$, is we expected to be the same as the distance obtained by the first approach), that is 1.

Example
What is the shortest distance from the origin, $O(0, 0)$ to the line $3x - y - 4 = 0$?

Solution
Again, there are two solution approaches for solving this problem.

Approach 1- We know that the distance from a point (x_0, y_0) to a line $ax + by + c = 0$, will be :

$$d = \frac{|ax_0 + by_0 + c|}{\sqrt{a^2 + b^2}}.$$

In this case, we are given the point $O(0, 0)$ and the line $3x - y - 4 = 0$. Hence, we have, $a = 3$, $b = -1$, $c = -4$, . So,

$$d = \frac{|ax_0 + by_0 + c|}{\sqrt{a^2 + b^2}}$$

$$d = \frac{|(3)(0) + (-1)(0) + (-4)|}{\sqrt{3^2 + (-1)^2}}$$ we plugged $a = 3$, $b = -1$, $c = -4$ and $(x_0, y_0) = (0, 0)$ into the equation

$$d = \frac{|0 + 0 - 4|}{\sqrt{9 + 1}}$$

$$d = \frac{|-4|}{\sqrt{10}}$$

$$d = \frac{4}{\sqrt{10}}$$

$$d = \frac{4}{\sqrt{10}} \times \frac{\sqrt{10}}{\sqrt{10}}$$

$$d = \frac{4\sqrt{10}}{10}$$

$$d = \frac{2 \times 2\sqrt{10}}{2 \times 5} = \frac{2\sqrt{10}}{5}.$$

Therefore, the shortest distance from point $O(0, 0)$ to the line $3x - y - 4 = 0$ is $d = \frac{2}{5}\sqrt{10}$.

Approach 2- Let us sketch the graph of the line $3x - y - 4 = 0$. To do so, we first need to plot two arbitrary points from the line. By letting $x = 0$, we find,

$3x - y - 4 = 0$

$3(0) - y - 4 = 0$ we plugged $x = 0$ into the equation

$-y - 4 = 0$

$-y - 4 + 4 = 0 + 4$ to isolate y, we added 4 to both sides

$-y = 4$

$(-1) \times (-y) = (-1) \times 4$ we multiplied both sides by (-1)

$y = -4.$

Therefore, one of the points on the line $3x - y - 4 = 0$ is $P(0, -4)$.

By letting $y = 0$, we will have:

$3x - y - 4 = 0$

$3x - (0) - 4 = 0$ we plugged $y = 0$ into the equation

$3x - 4 = 0$

$3x - 4 + 4 = 0 + 4$ to isolate x, we added 4 to both sides

$3x = 4$

$$\frac{1}{3} \times 3x = \frac{1}{3} \times 4$$ we multiplied both sides by the $\frac{1}{3}$

$$x = \frac{4}{3} = 1.33.$$

Therefore, the other point on the line $3x - y - 4 = 0$ is $Q(1.33, 0)$.

We now can sketch the graph of the line $3x - y - 4 = 0$, by plotting P and Q and connecting them. Here is the graph of the line $3x - y - 4 = 0$, and the graph of the line L_2, perpendicular to this line.

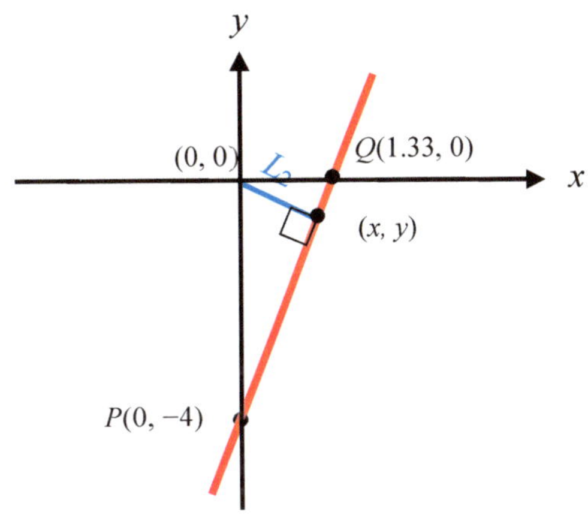

Next, to calculate the shortest distance from the origin, $O(0, 0)$ to the line $3x - y - 4 = 0$, we need to find the xy-coordinates of the point intersecting the line L_2 and the line $3x - y - 4 = 0$. However, for finding the xy-coordinates of this point, we need to find the equation of the line L_2. Given the origin, $O(0, 0)$ as one point on this line we need to find its slope first. Hence, let us find the slope, m_1 of the line $3x - y - 4 = 0$. To find m_1 we rearrange its equation to the form $y = mx + b$. So,

$$3x - y - 4 = 0.$$

$$3x - y - 4 - 3x + 4 = 0 - 3x + 4$$ to isolate y, we added $-3x + 4$ to both sides

$$-y = -3x + 4$$

$$(-1) \times (-y) = (-1) \times (-3x + 4)$$ we multiplied both sides by (-1)

$$y = 3x - 4.$$

Therefore, the slope, m_1, of the line $3x - y - 4 = 0$ is 3.

Since the line L_2 is perpendicular to the line $3x - y - 4 = 0$, its slope m_2 is $m_2 = -\frac{1}{\,} = -\frac{1}{3}$. Now, having the point $O(0, 0)$ and the slope $m_2 = -\frac{1}{3}$, we use the point-slope formula, $y - y_1 = m(x - x_1)$, to find the equation of the line L_2. So,

$$y - y_1 = m(x - x_1)$$

$$y - 0 = -\frac{1}{3}(x - 0)$$ we plugged $y_1 = 0$, $y_1 = 0$, and $m = -\frac{1}{3}$, into the equation

$$y = -\frac{1}{3}x.$$

Having the equation of the line L_2 as $y = -\frac{1}{3}x$ and the equation of the line $3x - y - 4 = 0$, we use the formula for finding the xy-coordinates for the point of intersection of the two lines (x, y). Rearranging the line $y = -\frac{1}{3}x$ as $\frac{1}{3}x + y = 0$, we have $a_1 = \frac{1}{3}, b_1 = 1, c_1 = 0,$ and the given line $3x - y - 4 = 0$ we have $a_2 = 3, b_2 = -1,$ and $c_2 = -4$.So,

$$(x, y) = (\frac{b_1c_2 - b_2c_1}{a_1b_2 - a_2b_1}, \frac{a_2c_1 - a_1c_2}{a_1b_2 - a_2b_1})$$

$$(x, y) = \left(\frac{1 \times (-4) - (-1)(0)}{\frac{1}{3}(-1) - 3(1)}, \frac{3(0) - \frac{1}{3}(-4)}{\frac{1}{3}(-1) - 3(1)}\right)$$

we plugged $a_1 = \frac{1}{3}, b_1 = 1, c_1 = 0, a_2 = 3,$ $b_2 = -1$ and $c_2 = -4$ into the equation

$$(x, y) = (\frac{-4 - 0}{-\frac{1}{3} - 3}, \frac{0 - \frac{4}{3}}{-\frac{1}{3} - 3})$$

$$(x, y) = (\frac{-4}{-\frac{1}{3} - 3 \times \frac{3}{3}}, \frac{\frac{4}{3}}{-\frac{1}{3} - 3 \times \frac{3}{3}})$$

to make a common denominator, we multiplied 3 by the $\frac{3}{3}$

$$(x, y) = (\frac{-4}{-\frac{1}{3} - \frac{9}{3}}, \frac{\frac{4}{3}}{-\frac{1}{3} - \frac{9}{3}})$$

$$(x, y) = (\frac{-4}{\frac{-1-9}{3}}, \frac{\frac{4}{3}}{\frac{-1-9}{3}})$$

$$(x, y) = (\frac{-4}{\frac{-10}{3}}, \frac{\frac{4}{3}}{\frac{-10}{3}})$$

$$(x, y) = (-4 \times \frac{-3}{10}, \frac{4}{3} \times \frac{-3}{10})$$

$$(x, y) = (\frac{12}{10}, \frac{-4}{10})$$

$$(x, y) = (\frac{6}{5}, -\frac{2}{5}).$$

Having the two points $(x, y) = (\frac{6}{5}, -\frac{2}{5})$, and $O(0, 0)$, we can use the distance formula between two points, $d = \sqrt{(x_1 - x_0)^2 + (y_1 - y_0)^2}$ to find the distance d between the two points with $x_1 = \frac{6}{5}, y_1 = -\frac{2}{5}$, $x_0 = 0, y_0 = 0$. So,

$$d = \sqrt{(x_1 - x_0)^2 + (y_1 - y_0)^2}$$

$$d = \sqrt{\left(\frac{6}{5} - 0\right)^2 + \left(-\frac{2}{5} - 0\right)^2}$$

we plugged $x_1 = \frac{6}{5}, y_1 = -\frac{2}{5}, x_0 = 0,$ and $y_0 = 0$ into the equation

$$d = \sqrt{\left(\frac{6}{5}\right)^2 + \left(-\frac{2}{5}\right)^2}$$

$$d = \sqrt{\frac{36}{25} + \frac{4}{25}}$$

$$d = \sqrt{\frac{36 + 4}{25}}$$

$$d = \sqrt{\frac{40}{25}}$$

$$d = \sqrt{\frac{4 \times 10}{25}}$$

$$d = \sqrt{\frac{2^2 \times 10}{5^2}}$$

$$d = \frac{2}{5}\sqrt{10}.$$

shortest distance from the origin, $O(0, 0)$ to the line $3x - y - 4 = 0$ is $d = \frac{2}{5}\sqrt{10}$.

Example

Find the distance between point P (1, 4) to the line $y = \frac{3}{4}x - 3$.

Solution

Approach 1 - The line $y = \frac{3}{4}x - 3$ can be written as $\frac{3}{4}x - y - 3 = 0$.

$$\frac{3}{4}x - y - 3 = 0$$

$$4 \times (\frac{3}{4}x - y - 3) = 4 \times 0$$ to make the coefficient of variable an integer value, we multiplied both sides by 4

$$3x - 4y - 12 = 0.$$

Now, we know that the distance from a point (x_0, y_0) to a line $ax + by + c = 0$, will be:

$$d = \frac{|ax_0 + by_0 + c|}{\sqrt{a^2 + b^2}}.$$

In this case, we are given the point P (1, 4) and the line $3x - 4y - 12 = 0$. Hence, we have, $a = 3$, $b = -4$, $c = -12$, . So,

$$d = \frac{|ax_0 + by_0 + c|}{\sqrt{a^2 + b^2}}$$

$$d = \frac{|(3)(1) + (-4)(4) + (-12)|}{\sqrt{3^2 + (-4)^2}}$$ we plugged $a = 3$, $b = -4$, $c = -12$ and $(x_0, y_0) = (1, 4)$ into the equation

$$d = \frac{|3 - 16 - 12|}{\sqrt{9 + 16}}$$

$$d = \frac{|3 - 28|}{\sqrt{25}}$$

$$d = \frac{|-25|}{\sqrt{25}}$$

$$d = \frac{25}{5} = 5.$$

Therefore, the shortest distance from point P (1, 4) to the line $3x - 4y - 12 = 0$ is $d = 5$.

Approach 2 - Let us sketch the graph of the line $y = \frac{3}{4}x - 3$. To do so, we need to plot two arbitrary points from the line. By letting $x = 0$, we find $y = -3$ and by letting $y = 0$, we will have:

$$y = \frac{3}{4}x - 3$$

$0 = \frac{3}{4}x - 3$

$0 + 3 = \frac{3}{4}x - 3 + 3$ to isolate $\frac{3}{4}x$, we added 3 to both sides

$3 = \frac{3}{4}x$

$\frac{4}{3} \times 3 = \frac{4}{3} \times \frac{3}{4}x$ to find x, we multiplied both sides by the reciprocal of x

$4 = x.$

Therefore, the two points on the line are (0, – 3) and (4, 0).

We now plot the point $P(1, 4)$ and we sketch the graph of the line by connecting the two these two points (0, – 3) and (4, 0). So,

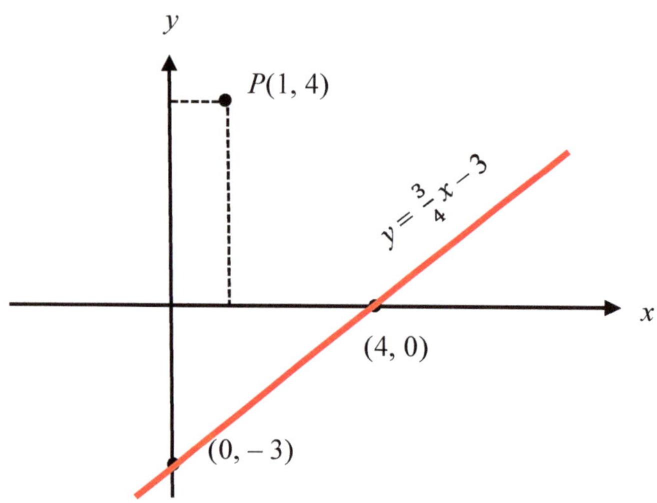

Next, we draw a line, L, that connects the point $P(1, 4)$ and intersects the line at a perpendicular angle at point H, as shown in the figure below.

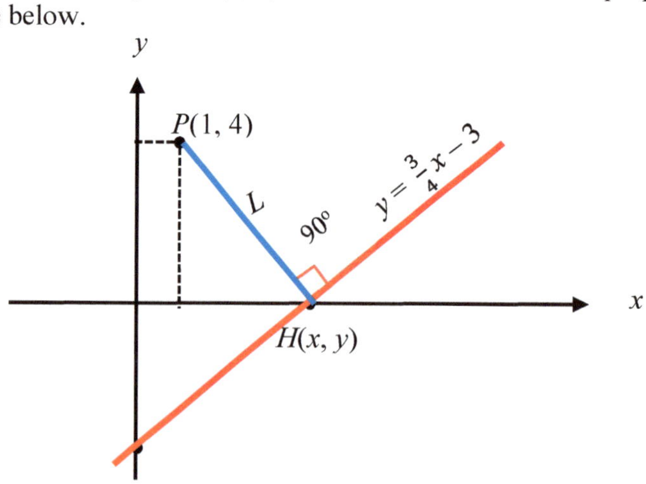

We now need to find the coordinates of H by first finding its equation. Since this line is perpendicular to the line $y = \frac{3}{4}x - 3$, it has a slop of $m = -\frac{1}{\frac{3}{4}} = -\frac{4}{3}$. Having the point $P(1, 4)$ and the slope $m = -\frac{4}{3}$, we simply use the slop-point formula to get its equation. So,

$y - y_1 = m(x - x_1)$

$y - 4 = (-\frac{4}{3})(x - 1)$ we plugged $y_1 = 4$, $x_1 = 1$, and the slope $m = -\frac{4}{3}$, into the equation

$y - 4 = -\frac{4}{3}x + (-\frac{4}{3})(-1)$

$y - 4 = -\frac{4}{3}x + \frac{4}{3}$

$y - 4 + 4 = -\frac{4}{3}x + \frac{4}{3} + 4$ to isolate y, we added 4 to both sides

$y = -\frac{4}{3}x + \frac{4}{3} + 4$

$y = -\frac{4}{3}x + \frac{4}{3} + 4 \times \frac{3}{3}$ to make a common denominator, we multiplied the last term in the right-side by $\frac{3}{3}$

$y = -\frac{4}{3}x + \frac{4}{3} + \frac{4 \times 3}{3}$

$y = -\frac{4}{3}x + \frac{4}{3} + \frac{12}{3}$

$y = -\frac{4}{3}x + \frac{4 + 12}{3}$

$y = -\frac{4}{3}x + \frac{16}{3}.$

Having the equation of the line L as $y = -\frac{4}{3}x + \frac{16}{3}$ and the equation of the line $y = \frac{3}{4}x - 3$, we use the formula for finding the xy-coordinates for the point of intersection of the two lines (x, y). Rearranging the line L: $y = -\frac{4}{3}x + \frac{16}{3}$ gives us $-\frac{4}{3}x - y + \frac{16}{3} = 0$ and rearranging the line $y = \frac{3}{4}x - 3$ we get $\frac{3}{4}x - y - 3 = 0$. Now, given the line $-\frac{4}{3}x - y + \frac{16}{3} = 0$ and the line $\frac{3}{4}x - y - 3 = 0$, we have $a_1 = -\frac{4}{3}, b_1 = -1, c_1 = \frac{16}{3}, a_2 = \frac{3}{4}, b_2 = -1$, and $c_2 = -3$. So,

$(x, y) = (\frac{b_1c_2 - b_2c_1}{a_1b_2 - a_2b_1}, \frac{a_2c_1 - a_1c_2}{a_1b_2 - a_2b_1})$

$(x, y) = (\frac{(-1)(-3) - (-1)(\frac{16}{3})}{(-\frac{4}{3})(-1) - (\frac{3}{4})(-1)}, \frac{(\frac{3}{4})(\frac{16}{3}) - (-\frac{4}{3})(-3)}{(-\frac{4}{3})(-1) - (\frac{3}{4})(-1)})$ we plugged $a_1 = -\frac{4}{3}, b_1 = -1, c_1 = \frac{16}{3}, a_2 = \frac{3}{4}$, $b_2 = -1$, and $c_2 = -3$. Into the equation

$(x, y) = (\frac{3 + \frac{16}{3}}{\frac{4}{3} + \frac{3}{4}}, \frac{\frac{16}{4} - \frac{12}{3}}{\frac{4}{3} + \frac{3}{4}})$

$(x, y) = (\frac{3 + \frac{16}{3}}{\frac{4}{3} + \frac{3}{4}}, \frac{4 - 4}{(\frac{4}{3} + \frac{3}{4})})$

$(x, y) = \left(\frac{3 + \frac{16}{3}}{\frac{4}{3} + \frac{3}{4}}, \frac{0}{(\frac{4}{3} + \frac{3}{4})}\right)$

$(x, y) = \left(\frac{3 + \frac{16}{3}}{\frac{25}{12}}, 0\right)$

$(x, y) = \left(\frac{3 \times \frac{3}{3} + \frac{16}{3}}{\frac{25}{12}},\ 0 \right)$ to make a common denominator, we multiplied 3 by the $\frac{3}{3}$

$(x, y) = \left(\frac{\frac{9}{3} + \frac{16}{3}}{\frac{25}{12}},\ 0 \right)$ again, to make a common denominator, we multiplied the two terms in denominator by the $\frac{4}{4}$, and $\frac{3}{3}$, respectoively

$(x, y) = \left(\frac{\frac{9+16}{3}}{\frac{25}{12}},\ 0 \right)$

$(x, y) = \left(\frac{\frac{25}{3}}{\frac{25}{12}},\ 0 \right)$

$(x, y) = (\frac{25}{3} \times \frac{12}{25}, 0)$

$(x, y) = (\frac{12}{3}, 0)$

$(x, y) = (4, 0)$.

Having the two points $H = (4, 0)$, and $P(1, 4)$, we can use the distance formula between two points, $d = \sqrt{(x_1 - x_0)^2 + (y_1 - y_0)^2}$ to find the distance d between the two points with $x_1 = 4, y_1 = 0, x_0 = 1$, and $y_0 = 4$. So,

$d = \sqrt{(x_1 - x_0)^2 + (y_1 - y_0)^2}$

$d = \sqrt{(4 - 1)^2 + (0 - 4)^2}$ we plugged $x_1 = 4, y_1 = 0, x_0 = 1$, and $y_0 = 4$ into the equation

$d = \sqrt{(3)^2 + (-4)^2}$

$d = \sqrt{9 + 16}$

$d = \sqrt{25}$

$d = \pm 5$.

The distance from a point to a line cannot have a negative value. So $d \neq -5$.

Therefore, the shortest distance from point P (1, 4) to the line $3x - 4y - 12 = 0$ is $d = 5$.

Example
Find the area of triangle ABC, as shown below, by finding its base and a height to that base.

Hint: The area A of a triangle is given by the formula $A = \frac{1}{2}hb$ where b is the base, and h is the height to the base of the triangle.

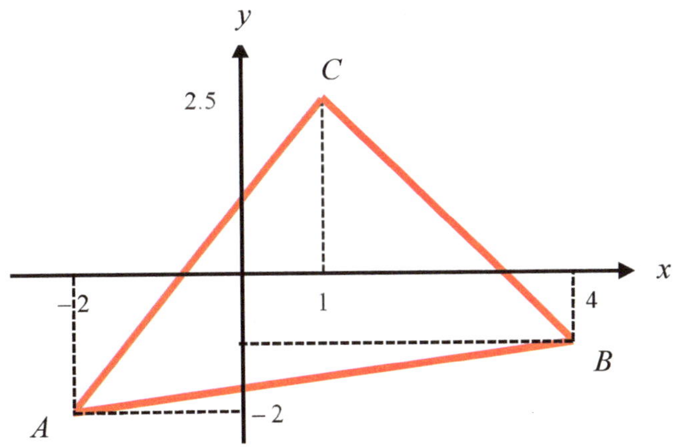

Solution
To find the area of the triangle, *ABC*, we need to find the length of one of the bases and the is the height to that base. Let us select the base *AB* with the two ends *A*(– 2, – 2) and *B*(4, – 1). Having the two points *A*(– 2, – 2) and *B*(4, – 1), we can use the distance formula $d = \sqrt{(x_1 - x_0)^2 + (y_1 - y_0)^2}$ to find the distance *d* between the two points with $x_1 = -2, y_1 = -2, x_0 = 4$, and $y_0 = -1$. So,

$$AB = \sqrt{(x_1 - x_0)^2 + (y_1 - y_0)^2}$$

$$AB = \sqrt{(-2-4)^2 + [-2-(-1)]^2}$$ we plugged $x_1 = -2, y_1 = -2, x_0 = 4$, and $y_0 = -1$ into the equation

$$AB = \sqrt{(-6)^2 + (-2+1)^2}$$

$$AB = \sqrt{(-6)^2 + (-1)^2}$$

$$AB = \sqrt{36+1}$$

$$AB = \sqrt{37}.$$

Now, we need to find the length of height to this base. We first sketch this height. Here is the graph of height *CH* = *h*.

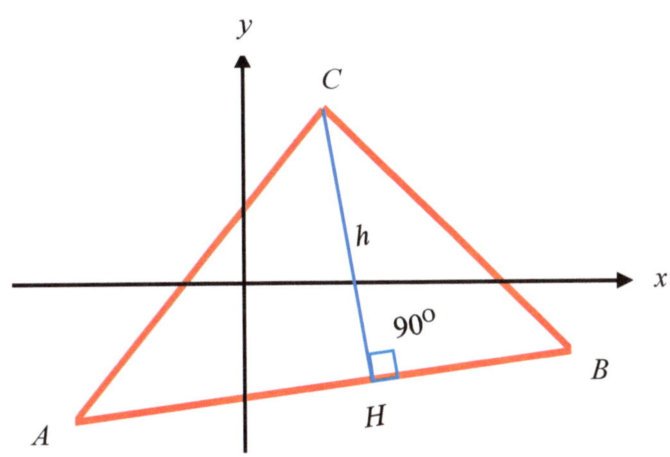

Next, we need to find the length of *CH*. We can use two approaches for finding the length of *CH*. The easier approach is the use of the distance from a point to a line formula, Let us use this approach and leave the other approach as an exercise for readers. However, to apply this formula we need to find the equation of the line AB. Given the two end points of AB, its equation can be found by the line equation from 2 points, $y - y_A = \frac{y_B - y_A}{x_B - x_A}(x - x_A)$.

Given $A(-2, -2)$ and $B(4, -1)$, equation of the line passing through these two points with $x_A = -2$, $y_A = = -2$, $x_B = 4$, and $y_B = -1$, is:

$$y - y_A = \frac{y_B - y_A}{x_B - x_A}(x - x_A)$$

$$y - (-2) = \frac{-1-(-2)}{4-(-2)}[x - (-2)]$$ we plugged $x_A = -2$, $y_A = = -2$, $x_B = 4$, and $y_B = -1$ into the equation

$$y + 2 = \frac{-1+2}{4+2}(x + 2)$$

$$y + 2 = \frac{1}{6}(x + 2)$$

$$y + 2 = \frac{1}{6}x + \frac{2}{6}$$

$$y + 2 = \frac{1}{6}x + \frac{1}{3}$$

$$y + 2 - 2 = \frac{1}{6}x + \frac{1}{3} - 2$$ to isolate y, we added -2 to both sides

$$y = \frac{1}{6}x + \frac{1}{3} - 2$$

$$y = \frac{1}{6}x + \frac{1}{3} - 2 \times \frac{3}{3}$$ to make a common denominator, we multiplied the last term in the right-side by $\frac{3}{3}$

$$y = \frac{1}{6}x + \frac{1}{3} - \frac{2 \times 3}{3}$$

$$y = \frac{1}{6}x + \frac{1}{3} - \frac{6}{3}$$

$$y = \frac{1}{6}x + \frac{1-6}{3}$$

$$y = \frac{1}{6}x + \frac{-5}{3}$$

$$6 \times y = 6 \times (\frac{1}{6}x + \frac{-5}{3})$$ to make integer coefficient, we multiplied both sides by 6

$$6y = x + \frac{(-5)(6)}{3})$$

$$6y = x + \frac{(-5)(2 \times 3)}{3}$$

$$x - 6y - 10 = 0$$ we rearranged terms.

Now, having point the point $C(1, 2.5)$, and the line $x - 6y - 10 = 0$, we use the point-distance formula, $d = \frac{|ax_0 + by_0 + c|}{\sqrt{a^2 + b^2}}$. In this case we are given the point $C(1, 2.5)$ and the line $x - 6y - 10 = 0$. Hence, we have, $d = h$, $a = 1$, $b = -6$, $c = -10$, . So,

$$h = \frac{|ax_0 + by_0 + c|}{\sqrt{a^2 + b^2}}$$

$$h = \frac{|(1)(1) + (-6)(2.5) + (-1\ \ )|}{\sqrt{1^2 + (-6)^2}}$$ we plugged $a = 1$, $b = -6$, $c = -10$ and $(x_0, y_0) = (1, 2.5)$ into the equation

$$h = \frac{|1 - 15 - 10|}{\sqrt{1 + 36}}$$

$$h = \frac{|1 - 25|}{\sqrt{37}}$$

$$h = \frac{|-24|}{\sqrt{37}}$$

$$h = \frac{24}{\sqrt{37}}.$$

Now that we obtained h, we can use the formula for a triangle area. We know that the area of a triangle, S, is half of the product of its height and the base length, that is $S = \frac{1}{2} \times h \times AB$. In this case we have the height $h = \frac{24}{\sqrt{37}}$ and the base $AB = \sqrt{37}$. So,

$$S_{ABC} = \frac{1}{2} \times h \times AB$$

$$S_{ABC} = \frac{1}{2} \times \frac{24}{\sqrt{37}} \times \sqrt{37}$$ we plugged $h = \frac{24}{\sqrt{37}}$ and $AB = \sqrt{37}$ into the equation

$$S_{ABC} = \frac{24 \times \sqrt{37}}{2 \times \sqrt{37}}$$

$$S_{ABC} = \frac{24}{2} =$$

Therefore, the area of the triangle ABC is 12.

Example

Find the area of parallelogram $ABCD$, as shown below, by finding its base and a height to that base. (Round area to nearest square unit).

Hint: The area A of a parallelogram is given by the formula $A = hb$ where b is the base, and h is the height to the base of the parallelogram.

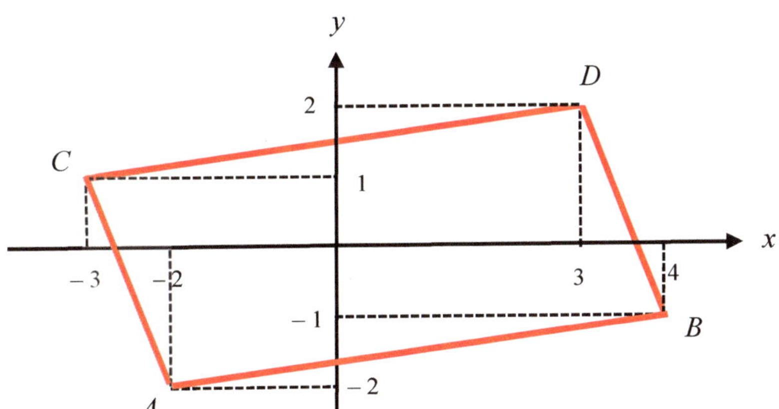

Solution

To find the area of parallelogram $ABCD$, we need to find the length of one of the bases and is the height to that base. Let us select the base AB with the two ends $A(-2, -2)$ and $B(4, -1)$. Having the two points $A(-2, -2)$ and $B(4, -1)$, we can use the distance formula $d = \sqrt{(x_1 - x_0)^2 + (y_1 - y_0)^2}$ to find the distance d between the two points with $x_1 = -2$, $y_1 = -2$, $x_0 = 4$ and $y_0 = -1$. So,

$$AB = \sqrt{(x_1 - x_0)^2 + (y_1 - y_0)^2}$$

$AB = \sqrt{(-2-4)^2 + [-2-(-1)]^2}$ we plugged $x_1 = -2$, $y_1 = -2$, $x_0 = 4$ and $y_0 = -1$ into the equation

$AB = \sqrt{(-6)^2 + (-2+1)^2}$

$AB = \sqrt{(-6)^2 + (-1)^2}$

$AB = \sqrt{36+1}$

$AB = \sqrt{37}.$

Therefore, the base of parallelogram is $AB = \sqrt{37}$.

Now, we need to find the length of height to this base. We first sketch this height. Here is the graph of height $DH = h$.

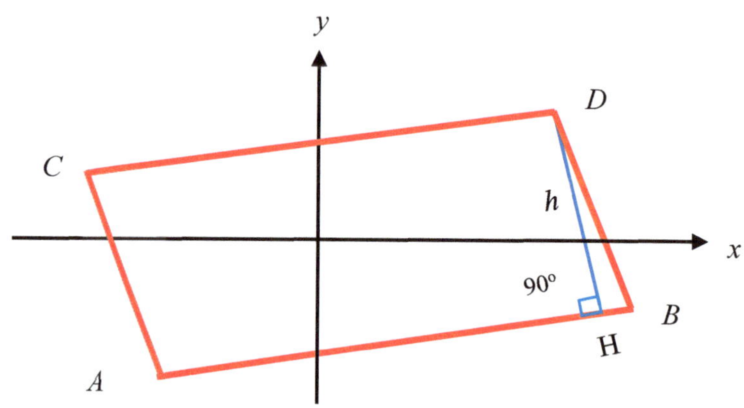

Next, we need to find the length of DH. To get $DH = h$, we need to use of the distance from a point to a line formula, To apply this formula, we need to find the equation of the line AB.

Given the two end points of AB, its equation can be found by the line equation from 2 points, $y - y_A = \frac{y_B - y_A}{x_B - x_A}(x - x_A)$. Given $A(-2, -2)$ and $B(4, -1)$, equation of the line passing through these two points with $x_A = -2$, $y_A = = -2$, $x_B = 4$, and $y_B = -1$, is:

$y - y_A = \frac{y_B - y_A}{x_B - x_A}(x - x_A)$

$y - (-2) = \frac{-1-(-2)}{4-(-2)}[x - (-2)]$ we plugged $x_A = -2$, $y_A = = -2$, $x_B = 4$, and $y_B = -1$ into the equation

$y + 2 = \frac{-1+2}{4+2}(x+2)$

$y + 2 = \frac{1}{6}(x+2)$

$y + 2 = \frac{1}{6}x + \frac{2}{6}$

$y + 2 = \frac{1}{6}x + \frac{1}{3}$

$y + 2 - 2 = \frac{1}{6}x + \frac{1}{3} - 2$ to isolate y, we added -2 to both sides

$y = \frac{1}{6}x + \frac{1}{3} - 2$

$y = \frac{1}{6}x + \frac{1}{3} - 2 \times \frac{3}{3}$ to make a common denominator, we multiplied the last term in the right-side by $\frac{3}{3}$

$y = \frac{1}{6}x + \frac{1}{3} - \frac{2 \times 3}{3}$

$y = \frac{1}{6}x + \frac{1}{3} - \frac{6}{3}$

$y = \frac{1}{6}x + \frac{1 - 6}{3}$

$y = \frac{1}{6}x + \frac{-5}{3}$

$6 \times y = 6 \times (\frac{1}{6}x + \frac{-5}{3})$ to make the integer coefficient, we multiplied both sides by 6.

$6y = x + \frac{(-5)(6)}{3})$

$6y = x + \frac{(-5)(2 \times 3)}{3}$

$x - 6y - 10 = 0$ we rearranged terms.

Now, having point the point D (3, 2), and the line $x - 6y - 10 = 0$, we use the point-distance formula, $d = \frac{|ax_0 + by_0 + c|}{\sqrt{a^2 + b^2}}$. In this case we are given the point D (3, 2) and the line $x - 6y - 10 = 0$. Hence, we have, $d = h$, $a = 1$, $b = -6$, $c = -10$, . So,

$h = \frac{|ax_0 + by_0 + c|}{\sqrt{a^2 + b^2}}$

$h = \frac{|(1)(3) + (-6)(2) + (-10)|}{\sqrt{1^2 + (-6)^2}}$ we plugged $a = 1$, $b = -6$, $c = -10$ and $(x_0, y_0) = (3, 2)$ into the equation

$h = \frac{|3 - 12 - 10|}{\sqrt{1 + 36}}$

$h = \frac{|3 - 22|}{\sqrt{37}}$

$h = \frac{|-19|}{\sqrt{37}}$

$h = \frac{19}{\sqrt{37}}$.

We know that the rea of a parallelogram S_{ABCD} is obtained by multiplying the base, AB, by its height h, that is $S = h \times AB$. In this case, we found the base $AB = \sqrt{37}$ and the height $h = \frac{19}{\sqrt{37}}$. So,

$S_{ABCD} = h \times AB$

$S_{ABCD} = \frac{19}{\sqrt{37}} \times \sqrt{37}$ we plugged $h = \frac{19}{\sqrt{37}}$, and $AB = \sqrt{37}$ into the equation

$S_{ABCD} = \frac{19 \times \sqrt{37}}{\sqrt{37}} = 19.$

Therefore, the area of the parallelogram $ABCD$ is $S_{ABCD} = 19$.

2.3. Shortest Distance Between Two Parallel Lines

The shortest distance between two parallel lines in the plane is the minimum distance between any two points lying on the lines. It equals the perpendicular distance from any point on one line to the other line. Because the lines are parallel, the perpendicular distance between them is a constant, so it does not matter which point is chosen to measure the distance.

Given the equations of two non-vertical parallel lines $y = mx + b_1$, and $y = mx + b_2$ the shortest distance d between them is: $d = \frac{|b_2 - b_1|}{\sqrt{m^2 + 1}}$.

When the equations of two lines are given by $d = \frac{|c_2 - c_1|}{\sqrt{c^2 + b^2}}$.

y

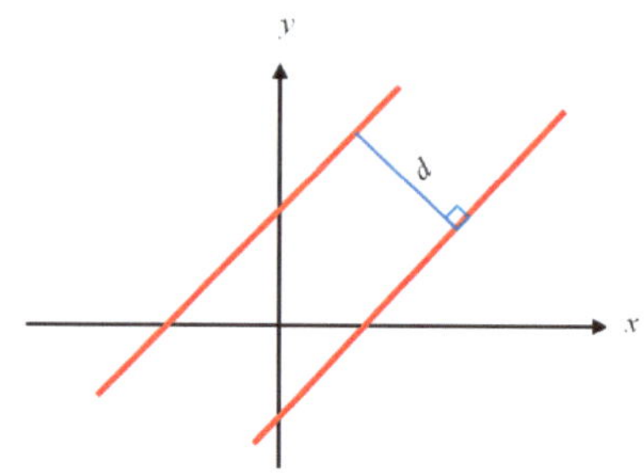

Example
Find the shortest distance between the two parallel lines $y = 5x + 11$ and $y = 5x + 7$.

Solution
We know the shortest distance, d, between two parallel lines with the equations $y = mx + b_1$, and $y = mx + b_2$ is given by $d = \frac{|b_2 - b_1|}{\sqrt{m^2 + 1}}$. In this case, $m = 5$, $b_1 = 11$, and $b_2 = 7$. So,

$$d = \frac{|b_2 - b_1|}{\sqrt{m^2 + 1}}$$

$$d = \frac{|7 - 11|}{\sqrt{5^2 + 1}}$$ we plugged $m = 5$, $b_1 = 11$, and $b_2 = 7$

$$d = \frac{|-4|}{\sqrt{25 + 1}}$$

$$d = \frac{|-4|}{\sqrt{26}}$$

$$d = \frac{4}{5.10} = 0.78.$$

Therefore, the shortest distance between the two parallel lines $y = 5x + 11$ and $y = 5x + 7$ is $d = 0.78$.

Example
Find the shortest distance between the two parallel lines $y = 3x + \frac{5}{7}$ and $y = 3x + \frac{52}{14}$.

Solution
We know the shortest distance, d, between two parallel lines with the equations $y = mx + b_1$, and $y = mx + b_2$ is given by $d = \frac{|b_2 - b_1|}{\sqrt{m^2 + 1}}$. In this case, $m = 3$, $b_1 = \frac{5}{7}$, and $b_2 = \frac{52}{14}$. So,

$$d = \frac{|b_2 - b_1|}{\sqrt{m^2 + 1}}$$

$d = \frac{\left|\frac{52}{14} - \frac{5}{7}\right|}{\sqrt{3^2 + 1}}$ we plugged $m = 3$, $b_1 = \frac{5}{7}$, and $b_2 = \frac{52}{14}$

$d = \frac{\left|\frac{52}{14} - \frac{5}{7} \times \frac{2}{2}\right|}{\sqrt{9 + 1}}$ to make a common denominator we multiplied the second term in the nominator by $\frac{2}{2}$.

$d = \frac{\left|\frac{52}{14} - \frac{10}{14}\right|}{\sqrt{10}}$

$d = \frac{\left|\frac{52 - 10}{14}\right|}{3.16}$

$d = \frac{\left|\frac{42}{14}\right|}{3.16}$

$d = \frac{\left|\frac{14 \times 3}{14}\right|}{3.16}$

$d = \frac{|3|}{3.16}$

$d = \frac{3}{3.16} = 0.95.$

Therefore, the shortest distance between the two parallel lines $y = 6x + \frac{5}{7}$ and $y = 6x + \frac{52}{14}$ is $d = 0.95$.

Example
Find the shortest distance between the two parallel lines $6x - 8y + 4 = 0$ and $6x - 8y - 16 = 0$.

Solution
We have learned that the shortest distance, d, between two parallel lines with the equations is given by $d = \frac{|c_2 - c_1|}{\sqrt{a^2 + b^2}}$. In this case, $a = 6$, $b = -8$, $c_1 = -4$, and $c_2 = -16$.

$d = \frac{|c_2 - c_1|}{\sqrt{a^2 + b^2}}$

$d = \frac{|-16 - 4|}{\sqrt{6^2 + (-8)^2}}$ we plugged $a = 6$, $b = -8$, $c_1 = -4$, and $c_2 = -16$ into the equation.

$d = \frac{|-20|}{\sqrt{36 + 64}}$

$d = \frac{20}{\sqrt{100}}$

$d = \frac{20}{\sqrt{10^2}}$

$d = \frac{10 \times 2}{10}$

$d = 2.$

Therefore, the shortest distance between the two given parallel lines $6x - 8y + 4 = 0$ and $6x - 8y - 16 = 0$ is $d = 2$.

2.3.1. Shortest Distance Between Two Vertical Parallel Lines

Given the equations of two vertical parallel lines $y = a_1$, and $y = a_2$, the shortest distance d between them is given as:

$$d = |\boldsymbol{a_2} - \boldsymbol{a_1}|.$$

2.3.2. Shortest Distance Between Two Horizontal Parallel Lines

Given the equations of two horizontal parallel lines $x = b_1$, and $x = b_2$ the shortest distance d between them is given as:

$$d = |\boldsymbol{b_2} - \boldsymbol{b_1}|.$$

Example
Find the shortest distance between the two vertical parallel lines $y = 7$, and $y = 3$.

Solution
We have learned given the equations of two vertical parallel lines $y = a_1$, and $y = a_2$ the shortest distance d between them is $d = |\boldsymbol{a_2} - \boldsymbol{a_1}|$. In this case, $a_1 = 7$ and $a_2 = 3$. So,

$d = |\boldsymbol{a_2} - \boldsymbol{a_1}|$

$d = |\mathbf{3 - 7}|$ we plugged $a_1 = 7$ and $a_2 = 3$ into the equation

$d = |\mathbf{-4}|$

$d = 4.$

Therefore, the shortest distance between the two vertical parallel lines $y = 7$, and $y = 3$ is $d = 4$.

Example
Find the shortest distance between the two horizontal parallel lines $x = 9$, and $x = 12$.

Solution
We have learned given the equations of two horizontal parallel lines $x = b_1$, and $x = b_2$ the shortest distance d between them is $d = |\boldsymbol{b_2} - \boldsymbol{b_1}|$. In this case, $b_1 = 9$ and $b_2 = 12$. So,

$d = |\boldsymbol{b_2} - \boldsymbol{b_1}|$

$d = |\mathbf{12 - 9}|$ we plugged $b_1 = 9$ and $b_2 = 12$ into the equation

$d = |\mathbf{-3}|$

$d = 3.$

Therefore, the shortest distance between the two horizontal parallel lines $x = 9$, and $x = 12$ is $d = 3$.

2.4. Exercises

1. Given the equations of two lines below, find the xy-coordinates for the point of their intersection.
   L_1: $4x - 3y - 15 = 0$.

   L_2: $-2x + 4y + 5 = 0$.

2. Given the equations of two lines below, find the xy-coordinates for the point of their intersection.
   L_1: $y = \frac{3}{7}x - \frac{12}{7}$.

   L_2: $x = 4y - 1$.

3. The slope of line L_1 is 3. Write the general equation of lines parallel to the given line.

4. Given two lines with the equations $5x + y = 6$ and $-5x + 3y = 10$, find whether they are parallel, perpendicular, or neither?

5. Write the equation of a line having the point (3, 1) and is perpendicular to $2y - 4x + 9 = 0$.

6. Write the equation for the line passing through point $P(2, 3)$ and is perpendicular to the line that its graph is given below.

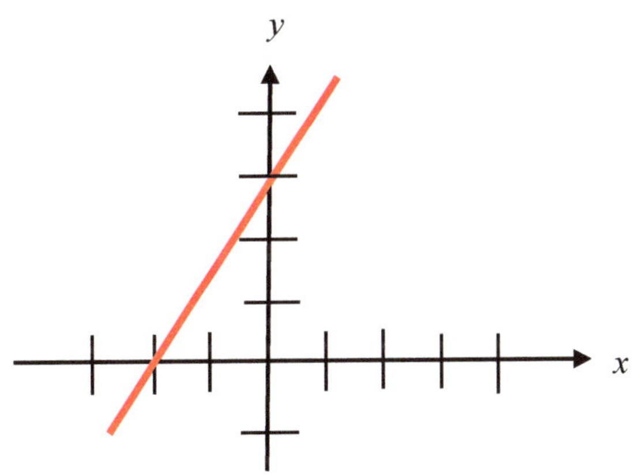

7. Write the equation for the line parallel to $y = 2x + 5$, and passes though the point (7, 1).

8. Find the equation of the line passing through point $P(-1, -1)$ and is parallel to the line from its graph below, and then sketch its graph.

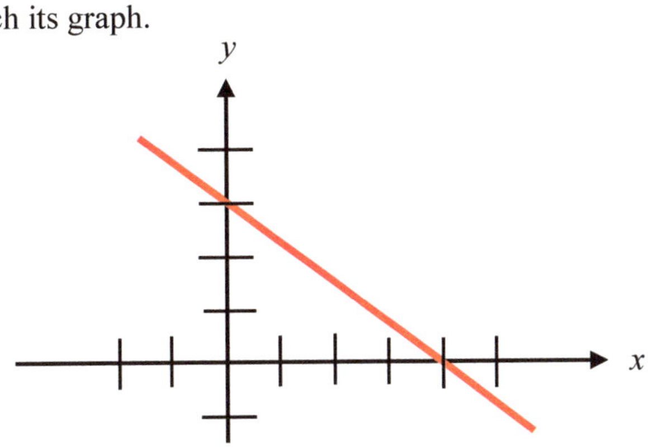

9. Find the two points of the form $P(x, -6)$ that have the same distance of 8 units from the point:
   a) $Q(2, 1)$.

b) $R(-3, 0)$.

10. Given a line with the equation $y = 6x - 5$, write the equation of the line parallel to this line and passing through $(-2, 8)$.

11. Given a line with the equation $3y = 2x + 4$, write the equation of the line perpendicular to this line and passing through $(1, -3)$.

12. Find the shortest distance between the line $4y - 3x = 1$ and the origin.

13. Find the shortest distance from point $(-3, -\frac{2}{3})$ to the line $y = 7$.

14. Find the shortest distance between point $(-4, \frac{1}{2})$ to the line $x = -\frac{3}{5}$.

15. Find the shortest distance between the line $y = 6x - 7$ and $(5, 2)$.

16. Find the shortest distance between each of the two parallel lines.

a) $y = 8x + 9$ and $y = 8x + 13$.

b) $y = -\frac{3}{7}x - \frac{5}{11}$ and $y = -\frac{3}{7}x + \frac{17}{22}$.

c) $3x - 2y + 15 = 0$ and $3x - 2y + 8 = 0$.

d) $\frac{2}{3}x - 5y + 6 = 0$ and $\frac{2}{3}x - 5y + \frac{5}{9} = 0$.

17. Find the shortest distance between each of the two vertical parallel lines.

a) $y = 13$ and $y = 9$

b) $y = \frac{8}{17}$ and $y = \frac{23}{34}$.

c) $y = -\frac{3}{5}$ and $y = 12$.

d) $y = \frac{4}{5}$ and $y = -7$.

e) $y = -3$ and $y = -5$.

18. Find the shortest distance between each of the two horizontal parallel lines.

a) $x = -1$ and $x = 8$.

b) $x = \frac{2}{5}$ and $x = -4$.

c) $x = -\frac{4}{9}$ and $x = -\frac{2}{3}$.

d) $x = 6$ and $x = -\frac{7}{9}$.

e) $x = -\frac{11}{6}$ and $x = 7$.

f) $x = -13$ and $x = -5$.

3. Common Graphs

The purpose of this section is to show you shapes of various **types of graphs** in order that you can become familiar with the types. For those which have more application and have a functional form, we will present a more detailed explanation in the later chapters (Chapter 5, Functions). Here is a listing of the topics that we will be looking at in this section.

- Circles.
- Ellipses.
- Hyperbolas.
- Parabolas.

3.1. Graph of Circles

A **circle** is formed when the plane is parallel to the base of the cone. Its intersection with the cone is therefore a set of points equidistant from a common point (the central axis of the cone), which meets the definition of a circle.

A **circle** can be determined by fixing a point, called the center, and a positive number, called the radius, as follows.

> **Definition of Circle**
> A circle with center (a, b) and radius $r > 0$ is the set of all points (x, y) in the plane whose distance to (a, b) is r.

The graph of a circle is illustrated below:

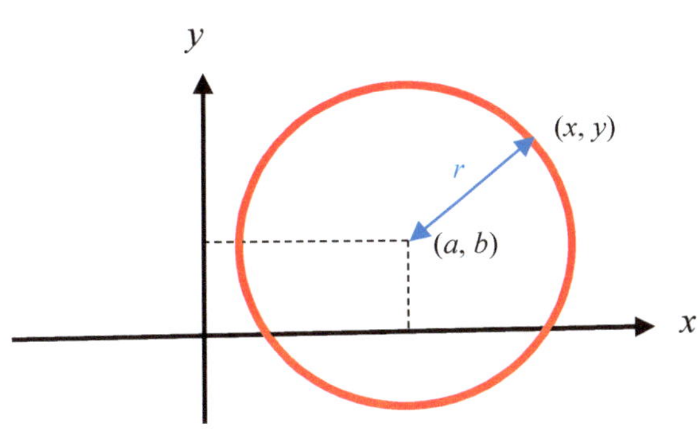

From the graph, we find that a point (x, y) is on the circle if and only if its distance to (a, b) is r. We express this relationship algebraically using the "distance formula", as:

$$r^2 = (x - a)^2 + (y - b)^2$$

$r = \sqrt{(x-a)^2 + (y-b)^2}$, since $r > 0$.

> **Standard Equation of a Circle**
>
> The equation of a circle with center (a, b) and radius $r > 0$ is
>
> $$(x-a)^2 + (y-b)^2 = r^2.$$

Example
Write the standard equation of the circle with center (5, – 6) and radius 4.

Solution
We have learned that the standard equation of the circle is, $(x-a)^2 + (y-b)^2 = r^2$. Here. $(a, b) = (5, -6)$ and $r = 4$, so

$$(x-5)^2 + (y-(-6))^2 = 4^2$$

$$(x-5)^2 + (y+6)^2 = 16.$$

Therefore, the standard equation of the circle is $(x-5)^2 + (y+6)^2 = 16$.

Example
Graph $(x-2)^2 + (y+3)^2 = 4$ and find the center and radius.

Solution
Considering the standard form of a circle, here for $x - a$ we have $x - 2$ or $a = 2$ and for $y - b$ we have $y + 3$ or $b = -3$. This tells us that our center is $(2, -3)$. Furthermore, $r^2 = 4$, so $r = 2$. Thus, we have a circle centered at $(2, -3)$ with a radius of 2. The graph will be given as:

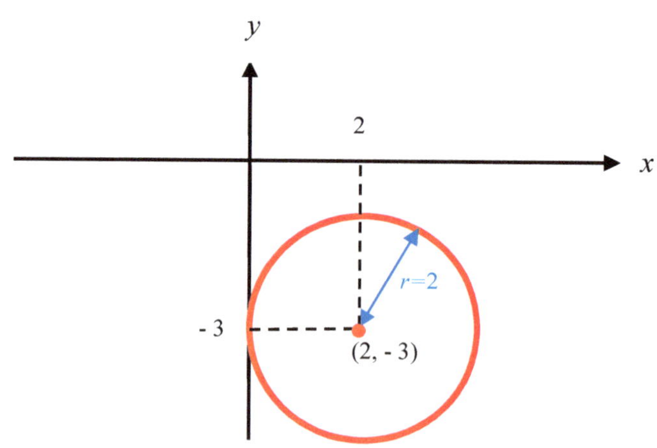

> **Unit Circle**
>
> The unit circle is the circle centered at (0, 0) with a radius of 1.
> The standard equation of the unit circle is:
>
> $$x^2 + y^2 = 1.$$

Example
Find the points on the unit circle with y-coordinate $\frac{1}{5}$.

Solution

We replace y with $\frac{1}{5}$ in the equation $x^2 + y^2 = 1$ to get

$$x^2 + (\frac{1}{5})^2 = 1$$

$$x^2 + \frac{1}{25} = 1$$

$$x^2 = -\frac{1}{25} + 1$$

we isolated the variable in one side and the constants on the other side of the equation

$$x^2 = \frac{-1 + 25}{25}$$

$$x^2 = \frac{24}{25}$$

$$x = \pm\sqrt{\frac{24}{25}}.$$

$$x = \pm\sqrt{\frac{4 \times 6}{5^2}}.$$

$$x = \pm\sqrt{\frac{2^2 \times 6}{5^2}}.$$

$$x = \pm\frac{2}{5}\sqrt{6}.$$

Therefore, the final answer is $(\frac{2}{5}\sqrt{6}, \frac{1}{5})$ and $(-\frac{2}{5}\sqrt{6}, \frac{1}{5})$.

3.2. Graph of Ellipse

An **ellipse** is a plane oval curve surrounding two focal points, F_1 and F_2, such that for the set of all points on the curve, the sum of the two distances to the focal points is a constant. Fundamentally, it generalizes a circle, which is the special type of ellipse in which the two focal points are the same.

Every ellipse has two axes of symmetry. The longer axis is called the major axis, and the shorter axis is called the minor axis. Each endpoint of the major axis is the vertex of the ellipse, and each endpoint of the minor axis is a co-vertex of the ellipse.

The center of an ellipse is the midpoint of both the major and minor axes. The semi-major axis, a, is one half of the major axis, and thus runs from the center, through a focus, and to the perimeter. The semi-minor axis, b, of an ellipse is a line segment that is at right angles with the semi-major axis and has one end at the center of the conic.

For the special case of a circle, the lengths of the semi-axes are both equal to the radius of the circle. The axes are perpendicular at the center. The foci always lie on the major axis, and the sum of the distances from the foci to any point on the ellipse (the constant sum) is greater than the distance between the foci. See figure below.

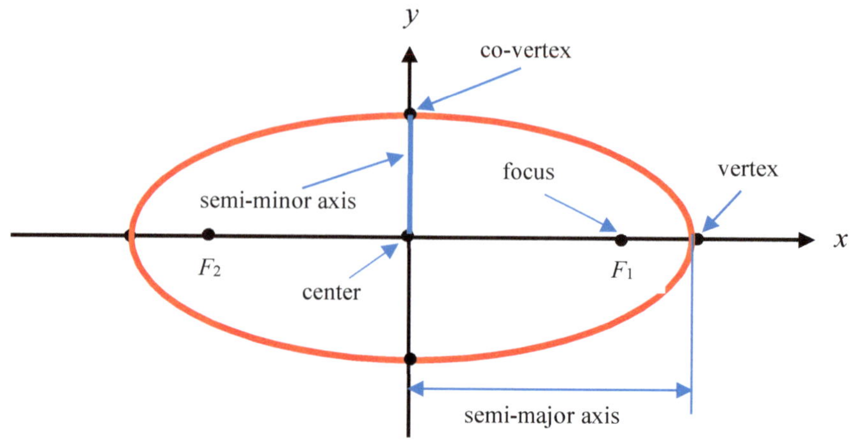

3.2.1. Equation of an Ellipse Centered at the Origin

For the simplicity of discussion, we restrict ellipses to those that are positioned vertically or horizontally in the coordinate plane. That is, the axes will either lie on or be parallel to the x– and y-axes. To work with horizontal and vertical ellipses in the coordinate plane, we consider two cases: those that are centered at the origin and those that are centered at a point other than the origin.

To derive the equation of an ellipse centered at the origin, we begin with the foci $(-c, 0)$ and $(c, 0)$. The ellipse is the set of all points (x, y) such that the sum of the distances from (x, y) to the foci is constant, as shown in the figure below.

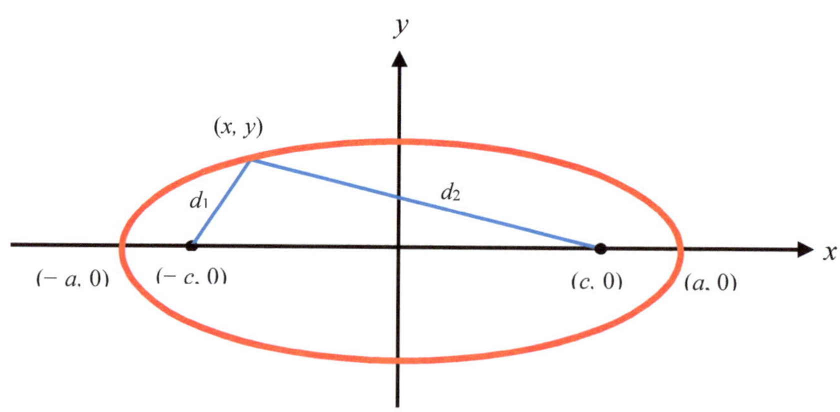

If $(a, 0)$ is a vertex of the ellipse, the distance from $(-c, 0)$ to $(a, 0)$ is $a - (-c) = a + c$. The distance from $(c, 0)$ to $(a, 0)$ is $a - c$. The sum of the distances from the foci to the vertex is:

$$(a + c) + (a - c) = 2a.$$

If (x, y) is a point on the ellipse, then we can define the following variables:

d_1 = the distance from $(-c, 0)$ to (x, y)

d_2 = the distance from $(c, 0)$ to (x, y).

Ellipse

Given two distinct points F_1 and F_2 in the plane and a fixed distance d, an ellipse is the set of all points (x, y) in the plane such that the sum of each of the distances from F_1 and F_2 to (x, y) is d. The points F_1 and F_2 are called the foci of the ellipse.

By the definition of an ellipse, $d_1 + d_2$ is constant for any point (x, y) on the ellipse. We know that the sum of these distances is $2a$ for the vertex $(a, 0)$. It follows that $d_1 + d_2 = 2a$ for any point on the ellipse.

The derivation of the standard form of the equation of an ellipse relies on this relationship and the distance formula. The derivation is beyond the scope of this book, but the equation is:

$$\frac{x^2}{b^2} + \frac{y^2}{a^2} = 1.$$

For an ellipse centered at the origin with its major axis on the x-axis.

Note if $a > b$, the ellipse is stretched further in the horizontal direction, and if $b > a$, the ellipse is stretched further in the vertical direction.

For an ellipse centered at the origin with its major axis on the y-axis the equation is:

$$\frac{x^2}{b^2} + \frac{y^2}{a^2} = 1.$$

3.2.2. How to Write the Equation of Ellipse, Vertices and Foci Centered at the Origin

Given the vertices and foci of an ellipse centered at the origin, we can follow the steps bellow for writing its equation in standard form.

1. Determine whether the major axis lies on the x-axis or y-axis.
   a) If the coordinates of the vertices and foci have the form $(\pm a, 0)$ and $(\pm c, 0)$ respectively, then the major axis is the x-axis. Use the standard form $\frac{x^2}{a^2} + \frac{y^2}{b^2} = 1$.
   b) If the coordinates of the vertices and foci have the form $(0, \pm a)$ and $(\pm c, 0)$, respectively, then the major axis is the y-axis. Use the standard form $\frac{x^2}{b^2} + \frac{y^2}{a^2} = 1$.
2. Use the equation $c^2 = a^2 - b^2$, along with the given coordinates of the vertices and foci, to solve for b^2.
3. Substitute the values for a^2 and b^2 into the standard form of the equation determined in Step 1.

Example
Write the equation of the ellipse that its vertices are located at $(\pm 9, 0)$ and its foci are located at $(\pm 6, 0)$.

Solution
As y values are zero, the foci are on the major x-axis. Thus, the equation will have the form:

$$\frac{x^2}{a^2} + \frac{y^2}{b^2} = 1.$$

The vertices are $(\pm 9, 0)$, so $a = 9$ and $a^2 = 81$. Also, the foci are $(\pm 6, 0)$, so $c = 6$ and $c^2 = 36$. We know that the vertices and foci are related by the equation $c^2 = a^2 - b^2$. Solving for b^2, we have:

$c^2 = a^2 - b^2$	
$36 = 81 - b^2$	we substituted $a^2 = 81$ and $c^2 = 36$ into the equation
$36 - 81 = 81 - b^2 - 81$	to isolate b, we added -81 to both sides
$-45 = -b^2$	
$(-1)(-45) = (-1)(-b^2)$	we multiplied both sides by -1
$b^2 = 45$.	

Now we need just substitute $a^2 = 81$ and $b^2 = 45$ into the standard form of the equation. The equation of the ellipse is:

$$\frac{x^2}{81} + \frac{y^2}{45} = 1.$$

Standard Forms of the Equation of an Ellipse with Center (0, 0)

The standard form of the equation of an ellipse with center (0, 0) and major axis on the x-axis is:

$$\frac{x^2}{a^2} + \frac{y^2}{b^2} = 1,$$

where,

- $a > b$.
- The length of the major axis is $2a$.
- The coordinates of the vertices are $(\pm a, 0)$.
- The length of the minor axis is $2b$.
- The coordinates of the co-vertices are $(0, \pm b)$.
- The coordinates of the foci are $(\pm c, 0)$, where $c^2 = a^2 - b^2$. See figure below.

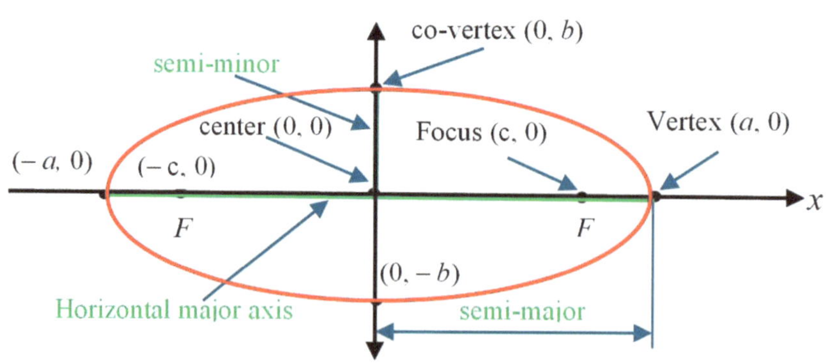

Major axis on the x-axis

Standard Forms of the Equation of an Ellipse with Center (0, 0)

The standard form of the equation of an ellipse with center (0, 0) and major axis on the y-axis is: $\frac{x^2}{b^2}+\frac{y^2}{a^2}=1$, where:

- $a > b$
- The length of the major axis is $2a$.
- The coordinates of the vertices are $(0, \pm a)$.
- The length of the minor axis is $2b$.
- The coordinates of the co-vertices are $(\pm b, 0)$.
- The coordinates of the foci are $(0, \pm c)$, where $c^2 = a^2 - b^2$. See figure below.

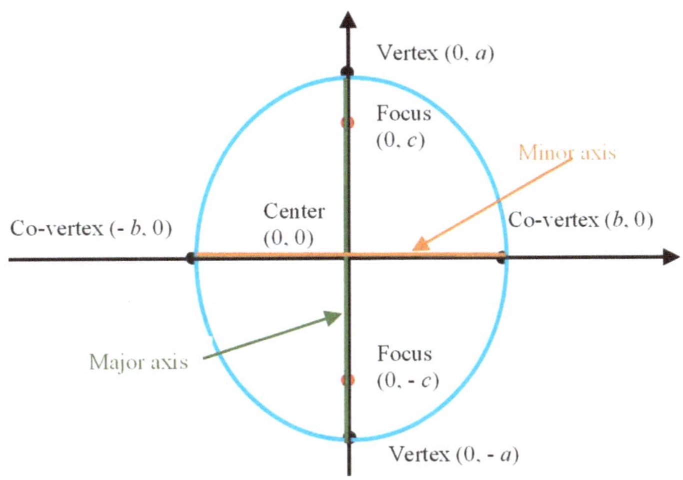

Major axis on the y-axis

Example

Write the standard form equation of the ellipse that its vertices are located at $(0, \pm 10)$ and its foci are located at $(0, \pm 7)$.

Solution

The foci are on the y-axis, since their x values are zero, so the major axis is the y-axis. Thus, the equation will have the form:

$$\frac{x^2}{b^2}+\frac{y^2}{a^2}=1.$$

The vertices are $(0, \pm 10)$, so $a = 10$ and $a^2 = 100$. The foci are $(0, \pm 7)$, so $c = 7$ and $c^2 = 49$.

We know that the vertices and foci are related by the equation $c^2 = a^2 - b^2$. Solving for b^2, we have:

$c^2 = a^2 - b^2$

$49 = 100 - b^2$ we substituted $a^2 = 100$ and $c^2 = 49$ into the equation.

$49 - 100 = 100 - b^2 - 100$ to isolate the variable, we added -100 to both sides

$-51 = -b^2$

$(-1)(-51) = (-1)(-b^2)$ we multiplied both sides by -1

$b^2 = 51.$

Now we need just substitute $a^2 = 100$ and $b^2 = 51$ into the standard form of the equation.

$$\frac{x^2}{51} + \frac{y^2}{100} = 1.$$

3.2.3. Equations of Ellipses Not Centered at the Origin

To get the formula for the ellipse centered at (h, k), we must use the transformations of functions. Since the transformations of functions will be presented in Chapter 5 of this book, here we just present the final form equations of ellipses, not centered at the origin, without a detailed explanation.

For positive unequal numbers a and b, the standard equation of an ellipse with center (h, k) is:

$$\frac{(x-h)^2}{a^2} + \frac{(y-k)^2}{b^2} = 1.$$

3.2.4. How to Write the Equation of Ellipse, Vertices and Foci Not Centered at the Origin

Given the vertices and foci of an ellipse centered at (h, k), we can follow the steps below for writing its equation in standard form.

1. Determine whether the major axis is parallel to the x-axis or y-axis.
    a. If the y-coordinates of the given vertices and foci are the same, then the major axis is parallel to the x-axis. Use the standard form $\frac{(x-h)^2}{a^2} + \frac{(y-k)^2}{b^2} = 1.$
    b. If the x-coordinates of the given vertices and foci are the same, then the major axis is parallel to the y-axis. Use the standard form $\frac{(x-h)^2}{b^2} + \frac{(y-k)^2}{a^2} = 1.$
2. Identify the center of the ellipse (h, k) using the midpoint formula and the given coordinates for the vertices.
3. Find a^2 by solving the length of the major axis, $2a$, it is the distance between the given vertices.
4. Find c^2 using h and k, found in Step 2, along with the given coordinates for the foci.
5. Solve for b^2 using the equation $c^2 = a^2 - b^2$.
6. Substitute the values for h, k, a^2, and b^2 into the standard form of the equation obtained in Step 1.

Example

Write the standard form equation of the ellipse with vertices $(1, -6)$ and $(1, 4)$ and foci $(1, -5)$ and $(1, 3)$.

Solution

The x-coordinates of the vertices and foci are the same, so the major axis is parallel to the y-axis. Thus, the equation of the ellipse will have the form:

$$\frac{(x-h)^2}{b^2} + \frac{(y-k)^2}{a^2} = 1.$$

First, we identify the center, (h, k). The center is halfway between the vertices, $(1, -6)$ and $(1, 4)$. Applying the midpoint formula, we have:

$$(h, k) = (\frac{1+1}{2}, \frac{4+(-6)}{2})$$

$$(h, k) = (\frac{2}{2}, \frac{-2}{2})$$

$$(h, k) = (1, -1).$$

Next, we find a^2. The length of the major axis, $2a$, is bounded by the vertices. We solve for a by finding the distance between the y-coordinates of the vertices.

$$2a = 4 - (-6)$$

$$2a = 10$$

$$a = 5$$

$$a^2 = 25.$$

Now we find c^2. The foci are given by $(h, k \pm c)$. So, $(h, k - c) = (1, -5)$ and $(h, k + c) = (1, 3)$. We substitute $k = -1$ using either of these points to solve for c.

$$k + c = 3$$

$$-1 + c = 3$$

$$c = 4$$

$$c^2 = 16.$$

Next, we solve for b^2 using the equation $c^2 = a^2 - b^2$.

$$16 = 25 - b^2$$

$$16 - 25 = -b^2$$

$$-b^2 = -9$$

$$(-1)(-b^2) = (-1)(-9)$$

$$b^2 = 9.$$

Finally, we substitute the values found for h, k, a^2, and b^2 into the standard form equation for an ellipse:

$$\frac{(x-h)^2}{b^2} + \frac{(y-k)^2}{a^2} = 1.$$

$$\frac{(x-1)^2}{9} + \frac{(y-(-1))^2}{25} = 1.$$

$$\frac{(x-1)^2}{9} + \frac{(y+1)^2}{25} = 1.$$

Example

Find the center, vertices, co-vertices, and foci of $\frac{(x+4)^2}{81} + \frac{(x-9)^2}{36} = 1$.

Solution

The center is $(-4, 9)$, $a = \sqrt{81} = 9$ and $b = \sqrt{36} = 6$. This results in a horizontal ellipse, where the major axis aligns parallel to the x-axis.

The vertices are $(-4 \pm 9, 9)$ or $(-13, 9)$ and $(5, 9)$. The co-vertices are $(-4, 9 \pm 6)$ or $(-4, 3)$ and $(-4, 15)$.

Using $c^2 = a^2 - b^2$ we can find c.

$c^2 = a^2 - b^2$

$c^2 = 81 - 36$

$c^2 = 45$

$c = \pm\sqrt{45}$

$c = 6.7$ or $c = -6.7$.

Therefore, the foci are $(-4 - 6.7, 9)$ and $(-4 + 6.7, 9)$ or $(-10.7, 9)$ and $(2.7, 9)$.

Standard Forms of the Equation of an Ellipse with Center (h, k)

The standard form of the equation of an ellipse with center (h, k) and major axis parallel to the x-axis is: $\frac{(x-h)^2}{a^2} + \frac{(y-k)^2}{b^2} = 1$,

where,

- $a > b$.
- The length of the major axis is $2a$.
- The coordinates of the vertices are $(h \pm a, k)$.
- The length of the minor axis is $2b$.
- The coordinates of the co-vertices are $(h, k \pm b)$.
- The coordinates of the foci are $(h \pm c, k)$, where $c^2 = a^2 - b^2$.

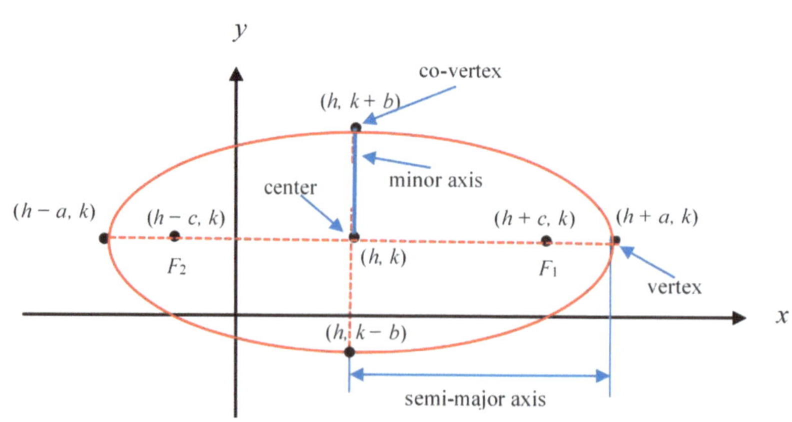

Major axis is parallel to x-axis.

Standard Forms of the Equation of an Ellipse with Center (h, k)

The standard form of the equation of an ellipse with center (h, k) and major axis parallel to the y-axis is: $\frac{(x-h)^2}{b^2} + \frac{(y-k)^2}{a^2} = 1$,

where,

- $a > b$.
- The length of the major axis is $2a$.
- The coordinates of the vertices are $(h, k \pm a)$.
- The length of the minor axis is $2b$.
- The coordinates of the co-vertices are $(h \pm b, k)$.
- The coordinates of the foci are $(h, k \pm c)$, where $c^2 = a^2 - b^2$.

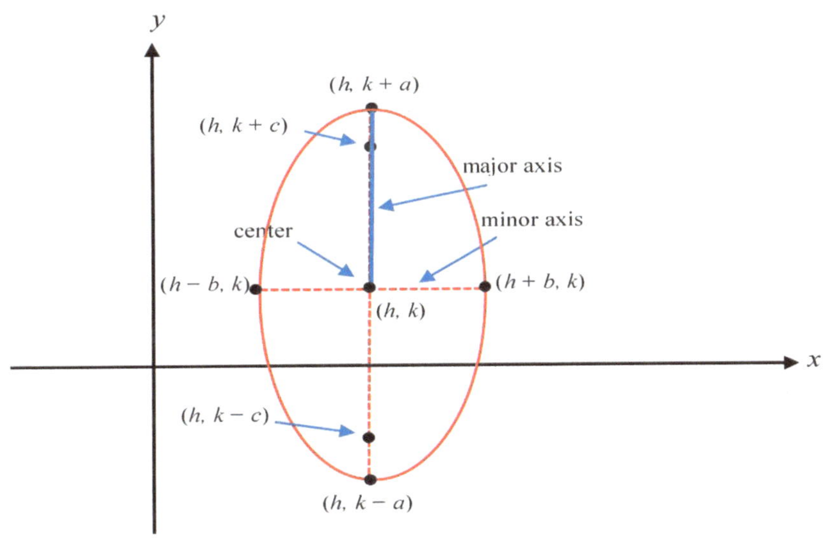

The major axis is parallel to y-axis.

3.2.5. Graphing Ellipses Centered at the Origin

We can graph an ellipse given its equation. To graph ellipses centered at the origin, we use the standard form $\frac{x^2}{a^2} + \frac{y^2}{b^2} = 1$, $a > b$ for horizontal ellipses and $\frac{x^2}{b^2} + \frac{y^2}{a^2} = 1$, $a > b$ for vertical ellipses.

Given the standard form of an equation for an ellipse centered at $(0, 0)$, we can sketch the graph completing the steps below:

1. Use the standard forms of the equations of an ellipse to determine the major axis, vertices, co-vertices, and foci.
   a) If the equation is in the form $\frac{x^2}{a^2} + \frac{y^2}{b^2} = 1$, where $a > b$, then
      - The major axis is the x-axis.

- The coordinates of the vertices are ($\pm a$, 0).
- The coordinates of the co-vertices are (0, $\pm b$).
- The coordinates of the foci are ($\pm c$, 0).

b) If the equation is in the form $\frac{x^2}{b^2} + \frac{y^2}{a^2} = 1$, where $a > b$, then:

- The major axis is the y-axis.
- The coordinates of the vertices are (0, $\pm a$).
- The coordinates of the co-vertices are ($\pm b$, 0).
- The coordinates of the foci are (0, $\pm c$).

2. Solve for c using the equation $c^2 = a^2 - b^2$.

3. Plot the center, vertices, co-vertices, and foci in the coordinate plane, and draw a smooth curve to form the ellipse.

Example

Graph the ellipse, $\frac{x^2}{9} + \frac{y^2}{16} = 1$. Identify and label the center, vertices, co-vertices, and foci.

Solution

First, we determine the position of the major axis. Since 16 > 9, the major axis is on the y-axis.

Therefore, the equation is in the form $\frac{x^2}{b^2} + \frac{y^2}{a^2} = 1$, where $b^2 = 9$ and $a^2 = 16$. It follows that:

- The center of the ellipse is (0, 0).
- The coordinates of the vertices are (0, $\pm a$).

  = (0, $\pm\sqrt{16}$)

  = (0, $\pm$ 4).
- The coordinates of the co-vertices are ($\pm b$, 0).

  = ($\pm\sqrt{9}$, 0)

  = ($\pm$ 3, 0).
- The coordinates of the foci are (0, $\pm c$), where $c^2 = a^2 - b^2$. Solving for c, we have:

  $c = \pm\sqrt{a^2 - b^2}$

  $c = \pm\sqrt{16 - 9}$ we plugged $a^2 = 16$ and $b^2 = 9$ into the equation

  $c = \pm\sqrt{7} = \pm 2.65$.

Therefore, the coordinates of the foci are (0, $\pm\sqrt{7}$) = (0. $\pm$ 2.65).

Next, we plot and label the center, vertices, co-vertices, and foci, and draw a smooth curve to form the ellipse. See

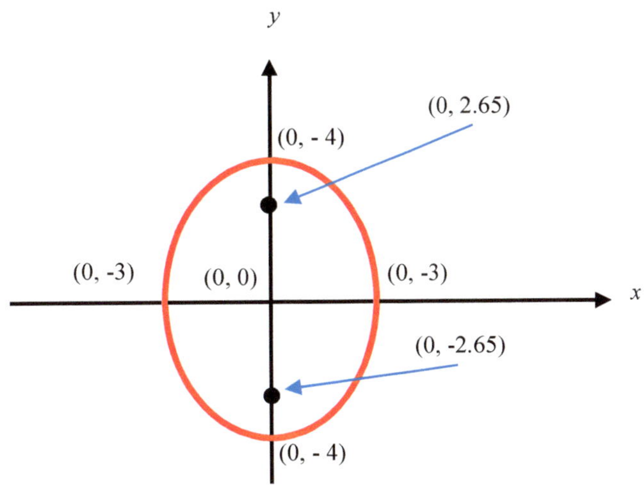

Example
Graph the ellipse given by the equation $x^2 + 4y^2 = 144$. Rewrite the equation in standard form and identify and label the center, vertices, co-vertices, and foci.

Solution
First, we rewrite the equation in standard form.

$x^2 + 4y^2 = 144$

$\frac{1}{144} \times (x^2 + 4y^2) = \frac{1}{144} \times 144$ — we multiplied both side of the equation by $\frac{1}{144}$, to change 144, the right-hand side of the equation, to 1

$\frac{x^2}{144} + \frac{4y^2}{144} = 1$

$\frac{x^2}{144} + \frac{y^2}{36} = 1.$

Next, we observe the position of the major axis. Since $144 > 36$, the major axis is on the x-axis. Therefore, the equation is in the form $\frac{x^2}{a^2} + \frac{y^2}{b^2} = 1$, where $a^2 = 144$ and $b^2 = 36$. It concludes that:

- The center of the ellipse is (0, 0).
- The coordinates of the vertices are $(\pm\, a, 0)$.

  $= (\pm \sqrt{144}, 0)$

  $= (\pm\, 12, 0)$.
- The coordinates of the co-vertices are $(0, \pm\, b)$.

  $= (0, \pm \sqrt{36})$

  $= (0, \pm .6)$.
- The coordinates of the foci are $(\pm c, 0)$, where $c^2 = a^2 - b^2$. Solving for c, we have:

  $c^2 = a^2 - b^2$

  $c^2 = 144 - 36$ — we substituted $a^2 = 144$ and $b^2 = 36$

  $c = \pm \sqrt{144 - 36}$

$c = \pm\sqrt{108} = \pm 10.39.$

Therefore, the coordinates of the foci are (±10.39, 0).

Next, we plot and label the center, vertices, co-vertices, and foci, and draw a smooth curve to form the ellipse. See the graph below:

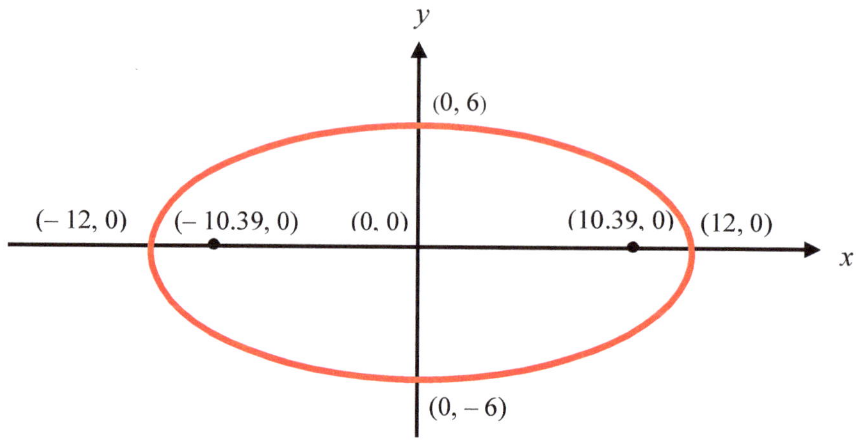

3.2.6. Graphing the Standard Form of an Equation for an Ellipse Not Centered at Origin

Provided the standard form of an equation for an ellipse centered at (h, k), the steps to sketch the graph are given below:

1. Use the standard forms of the equations of an ellipse to determine the center, position of the major axis, vertices, co-vertices, and foci.
   a) If the equation is in the form $\frac{(x-h)^2}{a^2} + \frac{(y-k)^2}{b^2} = 1$, where $a > b$, then:
      - The center is (h, k).
      - The major axis is parallel to the x-axis.
      - The coordinates of the vertices are $(h \pm a, k)$.
      - The coordinates of the co-vertices are $(h, k \pm b)$.
      - The coordinates of the foci are $(h \pm c, k)$.

   b) If the equation is in the form $\frac{(x-h)^2}{b^2} + \frac{(y-k)^2}{a^2} = 1$, where $a > b$, then:
      - The center is (h, k).
      - The major axis is parallel to the y-axis.
      - The coordinates of the vertices are $(h, k \pm a)$.
      - The coordinates of the co-vertices are $(h \pm b, k)$.
      - The coordinates of the foci are $(h, k \pm c)$.

2. Solve for c using the equation $c^2 = a^2 - b^2$.

3. Plot the center, vertices, co-vertices, and foci in the coordinate plane, and draw a smooth curve to form the ellipse.

Example

Graph the ellipse given by the equation, $\frac{(x-3)^2}{4}+\frac{(y-4)^2}{9}=1$. Identify and label the center, vertices, covertices, and foci.

Solution

First, we observe the position of the major axis. Since 9 > 4, the major axis is parallel to the y-axis. Therefore, the equation is in the form $\frac{(x-h)^2}{b^2}+\frac{(y-k)^2}{a^2}=1$, where $b^2 = 4$ and $a^2 = 9$. It concludes that:

- the center of the ellipse is $(h, k) = (3, 4)$
- the coordinates of the vertices are $(h, k \pm a) = (3, 4 \pm \sqrt{9}) = (3, 4 \pm 3)$, or (3, 1) and 3, 7)
- the coordinates of the co-vertices are $(h \pm b, k) = (3 \pm \sqrt{4}, 4) = (3 \pm 2, 4)$, or (1, 4) and (5, 4)
- the coordinates of the foci are $(h, k \pm c)$, where $c^2 = a^2 - b^2$. Solving for c, we have:

$c = \pm\sqrt{a^2 - b^2}$

$c = \pm\sqrt{9-4}$ we substituted $a^2 = 9$ and $b^2 = 4$ into the equation

$c = \pm\sqrt{5} = \pm 2.24$.

Therefore, the coordinates of the foci are $(3, 4+\sqrt{5})$ and $(3, 4-\sqrt{5})$ or (3, 6.26) and (3, 1.76).

Next, we plot and label the center, vertices, co-vertices, and foci, and draw a smooth curve to form the ellipse. See the graph below:

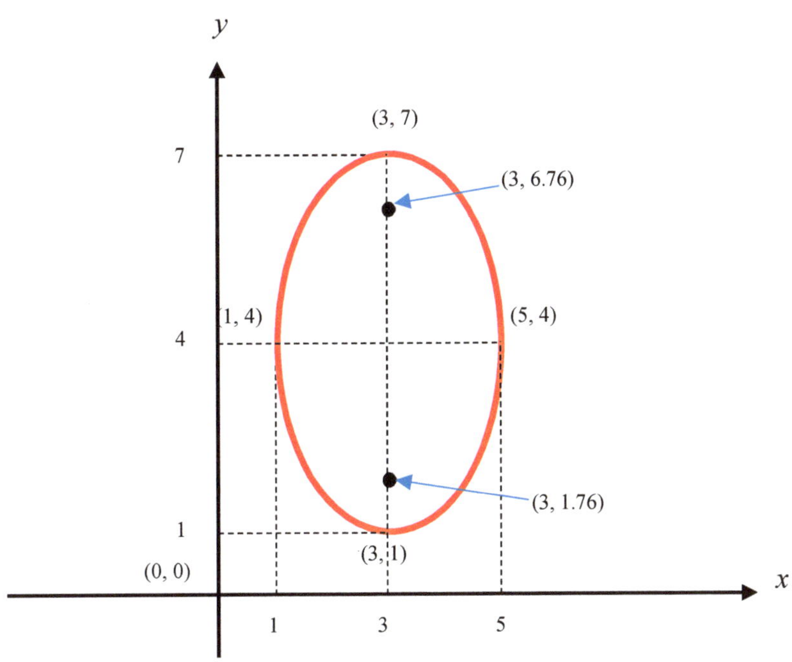

3.3. Hyperbolas

A **hyperbola** is formed when the plane is parallel to the cone's central axis, meaning it intersects both parts of the double cone. Hyperbolas have two branches, as well as these features: Asymptote lines these are two linear graphs that the curve of the hyperbola approaches, but never touches.

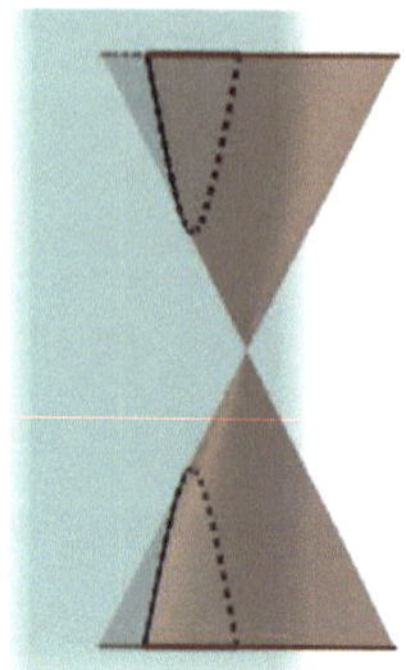

In the definition of an ellipse, we learned that an ellipse is a curve surrounding two focal points, F_1 and F_2, such that for the set of all points on the curve, the sum of the two distances, d, to the focal points is a constant. In this definition if we replace sum with subtract, the resulting curve will be a hyperbola.

3.3.1. Definition of Hyperbolas

Given two distinct points F_1 and F_2 in the plane and a fixed distance d, a hyperbola is the set of all points (x, y) such that for any point of the set, the absolute value of the difference of each of the distances from F_1 and F_2 to (x, y) is d. that is $d = PP$

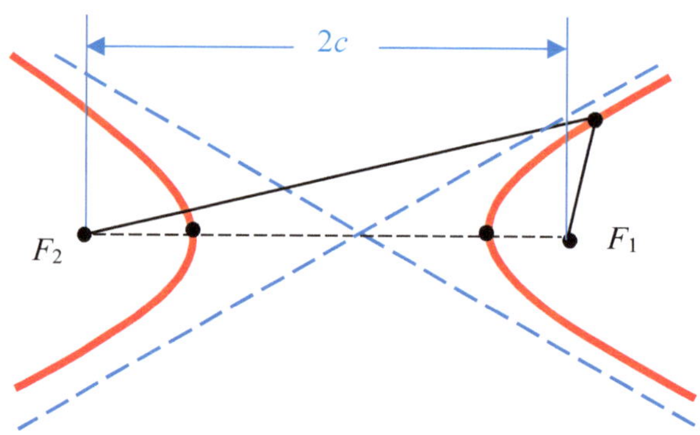

Considering the above figure, the midpoint of the line segment joining the foci is called the *center* of the hyperbola. The line through the foci is called the major axis or transverse axis. It contains the vertices , which have distance to the center.

The distance of the foci to the center is called the focal distance or linear eccentricity. The two blue dashed lines are called asymptotes which approach the branches of the hyperbola for large x and y values. Note that the hyperbola has two parts, which are called branches.

Before we derive the standard equation of the hyperbola, we need to discuss one further parameter, the conjugate axis of the hyperbola. The conjugate axis of a hyperbola is the line segment through the center which is perpendicular to the major or transverse axis and has the same length as the line segment through a vertex which connects the asymptotes.

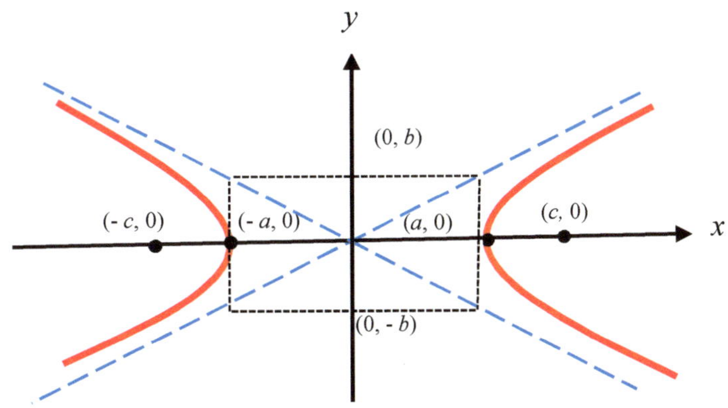

There are two basic forms of hyperbola. Here are examples of each. As was mentioned earlier, hyperbolas consist of two branches that open either up and down or right and left.

Also, just like parabola each of the pieces has a vertex. Note that they are not really parabolas, they just resemble parabolas.

Below we present a figure which illustrates the two standard forms of hyperbola. Figure (a) shows an open left and right hyperbola and Figure (b) reveals an open up and down hyperbola.

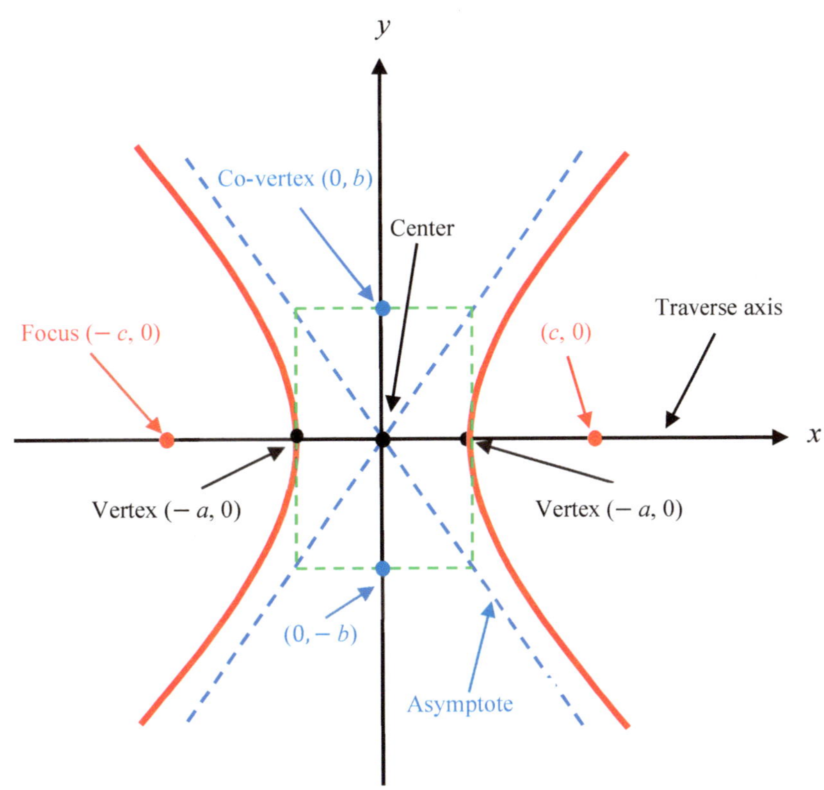

Figure (a) Open left and right hyperbola with the graph of $\frac{x^2}{a^2} - \frac{y^2}{b^2} = 1$.

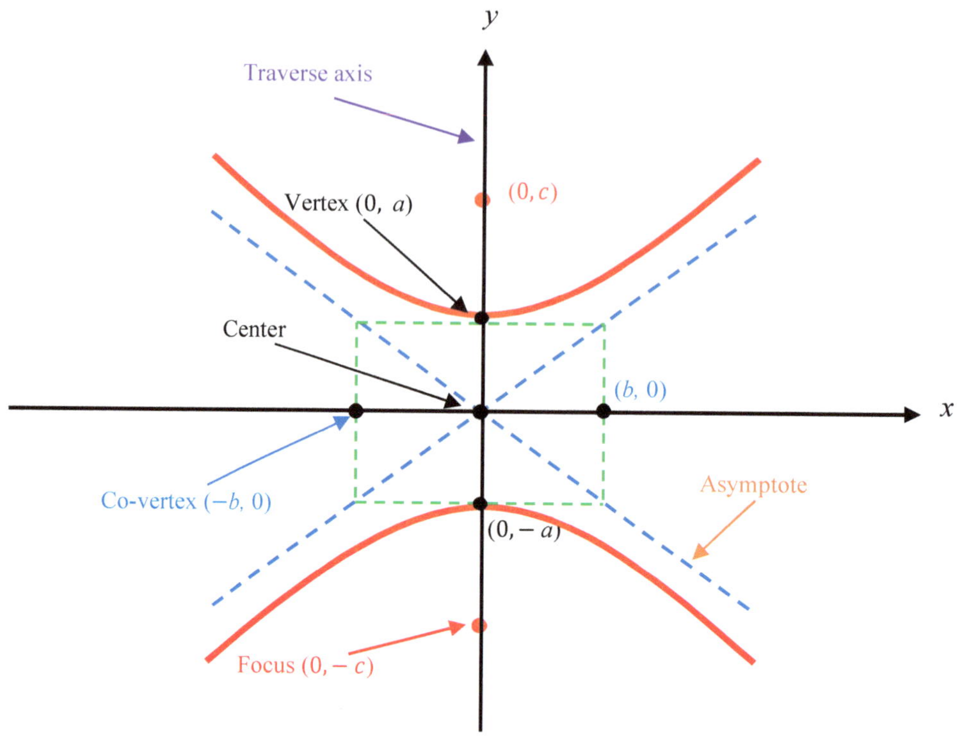

Figure (b) Open up and down hyperbola with the graph of $\frac{y^2}{a^2} - \frac{x^2}{b^2} = 1$.

Below is a table giving each hyperbola form as well as the information we can get from each one.

Form	$\frac{x^2}{a^2} - \frac{y^2}{b^2} = 1$	$\frac{y^2}{a^2} - \frac{x^2}{b^2} = 1$
Centers	(0, 0)	(0, 0)
Vertices	$(\pm a, 0)$	$(0, \pm a)$
Foci	$(\pm c, 0)$	$(0, \pm c)$
Form	$\frac{(x-h)^2}{a^2} - \frac{(y-k)^2}{b^2} = 1$	$\frac{(y-k)^2}{a^2} - \frac{(x-h)^2}{b^2} = 1$
Centers	(h, k)	(h, k)
Open	Opens left and right	Opens up and down
Vertices	$(h+a, k)$ and $(h-a, k)$	$(h, k+b)$ and $(h, k-b)$
Focuses	$(h+c, k)$ and $(h-c, k)$, where, $c^2 = a^2 + b^2$	$(h, k+c)$ and $(h, k-c)$ where, $c^2 = a^2 + b^2$
Equation of asymptotes	$y = k \pm \frac{b}{a}(x-h)$	$y = k \pm \frac{a}{b}(x-h)$

3.3.2. Finding the Center, Vertices, Foci, and Asymptotes of Hyperbola

Given a hyperbola with equations, $\frac{x^2}{a^2} - \frac{y^2}{b^2} = 1$ or $\frac{y^2}{a^2} - \frac{x^2}{b^2} = 1$, we will have:

For $\frac{x^2}{a^2} - \frac{y^2}{b^2} = 1$.

- The center is $C = (0, 0)$.
- The vertices are $V_2 = (-a, 0)$ and $V_1 = (a, 0)$.
- To find the focuses, we first need the distance from the center to the foci, $c^2 = a^2 + b^2$. Then $F_2 = (-c, 0)$ and $F_1 = (c, 0)$.
- The equations of the asymptotes are $y = \pm\frac{b}{a}x$.

For $\frac{y^2}{a^2} - \frac{x^2}{b^2} = 1$.

- The center is $C = (0, 0)$.
- The vertices are $V_2 = (0, -a)$ and $V_1 = (0, a)$.
- To find the foci, we first need the distance from the center to the foci, $c^2 = a^2 + b^2$. Then $F_2 = (0, -c)$ and $F_1 = (0, c)$.
- The equations of the asymptotes are $y = \pm\frac{a}{b}x$.

Example
Find the foci and asymptotes of the hyperbola $4y^2 - 9x^2 - 36 = 0$ and graph it.

Solution
We need to convert the hyperbola $4y^2 - 9x^2 - 36 = 0$ to a standard form, $\frac{x^2}{a^2} - \frac{y^2}{b^2} = 1$, or $\frac{y^2}{a^2} - \frac{x^2}{b^2}$. So,

$$4y^2 - 9x^2 - 36 = 0$$

$$\left(\frac{1}{36}\right)(4y^2 - 9x^2 - 36) = \left(\frac{1}{36}\right)(0)$$ we multiplied both sides by $\frac{1}{36}$

$$\frac{4}{36}y^2 - \frac{9}{36}x^2 - \frac{36}{36} = 0$$

$$\frac{4}{9\times 4}y^2 - \frac{9}{9\times 4}x^2 - 1 = 0$$

$$\frac{y^2}{9} - \frac{x^2}{4} - 1 = 0$$

$$\frac{y^2}{9} - \frac{x^2}{4} - 1 + 1 = 0 + 1$$ to isolate the variables in one side, we added 1 to both sides

$$\frac{y^2}{9} - \frac{x^2}{4} = 1.$$

The standard form equation of this hyperbola is $\frac{y^2}{9} - \frac{x^2}{4} = 1$. To find focuses of the hyperbola, we first need to compute the distance from the center, c, to the foci. We have $c^2 = a^2 + b^2$. In this case, $a^2 = 9$ and $b^2 = 4$. So,

$c^2 = a^2 + b^2$

$c^2 = 4 + 9$ we plugged $a^2 = 9$ and $b^2 = 4$ into the equation

$c^2 = 13$

$c = \pm\sqrt{13}.$

Then $F_2 = (0, -c) = (0, -\sqrt{13}) = (0,$ and $F_1 = (0, c) = (0, \sqrt{13})$.

Then $F_2 = (0, -c) = (0, -\sqrt{13})$, and $F_1 = (0, c) = (0, \sqrt{13})$. The asymptotes are $y = \pm\frac{a}{b}x = \pm\frac{3}{2}x$.

Here is the graph of $\frac{y^2}{9} - \frac{x^2}{4} = 1$.

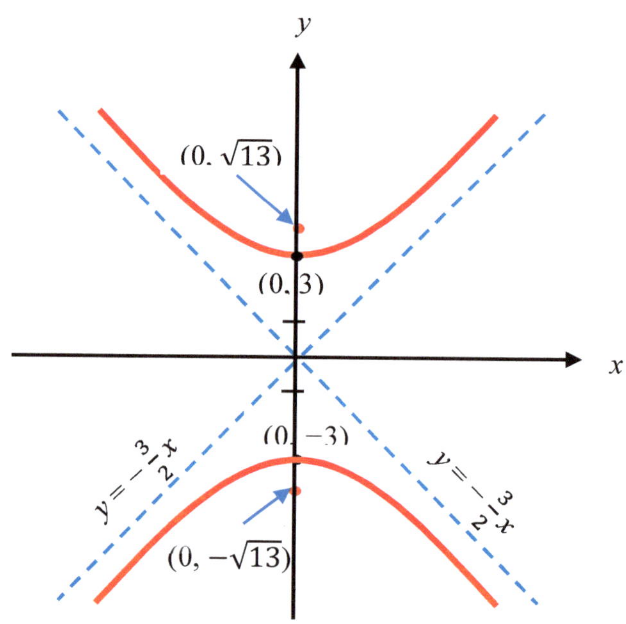

The above example provided an open up and down hyperbola. In this example, the asymptotes of the hyperbola crossed each other at the origin of xy-axis.

To illustrate the other version of a hyperbola, open left and right, as well as a hyperbola in which its center (the crossing point of the asymptote) is located at a point other than the origin of xy-axis, let us present the example below.

Example
Sketch the curve that presented by the equation $9x^2 - 4y^2 - 18x + 32y - 91 = 0$. Find coordinates of vertices and equations of asymptotes.

Solution:
To understand what this curve might look like, we must work towards a standard form. This is best accomplished by completing the square in the x terms and in the y terms.

$9x^2 - 4y^2 - 18x + 32y - 91 = 0.$

$9(x^2 - 2x) - 4(y^2 - 8y) - 91 = 0.$

$9(x^2 - 2x) - 9 + 9 - 4(y^2 - 8y) - 91 = 0.$ We added − 9 and + 9 into the equation

$9(x^2 - 2x + 1) - 9 - 4(y^2 - 8y) - 91 = 0.$

$9(x^2 - 2x + 1) - 9 - 4(y^2 - 8y) + 4(16) - 4(16) - 91 = 0.$ We added − 4(16) and + 4(16) into the equation

$9(x^2 - 2x + 1) - 9 - 4(y^2 - 8y + 16) - 4(16) - 91 = 0.$

$9(x - 1)^2 - 4(y - 4)^2 - 36 = 0.$

$9(x - 1)^2 - 4(y - 4)^2 = 36$

$\frac{1}{36} \times [9(x - 1)^2 - 4(y - 4)^2] = \frac{1}{36} \times 36$ we multiplied both side by $\frac{1}{36}$

$\frac{(x - 1)^2}{4} - \frac{(y - 4)^2}{9} = 1$

$\frac{(x - 1)^2}{2^2} - \frac{(y - 4)^2}{3^2} = 1.$

From this, we see that the curve is a hyperbola centered at (1, 4). When $y = 4$ we have:

$\frac{(x - 1)^2}{2^2} - \frac{(y - 4)^2}{3^2} = 1$

$\frac{(x - 1)^2}{2^2} - \frac{(4 - 4)^2}{3^2} = 1$ we plugged $y = 4$ into the equation

$\frac{(x - 1)^2}{2^2} - \frac{(0)^2}{3^2} = 1$

$\frac{(x - 1)^2}{2^2} = 1$

$(2^2)\frac{(x - 1)^2}{2^2} = 1(2^2)$ we multiplied both sides by (2^2)

$(x - 1)^2 = 2^2$

$(x - 1) = \pm 2.$

For $(x - 1) = 2$:

$x - 1 = 2$

$x - 1 + 1 = 2 + 1$ we added 1 to both sides

$x = 3.$

For $(x - 1) = - 2$:

$x - 1 = - 2$

$x - 1 + 1 = - 2 + 1$ we added 1 to both sides

$x = - 1.$

Thus, $x = 3$ or $x = - 1$.

Therefore, the coordinates of vertices are (3, 4) and (− 1, 4).

We learned, the equation of the asymptotes of a hyperbola are the lines $y = k \pm \frac{b}{a}(x - h)$. In this case, we have $k = 4$, $b = 3$, $a = 2$, and $h = 1$.

Using $y = k + \frac{b}{a}(x - h)$, we will have:

$y = k + \frac{b}{a}(x - h)$

$y = 4 + \frac{3}{2}(x - 1)$ we plugged $k = 4$, $b = 3$, $a = 2$, and $h = 1$ into the equation

$y = 4 + \frac{3}{2}x - \frac{3 \times 1}{2}$

$y = 4 + \frac{3}{2}x - \frac{3}{2}$

$y = 4 \times \frac{2}{2} + \frac{3}{2}x - \frac{3}{2}$

$y = \frac{8}{2} + \frac{3}{2}x - \frac{3}{2}$

$y = \frac{3}{2}x + \frac{8 - 3}{2}$

$y = \frac{3}{2}x + \frac{5}{2}$

$y = 1.5x + 2.5.$

Therefore, one of the equations for the asymptotes of the hyperbola is $y = 1.5x + 2.5$.

Now, using $y = k - \frac{b}{a}(x - h)$, we will have:

$y = k - \frac{b}{a}(x - h)$

$y = 4 - \frac{3}{2}(x - 1)$ we plugged $k = 4$, $b = 3$, $a = 2$, and $h = 1$ into the equation

$y = 4 - \frac{3}{2}x + \frac{3 \times 1}{2}$

$y = 4 - \frac{3}{2}x + \frac{3}{2}$

$y = 4 \times \frac{2}{2} - \frac{3}{2}x + \frac{3}{2}$

$y = \frac{8}{2} - \frac{3}{2}x + \frac{3}{2}$

$y = -\frac{3}{2}x + \frac{8 + 3}{2}$

$y = -\frac{3}{2}x + \frac{11}{2}$

$y = -1.5x + 5.5.$

Therefore, the other equation for the asymptotes of the hyperbola is $y = -1.5x + 5.5$.

We can now sketch the graph of the hyperbola. Here is the requested graph.

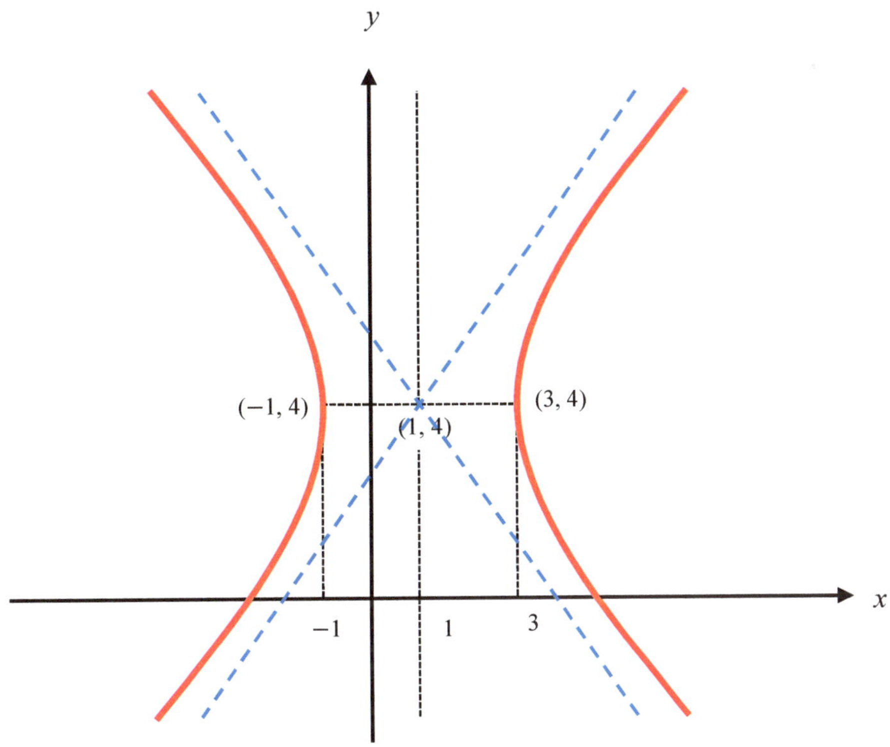

A **Parabola**, is an open curve, a conic section produced by the intersection of a right circular cone and a plane parallel to a line along the side of the cone.

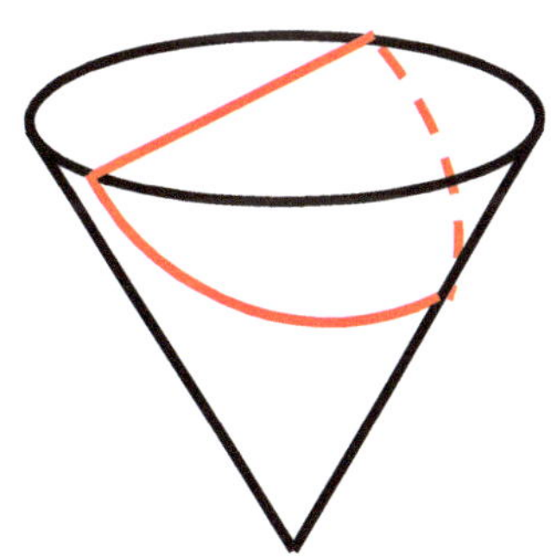

A parabola is a plane curve which is mirror-symmetrical and is approximately U-shaped. It involves a point (the focus) and a line (the directrix). The focus does not lie on the directrix. As a plane curve, it may be defined as the path of a point moving so that its distance from a fixed line (the directrix) is equal to its distance from a fixed point (the focus). In short, a parabola is defined as the set of points that are equidistant from both the directrix and the focus.

The line perpendicular to the directrix and passing through the focus is called the "axis of symmetry". This line splits the parabola through the middle. The point where the parabola intersects its axis of symmetry is called the "vertex" and is the point where the parabola is most sharply curved. Parabolas can open up, down, left, right, or in some other arbitrary direction.

Parabolas have the property that, if they are made of material that reflects light, then light that travels parallel to the axis of symmetry of a parabola and strikes its concave side is reflected to its focus, regardless of where on the parabola the reflection occurs. Conversely, light that originates from a point source at the focus is reflected into a parallel ("collimated") beam, leaving the parabola parallel to the axis of symmetry. The same effects occur with sound and other waves. This reflective property is the basis of many practical uses of parabolas. The advantages of this property are evidenced by the vast list of parabolic objects we use every day: satellite dishes, suspension bridges, telescopes, microphones, spotlights, and car headlights, to name a few.

3.4.1.Standard Parabola, Definition

A parabola is the path (locus) of a point moving so that its shortest distance from a fixed line, the directrix, is equal to its distance from a fixed point, the focus.

Below is the graph of a parabola.

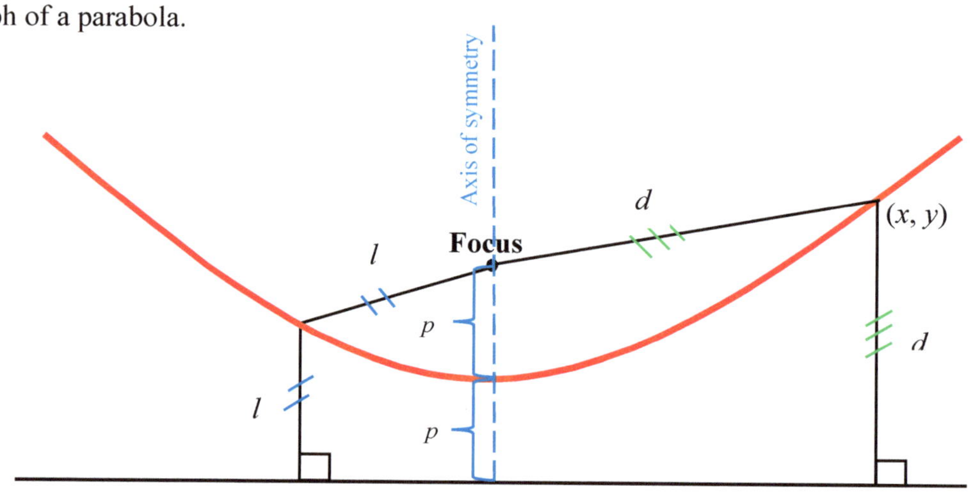

The specific distance from the vertex to the focus is traditionally labeled "*p*". Thus, the distance from the vertex to the directrix is also "*p*". The focus is a point which lies "inside" the parabola on the axis of symmetry. The directrix is a line that is perpendicular to the axis of symmetry and lies "outside" the parabola. For any point (x, y) on the parabola the distance from this point to focus d is the same as its shortest distance to directrix d.

Parabola, Definition
A parabola is a set of points that are equidistant from both the focus and the directrix (shortest distance or the perpendicular distance to directrix).

For any point on a parabola, the distance from that point to the focus is the same as the shortest distance from that point to the directrix. As we will discuss more thoroughly in Chapter 4, the equation of a parabola is a quadratic equation. Hence, it has a general form of $y = ax^2 + bx + c$.

We examine a parabola in relation to its focal point (focus) and directrix, we can determine more information about the parabola. We are now going to look more closely at the coefficient of the x^2 term to see what additional information it can tell us about the graph of the parabola.

3.4.3. Standard Forms of Parabola with Vertex at Origin (0, 0)

The table below summarizes the standard forms of parabola with a vertex at (0,0).

Axis symmetry	Equation	Focus	Directrix
x-axis	$y^2 = 4px$ or $x = \frac{1}{4p}y^2$	$(p, 0)$	$x = -p$
y-axis	$x^2 = 4py$ or $y = \frac{1}{4p}x^2$	$(0, p)$	$y = -p$

Here are graphs for the two parabolas, one with the axis symmetry along x-axis, and the other with the axis symmetry along y-axis.

Figure (a), illustrates a parabola with the axis symmetry along x-axis. Figure (b) depicts a parabola with the axis symmetry along y-axis.

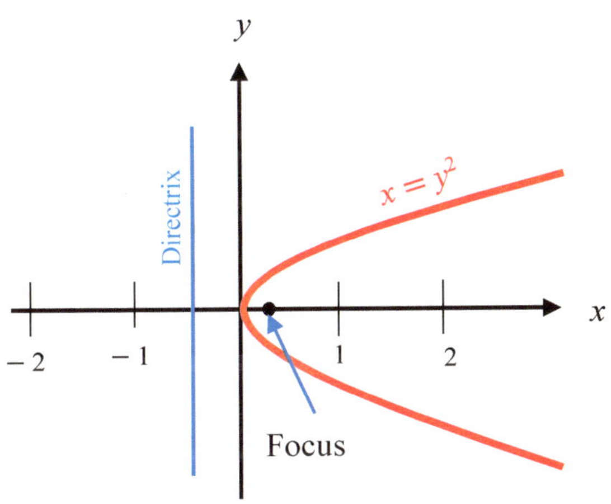

Figure (a). Axis symmetry is x-axis

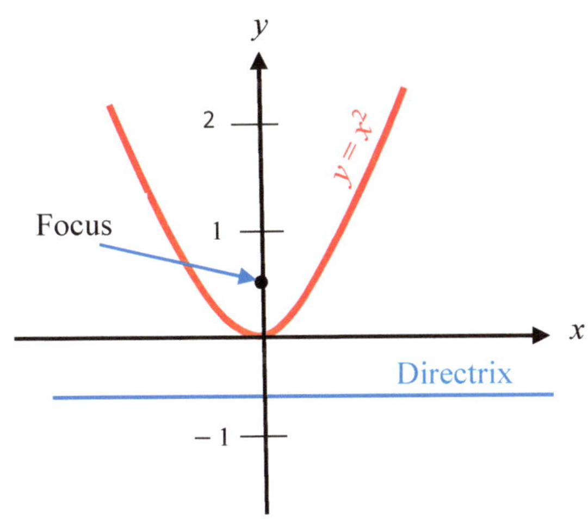

Figure (b). Axis symmetry is y-axis

3.4.4. Standard Forms of Parabola with Vertex at Origin (h, k)

The table below summarizes the standard forms of parabolas with a vertex at (h, k).

Axis symmetry parallel to	Equation	Focus	Directrix
x-axis	$(y - k)^2 = 4p(x - h)$ or $x = \frac{1}{4p}(y - k)^2 + h$	$(h + p, k)$	$x = h - p$
y-axis	$(x - h)^2 = 4p(y - k)$ or $y = \frac{1}{4p}(x - h)^2 + k$	$(h, k + p)$	$y = k - p$

Here are graphs for the two parabolas, one with the axis symmetry parallel to x-axis and the other with the axis symmetry parallel to y-axis Figure (a), illustrates a parabola with the axis symmetry parallel to x-axis.

Figure (a) depicts a parabola with its axis of symmetry parallel to the x-axis, while Figure (b) illustrates a parabola with its axis of symmetry aligned parallel to the y-axis..

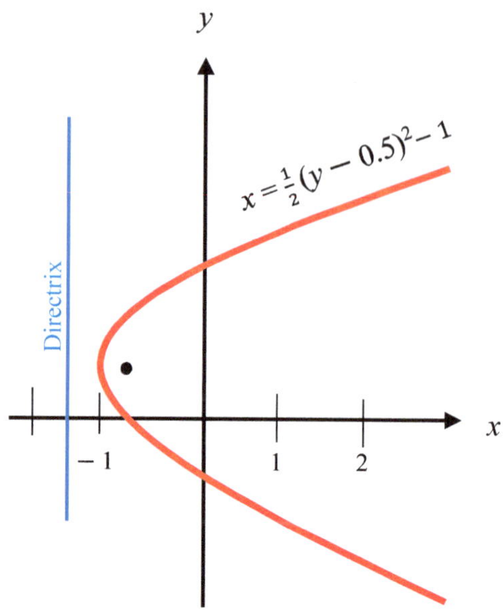

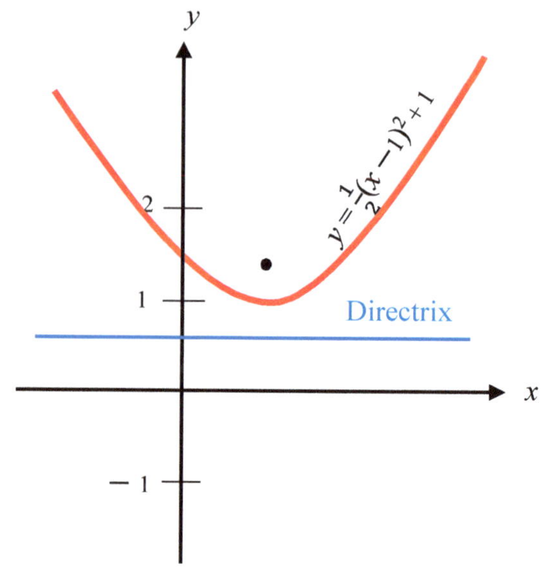

Figure (a) Axis symmetry is parallel to x-axis Figure (b) Axis symmetry is parallel to y-axis

Example
Given the parabola, $y = x^2$, determine p, coordinate of vertex, coordinate of focus, and the equation of directrix line. Additionally, sketch the graph and determine whether it is upward or downward.

Solution
The coefficient of x^2 is 1, so $\frac{1}{4p} = 1$, making $p = \frac{1}{4}$ (p is the distance from the vertex to the focus). The vertex is located at (0, 0), the focus is located at $(0, \frac{1}{4})$, and the directrix equation is $y = -\frac{1}{4}$.

The graph is shown below. As it is seen the parabola is open upward.

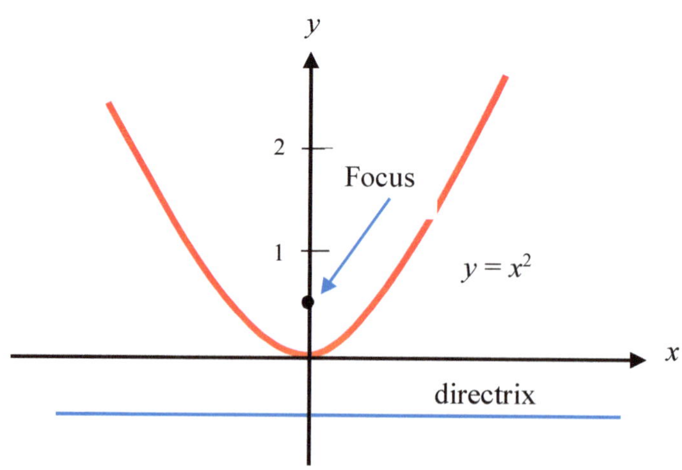

Example
Given the parabola, $y = (x - 2)^2 + 3$, determine p, coordinate of vertex, coordinate of focus, and the equation of directrix line. Additionally, sketch the graph and determine whether it is upward or downward.

Solution

Comparing the given equation with the equation of the previous example, we realize that it is a translation of the previous graph. Hence, the relationship between the parabola and its focus and directrix remains the same ($p = \frac{1}{4}$). So, with a vertex of (2, − 3), we have

$$(x-2)^2 = 4(\tfrac{1}{4})[y-(-3)]$$

$$(x-2)^2 = (\tfrac{4}{4})(y+3)$$

$$(x-2)^2 = y+3.$$

The focus is at $(2, -3+\frac{1}{4})$ or $(2, -2\frac{3}{4})$ and the directrix is $y = -3-\frac{1}{4}$ or $y = -3\frac{1}{4}$.

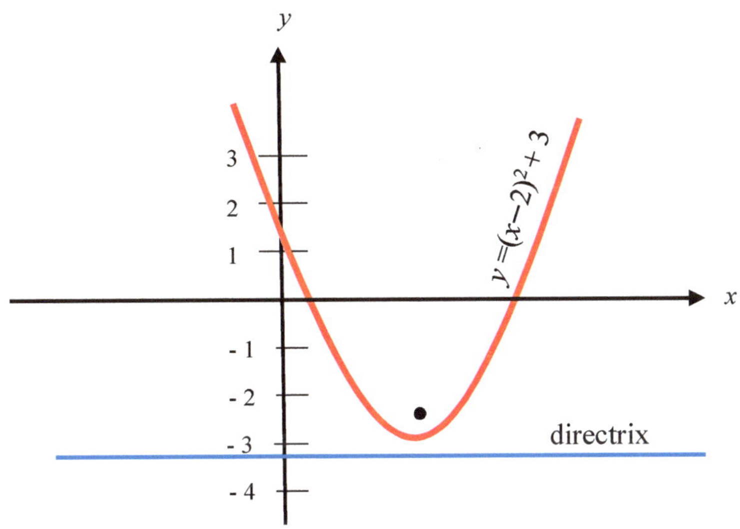

So far, we looked at examples in which parabolas of being either opening upward or opening downward. The axis of symmetry of those parabolas is parallel to the *y*-axis. We will now be looking at examples involving parabolas that open to the right or to the left, with their axis of symmetry parallel to the *x*-axis.**Example**

Given the parabola, $x = y^2$, determine the value of *p*, the coordinate of its vertex and its focus. Also, find the equation of directrix line, sketch the graph of the parabola, and determine whether it is open to the right or open to the left.

Solution

To find the value of *p* we compare the standard form of parabolas equation, $x = \frac{1}{4p}y^2$, with the given equation $x = y^2$. We realize $\frac{1}{4p} = 1$. So,

$$\frac{1}{4p} = 1$$

$$p(\frac{1}{4p}) = p(1) \qquad \text{we multiplied both sides by } p$$

$$\frac{p}{4p} = p$$

$$\frac{1}{4} = p.$$

Therefore, the value of p is $\frac{1}{4}$. The vertex is (0, 0), and the focus is $(p, 0) = (\frac{1}{4}, 0)$. The equation of directrix is $x = -p$.

Since $p = \frac{1}{4}$, the equation of directrix is $x = -\frac{1}{4}$.

The graph of the parabola, as below, shows it is open right.

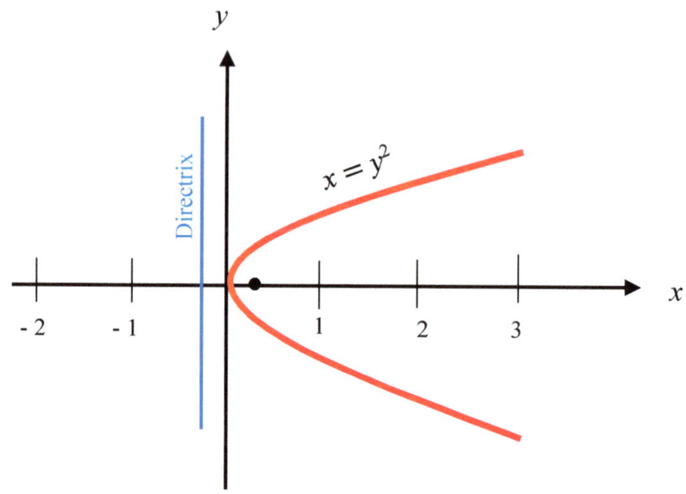

Given the parabola, $x = (y - 1)^2 - 2$, determine p, coordinate of vertex and focus. Also, find the equation of directrix line, sketch the graph, and determine whether it is open to the right or open to the left. Comparing the given equation, $x = (y - 1)^2 - 2$, with the parabola's standard equation, $x = \frac{1}{4p}(y - k)^2 + h$ we realize that $\frac{1}{4p} = 1$, and hence $p = \frac{1}{4}$.

Also, through this comparison, we find that the vertex of (h, k) in a xy- axis is located at the point with coordinate $(h = -2, k = 1) = (-2, 1)$.

For the parabola with a vertex at (h, k), the focus is $(h + p, k)$. Having $h = -2$, $p = \frac{1}{4}$, and $k = 1$ we find the focus is located at the point with coordinate $(-2 + \frac{1}{4}, 1) = (-1\frac{3}{4}, 1)$.

Similarly, for the parabola with a vertex at (h, k) and with axis of symmetry parallel to x-axis, the equation of the directrix is $x = h - p$. So,

$$x = h - p$$

$$x = -2 - \frac{1}{4}$$ we plugged $h = -2$ and $p = \frac{1}{4}$ into the equation

$$x = -2\frac{1}{4}.$$

To sketch the graph of parabola, $x = (y - 1)^2 - 2$, we know that its vertex is located at $(-2, 1)$. Adding more arbitrary points to the plot would be beneficial. So,

For $y = 0$, we will have:

$$x = (y - 1)^2 - 2$$

$x = (0 - 1)^2 - 2$ we plugged $y = 0$ into the equation

$x = (-1)^2 - 2$

$x = 1 - 2$

$x = -1.$

For $x = 0$, we will have:

$x = (y - 1)^2 - 2$

$0 = (y - 1)^2 - 2$ we plugged $x = 0$ into the equation

$0 + 2 = (y - 1)^2 - 2 + 2$ to isolate the variable we added 2 to both sides

$2 = (y - 1)^2$

$\sqrt{2} = \sqrt{(y - 1)^2}$ we took the square root of both sides

$\sqrt{2} = \pm (y - 1)$

$\sqrt{2} = (y - 1)$ or $\sqrt{2} = -(y - 1).$

For $\sqrt{2} = (y - 1)$:

$\sqrt{2} = y - 1$

$1 + \sqrt{2} = y - 1 + 1$ to isolate the variable we added 1 to both sides

$1 + 1.41 = y$

$y = 2.41.$

For $\sqrt{2} = -(y - 1)$:

$\sqrt{2} = -y + 1$

$-1 + \sqrt{2} = y + 1 - 1$ to isolate the variable we added -1 to both sides

$-1 + 1.41 = y$

$y = 0.41.$

Having the three points $(-2, 1)$, $(0, 1.41)$, and $(0, 0.41)$ on the graph, we can more easily sketch the graph of $x = (y - 1)^2 - 2$.

Here is the graph of parabola, $x = (y - 1)^2 - 2$.

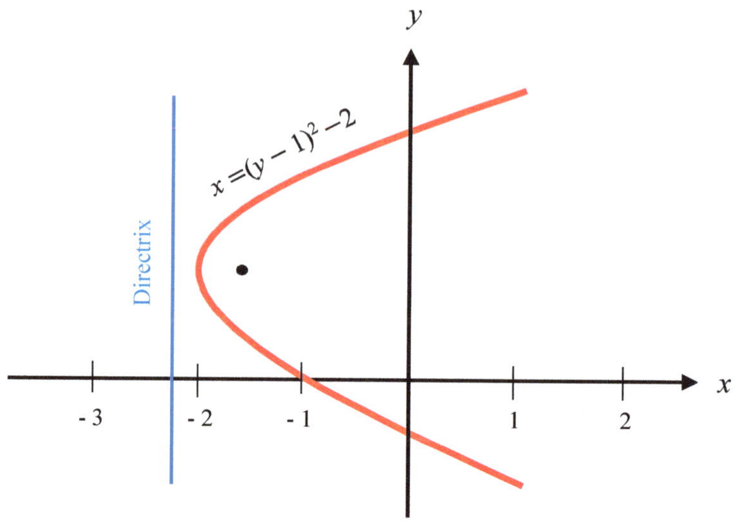

The table below summarizes parabolas information.

Vertical Parabola (up/down)		Horizontal Parabola (left/right)	
Axis of symmetry	parallel to y-axis	Axis of symmetry	parallel to x-axis
If $p > 0$	Open upward	If $p > 0$	Open to the right
If $p < 0$	Open downward	If $p < 0$	Open to the left

Vertical Parabola (up/down)		Horizontal Parabola (left-right)	
	Vertex (0, 0)		**Vertex (0, 0)**
Equation	$y = \frac{1}{4p}x^2$ or $x^2 = 4py$	Equation	$x = \frac{1}{4p}y^2$ or $y^2 = 4py$
Focus	$(0, p)$	Focus	$(0, p)$
Directrix	$y = -p$	Directrix	$x = -p$
	Vertex (h, k)		**Vertex (h, k)**
Equation	$y - k = \frac{1}{4p}(x-h)^2$ or $(x-h)^2 = 4p(y-k)$	Equation	$x - h = \frac{1}{4p}(y-k)^2$ or $(y-k)^2 = 4p(x-h)$
Focus	$(h, k+p)$	Focus	$(h+p, k)$
Directrix	$y = k - p$	Directrix	$x = h - p$

Example
Given $x^2 = 8y$, state whether the parabola opens upward, downward, right, or left, and state the coordinates of the focus.

Solution
Form: $x^2 = 4py$

$4p = 8$

$p = 2.$

The focal length is 2. Since this "form" squares x, and the value of $4p$ is positive, the parabola opens upward.

This form of parabola has its vertex at the origin, (0, 0). The focal length is 2 units, and the focus is located at = (0, 2).Given the parabola, $(x - 1)^2 = -16(y - 5)$, state whether the parabola opens upward, downward, right or left, and state the coordinates of the vertex, the focus, and the equation of the directrix.
Conic form of parabola equation is $(x - h)^2 = 4p(y - k)$, so,

$$4p = -16$$

$$p = -4.$$

Since $p < 0$ the parabola is open downward. The focal length is 4, and the parabola has its vertex at $(h, k) = (1, 5)$.

The focal length is 4 units. The focus is located at $(h, k + p) = (1, 5 - 4) = (1, 1)$. The directrix equation is $y = k - p = 5 - (-4) = 9$.

Example
Write the equation of a parabola with a vertex at the origin and a focus of (0, − 3).

Solution
The graph of the parabola is given below.

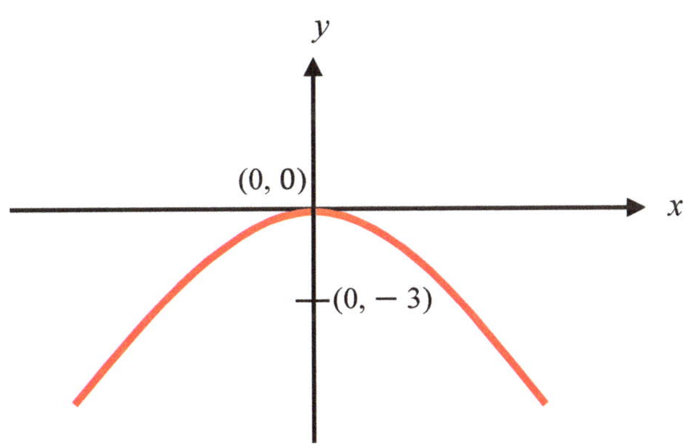

The vertex is located at (0, 0) and focus is located at (0, − 3). Focal length is $p = 3$ and the parabola is opening downward means negative p. Form of the equation of $x^2 = 4py$, we have:

$$x^2 = 4(-3)y$$

$x^2 = -12y$.Find the focus and directrix of the parabola with the equation given as: $x^2 - 4x + 4y + 16 = 0$.
You need to complete the square so the vertex, focus and directrix information will be visible.

$$x^2 - 4x + 4y + 16 = 0$$

$$x^2 - 4x = -4y - 16$$

$$x^2 - 4x + 4 = -4y - 16 - 4$$

$$(x - 2)^2 = -4y - 20$$

$$(x - 2)^2 = -4(y + 5).$$

- The vertex is (2, − 5).

- The x-squared term indicates the parabola opens upward or downward.
- The negative value of p indicates the parabola opens downward.
- The focal length, p, is: $4p = -4$; $p = -1$.
- The focus is at $= (2, -5 - 1) = (2, -6)$.
- The directrix is $y = k - p = -5 - (-1) = -5 + 1 = -4$.

Example
Write the equation of a parabola whose focus $(-2, 1)$ and whose directrix is $x = -4$.

Solution
Remember that the parabola opens "around" the focus, and the vertex is halfway between the focus and the directrix.

- Vertex: $(-3, 1) = (h, k)$.
- Opens to the right (around the focus).
- Focal length, $p = 1$.
- Form of the equation is, $(y - k)^2 = 4p(x - h)$.

  $(y - 1)^2 = 4(1)[x - (-3)]$.

  $(y - 1)^2 = 4(x + 3)$.

parabola

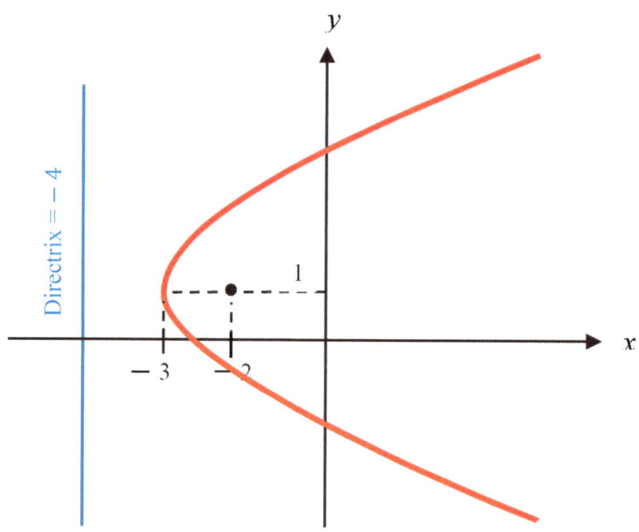

3.5. Exercises

1. Graph $(x-3)^2 + (y-4)^2 = 9$ and find the center and radius.

2. Write the standard equation of the circle with center $(-2, 7)$ and radius 8.

3. Find the points on the unit circle with y-coordinate $\frac{1}{4}$.

In Exercises 4 -7, find the standard equation of the circle and then graph it.

4. Center at (1, 2), and radius 2.

5. Center at (– 3, 1) and radius 2.

6. Center at (0, – 4) and radius 2.

7. Center at (– 2, 0) and radius 3.

In Exercises 5 - 13, complete the square to put the equation of circle into standard form. Identify the center and the radius or explain why the equation does not represent a circle.

Hint: note that the equation, $(x-a)^2 = x^2 - 2ax + b^2$, always holds for $a > 0$, as well as $(x+a)^2 = x^2 + 2ax + b^2$.

8. $x^2 - 6x + y^2 + 8y = -21$.

9. $x^2 + 10x = -y^2 + 8y - 32$.

10. $x^2 - 14x + y^2 + 12y + 60 = 0$.

11. $y^2 + x^2 = -71 - 2x + 16y$.

12. Write the standard form equation of the ellipse that its vertices are located at $(\pm 8, 0)$ and its foci are located at $(\pm 5, 0)$.

13. Write the standard form equation of the ellipse with vertices $(2, -7)$ and $(2, 3)$ and foci $(2, -4)$ and $(2, 2)$.

14. Graph the ellipse given by the equation, $\frac{x^2}{25} + \frac{y^2}{9} = 1$. Identify and label the center, vertices, co-vertices, and foci.

15. Graph the ellipse given by the equation, $\frac{(x-1)^2}{16} + \frac{(y-5)^2}{36} = 1$. Identify and label the center, vertices, covertices, and foci.

16. Sketch the curve that presented by the equation $2x^2 - 3y^2 - 4x + 18y + 16 = 0$. Find coordinates of vertices and equations of asymptotes.

17. Sketch the curve that presented by the equation $9x^2 - 4y^2 + 8y - 40 = 0$. Find coordinates of vertices and equations of asymptotes.

18. Sketch the curve that presented by the equation $3x^2 - 2y^2 - 12x + 24 = 0$. Find coordinates of vertices and equations of asymptotes.

19. Given the parabola, $16y = x^2$, determine p, coordinate of vertex, coordinate of focus, and the equation of directrix line. Additionally, sketch the graph and determine whether it is upward or downward.

20. Given $(x - 2)^2 = 3y$, state whether the parabola opens upward, downward, right, or left, and state the coordinates of the focus.

21. Given the parabola, $(y - 5)^2 = -5(x - 9)$, state whether the parabola opens upward, downward, right, or left, and state the coordinates of the vertex, the focus, and the equation of the directrix.

22. Write the equation of a parabola with a vertex at the origin and a focus of $(0, -2)$.Find the focus and directrix of the parabola with the equation given as: $x^2 - 12x + 4y - 4 = 0$.

24. Given the parabola, $(x - 7)^2 = -4(y - 6)$, state whether the parabola opens upward, downward, right or left, and state the coordinates of the vertex, the focus, and the equation of the directrix.

25. Write the equation of a parabola whose focus $(4, -2)$ and whose directrix is $x = -3$.

4. Real - World Applications of Principle of Analytic Geometry

Analytic geometry, also referred to as coordinate geometry or Cartesian geometry, involves studying geometry using a coordinate system. Let us delve into some intriguing real-world applications of this mathematical field.

4.1. Example: Architectural Design of University Entrance

An architectural design for a university entrance is given below. It is an imaginary image of an open book that shows the value of reading, education, and research. In this design, there are two quarter-circles, three vertical-lines, and four non-vertical straight lines. Given the *xy*-coordinates, of points:

a) Find the equations of the vertical lines.

b) Find the equations of the non-vertical lines.

c) Find the equations of the circles.

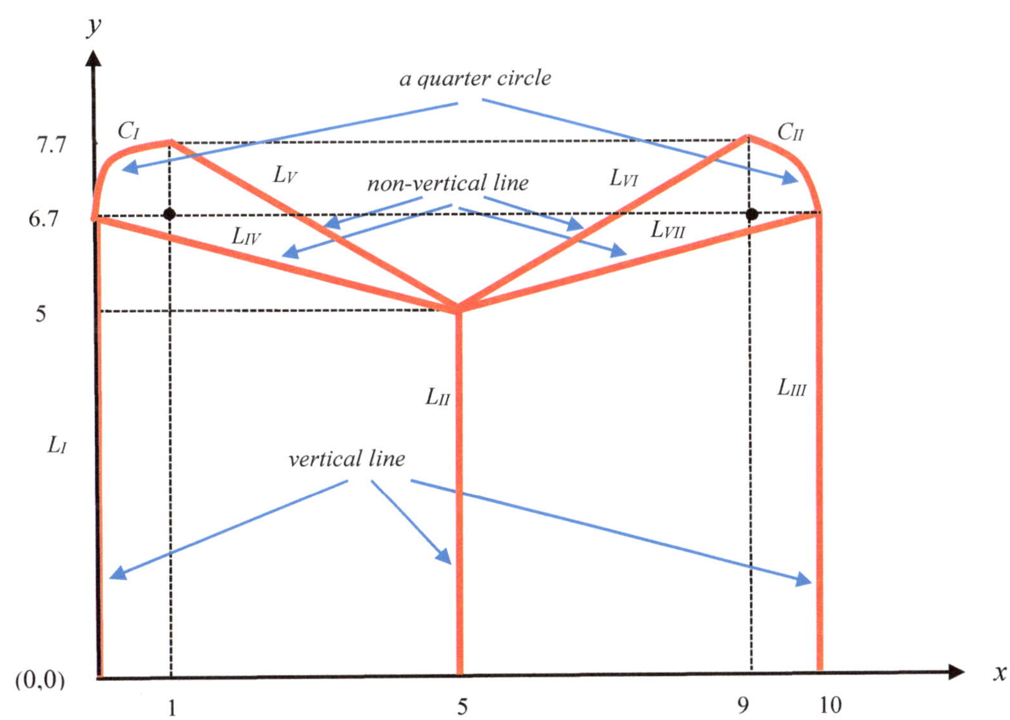

Solution

We have learned to find equations of non-vertical line, straight lines, and circles. So,

a) Find the equations of the vertical lines, we know vertical lines have the general form $x = b$. we have three vertical-lines L_I with $b = 0$, L_{II} with $b = 5$ and L_{III} with $b = 10$.

For L_I, the equation of the line is $x = 0$.

For L_{II}, the equation of the line is $x = 5$.

For L_{III}, the equation of the line is $x = 10$.

b) Find the equations of the non-vertical lines, we learned they can be obtained by constructing a table of values. For L_{IV}, a table of values is:

x	y	Ordered pair
0	6.7	(0, 6.7)
5	5	(5, 5)

The equation of the line L_{IV} is obtained by the "line equation from 2 points" which suggests: the equation of the line passing through two points (x_A, y_A) and (x_B, y_B) is given by $y - y_A = \frac{y_B - y_A}{x_B - x_A}(x - x_A)$. In this case, $x_A = 0$, $y_A = 6.7$, $x_B = 5$, and $y_B = 5$. So,

$$y - y_A = \frac{y_B - y_A}{x_B - x_A}(x - x_A)$$

$$y - 6.7 = \frac{5 - 6.7}{5 - 0}(x - 0)$$ we plugged $x_A = 0$, $y_A = 6.7$, $x_B = 5$, and $y_B = 5$ into the equation

$$y - 6.7 = \frac{-1.7}{5}x$$

$$y - 6.7 = -0.34\,x$$

$$y - 6.7 + 6.7 = -0.34\,x + 6.7$$ to isolate y, we added 6.7 to both sides

$$y = -0.34\,x + 6.7.$$

Therefore, the equation of the line L_{IV} is $y = -0.34\,x + 6.7$.

For L_V, a table of values is:

x	y	Ordered pair
1	7.7	(1, 7.7)
5	5	(5, 5)

Again, using the "line equation from 2 points", $y - y_A = \frac{y_B - y_A}{x_B - x_A}(x - x_A)$, where, in this case, $x_A = 1$, $y_A = 7.7$, $x_B = 5$, and $y_B = 5$. So,

$$y - y_A = \frac{y_B - y_A}{x_B - x_A}(x - x_A)$$

$$y - 7.7 = \frac{5 - 7.7}{5 - 1}(x - 1)$$ we plugged $x_A = 1$, $y_A = 7.7$, $x_B = 5$, and $y_B = 5$ into the equation

$$y - 7.7 = \frac{-2.7}{4}(x - 1)$$

$$y - 7.7 = \frac{-2.7}{4}x - \frac{-2.7}{4}$$

$$y - 7.7 = -0.675\,x + 0.675$$

$$y - 7.7 + 7.7 = -0.675\,x + 0.675 + 7.7$$ to isolate y, we added 7.7 to both sides

$$y = -0.675\,x + 8.375.$$

Therefore, the equation of the line L_V is $y = -0.675\,x + 8.375$.

For L_{VI}, a table of values is:

x	y	Ordered pair
5	5	(5, 5)
9	7.7	(9, 7.7)

Again, using the "line equation from 2 points", $y - y_A = \frac{y_B - y_A}{x_B - x_A}(x - x_A)$, where, in this case, $x_A = 5$, $y_A = 5$, $x_B = 9$, and $y_B = 7.7$. So,

$$y - y_A = \frac{y_B - y_A}{x_B - x_A}(x - x_A)$$

$$y - 5 = \frac{7.7 - 5}{9 - 5}(x - 5)$$ we plugged $x_A = 5$, $y_A = 5$, $x_B = 9$, and $y_B = 7.7$ into the equation

$$y - 5 = \frac{2.7}{4}(x - 5)$$

$$y - 5 = \frac{2.7}{4}x - \frac{2.7 \times 5}{4}$$

$$y - 5 = 0.675\,x - 3.375$$

$$y - 5 + 5 = 0.675\,x - 3.375 + 5$$ to isolate y, we added 5 to both sides.

$$y = 0.675\,x + 1.625.$$

Therefore, the equation of the line L_{VI} is $y = 0.675\,x + 1.625$.

For L_{VII}, a table of values is:

x	y	Order pair
5	5	(5, 5)
10	6.7	(10, 6.7)

Again, using the "line equation from 2 points", $y - y_A = \frac{y_B - y_A}{x_B - x_A}(x - x_A)$, where, in this case, $x_A = 5$, $y_A = 5$, $x_B = 10$, and $y_B = 6.7$. So,

$$y - y_A = \frac{y_B - y_A}{x_B - x_A}(x - x_A)$$

$$y - 5 = \frac{6.7 - 5}{10 - 5}(x - 5)$$ we plugged $x_A = 5$, $y_A = 5$, $x_B = 9$, and $y_B = 7.7$ into the equation

$$y - 5 = \frac{1.7}{5}(x - 5)$$

$$y - 5 = \frac{1.7}{5}x - \frac{1.7 \times 5}{5}$$

$$y - 5 = 0.34\,x - 1.7$$

$$y - 5 + 5 = 0.34\,x - 1.7 + 5$$ to isolate y, we added 5 to both sides

$$y = 0.34\,x + 3.3.$$

Therefore, the equation of the line L_{VII} is $y = 0.34\,x + 3.3$.

c) To find the equation of a circle, we need to find its radius. In this case the left-end point of a quarter circle C_I is located at (0, 6.7) and its center is located at (1, 6.7). For finding the length of the radius r_1 for circle C_I, we simply need to find the absolute value of their x coordinates difference. Hence, $r_1 = |1 - 0| = 1$ since they have the same y coordinate.

Similarly, the right-end point of a quarter circle C_{II} is located at (10, 6.7) and its center is located at (9, 6.7). For finding the length of the radius r_2 for circle C_{II}, we simply need to find $r_2 = |10 - 9| = 1$, again since they have the same y coordinate.

We know the standard equation of a circle with center (a, b) and radius $r > 0$ is $(x - a)^2 + (y - b)^2 = r^2$. Using this formula, we have:

For a quarter circle C_I, with the center (1, 6.7) and the radius $r_1 = 1$,

$(x - a)^2 + (y - b)^2 = r^2$

$(x - 1)^2 + (y - 6.7)^2 = 1^2$ we plugged $a = 1$, $b = 6.7$, and $r = 1$ into the equation

$(x - 1)^2 + (y - 6.7)^2 = 1.$

Therefore, the equation of the circle C_I is $(x - 1)^2 + (y - 6.7)^2 = 1$.

For a quarter circle C_I, with the center (9, 6.7) and the radius $r_1 = 1$,

$(x - a)^2 + (y - b)^2 = r^2$

$(x - 9)^2 + (y - 6.7)^2 = 1^2$ we plugged $a = 9$, $b = 6.7$, and $r = 1$ into the equation

$(x - 9)^2 + (y - 6.7)^2 = 1.$

Therefore, the equation of the circle C_{II} is $(x - 9)^2 + (y - 6.7)^2 = 1$.

4.2. Example: Depreciation and Appreciation Rate

A Mustang was purchased in 1980 for \$8,000 and depreciated in value over time until it was sold in 1992 for \$2,000. In the year 2020 this 40-year-old car was considered as a classic car, so it is sold at San Jose auction for \$44,000.

a) Graph the value of the car over time.

b) Determine the rate at which the car depreciated in value from 1980 to 1992.

c) Determine the rate at which the car appreciated in value from 19925 to 2020.

Solution

Recall that the car's value depends on its age and that the slope measures the rate in dollars per year. So,

a) Sketch the graph of car value over time, let us consider a xy-axis in which the x-axis represents the time, and the y-axis illustrates the car value.

Remember that we need two points for graphing a straight line. Thus, for the time span between 1980 to 1992, let us apply the method of constructing a table value to get two points. Given the data, a table of values is:

year	Car value	Ordered pair
1980	\$8,000	(1980, 8,000)
1992	\$2,000	(1992, 2,000)

For the time span between 1992 to 2020, the table values are:

year	Car value	Ordered pair
1992	\$2,000	(1992, 2,000)
2020	\$44,000	(1992, 44,000)

Using these tables, the graphs for the value of the car over time are illustrated below. In this figure the red line is the depreciation trend, and the blue line is appreciation trend.

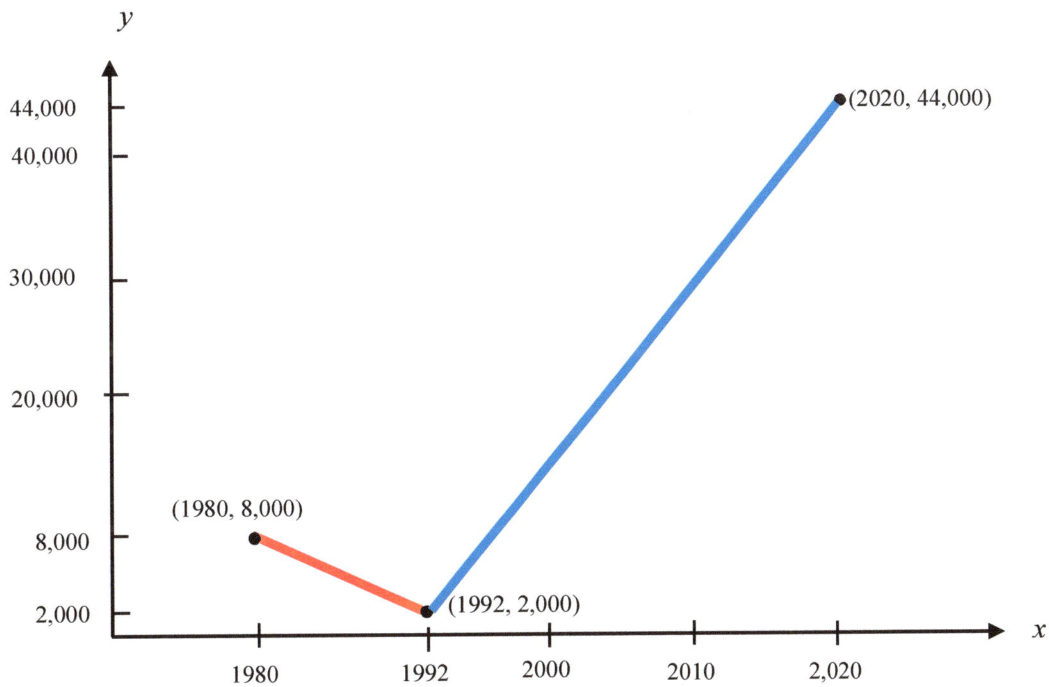

b) The depreciation rate is just the slope of the line segment depicting the car value for the first 12 years. Using the formula for the slope of a line, $m = \frac{y_2 - y_1}{x_2 - x_1}$, we can find the rate at which the car depreciated in value from the year 1980 to 1992. So,

$$m = \frac{y_2 - y_1}{x_2 - x_1}$$

$$m = \frac{2{,}000 - 8{,}000}{1992 - 1980}$$

we plugged $y_2 = 2{,}000$, $y_1 = 8{,}000$, $x_2 = 1992$, and $x_1 = 1980$ into the equation

$$m = \frac{-6{,}000}{12} = -500.$$

Therefore, the value of the car depreciated \$500 per year from 1980 to 1992.

c) Similarly, the appreciation rate is the slope of the line segment depicting the car value for the next 28 years, from the year 1992 to the year 2020. Again, using the formula for the slope of a line, $m = \frac{y_2 - y_1}{x_2 - x_1}$, we can find the rate at which the car appreciated in value from 1992 to 2020. So,

$$m' = \frac{y_2 - y_1}{x_2 - x_1}$$

$$m' = \frac{44{,}000 - 2{,}000}{2020 - 1992}$$ we plugged $y_2 = 44{,}000$, $y_1 = 2{,}000$, $x_2 = 2020$, and $x_1 = 1992$ into the equation

$$m' = \frac{42{,}000}{28} = 1{,}500.$$

Therefore, the value of the car appreciated \$1,500 per year from 1992 to 2020.

4.4. Example: Center of Mass

Suppose that two $1m$ by $1m$ squares are cut from a square piece of plywood which originally had sides of length $4m$, as shown below. If mass the plywood is 2 per square meter, what is the center of mass of the resulting shape?

Hint: the area, A, of squares with side s is $A = s^2$, and the area of rectangles with sides of w and l is $A = wl$.

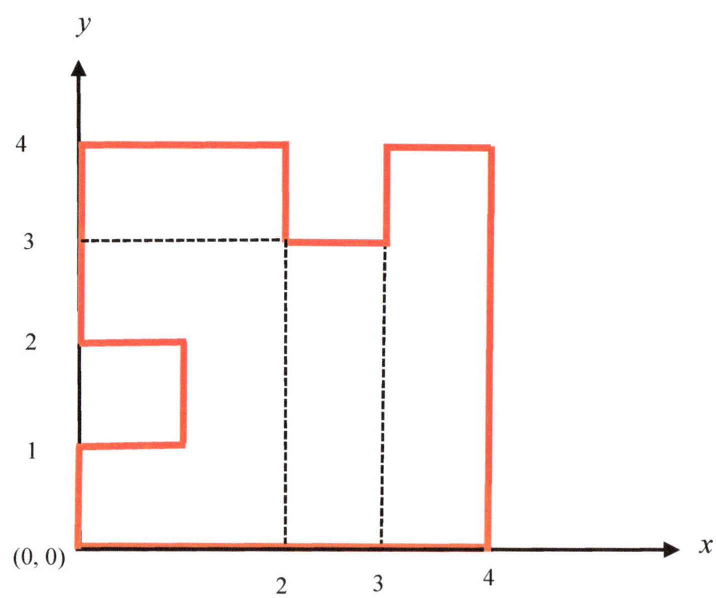

Solution

We know the center of mass (COM) is a position defined as the average position of all the parts of the system, weighted according to their masses. In other words, the position of COM is mass weighted average of the positions of components.

To find the COM separately for the components along each axis,

For object positions along the x axis:

$$\text{COM}_x = \frac{m_1x_1 + m_2x_2 + m_3x_3 + \cdots}{m_1 + m_2 + m_3 + \cdots}.$$

And similarly, for objects positions along the y axis:

$$\text{COM}_y = \frac{m_1y_1 + m_2y_2 + m_3y_3 + \cdots}{m_1 + m_2 + m_3 + \cdots}.$$

where, m_i is the mass of component i, x_i is the x coordinate of component i, and y_i is the y coordinate of component i.

To use the above concept, we need to divide the given shape into two squares and three rectangles. For the two squares, one has sides of length $2m$ and the other has sides of length $1m$. For the three rectangles, we have one with $1m$ by $2m$, the second with $1m$ by $3m$, and the third with $1m$ by $4m$. Note that this is not the only way to divide the shape into squares and rectangles, and any of the other ways will work just as well. We then find the centers of the squares and rectangles. The center of square (rectangle) is the midpoint of the diagonal end points of square (rectangle). Mathematically, the center of square (rectangle), $C(x, y)$ for a square (rectangle) with end pints of (x_1, y_1) and (x_2, y_2) is $\left(\frac{x_2 + x_1}{2}, \frac{y_2 + y_1}{2}\right)$. Hence, for the $2m$ by $2m$ square with the end points of (0, 2) and (2, 4) the center is,

$$C_1\left(\frac{x_2 + x_1}{2}, \frac{y_2 + y_1}{2}\right)$$

$$C_1\left(\frac{2 + 0}{2}, \frac{4 + 2}{2}\right)$$ we plugged $x_1 = 0$, $y_1 = 2$, $x_2 = 2$ and $y_2 = 4$ into the equation

$$C_1\left(\frac{2}{2}, \frac{6}{2}\right)$$

$$C_1(1, 3).$$

For the $1m$-by-1m square with the end points of (1, 1) and (2, 2) the center is,

$$C_2\left(\frac{x_2 + x_1}{2}, \frac{y_2 + y_1}{2}\right)$$

$$C_2\left(\frac{2 + 1}{2}, \frac{2 + 1}{2}\right)$$ we plugged $x_1 = 1$, $y_1 = 1$, $x_2 = 2$ and $y_2 = 2$ into the equation

$$C_2\left(\frac{3}{2}, \frac{3}{2}\right)$$

$$C_2(1.5, 1.5).$$

For the $1m$ by $2m$ rectangle with the end points of (0, 0) and (2, 1) the center is,

$$C_3\left(\frac{x_2 + x_1}{2}, \frac{y_2 + y_1}{2}\right)$$

$$C_3\left(\frac{2 + 0}{2}, \frac{1 + 0}{2}\right)$$ we plugged $x_1 = 0$, $y_1 = 0$, $x_2 = 2$ and $y_2 = 1$ into the equation

$$C_3\left(\frac{2}{2}, \frac{1}{2}\right) = (1, 05).$$

For the $1m$ by $3m$ rectangle with the end points of (2, 0) and (3, 3) the center is,

$$C_4\left(\frac{x_2 + x_1}{2}, \frac{y_2 + y_1}{2}\right)$$

$$C_4\left(\frac{3 + 2}{2}, \frac{3 + 0}{2}\right)$$ we plugged $x_1 = 2$, $y_1 = 0$, $x_2 = 3$ and $y_2 = 3$ into the equation

$$C_4\left(\frac{5}{2}, \frac{3}{2}\right) = (2.5, 1.5).$$

For the $1m$ by $4m$ rectangle with the end points of (3, 0) and (4, 4) the center is,

$$C_5\left(\frac{x_2+x_1}{2}, \frac{y_2+y_1}{2}\right)$$

$$C_5\left(\frac{4+3}{2}, \frac{4+0}{2}\right)$$

we plugged $x_1 = 3$, $y_1 = 0$, $x_2 = 4$ and $y_2 = 4$ into the equation

$$C_5\left(\frac{7}{2}, \frac{4}{2}\right) = (3.5, 2).$$

We now can plot all the centers on a *xy*-coordinate, as shown below.

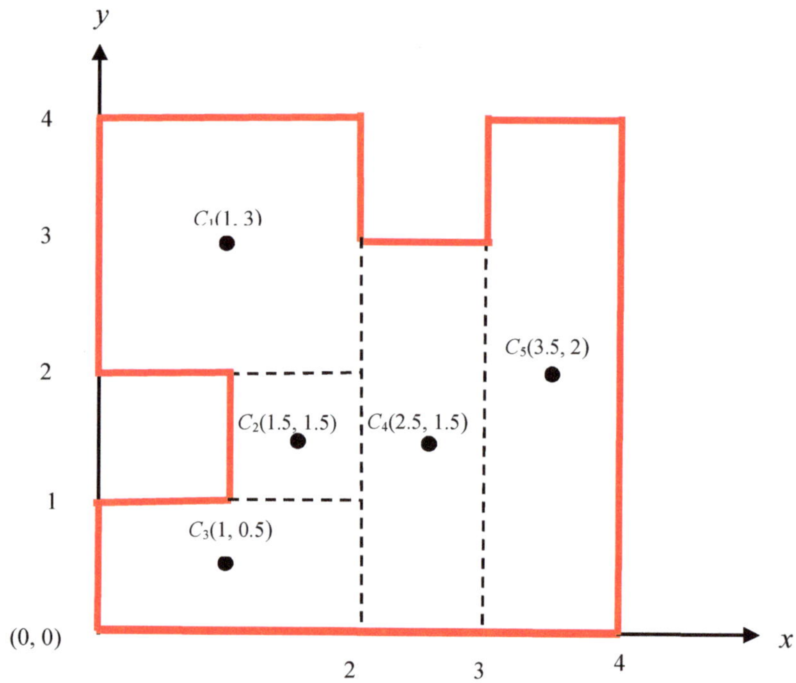

To find the center of mass of the given shape, we just need to determine the area of the divided shapes. Given the area, *A*, of a square with side *s* is $A = s^2$, and the area of the area of rectangles with sides of *w* and *l* is $A = wl$. We will have:

The area for the 2*m*-by-2m square is $A_1 = 2^2 = 4m$.

The area for the 1*m*-by-1m square is $A_2 = 1^2 = 1m$.

The area for the 1*m*-by-2*m* rectangle is $A_3 = 1 \times 2 = 2m$.

The area for the 1*m*-by-3*m* rectangle is $A_4 = 1 \times 3 = 3m$.

The area for the 1*m*-by-4*m* rectangle is $A_5 = 1 \times 4 = 4m$.

Now we can find the COM separately for components along each axis,

For object positions along the *x* axis with $m_1 = 4 \times 2 = 8$, $x_1 = 1$, $m_2 = 1 \times 2 = 2$, $x_2 = 1.5$, $m_3 = 2 \times 2 = 4$, $x_3 = 1$, $m_4 = 3 \times 2 = 6$, $x_4 = 2.5$, $m_5 = 4 \times 2 = 8$, $x_5 = 3.5$, we will have:

$$\text{COM}_x = \frac{m_1x_1 + m_2x_2 + m_3x_3 + \cdots}{m_1 + m_2 + m_3 + \cdots}$$

$$\text{COM}_x = \frac{8 \times 1 + 2 \times 1.5 + 4 \times 1 + 6 \times 2.5 + 8 \times 3.5}{8 + 2 + 4 + 6 + 8}$$

we plugged $m_1 = 8$, $x_1 = 1$, $m_2 = 2$, $x_2 = 1.5$, $m_3 = 4$, $x_3 = 1$, $m_4 = 6$, $x_4 = 2.5$, $m_5 = 8$, $x_5 = 3.5$ into the equation

$$\text{COM}_x = \frac{8 + 3 + 4 + 15 + 28}{28}$$

$$\text{COM}_x = \frac{58}{28} = 2.07.$$

For object positions along the x axis with $m_1 = 8$, $y_1 = 3$, $m_2 = 2$, $y_2 = 1.5$, $m_3 = 4$, $y_3 = 0.5$, $y_4 = 1.5$, $m_4 = 6$, $y_5 = 2$, and $m_5 = 8$, we will have:

$$\text{COM}_y = \frac{m_1 y_1 + m_2 y_2 + m_3 y_3 + \cdots}{m_1 + m_2 + m_3 + \cdots}$$

$$\text{COM}_y = \frac{8 \times 3 + 2 \times 1.5 + 4 \times 0.5 + 6 \times 1.5 + 8 \times 2}{8 + 2 + 4 + 6 + 8}$$

we plugged $m_1 = 8$, $y_1 = 3$, $m_2 = 2$, $y_2 = 1.5$, $m_3 = 4$, $y_3 = 0.5$, $y_4 = 1.5$, $m_4 = 6$, $y_5 = 2$, and $m_5 = 8$ into the equation

$$\text{COM}_y = \frac{24 + 3 + 2 + 9 + 16}{28}$$

$$\text{COM}_y = \frac{54}{28} = 1.93.$$

Therefore, the center of mass of the given shape is ($2.07m$, $1.93m$).

4.5. Example: Tunnel Entrance Coating

A schematic figure of a tunnel entrance surface is drawn on a xy-axis as below. As it is seen the tunnel entrance surface consists of an isosceles trapezoid with 36 feet and the 42 feet bases and the 21 feet height from which a semi-circle with the radius of 18 feet has been cut off.

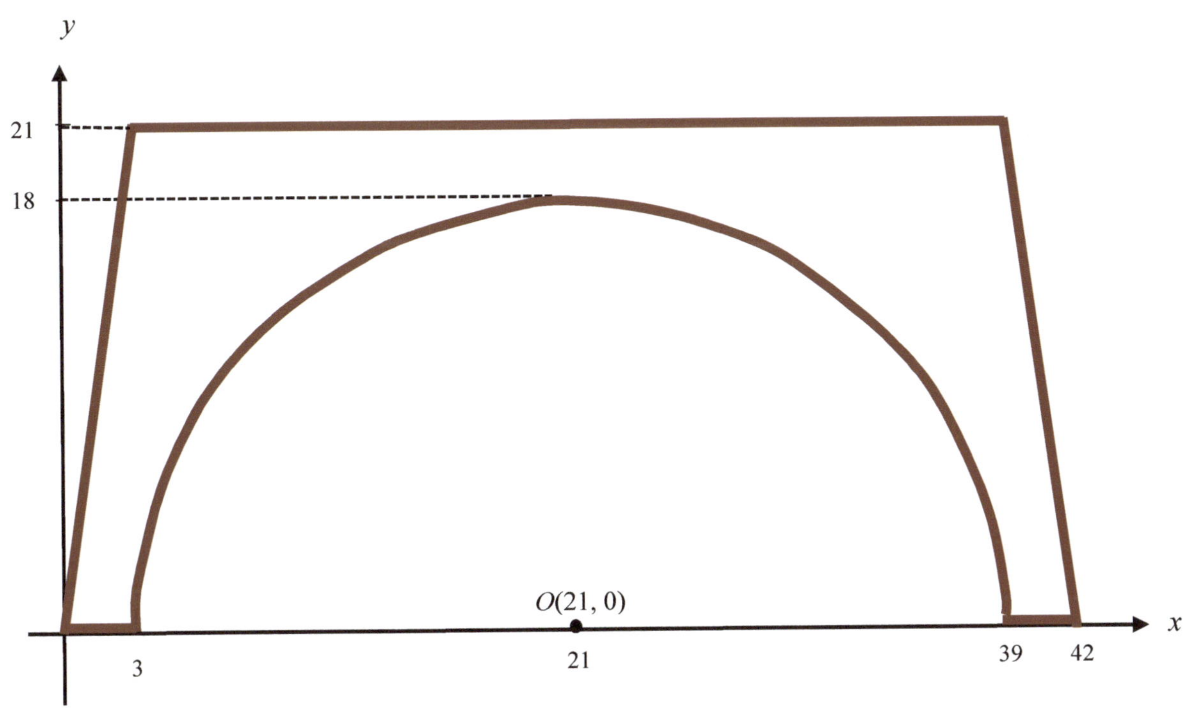

The surface needs to be coated with a special type of anti-freeze and anti-corrosion coating substance. A contractor has been hired for the coating project and their accountant has been asked to estimate the total project's costs.

The accountant consolidated the total project costs as only two types of costs: the cost of material, and the manpower costs. The coating material cost is estimated to be $285 per square foot, and the manpower cost for this project is estimated to be $75 per square foot. Determine the amount that accountant will come up for the total project's costs.
Hint:

- The area, A, of a circle is pi, π, times the radius squared r^2 or $A = \pi r^2$ ($\pi = 3.14$).
- The area, A, of a trapezoid is half of its height, h, times sum of its bases, $a + b$, or $A = \frac{1}{2}h(a + b)$.

Solution
To find the total costs of the coating project, we first need to calculate the surface area of the tunnel entrance.

The surface area of the tunnel can be calculated by dividing its surface to the two sub-surfaces S_1, and S_2, as shown below. As it is seen, S_1 is an isosceles trapezoid with the shorter base of 36 feet and the longer base of 42 feet and the height of 21 feet. The sub-surface S_2 is a semi-circle with a radius of 18 feet. Having these sub-surfaces, the surface area for the tunnel, A_{Tl}, can be obtained by the area of sub-surface of S_1, the area of the trapezoid, A_{Tr}, minus the area of sub-surfaces S_2, the area of the simi-circle, A_{Cr},. Mathematically, $A_{Tl} = A_{Tr} - A_{Cr}$.

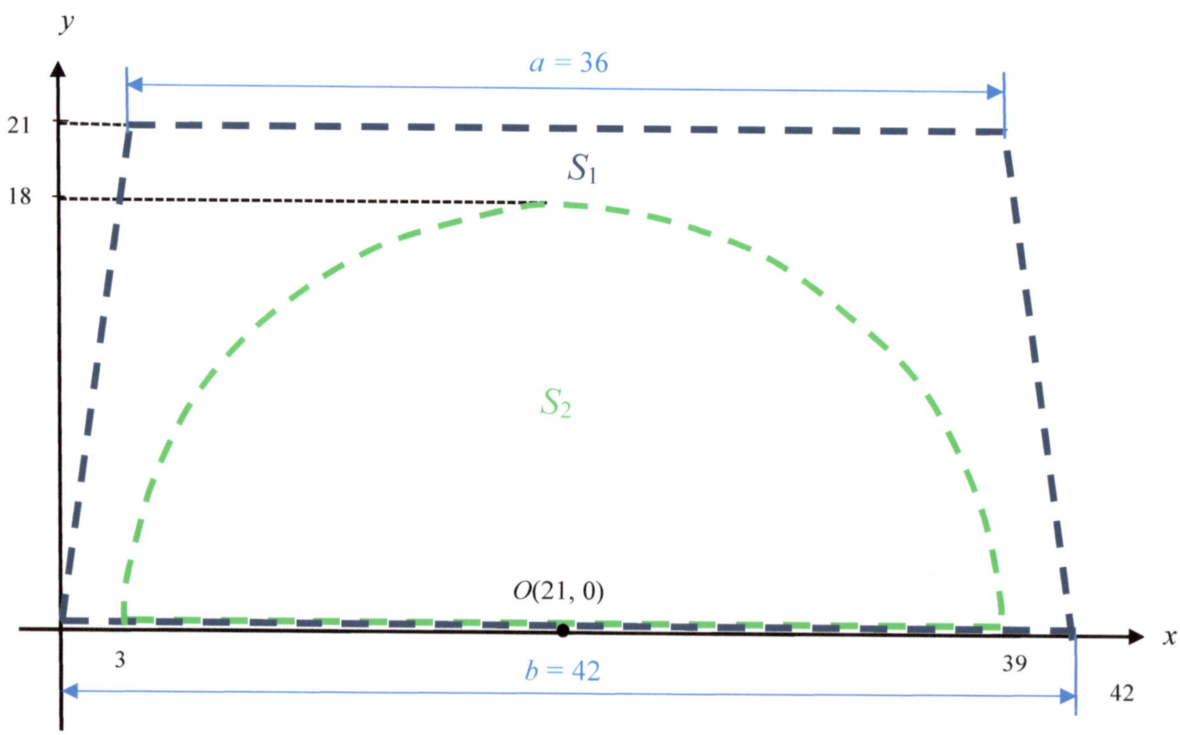

We are given, the area of a circle is pi, π, times the radius squared r^2 or $A = \pi r^2$, hence half of this area will be $A' = \frac{1}{2}\pi r^2$.

Also, the area of a trapezoid is half of its height, h, times sum of its bases, $a + b$, or $A = \frac{1}{2}h(a + b)$. In this case, we have a circle with a radius of 18 feet. So,

$$A_{Cr} = A' = \frac{1}{2}\pi r^2$$

$A_{Cr} = \frac{1}{2} \times 3.14 \times 18^2$ we plugged $\pi = 3.14$ and $r = 18$

$A_{Cr} = \frac{1}{2} \times 3.14 \times 324$

$A_{Cr} = 508.68.$

We have a trapezoid with the bases of $a = 36$ feet and $b = 42$ feet, and the height $h = 21$ feet. So,

$A_{Tr} = \frac{1}{2} h(a + b)$

$A_{Tr} = \frac{1}{2} \times 21(36 + 42)$ we plugged $a = 36$, $b = 42$, and $h = 21$

$A_{Tr} = \frac{21}{2}(78)$

$A_{Tr} = \frac{21 \times 78}{2}$

$A_{Tr} = 21 \times 39$

$A_{Tr} = 819.$

Now, by obtaining the semi-circle area, $A_{Cr} = 508.68$, and the trapezoid $A_{Tr} = 819$. We will have.

$A_{Tl} = A_{Tr} - A_{Cr}.$

$A_{Tl} = 819 - 508.68$ we plugged $A_{Tr} = 819$ and $A_{Cr} = 508.68$ into the equation.

$A_{Tl} = 310.32.$

Therefore, the surface area of the tunnel is $A_{Tl} = 310.32$.

We now need to calculate the total project's costs. Given the material costs of \$285 per square foot and the manpower costs of \$75 per square foot, the total costs for one square foot will be $C_f = \$285 + \$75 = \$360$. Multiplying costs of one square foot by the surface area of the tunnel, we get the total costs of the coating project. So:

The total project's costs = tunnel surface area × costs per square foot.

The total project's costs = $A_{Tl} \times C_f$

The total project's costs = $310.32 ft^2 \times \$360/ft^2$ we plugged $A_{Tl} = 310.32\, ft^2$ and $C_f = \$360/ft^2$ into the equation

The total project's costs = \$111,715.20.

Therefore, the total project's costs are \$111,715.20.

4.6. Example: Design of a Wheelchair Ramp

Mr. Carlson starts working as a life insurance agent. He is planning to convert his house patio to the insurance selling working office. The patio is 18 inches above the ground with three 6" stairs, and he must build a wheelchair ramp for his disabled clients. He has realized that the Americans Disabilities Act (ADA) guidelines, requires that all business and public use ramps adhere to a 1:12 slope ratio, meaning that for every inch of vertical rise there must be 12 horizontal inches of ramp.

Sketch a graph, showing the dimensions, for the wheelchair ramp.

Solution
To find the dimensions of the wheelchair ramp, let us present the graph of the ramp in a *xy*-coordinate system. Here is the preliminary sketch of the wheelchair ram in which we need to find *l* and *d* values.

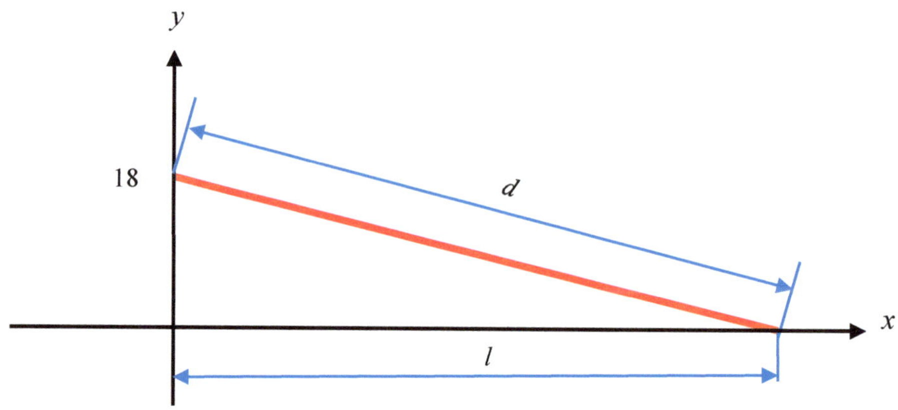

As it is seen, the wheelchair ramp is a straight line in the *xy*-axis. Using the slope-intercept form of a line, $y = mx + b$, in which m is the slope and b is the y intercept, we can find m by the raise over run concept, as it is shown in the figure below.

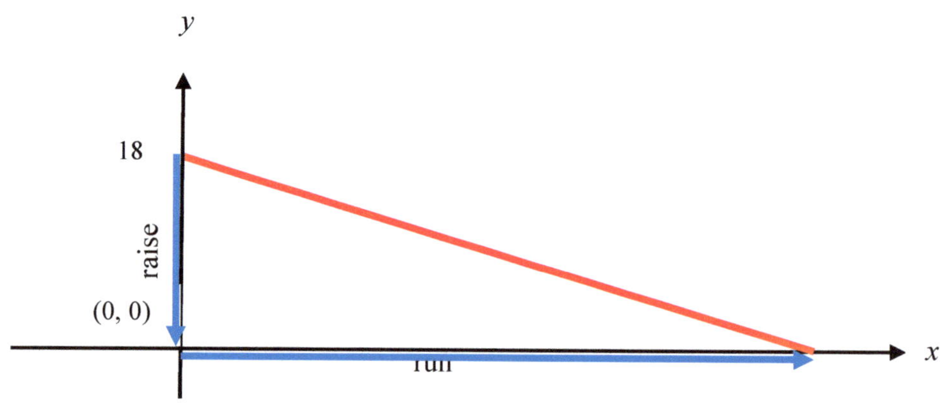

In this figure, to find the length of raise we started from point (0, 18) and moved 18 units down. Hence the length of raise is -18. We next moved from the point (0, 0) and moved l units to the right. Hence the length of run is l. we therefore found $m = \frac{-18}{l} = -\frac{18}{l}$.

The ADA guidelines require 1:12 for the slope ratio for the public ramps. This implies for the raise with the absolute value of 18 we need to have a run that is at least 12 times 18 or run $= l = 12 \times 18 = 216$ inches and $m = -\frac{18}{l} = -\frac{18}{216} = -\frac{18}{12\times18} = -\frac{1}{12}$. In this case it is interesting that the end point of run is the y intercept of the line, and thus, the y intercept is $b(216, 0)$.

Now we write the line equation using the slope-intercept form of a line, $y = mx + b$, so,

$$y = mx + b$$

$y = -\frac{1}{12}x + 216$ we plugged $m = -\frac{1}{12}$, and $b = 216$ into the equation.

To find the length of the ramp, d, we have the two-end point (0, 18) and (216, 0) of the line segment, and hence we can use the distance between tow pints formula, $d = \sqrt{(x_1 - x_0)^2 + (y_1 - y_0)^2}$. For $x_0 = 0$, $y_0 = 18$, $x_1 = 216$, and $y_1 = 0$, we will have:

$$d = \sqrt{(x_1 - x_0)^2 + (y_1 - y_0)^2}$$

$$d = \sqrt{(216 - 0)^2 + (0 - 18)^2}$$ we plugged $x_0 = 0$, $y_0 = 18$, $x_1 = 216$, and $y_1 = 0$ into the equation.

$$d = \sqrt{(216)^2 + (-18)^2}$$

$$d = \sqrt{46{,}656 + 324}$$

$$d = \sqrt{46{,}980}$$

$$d = 216.75.$$

Therefore, the length of the wheelchair ramp must be at least 216.75". We know that 1 foot is 12 inches. Hence the length of the ramp is 216.75" divided by 12 or 18.06' which can be rounded up to 18.1'.

Finally, we can present the graph of the wheelchair ramp, showing the dimensions for the wheelchair ramp. Here is the graph of the wheelchair ramp.

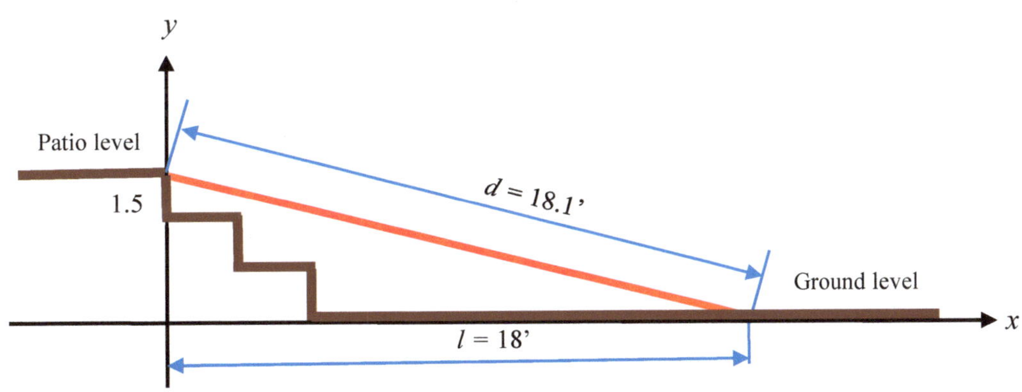

4.7. Example: Home Décor Wallpaper

Jenny is redesigning one of the wall's decorations in her house. She selected one of the highest quality papers to ensure a vivid print with a finishing cost of $35 per square foot. The figure below shows the dimensions of the wall in a *xy*-coordinate in which the curvature is a half of an ellipse with its major and minor axes crossing each other at the center point $P(7.5, 6)$. Calculate the total cost of her project.

H*int:* The area of rectangle = length × width. Also, the area, A, of an ellipse can be found by $A = \pi ab$, where a is the distance from the center to a vertex (the semi major axis of the ellipse), b is the distance from the center to a co-vertex (the semi minor axis of the ellipse) and $\pi = 3.14$.

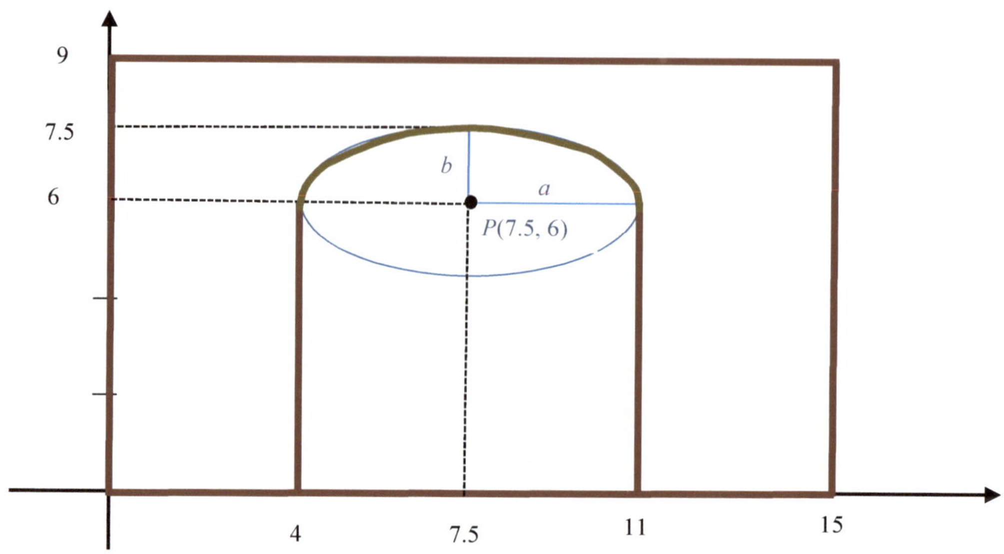

Solution
To calculate the total cost of Jenny's project, we first need to find the surface area of the wall. The area of the wall can be calculated by dividing its surface to the sub-surfaces S_1, S_2, and S_3, as shown below.

As it is seen, S_1 and S_3 are two 4 feet by 9 feet rectangles and S_3 is 3 feet by 7 feet rectangle from which half of an ellipse has been cut out. Having these three surfaces, the area of the wall is the sum of the areas of S_1, S_2, and S_3.

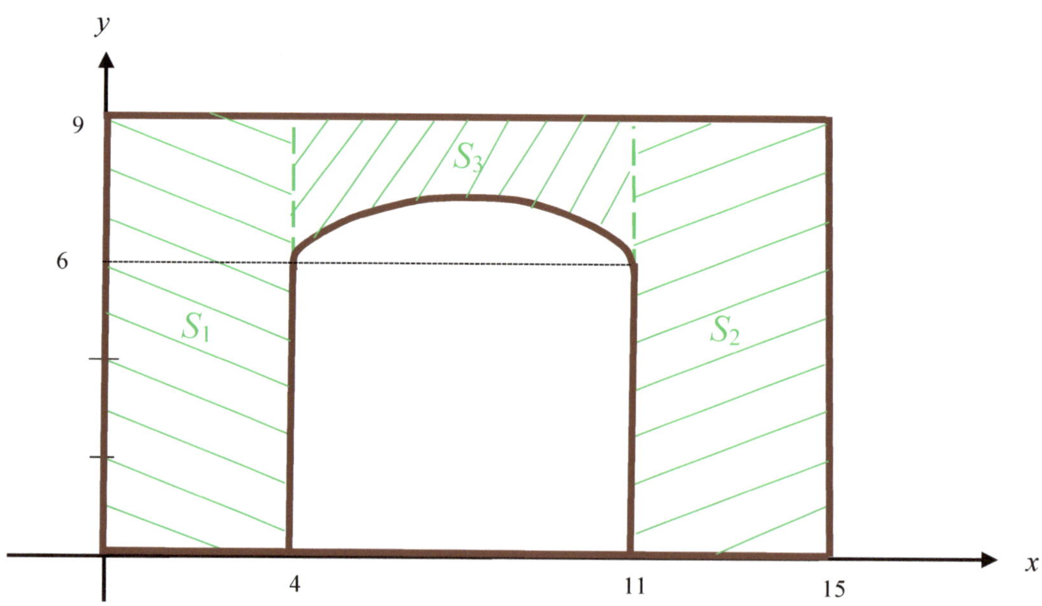

Given the area of a rectangle is the product of its length and its width, and the area of an ellipse is equal to the pi times its semi major axis times, a, its semi minor axis, b, we can calculate the area of the wall. Referring to the wall figure, it is seen that the semi major axis a is a line segment with two end points of (7.5, 6) and (11, 6). Also, semi minor axis b is a line segment with two end points of (7.5, 6) and (7.5, 7.5). We simply can find the length of a, and the length of b using the distance formula between two points, $d = \sqrt{(x_1 - x_0)^2 + (y_1 - y_0)^2}$. So,

To find *a,* we will have:

$$d = \sqrt{(x_1 - x_0)^2 + (y_1 - y_0)^2}$$

$$a = \sqrt{(11 - 7.5)^2 + (6 - 6)^2}$$

we plugged $x_1 = 11, x_0 = 7.5, y_1 = 6$, and $y_0 = 6$ into the equation

$$a = \sqrt{(3.5)^2 + (0)^2}$$

$$a = \sqrt{(3.5)^2}$$

$$a = \pm 3.5.$$

Since a is the length of a line segment, it cannot be negative, and hence a is equal to 3.5.

To find *b,* we will have:

$$d = \sqrt{(x_1 - x_0)^2 + (y_1 - y_0)^2}$$

$$b = \sqrt{(7.5 - 7.5)^2 + (7.5 - 6)^2}$$

we plugged $x_1 = 5.5, x_0 = 7.5, y_1 = 7.5$, and $y_0 = 6$ into the equation

$$b = \sqrt{(0)^2 + (1.5)^2}$$

$$b = \sqrt{(1.5)^2}$$

$$b = \pm 1.5.$$

Since b is the length of a line segment, it cannot be negative, and hence $b = 1.5$.

Now, we can find the area of the wall.

The area of S_1 = length × width = 9 × 4 = 36 square feet

The area of S_2 = length × width = 9 × 4 = 36 square feet

The area of S_3 = the area of 3 feet by 7 feet rectangle – one half of the area of the ellipse

The area of $S_3 = 7 \times 3 - \frac{1}{2}\pi\,(3.5)(1.5)$

The area of $S_3 = 21 - \frac{1}{2}(3.14)(3.5)(1.5)$

The area of $S_3 = 21 - 8.2425$

The area of $S_3 = 12.7575$

The area of the wall = area of S_1 + area of S_2 + area of S_3

The area of the wall = 36 + 36 + 12.7575 = $84.7575 ft^2$.

To calculate the total cost of Jenny's project, we need to multiply the area by the unit cost of the wallpaper installation. So,

The total cost of Jenny's project = the wall area × the unit cost of the wallpaper installation

The total cost of Jenny's project = $84.7575 ft^2 \times \$35/ft^2$

The total cost of Jenny's project = \$2,966.51.

Therefore, Jenny must spend \$2,966.51 for the wallpaper installation.

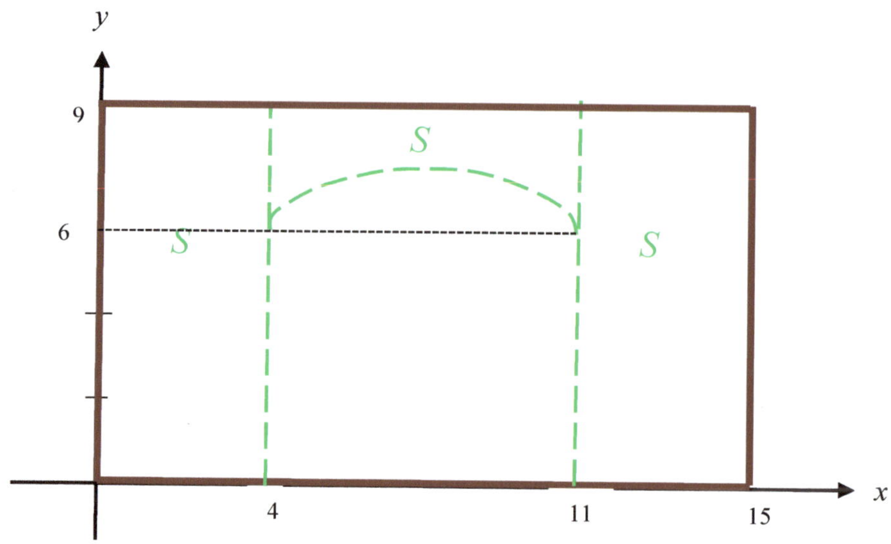

4.8. Example: Hanging a Painting Frame

Susan has recently purchased a framed Picasso painting. She wants to find a proper position for hanging this 9 square feet frame (3 by 3) between two other existing frames on her gust room wall. To find a correct spot, she drew the sketch of the place of the existing frames on a *xy*-axis as shown below. As it is seen she drew tow rectangles *ABCD* and *EFGH* to represent them. Then she drew two lines segment *BG* and *DE* and is planning to locate the center of the new frame on the intersection of these two tines *O*(*x*, *y*). Find the *xy*-coordinate of the two-line segments.

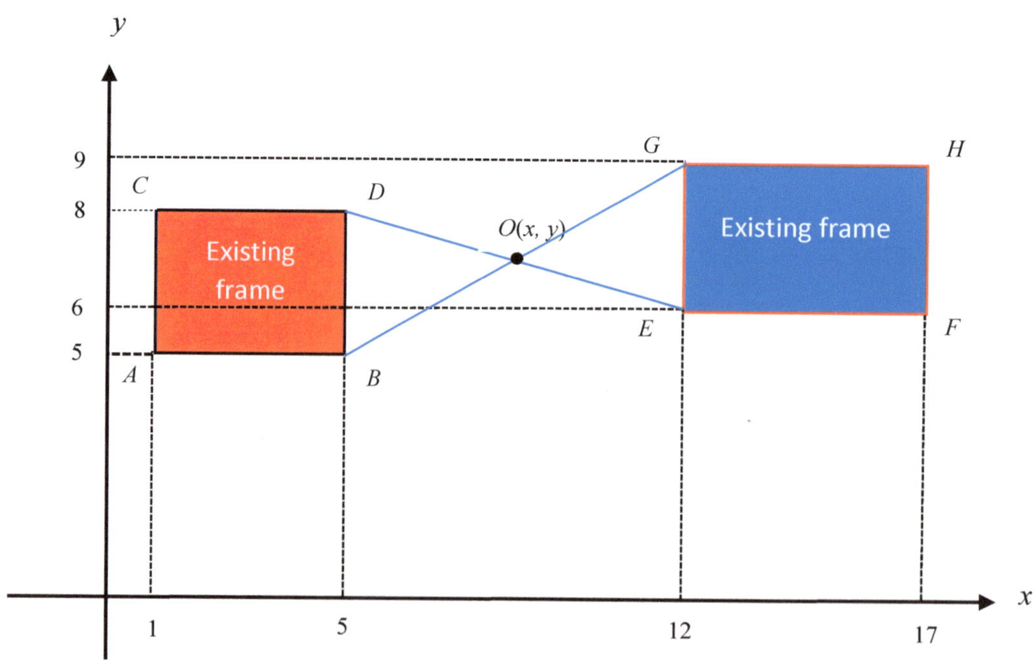

Solution

Given the two-line segments *BG* and *DE*, we can take their end-point coordinates from the above figure. Letting x_B and y_B to be the coordinates of the point *B* and x_G and y_G to be the coordinates of the point *G*, we will have $x_B = 5$, $y_B = 5$, $x_G = 12$, and $y_G = 9$. Similarly, letting x_D and y_D to be the coordinates of the point D and x_E and y_E to be the coordinates of the point E, we will have $x_D = 5$, $y_D = 8$, $x_E = 12$, and $y_E = 6$.

Now there are two approaches to find the coordinate of $O(x, y)$.

Approach 1:

In this approach we first need to find the equations for the two lines BG and DE and then we can find the point of their intersection. We have learned the equation of the line passing through two points (x_B, y_B) and (x_G, y_G) is given by, $y - y_G = \frac{y_B - y_G}{x_B - x_G}(x - x_G)$. So,

For the line BG with the two points having the coordinates $x_B = 5$, $y_B = 5$, $x_G = 12$, and $y_G = 9$, we will have:

$$y - y_G = \frac{y_B - y_G}{x_B - x_G}(x - x_G)$$

$$y - 9 = \frac{5-9}{5-12}(x - 12)$$ we plugged $x_B = 5$, $y_B = 5$, $x_G = 12$, and $y_G = 9$ into the equation

$$y - 9 = \frac{-4}{-7}(x - 12)$$

$$y - 9 = \frac{4}{7}(x - 12)$$

$$y - 9 = \frac{4}{7}x - \frac{4 \times 12}{7}$$

$$y - 9 = \frac{4}{7}x - \frac{48}{7}$$

$$y - 9 + 9 = \frac{4}{7}x - \frac{48}{7} + 9$$ to isolate y, we added 9 to both sides

$$y = \frac{4}{7}x - \frac{48}{7} + 9$$

$$y = \frac{4}{7}x - \frac{48}{7} + 9 \times \frac{7}{7}$$ to make a common denominator, we multiplied the last term in the right-side by $\frac{7}{7}$

$$y = \frac{4}{7}x - \frac{48}{7} + \frac{63}{7}$$

$$y = \frac{4}{7}x + \frac{-48+63}{7}$$

$$y = \frac{4}{7}x + \frac{15}{7}$$

$$7 \times y = 7\left(\frac{4}{7}x + \frac{15}{7}\right)$$ to convert the fractions to integers, we multiplied both sides by 7

$7y = 4x + 15$.

Therefore, the equation of the line BG is $7y = 4x + 15$.

For the line DE with the two points having the coordinates $x_D = 5$, $y_D = 8$, $x_E = 12$, and $y_E = 6$, we will have:

$$y - y_E = \frac{y_D - y_E}{x_D - x_E}(x - x_E)$$

$$y - 6 = \frac{8-6}{5-12}(x - 12)$$ we plugged $x_D = 5$, $y_D = 8$, $x_E = 12$, and $y_E = 6$ into the equation

$$y - 6 = \frac{2}{-7}(x - 12)$$

$$y - 6 = -\frac{2}{7}(x - 12)$$

$$y - 6 = -\frac{2}{7}x + \frac{2 \times 12}{7}$$

$$y - 6 = -\frac{2}{7}x + \frac{24}{7}$$

$y - 6 + 6 = -\frac{2}{7}x + \frac{24}{7} + 6$ to isolate y, we added 6 to both sides

$y = -\frac{2}{7}x + \frac{24}{7} + 6$

$y = -\frac{2}{7}x + \frac{24}{7} + 6 \times \frac{7}{7}$ to make a common denominator, we multiplied the last term in the right-side by $\frac{7}{7}$

$y = -\frac{2}{7}x + \frac{24}{7} + \frac{42}{7}$

$y = -\frac{2}{7}x + \frac{24 + 42}{7}$

$y = -\frac{2}{7}x + \frac{66}{7}$

$7 \times y = 7(-\frac{2}{7}x + \frac{66}{7})$ to convert the fractions to integers, we multiplied both sides by 7

$7y = -2x + 66.$

Therefore, the equation of the line BG is $7y = -2x + 66$.

We also learned the point of intersection, (x, y), of two lines $a_1x_1 + b_1y_1 + c_1 = 0$ and $a_2x_2 + b_2y_2 + c_2 = 0$ is obtained by $(x, y) = \left(\frac{b_1c_2 - b_2c_1}{a_1b_2 - a_2b_1}, \frac{a_2c_1 - a_1c_2}{a_1b_2 - a_2b_1}\right)$. In this case we are given $7y = 4x + 15$ and $7y = -2x + 66$, or $4x - 7y + 15 = 0$ and $-2x - 7y + 66 = 0$ with $a_1 = 4$ $b_1 = -7$, $c_1 = 15$, $a_2 = -2$ $b_2 = -7$, and $c_2 = 66$, we will have:

$$O(x, y) = \left(\frac{b_1c_2 - b_2c_1}{a_1b_2 - a_2b_1}, \frac{a_2c_1 - a_1c_2}{a_1b_2 - a_2b_1}\right)$$

Example: Home Décor Wallpaper

$O(x, y) = \left(\frac{(-7)(66) - (-7)(15)}{(4)(-7) - (-2)(-7)}, \frac{(-2)(15) - (4)(66)}{(4)(-7) - (-2)(-7)}\right)$ we plugged $a_1 = 4$ $b_1 = -7$, $c_1 = 15$, $a_2 = -2$ $b_2 = -7$, and $c_2 = 66$ into the equation

$O(x, y) = \left(\frac{-462 - (-105)}{-28 - (14)}, \frac{-30 - (264)}{-28 - (14)}\right)$

$O(x, y) = \left(\frac{-462 + 105}{-28 - (14)}, \frac{-30 - (264)}{-28 - (14)}\right)$

$O(x, y) = \left(\frac{-357}{-42}, \frac{-294}{-42}\right)$

$O(x, y) = \left(\frac{357}{42}, \frac{294}{42}\right)$

$O(x, y) = (8.5, 7).$

Approach 2:

This approach is easier than approach 1. Referring to the above figure, we realize that, the point $O(x, y)$ is in the midpoint of the line segment BG or the line segment DE. We have learned given two points, $B(x_1, y_1)$ and $G(x_2, y_2)$, the midpoint M of B and G is $M = (\frac{x_B + x_G}{2}, \frac{y_B + y_G}{2})$. In this case the two points having the coordinates $x_B = 5$, $y_B = 5$, $x_G = 12$, and $y_G = 9$, we will have.

$M = O(x, y) = (\frac{x_B + x_G}{2}, \frac{y_B + y_G}{2})$

$O(x, y) = (\frac{5 + 12}{2}, \frac{5 + 9}{2})$ we plugged $x_B = 5$, $y_B = 5$, $x_G = 12$, and $y_G = 9$ into the equation

$O(x, y) = (\frac{17}{2}, \frac{14}{2})$

$O(x, y) = (8.5, 7)$.

Similarly considering the line segment DE, with $x_D = 5$, $y_D = 8$, $x_E = 12$, and $y_E = 6$, we will have.

$$M = O(x, y) = (\frac{x_D + x_E}{2}, \frac{y_D + y_E}{2})$$

$$O(x, y) = (\frac{5 + 12}{2}, \frac{8 + 6}{2})$$ we plugged $x_D = 5$, $y_D = 8$, $x_E = 12$, and $y_E = 6$ into the equation

$$O(x, y) = (\frac{17}{2}, \frac{14}{2})$$

$$O(x, y) = (8.5, 7).$$

Now that we have found the coordinate of the point $O(x, y)$, we can sketch the new frame on the above xy-axis. Given the new frame is a 3 by 3 rectangle and its center is on $O(x, y) = (8.5, 7)$, by adding half of the length of its width to x-coordinate of the center we find the location for the right side of the frame, and by subtraction half of the length of its width from x-coordinate of the center, we find the location for the left side of the frame.

That is, the frame right side is at $x = 8.5 + \frac{1}{2} \times 3 = 8.5 + 1.5 = 10$, the frame right side is at $x = 8.5 - \frac{1}{2} \times 3 = 8.5 - 1.5 = 7$. Similarly, for the frame upper side and lower side we need to perform the same operations on the y-coordinate of the frame center. That is, the frame upper side is at $y = 7 + \frac{1}{2} \times 3 = 7 + 1.5 = 8.5$, and the frame lower side is at $x = 7 - \frac{1}{2} \times 3 = 7 - 1.5 = 5.5$.

Here is the figure illustration of the location of the new frame.

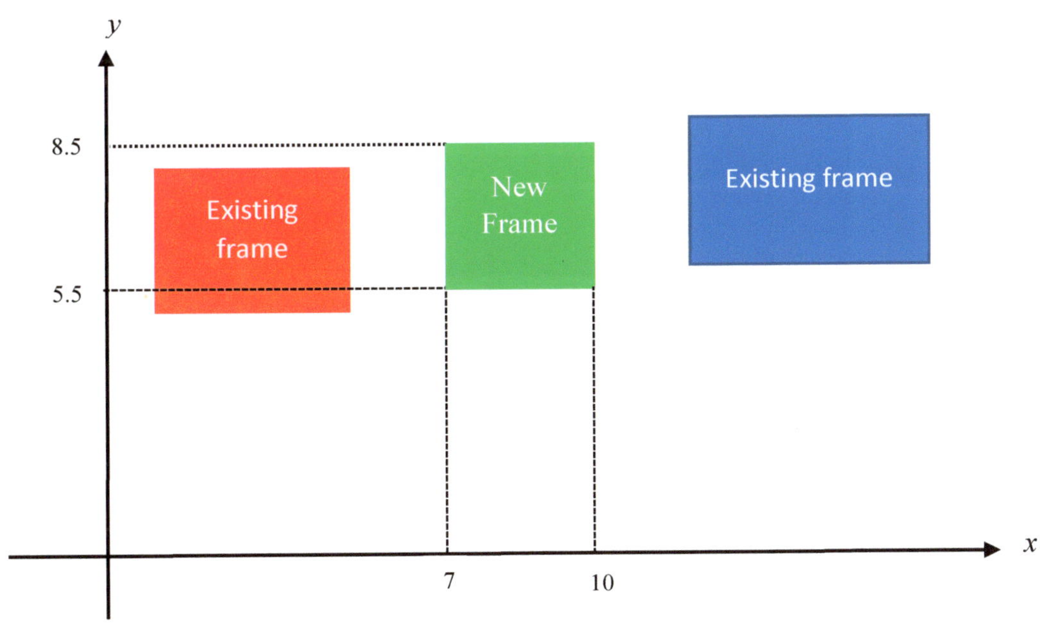

4.8. Exercises

1. Sheldon has a checking account and uses checks to pay his apartment monthly rent. On January 1st, the balance on his checking account was \$15,000. By the end of September, Sheldon's account statement showed that his tenant cashed 9 checks and the initial balance of his account has reduced to the current balance of \$7,800. Assuming there was no other transaction on the Sheldon's checking account during the above period:
   a) Graph his 9-month account transactions on an *xy*-axis.
   b) Determine the rate of his monthly rent.

2. For every kilometer descended into the ocean, the water pressure is approximately 0.72 tons per square inch.
   a) Find the water pressure for points 3, 5, 7, and 9 kilometers below sea level.
   b) Plot the ordered pairs (pressure, sea level depth) for the above points.
   c) Graph the water pressures in terms of sea level depths for the above points on the coordinate plane.

3. A tank with capacity of holding 12 liters of water is empty at 7.00*AM* in June 2021. It is filled up by a tap at 1.00*PM* on the same day. The tank is emptied by the other taps in 10 hours in the same day.
   a) Graph the tank water levels in terms of time on the coordinate plane.
   b) Determine the rate at which the tank is filled from 7.00*AM* to 1.00*PM*.
   c) Determine the rate at which the tank is emptied in the same day.

4. Suppose that two $1m$ by $1m$ squares and a $2m$ diameter circle are cut from a square piece of plywood which originally had sides of length $4m$ by $6m$, as shown below. If the plywood mass is 3 per square meter, what hat is the center of mass of the resulting shape?

   Hint: the area, A, of squares with side s is $A = s^2$, the area of rectangles with sides of w and l is $A = wl$, and the area of circles with diameter d is $A = \frac{1}{4}\pi d^2$.

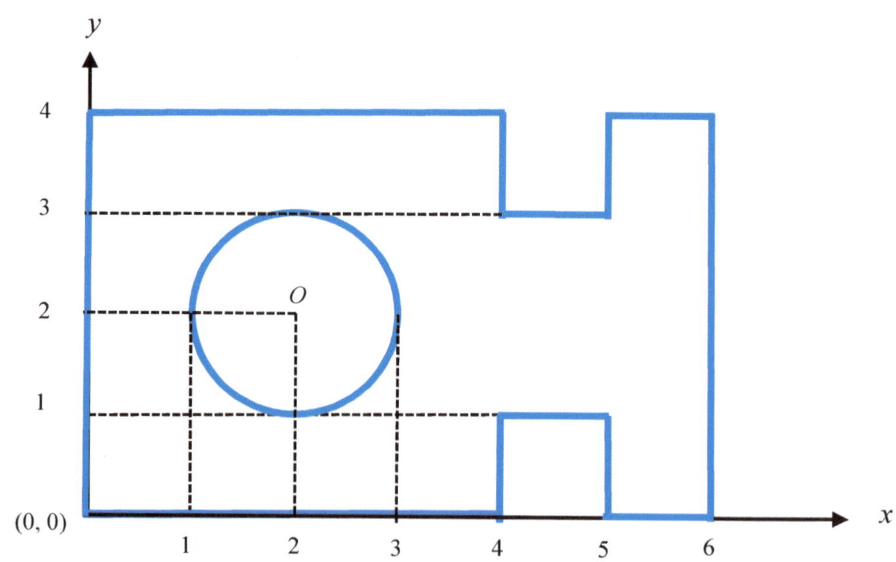

5. Mr. Douglass runs a convenience store in a small town. He has \$480 in his safe on Monday of the first week in March. He paid \$2,170 to the store suppliers plus other expenses on Friday night. The store revenue for selling items was \$3,150 by the end of the week (Sunday night).
   a) Considering the payments as the total expenses of the week, determine the gross profit of the week (profit is the store revenues minus the store expenses).
   b) Determine the safe balance at the end of the week (Sunday night).
   c) Determine the average daily expenses of the store for this week.
   d) Determine the average daily revenue of the store for this week.
   e) Determine the average daily profit of the store for this week.

6. A horseshoe tunnel entrance surface is drawn on a xy-axis as below. As it is seen, the tunnel entrance surface consists of an isosceles trapezoid with the 36 feet and the 42 feet bases and the 22 feet height from which a 24 by 11 rectangle and half of an ellipse with the semi major axis of $r_1 = 12$ feet and the semi minor axis of $r_2 = 8$ has been cut off from its surface.

7. The surface needs to be coated with a special type of coating material. The coating material cost is estimated to be \$190 per square foot and the manpower cost for this project is estimated to be \$80 per square foot. Determine the total costs for the surface coating.

Hint: The area of a trapezoid is $A = \frac{1}{2}h(a + b)$, where a and b are its bass' lengths and h is its height. The area of a rectangles with sides of w and l is $A = wl$. The area of an ellipse is $A = \pi r_1 r_2$, where r_1 and r_2 are the semi major axis and semi minor axis lengths of the ellipse, respectively.

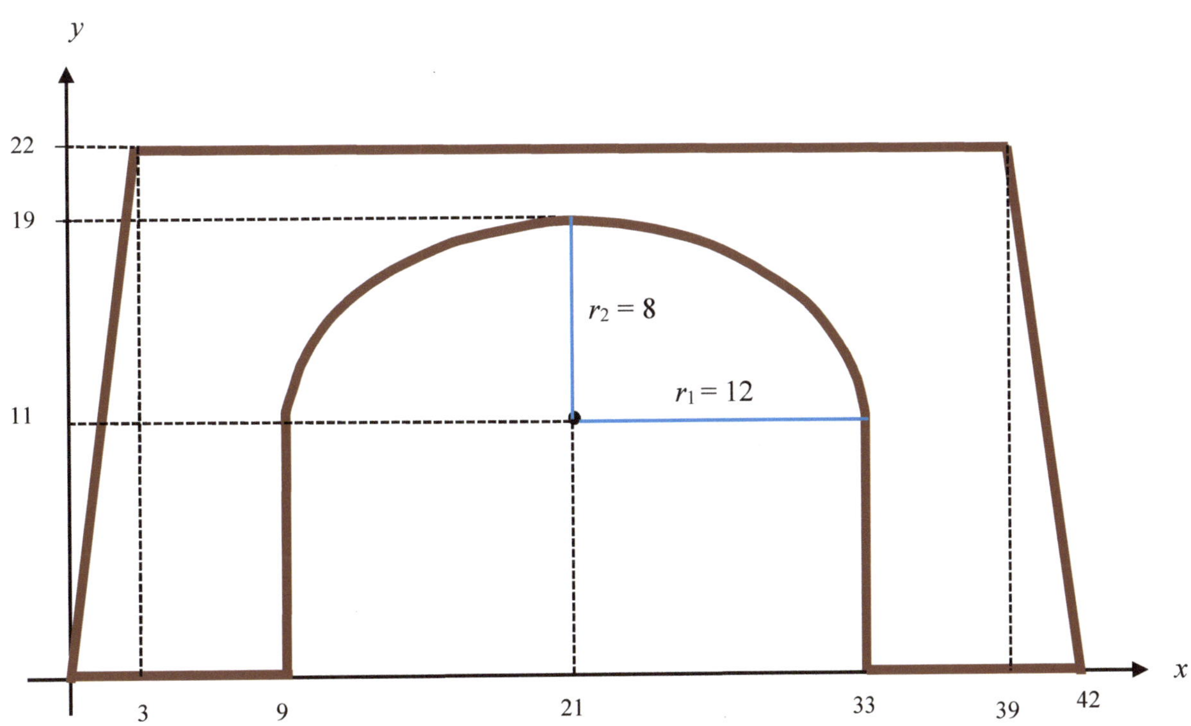

8. Consider the tunnel surface coating project of exercise 6, in which tunnel entrance surface is slightly different. It is graphed as the figure below. As it is seen the tunnel entrance surface consists of 40 feet by 28 feet rectangle from which an isosceles trapezoid with 16 feet and 28 feet bases and the height of 18 feet and on its top a semi-circle with the diameter of 16 feet is cut off. The coating material cost is estimated to be \$210 per square foot and the manpower cost for this project is estimated to be \$74 per square foot. Determine the total costs for the surface coating.

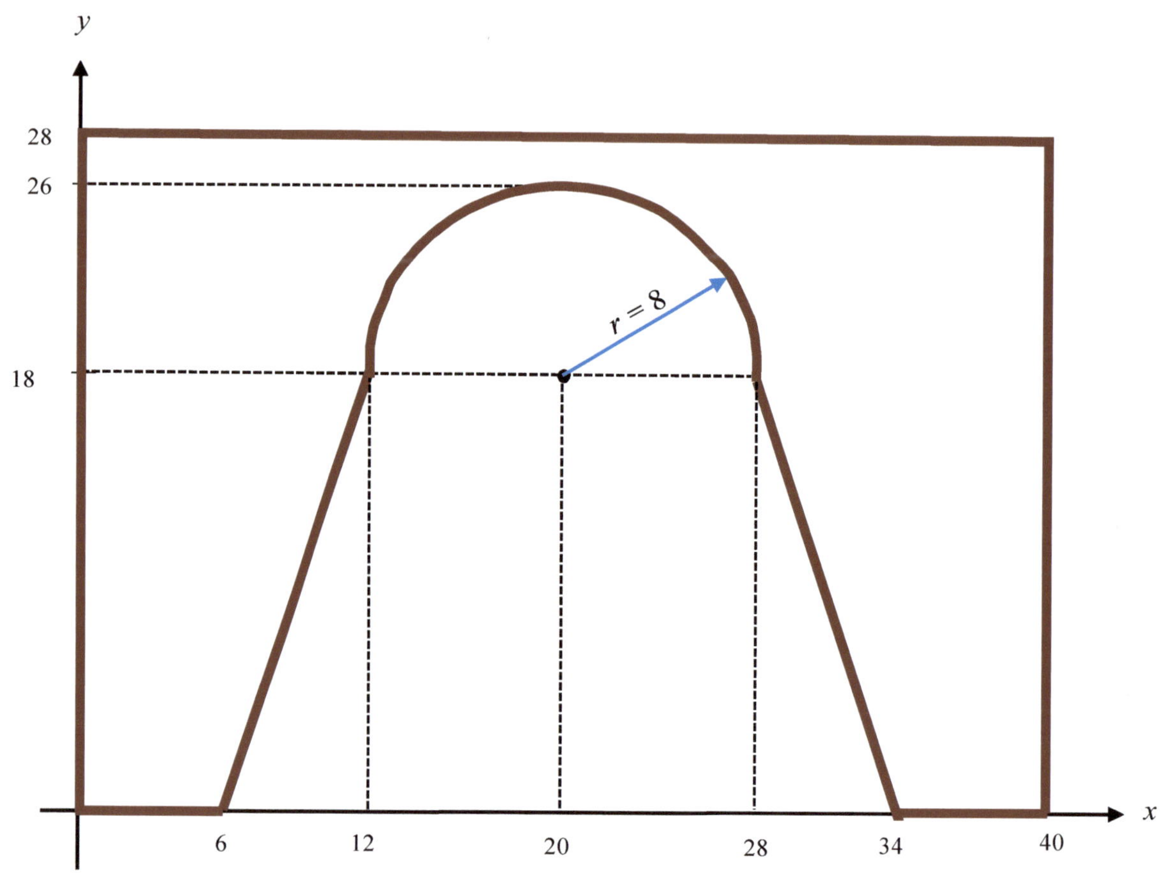

9. It is known that the boiling point of water is 100 degrees centigrade at sea level. It is also known at higher elevations, water boils at a lower temperature. The density of the air will cause that boiling point to decrease as the altitude increases. The table below illustrates the relation between the above sea altitude and the water boiling temperature.

Altitude (m)	Temperature (°C)
0	100
300	99
600	98
900	97
1200	96
1500	95
1800	94
2100	93

2400	92
2700	91
3000	90

a) Find various slopes using the above data points to verify that the relationship between the altitude and the temperature seems to be linear.

b) Find an equation relating the two variables using the two-point formula.

c) Find the slope-intercept form of the line equation relating the two variables.

10. Two rectangular painting frames, one with 2 feet by 3 feet sides and the other with 2 feet by 3.5 feet sides are hung on a wall as illustrated on a *xy*-axis below. A third new oval painting frame with the major axis of 3 feet and minor axis of 2 feet must be hung between these two existing frames, where its center to be located at point $C(x, y)$. Find x, and y.

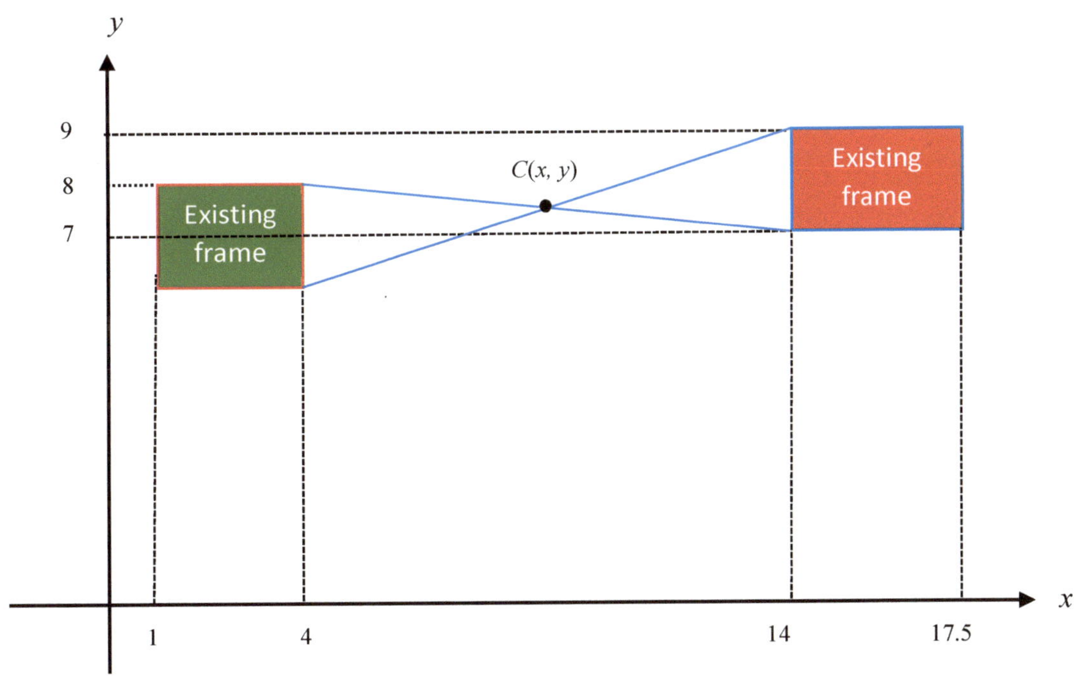

11. You have two coupons that could be used on an item. One is for $25 off. Another is for 15 percent off. They could be applied in either order. How could an equation be written to express these situations, and which would be the best way to apply the coupons?

12. Michael is planning to build a wheelchair ramp for his newly purchased office building. The side-way in front of the building is 30 inches high. He wants to comply with the Americans Disabilities Act (ADA) guidelines, which requires that all business and public use ramps adhere to a 1:12 slope ratio. Using the exact ratio requirement, sketch a graph, showing the dimensions, for the wheelchair ramp.

13. Sue is getting a new carpet for the guestroom of her house. The carpet installers charge $1.10 per square foot. The floor plan of the guestroom, 40 feet by 23 feet rectangle, is sketched on a *xy*-coordinates as shown in the figure below. In her guestroom she has a bar formed by 6 feet by 2 feet rectangle with an attached 2 feet radius semicircle, a regular hexagon aquarium with 3 feet sides, and

an isosceles trapezoid stone tile in front of fireplace with the bases of 6 feet and 4 feet and the height of 3 feet. Find the total costs of the floor carpeting.

Hint: the area, A, of a regular hexagon with side s is $A = \frac{3\sqrt{3}}{2}s^2$.

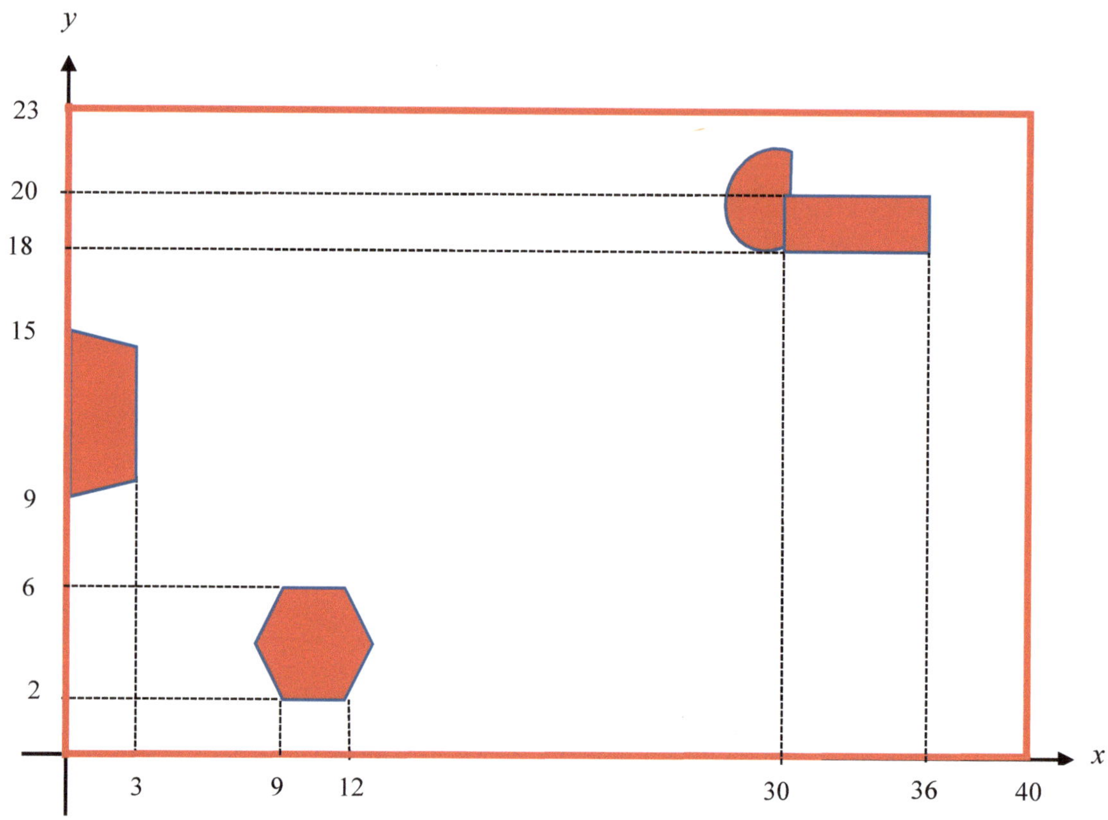

14. Consider exercise 12 above in which Sue has a liquor bar cabinet of 1 foot by 4 feet rectangle attached to another 6 feet by 2 feet rectangle, an oval aquarium with 3 feet semi-major axis and 2 feet minor-semi axis, and an isosceles trapezoid stone tile in front of fireplace with the bases of 6 feet and 4 feet and the height of 3 feet. Find the total costs of the floor carpeting.

15. A business collects some amount of money from sales and pays its expenses out of this revenue. The profit equation, P, is determined by subtracting the cost equation C from the revenue equation R. Mathematically, the profit equation is equation $P= R - C$. Suppose the business is selling an item for \$400. Also, the cost to produce each item is \$250. Additionally, there is a fixed cost of \$3,400 per month, for rent, utilities, equipment maintenance, etc.
    a) Write the cost equation for the month.
    b) Write the revenue equation for the month.
    c) Write the profit equation for the month.
    d) Graph the cost equation for the month.
    e) Graph the revenue equation for the month.
    f) Graph the profit equation for the month.

5. Multiple Choice Questions

1. Line L is parallel to y-axis and passes through the point $P(5, 7)$. What is the slope of L and its x-intercept?

   A. $m = \infty, x = (7, 0)$.
   B. $m = 0, x = (5, 0)$.
   C. $m = \infty, x = (5, 0)$.
   D. $m = 0, x = (7, 0)$.

2. Graph of line L is shown below. What could be its equation?

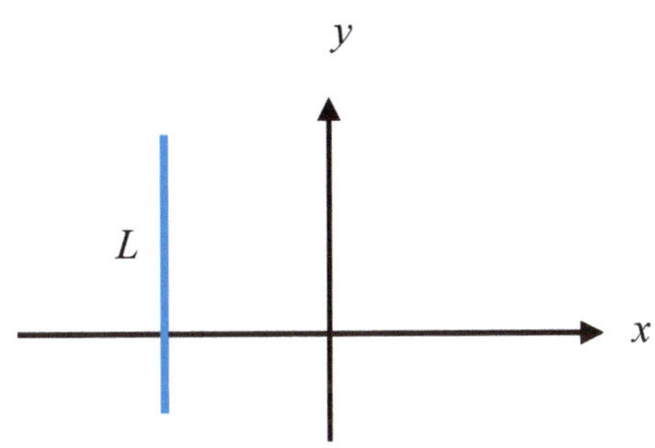

   A. $y = -3 + x$.
   B. $x + 3 = y$.
   C. $x + y = -3$.
   D. $x + y = -3 + y$.

3. Points P and Q are plotted below. Which option represents the point P and the point and Q?

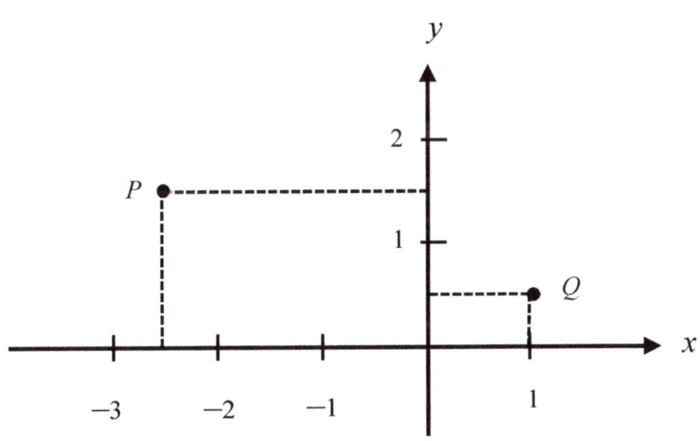

   A. Points $P\left(\frac{5}{2}, -\frac{3}{2}\right)$ and $Q\left(1, -\frac{1}{2}\right)$.
   B. The point $P\left(-\frac{5}{2}, \frac{3}{2}\right)$ and $Q\left(1, \frac{1}{2}\right)$.
   C. The point $P\left(-\frac{5}{2}, -\frac{3}{2}\right)$ and $Q\left(-1, \frac{1}{2}\right)$.
   D. The point $P\left(\frac{5}{2}, \frac{3}{2}\right)$ and $Q\left(-1, -\frac{1}{2}\right)$.

4. Given points $P(-3)$, $Q(3)$, and $R(-1)$ on a line, which statement is correct?

   A. The distance between P and Q is 6, the distance between R and Q is 3.
   B. The distance between P and Q is 6, the distance between R and Q is 4.
   C. The distance between P and Q is 3, the distance between R and Q is 6.
   D. The distance between P and Q is 4, the distance between R and Q is 6.

5. Given points $P(-4)$ and $Q(6)$ on a line, the midpoint of the line segment PQ is.

   A. The midpoint is $M(-2)$.
   B. The midpoint is $M(1)$.
   C. The midpoint is $M(2)$.
   D. The midpoint is $M(-1)$.

6. Line L_1 contains points (6, 5) and (7, 2). If L_2 is perpendicular to line L_1, then which of the following could be the equation of line L_2?

   A. $2x - 6y = 9$.
   B. $\frac{1}{3}x - 5y = 7$.
   C. $2x + 6y = 9$.
   D. $-\frac{1}{3}x - 5y = 7$.

7. Line L_1 contains points (2, 8) and (4, 4). If L_2 is parallel to line L_1, then which of the following could be the equation of line L_2?

   A. $5x + 10y = 13$.
   B. $-2x - 9y = 17$.
   C. $5x - 10y = 13$.
   D. $2x - 9y = 17$.

8. The graph below could represent which of the following equations?

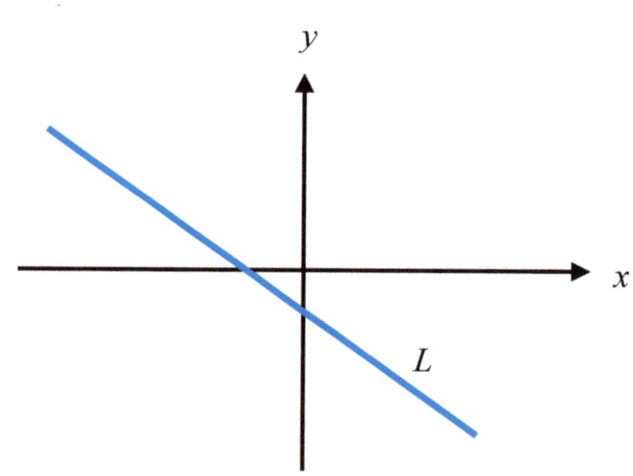

   A. $3x - 4y = -3$
   B. $-3x + 4y = -3$
   C. $3x - 4y = 3$
   D. $-3x - 4y = 3$

9. The figure below shows the straight-line depreciation, D, of a sedan car over the first four years of its use. According to the figure, what is the absolute value of the average rate of change in dollars per year of the value of the car over the four-year period?

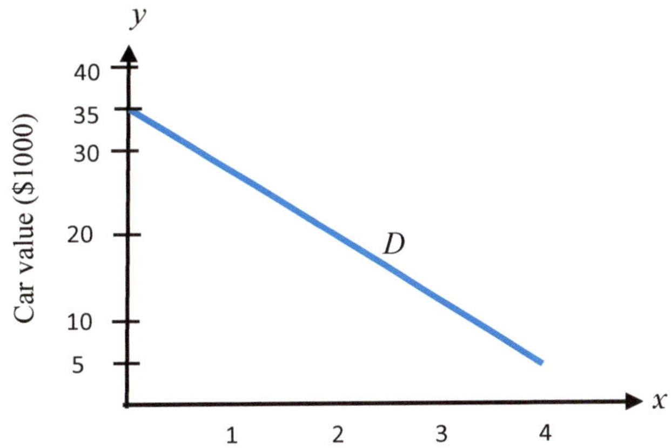

A. $8,500
B. $8,000
C. $7,500
D. $7,000

10. If the equation of the circle shown below is written in the form $(x-a)^2 + (y-b)^2 = r^2$, what is the value of $ab - r$?

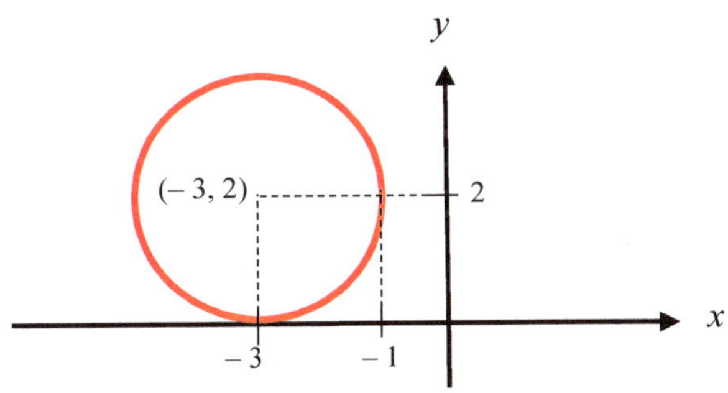

A. – 4
B. – 6
C. – 8
D. – 10

11. The equation of a vertical line, *L*, is $x = 5$, and the equation of a horizontal line, *H*, is $y = 3$, at what point *L* and *H* intersect.

A. (5, 3)
B. (3, 0)
C. (3, 5)
D. (5, 0)

12. Given the ellipse shown below, with the semi major axis r_1 and the semi minor axis r_2 what is the value of $3r_1 - 5r_2$?

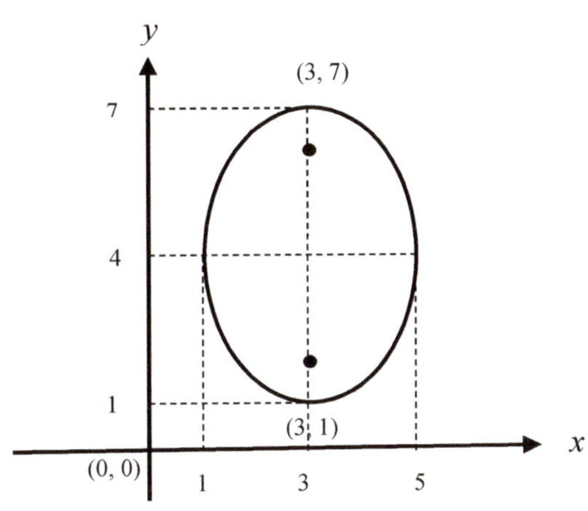

A. – 4.
B. – 3.
C. – 2.
D. – 1.

13. Given the equation of the line $5x + 8y = 24$, find the intercepts.

A. The x-intercept is (0, 3) and the y-intercept is (4.8, 0).
B. The x-intercept is (4.8, 0) and the y-intercept is (0, 3).
C. The x-intercept is (5, 0) and the y-intercept is (0, 8).
D. The x-intercept is (0, 8) and the y-intercept is (5, 0).

14. Given a hyperbola with the graph of $\frac{y^2}{a^2} - \frac{x^2}{b^2} = 1$, then,

A. It is open up and down with vertexes are at $(0, -a)$ and $(0, a)$
B. It is open right and left with vertexes are at $(-a, 0)$ and $(a, 0)$
C. It is open up and down with vertexes are at $(0, -b)$ and $(0, b)$
D. It is open right and left with vertexes are at $(-b, 0)$ and $(b, 0)$

15. Suppose Tony starts a manufacturing company to produce study lamps. His production cost is a linear function C where $C(x)$ is the cost for x lamps produced in each month. The cost Tony incurs is the sum of these two costs, represented by $C(x) = 1420 + ax$, in which he incurs a fixed cost of $1,420 per month for the overhead, which includes his office rent. In this cost function, what would "a" represent?

A. It represents the variable cost for producing a lamp
B. It represents the unit overhead cost.
C. It represents the total costs for production of the lamps in one month
D. It represents the daily costs of overhead and rent.

16. Given the probole $y = 2(x - h)^2 + k$, then,

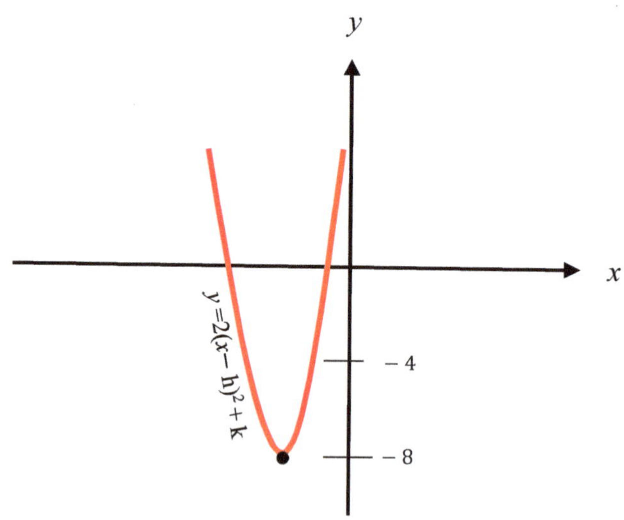

A. It is open upward with $h = 3$ and $k = 8$.
B. It is open upward with $h = -3$ and $k = - 8$).
C. It is open upward with $h = 3$ and $k = -8$.
D. It is open upward with $h = -3$ and $k = 8$

17. What is the slop of the line graphed in the figure below?

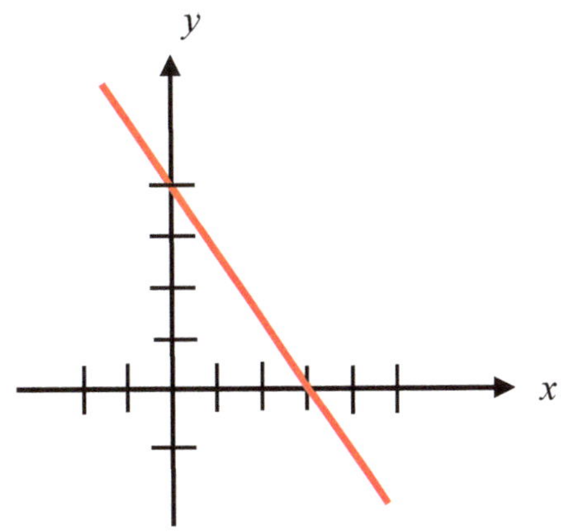

A. $\frac{3}{4}$.

B. $\frac{4}{3}$.

C. $-\frac{3}{4}$.

D. $-\frac{4}{3}$.

18. Suppose the equation of the line in the figure below is written in the form $\frac{y}{x} = m$. Which of the following could be the value of m?

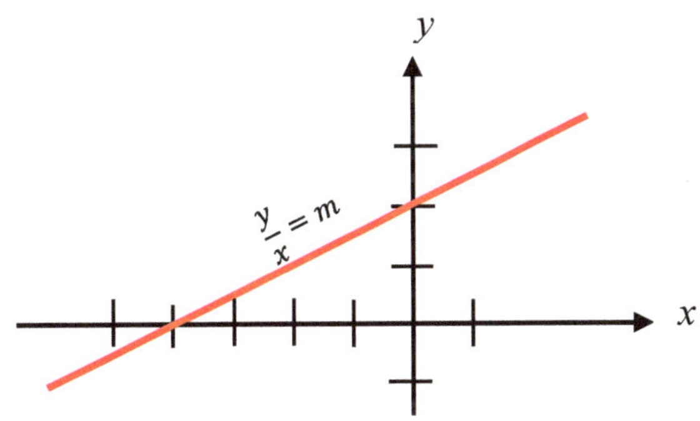

A. $-\frac{1}{2}$.

B. 2.

C. $\frac{1}{2}$.

D. -2.

19. What is the shortest distance between the two parallel lines $y = 3x$ and $y = 3x + 10$?

A. $\frac{10}{\sqrt{3}}$.

B. $0.3\sqrt{10}$.

C. $\frac{0.3}{\sqrt{10}}$.

D. $\sqrt{10}$.

20. The triangle ABC is plotted in xy-axis as shown in figure below. What is the area of the triangle? Hint: use the side AC as the base.

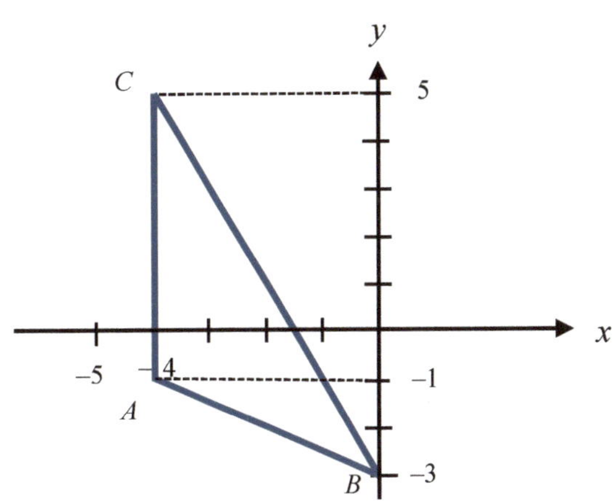

A. 24.

B. 12.

C. 10.

D. 18.

21. What is the shortest distance from the origin, $O(0, 0)$ to the line $x - y - 6 = 0$?

A. $3\sqrt{2}$.

B. $\frac{3}{\sqrt{2}}$.

C. $2\sqrt{3}$.

D. $\frac{2}{\sqrt{3}}$.

22. Suppose we know that the slope of line L_1 is $-\frac{3}{5}$. What are the general equations of lines perpendicular to the given line?

A. $-\frac{3}{5}x + y = c$

B. $\frac{3}{5}x + y = c$

C. $\frac{5}{3}x + y = c$

D. $-\frac{5}{3}x + y = c$

23. Which of the following is a possible equation for a circle that is tangent to both the x-axis and the line $x = 3$?

A. $(x - 3)^2 + (y - 3)^2 = 9$

B. $(x + 3)^2 + (y + 3)^2 = 9$

C. $(x - 6)^2 + (y - 3)^2 = 9$

D. $(x + 3)^2 + (y - 3)^2 = 9$

24. The graph of the line of L_1 is shown below in the xy-plane. The three points $P(0, 3)$, $Q(4a, a)$, and $R(8a, -a)$ are on the line L_1. If a is a positive number, what are the coordinates of point R?

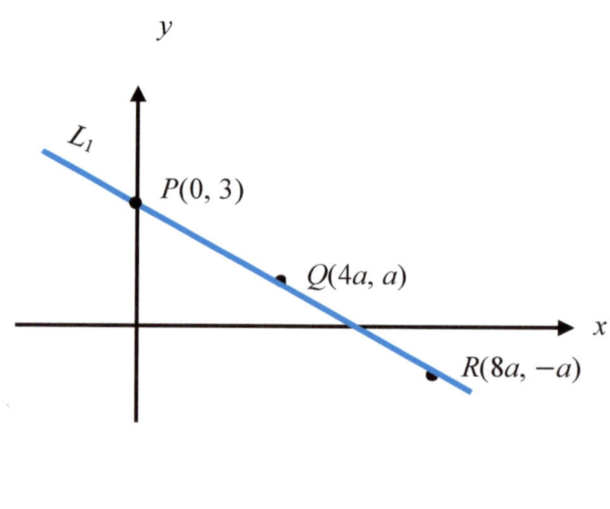

A. $R(4, -0.5)$

B. $R(4, -1)$

C. $R(8, -0.5)$

D. $R(8, -1)$

25. The equation for the line L_1 is written as $y = \frac{2}{3}x + 6$. If line L_2 is perpendicular to this line, which of the following could be the equation of L_2?

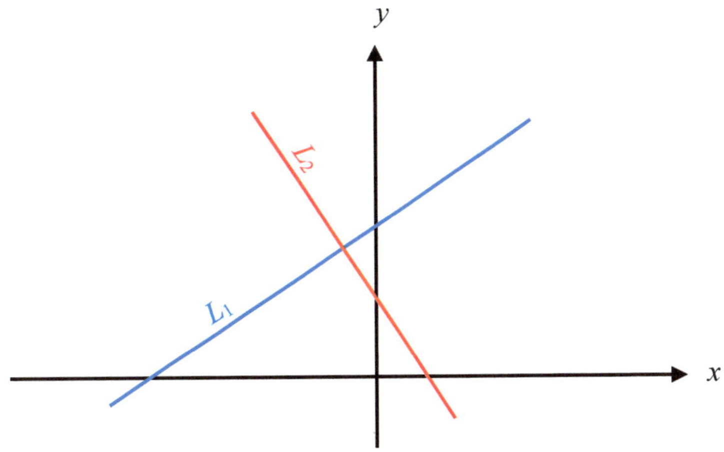

A. $L_2: y = -\frac{3}{2}x + 3$

B. $L_2: y = \frac{3}{2}x + 3$

C. $L_2: y = -\frac{2}{3}x + 3$

D. $L_2: y = \frac{2}{3}x + 3$

26. In the following figure, the graph of line L is shown. Which of the following could be the equation of L?

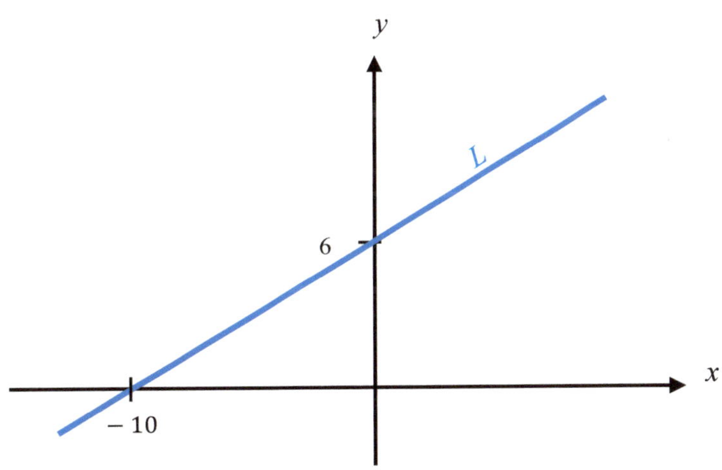

A. $L: y = -\frac{3}{5}x + 6$

B. $L: y = \frac{3}{5}x - 6$

C. $L: y = -\frac{3}{5}x - 6$

D. $L: y = \frac{3}{5}x + 6$

27. The price of an item that costs \$65 in 2020 continually increases by \$5 per year. The current price in dollars, C, of the item can be represented by the equation $C = 5t + 20$, where t is the number of years since the item was first manufactured. Which of the following best describes the meaning of the number 20 in the equation?

A. It is the price of the item in 2013
B. It is the price of the item in 2012
C. It is the price of the item in 2011
D. It is the annual increase in the price of the item

28. If a line that passes through the point $P(k, 3k)$ and $(k + 2, 4)$ has a slope of $m = 1$, what is the value of k?

A. $-\frac{2}{3}$.
B. $\frac{3}{2}$.
C. $-\frac{3}{2}$
D. $\frac{2}{3}$

29. In the figure below, the graph of the line L is shown. Where will line L intersect the x-axis?

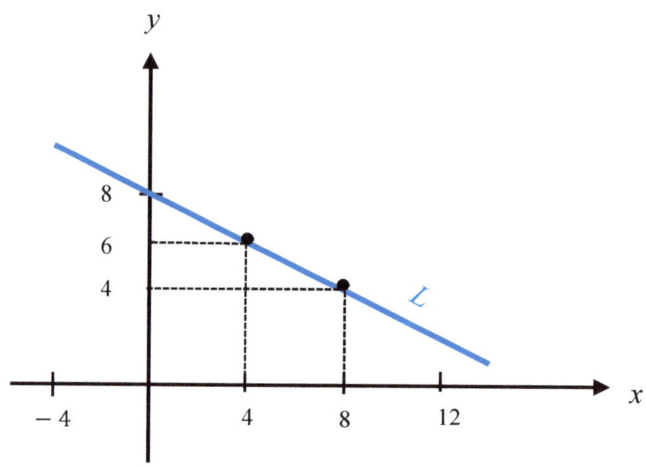

A. 17.
B. 16.
C. 15
D. 14

30. Given the following line equations, which one will not cross the y-axis when graphed on a coordinate plane?

A. $17x - 6(\frac{3}{5}y + 2x) = \frac{7}{15} - 4y.$
B. $12x - 10(\frac{3}{5}y + x) = 13 - 6y.$
C. $11x - 4(\frac{6}{7}y + 9) = -7y.$
D. $19(\frac{11}{38}y + \frac{2}{7}x) = 8x - 3y.$

31. Given the following line equations, which one will not cross the x-axis when graphed on a coordinate plane?

A. $\frac{2}{9}y + 5x - 6(1 + 2x) = 0.$

B. $7x - 5(\frac{3}{5}y + 3x) = 8 - 5y.$

C. $11y - 8(\frac{3}{14}x + 1) = 15y - \frac{12}{7}x + 5.$

D. $9(\frac{5}{7}y + \frac{2}{3}x) = 11x + 3.$

32. The circle shown is given by the equation $(x + 3)^2 + (y - 2)^2 = 16$. What is the shortest distance from point A to the point B?

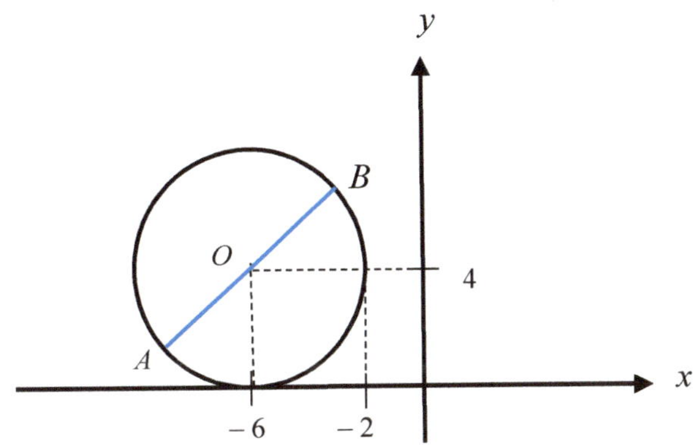

A. 16.
B. 12.
C. 10
D. 8

33. If line L_1, shown below, is reflected graph of the line L_2 over the x-axis, what would be the slope of the line L_2?

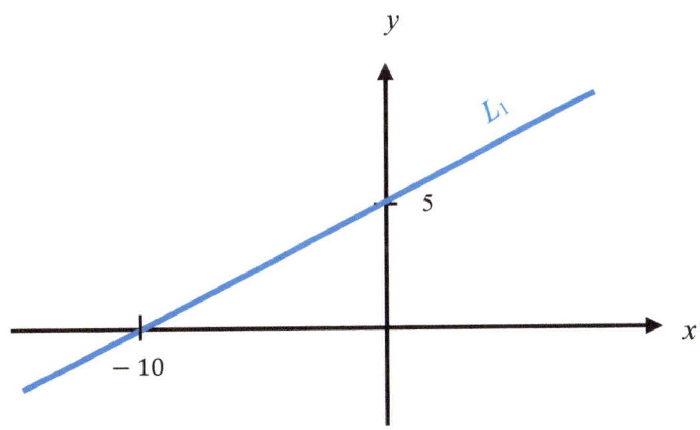

A. $-2.$
B. $\frac{1}{2}.$
C. 2
D. $-\frac{1}{2}.$

34. Which of the following lines contains the point equidistant from the points $P(0, 6)$ and $Q(10, 0)$ in the xy-plane?

A. $y = \frac{1}{2}x - \frac{1}{2}.$

B. $y = \frac{1}{2}x + \frac{1}{2}.$

C. $6y = 5x - 10$

D. $5y = 6x - 10.$

35. Given the trapezoid $ABCD$ shown below, what is its area? Hint: the area of a trapezoid is given by the formula $A = \frac{1}{2}h(a + b)$, where h is the height and a and b are the bases.

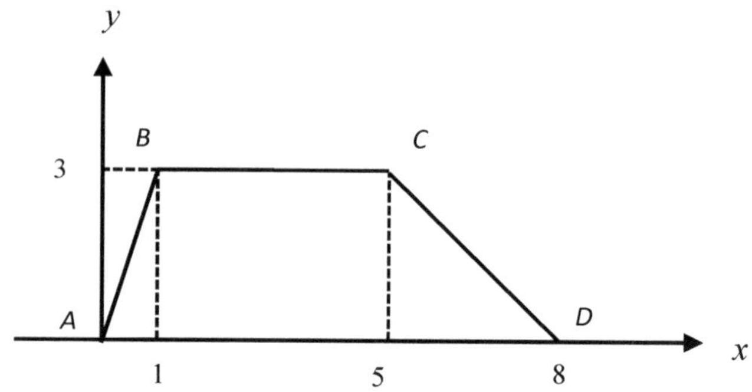

A. 36.
B. 24.
C. 18.
D. 16

36. Which point lies in quadrant III?

A. $(-11,\ 8)$.
B. $(5, -10)$.
C. $(-2, -9)$.
D. $(7,\ 2)$.

37. The equation of a line passing through the origin and perpendicular to the line $5x - 2y + 6 = 0$ is

A. $5x - 2y + 6 = 0$.
B. $2x + 5y + 6 = 0$
C. $5x + 2y + 6 = 0$
D. $2x + 5y = 0$

38. Let P be the point $(-6, 4)$. Find the points which are symmetric to P about the origin.

A. $(-6,\ -4)$.
B. $(6,\ 4)$
C. $(-6,\ 4)$
D. $(6,\ -4)$.

39. What is the equation of the line with a y-intercept of 13 and a slope of 4?

A. $y = 13x + 4$.
B. $y = 4x + 13$.
C. $y = 4x$.
D. $y = 13x$.

40. The slope of the line joining $(5, 6)$, $(9, a)$ is $-\frac{1}{4}$. The value of 'a' is.

A. 4.
B. 5.
C. 6.
D. 7

41. How would the graph of the relation $y = 5x - 8$ change if the 5 and – 8 were both doubled? The graph would be:

A. Less steep and have a higher y-intercept
B. Steeper and have a higher y-intercept
C. Steeper and have a lower y-intercept
D. Less steep and have a lower y-intercept

42. Consider the line $y = \frac{3}{5}x - \frac{3}{5}$. Which of the following statements about the graph of this line is NOT true?

A. The graph cross x-axis at (1, 0).
B. The slop of the line is $\frac{3}{5}$.
C. The graph cross y-axis at $(0, -\frac{3}{5})$.
D. For a raise of 5 the run is 3.

43. The equation of the line that passes through the points P(–1, 4) and Q(2, – 5) is.

A. $y = 3x - 1$.
B. $y = -3x + 1$.
C. $y = 3x + 1$.
D. $y = -3x - 1$.

44. Given the equation of a line, $4y - 3x - 16 = 0$, rearrange it in the form $y = mx + b$.

A. $y = \frac{3}{4}x + 4$.
B. $y = -\frac{3}{4}x + 4$.
C. $y = \frac{3}{4}x - 4$.
D. $y = -\frac{3}{4}x - 4$.

45. If (7, 8), (5, a) and (8, 7) are on a line, then the value of p is.

A. 6.
B. 8.
C. 10.
D. 12

46. If slope of the line AB is $\sqrt{2}$, then slope of the perpendicular bisector of AB is.

A. $\sqrt{2}$
B. $-\frac{1}{2}\sqrt{2}.$
C. $-\sqrt{2}$
D. $\frac{1}{2}\sqrt{2}$

47. The point of intersection of $2x - y = 5$ and $x + y = 7$ is

A. (2, 5).
B. (3, 4).
C. (6, 1).
D. (0, 7).

48. The slope of the line which is perpendicular to a line joining the points (0, 0) and (– 7, 7) is.

A. 1.
B. – 1
C. $\frac{1}{7}.$
D. $-\frac{1}{7}.$

49. Consider three straight lines; (I) L_1: $11y = 12x + 8$, (II) L_2: $12y = 11x - 9$, (III) L_3: $12x + 11y - 7 = 0$, which of the following statement is true?

A. L_1 and L_2 are perpendicular.
B. L_1 and L_3 are perpendicular.
C. L_2 and L_3 are parallel.
D. L_2 and L_3 are perpendicular.

50. What is the length of the line segment that has endpoints at (– 5, 4), (– 8, 0)?

A. $\sqrt{17}.$
B. $\frac{1}{2}\sqrt{17}.$
C. $\frac{1}{2}\sqrt{13}.$
D. 5.

51. A line intersects the x-axis and y-axis at the points P and Q, respectively. If (–3, 2) is the mid - point of PQ, then the coordinates of P and Q are, respectively.

A. (– 3, 0), (0, 2).
B. (6, 0), (0, – 4).
C. (0, 4), (– 6, 0).
D. (– 6, 0), (0, 4).

52. What is the equation of the line passes through the two points $P(3, 0)$ and $Q(-9, 2)$?

A. $3y = 3x - 1$.
B. $3y = -6x + 1$.
C. $6y = x + 3$.
D. $6y = -x + 3$.

53. Given the isosceles trapezoid $ABCD$ shown below, which of the following could be the equation of one of the sides?

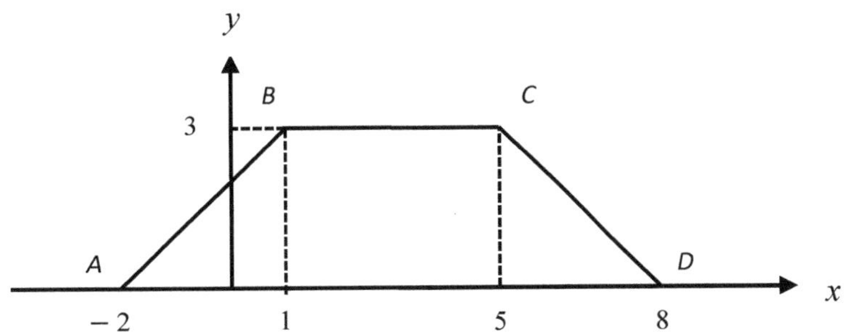

A. $y = x - 8$.
B. $3y = -3x + 2$.
C. $3y = x - 8$.
D. $y = x + 2$.

54. The three vertices of a square are at points (0, 0), (5, 0), and (0, 5). What are the coordinates of the fourth vertex?

A. $(-5, -5)$.
B. $(5, -5)$.
C. $(5, 5)$.
D. $(-5, 5)$.

55. The distance of the point $P(3, 4)$ from the x-axis is.

A. 3.
B. 4.
C. 5.
D. 1.

56. The distance between the points $P(3, 0)$ and $Q(-6, 0)$ is.

A. 9.
B. 8.
C. 7.
D. 3.

57. Given a rectangle $OABC$ whose three vertices are $O(0, 0)$, $A(4, 0)$, and B(0, 3), what is the length of its diagonal?

A. 6.
B. 5.
C. 4.
D. 3.

58. Given a right triangle *OCD*, with $\angle O = 90°$ whose three vertices are *O*(0, 0), *C*(5, 0), and *D*(0, 6), what is the area of the triangle?

A. 18.
B. 16.
C. 15.
D. 12.

59. Given a parallelogram *OABC*, whose three vertices are *O*(0, 0), *A*(4, 0), *B*(2, 3), and *C*(6, 3), what is the area of the parallelogram?

A. 18.
B. 16.
C. 14.
D. 12.

60. Given a rhombus *ABCD*, whose three vertices are *A*(– 1, 2), *B*(4, 0), *C*(9, 2), and *D*(4, 4), what is the area of the rhombus? Hint: the area, *A*, of a rhombus is $A = \frac{1}{2}d_1 d_2$, where d_1 is the length of diagonal 1, and d_2 is the length of diagonal 2.

A. 20.
B. 18.
C. 16.
D. 14.

5.1. Key to the Multiple-Choice Questions

Question No.	Answer	Question No.	Answer	Question No.	Answer
1	C	21	A	41	C
2	D	22	D	42	D
3	B	23	C	43	B
4	B	24	D	44	A
5	B	25	A	45	C
6	A	26	D	46	B
7	A	27	C	47	B
8	D	28	D	48	A
9	C	29	B	49	D
10	C	30	B	50	D
11	C	31	C	51	D
12	D	32	D	52	D
13	B	33	A	53	D
14	A	34	B	54	C
15	A	35	C	55	B
16	B	36	C	56	A
17	D	37	B	57	B
18	C	38	D	58	C
19	D	39	B	59	D
20	B	40	B	60	A

Goodbyes are only for those who love with their eyes.
Because for those who love with heart and soul,
there is no such thing as separation.

Rumi

Chapter 3

Equalities and Inequalities in One Variable

Contents

Introduction

This chapter is dedicated to fundamental principles in algebra, specifically concentrating on the examination of equations and inequalities concerning a solitary variable. Proficiency in solving such equations and inequalities is essential in algebra, serving as the cornerstone of problem-solving abilities. We will explore different types of equations and inequalities involving just one variable, along with strategies to effectively solve them. Furthermore, we'll delve into practical instances demonstrating the application of these concepts through illustrative examples.

Prior to delving into the complexities of solving equations and inequalities, it's vital to comprehend algebraic expressions and the principles governing their operations. By mastering these foundational concepts, we'll enhance our ability to navigate through subsequent topics with clarity and confidence.

Here is a brief listing of the material covered in this chapter:

- Algebraic Expressions
    - Algebraic Expressions with Exponents
    - Rational algebraic Expressions
    - Algebraic Expressions Involving Radicals
    - Algebraic Expressions Involving Logarithms
- Interpolation and Extrapolation
- Linear Equations and Solution Methods
- Linear Inequalities and Solution Methods
- Compounded Inequalities
- Absolute Value Equalities and Solution Methods
- Absolute Value Inequalities and Solution Methods
- Polynomial Equations and Solution Methods
- Polynomial Inequalities and Solution Methods
- Rational Equations and Solution Methods
- Rational Inequalities and Solution Methods
- Radical Equations and Solution Methods
- Radical Inequalities and Solution Methods
- Logarithmic Equations and Solution Methods
- Logarithmic Inequalities and Solution Methods
- Real-Life Applications of Equalities and Inequalities.

1. Algebraic Expressions

An **algebraic expression** is an expression which is made up of variables and constants and coefficient along with algebraic operations (addition, subtraction, etc.) The combination of these three (as terms) is said to be an expression. Expressions are made up of terms. They are also termed algebraic equations. Two of its examples include $8x + 3y - 9$, $3x - 12 = 0$, and $6x^2 - 4xy + 7$. First, we need to define some vocabulary.

First, we need to define some terminologies. These words are especially important since we don't want to confuse terms with factors.

Vocabulary

Variable

A Variable is a symbol for a number we do not know yet. It is usually a letter like x or y.

- Variables are letters that represent an unknown or generic real number.
- A variable could be any number. All you know is that it is some number.
- Sometimes with restrictions, such as a member of a certain set, or the set of values that makes an equation true.
- A letter that stands for a physical quantity such as weight, distance, time, etc.

Constants

A constant is a number on its own, or sometimes a letter like a, b. or c stands for a fixed number.

- Fixed values, like 2, 3 or 9.
- Can also be represented by letters such as: a, b, c, k.

Terms

Terms are parts of the expression that are added or subtracted. A term can be a signed number, a variable, or a constant multiplied by a variable or variables. Each term in an algebraic expression is separated by a "+" sign or "–" sign.

$$5x^2 + 7x - y^2 + 6$$

4 Terms

Factors

Factors are quantities that are multiplied together.

Coefficients

Coefficients are constant factors that multiply a variable or powers of a variable. For example, consider the two middle terms; $7x$ and $-y^2$, the first term, $7x$, has 2 factors, namely 7 and x, and the second term, $-y^2$, has 3 factors, namely -1 and two factors of y. In the latter case, we say that -1 is the coefficient of y^2.

$$5x^2 + 7 \times x - y^2 + 6$$

Two Factors

$$5x^2 + 7x - 3 \times y^2 + 6$$

Three Factors

The last term is a factor all by itself because it is not multiplying anything.

$$5x^2 + 7x - v^2 + 6$$

One Factor

There are 6 main types of algebraic expressions. These are:

1. Monomial expression.
2. Binomial expression.
3. Trinomial expression.
4. Linear polynomial.
5. Quadratic polynomial.
6. Cubic polynomial.

Let us have a look at them one by one:

Monomial Expression–A polynomial with a single term is known as a monomial. Some examples of monomial expressions are: $3x$, $5x^4$, $7xy^2z$, etc.

Binomial Expression – A polynomial with two terms is known as a binomial. Some examples of binomial expressions are: $4x + 9y^3$, $11x^5 - 7y^2$, $15x^2yz^3 + 5$, etc.

Trinomial Expression – A polynomial with three terms is known as a trinomial. Some examples of trinomial expressions are: $6x^2 - 3xy + 14xy^3$, $11x^5 - 18xy^4 + 5x^5yz^2$, $13x^7y^4v^2 + 22 - x$, etc.

Linear Polynomial – A polynomial of degree 1 is called a linear polynomial. In other words, in a linear polynomial, the highest exponent of the variable is one. Some examples of linear polynomials are: $4y - 3$, $12x + 23$, $14u - 45$, etc.

Quadratic Polynomial – A polynomial of degree 2 is called a quadratic polynomial. This means, in a quadratic polynomial, the highest exponent of the variable is two. Some examples of the quadratic polynomial are: $5x^2 + 9x + 24$, $16x^2 - 33$, $19x^2 + 3x - 43$, etc.

Cubic Polynomial – A polynomial of degree 3 is called a cubic polynomial. Here, the highest exponent of the variable is three. Some examples of the cubic polynomial are: $4x^3 + 9x^2 + 24$, $8y^3 + 7y^2 + 9y - 24$, $z^3 + 13z^2 + 27$, $29x^3 + 6x$, etc.

Now that you are familiar with the types of algebraic expressions, let us have a look at how to simplify them.

1.1. Simplifying Algebraic Expressions

By simplifying an algebraic expression, we mean rewriting it in the most compact or efficient form without changing the value of the expression. The simplification is conducted by combining the like terms.

The main rule here is "only like terms can be added together". By combining the like terms, we mean those terms which contain the same variable with the same power but may have different coefficients. Below are some examples of like terms and NOT like terms.

$8x, -5x, 11x, x$	Like terms.
$7x^2, 3x^2, -x^2, x^2$	Like terms.
$k^3t^2, 3k^3t^2, -4k^2t^3, 6k^3t^2$	NOT like terms.
$12xy^3, -2xy^3, xy^3, -xy^3$	Like terms.

$dx, -6x^2, x^2$ NOT like terms.

Like Terms

Two terms are called like terms if they have identical variable parts, which means that the terms must contain the same variables raised to the same exponents.

For combining the like terms, we first need to look at the properties of algebraic expressions. As we presented the properties of real numbers in Chapter 1, here we will look at those properties applied to algebraic expressions.

Definition: Associative Property

The associative property states that you can add or multiply regardless of how the numbers are grouped. By 'grouped' we mean 'how you use parenthesis'. In other words, if you are adding or multiplying, it does not matter where you put the parenthesis.

Associative Property of Addition

Let a, b, and c be any numbers. Then:

$$a + (b + c) = (a + b) + c.$$

Associative Property of Multiplication.

Let a, b, and c be any numbers. Then:

$$a \times (b \times c) = (a \times b) \times c.$$

Example

Simplify, the expression $-5a + (7a - 4b) + 12b$.

Solution

We can use the associative property of addition to regroup.

$-5a + (7a - 4b) + 12b = (-5a + 7a) + (-4b + 12b)$ The associative property of addition

$-5a + (7a - 4b) + 12b = 2a + 8b.$

Therefore, $-5a + (7a - 4b) + 12b = 2a + 8b$.

Example

Simplify, the expression $-5(-4xy^2)$.

Solution

We can use the associative property of multiplication to regroup, first multiplying -5 and -4.

$-5(-4xy^2) = [(-5) \times (-4)]\, xy^2.$ The associative property of multiplication

$-5(-4xy^2) = 20\, xy^2.$

Therefore, $-5(-4xy^2) = 20\, xy^2$.

The Distributive Property

Let a, b, and c be any numbers. Then:

$$a \times (b + c) = (a \times b) + (a \times c).$$

Example

Use the distributive property to expand $3(-4x + 9)$.

Solution

First distribute the 3 times each term in parentheses. Then simplify.

$3(-4x + 9) = 3\times(-4x) + 3\times(9)$ we used the distributive property

$3(-4x + 9) = -12x + 27.$

Therefore, $3(-4x + 9) = -12x + 27$.

Let us now present some examples of simplifying algebraic expressions.

Example

Simplify $-9w^2 + 13w^2$.

Solution

These are like terms. We just add the coefficients -9 and 13, and we get:

$-9w^2 + 13w^2$

$-9w^2 + 13w^2 = (-9 + 13)w^2$

$-9w^2 + 13w^2 = 4w^2.$

Therefore, $-9w^2 + 13w^2 = 4w^2$.

Example

Simplify $-3xy^2 - 11xy^2$.

Solution

These are like terms. We just add the coefficients -3 and -11, and we get.

$-3xy^2 - 11xy^2.$

$-3xy^2 - 11xy^2 = (-3 - 11)\ xy^2$

$-3xy^2 - 11xy^2 = -14\ xy^2.$

Therefore, $-3xy^2 - 11xy^2 = 14\ xy^2$

Example

Simplify $8xu^3y^2 - 5xu^3y^2 + 14xu^3y^2$.

Solution

These are like terms. We just add the coefficients 8, -5 and 14, and we get.

$8xu^3y^2 - 5xu^3y^2 + 14xu^3y^2$

$8xu^3y^2 - 5xu^3y^2 + 14xu^3y^2 = (8 - 5 + 14)\ xu^3y^2$

$8xu^3y^2 - 5xu^3y^2 + 14xu^3y^2 = (17)\ xu^3y^2.$

Therefore, $8xu^3y^2 - 5xu^3y^2 + 14xu^3y^2 = 17xu^3y^2$.

1.2. Algebraic Equations Zero Factor Principle

All that the property says is that if a product of two or more terms is zero, then at least one of the terms had to be zero to start. Notice that this property will only work if the product is equal to zero. Consider the following product:

$$(a_1x_1 + b_1)\,(a_2x_2 + b_2)\,(a_3x_3 + b_3) \cdots (a_nx_n + b_n) = 0.$$

Then,

$$(a_1x_1 + b_1) = 0.$$

$$(a_2x_2 + b_2) = 0.$$

$$(a_3x_3 + b_3) = 0.$$

$$\vdots$$

$$(a_nx_n + b_n) = 0.$$

1.3. Algebraic Expressions with Exponents

Let us present some definitions.

Powers - A product in which the factors are identical is called a **power** of that factor. For instance, $16 = 2 \times 2 \times 2 \times 2$, so 16 is the fourth power of 2.

Exponent - The **exponent** indicates how many times the base is being multiplied by itself. When you use an exponent to write an expression you are using exponential form.

1.3.1. Exponential Expressions Definition

Exponential Expressions are just a way to write powers in short form. The exponent indicates the number of times the base is used as a factor. So, in the case of 16 it can be written as $2 \times 2 \times 2 \times 2 \times 2 = 2^4$, where 2 is the "base" and 4 is the "exponent". We read this expression as "two to the fourth power". In general, we will have that $a^n = a \times a \times a \times \cdots \times a$ (n times), where a denotes the base and n denotes the exponent, and it is interpreted as "a to the nth power."

1.3.2. Exponent Rules

The rules of exponents, also known as the "exponent rules", are some of the rules about algebra that we need to be familiar with. Assuming a and b are nonzero real numbers, the exponent rules along with a simple example are illustrated in table below.

Rule No.	Property	Formula	Example
1	Zero property	$a^0 = 1$	$(x - 3y)^0 = 1$
2	Negative property	$a^{-n} = \dfrac{1}{a^n}$	$u^{-4} = \dfrac{1}{u^4}$
3	Product property	$(a^m)(a^n) = a^{m+n}$	$(k)^3(k^5) = k^8$

4	Quotient property	$\frac{a^m}{a^n} = a^{m-n}$	$\frac{c^9}{c^6} = c^3$
5	Power of power property	$(a^m)^n = a^{mn}$	$(v^2)^7 = v^{14}$
6	Power of product property	$(ab)^n = a^n b^n$	$(xy)^6 = x^6 y^6$
7	Power of quotient property	$\left(\frac{a}{b}\right)^n = \frac{a^n}{b^n}$	$\left(\frac{w}{u}\right)^8 = \frac{w^8}{u^8}$
8	One to one property	$a^S = a^T$ if and only if $S = T$	$x^2 = x^u$, $u = 2$.

Example

Simplify each exponential expression.

a) $5(a^3b^{-2})^4$.

b) $15^4 \times 15^{-6} \times 15^{-3} \times 15^8$.

c) $\frac{3x^4 y^3 z^8}{9x^2 y^7 z^6}$.

d) $(5t^3)^4(2t^2)^5(t^{-7})$.

e) $\frac{(a-2b)^6}{(a-2b)^4(2c)^4}$.

f) $\frac{(abc)^5}{a^2b^7c}$.

Solution

a) $5(a^3b^{-2})^4$.
Using the power of power property, we can rewrite.

$$5(a^3b^{-2})^4 = 5(a^{12}b^{-8})$$
$$5(a^3b^{-2})^4 = 5a^{12}b^{-8}.$$

b) $15^4 \times 15^{-6} \times 15^{-3} \times 15^8$.
The base of terms is 15, so we use the product property to rewrite the expression. So,

$$15^4 \times 15^{-6} \times 15^{-3} \times 15^8 = 15^{(4+(-6)+(-3)+8)}.$$

$$15^4 \times 15^{-6} \times 15^{-3} \times 15^8 = 15^{(4-12-3+8)}.$$

$$15^4 \times 15^{-6} \times 15^{-3} \times 15^8 = 15^{-3}.$$

$$15^4 \times 15^{-6} \times 15^{-3} \times 15^8 = \frac{1}{15^3}.$$ We applied the negative property

$$15^4 \times 15^{-6} \times 15^{-3} \times 15^8 = \frac{1}{3{,}375}.$$

c) $\frac{3x^4 y^3 z^8}{9x^2 y^7 z^6}$.

$$\frac{3x^4 y^3 z^8}{9x^2 y^7 z^6} = \frac{3x^4 y^3 z^8}{3^2 x^2 y^7 z^6}$$

$$\frac{3x^4 y^3 z^8}{9x^2 y^7 z^6} = (3)^{-1}x^2y^{-4}z^2$$ we applied the quotient property

$$\frac{3x^4 y^3 z^8}{9x^2 y^7 z^6} = \frac{1}{3}x^2\frac{1}{y^4}z^2$$ we applied the negative property

$$\frac{3x^4 y^3 z^8}{9x^2 y^7 z^6} = \frac{x^2z^2}{3y^4}.$$

d) $(5t^3)^4(2t^2)^5(t^{-7})$.

$$(5t^3)^4(2t^2)^5(t^{-7}) = (5^4t^{12})(2^5t^{10})(t^{-7})$$ we applied the power of power property.

$$(5t^3)^4(2t^2)^5(t^{-7}) = (5^4t^{12})(2^5t^{10})(\frac{1}{t^7})$$ we applied the negative property

$$(5t^3)^4(2t^2)^5(t^{-7}) = (625t^{12})(32t^{10})(\frac{1}{t^7})$$

$$(5t^3)^4(2t^2)^5(t^{-7}) = (20{,}000t^{22})(\frac{1}{t^7})$$ we applied the product property

$$(5t^3)^4(2t^2)^5(t^{-7}) = \frac{20{,}000t^{22}}{t^7}$$

$$(5t^3)^4(2t^2)^5(t^{-7}) = 20{,}000t^{15}$$ we applied the quotient property.

e) $\frac{(a-2b)^6}{(a-2b)^4(2c)^4}$.

$$\frac{(a-2b)^6}{(a-2b)^4(2c)^4} = \frac{(a-2b)^2}{(2c)^4}$$ we applied the power of quotient property

$$\frac{(a-2b)^6}{(a-2b)^4(2c)^4} = \frac{(a-2b)^2}{(2^4)(c)^4}$$ we applied the power of product property

$$\frac{(a-2b)^6}{(a-2b)^4(2c)^4} = \frac{(a-2b)^2}{16c^4}$$

$$\frac{(a-2b)^6}{(a-2b)^4(2c)^4} = \frac{1}{16} \times (a-2b)^2(c)^{-4}$$ we applied the negative property

$$\frac{(a-2b)^6}{(a-2b)^4(2c)^4} = \frac{(a-2b)^2c^{-4}}{16}.$$

f) $\frac{(abc)^5}{a^2b^7c}$.

$$\frac{(abc)^5}{a^2b^7c} = \frac{a^5\,b^5c^5}{a^2b^7c}$$ we applied the power of product property

$$\frac{(abc)^5}{a^2b^7c} = a^3b^{-2}c^4$$ we applied the power of quotient property

$$\frac{(abc)^5}{a^2b^7c} = \frac{a^3c^4}{b^2}$$ we applied the negative property.

1.3.3. Exponential Expressions, One-to-One Property

The one-to-one property of exponential expressions implies that, for any real numbers a, S, and T, where $a > 0$, $a \neq 1$, $a^S = a^T$ if and only if $S = T$.

1.4. Rational Algebraic Expressions

A **rational algebraic expression** is simply a quotient of two polynomials. Or in other words, it is a fraction whose numerator and denominator are polynomials e.g., $\frac{1}{x}, \frac{5}{3x+2}, \frac{x-2}{3x+5}$.

1.4.1. Operations on Rational Expressions

Operations on rational expression consists of adding, subtracting, multiplying, and dividing. Here, we will look at each of these operations.

1.4.1.1. Adding and Subtracting of Rational Expressions

These are the steps in adding and subtracting rational expressions:

Step 1- Find the lowest common denominator.

Step 2- Rewrite each fraction, using the lowest common denominator.

Step 3- Add or subtract the numerators and place the answer over the common denominator.

Step 4- Simplify the expression.

Step 5 - Identify any excluded values (any x value's that would make the denominator 0).

To find the lowest common denominator, you must first find the least common multiple for the given denominators. That is the smallest value that is a multiple of both original denominators. If the denominators do not have any common multiples, they can be multiplied by each other to get the common denominator.

Least Common Multiple (LCM) of an Algebraic Expression

The least common multiple (LCM) of two or more expressions is the smallest (or simplest) expression that is divisible by each of these expressions. It is found by first factoring completely each of the given expressions then use these factors to write the LCM. Detailed examples are shown below.

How to Find the Least Common Multiple of an Algebraic Expression

To find the least common multiple of given expressions, we proceed as follows:

- Express each of the expressions as the product of its factors including prime factors where applicable.
- Find the product of each factor with the highest power which occurs in the given expressions.
- This product is LCM.

Example
Find the lowest common multiple of the expressions $8a^2b^3c$, $12a\,b^4c^2$.

Solution
We first factor into the given expressions.

$$8a^2b^4c = 2^3 \times a^2 \times b^3 \times c$$

$$12abc^2 = 2^2 \times 3 \times a \times b^4 \times c^2.$$

Then the least common multiple will be:

$$\text{LCM} = 2^3 \times 3 \times a^2 \times b^4 \times c^2$$

$$\text{LCM} = 8 \times 3 \times a^2 \times b^4 \times c^2$$

$$\text{LCM} = 24a^2b^4c^2.$$

Example

Find the lowest common multiple of the expressions $9kt^4l^2$, $18kt^5l$, $6k^6t^3l$, $12k^2t^7l^4$.

Solution

We first factor into the given expressions.

$$9kt^4l^2 = 3^2 \times k \times t^4 \times l^2$$

$$18kt^5l = 2 \times 3^2 \times k \times t^5 \times l$$

$$6k^6t^3l = 2 \times 3 \times k^6 \times t^3 \times l$$

$$12k^2t^7l^4 = 2^2 \times 3 \times k^2 \times t^7 \times l^4.$$

Then the least common multiple will be:

$$\text{LCM} = 2^2 \times 3^2 \times k^6 \times t^7 \times l^4$$

$$\text{LCM} = 4 \times 9 \times k^6 \times t^7 \times l^4$$

$$\text{LCM} = 36k^6t^7l^4.$$

Example

Find the lowest common multiple of the expressions:

$$25(x - 8)^2 (x + 3)^2,\ 5(x + 3)^4(x - 8)^3,\ 50(x + 3)(x - 8),\ 10(x + 3)^5.$$

Solution

We first factor into the given expressions.

$$25(x - 8)^2 (x + 3)^2 = 5^2(x - 8)^2 \times (x + 3)^2$$

$$5(x + 3)^4(x - 8)^3 = 5 \times (x + 3)^2 \times (x + 3)^2 \times (x - 8)^3$$

$$50(x + 3)(x - 8) = 2 \times 5^2 \times (x + 3) \times (x - 8)$$

$$10(x + 3)^5 = 2 \times 5 \times (x + 3)^2 \times (x + 3)^3.$$

Then the least common multiple will be:

$$\text{LCM} = 2 \times 5^2 \times (x - 8)^3 \times (x + 3)^2 \times (x + 3)^3$$

$$\text{LCM} = 2 \times 25 \times (x - 8)^3 \times (x + 3)^2 \times (x + 3)^3$$

$$\text{LCM} = 50(x - 8)^3(x + 3)^5.$$

Example

Simplify $\frac{7}{13}x^4uv^2 - \frac{9}{26}x^4uv^2 + 8\,x^4uv^2$.

Solution

These are like terms. We just add the coefficients $\frac{7}{13}$, $-\frac{9}{26}$ and 8, we get,

$$\frac{7}{13}x^4uv^2 - \frac{9}{26}x^4uv^2 + 8\,x^4uv^2$$

$$\frac{7}{13}x^4uv^2 - \frac{9}{26}x^4uv^2 + 8\,x^4uv^2 = (\frac{7}{13} - \frac{9}{26} + 8)\,x^4uv^2$$

$\frac{7}{13}x^4uv^2 - \frac{9}{26}x^4uv^2 + \mathbf{8}\, x^4uv^2 = (\frac{7}{13} \times \frac{2}{2} - \frac{9}{26} + \mathbf{8} \times \frac{26}{26})\, x^4uv^2$

to make each fraction with the same LCD, we multiplied the nominator and the denominator of each fraction by a proper number

$\frac{7}{13}x^4uv^2 - \frac{9}{26}x^4uv^2 + \mathbf{8}\, x^4uv^2 = (\frac{14}{26} - \frac{9}{26} + \frac{208}{26})\, x^4uv^2$

$\frac{7}{13}x^4uv^2 - \frac{9}{26}x^4uv^2 + \mathbf{8}\, x^4uv^2 = (\frac{213}{26})\, x^4uv^2.$

Therefore, $\frac{7}{13}x^4uv^2 - \frac{9}{26}x^4uv^2 + \mathbf{8}\, x^4uv^2 = \frac{213}{26}x^4uv^2.$

Example

Simplify $\frac{1}{ab} - \frac{5}{a^2b} + 3a - \frac{3}{ab^2}.$

Solution

We need first to find a proper LCD. A proper LCD is $\frac{3}{a^2b^2}$. So,

$\frac{1}{ab} - \frac{5}{a^2b} + 3a - \frac{3}{ab^2}$

$\frac{1}{ab} - \frac{5}{a^2b} + 3a - \frac{3}{ab^2} = \frac{1}{ab} \times \frac{ab}{ab} - \frac{5}{a^2b} \times \frac{b}{b} + 3a - \frac{3}{ab^2} \times \frac{a}{a}$

to make each fraction with the same LCD, we multiplied the nominator and the denominator of each fraction by a proper expression

$\frac{1}{ab} - \frac{5}{a^2b} + 3a - \frac{3}{ab^2} = \frac{ab}{a^2b^2} - \frac{5b}{a^2b^2} + 3a - \frac{3a}{a^2b^2}$

$\frac{1}{ab} - \frac{5}{a^2b} + 3a - \frac{3}{ab^2} = \frac{ab - 5b - 3a}{a^2b^2} + 3a$

$\frac{1}{ab} - \frac{5}{a^2b} + 3a - \frac{3}{ab^2} = \frac{ab - 5b - 3a}{a^2b^2} + 3a \times \frac{a^2b^2}{a^2b^2}$

to have the same LCD, we multiplied the last term by $\frac{a^2b^2}{a^2b^2}$

$\frac{1}{ab} - \frac{5}{a^2b} + 3a - \frac{3}{ab^2} = \frac{ab - 5b - 3a}{a^2b^2} + \frac{3a^3b^2}{a^2b^2}$

$\frac{1}{ab} - \frac{5}{a^2b} + 3a - \frac{3}{ab^2} = \frac{ab - 5b - 3a + 3a^3b^2}{a^2b^2}.$

1.4.1.2. Multiplication of Rational Expressions

Multiplying rational expressions is like multiplying any factions. These are the steps in multiplying rational expressions:

Step 1- Multiply the nominators and denominators.

Step 2- Simplify the expression.

Step 3 - Identify any excluded values (any x value's that would make the denominator 0).

Example

Simplify $\frac{(8x - 1)}{3} \times \frac{7}{2(8x - 1)^2}.$

Solution

We simply multiply the nominators and denominators. So,

$$\frac{(8x-1)}{3} \times \frac{7}{2(8x-1)^2} = \frac{(8x-1)\times(7)}{3 \times 2(8x-1)^2}$$

$$\frac{(8x-1)}{3} \times \frac{7}{2(8x-1)^2} = \frac{7}{6(8x-1)}$$ we canceled out factors, not terms.

To identify the excluded value, we need first to put the denominator equal to zero. So,

$6(8x - 1) = 0$

$(8x - 1) = 0$

$8x = 1$ we isolated the variable in one side and the constant term on the other side

$\frac{1}{8} \times 8x = \frac{1}{8} \times 1$ we multiplied both side by the reciprocal of $8x$

$x = \frac{1}{8}$.

Therefore, the simplification of $\frac{(8x-1)}{3} \times \frac{7}{2(8x-1)^2}$ is $\frac{7}{6(8x-1)}$, where, $x \neq \frac{1}{8}$.

1.4.1.3. Dividing Rational Expressions

Dividing rational expressions is like dividing any factions. These are the steps in dividing two rational expressions:

Step 1- Flip the second term so that you are multiplying by the reciprocal.

Step 2- Simplify the expression, factor, and cancel, if necessary.

Step 3 - Multiply the nominators and denominators.

Step 4 - Simplify further if necessary.

Step 5 - Identify any excluded values (any x value's that would make the denominator 0).

Example

Simplify $\frac{(11x-4)}{3} \div \frac{(3x+6)}{9}$.

Solution

We need first to flip the second term and then multiply the first term by the resulting second term. So,

$$\frac{(11x-4)}{3} \div \frac{(3x+6)}{9} = \frac{(11x-4)}{3} \times \frac{9}{(3x+6)}$$ we flipped the second term and changed symbol "÷ " to symbol "×".

$$\frac{(11x-4)}{3} \div \frac{(3x+6)}{9} = \frac{(11x-4)(9)}{3\times(3x+6)}$$

$$\frac{(11x-4)}{3} \div \frac{(3x+6)}{9} = \frac{99x-36}{9x+18}.$$

To identify the excluded value, we need first to put the denominator equal to zero. So,

$9x + 18 = 0$

$9x = -18$ we isolated the variable in one side and the constant term on the other side

$\frac{1}{9} \times 9x = \frac{1}{9} \times (-18)$ we multiplied both side by the reciprocal of $9x$

$x = \frac{-1}{9} = -2.$

Therefore, the simplification of $\frac{(11x - 4)}{3} \div \frac{(3x + 6)}{9}$ is $\frac{99x - 36}{9x + 18}$, where, $x \neq -2$.

1.5. Evaluating Algebraic Expressions

To evaluate an algebraic expression, you must substitute a number for each variable and perform the arithmetic operations.

Example
Given the expression $7x + 9xy - 3x^2$, evaluate it for $x = -2$ and $y = 4$.

Solution
We replace x with -2 and y with 4 to evaluate the expression.

$7x + 9xy - 3x^2$

$= 7(-2) + 9(-2)(4) - 3(-2)^2$ — we substituted $x = -2$ and $y = 4$ into the equation

$= -14 - 72 - 3(4)$

$= -14 - 72 - 12$

$= 98.$

Therefore, the expression $7x + 9xy - 3x^2$, evaluated by $x = -2$ and $y = 4$, is 98.

Example
Given the expression $5ab - \frac{3}{a^2b^3} - 2ab^2 + \frac{3}{a^2b^2} + \frac{1}{(a+b)^3} - 4$, evaluate it for $a = 1$ and $b = -2$.

Solution
We replace a with 1 and b with -2 to evaluate the expression.

$5ab - \frac{3}{a^2b^3} - 2ab^2 + \frac{3}{a^2b^2} + \frac{1}{(a+b)^3} - 4$

$= 5 \times 1 \times (-2) - \frac{3}{(1)^2(-2)^3} - 2 \times 1 \times (-2)^2 + \frac{3}{(1)^2(-2)^2} + \frac{1}{(1+(-2))^3} - 4$ — we substituted $a = 1$ and $b = -2$ into the equation

$= -10 - \frac{3}{1 \times (-8)} - 2 \times 1 \times (4) + \frac{3}{1 \times (4)} + \frac{1}{(-1)^3} - 4$

$= -10 - \frac{3}{-8} - 8 + \frac{3}{4} + \frac{1}{-1} - 4$

$= -10 + \frac{3}{8} - 8 + \frac{3}{4} - 1 - 4$

$= -23 + \frac{3}{8} + \frac{3}{4}$

$= -23 + \frac{3}{8} + \frac{3}{4} \times \frac{2}{2}$ — to make the same LCD, we multiplied the nominator and the denominator of the second fraction by $\frac{2}{2}$

$= -23 + \frac{3}{8} + \frac{6}{8}$

$= -23 + \frac{3 + 6}{8}$

$= -23 + \frac{9}{8}$

$= -23 \times \frac{8}{8} + \frac{9}{8}$ to make the same LCD, we multiplied the nominator and the denominator of the fraction by $\frac{8}{8}$

$= -\frac{184}{8} + \frac{9}{8}$

$= -\frac{175}{8}$.

Therefore, the expression $5ab - \frac{3}{a^2b^3} - 2ab^2 + \frac{3}{a^2b^2} + \frac{1}{(a+b)^3} - 4$, evaluated by $a = 1$ and $b = -2$, is $-\frac{175}{8}$.

1.6. Algebraic Expressions Involving Radicals

An algebraic expression with roots is called a **radical expression**. In general terms, if a is a positive real number, then the square root of a is a number that, when multiplied by itself, gives a. The square root could be positive or negative because multiplying two negative numbers gives a positive number.

The principal square root is the nonnegative number that when multiplied by itself equals a. For example, 5 is a square root of 25, because $5^2 = 25$. Since $(-5)^2 = 25$, we can say that -5 is a square root of 25 as well.

Every positive real number has two square roots, one positive and one negative. For this reason, we use the radical sign $\sqrt{\ }$ to denote the principal (nonnegative) square root and a negative sign in front of the radical $-\sqrt{\ }$ to denote the negative square root.

$\sqrt{25} = 5$ positive square root of 25

$-\sqrt{25} = -5$ negative square root of 25.

Zero is the only real number with one square root.

$\sqrt{0} = 0$ because $0^2 = 0$.

When we have an expression containing a radical ($\sqrt{\ }$) symbol, some people mistakenly call this a 'square root' symbol. However, it can also be used to describe a cube root, a fourth root, or higher. When the radical symbol is used to denote any root other than a square root, there will be a superscript number in the 'V'-shaped part of the symbol.

For example, $\sqrt[3]{125}$ means to find the cube root of 125. If there is no superscript number, the radical expression is calling for the square root. Generally, the symbol $\sqrt[n]{x}$ used to indicate a root is called a radical, or sometimes a surd. The expression $\sqrt[n]{x}$ is therefore read "x radical n," or "the nth root of x." In the radical symbol, the quantity under the vinculum is called the radicand, and the quantity n written to the left is called the index. When n is an even number, the radical is called an even radical.

Notice that if n is an odd number, then any real number a has an nth root. If n is an even number, then only a positive real number a has an nth root.

When we are given a problem in radical form, we may have an easier time if we rewrite it by using **rational exponents** (exponents that are fractions). We can rewrite every radical as an exponent by using the following property.

- The top number in the resulting rational exponent tells us the power.
- The bottom number tells us the root we are taking.

Algebraically, we will have:

$$x^{\frac{m}{n}} = \sqrt[n]{x^m} = \left(\sqrt[n]{x}\right)^m.$$

Example

Rewrite each of the followings:

a) $7^{\frac{5}{4}}$.

b) $\left(\sqrt[3]{t}\right)^8$.

c) $\sqrt[6]{a^7}$.

Solution

a) We learned that the rational exponent can be rewritten as the radical form. So,

$$7^{\frac{5}{4}} = \left(\sqrt[4]{x}\right)^5 = \sqrt[4]{x^5}.$$

b) We learned that radical expression can be rewritten as the rational exponent. So,

$$\left(\sqrt[3]{t}\right)^8 = \sqrt[3]{t^8} = t^{\frac{8}{3}}.$$

c) We learned that radical expression can be rewritten as the rational exponent. So,

$$\sqrt[6]{a^7} = \left(\sqrt[6]{a}\right)^7 = a^{\frac{7}{6}}.$$

1.6.1. Arithmetic of Algebraic Expressions Involving Radicals

We will look carefully at how to work with radicals by way of arithmetic operations (addition, subtraction, division, and multiplication). To simplify a square root, we rewrite it such that there are no perfect squares in the radicand.

There are several properties of square roots that allow us to simplify complicated radical expressions. The first rule we will look at is the product rule for simplifying square roots, which allows us to separate the square root of a product of two numbers into the product of two separate rational expressions.

1.6.1.1. The Product Rule of Square Roots

If a and b are nonnegative, the square root of the product ab is equal to the product of the square roots of a and b.

$$\sqrt{ab} = \sqrt{a} \times \sqrt{b}.$$

How to Use the Product Rule to Simplify Square Roots

1. Factor any perfect squares from the radicand.
2. Write the radical expression as a product of radical expressions.
3. Simplify.

Example
Simplify the radical expression.

a) $\sqrt{800}$.
b) $\sqrt{144a^7b^6c}$.

Solution
a) $\sqrt{800}$.

$\sqrt{800} = \sqrt{2 \times 4 \times 100}$

$\sqrt{800} = \sqrt{2} \times \sqrt{4} \times \sqrt{100}$ — we wrote radical expression as product of radical expressions

$\sqrt{800} = \sqrt{2} \times 2 \times 10$ — we factored the perfect square from radicand

$\sqrt{800} = 20\sqrt{2}$.

b) $\sqrt{144a^7b^6c}$.

$\sqrt{144a^7b^6c} = \sqrt{144 \times a \times a^2 \times a^2 \times a^2 \times b^2 \times b^2 \times b^2 \times c}$

$\sqrt{144a^7b^6c} = \sqrt{144} \times \sqrt{a} \times \sqrt{a^2} \times \sqrt{a^2} \times \sqrt{a^2} \times \sqrt{b^2} \times \sqrt{b^2} \times \sqrt{b^2} \times \sqrt{c}$ — we wrote the radical expression as product of radical expressions

$\sqrt{144a^7b^6c} = \sqrt{16 \times 9} \times \sqrt{a} \times \sqrt{a^2} \times \sqrt{a^2} \times \sqrt{a^2} \times \sqrt{b^2} \times \sqrt{b^2} \times \sqrt{b^2} \times \sqrt{c}$

$\sqrt{144a^7b^6c} = \sqrt{4 \times 4 \times 9} \times \sqrt{a} \times \sqrt{a^2} \times \sqrt{a^2} \times \sqrt{a^2} \times \sqrt{b^2} \times \sqrt{b^2} \times \sqrt{b^2} \times \sqrt{c}$

$\sqrt{144a^7b^6c} = \sqrt{2^2 \times 2^2 \times 3^2} \times \sqrt{a} \times \sqrt{a^2} \times \sqrt{a^2} \times \sqrt{a^2} \times \sqrt{b^2} \times \sqrt{b^2} \times \sqrt{b^2} \times \sqrt{c}$.

Again, writing a radical expression as product of radical expressions, we will have:

$\sqrt{144a^7b^6c} = \sqrt{2^2} \times \sqrt{2^2} \times \sqrt{3^2} \times \sqrt{a} \times \sqrt{a^2} \times \sqrt{a^2} \times \sqrt{a^2} \times \sqrt{b^2} \times \sqrt{b^2} \times \sqrt{b^2} \times \sqrt{c}$

$\sqrt{144a^7b^6c} = 2 \times 2 \times 3 \times \sqrt{a} \times a \times a \times a \times b \times b \times b \times \sqrt{c}$ — we factored the perfect squares from radicand

$\sqrt{144a^7b^6c} = 12 \times a^3 \times b^3 \times \sqrt{a \times c}$

$\sqrt{144a^7b^6c} = 12a^3b^3\sqrt{ac}.$

1.6.1.2. Using the Quotient Rule to Simplify Square Roots

Just as we can rewrite the square root of a product as a product of square roots, we can also rewrite the square root of a quotient as a quotient of square roots. This is the quotient rule for simplifying square roots. Usually, it can be helpful to separate the numerator and denominator of a fraction under a radical so that we can take their square roots separately. For example, we can rewrite $\sqrt{\frac{7}{11}}$ as $\frac{\sqrt{7}}{\sqrt{11}}$.

The square root of the quotient $\frac{a}{b}$ is equal to the quotient of the square roots of a and b, where $b \neq 0$.

$$\sqrt{\frac{a}{b}} = \frac{\sqrt{a}}{\sqrt{b}}.$$

Steps to Use the Quotient Rule to Simplify Square Roots

1. Write the radical expression as the quotient of two radical expressions.
2. Simplify the numerator and denominator.

Example

Simplify the radical expression. $\sqrt{\frac{81}{16}}$.

Solution

$$\sqrt{\frac{81}{16}}$$

$$\sqrt{\frac{81}{16}} = \frac{\sqrt{81}}{\sqrt{16}}$$ we wrote the radical expression as quotient of two radical expressions

$$\sqrt{\frac{81}{16}} = \frac{\sqrt{9^2}}{\sqrt{4^2}}$$

$$\sqrt{\frac{81}{16}} = \frac{9}{4}$$ we factored the perfect square from radicands.

Example

Simplify $\sqrt{\frac{2x^2}{9y^4}}$.

Solution

$$\sqrt{\frac{2x^2}{9y^4}}$$

$$\sqrt{\frac{2x^2}{9y^4}} = \frac{\sqrt{2x^2}}{\sqrt{9y^4}}$$ we wrote the radical expression as quotient of two radical expressions

$$\sqrt{\frac{2x^2}{9y^4}} = \frac{x\sqrt{2}}{3y^2}$$ we factored the perfect squares from radicands.

Note: We do not need the absolute value signs for y^2 because that term will always be nonnegative.

Example

Simplify the radical expression $\frac{\sqrt{234x^{11}y}}{\sqrt{26x^7y}}$.

Solution

$$\frac{\sqrt{234x^{11}y}}{\sqrt{26x^7y}}$$

$$\frac{\sqrt{234x^{11}y}}{\sqrt{26x^7y}} = \sqrt{\frac{234x^{11}y}{26x^7y}}$$ we combined numerator and denominator into one radical expression

$$\frac{\sqrt{234x^{11}y}}{\sqrt{26x^7y}} = \sqrt{9x^4}$$

$$\frac{\sqrt{234x^{11}y}}{\sqrt{26x^7y}} = 3x^2$$ we factored the perfect squares from radicands.

1.6.1.3. Adding and Subtracting Square Roots

We can add or subtract radical expressions only when they have the same radicand and when they have the same radical type such as square roots. For example, the sum of $\sqrt{2}$ and $5\sqrt{2}$ is $6\sqrt{2}$.

However, it is often possible to simplify radical expressions, and that may change the radicand. The radical expression $\sqrt{50} = \sqrt{2 \times 25}$ can be written with a 2 in the radicand, as $5\sqrt{2}$, so $\sqrt{2} + \sqrt{50} = \sqrt{2} + 5\sqrt{2} = 6\sqrt{2}$. Similarly, the subtraction of $\sqrt{2}$ and $5\sqrt{2}$ is $-5\sqrt{2}$.

How to Use the Adding/Subtracting Rule to Simplify Square Roots

1. Simplify each radical expression.
2. Add or subtract expressions with equal radicands.

Example

Add $6\sqrt{28}$ and $2\sqrt{7}$.

Solution

We can rewrite $6\sqrt{28}$ as $6\sqrt{7 \times 4}$. According to the product rule, this becomes $6\sqrt{7} \times \sqrt{4}$. The square root of $\sqrt{4}$ is 2, so the expression becomes $6 \times 2 \times \sqrt{7}$ which is $12\sqrt{7}$. Now we can add the terms having the same radicand. So,

$$6\sqrt{28} + 2\sqrt{7}. = 12\sqrt{7} + 2\sqrt{7}$$

$$6\sqrt{28} + 2\sqrt{7}. = 14\sqrt{7}.$$

Example

Subtract $15\sqrt{45a^3b^5c^3}$ and $13\sqrt{5a^3b^5c^3}$.

Solution

Rewrite each term so they have equal radicands. First,

$$15\sqrt{45a^3b^5c^3}$$

$$15\sqrt{45a^3b^5c^3} = 15\sqrt{5 \times 9 \times a^3b^5c^3}$$

$$15\sqrt{45a^3b^5c^3} = 15\times \sqrt{5} \times \sqrt{9} \times \sqrt{a^3}\times\sqrt{b^5} \times \sqrt{c^3}$$ we wrote the radical expression as product of radical expressions

$$15\sqrt{45a^3b^5c^3} = 15\times \sqrt{5} \times \sqrt{3^2} \times \sqrt{a^3}\times\sqrt{b^5} \times \sqrt{c^3}$$

$$15\sqrt{45a^3b^5c^3} = 15\times \sqrt{5} \times \sqrt{3^2} \times \sqrt{a \times a^2}\times\sqrt{b \times b^2 \times b^2} \times \sqrt{c \times c^2}$$

$$15\sqrt{45a^3b^5c^3} = 15\times \sqrt{5} \times 3 \times a \times \sqrt{a} \times b \times b \times \sqrt{b} \times c \times \sqrt{c}$$ we factored the perfect squares from radicands

$$15\sqrt{45a^3b^5c^3} = 45ab^2c \times \sqrt{5} \times \sqrt{a} \times \sqrt{b} \times \sqrt{c}$$

$$15\sqrt{45a^3b^5c^3} = 45ab^2c\sqrt{5abc}.$$

Now, using similar operations, we can rewrite the other term. So,

$$13\sqrt{5a^3b^5c^3}$$

$$13\sqrt{5a^3b^5c^3} = 13\times \sqrt{5 \times a \times a^2 \times b \times b^4 \times c \times c^2}$$

$$13\sqrt{5a^3b^5c^3} = 13\times \sqrt{5} \times a \times \sqrt{a} \times b^2 \times \sqrt{b} \times c \times \sqrt{c}$$ we factored the perfect squares from radicands

$$13\sqrt{5a^3b^5c^3} = 13ab^2c\sqrt{5abc}.$$

Here, we have two expressions with equal radicands, so we can proceed the subtracting operation.

$$15\sqrt{45a^3b^5c^3} - 13\sqrt{5a^3b^5c^3}$$

$$15\sqrt{45a^3b^5c^3} - 13\sqrt{5a^3b^5c^3} = 45ab^2c\sqrt{5abc} - 13ab^2c\sqrt{5abc}.$$

$$15\sqrt{45a^3b^5c^3} - 13\sqrt{5a^3b^5c^3} = 32ab^2c\sqrt{5abc}.$$

Therefore, $15\sqrt{45a^3b^5c^3}$ minus $13\sqrt{5a^3b^5c^3}$ is $32ab^2c\sqrt{5abc}$.

1.7. Algebraic Expressions Involving Logarithms

Properties of Logarithms - Here, we propose some of the properties of logarithms and natural logarithms, which will later help us to solve both exponential and logarithmic equations.

Properties of logarithms and natural logarithms can be summarized as below:

- If $log_b x = a$, then $x = b^a$.
- If $ln x = a$, then $x = e^a$.
- $log_b 1 = 0$.
- $ln 1 = 0$.
- $log_b b = 1$.
- $ln\, e = 1$.

- $log_b\, a^x = x\, log_b\, a.$
- $ln\, a^x = x\, ln\, a.$
- $log_b\,(K \times H) = log_b\, K + log_b\, H.$
- $ln(K \times H) = ln\, K + ln\, H.$
- $log_b\,(\frac{K}{H}) = log_b\, K - log_b\, H.$
- $ln\,(\frac{K}{H}) = ln\, K - ln\, H.$

One-to-One Property of Logarithmic and Natural Logarithmic Expressions

- If there is an equation having logarithms, there is a property that implies: if $log_b\, K = log_b\, H$, then $K = H$.
- If there is an equation having natural logarithms, is a property that implies: if $ln\, K = ln\, H$, then $K = H$.

Example
Evaluate the following expressions:

a) $23 log_7\, 1.$
b) $8 log_9\, 9.$
c) $log_{12}\, 12^9.$
d) $13 ln\, 1.$
e) $ln\, e^3.$

Solution
Using some of the above properties, we have:

a) $23 log_7\, 1 = 23 \times 0 = 0.$
b) $8 log_9\, 1 = 8 \times 0 = 0.$
c) $log_{12}\, 12^9 = 9 \times log_{12}\, 12 = 9 \times 1 = 9.$
d) $13 ln\, 1 = 13 \times 0 = 0.$
e) $ln\, e^3 = 3 \times\, ln\, e = 3 \times 1 = 3.$

Example
What is the value of x when:

a) $log x = 2.$
b) $log_3 x = 5.$
c) $log_7 x = 4.$
d) $ln\, e^x = 3.$

Solution
Using some of the above properties, we have:

a) $\log x = 2$, since the base is 10, $x = 10^2 = 100$.

b) $\log_3 x = 5$, then $x = 3^5 = 243$.

c) $\log_7 x = 4$, then $x = 7^4 = 2401$.

d) $\ln e^x = 3$, then $x \ln e = 3$, since $\ln e = 1$, then $x = 3$, or $e^x = e^3$, since the base are equal, then the exponent must be equal, then $x = 3$.

Example
Expand the following expressions.

a) $\log 9x$.

b) $\log_3 xyz$.

c) $\ln xy$.

d) $\ln 5xyz$.

Solution
Using some of the above properties, we have:

a) $\log 9x = \log 9 + \log x$.

b) $\log_3 xyz = lg_3 x + \log_3 y + \log_3 z$.

c) $\ln xy = \ln x + \ln y$.

d) $\ln 5xyz = \ln 5 + \ln x + \ln y + \ln z$.

Example
Expand the following expressions.

a) $\log \frac{x}{y}$.

b) $\log_6 \frac{x}{y} z$.

c) $\ln \frac{7x}{y}$.

d) $\ln \frac{4xy}{z}$.

Solution
Using some of the above properties, we have:

a) $\log \frac{x}{y} = \log x - \log y$.

b) $\log_6 \frac{x}{y} z = \log_6 x - \log_6 y + \log_6 z$.

c) $\ln \frac{7x}{y} = \ln 7 + \ln x - \ln y$.

d) $\ln \frac{4xy}{z} = \ln 4 + \ln x + \ln y - \ln z$.

1.8. Exercises

1. Simplify each expression and state which property is applied.

   a) $-12(a + 3) + (4a - 9b) + 14b$.

   b) $3x - 12y(x + 2) + 3x(6x - 8y) + 11(y + 3) + 13x$.

   c) $-5b(a + 11) + 2a(13a - 7b) + 15b + 12(a + 2b)$.

   d) $5(k + \frac{2}{5}) + \frac{3}{7}(2k - 7t) + 14k + 16kt + k(2t - \frac{1}{2}), + 4t$.

   e) $-6x(\frac{19}{12y} + 3) + \frac{2}{3y}(4x - 8) + 17\frac{1}{6y} + 13x$.

   f) $4(\frac{1}{2ab} + 2b) + \frac{2a}{5}(\frac{1}{3ab} - 8a) + 9\left(\frac{1}{10a} + b\right) + 16ab$.

2. Given the set of terms below, designate which one contains like terms.

   a) $\{3xyz, -xyz, -5xyz, 12\,xyz\}$.

   b) $\{4a(b^2c), 5ab^2c, b^2ca, acb^2, 8c(b^2a) - 5acb^2\}$.

   c) $\{k(3t^2), 7t(3tk), kt^2(9), -kt^2\}$.

   d) $\{3uvw, -uwv, -5uvw^2, 14uwv\}$.

   e) $\{x(y + z), -xy - xz, 5xy + 5xz\}$.

   f) $\{-b(a^2c^2), 22c^2ba^2, b^2ca, 3a^2cb^2, -5\,a^2bc^2\}$.

3. Given the expression $5x + 8xy + 3y - 11y^2 + -11xy^2 + 3x^2$, evaluate it for $x = -1$ and $y = 3$.

4. Given the fractional expression $\frac{7}{2ab} + 4ab + \frac{3}{2b} - 11a^2 + \frac{3}{2a^2b} - 6ab^2 + 18b^2$, evaluate it for $a = 2$ and $b = -1$.

5. Simplify the given expressions.

   a) $7x + 3x^2y^3 + 9(x^2y^3 - xy^3) + 5(y^3x + x)$.

   b) $ab^2c^2 + 2a^3b^2 - 8abc - 5ab^2c^2 - 2b^2a^3 + 2c(ab + ab^2c)$.

   c) $34x^3 - 19x(x^2 - 3) + 11x + 42x^3 - 25x + 17x - 44x^3$.

   d) $4wu^2v^2 + 5w^3u^2 - 7w^3u^2 + 12w(uv - w^2u^2) - 5wu^2v^2 - 9u(w^2v^3 + 2w^2v^3) + 3w(uv + u^2v^2) - 6wu^2v^2$.

6. Simplify each expression.

   a) $\frac{9(x+1)}{x^3y^2} - \frac{3}{x^2y} + 8y - \frac{7(y-2)}{xy^2}$.

   b) $\frac{10k}{k^3tl^2} + \frac{2(t+11)}{t^3l^2} - \frac{3}{k^2l}$.

   c) $\frac{12}{a^3b^2} - \frac{1}{a^2b^3} + 2ab - \frac{7(b-4)}{ab^5}$.

7. Simplify each expression.

   a) $\frac{2(x+1)}{5(x-3)} \times \frac{3(x-3)^2}{x} \times \frac{6}{y}$.

b) $\frac{13k^3}{tl^2} \times \frac{3(t+8)}{t^2k^2} \times \frac{3tl}{k^2}$

c) $5 \times \frac{(3-a)}{b^2} \times \frac{(b-4)}{a^2} \times 2a^3b^4$.

8. Simplify each expression.

a) $\frac{4(x+3)}{(y-2)^2} \div \frac{3(2x+6)^2}{xy}$.

b) $\frac{\frac{15}{k}}{t^2} \div \frac{6(t+3)}{t^3}$

c) $\frac{12a^2}{\frac{4}{ab}} \div \frac{\frac{5b^2}{3a}}{4a^2b^2}$.

d) $\frac{4x}{(y-2)^2}(x+2y) \div \frac{x^2(4x+8y)^2}{3(2x+6)^2}$.

9. Simplify each expression.

a) $\frac{4x}{(x+5y)^3} \times \frac{3(2x+10y)^2}{3y} - \frac{3}{(x+5y)^2}$.

b) $\frac{2(k-3t)}{8k^2t^3} \times \frac{12k^3t^2}{3(4k^2-6t)} + \frac{kt}{2(2k^2-3t)^2}$.

c) $\frac{2}{(4w-5u)^2} - \frac{11w^3u}{(8w-10u)} \times \frac{\frac{4}{3}(12w-15u)^3}{22w^2u^2}$.

d) $\frac{3ac^3}{(2a-3b+c)^2} + \frac{11a^2bc}{(6a-9b+3c)} \times \frac{(2a-3b+c)^3}{33a^2b^2c^3}$.

10. Simplify each expression.

a) $\frac{-4xy^2z}{(2x+8y)(x+4y)^2} + \frac{3(2x-z)^2}{(2x+8y)(x+4y)\times 2^2} - \frac{2yz}{(2x+8y)(x+4y)} \div \frac{6(2x-z)}{(2x+8y)(x+4y)^2}$.

b) $\frac{(5w+15u)}{7w^2u^2} \times \frac{13}{(3w+9u)} \div \frac{\frac{2}{w^2u}}{\frac{(w+3u)^2}{wu^2}}$.

e) $\frac{\frac{2}{w^2u}}{\frac{(w+3u)^2}{wu^2}} \div \frac{(5w+15u)}{7w^2u^2} \times \frac{13}{(3w+9u)}$.

11. Simplify the radical expression.

a) $\sqrt{32{,}400}$.

b) $\sqrt{1{,}764w^9v^4u^7}$.

c) $\frac{\sqrt{225k^8t^3l^5}}{\sqrt{81k^7t^4}}$.

12. Simplify $(\sqrt{50a^5b^3c^7})(\sqrt{128a^{11}b^5c^{-3}})$.

13. Simplify $\frac{\sqrt{288x^{13}y^5z^3}}{\sqrt{36x^9yz^2}} \times \frac{\sqrt{16x^3y^5z^5}}{\sqrt{36\ {}^9y^2z^7}}$.

14. Simplify $\frac{\sqrt{32k^7(4h-2l)}}{\sqrt{36(k-5h)^2hl^2}} \div \frac{\sqrt{27k^3(4h-2l)^3}}{\sqrt{9(k-5h)^3l^5}}$.

15. Add radicals.
    a) $3\sqrt{\mathbf{32}}$, $5\sqrt{\mathbf{128}}$. and $5\sqrt{\mathbf{8}}$.
    b) $2\sqrt{xy^5}$, $9\sqrt{x^3y}$, $8\sqrt{x^5y^3}$, and $7\sqrt{xy^3}$.
    c) $11\sqrt{a^3(2+b)}$, $4\sqrt{a^5(2+b)^3}$, $3\sqrt{a(2+b)}$.
    d) $8u\sqrt{(2u-3v)^2}$, $(3u-2)\sqrt{(2u-3v)^3}$, and $(5u-1)\sqrt{4u-9v}$.

16. Subtract radicals.
    a) $6\sqrt{(2a-3b)^3}$, and $(5a+3b-2)\sqrt{2a-3b}$.
    b) $14x\sqrt{28x^3y^7z^3}$, and $19\sqrt{343x^5y^7z^3}$.
    c) $8\sqrt{k^5(2h-3l)^3}$, and $(5h+3k-5)\sqrt{k^3(2h-3l)}$.

17. Evaluate the following expressions:
    a) $6log_{15}1$.
    b) $17log_4 4$.
    c) $9log_8 8^3$.
    d) $19ln\,1$.
    e) $10ln\,e^{14}$.

18. What is the value of x when:
    a) $3log x = 9$.
    b) $log_4 x = 2$.
    c) $4log_6 x = 16$.
    d) $ln\,e^{2x} = 8$.
    e) $ln\,e^{(x-1)} = 16$.

19. Expand the following expressions:
    a) $log7x$.
    b) $log_5 abcd$.
    c) $ln8khl$.
    d) $ln\,xyz^2$.

20. Expand the following expressions:
    a) $log\frac{a}{4b}$.
    b) $log_6\frac{5u}{v}w$.
    c) $ln\frac{12kh}{5l}$.

d) $ln \frac{4x^3y}{5z} \times \frac{4xy^2}{z^2}$.

2. Interpolation and Extrapolation

Interpolation and extrapolation, although frequently encountered in mathematics and often confused due to their similar contexts, represent distinct concepts.

- **Interpolation:** Interpolation entails estimating or predicting values within the range of known data points. It serves to fill in gaps between existing data points or to estimate values within a given data set. Interpolation operates under the assumption that the relationship between data points remains consistent within the observed range.

- **Extrapolation:** Extrapolation involves predicting or estimating values that extend beyond the range of known data points. It projects the trend or pattern observed in existing data to make predictions outside the observed range. However, extrapolation can be less reliable than interpolation because it assumes that the observed trend continues indefinitely beyond the known data range.

In essence, interpolation deals with values inside the range of observed data, while extrapolation deals with values outside this range. It's crucial to exercise caution when extrapolating, as the accuracy of predictions can diminish significantly when venturing beyond the scope of the available data.

2. 1. Interpolation

Tabular presentation of data is a common technique for displaying information. Data, often gathered through experiments or sampling, typically represents values of a function for a limited number of independent variable values, resulting in fragmented data. Interpolation becomes necessary to estimate the function's value for intermediate independent variable values.

Outside the tabulated data, there may exist values that lie between the known data points, which require estimation through interpolation. Interpolation functions by utilizing known data to predict values at unknown points. For instance, if it took 15 minutes to walk 1 kilometer on Saturday and 2 hours to walk 8 kilometers on Sunday, interpolation could estimate that it would take 1 hour to walk 4 kilometers.

To illustrate how interpolation functions, consider the following example where a table provides some values of an unknown function, $f(x)$.

x	$f(x)$
1	7
2	3
3	2
4	1
5	4
6	5
7	6
8	8

The graph of the above data is presented below:

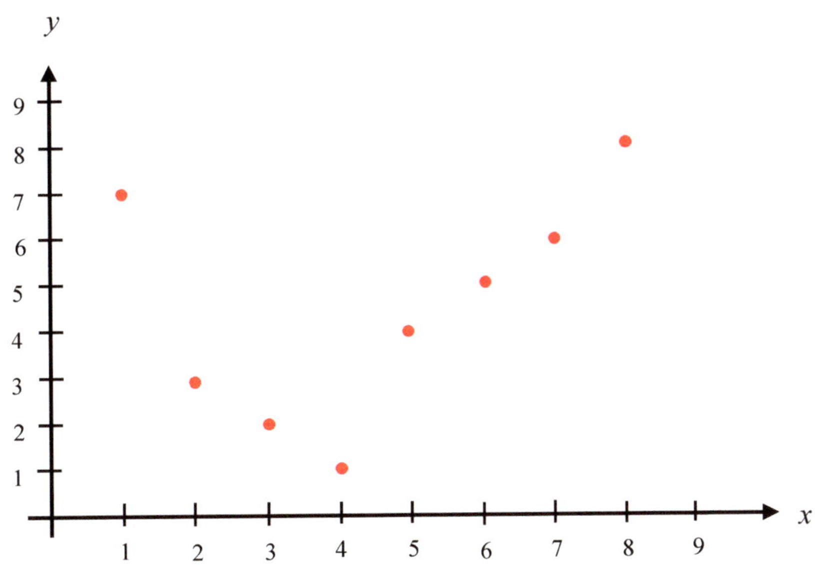

Interpolation gives a means of estimating the function values at intermediate points, such as $f(1.5)$, $f(3.6)$ or $f(6.2)$.

> **Interpolation**
> **Interpolation** is a type of estimation of finding new data points between two known values on a line or a curve.

2.1.1. Interpolation Methods

There are different types of interpolation methods. The methods differ according to their precision, cost, number of data points needed, and smoothness of the resulting interpolant function. Blows are different interpolation methods:

- **Linear Interpolation Method** – Linear interpolation is the simplest methods of estimation. This method applies a distinct linear polynomial between each pair of data points for curves, or within the sets of three points for surfaces.
- **Nearest Neighbor Interpolation Method** (also known as proximal interpolation method) –This method inserts the value of an interpolated point to the value of the most adjacent data point. Therefore, this method does not produce any new data points.
- **Polynomial Interpolation Method** - Polynomial interpolation is a generalization of linear interpolation. Note that the linear interpolant is a linear function. We now replace this interpolant with a polynomial of higher degree.
- **Spline Interpolation Method** - Spline interpolation uses low-degree polynomials in each of the intervals and chooses the polynomial pieces such that they fit smoothly together. The resulting function is called a spline.

Choosing one method versus another is a subject to both practicality and usefulness. As we will see, linear interpolation uses only standard arithmetic operations, and therefore may be easier to use in certain cases. No matter which method we choose we should always remember that it is an estimate only and the smaller the interval, the more reliable the estimate.

Among the above-mentioned methods, the simplest interpolation method is to locate the nearest data value (the nearest neighbor interpolation) and assign the same value. However, through this method we will not get any new data points.

In simple problems, this method is unlikely to be used, as linear interpolation is almost as easy. Hence, in this manuscript we limit the discussion of the interpolation to the linear method only.

In the next section we will look at this method.

2.12. Linear Interpolation

Linear interpolation is the simplest method that is used for estimating the value of a function between any two known values. Mathematically, it is a method of curve fitting using linear polynomials to construct new data points within the range of a discrete set of known data points.

Consider the above example in which a table provided some values of unknown function $f(x)$. Suposse we want to estimate $f(1.5)$. Since 1.5 is midway between 1 and 2, it is reasonable to take $f(1.5)$ midway between $f(1) = 7$ and $f(2) = 3$, which yields 5 (the midpoint of 7 and 3 is obtained by: $\frac{7+3}{2} = \frac{10}{2} = 5$).

Graphically, linear interpolation means drawing a secant line between the points that you know and finding the point on the line that corresponds with the value you want to know about. More precisely, in linear interpolation, the estimated point is assumed to lie on the line joining the nearest points to the left and right. Again, considering the above example, the visualization of estimating $f(1.5)$ is:

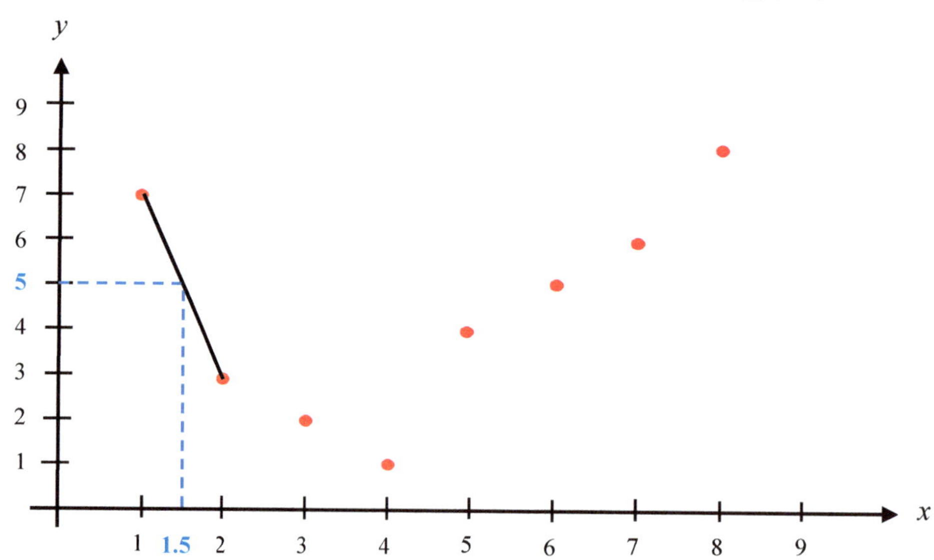

As it is seen in the above graph, by selecting point 1.5 on the x-axis it provides $f(1.5) = 5$ on the y-axis.

2.1.3. Linear Interpolation Formula

Generally, linear interpolation takes two data points, (x_a, y_a) and (x_b, y_b), and the interpolant is given by:

$$y = y_a + \frac{y_b - y_a}{x_b - x_a}(x - x_a) \text{ at the point } (x, y)$$

or

$$\frac{y - y_a}{y_b - y_a} = \frac{x - x_a}{x_b - x_a}$$

$$\frac{y - y_a}{x - x_a} = \frac{y_b - y_a}{x_b - x_a}.$$

This formula is using coordinates of two given values, (x_a, y_a) and (x_b, y_b), to find the best fit curve as a straight line. Then this will give any required value of y at a known value of x, see the graph below:

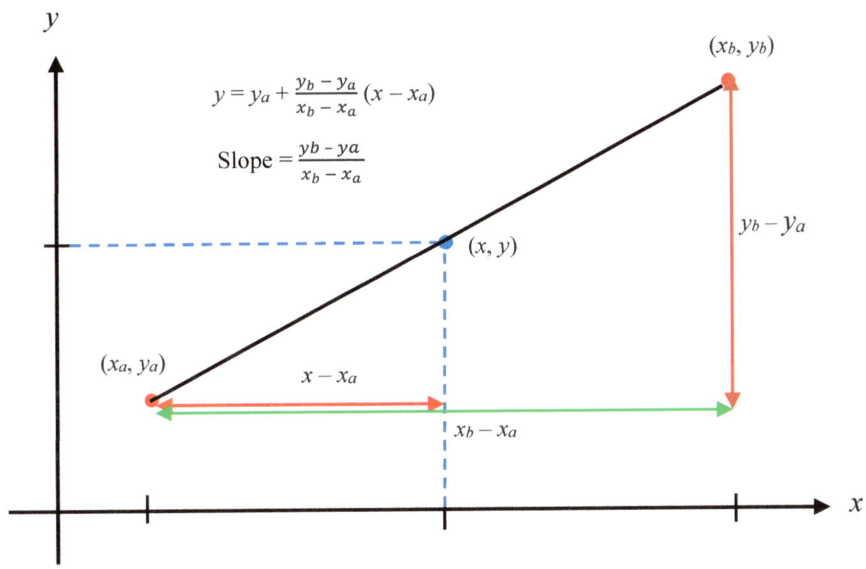

Example:
Find the value of y at $x = 3$ given some set of values (1, 3), (5, 9) using the linear interpolation.

Solution
We learned that given values, (x_a, y_a) and (x_b, y_b), the linear interpolant formula for finding y at a point x, is given by $y = y_a + \frac{y_b - y_a}{x_b - x_a}(x - x_a)$. In this case $x_a = 1$, $y_a = 3$, $x_b = 5$, $y_b = 9$, and $x = 3$. So,

$$y = y_a + \frac{y_b - y_a}{x_b - x_a}(x - x_a)$$

$$y = 3 + \frac{9 - 3}{5 - 1}(3 - 1)$$

we plugged $x_a = 1$, $y_a = 3$, $x_b = 5$, $y_b = 9$, and $x = 3$ into the equation

$$y = 3 + \frac{6}{4}(2)$$

$$y = 3 + \frac{6 \times 2}{2 \times 2}$$

$$y = 3 + \frac{3 \times 2 \times 2}{2 \times 2}$$

$$y = 3 + 3 = 6.$$

Therefore, we obtained $y = 6$.

Example
Calculate y when x value is 70 from the given set of data below using linear interpolation.

x	$f(x)$
20	340
40	380
60	440
80	490
100	520
120	590

Solution

We know that the value of y can be derived when x is 70 using linear interpolation from the formula given as: $y = y_a + \frac{y_b - y_a}{x_b - x_a}(x - x_a)$. In this case we can use the two points low and up of $x = 70$ such as $x_a = 60$, $y_a = 440$, $x_b = 80$, and $y_b = 490$. So,

$$y = y_a + \frac{y_b - y_a}{x_b - x_a}(x - x_a)$$

$$y = 60 + \frac{490 - 440}{80 - 60}(70 - 60)$$

we plugged $x_a = 60$, $y_a = 440$, $x_b = 80$, $y_b = 490$, and $x = 70$ into the equation

$$y = 60 + \frac{50}{20}(10)$$

$$y = 60 + \frac{50 \times 10}{2 \times 10}$$

$$y = 60 + \frac{25 \times 2 \times 10}{2 \times 10}$$

$$y = 60 + 25 = 85.$$

Therefore, we found $y = 85$ when x value is 70.

Example

The number of bacteria per unit volume existing in an incubation after x hours is presented in the following table. Find the number of bacteria per volume for the time of 4.5 hours, using linear interpolation.

Hour	Bacteria/unit volume
0	0
1	35
2	66
3	98
4	144
5	256
6	321

Solution

The reference table does not establish a value that indicates the number of bacteria per volume for the time of 4.5 hours. However, we do have higher and lower values corresponding to a time of 4 and 5 hours, respectively. We learned that having two values such as, (x_a, y_a) and (x_b, y_b), the linear interpolant formula for finding y at a point x, is given by: $y = y_a + \frac{y_b - y_a}{x_b - x_a}(x - x_a)$. In this case, $x_a = 4$, $y_a = 144$, $x_b = 5$, $y_b = 256$ and $x = 4.5$. So,

$$y = y_a + \frac{y_b - y_a}{x_b - x_a}(x - x_a)$$

$$y = 144 + \frac{256 - 144}{5 - 4}(4.5 - 4)$$

we plugged $x_a = 4$, $y_a = 144$, $x_b = 5$, $y_b = 256$ and $x = 4.5$ into the equation

$$y = 144 + \frac{112}{1}(0.5)$$

$$y = 144 + 56 = 200.$$

Therefore, we found the number of bacteria per volume for the time of 4.5 hours equal to 200 units, which represents an intermediate level between the volume of bacteria existing at 4 and 5 hours.

2.2. Extrapolation

Extrapolation is another type of estimation. Extrapolation is another process of estimation, beyond the distinct range of the known variable or observation, based on its relationship with another variable. In a general sense, to extrapolate is to infer something that is not explicitly stated from existing information. It is an estimation method that works beyond the original observation range of the value of a variable based on its relationship with another variable. The action of estimating or concluding something by assuming that existing trends will continue, or a current method will remain applicable. Although we do not have a reason to believe that patterns will continue outside of an interval that we have known values for, it would be essential to assume it follows the same pattern to apply extrapolation method.

Although interpolation and extrapolation sound similar and are both methods of estimating hypothetical values, they have different purposes and work well in different scenarios. Interpolation is used to predict values that exist within a data set, and extrapolation is used to predict values that fall outside of a data set and use known values to predict unknown values. Often, interpolation is more reliable than extrapolation, but both types of prediction can be valuable for different purposes.

Extrapolation methods are a vital component in mathematics. It also has its branches in science, statistics, sociology, psychology, and other fields of study, with some categorical data.

2.2.1. Extrapolation Methods

Extrapolation methods are of three types, linear, conic, and polynomial extrapolation. Given below is a brief description of these methods:

- **Linear Extrapolation Method:** Linear extrapolation means drawing a tangent line at the end of the known data and extending it beyond that limit. For any linear function, linear extrapolation provides a good result when the point to be predicted is not too far from the given data.

- **Conic Extrapolation Method:** In a conic extrapolation, a conic section is created with the help of five points nearer to the end of the given data. The conic section will curve back on itself if it is a circle or ellipse. But for parabola or hyperbola, the curve will not back on itself because it is relative to the x-axis.

- **Polynomial Extrapolation Method:** In a polynomial extrapolation, a polynomial curve is created with the help of entire known data or near the endpoints. The resulting curve will then be extended beyond the end of the known data.

Typically, the quality of a particular method of extrapolation is limited by the assumptions about the function made by the method. If the method assumes the data are smooth, then a non-smooth function will be poorly extrapolated.

As for the interpolation, in this manuscript we limit the discussion of the extrapolation to the linear method only. Next, we will look at this the linear extrapolation method.

2.2.2. Linear Extrapolation

The extrapolation method (i.e., curve fitting) is a simplistic model that uses past gross population trends to project future population levels. The defining characteristics of trend of linear extrapolation are that future values of any variable are determined solely by its historical values.

2.2.3. Linear Extrapolation Formula

Let us denote the two data points nearest the point x to be extrapolated as (x_a, y_a) and (x_b, y_b). Then the linear extrapolation formula gives the function:

$$y(x) = y_a + \frac{y_b - y_a}{x_b - x_a}(x - x_a).$$

Linear extrapolation formula has some advantages and disadvantages. Its advantages are low data requirements, very easy methodology and low resource requirements. Its disadvantages are, the use of only aggregated data, and the assumption that past trends will predict future trends.

Example:
Suppose Jennifer is willing to estimate the amount of money needed to pay for gas during the next year, based on the average annual distance she is travelling each year. She has been laboriously tracking her car's efficiency for the last four years. So, she tabulated the following data from her computer. Find the cost of gas if she wants to travel 21,500 miles in the coming year.

year	Milage (miles/year)	Cost of gas ($/year)
1	18,678	3,688.58
2	22,930	4.403.89
3	19,743	3,874.53
4	20,986	4,119.24

Solution
We learned the linear extrapolation formula to extrapolate y of x with its two nearest data points (x_a, y_a) and (x_b, y_b), is $y(x) = y_a + \frac{y_b - y_a}{x_b - x_a}(x - x_a)$. In this case, we can select $(x_a, y_a) = (19.743, 3,874.53)$ and $(x_b, y_b) = (20,986, 4,119.24)$ to find $f(x)$ for $x = 21,500$. So,

$$y(x) = y_a + \frac{y_b - y_a}{x_b - x_a}(x - x_a)$$

$y(21{,}500) = 3{,}874.53 + \frac{4{,}119.24 - 3{,}874.53}{20{,}986 - 19.743}(21{,}500 - 19{,}743)$

we plugged $x_a = 19.743$, $y_a = 3{,}874.53$ $x_b = 20{,}986$, $y_b = 4{,}119.24$ and $x = 21{,}500$ into the equation

$y(21{,}500) = 3{,}874.53 + \frac{244.71}{1{,}243}(1{,}757)$

$y(21{,}500) = 3{,}874.53 + \frac{244.71 \times 1{,}757}{1{,}243}$

$y(21{,}500) = 3{,}874.53 + \frac{429{,}955.47}{1{,}243}$

$y(21{,}500) = 3{,}874.53 + 345.90 = 4{,}220.43.$

Therefore, the cost of gas for Jenifer to travel 21,500 miles in the coming year is estimated as $4,220.43.

Example:
Charles is reviewing details of sales and profits of his corporation. He is anxious to find out the profits of his corporation when the sales figures reach $8,000,000. Calculate his corporation profits based on the given data:

Sales ($×1000)	Profits ($×1000)
1,000	160
2,000	360
3,000	570
4,000	800
5,000	1,100
6,000	1,400

Solution
We know that if the two data points nearest the point x to be extrapolated are (x_a, y_a) and (x_b, y_b), then the linear extrapolation formula is given by, $y(x) = y_a + \frac{y_b - y_a}{x_b - x_a}(x - x_a)$.
In this case we can select $(x_a, y_a) = (5{,}000, 1{,}100)$ and $(x_b, y_b) = (6{,}000, 1{,}400)$ to find $f(x)$ for $x = 8{,}000{,}000$ (8,000 $K). So,

$y(x) = y_a + \frac{y_b - y_a}{x_b - x_a}(x - x_a)$

$y(8{,}000) = 1{,}100 + \frac{1{,}400 - 1{,}100}{6{,}000 - 5{,}000}(8{,}000 - 5{,}000)$

we plugged $x_a = 5{,}000$, $y_a = 1{,}100$, $x_b = 6{,}000$, $y_b = 1{,}400$, and $x = 8{,}000$ into the equation

$y(8{,}000) = 1{,}100 + \frac{300}{1{,}000}(3{,}000)$

$y(8{,}000) = 1{,}100 + \frac{300 \times 3{,}000}{1{,}000}$

$y(8{,}000) = 1{,}100 + \frac{300 \times 3 \times 1{,}000}{1{,}000}$

$y(8{,}000) = 1{,}100 + 900 = 2{,}000.$

Therefore, the profits of Charles corporation when the sales figure reaches \$8,000,000 are estimated as \$2,000,000.

Example:
Consider the data below. Use extrapolation with a linear function to estimate the value of y for $x = 1.436$.

x	y
– 0.800	0.102
– 0.437	0.147
– 0.018	0.175
0.122	0.193
0.301	0.209
0.509	0.242
0.639	0.276
0.818	0.292
1.008	0.316

Solution
Again, we know that the linear extrapolation formula for estimating y values for x using the nearest given the two points data (x_a, y_a) and (x_b, y_b) is $y(x) = y_a + \frac{y_b - y_a}{x_b - x_a}(x - x_a)$. In this case we can select the two last points of data $(x_a, y_a) = (0.818, 0.292)$ and $(x_b, y_b) = (1.008, 0.316)$ to find y for $x = 1.436$. So,

$$y(x) = y_a + \frac{y_b - y_a}{x_b - x_a}(x - x_a)$$

$$y(1.436) = 0.292 + \frac{0.316 - 0.292}{1.008 - 0.818}(1.436 - 0.818)$$

we plugged $x_a = 0.818$, $y_a = 0.292$, $x_b = 1.008$, $y_b = 0.316$, and $x = 1.436$ into the equation

$$y(1.436) = 0.292 + \frac{0.024}{0.190}(0.618)$$

$$y(1.436) = 0.292 + \frac{0.024 \times 0.618}{0.190}$$

$$y(1.436) = 0.292 + \frac{0.014832}{0.190}$$

$$y(1.436) = 0.292 + 0.078 = 0.370.$$

Therefore, the estimated value of y for $x = 1.436$ is $y = 0.370$.

2.3. Exercises

1. Find the value of y at $x = 4.2$ given the following data, using linear interpolation.

x	y
0	1.28
1	1.56
2	1.86
3	1.13
4	1.47
5	1.89
6	2.21

2. Find the value of y at $x = 2.6$ given the following linear graph, using linear interpolation.

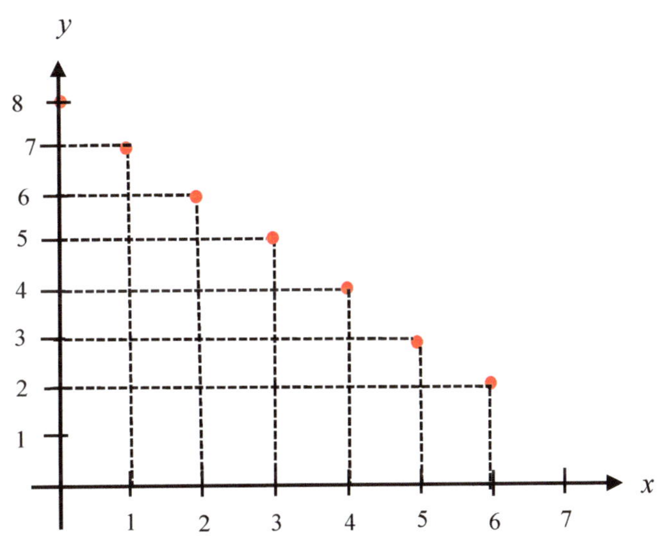

3. Find the value of y at $x = 6.5$ given the following nonlinear graph, using linear interpolation.

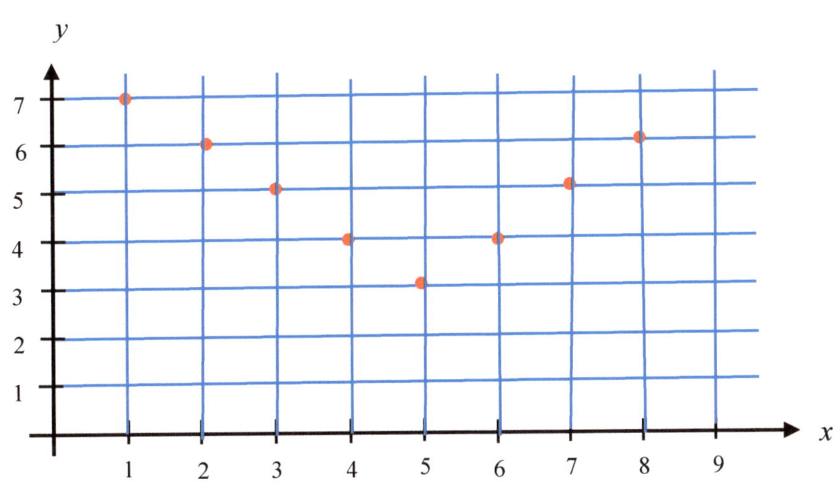

4. Find the value of y at $x = 9.2$ given some set of values (8, 5.88), (10, 7.94) using linear interpolation.

5. Find the value of y at $x = 8$ given some set of values (4, 1), (12, 12) using linear interpolation.

6. Find the value of y at $x = 1.3$ given some set of values (0, 6), (2, 14) using linear interpolation.

7. In the example that Jennifer wanted to estimate the amount of money needed to pay for gas during the next year, find the cost of gas if she wants to travel 22,000 miles in the coming year.

8. In the example that Charles was reviewing details of sales and profits of his corporation, suppose he is anxious to find out the cost of his products when the production figure reaches 800,000 units per year. Calculate product cost based on the given data:

Production quantity	Product cost ($)
100,000	45,000
200,000	84,000
300,000	114,000
400,000	144,000
500,000	165,000
600,000	186,000

9. Find the value of y at $x = 4.5$ given some set of values (1, 766), (3, 584), using the linear extrapolation method.

10. Find the value of y at $x = 24$ given some set of values (18, 1.06), (21, 13.87), using the linear extrapolation method.

11. Find the value of y at $x = 9$ given the following nonlinear graph, using linear extrapolation.

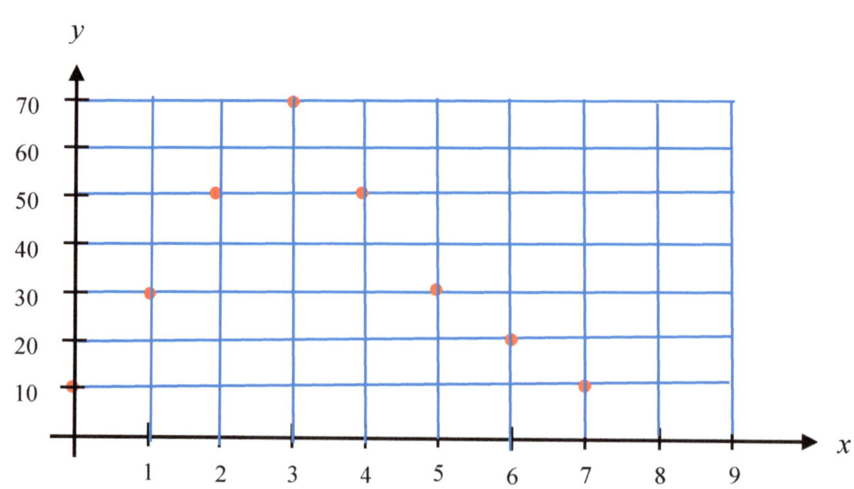

12. Find the value of y at $x = -8.5$ given some set of values $(-5, 18)$, $(-7, 22)$, using the linear extrapolation method.

13. Find the value of y at $x = -38$ given some set of values $(-34, 1.68)$, $(-36, 2.12)$, using the linear extrapolation method.

14. Consider the following data, use extrapolation with a linear function to estimate the value of y for $x = 7.54$.

x	y
0	42.66
1	41.23
2	40.08
3	39.14
4	38.24
5	37.37
6	36.52

3. Linear Equation in One Variable

This section focuses on linear equations in one variable, which represent straight lines and are written with only the variable raised to the power of one. Such equations take the form $ax + b = 0$, where a and b are real numbers, and a cannot be zero. For instance, the equation $3x + 2 = 14$ exemplifies a linear equation with a single variable. Consequently, such equations possess only one solution, like in this case, $x = 4$. Conversely, linear equations in two or more variables may entail multiple solutions. This section delves into the definition, methods for finding intercepts and solutions, along with some examples, pertaining to linear equations in one variable.

3.1. Solving Linear Equation

To solve a linear equation of one variable, is to find the x value to satisfy the equation. There are two basic methods for solving linear equations. The first method is graphing by plotting points. The second method is the algebraic method.

In the graphical method, we simply apply the rule of finding the x intercept. In the algebraic method, we find the intercepts and perform algebraic operations.

Finding the intercepts are the core steps of both methods. So, let us first explain the procedure of finding the intercepts.

3.2. Given a Linear Equation, Finding the Intercepts

Given an equation for a line, the intercepts of the line are points at which the line intercepts, or crosses, the horizontal and vertical axes. The x–intercept is the point at which the graph crosses the x–axis. At this point, the y–coordinate is zero. The y–intercept is the point at which the graph crosses the y–axis. At this point, the x–coordinate is zero.

Based on these facts, to find the x–intercept, we must set y equal to zero and solve the equation to get x value. Using the same method, for determining the y–intercept, we must set x equal to zero and solve the equation to get y value.

Example
Given $y = 5x - 3$, find the x–intercept and the y intercept.

Solution

a) To find y–intercept, set $x = 0$ and solve $y = 5x - 3$ to get y:
$$y = 5(0) - 3 = 3.$$

b) To find x–intercept, set $y = 0$ and solve $y = 5x - 3$ to get x:
$$0 = 5(x) - 3$$
$$3 = 5(x)$$
$$x = \frac{3}{5}.$$

Example
Given $y = 15x - 5$, find the x–intercept and the y intercept.

Solution

a) To find y–intercept, set $x = 0$ and solve $y = 15x - 5$ to get y:
$y = 15(0) - 5 = -5.$

b) To find x–intercept, set $y = 0$ and solve $y = 15x - 5$ to get x:
$0 = 15(x) - 5$

$5 = 15(x)$

$x = \frac{5}{15} = \frac{1}{3}.$

> **Given a line equation, finding the intercepts:**
>
> 1. Find the x-intercept by setting $y = 0$ and solving for x.
> 2. Find the y-intercept by setting $x = 0$ and solving for y.

3.3. Given a Linear Equation, Graphing by Plotting Points

To graph an equation by plotting points, we can execute the following steps:

1. Make a table with a first column labeled x, a second column labeled with the equation, and a third column listing the resulted ordered pairs.
2. Enter x–values down the first column using positive and negative values. To make the graphing simpler, select the x–values in numerical order.
3. Select x–values so that they yield y–values.
4. Plot the ordered pairs and connect the points to form a line.

Example
Graph the equation $y = -2x - 1$ by plotting points.

Solution
Using the above instruction, we must first construct a table.

x–value	$y = -2x - 1$	(x, y)
-3	$y = -2(-3) - 1$	$(-3. 5)$
-2	$y = -2(-2) - 1$	$(-2. 3)$
-1	$y = -2(-1) - 1$	$(-1. 1)$
0	$y = -2(0) - 1$	$(0. -1)$
1	$y = -2(1) - 1$	$(1, -3)$
2	$y = -2(2) - 1$	$(1, -5)$

Next, we plot the ordered pairs and connect the points to get the line.

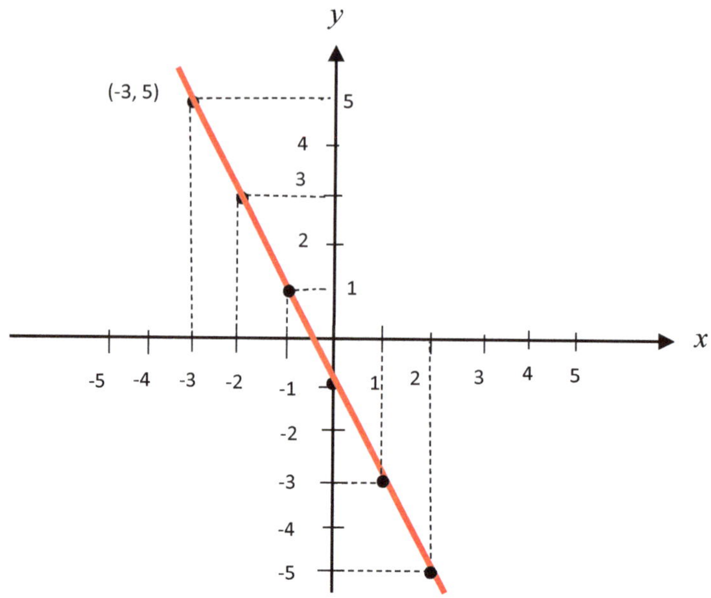

3.4. Solving a Linear Equation, Graphical Method

To solve a linear equation of one variable is to find the x value to satisfy the equation. In graphical method, we simply apply the rule of finding the x intercept.

Example

Solve the equation graphically.

a) $x - 4 = 0$.
b) $5x + 10 = 0$.

Solution

a) The graph of $x - 4 = 0$ is depicted below.

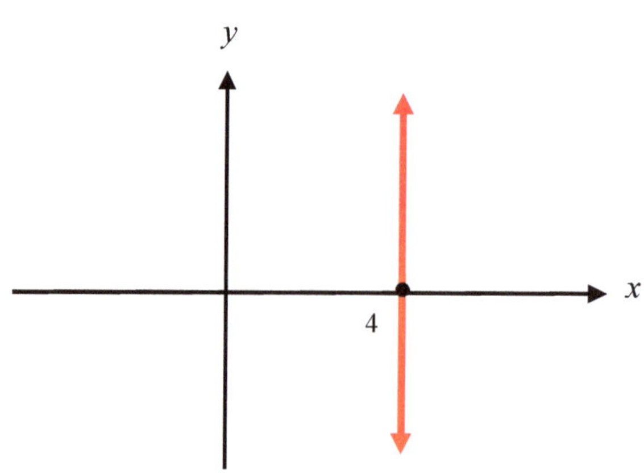

We must perform operations to find the x intercept.

$x - 4 = 0$

$x - 4 + 4 = 0 + 4$ — we added 4 to both sides of equation to isolate x in one side of the equation

$x + 0 = 4$

$x = 4.$

b) The graph of $5x + 10 = 0$ is depicted below.

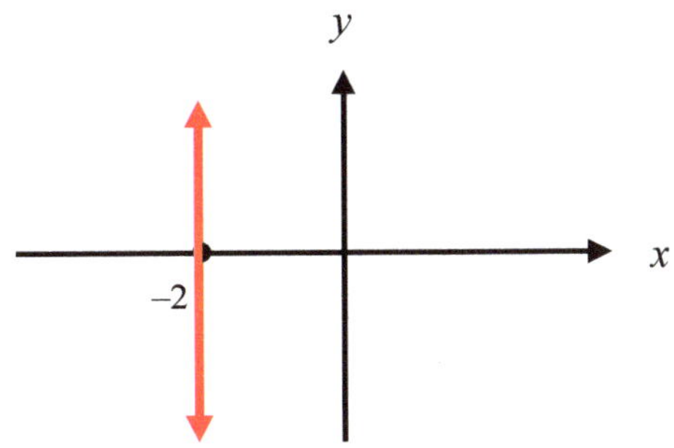

$5x + 10 = 0$

$5x + 10 - 10 = -10$ — we added – 10 to both sides of equation to isolate x in one side of the equation

$5x = -10$

$\frac{1}{5} \times 5x = \frac{1}{5} \times (-10)$ — we multiplied both sides of the equation by the reciprocal of the x coefficient

$x = -2.$

Example

Solve the equation graphically.

a) $y = x - 8.$

b) $y = 4x + 8.$

c) $y = -5x - 10.$

Solution

a) The graph of $y = x - 8$ is depicted below.

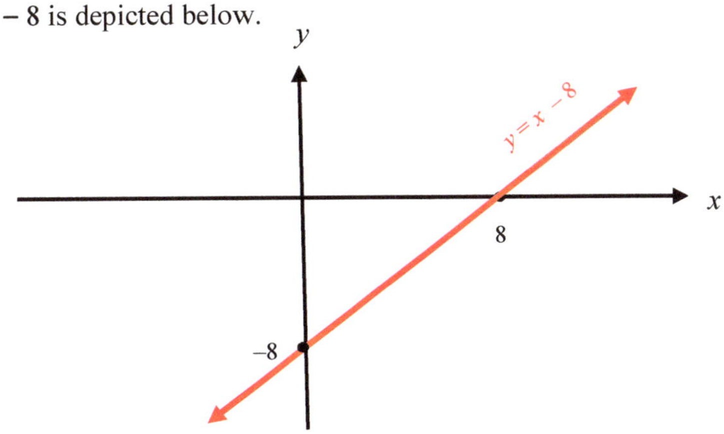

We simply apply the rule of finding the x intercept, $y = 0$. We must then perform operations to find x intercept.

$y = x - 8$

$x - 8 = 0$

$x - 8 + 8 = 0 + 8$ we added 8 to both sides of equation to isolate x in one side of the equation

$x + 0 = 8$

$x = 8.$

b) The graph of $y = 4x + 8$ is depicted below.

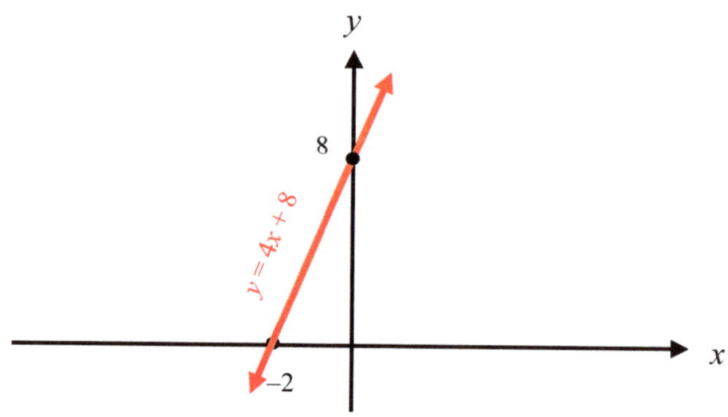

$y = 4x + 8$

$4x + 8 = 0$

$4x + 8 - 8 = -8$ we added – 8 to both sides of equation to isolate x in one side of the equation

$4x = -8$

$\frac{1}{4} \times 4x = \frac{1}{4} \times (-8)$ we multiplied both sides of the equation by the reciprocal of the x coefficient

$x = -2.$

c) The graph of $y = -5x - 10$ is depicted below.

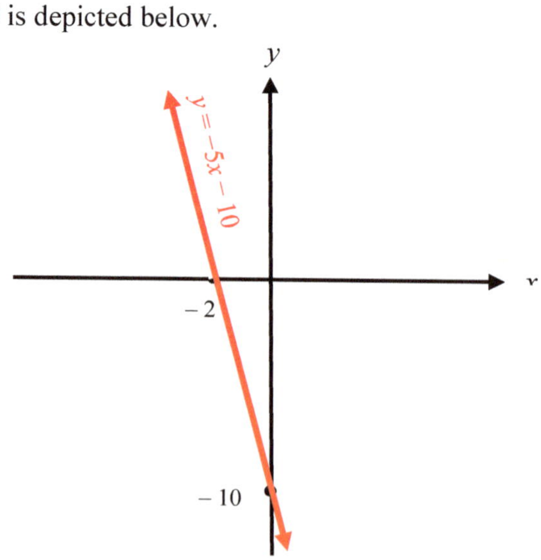

$y = -5x - 10$

$-5x - 10 = 0$

$-5x - 10 + 10 = 0 + 10$ — we added 10 to both sides of the equation to isolate x in one side of the equation

$-5x = 10$

$-\frac{1}{5} \times (-5x) = -\frac{1}{5} \times 10$ — we multiplied both sides of the equation by the reciprocal of the x coefficient

$x = -2.$

3.5. Solving Linear Equations, Algebraic Method

As we learned, to solve a linear equation of one variable, is to find the x value to satisfy the equation. The following steps are used to manipulate an equation and isolate the unknown variable, so that the last line we get the value of x, where x is the unknown.

1. We may add, subtract, multiply, or divide (a non–zero value) an equation by a number or an expression if we do the same thing to both sides of the equal sign.
2. Apply the distributive property as needed: $a(b + c) = ab + ac$.
3. Isolate the variable on one side of the equation and the constant on the other side of equation.
4. When the variable is multiplied by a coefficient in the final stage, multiply both sides of the equation by the reciprocal of the coefficient.

Example
Solve the equation, $4(x - 3) + 12 = 15 - 5(x + 6)$.

Solution
Applying algebraic operations,

$4(x - 3) + 12 = 15 - 5(x + 6)$

$4x - 12 + 12 = 15 - 5x - 30$ — we applied the distribution property

$4x = -5x - 15$ — we combined the like terms

$4x + 5x = -15$ — we placed the x–terms on one side of the equation

$9x = -15$ — we simplified

$\frac{1}{9} \times 9x = \frac{1}{9} \times (-15)$ — we multiplied both sides by $\frac{1}{9}$

$x = -\frac{15}{9} = -\frac{5}{3}.$

3.6. Linear Inequalities

In mathematics, inequality denotes a mathematical expression where both sides are not equal. It signifies a comparison between two expressions or numbers where one is not equal to the other. In such cases, the equal sign "=" is replaced by inequality symbols such as greater than (">"), less than ("<"), greater than or equal to (≥), less than or equal to (≤), or not equal to (≠). Various types of inequalities in mathematics include polynomial inequalities, rational inequalities, and absolute value inequalities.

The symbols ">" and "<" represent strict inequalities, indicating a clear distinction between values. On the other hand, "≥" and "≤" denote slack inequalities, allowing for the possibility of equality between the expressions.

Although linear inequalities resemble linear equations in appearance, they differ in the symbol used to relate the two expressions, signifying a comparison rather than equality.

Symbols of Inequality
A linear inequality contains one of the symbols of inequality:

- "<" : less than
- ">" : greater than
- "≤" : less than or equal to
- "≥" : greater than or equal to.

3.7. Properties of Inequalities

Working with inequalities would be like but not exactly as working with equalities. To describe the differences, we will introduce two properties which are applicable inequalities. These properties are the addition property and the multiplication property.

Here are properties of inequalities:

- **Addition Property**– for any real numbers a, b, and c, if $a < b$, then $a + c < b + c$.
- **Multiplication Property**– for any real numbers a, b, and c, if $a < (>)\ b$ and $c > 0$, then $ac < (>)\ bc$. If $a < (>)\ b$ and $c < 0$, then $ac > (<)\ bc$.

These properties apply to the symbol less–than–and–equal, ≤, and the symbol greater–than–and–equal, ≥, as well.

Note that when we multiply or divide by a negative number, the inequality symbol is reversed.

Properties of Inequalities

Addition Property	If $a < b$, then $a + c < b + c$.
Multiplication Property	If $a < b$ and $c > 0$, then $ac < bc$.
	If $a < b$ and $c < 0$, then $ac > bc$.

The above properties also apply to $a \leq b$, and $a \geq b$.

Example
Illustrate the addition property for inequalities by solving each of the following:

a) $x - 13 < 4$.

b) $8 \geq x - 4$.

c) $x + 11 > 7$.

Solution

a) By the addition property for inequalities, adding or subtracting the same number on both sides does not change the inequality, then:

$x - 13 < 4$

$x - 13 + 13 < 4 + 13$ we added 13 to both sides

$x < 17.$

b) $8 \geq x - 4$:

$8 + 4 \geq x - 4 + 4$ we added 4 to both sides

$8 + 4 \geq x + 0$

$x \leq 12.$

c) $x + 11 > 7$:

$x + 11 + (-11) > 7 + (-11)$ we added (– 11) to both sides

$x + 11 - 11 > 7 - 11$

$x + 0 > 7 - 11$

$x > -4.$

Example
Illustrate the multiplication property for inequalities by solving each of the following:

a) $4x > 12.$

b) $-3x - 4 \leq 5.$

c) $9x + 8 > 35.$

Solution

a) $4x > 12$:

$\frac{1}{4} \times 4x > \frac{1}{4} \times 12$ we multiplied each side by $\frac{1}{4}$

$x > 3.$

b) $-3x - 4 \leq 5$:

$-3x - 4 + 4 \leq 5 + 4$ we added 4 to both sides

$-3x \leq 9$

$-3x \times (-\frac{1}{3}) \geq 9 \times (-\frac{1}{3})$ we multiplied by $-\frac{1}{3}$, the symbol was changed to "≥"due to negative multiplication

$x \geq -3.$

c) $9x + 8 > 35$:

$9x + 8 + (-8) > 35 + (-8)$ we added (– 8) to both sides

$9x > 27$

$9x \times \frac{1}{9} > 27 \times \frac{1}{9}$ we multiplied each side by $\frac{1}{9}$

$x > 3.$

3.8. Solving Inequalities in One Variable

As mentioned earlier, working with inequalities is somewhat like working with equalities. That is; we can perform the same operations on both sides of an inequality, just as we do with equations, combining similar terms and perform operations.

To solve an inequality, we isolate the variable and solve to get a range for the variable.

Example
Solve the inequality: $21 - 2x \geq 3x - 9$.

Solution

$21 - 2x \geq 3x - 9$

$-3x - 2x \geq -21 - 9$ — we moved variable terms to one side of the inequality and constant terms to the other side of the inequality

$-5x \geq -30$

$-5x \times (-\frac{1}{5}) \leq -30 \times (-\frac{1}{5})$ — we multiplied each side by $-\frac{1}{5}$, the symbol was changed to "$\leq$"

$x \leq 6$.

Example
Solve the inequality: $1 - \frac{1}{5}x \geq \frac{1}{15}x - 3$.

Solution

$1 - \frac{1}{5}x \geq \frac{1}{15}x - 3$

$-\frac{1}{5}x - \frac{1}{15}x \geq -3 - 1$ — we moved variable terms to one side of the inequality and constant terms to the other side of the inequality

$-\frac{3}{15}x - \frac{1}{15}x \geq -3 - 1$ — we wrote fractions with a common denominator

$-\frac{4}{5}x \geq -4$

$-\frac{4}{5}x \times (-\frac{4}{5}) \geq -4 \times (-\frac{4}{5})$ — we multiplied each side by $-\frac{5}{4}$, the symbol was changed to "$\leq$"

$x \leq 5$.

3.9. Graphing Linear Inequalities

When you are graphing inequalities, you will graph the ordinary linear functions. The difference is that the solution to the inequality is not the drawn line but the area of the coordinate plane that satisfies the inequality.

Given linear inequalities, $y > mx + b$, the first thing is to make sure that variable y is by itself on the left side of the inequality symbol, which is the case in this problem. The next step is to graph the boundary line by momentarily changing the inequality symbol to equality symbol. In other words, the steps to graph a linear inequality are.

1. Reorder the equation so "y" is on the left and everything else on the right.
2. Plot the "$y =$" line.
3. Shade above the plotted line for a "greater than" ($y >$ or $y \geq$) or below the plotted line for a "less than" ($y <$ or $y \leq$).

Example
Graph the linear inequality $y \geq \frac{1}{2}x - 1$.

Solution
As mentioned in the above steps, we need first to change the inequality symbol to equality symbol. So,

$$y \geq \frac{1}{2}x - 1$$

$$y = \frac{1}{2}x - 1$$ we changed the greater symbol, "≥", to equality symbol "=".

Next, we graph the line $y = \frac{1}{2}x - 1$, in the xy axis. In the figure below, the plotted line presents the graph of line $y = \frac{1}{2}x - 1$. Observe that this line cuts the plane into two half–planes. All points on one side of this line (including the line itself) satisfy $y \geq \frac{1}{2}x - 1$, whereas on the other side, $y \leq \frac{1}{2}x - 1$. Note that $(x_1, x_2) = (0, 0)$ satisfies $y \geq \frac{1}{2}x - 1$, and that (0, 0) lies in the upper–left direction of the line. Therefore, the solution of the inequality, $y \geq \frac{1}{2}x - 1$ is the set points on the upper–left portion of the plane, indicted by two arrows.

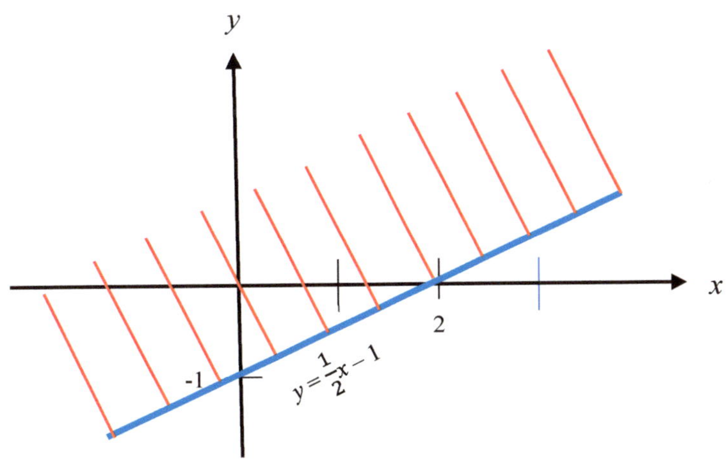

Example
Graph the linear inequality $y \leq -3x + 6$.

Solution
Using the steps of graphing linear inequalities, we need first to change the inequality symbol to equality symbol. So,

$$y \leq -3x + 6$$

$y = -3x + 6$ we changed the greater symbol, "≤", to equality symbol "=".

Next, we graph the line $y = -3x + 6$, in the xy axis. In figure below, the plotted line presents the graph of line $y = -3x + 6$. Observe that this line cuts the plane into two half–planes. All points on one side of this line (including the line itself) satisfy $y \leq -3x + 6$, whereas on the other side, $y \geq -3x + 6$. Note that $(x_1, x_2) = (0, 0)$ satisfies $y \leq -3x + 6$, and that (0, 0) lies in the lower–left direction of the line. Therefore, the solution of the inequality, $y \leq -3x + 6$ is the set points on the lower–left portion of the plane, indicated by two arrows.

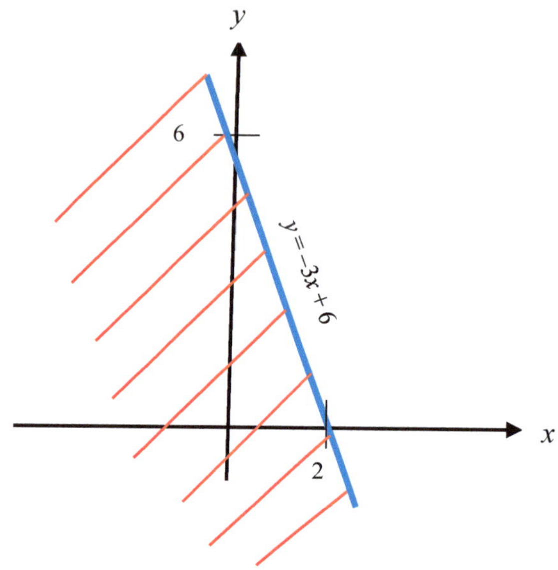

Example
Graph y ≥ 3.

Solution
Like what we have done in the two previous examples, we first plot the line y = 3.

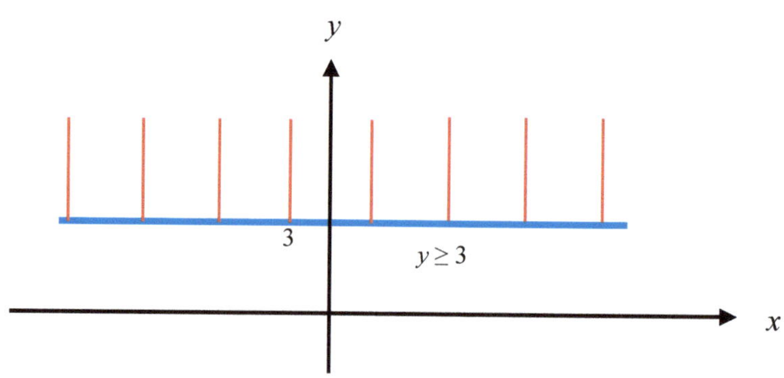

Here we observe that this line cuts the plane into two half–planes, one is on the upper side and the other is on the lower side of the line. All points on one side of this line satisfy y ≥ 3, whereas on the other side, y ≤ 3. It is seen that that $(x_1, x_2) = (0, 0)$ satisfies $y \leq 3$, so we are interested in the other side.

Therefore, the solution of the inequality, y ≥ 3 is the set points on the upper portion of the plane, indicated by two arrows.

Example
Graph $x \leq 5$.

Solution
Like what we have done in the two previous examples, we first plot the line $x = 5$. Here we observe that this line cuts the plane into two half–planes, one is on the left side and the other is the right side of the line.

All points on one side of this line satisfy $x \leq 5$, whereas on the other side, $x \geq 5$. It is seen that that $(x_1, x_2) = (0, 0)$ satisfies $x \leq 5$. Therefore, the solution of the inequality, $x \leq 5$ is the set points on the left portion of the plane, indicated by two arrows.

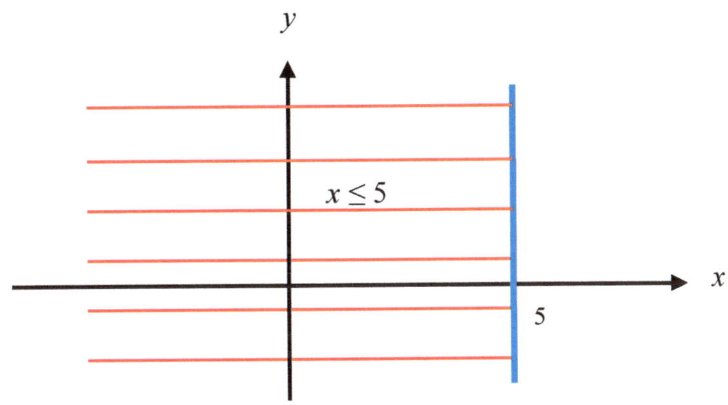

3.10.1. Evaluating Inequality Expressions

To evaluate an inequality expression, you need to substitute a number for the variable and perform the arithmetic operations.

Example
Evaluate $y < 9x - 12$ at:

a) $x = 22$.

b) $x = -4$.

c) $x = 0$.

Solution

a) We just put the value of $x = 22$ into the inequality. So,

$y < 9x - 12$

$y < 9\,(22) - 12$ we substituted $x = 22$ into the inequality

$y < 198 - 12$

$y < 188$.

b) Like above we just put the value of $x = -4$ into the inequality. So,

$y < 9x - 12$

$y < 9\,(-4) - 12$ we substituted $x = -4$ into the inequality

$y < -36 - 12$

$y < -48$.

c) Again, we just put the value of $x = 0$ into the inequality. So,

$y < 9x - 12$

$y < 9\,(0) - 12$ — we substituted $x = 0$ into the inequality

$y < -12.$

3.10.2. Compounded Inequalities

A compound inequality comprises two inequalities in one statement. A compound inequality contains at least two inequalities that are separated by either "and" or "or".

Inequalities Joined by And- When two inequalities are joined by the word **and**, the solution of the compound inequality occurs when **both** inequalities are true at the same time. It is the overlap, or intersection, of the solutions for each inequality. As an example, $-3 < x \leq 7$ is a compounded inequality which implies, $-3 < x$ and $x \leq 7$.

Inequalities Joined by Or - When the two inequalities are joined by the word **or**, the solution of the compound inequality occurs when **either** of the inequalities is true. The solution is the combination, or union, of the two individual solutions.

Example

If $-\frac{22}{7} < -3x + 7 < -\frac{17}{8}$, what is the smallest possible integer value t by which $9x - 21 < t$?

Solution

We can first rewrite the compounded inequality $-\frac{22}{7} < -3x + 7 < -\frac{17}{8}$ as two inequalities, $-\frac{22}{7} < -3x + 7$ and $-3x + 7 < -\frac{17}{8}$. Considering the first one we have:

$-\frac{22}{7} < -3x + 7.$

It can be observed that $9x - 21$ is minus three times of $-3x + 7$. So, we need to multiply both sides by minus three. Then we have

$-3\times(-\frac{22}{7}) > -3\times(-3x + 7)$ — we multiplied both sides by -3, and we changed the less that symbol "<" to greater symbol ">"

$\frac{66}{7} > 9x + -21$

$9.43 > 9x - 21.$

The smallest integer value that the left side can have is $10 > 9.43 > 9x - 21$. So, the smallest possible integer value of t is 10.

3.11. Solving and Graphing Compounded Inequalities Joined by And

Steps to solve compounded inequalities combined by the word and is.

1. Solve each part of the inequality separately.
2. Graph them on the number line.

Let us look at an example.

Example
Solve and graph the compound inequalities $3x - 4 < 8$ and $3x - 4 \geq 5$, or equivalently, $5 \leq 3x - 4 < 8$.

Solution
First, we solve $3x - 4 < 8$.

$3x - 4 < 8$

$3x - 4 + 4 < 8 + 4$ we added 4 to both sides of equation

$3x < 12$

$3x \times \frac{1}{3} < 12 \times \frac{1}{3}$ we multiplied $\frac{1}{3}$ to both sides of equation

$x < 4.$

Next, we solve $3x - 4 \geq 5$.

$3x - 4 \geq 5.$

$3x - 4 + 4 \geq 5 + 4$ we added 4 to both sides of equation

$3x \geq 9$

$3x \times \frac{1}{3} \geq 9 \times \frac{1}{3}$ we multiplied $\frac{1}{3}$ to both sides of equation

$x \geq 3.$

The solution of $3x - 4 < 8$ and $3x - 4 \geq 5$, is $x < 4$ and $x \geq 3$. Or equivalently $3 \leq x < 4$.

The number line graph of $3x - 4 < 8$ is:

The number line graph of $3x - 4 \geq 5$ is:

The number line graph of $3x - 4 < 8$ and $3x - 4 \geq 5$ is:

3.12. Solving and Graphing Compounded Inequalities Joined by Or

Let's take a closer look at a **compound inequality** that uses or to combine two inequalities. Steps to solve compounded inequalities combined by the word or is.

1. Solve each part of the inequality separately.

2. Graph them on the number line.

Let us look at an example.

Example
Solve and graph the compound inequalities $5x - 3 < 7$ or $2x - 9 > 3$.

Solution
First, we solve $5x - 3 < 7$.

$5x - 3 < 7$

$5x - 3 + 3 < 7 + 3$ we added 3 to both sides of equation

$5x < 10$

$5x \times \frac{1}{5} < 10 \times \frac{1}{5}$ we multiplied $\frac{1}{5}$ to both sides of equation

$x < 2.$

Next, we solve $2x - 9 > 3$.

$2x - 9 > 3$

$2x - 9 + 9 > 3 + 9$ we added 9 to both sides of equation

$2x > 12$

$2x \times \frac{1}{2} > 12 \times \frac{1}{2}$ we multiplied $\frac{1}{2}$ to both sides of equation

$x > 6.$

The solution of $5x - 3 < 7$ or $2x - 9 > 3$, is $x < 2$ or $x > 6$.

The number line graph of $5x - 3 < 7$ is:

The number line graph of $2x - 9 > 3$ is:

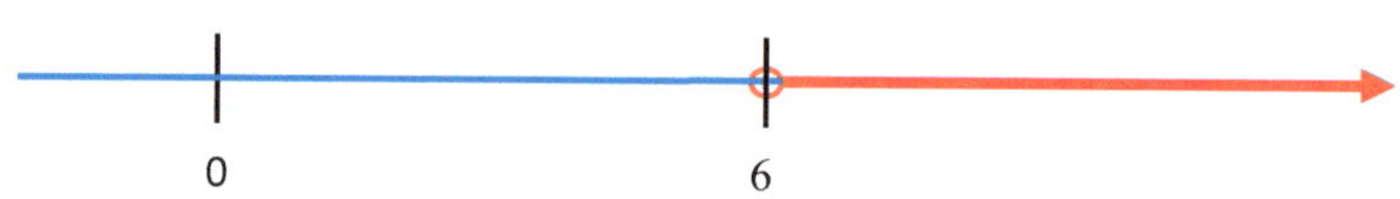

The number line graph of $5x - 3 < 7$ or $2x - 9 > 3$ is:

Referring to the above graph, it has an open blue circle on 6 and a blue arrow to the right and another open red circle at 2 and a red arrow to the left. In fact, the only parts of the number line that are not a solution to this compound inequality are points 2 and 6 and all the points in between these values, or equivalently, the range $2 \le x \le 6$ is not part of the solution. Everything else on the graph of number line is a solution to this compound inequality.

3.13. Exercises

1. For each of the equations below, find the x–intercept and the y–intercept without graphing. Write the coordinates of each intercept.
   a) $y = -3x + 6$.
   b) $4y = 2x - 1$.
   c) $3x - 2y = 6$.
   d) $4x - 3 = 2y$.

2. For each of the equations below, solve the equation for y in terms of x.
   a) $4x + 2y = 8$.
   b) $3x - 2y = 6$.
   c) $2x = 5 - 3y$.
   d) $x - 2y = 7$.
   e) $5y + 4 = 10x$.
   f) $5x + 2y = 0$.

3. Graph the equation $y = -5x - 1$ by plotting points.

4. Solve the equation graphically.
   a) $7x - 14 = 0$.
   b) $2y + 6 = 0$.
   c) $y = 4x - 8$.

5. Solve the equation algebraically.
   a) $5(3x - 1) = 14 - 2(x + 3)$.
   b) $6(x - 4) + 10 = 16 - 3(x + 2)$.
   c) $18 - 5x = 3(x - 1)$.
   d) $7(x - 4) = 6x$.

6. Solve the equation algebraically.
   a) $\frac{3}{x} - \frac{1}{4} = 1 + \frac{1}{x}$.
   b) $\frac{2}{x+1} - \frac{1}{3(x+1)} = \frac{1}{8}$.
   c) $\frac{2}{x} - \frac{1}{4} = 0$.
   d) $\frac{7}{2(x-6)} - \frac{1}{4} + \frac{1}{x-6} = 0$.

7. There are two car rental companies that offer different rental deals. Company A charges \$45 for initial fees and \$15 per day. Company B charges \$8 for initial fees and \$27 per day. Determine which company offers the better plan:
   a) If you want to rent a car for 3 days.

b) If you want to rent a car for 5 days.

c) If you want to rent a car for 7 days.

Hint: you may use the equation $C = IF + rx$, where C is the total rental cost, IF is the initial fees, r is the daily rental charge, and x is the number of days.

8. Illustrate the addition property for inequalities by solving each of the following:
   a) $x - 11 < 3$.
   b) $9 \geq x - 5$.
   c) $x + 13 > 8$.
   d) $-5 + 3x \leq 12$.
   e) $18 - 8x \leq 2$.

9. The formula $C = 0.80x + 1200$ is the cost of goods made for a small manufacturing firm, and $R = 3.80x$ is the revenue for selling its goods, where x is the number of products.
   a) Find the number of products that will keep the revenue greater than cost.
   b) Graph the lines for the cost and the revenues.

10. Illustrate the multiplication property for inequalities by solving each of the following:
    a) $\frac{7}{9} + 2x > \frac{1}{3}$.
    b) $-6x - 10 \leq 2$.
    c) $11x + 19 > -14$.
    d) $\frac{1}{5}x + 4 \geq -3$.

11. Solve each inequality.
    a) $7 - 3x \geq 5x - 11$.
    b) $3 + \frac{5}{6}x \geq \frac{4}{9}x - \frac{4}{7}$.
    c) $\frac{1}{3} - 8x \geq -5x - \frac{5}{12}$.
    d) $8 - \frac{1}{3}x \geq 3x - \frac{3}{4}$

12. Graph each of the linear inequality.
    a) $0 \geq \frac{1}{8}x - 3$.
    b) $5y - 15 < 0$.
    c) $y \leq 2x - \frac{1}{2}$.

13. Evaluate $5y < 7x - 14$ at
    a) $x = 13$.
    b) $x = -3$
    c) $x = 0$.

14. Evaluate $-12 \leq -2x + 4 < 27$ at

a) $x = 8$.

b) $x = -2$.

c) $x = 0$.

d) $x = -\frac{1}{2}$.

15. Solve and graph each of the compound inequalities.

a) $x > -3$ and $x \leq 2$.

b) $x \leq 1$ and $x > -1$.

c) $5x - 1 < 24$ and $4x \geq 16$.

d) $6x - 5 < 37$ and $4x - 6 \geq 10$.

16. Solve and graph each of the compound inequalities.

a) $x < -2$ or $x \geq 3$.

b) $x \leq 0$ or $x > 5$.

c) $7x < 14$ or $8x \geq 32$.

d) $2x - 3 \leq -9$ or $8x - 5 > 11$.

17. Solve and graph each of the compound inequalities.

a) $3 < x - 5 < 4$.

b) $\frac{5}{3} \leq \frac{3}{4}x + \frac{1}{6} \leq \frac{29}{12}$.

c) $-\frac{2}{3} < x - \frac{2}{3} \leq \frac{1}{3}$.

d) $-1 \leq \frac{2}{3}x - 1 < 1$.

4. Other Types of Equalities and Inequalities

In addition to solving linear equations and inequalities through various methods, we'll explore a range of other equation and inequality types in this section. These include equations and inequalities involving absolute values, rational expressions, polynomials, radicals, and more. Despite the diversity of equation types, the fundamental principles of algebra remain consistent in solving them.

Regardless of the specific equation form, solving equations and inequalities relies on applying basic algebraic rules and techniques. By mastering these foundational principles, we can effectively tackle a wide array of mathematical problems across different equation types.

4.1. Absolute Value Equalities and Inequalities

In this section, we will discuss the equations and inequalities involving absolute value. First, we look at definition and properties of absolute value.

Absolute value definition - The absolute value of a real number x, denoted $|x|$, is the non–negative value of x regardless of its sign. If x is positive, and $|x| = -x$ if x is negative(in which case $-x$ is positive) , and $|0| = 0$. For example, the absolute value of 2 is 2, and the absolute value of -2 is $-(-2) = 2$.

Absolute Value Definition
The absolute value of a real number x, denoted $|x|$, is:

$$|x| = \begin{cases} x & \text{if } x \geq 0 \\ -x & \text{if } x < 0 \end{cases}.$$

Using the number line, the absolute value of a number may view as the distance, on the number line, of that number from zero. For example, to find find $|3|$ we use the rule $|x| = x$ so $|3| = 3$. Similarly, to find find $|-3|$ we use the rule $|x| = -x$ so $|-3| = -(-3) = 3$. Before we start explaining absolute value functions, let us present the properties of absolute value.

Properties of Absolute Value:
Let a, b and x be real numbers and let n be an integer, we then have:

- Product Rule: $|a \times b| = |a| \times |b|$
- Power Rule: $|a^n| = |a|^n$ whenever a^n is defined
- Quotient Rule: $|\frac{a}{b}| = \frac{|a|}{|b|}$, provided $b \neq 0$.

Equality Properties:

- $|x| = 0$ if and only if $x = 0$.
- For $c > 0$, $|x| = c$ if and only if $x = c$ or $-x = c$.
- For $c < 0$, $|x| = c$ has no solution.

4.1.1. Solving Equations Involving Absolute Value

For solving equations involving absolute value, we use the above properties, we then perform similar operations as we described on solving linear equations. To do so:

- Isolate the variable on one side of the equation.

- When the variable is multiplied by a coefficient in the final stage, multiply both sides of the equation by the reciprocal of the coefficient.

> **Absolute Value Equation**
> An absolute value equation in the form $|ax + b| = c$ has the following properties:
> - If $c < 0$, $|ax + b| = c$ has no solution.
> - If $c = 0$, $|ax + b| = c$ has one solution.
> - If $c > 0$, $|ax + b| = c$ has two solutions.

Example
Solve each equation.

a) $|4x - 2| = 10$.

b) $-|2x + 7| = 3$.

c) $-|4x + 3| = -7$.

Solution.

a) The equation $|4x - 2| = 10$ is of the form $|x| = c$, so by the equality properties, for $c > 0$, the equation is equivalent to $4x - 2 = 10$ or $-(4x - 2) = 10$. Solving the former, we have:

$4x - 2 = 10$

$4x = 10 + 2$ we isolated the variable on one side of the equation

$4x = 12$ we added 10 to 2

$\frac{1}{4} \times 4x = \frac{1}{4} \times 12$ we multiplied both sides of equation by the reciprocal of the variable coefficient

$x = 3$.

Now solving the second we have:

$-(4x - 2) = 10$

$-4x + 2 = 10$

$-4x = 10 - 2$ we isolated the variable on one side of the equation

$-4x = 8$ we added 10 to -2

$-\frac{1}{4} \times (-4x) = -\frac{1}{4} \times 8$ we multiplied both sides of equation by the reciprocal of the variable coefficient

$x = -2$.

Therefore the $|4x - 2| = 10$ has two answers, $x = 3$, and $x = -2$.

b) The equation $-|2x + 7| = 3$, can be rewritten as $|2x + 7| = -3$ by multiplying both sides by -1. Now it is of the form $|x| = -c$, so by the equality properties, for $c < 0$, has no solution.

c) The equation $-|4x + 3| = -7$ can be rewritten as $|4x + 3| = 7$ by multiplying both sides by -1. We now have an equation which is of the form $|x| = c$, so by the equality properties, for $c > 0$, the equation is equivalent to $4x + 3 = 7$ or $-(4x + 3) = 7$. Solving the former, we have:

$4x + 3 = 7$

$4x = 7 - 3$ — we isolated the variable on one side of the equation

$4x = 4$ — we added 7 to – 3

$\frac{1}{4} \times 4x = \frac{1}{4} \times 4$ — we multiplied both sides of equation by the reciprocal of the variable coefficient

$x = 1.$

Now solving the second we have:

$-(4x + 3) = 7$

$-4x - 3 = 7$

$-4x = 7 + 3$ — we isolated the variable on one side of the equation

$-4x = 10$ — we added 7 to 3

$-\frac{1}{4} \times (-4x) = -\frac{1}{4} \times 10$ — we multiplied both sides of equation by the reciprocal of the variable coefficient

$x = -\frac{10}{4} = -2.5.$

Therefore the $-|4x + 3| = -7$ has two answers, $x = 1$, and $x = -2.5$.

Visualizing absolute value functions might be helpful is solving the equations involving absolute value expressions. Here we look at the graph of absolute value functions.

4.1.2. Graphing Absolute Value Functions

The general form of an absolute value function is $f(x)=a|x - b| + c$. Absolute value functions will be explored in more detail in chapter 5 of this text. However, we will present it here in a condensed discussion.

The absolute value function is classified as a pairwise function. Conceivably, the most significant feature of the absolute value graph is the corner point at which the graph changes direction.

To illustrate this point, consider the basic form $f(x) = |x|$, in which the origin of the axes is the corner point, as the following graph shows:

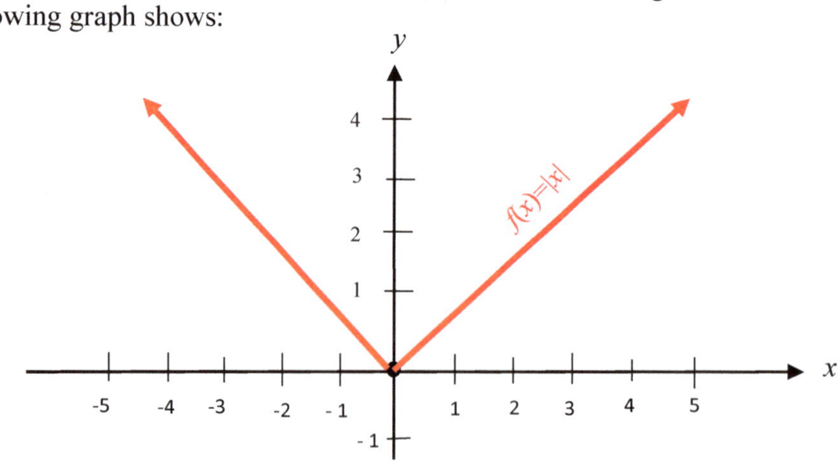

Now, let us explore a more problematic absolute value function, $f(x)=a|x-b|+c$. In this function, the graph of $f(x)=|x|$ has been shifted right b units, vertically stretched by a factor of a, and shifted up c units (see the graph presented below). This means that for this transformed function, the corner point is located at point (b, c).

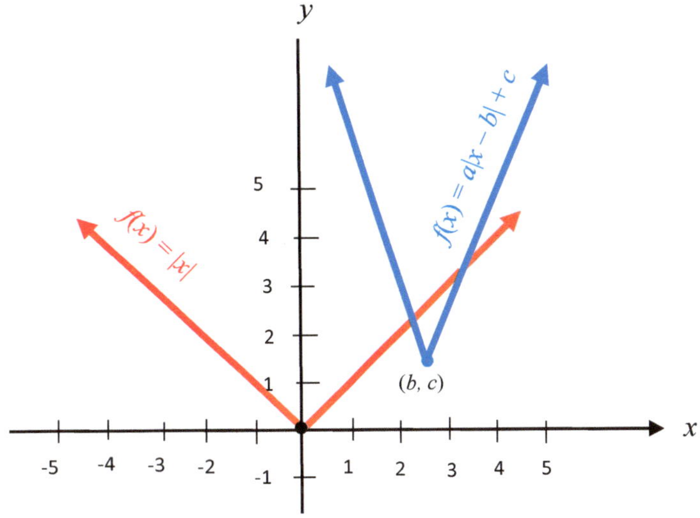

Example

Graph $f(x) = 2|x-4| - 3$.

Solution

As we have learned, the graph of $f(x) = 2|x-4| - 3$ ($f(x)=a|x-b|+c$) must be shifted right $b = 4$ units, vertically stretched by a factor of $a = 2$, and shifted down $c = -3$ units. Note that in the general formula $c = +3$.

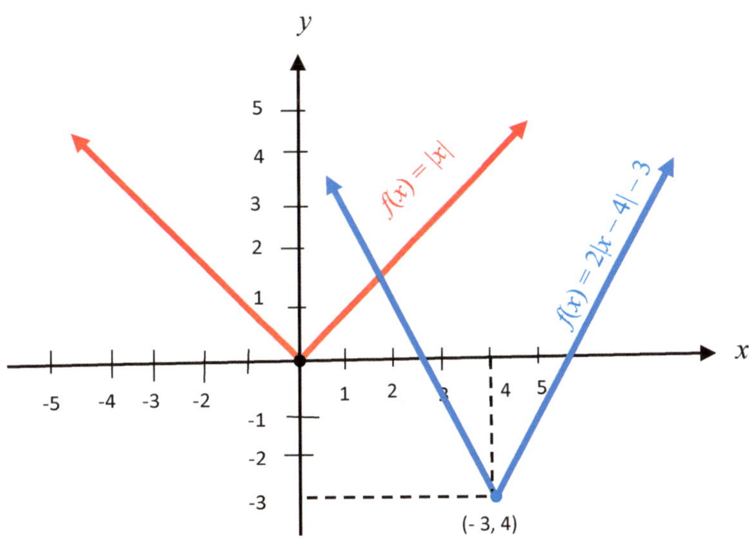

4.1.3. Solving Absolute Value Inequalities

An absolute value inequality is a statement of the form $|a| < b$, $|a| \leq b$, $|a| > b$, or $|a| \geq b$. In absolute value inequality however, there exists an algebraic expression dependent on a variable x, in which solving the inequality means finding the set of all x–values that satisfy the inequality.

There are two basic approaches to solving absolute value inequalities, algebraic approach, and graphical approach. The advantage of the graphical approach is we can visualize the solution by interpreting the graphs of absolute value equations. However, in graphical approaches, more precise solutions are occasionally difficult to read from a graph. While the advantage of the algebraic approach is that solutions are exact. Let us consider a simple form of the absolute value inequality, $|x| = 3$. Using algebraic approach, we have:

$|x| = 3$, then either $x = 3$, if $x > 0$, or $-x = 3$ which implies $x = -3$.

Therefore, $|x| = 3$ has two answers, $x = 3$, and $x = -3$, which this solution can be presented in a simplified form of $x = \pm 3$.

Now using graphical approach, we have learned earlier that $|x| = 9$ means that the distances from zero, is 9 unite in both directions of a number line, as it is illustrated below:

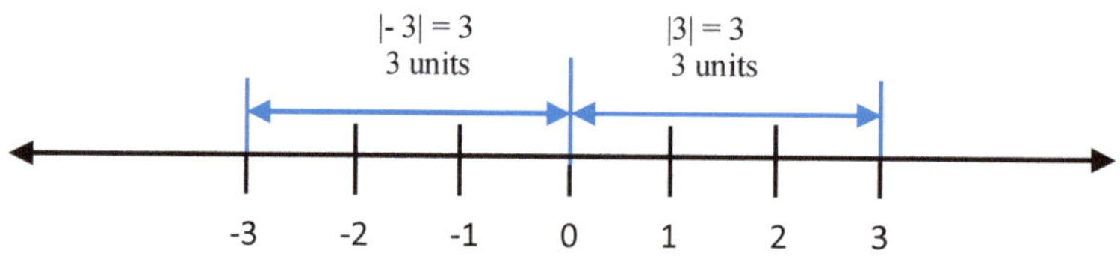

Therefore, the solution of $|x| = 3$ is ± 3.

Example

Solve $|9x - 13| \leq 5$, using the algebraic approach and graph the solution.

Solution

$|9x - 13| \leq 5$

$9x - 13 \leq 5$ — for $9x - 13 > 0$

$9x \leq 5 + 13$ — we isolated the variable on one side of the inequality

$9x \leq 18$ — we added 13 to 5

$\frac{1}{9} \times 9x \leq \frac{1}{9} \times 18$ — we multiplied both sides of the inequality by the reciprocal of the variable coefficient since it is positive the symbol did not change

$x \leq 2.$

Now consider $|9x - 13| \leq 5$ for $9x - 13 < 0$:

$|9x - 13| \leq 5$

$-(9x - 13) \leq 5$ — for $9x - 13 < 0$

$-9x + 13 \leq 5$

$-9x \leq 5 - 13$ — we isolated the variable on one side of the inequality

$-9x \leq -8$ — we added -13 to 5

$-\frac{1}{9} \times (-9x) \geq -\frac{1}{9} \times (-8)$ — we multiplied both sides of the inequality by the reciprocal of the variable coefficient, since it is negative the symbol was changed

$x \geq \frac{8}{9}$.

Therefore, the solution of $|9x - 13| = 5$ is $\frac{8}{9} \leq x \leq 2$.

To graph $|9x - 13| \leq 5$, we have $9x - 13 \leq 5$, for $9x - 13 > 0$, and $-(9x - 13) \leq 5$ (0r $9x - 13 \geq -5$ for $9x - 13 < 0$. Writing both inequality inequalities in form of the equivalent compound inequality, we have:

$$-5 \leq 9x - 13 \leq 5$$

$$-5 + 13 \leq 9x \leq 5 + 13$$

$$8 \leq 9x \leq 18$$

$$\frac{1}{9} \times 8 \leq \frac{1}{9} \times 9x \leq \frac{1}{9} \times 18$$

$$\frac{8}{9} \leq x \leq 2.$$

The graph of $|9x - 13| \leq 5$ is below:

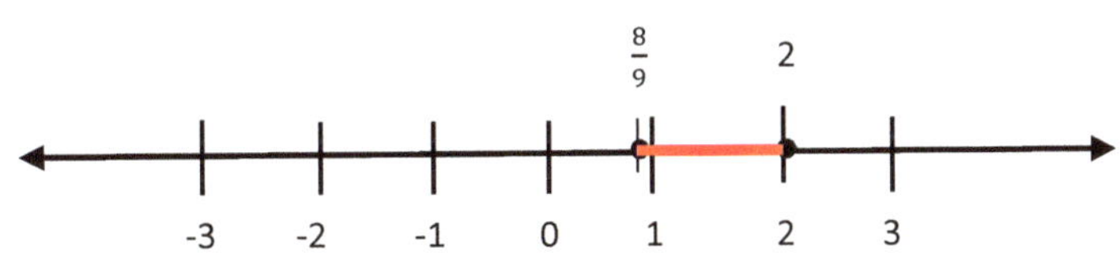

4.2. Polynomial Equations

A polynomial is an expression consisting of variables and coefficients, that involves only the operations of addition, subtraction, multiplication, and non–negative integer exponents of variables. Polynomials appear in many areas of mathematics and science.

For example, astrophysicists use them to calculate a star's velocity and distance from another object in space. Polynomial equations are also important in determining pressure in applications of fluid dynamics. Chemists use polynomials to determine the composition of certain compounds and molecules, and they are central to statistics.

Statistical formulas use polynomials to ascertain future values of animal birth and death rates, monetary flow, and population growth.

4.2.1. Standard Form of Polynomial Equations

Let n be a non–negative integer. The standard form of a polynomial equation is an expression in the form of:

$$a_n x^n + a_{n-1} x^{n-1} + \cdots + a_i x^i + \cdots + a_2 x^2 + a_1 x + a_0 = 0.$$

This is called the general form of a polynomial equation. Each a_i is a coefficient and can be any real number other than zero. Each term $a_i x^i$ is a term of a polynomial equation.

4.2.2. Terminology of Polynomial Equations

We often rearrange polynomials so that the powers are descending. When a polynomial is written in this way, we say that it is in general form.

Leading coefficient Degree

$$a_n x^n + \cdots + a_2 x^2 + a_1 x + a_0 = 0$$

Leading term

Polynomial Classifications

The exponent on a variable in a term is called the degree of that variable in that term; the degree of the term is the sum of the degrees of the variables in that term, and the degree of a polynomial is the largest degree of any one term with nonzero coefficient.

A polynomial with no variables or a polynomial of degree zero is a constant polynomial or simply a constant. Polynomials of degree one, two or three are respectively **linear polynomials**, **quadratic polynomials**, and **cubic polynomials**. In the case of polynomials in more than one variable, a polynomial is called **homogeneous polynomial** of degree n if all its non–zero terms have degree n.

The commutative law of addition can be used to rearrange terms into any preferred order. In polynomials with one variable, the terms are usually ordered according to degree, either in "descending powers of x", with the term of largest degree first, or in "ascending powers of x".

Two terms with the same indeterminates raised to the same powers are called "similar terms" or "like terms", and they can be combined, using the distributive law, into a single term whose coefficient is the sum of the coefficients of the terms that were combined. It may happen that this makes the coefficient 0.

Polynomials can be classified by the number of terms with nonzero coefficients, so that a one–term polynomial is called a monomial, a two–term polynomial is called a binomial, and a three–term polynomial is called a trinomial. The term "quadrinomial" is occasionally used for a four–term polynomial.

A polynomial in one variable is called a univariate polynomial, a polynomial in more than one variable is called a multivariate polynomial. A polynomial with two variables is called a bivariate polynomial.

4.2.3. Evaluating Polynomials Expressions

To evaluate a polynomial expression, we need to substitute a number for the variable and perform the arithmetic operations.

Example

Given the polynomial $2x^4 + 5x^3 - 4x - 6$, evaluate it at $x = -1$.

Solution

We simply need to plug $x = -1$ into the polynomial expression. So,

$2x^4 + 5x^3 - 4x - 6$

$2 \times (-1)^4 + 5 \times (-1)^3 - 4 \times (-1) - 6$ we substituted $x = -1$ into the polynomial

$2 \times (1) + 5 \times (-1) - 4 \times (-1) - 6$

$2 - 5 + 4 - 6$

$-5.$

Therefore, the value of the polynomial at $x = -1$ is -5.

4.2.4. Solving Polynomial Equations

When considering equations, the variables of polynomials are also called unknowns, and the solutions are the possible values of the unknowns for which the equality is true (in general more than one solution may exist).

Considering a polynomial $P = a_n x^n + a_{n-1} x^{n-1} + \cdots + a_i x^i + \cdots + a_2 x^2 + a_1 x + a_0$, then $P = 0$ is the polynomial equation. The solutions of this equation are called the **roots** of the polynomial equation. A number a is a root of a polynomial P if and only if the linear polynomial $x - a$ divides P, that is if there is another polynomial Q such that $P = (x - a)$ Q.

It may happen that $x - a$ divides P more than once: if $(x - a)^2$ divides P then a is called a multiple root of P, and otherwise a is called a simple root of P.

If P is a nonzero polynomial, there is a highest power m such that $(x - a)^m$ divides P, which is called the multiplicity of the root a in P. When P is the zero polynomial, the corresponding polynomial equation is trivial, and this case is usually excluded when considering roots, as, with the above definitions, every number is a root of the zero polynomial, with an undefined multiplicity. With this exception made, the number of roots of P, even counted with their respective multiplicities, cannot exceed the degree of P.

A detailed method of solving polynomial equations with three terms will be explained in the next chapter (chapter 4) of this book. For higher-degree equations, methods become more complicated and the general solution algorithms are beyond the scope of this book. However, some preliminary solution methods will be also explained in the next chapter.

We have not discussed graphing polynomials/polynomial functions yet; however, as we will see this later, the graphs of polynomials are nice smooth functions that have no breaks in them. This means that as one moving across the number line (in any direction) if the value of the polynomial changes sign (say from positive to negative) then it must go through zero or x intercept. To illustrate this, let us consider a polynomial function $y = f(x) = a_n x^n + a_{n-1} x^{n-1} + \cdots + a_i x^i + \cdots + a_2 x^2 + a_1 x + a_0$. We just show that the roots of the polynomial considering xy coordinate is the x intercept of the graph of the polynomial.

Referring to the figure below, the roots of the polynomial are $x = -3$, $x = -1$, $x = 1$, and $x = 3$. As it is seen, at each of these points, the polynomial changes sign. That is the value of the polynomial, $f(x)$, is positive for $x < -3$, is negetive for $-3 < x < -1$, is positive for $-1 < x < 1$, is negetive for $1 < x < 3$, and becomes positive after $x = 3$(for $x > 3$). This concept is a base for discussing polynomial inequalities.

x	y
3	0
2	− 9
1	0
0	9
− 1	0
− 2	− 9
− 3	0

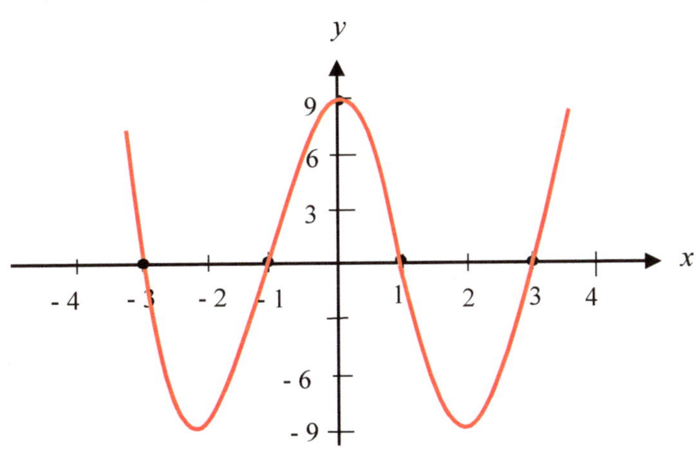

4.2.5. Polynomials Involving Inequalities

As we discussed before, in solving the linear inequalities, we were looking for all the values of the variable that will make the inequality true. This means that our solution will almost certainly involve inequalities as well. For polynomial inequalities, the process that we are going to go through will give the answers in that form. Since procedures for solving polynomial equations have not discussed yet, here we just consider those which can solved using factoring.

4.2.6. Solving Polynomials Involving Inequality

Considering the factorable polynomials, a procedure for solving polynomials involving inequality is given in the steps below.

Step 1. Move all terms in one side of the inequity leaving zero on the other side.

Step 2. Factor the polynomial.

Step 3. Determine the zeros of the polynomial, the x intercepts. Note that if the value of the polynomial changes sign, then it must go through zero. Also, note that the number line is divided into regions each specified by the two consecutive zero points.

Step 4. Graph a number line comprising the obtained zeros (the points from the previous step). Pick an arbitrary point from each region (pick an easy number to work with, this is our test point) and plug it into the polynomial to determine the sign of the polynomial at that region.

Step 5. Write down the answer because if the test point from a region satisfies the inequality, then that region is part of the solution. Contrary, if the test point does not satisfy the inequality, then that region is not part of the solution.

Example
Solve $x^3 + x^2 \leq 6x$.

Solution
We first need to move all terms in one side of the inequity leaving zero on the other side, so,

$$x^3 + x^2 \leq 6x$$

$$x^3 + x^2 - 6x \leq 0.$$

Hence, we want to solve $x^3 + x^2 - 6x \leq 0$. Next, we proceed to factoring. So,

$$x^3 + x^2 - 6x \leq 0$$

$$x(x^2 + x - 6) \leq 0 \qquad \text{we factored out } x$$

$$x(x - 2)\,(x + 3) \leq 0 \qquad \text{we perform further factoring.}$$

Now, to obtain zeros, we convert the inequality $x(x - 2)\,(x + 3) \leq 0$ to an equality.

$$x(x - 2)\,(x + 3) = 0.$$

Having three product terms implies that:

$$x = 0,\ (x - 2) = 0,\ \text{or}\ (x + 3) = 0.$$

The zeros are:

$$x = 0.$$

$(x - 2) = 0$

$x - 2 + 2 = 0 + 2$ we added 2 in both sides of the equation

$x = 2.$

$(x + 3) = 0$

$x + 3 - 3 = 0 - 3$ we added – 3 in both sides of the equation

$x = -3.$

Therefore, the zeros of $x (x - 2) (x + 3) = 0$, are $x = 0$, $x = 2$, and $x = -3$.

We now proceed to graph a number line and put zero points on it. The figure below illustrates the number line and the divided four regions resulted by the zero points. Let us denote the value of the polynomial $x^3 + x^2 - 6x$, as $f(x)$, and by picking test points $x = -4$ in region *I*, $x = -1$ in region *II*, $x = 1$ in region *III*, and $x = 4$ in region *IV*, we can find the sign of $f(x)$ in each region.

$f(x) = x^3 + x^2 - 6x$, at $x = -4$, is $-24 < 0$.

$f(x) = x^3 + x^2 - 6x$, at $x = -1$, is $6 > 0$.

$f(x) = x^3 + x^2 - 6x$, at $x = 1$, is $-4 < 0$.

$f(x) = x^3 + x^2 - 6x$, at $x = 4$, is $56 > 0$.

The solution of $x^3 + x^2 \leq 6x$ or equivalently $f(x) = x^3 + x^2 - 6x \leq 0$ is in those regions in which $f(x)$ has a negative value. Therefore, the solutions are $x \leq -3$, and $0 \leq x \leq 2$.

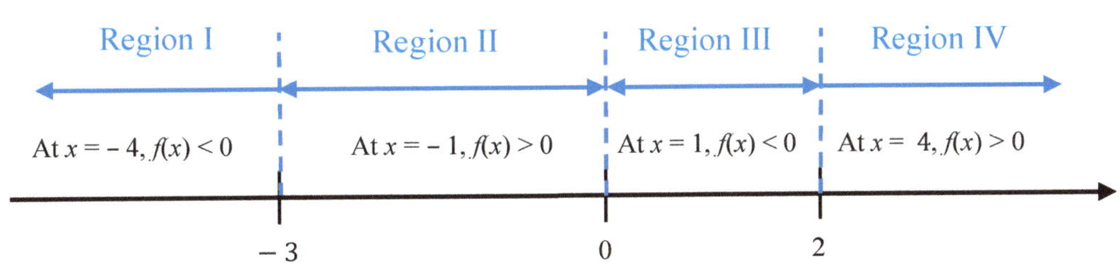

4.3. Rational Equations and Inequality

Rational equations definition - A **rational equation** contains at least one rational expression where the variable appears in at least one of the denominators.

4.3.1. Solving a Rational Equation

The steps to solve a rational equation are:

1. Find the common denominator.
2. Multiply everything by the common denominator.
3. Simplify.
4. Check the answer(s) to make sure there is not an extraneous solution.

Example

Solve $\frac{5}{x} - \frac{1}{3} = \frac{1}{x}$.

Solution

$\frac{5}{x} - \frac{1}{3} = \frac{1}{x}$

$\frac{5}{x} \times \frac{3}{3} - \frac{1}{3} \times \frac{x}{x} = \frac{1}{x}$ — we multiplied the first term by $\frac{3}{3}$ and the second term by $\frac{x}{x}$, to build up the denominator to the LCD

$\frac{15}{3x} - \frac{x}{3x} = \frac{1}{x}$

$3x \times (\frac{15}{3x} - \frac{x}{3x}) = 3x(\frac{1}{x})$ — we multiplied both sides by the LCD, $3x$

$15 - x = 3$

$15 - x - 15 = 3 - 15$ — we added -15 to both sides of the equation

$-x = -12$

$x = 12.$

We now check our answer by substituting 12 for x to see if we obtain a true statement.

$\frac{5}{x} - \frac{1}{3} = \frac{1}{x}$

$\frac{5}{12} - \frac{1}{3} \overset{?}{\Leftrightarrow} \frac{1}{12}$ — we substituted x by 12 into the equation

$\frac{5}{12} - \frac{1}{3} \times \frac{4}{4} \overset{?}{\Leftrightarrow} \frac{1}{12}$ — we multiplied the second term of the left side by $\frac{4}{4}$ to build up the denominator to the LCD

$\frac{5}{12} - \frac{4}{12} \overset{?}{\Leftrightarrow} \frac{1}{12}$

$\frac{5-4}{12} \overset{?}{\Leftrightarrow} \frac{1}{12}$

$\frac{1}{12} = \frac{1}{12}$ — we obtained a true statement.

Therefore, the solution of $\frac{5}{x} - \frac{1}{3} = \frac{1}{x}$ is $x = 12$.

4.3.2. Rational Inequalities

A **rational inequality** is an inequality which contains a variable in the denominator.

4.3.3. Solving Rational Inequalities

In solving rational equations, the first step was to clear out the denominators by multiplying by the least common denominator. That will not work for inequalities. Since we do not know the value of x we cannot multiply both sides by anything that contains an x.

As we mentioned before, when we multiply or divide inequalities by a negative number, we must reverse the inequality sign. However, since we do not know the value of x we do not know if the denominator is positive or negative and so we will not know if we need to switch the direction of the inequality or not. So, we need to leave the rational expression in the inequality.

The basic procedure of solving rational inequalities is like procedure of solving polynomial inequalities. Step below present a method for solving rational inequalities.

Step 1. Move all terms to one side of the inequity leaving zero on the other side.

Step 2. Factor the numerator and denominator.

Step 3. Determine the zeros of the numerator and the denominator. Then zeros are the numbers where the rational expression may change sign. So, we will build a number line using these points to define ranges out of which to pick test points just like we did with polynomial inequalities.

Step 4. Graph a number line comprising the obtained zeros (the points from the previous step). Pick an arbitrary point from each region (pick an easy number to work with, this is our test point) and plug it into the inequality expression to determine its sign at that region.

Step 5. Write down the answer by because; if the test point from a region satisfies the inequality, then that region is part of the solution. Contrary, if the test point does not satisfy the inequality, then that region is not part of the solution.

Example

Solve $\frac{x+3}{2x-4} \leq 0$.

Solution

Based on the above steps, the first step is to get a zero on one side. This has already been done here. The next step is to factor the numerator and denominator. This has already been done likewise. The next step is to determine the zeros of numerator and the denominator. So,

$x + 3 = 0, x = -3.$

$2x - 4 = 0, x = 4.$

Note that there is another reason for finding the value of x that make the denominator zero. Having a rational expression, we need to avoid dividing it by zero. Hence, determining at what value of x the denominator becomes zero, gives us the values of x to avoid doing so. Here the number line of this case:

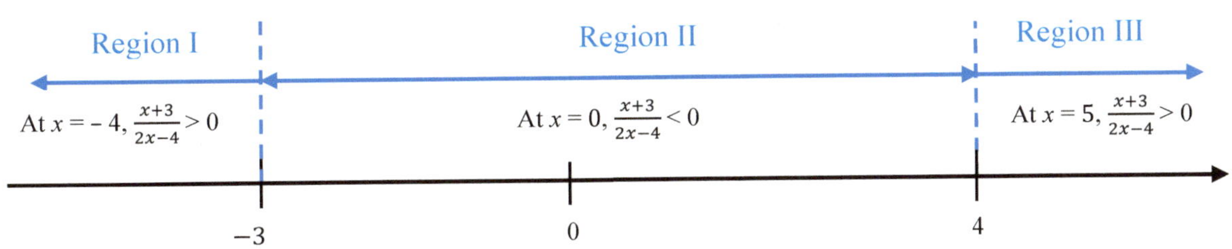

As is seen, the zeros resulted in three distinct regions on the number line. Picking test points $x = -4$, $x = 0$, and $x = 5$ in region I, in region II, and in region III, respectively, we will have:

$\frac{x+3}{2x-4}$, at $x = -4$, is $\frac{-4+3}{2\times(-4)-4} = \frac{-1}{-8-4} = \frac{-1}{-12} = \frac{1}{12} > 0.$

$\frac{x+3}{2x-4}$, at $x = 0$, is $\frac{0+3}{2\times(0)-4} = \frac{3}{-4} = -\frac{3}{4} < 0.$

$\frac{x+3}{2x-4}$, at $x = 5$, is $\frac{5+3}{2\times(5)-4} = \frac{8}{6} = \frac{4}{3} > 0.$

Therefore, the solution of $\frac{x+3}{2x-4} \leq 0$ is $-3 \leq x \leq 4$.

4.4. Equations and Inequality Involving Radicals

Radical equations - An equation that contains one or more radical signs, in which a variable appears under a radical sign, is called a radical equation (also known as irrational).

Radicand - The variable expression sits underneath the radical and is called a radicand.

Negative radicands property -The negative property of radicand implies, give the radical $\sqrt[n]{x}$, the radicand can NOT be negetive when n is even. In other words, the radicand cannot have a calculated final value that is negative when the index of the radical is an even number.

4.4.1. Solving Radical Equations

Given a radical equation, steps to solve radical equations are presented below.

Step 1. Isolate the radical expression on one side of the equation and the other terms on the others side.

Step 2. To eliminate the radical symbol, raise both sides of the equation to the nth power, if the radical is an nth root radical. In other words, If the radical is a square root, then square both sides of the equation. If it is a cube root, then raise both sides of the equation to the third power, etc.

Step 3. Isolate the variable on one side of the equation and the other terms on the others side, if necessary.

Step 4. Solve the remaining equation.

Step 5. Confirm solutions by substituting them into the original equation.

Example

Solve the equation $\sqrt{9x-23} = 2$.

Solution

Since the radical is already isolated on the left side of the equation, we need to proceed to square both sides. So,

$$\sqrt{9x-23} = 2$$

$$\left(\sqrt{9x-23}\right)^2 = 2^2$$

$$9x - 23 = 4$$

$$9x = 4 + 23$$ we isolated the variable on one side of the equation and the other terms on the others side

$$9x = 27$$

$$\frac{1}{9} \times 9x = \frac{1}{9} \times 27$$ we multiplied both side by the reciprocal of $9x$

$$x = 3.$$

Therefore, the solution of $\sqrt{9x-23} = 2$, is $x = 3$.

Now, the last step is to check the solution. So,

$\sqrt{9x - 23} = 2$

$\sqrt{9(3) - 23} \overset{?}{\Leftrightarrow} 2$ we replaced the variable with 3

$\sqrt{27 - 23} \overset{?}{\Leftrightarrow} 2$

$\sqrt{4} \overset{?}{\Leftrightarrow} 2$

$2 = 2$ true.

Example

Solve $\sqrt{4x + 12} - \sqrt{2x + 6} = 0$.

Solution

First, we isolate one radical, eliminate it, and then isolate the second radical.

$\sqrt{4x + 12} - \sqrt{2x + 6} = 0$

$\sqrt{4x + 12} - \sqrt{2x + 6} + \sqrt{2x + 6} = \sqrt{2x + 6}$ we added $\sqrt{2x + 6}$ to both sides

$\sqrt{4x + 12} = \sqrt{2x + 6}$

$\left(\sqrt{4x + 12}\right)^2 = \left(\sqrt{2x + 6}\right)^2$ we raised both sides to the power of two

$4x + 12 = 2x + 6$

$4x - 2x = -12 + 6$ we isolated the variables on one side and the constants terms on the other side.

$2x = -6$

$\frac{1}{2} \times 2x = \frac{1}{2} \times (-6)$ we multiplied both sides by the reciprocal of $2x$

$x = -\frac{6}{2}$

$x = -3$.

Therefore, the solution of $\sqrt{4x + 12} + \sqrt{2x + 6} = 0$, is $x = -3$.

Now, the last step is to check the solution. So,

$\sqrt{4x + 12} + \sqrt{2x + 6} = 0$

$\sqrt{4(-3) + 12} + \sqrt{2(-3) + 6} \overset{?}{\Leftrightarrow} 0$ we replaced x with $= -3$

$\sqrt{-12 + 12} + \sqrt{2(-3) - 6} \overset{?}{\Leftrightarrow} 0$

$0 = 0$ true.

4.4.2. Radical Inequalities

A **radical inequality** is an inequality that contains a variable within a radical. This means that the variable expression sits underneath the radical and is called a radicand.

4.4.3. Solving Radical Inequalities

We cannot mathematically solve the problem with the radical symbol in it. It must be removed. So, do not forget that we have this property that tells us how to do that.

$$(\sqrt[n]{x})^n = x.$$

This means that we take the index value and use this same value as an exponent. In doing so, it cancels out the radical symbol and leaves the radicand variable expression by itself. What is under an even radical must be positive (domain restriction); we must create another inequality and set what is under the even radical to ≥ 0. We then solve for x and take the intersection of both solutions. The reason we take the intersection of the two solutions is because both must work.

Here are some examples; note that we just raise each side to the root to get rid of it. We can do that in these examples since we know the sign of the values on both sides. It gets trickier when we do not know the sign of one of the sides.

Example
Solve $\sqrt{x+2} \leq 3$.

Solution
We need to solve two inequalities, one is the original inequality since our x must work in the original, the other is the domain restriction inequality since it must also work so anything under the even radical is positive. So,

$$\sqrt{x+2} \leq 3 \text{ and } x+2 \geq 0$$

For $\sqrt{x+2} \leq 3$, we will have.

$$\sqrt{x+2} \leq 3$$

$$\left(\sqrt{x+2}\right)^2 \leq 3^2$$ to get rid of the square roots, we square each side, and we can leave the inequality signs the same since we are multiplying by positive numbers

$$x+2 \leq 9$$

$$x+2-2 \leq -2+9$$ we added -2 to both sides

$$x \leq 7.$$

For $x+2 \geq 0$, we will have.

$$x+2 \geq 0$$

$$x+2-2 \geq 0-2$$ we added -2 to both sides

$$x \geq -2.$$

Check: To verify our solution, we need first convert the inequality to an equality by just changing the $\leq$ sign to $=$ sign. Then we need to plug them separately into the inequality. So,

For $x \leq 7$, we need to check $x = 7$.

$$\sqrt{x+2} \leq 3$$

$\sqrt{x+2} = 3$ — we converted the inequality to an equality

$\sqrt{7+2} \overset{?}{\Leftrightarrow} 3$ — we plugged $x = 7$ into the equality

$\sqrt{9} \overset{?}{\Leftrightarrow} 3$

$\left(\sqrt{9}\right)^2 \overset{?}{\Leftrightarrow} 3^2$ — we squared both sides

$9 = 9$ — true, $x = 7$ is a correct answer.

For ≥ -2, we need to check $x = -2$.

$\sqrt{x+2} \geq 0$

$\sqrt{x+2} = 0$ — we converted the inequality to an equality

$\sqrt{-2+2} \overset{?}{\Leftrightarrow} 0$ — we plugged $x = -2$ into the equality

$\sqrt{0} \overset{?}{\Leftrightarrow} 0$

$\left(\sqrt{0}\right)^2 \overset{?}{\Leftrightarrow} 0^2$ — we squared both sides

$0 = 0$ — true, $x = 0$ is a correct answer.

Therefore, our solutions are $x \leq 7$ and $x \geq -2$, or $[-2, 7]$. We also need to try numbers outside our solution (like $x = -3$ and $x = 14$) and see that they do not work.

To check $x = -3$.

$\sqrt{x+2} \geq 0$

$\sqrt{-3+2} \overset{?}{\Leftrightarrow} 0$ — we plugged $x = -3$ into the inequality

$\sqrt{-1} \overset{?}{\Leftrightarrow} 0$ — we realize, that -1 is under an even radical is negative.

Therefore, any $x < -2$ cannot be our solution.

To check $x = 14$.

$\sqrt{x+2} \leq 3$

$\sqrt{14+2} \overset{?}{\Leftrightarrow} 3$ — we plugged $x = 14$ into the inequality

$\sqrt{16} \overset{?}{\Leftrightarrow} 3$

$\left(\sqrt{16}\right)^2 \overset{?}{\Leftrightarrow} 3^2$ — we squared both sides

$16 \overset{?}{\Leftrightarrow} 9$

$16 > 9$ — we realize that the inequality is not hold.

Therefore, any $x > 7$ cannot be our solution.

Example

Solve $\sqrt[3]{5x-18} \leq 3$.

Solution
We first need to remove the radical by taking the index value equal to three and raise both side to the power of this value. So,

$\sqrt[3]{5x-18} \leq 3$

$(\sqrt[3]{5x-18})^3 \leq (3)^3$ we raised both side to the power of 3

$5x - 18 \leq 27$

$5x - 18 + 18 \leq 27 + 18$ we added 18 to both sides of the expression

$5x \leq 45$

$\frac{1}{5} \times 5x \leq \frac{1}{5} \times 45$ we multiplied both sides by the reciprocal of $5x$

$x \leq 9$.

Therefore, the solution is $x \leq 9$.

Next, we will check the solution. So,

$\sqrt[3]{5x-18} \leq 3$

$\sqrt[3]{5(9)-18} \overset{?}{\Leftrightarrow} 3$ we replace $x = 9$ into the expression

$\sqrt[3]{45-18} \overset{?}{\Leftrightarrow} 3$

$\sqrt[3]{27} \overset{?}{\Leftrightarrow} 3$

$\sqrt[3]{3^3} \overset{?}{\Leftrightarrow} 3$

$3 \leq 3$ true.

Example
Solve $\sqrt[4]{5x-11} \geq \sqrt[4]{3x+13}$.

Solution
We need to solve three inequalities, one is the original inequality since our x must work in the original, the two others are the domain restriction inequality. So,

a) $\sqrt[4]{5x-11} \geq \sqrt[4]{3x+13}$
b) $\sqrt[4]{5x-11} \geq 0$
c) $\sqrt[4]{3x+13} \geq 0$.

a) $\sqrt[4]{5x-11} \geq \sqrt[4]{3x+13}$.
We need to remove the radical by taking the index value equal to 4 and raise both side to the power of this value. So,

$\sqrt[4]{5x-11} \geq \sqrt[4]{3x+13}$

$(\sqrt[4]{5x-11})^4 \geq (\sqrt[4]{3x+13})^4$ we raised both sides to the power of 4

$5x - 11 \geq 3x + 13$

$5x - 3x \geq 11 + 13$ we isolated the variables in one side and the constant terms on the other side

$2x \geq 24$

$\frac{1}{2} \times 2x \geq \frac{1}{2} \times 24$ we multiplied both sides by the reciprocal of $2x$

$x \geq 12.$

b) $\sqrt[4]{5x-11} \geq 0.$
Again, we need to remove the radical by taking the index value of equal to 4 and raise both sides to the power of this value. So,

$\sqrt[4]{5x-11} \geq 0$

$(\sqrt[4]{5x-11})^4 \geq (0)^4$ we raised both side to the power of 4

$5x - 11 \geq 0$

$5x - 11 + 11 \geq 0 + 11$ we added 11 to both sides

$5x \geq 11$

$\frac{1}{5} \times 5x \geq \frac{1}{5} \times 11$ we multiplied both sides by the reciprocal of $5x$

$x \geq \frac{11}{5}.$

c) $\sqrt[4]{3x+13} \geq 0.$
Likewise, we need to remove the radical by taking the index value of equal to 4 and raise both side to the power of this value. So,

$\sqrt[4]{3x+13} \geq 0$

$(\sqrt[4]{3x+13})^4 \geq (0)^4$ we raised both sides to the power of 4

$3x + 13 \geq 0$

$3x + 13 - 13 \geq 0 - 13$ we added – 13 to both sides

$3x \geq -13$

$\frac{1}{3} \times 3x \geq \frac{1}{3} \times (-13)$ we multiplied both sides by the reciprocal of $3x$

$x \geq -\frac{13}{3}.$

Next, we will check the solutions separately. So,

For $x = 12$, we will have:

$\sqrt[4]{5x-11} \geq \sqrt[4]{3x+13}$

$\sqrt[4]{5(12) - 11} \overset{?}{\Leftrightarrow} \sqrt[4]{3(12) + 13}$ we replace $x = 12$ into the expression

$\sqrt[4]{60 - 11} \overset{?}{\Leftrightarrow} \sqrt[4]{36 + 13}$

$\sqrt[4]{49} \geq \sqrt[4]{49}$ true.

For $x = \frac{11}{5}$, we will have:

$\sqrt[4]{5x - 11} \geq 0$

$\sqrt[4]{5 \times \frac{11}{5} - 11} \overset{?}{\Leftrightarrow} 0$ we replace $x = \frac{11}{5}$ into the expression

$\sqrt[4]{11 - 11} \overset{?}{\Leftrightarrow} 0$

$\sqrt[4]{0} \overset{?}{\Leftrightarrow} 0$

$0 = 0$ true.

For $x = -\frac{13}{3}$, we will have:

$\sqrt[4]{3x + 13} \geq 0$

$\sqrt[4]{3 \times \left(-\frac{13}{3}\right) + 13} \overset{?}{\Leftrightarrow} 0$ we replace $x = -\frac{13}{3}$ into the expression

$\sqrt[4]{-13 + 13} \overset{?}{\Leftrightarrow} 0$

$\sqrt[4]{0} \overset{?}{\Leftrightarrow} 0$

$0 = 0$ true.

The solutions we obtained are, $x \geq 12$, $x \geq \frac{11}{5}$, and $x \geq -\frac{13}{3}$. Therefore, the final solution is intersection of these three solutions which is $x \geq 12$.

Now, we need to try numbers outside our solution (like $x = 11$) and see that they do not work.

To check $x = 11$, we will have.

$\sqrt[4]{5x - 11} \geq \sqrt[4]{3x + 13}$

$\sqrt[4]{5 \times 11 - 11} \overset{?}{\Leftrightarrow} \sqrt[4]{3 \times 11 + 13}$ we replace $x = 11$ into the expression

$\sqrt[4]{55 - 11} \overset{?}{\Leftrightarrow} \sqrt[4]{33 + 13}$

$\sqrt[4]{44} \overset{?}{\Leftrightarrow} \sqrt[4]{46}$

$44 \overset{?}{\Leftrightarrow} 46$ we raised both sides to the power of 4

$44 < 46$ we realize that inequality is not hold.

Therefore, any $x < 12$ cannot be our solution.

4.5. Equations Involving Exponent Expressions

Exponential equations involve exponents, where the power of a number (base) indicates how many times the base is multiplied. However, when the exponent itself is a variable within an equation, it leads to what is known as an exponential equation. Solving exponential equations often requires leveraging the relationship between exponents and logarithms.

Understanding exponential equations entails grasping how to solve them when the bases are either identical or different. This involves employing techniques to manipulate the equations to isolate the variable exponent and solve for its value. Solved examples and practice questions can aid in reinforcing these concepts and honing problem-solving skills.

Exponential Equations, Definition - Exponential equations are equations in which variables occur as exponents.

4.5.1. Solving Exponential Equations, Identical Bases

The first technique we will introduce for solving exponential equations involves two functions with identical bases. Bear in mind that the one-to-one property of exponential functions tells us that, for any real numbers *a*, *S*, and *T*, where $a > 0$, $a \neq 1$, $a^S = a^T$ if and only if $S = T$. In other words, when an exponential equation has the same base on each side, the exponents must be equal. This also applies when the exponents are algebraic expressions. Therefore, we can solve many exponential equations by using the rules of exponents to rewrite each side as a power with the same base. Then we use the fact that exponential functions are one-to-one to set the exponents equal and solve for the unknown.

Example

Solve $6^{3x-5} = \frac{6^{2x}}{36}$.

Solution

We need first to rewrite the equation so that both sides have the same base. To do so, we need to use the negative property of exponents to rewrite the right side of the equation. So,

$$6^{3x-5} = \frac{6^{2x}}{36}$$

$$6^{3x-5} = \frac{6^{2x}}{6^2}$$

$$6^{3x-5} = (6^{2x})(6^{-2})$$ we used the negative property of the exponent

$$6^{3x-5} = 6^{2x+(-2)} = 6^{2x-2}$$ we used the product property of the exponent.

Now, both sides have the common base 6. Then we apply the one-to-one property of exponents by setting the exponents equal. So,

$$6^{3x-5} = 6^{2x-2}$$

$$3x - 5 = 2x - 2$$ we used one to one property of the exponent

$$3x - 2x = 5 - 2$$ we isolated *x* in one side and constants in the other side

$$x = 3.$$

Therefore, the solution of $6^{3x-5} = \frac{6^{2x}}{36}$ is $x = 3$.

Check: to verify our answer, we need to plug $x = 3$ into the original equation $6^{3x-5} = \frac{6^{2x}}{36}$. So,

$$6^{3x-5} = \frac{6^{2x}}{36}$$

$$6^{3(3)-5} \overset{?}{\Leftrightarrow} \frac{6^{2(3)}}{36}$$ we plugged $x = 3$ into the equation

$$6^{9-5} \overset{?}{\Leftrightarrow} \frac{6^6}{6^2}$$

$$6^4 \overset{?}{\Leftrightarrow} 6^6(6^{-2})$$ we used the negative property of the exponent

$$6^4 \overset{?}{\Leftrightarrow} 6^{6+(-2)}$$ we used the product property of the exponent

$$6^4 \overset{?}{\Leftrightarrow} 6^{6-2}$$

$$6^4 = 6^4$$ true, both sides are the same.

Example
Solve $5^{7x-3} = \sqrt{5}$.

Solution
We need to make both sides have the common base 5. So,

$$5^{7x-3} = \sqrt{5}$$

$$5^{7x-3} = 5^{\frac{1}{2}}$$ we applied the rule of converting a radical to a rational exponent

$$7x - 3 = \frac{1}{2}$$ we used the one-to-one property of exponents

$$7x - 3 + 3 = \frac{1}{2} + 3$$ we added 3 to both sides

$$7x = \frac{1}{2} + 3 \times \frac{2}{2}$$ we multiplied 3 by $\frac{2}{2}$ to have a common denominator in the right side of the equation

$$7x = \frac{1}{2} + \frac{6}{2}$$

$$7x = \frac{7}{2}$$

$$\frac{1}{7} \times 7x = \frac{1}{7} \times \frac{7}{2}$$ we multiplied both sides by the reciprocal of the variable

$$x = \frac{1}{2}.$$

Therefore, the solution of $5^{7x-3} = \sqrt{5}$ is $x = \frac{1}{2}$.

Check: to verify our answer, we need to plug $x = \frac{1}{2}$ into the original equation $5^{7x-3} = \sqrt{5}$. So,

$$5^{7x-3} = \sqrt{5}$$

$$5^{7\left(\frac{1}{2}\right)-3} \overset{?}{\Leftrightarrow} \sqrt{5}$$ we plugged $x = \frac{1}{2}$ into the equation

$5^{\frac{7}{2} - 3} \overset{?}{\Leftrightarrow} \sqrt{5}$

$5^{\frac{7}{2} - (3 \times \frac{2}{2})} \overset{?}{\Leftrightarrow} \sqrt{5}$ — we multiplied 3 by $\frac{2}{2}$ to have a common denominator in the exponent of the right side

$5^{\frac{7}{2} - \frac{6}{2}} \overset{?}{\Leftrightarrow} \sqrt{5}$

$5^{\frac{7-6}{2}} \overset{?}{\Leftrightarrow} \sqrt{5}$

$5^{\frac{1}{2}} \overset{?}{\Leftrightarrow} \sqrt{5}$

$5^{\frac{1}{2}} = 5^{\frac{1}{2}}$ — we applied the rule of converting a radical to a rational exponent.

Since both sides are the same the answer must be correct.

4.5.2. Solving Exponential Equations, Unlike Bases

The following steps describe the solution for the exponential equations.

Step 1. Rewrite each side in the equation as a power with a common base.

Step 2. Use the rules of exponents to simplify, if necessary, so that the resulting equation becomes the form of $a^S = aT$.

Step 3. Use the one-to-one property to set the exponents equal.

Step 4. Solve the resulting equation, $S = T$, for the unknown.

Example

Solve $27^{x+3} = 9^{x+5}$ by rewriting it to have a common base.

Solution

We need to rewrite the equation to have a common base. So,

$27^{x+3} = 9^{x+5}$

$(3^3)^{x+3} = (3^2)^{x+5}$

$(3^3)^{x+3} = (3^2)^{x+5}$

$3^{3(x+3)} = 3^{2(x+5)}$ — we applied the power of power property

$3^{3x+9} = 3^{(2x+10)}$

$3x + 9 = 2x + 10$

$3x - 2x = -9 + 10$ — we isolated the variables in one side and the constant terms in the other side

$x = 1.$

Therefore, the solution of $27^{x+3} = 9^{x+5}$ is $x = 1$.

Check: to verify our answer, we need to plug $x = 1$ into the original equation $27^{x+3} = 9^{x+5}$. So,

$27^{x+3} = 9^{x+5}$

$27^{1+3} \overset{?}{\Leftrightarrow} 9^{1+5}$ — we plugged $x = 1$ into the equation

$27^{4} \overset{?}{\Leftrightarrow} 9^{6}$

$(3^3)^4 \overset{?}{\Leftrightarrow} (3^2)^6$

$3^{12} = 3^{12}$ — we applied the power of power property.

Since both sides are the same the answer must be correct.

1.5.3. Solving Exponential Equations, a Common Base Cannot be Found

The following steps describe the solution for the exponential equations in which the common base cannot be found.

Step 1. Apply the logarithm to both sides of the equation.

- If one of the terms in the equation has base 10, use the common logarithm.
- If none of the terms in the equation has base 10, use the natural logarithm.

Step 2. Use the rules of logarithms to solve for the unknown.

Example

Solve $7^{x+2} = 9^{x-3}$.

Solution

For solving $7^{x+2} = 9^{x}$, there is no easy way to get the powers to have the same base. So, we need to use the natural logarithm. Hence

$ln7^{x+2} = ln9^{x}$ — we took *ln* of both sides

$(x + 2)ln7 = xln9$ — we used the power rule for logarithms

$xln7 + 2ln7 = xln9$ — we used the distributive property

$xln7 - xln9 = -2ln7$ — we isolated the terms containing variable in one side and constant on the other side

$x(\ln 7 - ln9) = -2ln7$ — we factored out x on the left side

$xln(\frac{7}{9}) = -2ln7$ — we used the properties of logarithms

$\frac{1}{ln(\frac{7}{9})} \times xln(\frac{7}{9}) = \frac{1}{ln(\frac{7}{9})} \times (-2ln7)$ — we multiplied both sides by the reciprocal coefficient of the variable x

$x = \frac{-2ln7}{ln(\frac{7}{9})}$.

1.5.4. Exponential Equations Containing Base *e*

A common type of exponential equation is one with base *e*. When we have an equation with a base *e* on either side, we can use the **natural logarithm** to solve it.

1.5.5. Solving an Equation of the Form $y = Ae^{kt}$

The steps below can be used to solve $y = Ae^{kt}$.

1. Divide both sides of the equation by A.
2. Apply the natural logarithm to both sides of the equation.
3. Divide both sides of the equation by k.

Example
Solve $8e^{5t} = 64$.

Solution
As we have learned we need to divide both sides of the equation by 8, the coefficient of $8e^{5t}$. So,

$8e^{5t} = 64$

$\frac{1}{8} \times 8e^{5t} = \frac{1}{8} \times 64$ — we multiplied both sides by $\frac{1}{8}$, the coefficient of $8e^{5t}$

$e^{5t} = \frac{64}{8}$

$e^{5t} = 8.$

We now need to apply the natural logarithm to both sides of the equation. So,

$e^{5t} = 8$

$lne^{5t} = ln8$

$5t\,(lne) = ln8.$

We also learned that $ln\ e = 1$. So,

$5t\,(1) = ln8$

$5t = ln8$

$\frac{1}{5} \times 5t = \frac{1}{5} \times ln8$ — we multiplied both sides by $\frac{1}{5}$, the coefficient of the variable

$t = \frac{ln}{5}.$

Therefore, the solution of $8e^{5t} = 64$ is $t = \frac{ln8}{5}$.

Check: We want to verify our answer. We need to plug $t = \frac{ln8}{5}$ into the original equation $8e^{5t} = 64$. So,

$8e^{5t} = 64$

$8e^{5(\frac{ln}{5})} \overset{?}{\Leftrightarrow} 64$ — we plugged $t = \frac{ln8}{5}$ into the equation

$8e^{ln} \overset{?}{\Leftrightarrow} 64$

$\frac{1}{8} \times 8e^{ln} \overset{?}{\Leftrightarrow} \frac{1}{8} \times 64$ — we multiplied both sides by $\frac{1}{8}$, the coefficient of $8e^{ln8}$

$e^{ln8} \overset{?}{\Leftrightarrow} \frac{64}{8}$

$e^{ln8} = 8$

$ln(e^{ln8}) \overset{?}{\Leftrightarrow} ln8$ — we applied the natural logarithm to both sides of the equation

$ln8 \, ln\, e \overset{?}{\Leftrightarrow} ln8$ — we used the property of natural logarithms.

We have learned that $ln\ e = 1$. So,

$ln8\ (1) \overset{?}{\Leftrightarrow} ln8$

$ln8 = ln8$ — true, both sides are the same.

1.6. Equations Involving Logarithms

Logarithmic Equations, Definition - A logarithmic equation is an expression that involves the logarithm which holds a variable. The purpose of solving logarithmic equation is to find the value of the unknown variable.

4.6.1. Solving Exponential Equations Using Logarithms

We have mentioned that to solve exponential equations, we must write both sides of the equation with the same base. However, it is not always possible or convenient to write expressions with the same base. In that case we often take the common logarithm or natural logarithm of both sides once the exponential is isolated.

Example

Solove $3e^{(x-3)} = 21$.

Solution

$3e^{(x-3)} = 21$

$e^{(x-3)} = 7$ — we divided both side by 3

$ln\ e^{(x-3)} = ln\ 7$ — we took natural algorithm of both sides

$(x - 3)\ ln\ e = ln\ 7$

$(x - 3) = ln\ 7$ — we know that $lne = 1$

$x = 3 + 1.94591 = 4.94591$.

4.6.2. Solving Logarithmic Equations

When we talked about exponential equations, we introduced the number e. Just as e was a base for an exponential equation, it can be used as a base for logarithmic functions too. The logarithmic equation with base e is called the natural logarithmic equation.

Common Logarithmic Equation

The equation $y = log\, x$ is the common logarithmic equation with base 10, where $x > 0$. $y = log\, x$ is equivalent to $x = 10^y$.

Also, the equation $y = log_a\, x$ is a logarithmic equation with base a, where $x > 0$ and $a \neq 1$. $y = log_a\, x$ is equivalent to $x = a^y$.

To solve logarithmic equations, one method is to change the equation to exponential form and then solve the exponential equation as we presented it before.

Example
Solve $3log_4\, x = log_4\, 27$.

Solution

$3log_4\, x = log_4\, 27$

$log_4\, x^3 = log_4\, 27$ we applied the power property

$x^3 = 27$ we applied one-to-one property

$x^3 = 3^3$

$x = 3$.

Example
Solve $log_5\,(2x - 7) = log_5\,(x + 13)$.

Solution

$log_5\,(2x - 7) = log_5\,(x + 13)$

$(2x - 7) = (x + 13)$ we applied one-to-one property

$2x - x = 7 + 13$

$x = 20$.

4.6.3. Natural Logarithmic Equation

The equation $y = ln\, x$ is the natural logarithmic equation with base e, where $x > 0$. $y = ln\, x$ is equivalent to $x = e^y$ When the base of the logarithm equation is 10, we call it the common logarithmic equation and the base is not commonly shown. Hence, if the base a of a logarithm is not shown, we assume it is 10.

Example
Solve $2ln\,(2x + 3) - 2\, ln\,(2x + 3) = ln\,(x + 12)$.

Solution

$2ln\,(2x + 3) - ln\,(2x + 3) = ln\,(x + 12)$

$ln\, \frac{(2x + 3)^2}{(2x + 3)} = ln\,(x + 12)$ we applied the power property

$ln\,(2x + 3) = ln\,(x + 12)$

$(2x + 3) = (x + 12)$ we applied one-to-one property

$2x - x = -3 + 12$

$x = 9.$

We now present some more examples of equations involving common logarithms, logarithms with base a, and natural logarithms.

Example
Solve each logarithmic equation.

a) $\log x - 2 = 0.$

b) $\log_x 64 - 3 = 0.$

c) $\ln (x - 5) - 12 = 0.$

Solution

a) We learned, $y = \log x$ is equivalent to $x = 10^y$. So,

$\log x - 2 = 0$

$\log x - 2 + 2 = 0 + 2$ we added 2 to both sides

$\log x = 2$

$x = 10^2$ we used the common logarithmic equation rule

$x = 100.$

Therefore, the solution to $\log x - 2 = 0$ is $x = 100$.

Check: to verify the answer, we need to plug $x = 100$ into the equation. So,

$\log x - 2 = 0$

$\log 100 - 2 \overset{?}{\Leftrightarrow} 0$

$\log 10^2 - 2 \overset{?}{\Leftrightarrow} 0$

$2\log 10 - 2 \overset{?}{\Leftrightarrow} 0$ we used the property of logarithm.

By property of logarithms, we have learned, $\log 10 = 1$, So,

$2\,(1) - 2 \overset{?}{\Leftrightarrow} 0.$

$2 - 2 \overset{?}{\Leftrightarrow} 0$

$0 = 0$ true, both sides are the same.

b) We learned, $y = \log_a x$ is equivalent to $x = a^y$. So

$\log_x 64 - 3 = 0$

$\log_x 64 - 3 + 3 = 0 + 3$ we added 2 to both sides.

$\log_x 64 = 3$

$64 = x^3$ we applied the logarithmic equation rule

$4^3 = x^3$

$x = 4$ we used the one-to-one exponent rule.

Therefore, the solution to $log_x\, 64 - 3 = 0$ is $x = 4$.

Check: to verify the answer, we need to plug $x = 4$ into the equation. So,

$log_x\, 64 - 3 = 0$

$log_4\, 64 - 3 \overset{?}{\Leftrightarrow} 0$ we plugged $x = 4$ into the equation

$log_4\, 4^3 - 3 \overset{?}{\Leftrightarrow} 0$

$3\ log_4\, 4 - 3 \overset{?}{\Leftrightarrow} 0$ we used the property of logarithm.

By property of logarithms, we have learned, $log_4\, 4 = 1$, So,

$3\ (1) - 3 \overset{?}{\Leftrightarrow} 0.$

$3 - 3 \overset{?}{\Leftrightarrow} 0$

$0 = 0$ True, both sides are the same.

c) We learned, $y = ln\, x$ is equivalent to $x = e^y$. So

$ln\, (x - 5) - 12 = 0$

$ln\, (x - 5) - 12 + 12 = 0 + 12$ we added 2 to both sides

$ln\, (x - 5) = 12$

$(x - 5) = e^{12}$

$x - 5 + 5 = e^{12} + 5$ we added 5 to both sides

$x = e^{12} + 5.$

Therefore, the solution to $ln\, (x - 5) - 12 = 0$ is $x = e^{12} + 5$.

Check: to verify the answer, we need to plug $x = e^{12} + 5$ into the equation. So,

$ln\, (x - 5) - 12 = 0$

$ln\, (e^{12} + 5 - 5) - 12 \overset{?}{\Leftrightarrow} 0$ we plugged $x = e^{12} + 5$ into the equation

$ln\, (e^{12}) - 12 \overset{?}{\Leftrightarrow} 0$

$12 ln\, e - 12 \overset{?}{\Leftrightarrow} 0$ we used the property of natural logarithm.

By property of natural logarithms, we have learned, $ln\, e = 1$, So,

$12\ (1) - 12 \overset{?}{\Leftrightarrow} 0.$

$12 - 12 \overset{?}{\Leftrightarrow} 0$

$0 = 0.$ True, both sides are the same.

4.7. Exercises

1. Solve each equation.
   a) $|6x - 4| = 14$.
   b) $-|3x + 8| = 16$.
   c) $-|\frac{5}{7}x + \frac{3}{14}| = -13$.
   d) $-|\frac{3}{2}x + 4| + |x - 5| = -12$.

2. Solve each inequality, using the algebraic approach and graph the solution.
   a) $|3x| \leq 9$.
   b) $|2x - 10| \leq 4$.
   c) $|5x - \frac{7}{2}| > 12$.
   d) $|3x - 5| + |x| \leq 7$.

3. Graph $f(x) = 3|x - 7| - 5$.

4. Given the polynomial $6x^5 + 3x^3 - 2x^2 - 7x - 14$, evaluate it at:
   a) $x = 0$.
   b) $x = 2$.
   c) $x = -3$.
   d) $x = \frac{2}{3}$.

5. The graph of a polynomial in xy-axis with degree of 5 is given below. As we have shown earlier, at each of the x-intercepts the polynomial changes sign. For the given polynomial,
a) Find the x intercepts.
b) Write the inequities for each rage of x, with positive and/or negative values of the polynomial.

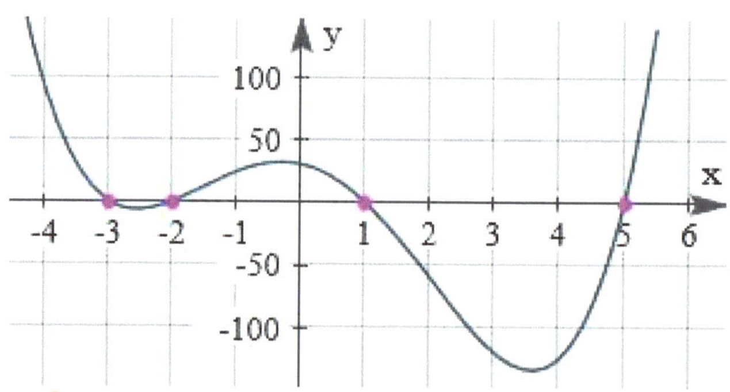

6. Solve each polynomial inequality.
   a) $x^2 + 5x - 14 \leq 0$.
   b) $x^3 + 2x^2 \leq 15x$.
   c) $x^4 + 2x^3 - 8x^2 > 0$.

7. Given the expression $\frac{(x-12)}{(14x-7)} - \frac{5(6x^2-1)}{(8x-19)^2} + \frac{1}{(3x^3-11)}$, evaluate it at $x = -2$.

8. Solve each rational equality.

a) $\frac{2}{3x} - \frac{1}{9} = \frac{5}{6x}$.

b) $\frac{3}{5x} - 4 = \frac{1}{3x}$.

c) $\frac{5}{(x+2)} - \frac{3}{(x+2)^2} + \frac{x}{4(x+2)^2} = \frac{7}{4(x+2)}$.

d) $\frac{1}{(14x-21)} - \frac{1}{(2x-3)^2} + \frac{1}{(10x-15)} - \frac{1}{35(2x-3)^2} = 0$.

9. Given $\frac{2(x^3-12)}{(5x^2+8)} \leq \frac{2(x^3+1)}{(3x-1)^2} \times \frac{(9x-3)}{(4x^3+5)}$, evaluate it at $x = -1$.

10. Solve each rational inequality.

a) $\frac{2x+4}{5x-11} \geq 0$.

b) $\frac{3x-9}{x-2} \leq 0$.

c) $\frac{4x+7}{2x-1} + \frac{2x+7}{3(2x-1)} > 0$.

d) $\frac{3x-3}{6(2x+1)} + \frac{4x-1}{(2x+1)} < 0$.

11. In a gas cylinder the volume for a certain gas is given by $V = \frac{1}{0.03}$, where P is the gas pressure. Find the set of volume values when the gas pressure varies between 32 and 41.

12. Evaluate $2\sqrt[3]{x+23} - \frac{5\sqrt{(2x^2+6)^2}}{\sqrt[4]{8(x^2+16)}} - 2\frac{4\sqrt{(2x-1)^2}}{\sqrt[3]{2x}}$ at $x = 4$.

13. Solve each equation.

a) $\sqrt{7x-12} = 3$.

b) $\sqrt[3]{6x-16} = 2$.

c) $9\sqrt{5x+8} - 3\sqrt{20x+32} = 0$.

d) $5\sqrt{3x+7} - \frac{4\sqrt{(2x+6)^2}}{\sqrt{12x+28}} + 2\sqrt{27x+63} = 0$.

14. Evaluate $7\sqrt[3]{9x+18} - \frac{3\sqrt{(3x^2+5)^2}}{\sqrt[3]{3x^2+24)}} \geq 2\frac{4\sqrt{(2x-1)(x^2+3)}}{\sqrt[3]{8x^2}}$ at $x = 1$

15. Solve each inequality.

a) $\sqrt{4x+5} < 12$.

b) $\sqrt[3]{2x+16} \leq 4$.

c) $\sqrt[4]{7x-10} \geq \sqrt[4]{3x+12}$.

16. Solve each equation.

a) $4^{x^2} = 16$.

b) $3^{(x^2-1)} = 27$.

c) $5^{x^2 - 2} = \frac{1}{5}$.

d) $27^{(u + 3)} = 9$.

e) $e^{3x} \cdot e^{6} = e^{12}$.

f) $\frac{e^{2x}}{e^{4x}} = e^{4}$.

17. Solve each of the exponent equations.

a) $e^{9t} = 81$.

b) $5e^{6t} = 125$.

c) $3e^{(14t + 7)} = 63$.

18. Use the property of logarithmic function which implies $a^b = c$ if and only if $log_a c = b$ and rewrite the given exponential equations as logarithmic equations and vice versa.

a) $2^4 = 16$.

b) $6^2 = \frac{1}{36}$.

c) $(\frac{1}{5})^{-2} = 25$.

d) $log100 = 2$.

e) $log_3 27 = 3$.

f) $log_4 \frac{1}{16} = -2$.

g) $ln\,(e^2) = 2$.

h) $ln\sqrt{e} = \frac{1}{2}$.

19. Solve logarithmic equations.

a) $log_{31} x = 3log_{31} 2$.

b) $2log_7 x = log_7 25$.

c) $2ln\,x - ln\,2x + ln\,1 = 0$.

d) $\frac{1}{3}e^{x} = 3$.

e) $6^{(5x - 2)} = 216$

f) $e^{(2x + 1)} = 7$.

20. Solove $3ln\,(2x + 3) - 2ln\,(2x + 3) = ln\,(x + 12)$.

5. Real-World Applications of Equalities and Inequalities

Real- world situations often lend themselves to being modeled through linear equations and inequalities. Take, for example, a car rental agency where the rental cost is influenced by both the mileage driven and a fixed daily fee. To formulate an equation for such scenarios, we gather known data and define the unknown quantity as a variable. Verbal descriptions are then translated into mathematical expressions.

In forthcoming scenarios, we will explore various real-world applications that illustrate the concepts of equality and inequality, demonstrating how mathematical models can be applied to practical situations.

5.1 Example: Simple Interest Rate

Suppose you put \$10,000.00 in a savings account that pays simple interest at a rate of 11% per annum. Consider that the amount of money in a savings account S depends on the principal P, the interest rate r, and the time in years' t that the money is invested. The equation $S = P(1 + rt)$ shows the relationship among the variables for an account earning simple interest.

a) How much interest do you get in 5 years?

b) How much interest do you get for 2 and a half years?

c) How many years will it take for the savings to exceed \$14,000? (Round to the nearest tenth of a year).

Solution

To answer these questions, we need to first write an equation. Letting x to be the number of years, the interest you get for x years is:

Earned Interest = Interest Rate $\times\ x \times$ The Initial Saving Amount.

Using this equality equation, we have:

a) For 5 years the interest you get is:
Earned Interest = Interest Rate $\times\ x \times$ The Initial Saving Amount

Earned Interest = $0.11\times5\times\$10,000.00 = \$5,500.00$.

b) For 2.5 years interest you get is:
Earned Interest = Interest Rate $\times\ x \times$ The Initial Saving Amount

Earned Interest $=0.11\times2.5\times\$10,000.00 = \$2,750$.

c) The equation relating the variables for an account earning simple interest is $S = P(1 + rt)$. The savings to exceed \$14,000 implies $S > \$14,000$ or $P(1 + rt) > \$14,000$. So,

$\$10,000.00(1 + 0.11t) > \$14,000$ — we substituted $P = \$10,000$ and $r = 0.11$ into the inequality

$10,000.00 + 10,000.00(0.11t) > \$14,000$ — we applied the distributive rule

$10,000.00 + 1,100t > \$14,000$

$10,000.00 + 1,100t - 10,000 > \$14,000 - 10,000$ — we added $-10,000$ to both sides.

$1,100t > 4,000$

$\frac{1}{1,100} \times 1,100t > \frac{1}{1,100} \times 4,000$ we multiplied both sides by the reciprocal of the coefficient of the variable

$t > \frac{4,000}{1,100} = 3.64.$

Therefore, the number of years taken for the saving to exceed \$14,000 is 3.7 (the rounded value to nearest tenth in usual method must be 3.6, but in this case, we need the savings to exceed \$14,000, so we must round up the number of years. Thus, the correct answer is 3.7 years.

5.2. Example: Compounded Interest Rate

Suppose you put \$5,000 in a savings account that pays compounded interest at a rate of 8.6 % per annum. Consider that compound interest, or 'interest on interest', is calculated with the compound interest formula. The formula for compound interest is $S(1+\frac{r}{n})^{nt}$, where S is the initial saving, r is the interest rate, n is the number of times interest is compounded per period and t is the number of time periods. Using this formula, we can find the mount of saving S_t at end of year t by $S_t = S(1+\frac{r}{n})^{nt}$.

a) How much interest do you get for 4 years?

b) How much interest do you get for 6 months?

c) What is the amount of saving at end of second year?

Solution

a) Leaving the total amount for 6 years, your interest earned at end of year six is,

$$\text{interest} = (1+0.086)^4 \times 5,000 = 6,954.87.$$

b) We first need to convert 6 months to its equivalent portion of a year. Knowing 6 months is $\frac{1}{2}$ of a year, leaving the total amount for 6 months, your interest earned at end of month four is:

$$\text{interest} = (1+0.086)^{\frac{1}{2}} \times 5,000 = 5,210.57.$$

c) The formula for finding the amount of saving at end second year t, where is $S_t = S(1+\frac{r}{n})^{nt}$. Since in this case, the compounded interest rate is per annum, it means that the number of times interest is compounded per period of a year is once. That is $n = 1$. So,

$S_t = S(1+\frac{r}{n})^{nt}$

$S_t = S(1+\frac{r}{1})^{1t} = S(1+r)^t$

$S_2 = \$5,000(1+0.086)^2$ we substituted $S = \$5,000$, $r = 0.086$, and $t = 2$ into the equation

$S_2 = \$5,000(1.086)^2$

$S_2 = \$5,000(1.179) = 5,896.98.$

Therefore, the amount of saving at end of second year is 5,896.98.

5.3. Example: Production Financial Analysis

A house appliances company manufactures four models of blender. Each blender requires a certain number of raw materials; plastic composite; and steel to make. The required and available amounts of materials, the unit production cost and the unit selling price are presented in the table below.

	Blender 1	Blender 2	Blender 3	Blender 4	Total available/week
Plastic composite	2.1 lbs.	3.4 lbs.	5.8 lbs.	8.1 lbs.	7,300 lbs.
Steel	3.6 lbs.	7.8 lbs.	6.2 lbs.	5.9 lbs.	8,750 lbs.
Unit production cost	\$28	\$36	\$32	\$44	
Unit price	\$43	\$59	\$48	\$78	

The company intended to quantify the following in forms of equalities and inequalities.

a) Inequalities for availability of raw materials.

b) The equality for the total production costs/week.

c) The equality for the total incomes.

d) The equality for the total profit.

Solution

To answer the requested questions, we need first to define variables as below.

x_1 = the number of blenders of type 1 produced, where $x_1 \geq 0$.

x_2 = the number of blenders of type 2 produced, where $x_2 \geq 0$.

x_3 = the number of blenders of type 3 produced, where $x_3 \geq 0$.

x_4 = the number of blenders of type 4 produced, where $x_4 \geq 0$.

a) Referring to the above table:
   - To produce 1 unit of blender 1 we need 2.1*lbs*. of plastic composite.
   - To produce 1 unit of blender 2 we need 3.4*lbs*. of plastic composite.
   - To produce 1 unit of blender 3 we need 5.8*lbs*. of plastic composite.
   - To produce 1 unit of blender 4 we need 8.1*lbs*. of plastic composite.

   Since the total available of plastic composite is 7,300*lbs*, the inequality below enforces that the total intake of plastic composite does not exceed its total available is:

   $$2.1x_1 + 3.4x_2 + 5.8x_3 + 8.1x_4 \leq 7{,}300 lbs$$

   $$x_1, x_2, x_3, \text{ and } x_4 \geq 0.$$

   Similarly, referring to the above table,
   - To produce 1 unit of blender 1 we need 3.6*lbs*. of steel.

- To produce 1 unit of blender 2 we need 7.8*lbs*. of steel.
- To produce 1 unit of blender 3 we need 6.2*lbs*. of steel.
- To produce 1 unit of blender 4 we need 5.9*lbs*. of steel.

Since the total available of steel is 8,750 *lbs*. The inequality below enforces that the total intake of steel does not exceed its total available is:

$$3.6x_1 + 7.8x_2 + 6.2x_3 + 5.9x_4 \leq 8{,}750lbs$$

$$x_1, x_2, x_3, \text{ and } x_4 \geq 0.$$

b) Referring to the above table:
- To produce 1 unit of blender 1 we need to spend \$28.
- To produce 1 unit of blender 2 we need to spend \$36.
- To produce 1 unit of blender 3 we need to spend \$32.
- To produce 1 unit of blender 4 we need we need to spend \$44.

The equality below determines the total production costs.

$$\text{Total Production Cost} = \$28x_1 + \$36x_2 + \$32x_3 + \$44x_4.$$

c) Referring to the above table:
- The company earns \$43 for selling one unit of blender 1.
- The company earns \$59 for selling one unit of blender 2.
- The company earns \$48 for selling one unit of blender 3.
- The company earns \$78 for selling one unit of blender 4.

The equality below determines the total incomes.

$$\text{Total Incomes} = \$43x_1 + \$59x_2 + \$48x_3 + \$78x_4.$$

d) To write the equality for the total profit, we must subtract the total production cost from the total income. Using the obtained equalities for the total production cost and the total incomes, we have:

$$\text{Total Profit} = \text{Total Incomes} - \text{Total Production Cost}.$$

$$\text{Total Profit} = (\$43x_1 + \$59x_2 + \$48x_3 + \$78x_4) - (\$28x_1 + \$36x_2 + \$32x_3 + \$44x_4).$$

$$\text{Total Profit} = (\$43 - \$28)x_1 + (\$59 - \$36)x_2 + (\$48 - \$32)x_3 + (\$78 - \$44)x_4.$$

$$\text{Total Profit} = \$15x_1 + \$23x_2 + \$16x_3 + \$34x_4.$$

In the previous example, we looked at a company producing four products. If we consider a situation in which there is a company producing two products, we can sketch inequalities in a two–dimensional coordinate as shown in example below.

5.4. Example: Analyzing Production Operations

Consider a company producing two products, A and B. Each product requires certain number of raw materials, machine hours and man-hours to make. The required and available amounts of raw materials, machine hours and man-hours are presented in table below.

	Product A	Product B	Total available/week
Raw material	$54lb.$	$68lbs.$	$940lbs.$
Machine hour	58 hours	42 hours	484 hours
Man-hour	48 hours	36 hours	242 hours

Given the information above, provide the requests below.

a) Write inequalities regarding the raw material, machine hours and manhours.

b) Plot the inequalities and designate the feasible space (a feasible space, or solution space is the set of all possible points that satisfy the problem's requirements).

Solution

To answer the requested questions, we need first to define variables as below.

x_1 = the number of product A to be produced, where $x_1 \geq 0$.

x_2 = the number of product B to be produced, where $x_2 \geq 0$.

a) Referring to the above table,
   - To produce 1 unit of product A, we need 54 lbs. of raw material.
   - To produce 1 unit of product B, we need 68 lbs. of raw material.

   Since the total available of raw material is $940lbs$, the inequality below enforces that the total intake of raw material does not exceed its total available is:

   $54x_1 + 68x_2 \leq 940lbs$

   $x_1, \geq 0$

   $x_2, \geq 0.$

   Similarly, referring to the above table,
   - To produce 1 unit of product A, we need 58 machine hours.
   - To produce 1 unit of product B, we need 42 machine hours.

   Since the total available of machine hours 484 hours, the inequality below enforces that the total manhours do not exceed the total available is:

   $58x_1 + 42x_2 \leq 484$ hours

   $x_1, \geq 0$

   $x_2, \geq 0.$

   Again, referring to the above table:
   - To produce 1 unit of product A, we need 48 manhours.

- To produce 1 unit of product B, we need 36 manhours.

Since the total available manhours 242 hours, the inequality below enforces that the total machine hours do not exceed its total available is:

$48x_1 + 36x_2 \leq 242$ hours.

$x_1, \geq 0.$

$x_2, \geq 0.$

b) To illustrate the region satisfying the constraints inequalities and constructing the feasible space, we examine the constraints one at a time. Let us first start with the constraint of raw material, $54x_1 + 68x_2 \leq 940$. The common approach is to replace this inequality by equality.

Considering $54x_1 + 68x_2 = 940$, it is easily seen that the set of points satisfying this equality corresponds to a straight line in the two–dimensional plane with x_1 axis and x_2 axis.

To plot this line, we need to identify the coordinates of two distinct points on this line. The standard approach for doing this is to set, alternatingly, one of the two variables to the value 0 and then solve the resulting equation with one unknown to obtain the corresponding value of the other variable. Thus, if we let $x_1 = 0$, then x_2 must equal to 13.82. This yields the point (0, 13.82). Similarly, by setting $x_2 = 0$, we obtain the point (17.41, 0). The resulting line that passes through these two points is shown in the figure below.

Observe that this line cuts the plane into two half–planes. All points on one side of this line (including the line itself) satisfy $54x_1 + 68x_2 \leq 940$, whereas on the other side, $54x_1 + 68x_2 \geq 940$.

Note that $(x_1, x_2) = (0, 0)$ satisfies $54x_1 + 68x_2 \leq 940$, and that (0, 0) lies in the lower–left direction of the line. Therefore, we are only interested in the lower–left portion of the plane, indicated by two arrows. Additionally, we have two more constraints presenting $x_1 \geq 0$ and $x_2 \geq 0$. It can be easily observed that the upper half of plane satisfies $x_1 \geq 0$ shown by two arrows upward, and the right half of plane satisfies $x_2 \geq 0$, show by two arrows directing the right half of plane.

In summary, after discarding points that violate the raw material constraint as well as discarding points that violate $x_1 \geq 0$ and $x_2 \geq 0$, we are left with the triangle with line in one side and x_1 and x_2 representing the horizonal axis and the vertical axis of the plane, the shaded triangle.

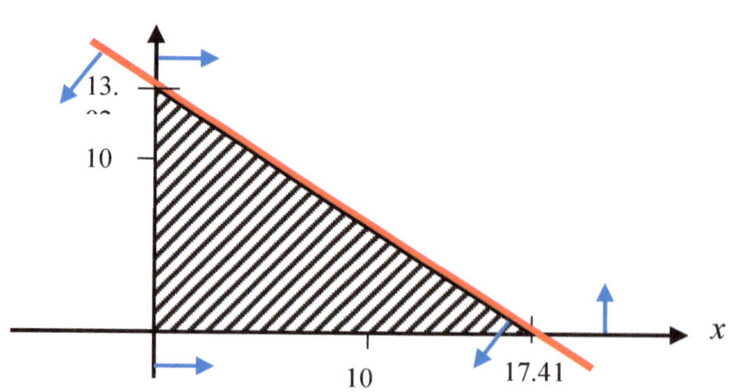

We can now, use the same procedure to plot the machine hour constraint and its feasible region.

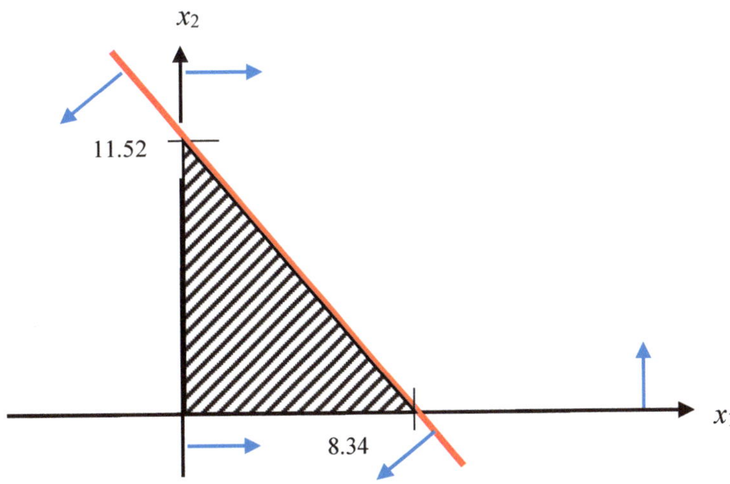

Similarly, we can now use the same procedure to plot the manhour constraint and its feasible region.

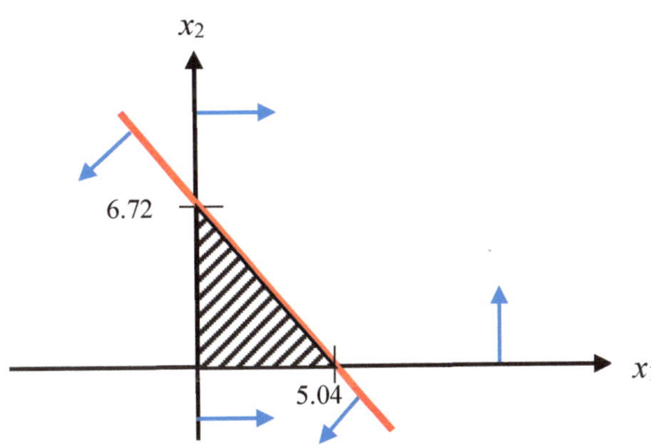

We now can combine the three figures to find the feasible spaces, satisfying all constraints simultaneously.

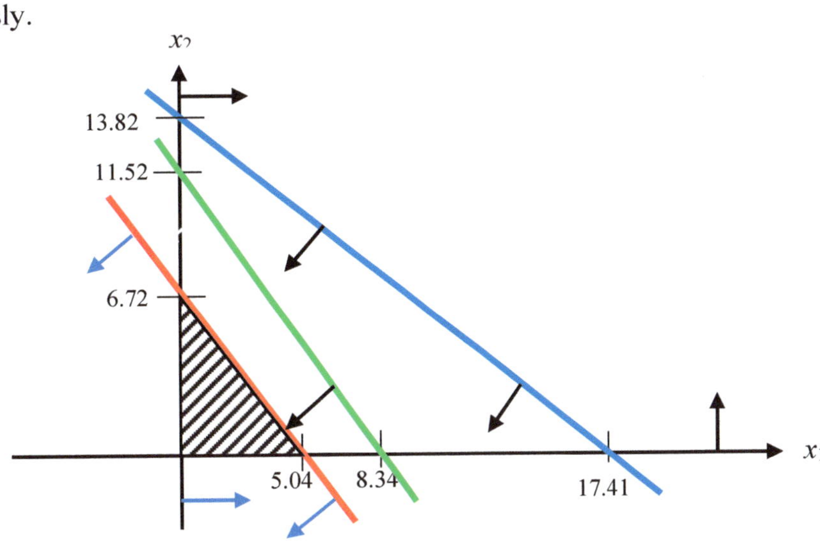

The final feasible region satisfying the three constraints simultaneously is the shaded triangle. Referring to the above figure, we realize that the two constraints, the raw material constraint $54x_1 + 68x_2 \leq 940$ and the machine hour constraint $58x_1 + 42x_2 \leq 484$, have NO effect in determining feasible region. These types of constraints are commonly called redundant constraints.

5.5. Example: Numbers

The sum of four consecutive even integers is 180. What are these numbers?

Solution

Denoting the first number as x, we will have:

The first number $= x$

The Second number $= x + 2$

The third number $= x + 4$

The fourth number $= x + 6$.

Given the sum of the four numbers is 180, we can write,

$(x) + (x + 2) + (x + 4) + (x + 6) = 180$

$x + x + 2 + x + 4 + x + 6 = 180$

$4x + 12 = 180$ — we combined the like terms

$4x = 180 - 12$ — we isolated the variable in one side and constant on the other side

$4x = 168$

$\frac{1}{4} \times 4x = \frac{1}{4} \times 168$ — we multiplied both side by the reciprocal of $4x$

$x = \frac{168}{4} = 42.$

Therefore, the first number is 42. So,

The first number $= 42$.

The second number $= x + 2 = 44$.

The third number $= x + 4 = 46$.

The fourth number $= x + 6 = 48$.

Note that:

$42 + 44 + 46 + 48 = 180.$

5.6. Example: Geometry, Angeles

Two angles of a triangle are the same size, angle 1 and angle 2. The third angle is 3 times as large as one of the other angles, either angle 1 or angle 2. What are the measures of the angles (note that the sum of angles in triangles is 180°)?

Solution

Denoting the first angle as F, the second angle as S, and the third one as T, and letting the variable x to represent the measure of the first angle, we have:

The measure of $F = x$.

The measure of $S = x$.

The measure of $T = 3x$.

Since the sum of angles in triangles is 180°, we will have:

the measure of F + the measure of S + the measure of T = 180°

$x + x + 3x = 180^\circ$

$5x = 180^\circ$ — we combined the like terms

$\frac{1}{5} \times 5x = \frac{1}{5} \times 180^\circ$ — we multiplied both side by the reciprocal of $5x$

$x = \frac{180^0}{5} = 36^\circ$.

Therefore, the first angle is 36°. So,

The measure of $F = x = 36^\circ$

The measure of $S = x = 36^\circ$.

The measure of $T = 3x = 3 \times 36^\circ = 108^\circ$.

Note that,

$36^\circ + 36^\circ + 108^\circ = 180^\circ$.

5.7. Example: Geometry, Areas

Suppose we have a rectangle, and we know that its total area is 25 centimeters squared (cm^2). If side A equals 5 cm^2 and side B equals $\sqrt{5x}$ cm^2, what is the value of x?

Note that, to find the area of a rectangle, we multiply the length of the rectangle by the width of the rectangle.

Solution

We have learned, the area S of a rectangle is the multiplications of its two sides. So,

$S = A \times B$

$25 = 5 \times \sqrt{5x}$ — we substituted $S = 25$, $A = 5$ and $B = \sqrt{5x}$ into the equation.

$(25)^2 = (5\sqrt{5x})^2$ — we raised both sides to the power of 2

$625 = 25 \times 5x$

$625 = 25 \times 5x$

$625 = 125x$

$\frac{1}{125} \times 625 = \frac{1}{125} \times 125x$ — we multiplied both sides by the reciprocal of coefficient of the variable

$\frac{1}{125} \times 625 = \frac{1}{125} \times 125x$

$5 = x$.

Therefore, the value of x is $5cm$.

5.8. Example: Unit Conversion

For a day in September, the temperatures in Oxnard, California ranged from $10°C$ to $20°C$. The formula relating Celsius temperatures to Fahrenheit temperatures is given by $C = \frac{5}{9}\ (F - 32)$.

Solution

Using the conversion formula, to convert the lower range of temperature we have:

$C = \frac{5}{9}(F - 32)$

$10 = \frac{5}{9}(F - 32)$ — we substituted $C = 10$ into the equation

$10 = \frac{5}{9}F - \frac{5}{9} \times 32$

$9 \times 10 = 9 \times (\frac{5}{9}F - \frac{5}{9} \times 32)$ — we multiplied both sides by 9 to convert the fraction to an integer number

$90 = 5F - 5 \times 32$

$90 = 5F - 160$

$-5F = -90 - 160$ — we multiplied both sides by -1 to change the negetive sign to a positive sign

$5F = 90 + 160$ — we isolated the variable in one side and the constant terms in the other side

$5F = 250$

$\frac{1}{5} \times 5F = \frac{1}{5} \times 250$ — we multiplied both side by the reciprocal of the coefficient of $5F$

$F = 50.$

Now, applying similar operations as the above, we can convert the upper limit temperature to its equivalent Fahrenheit. So,

$C = \frac{5}{9}(F - 32)$

$20 = \frac{5}{9}(F - 32)$ — we substituted $C = 20$ into the equation.

By the same operations we reach to:

$F = 68.$

Therefore, the inequality is converted to a range of $50°C$ to $68°C$.

5.9. Example: Breakeven Point Analysis

Suppose a company is expecting to sell 10,000 units at a price of $10 each. The variable cost associated with the product is $5 per unit and the fixed cost is $15,000 per year. Find the break-even point for the given company.

Solution

The breakeven point is the production level where total revenues equal total expenses. The breakeven point for the given company can be calculated either in quantity terms or in dollar terms.

To calculate the breakeven point, in quantity terms, we must divide the total fixed cost by the contribution per unit.

- Here, the selling price per unit = \$10.
- Variable costs per unit = \$5.
- Contribution per unit = \$10 – \$5 = \$5.
- Hence the breakeven point (quantity) = $\frac{\$15000}{\$5}$ = 3,000 units.

Therefore, the breakeven point is 3,000 units. It means by selling up to 3000 units the company will overcome its fixed costs only. Selling quantity beyond 3000 will help in earning a profit which will be equal to the contribution per unit for every additional unit sold beyond 3000.

To calculate the breakeven point, in dollar terms, we will divide the total fixed cost by the contribution margin ratio.

- Contribution per unit = \$5.
- Selling price per unit = \$10.
- So, contribution margin ratio = $\frac{\$5}{\$10}$ = 0.5.
- Hence, breakeven sales (\$) = $\frac{\$15000}{0.5}$ = \$30,000.

Therefore, the breakeven sales = \$30,000. It means by selling up to a sales value of \$30,000, the company will be at breakeven point and will overcome its fixed cost only and will earn profit beyond the sales value of beyond \$30,000.

The chart of breakeven helps in portraying the company's ability to earn a profit with the present cost structure. To prepare the chart, on the vertical axis, we plot the revenue, variable cost, and the fixed costs of the company and on the horizontal axis, we plot the production volume. Figure below illustrates the breakeven chart of the above example.

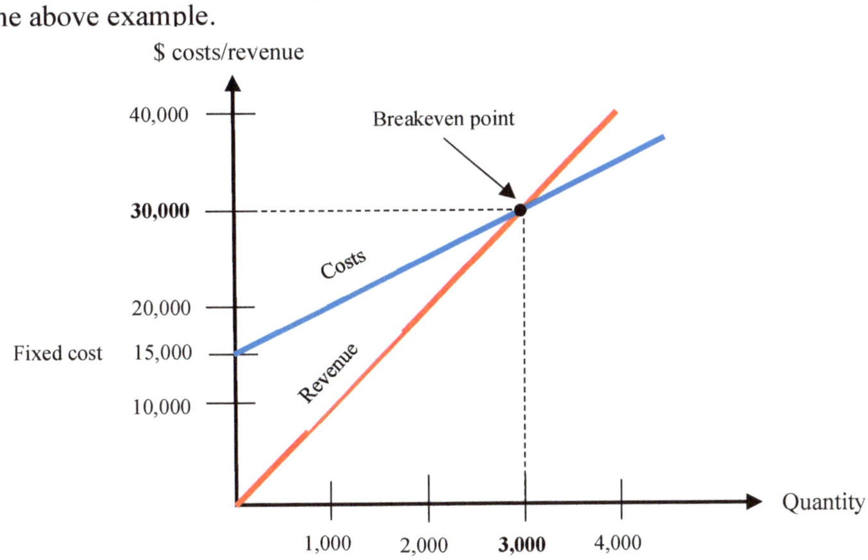

5.10. Example: Working Hours to Earn Money

Shery's best friend is having a wedding in Ventura California and the event will last 2 days. Shery has \$350 in savings and can earn \$18 an hour babysitting. She expects to pay \$310 airfare, \$286 for food and entertainment and \$56 per night for her share of a hotel room. How many hours must she babysit to have enough money to pay for the trip?

Solution

Let us define h as the number of hours Shery must babysit. Her total expenses, TE, for the event are the cost of airfare plus the cost of food and entertainment and the hotel bill. So,

TE = the f airfare cost + the food and entertainment costs + her share of two nights' hotel bill.

$TE = \$310 + \$286 + \$56(2).$

Her total available money, AM, is the amount of savings plus her earnings from the babysitting. So,

AM = the amount of savings + the babysitting income.

$AM = \$350 + \$18(h).$

Now, her expenses must be less than or equal to the total available money, $TE \leq AM$. So,

For $TE \leq AM$, we have:

$\$310 + \$286 + \$56(2) \leq \$350 + \$18(h)$ — we substituted TE and AM into the inequality

$\$310 + \$286 + \$112 \leq \$350 + \$18h$

$\$708 \leq \$350 + \$18h$

$\$708 - \$350 \leq \$350 + \$18h - \$350$ — we added − \$350 to both sides

$\$358 \leq \$18h$

$\frac{1}{\$18} \times \$358 \leq \frac{1}{\$18} \times \$18h$ — we multiplied both sides of the inequality by the reciprocal of \$18h

$19.89 \leq h.$

Therefore, she needs to babysit for at least 20 hours.

5.11. Example: Distance, Rate and Time

Two runners start from opposite ends of a 12-kilometer (km) course running towards each other. One runner has a speed of 6.4km per hour, and the other is running at a rate of 7.1km per hour. After how much time will the runners meet?

Solution

Let us first write the equation relating distance rate and time. The equation $d = st$, presents this relation, in which d denotes distance, s denotes speed rate and t denotes time. In this case, denoting d_1 as the distance the first runner has run and d_2 as the distance the second runner has run, we have:

$d = st$

$d_1 = 6.4t$

$d_2 = 7.2t.$

Since the total distance both runners have run, when they meet, is equal to the course running distance. Hence,

$d_1 + d_2 = 12$

$6.4t + 7.2t = 12$ — we substituted $d_1 = 6.4t$ and $d_2 = 7.3t$ into the equation

$13.6t = 12$

$\frac{1}{13.6} \times 13.6t = \frac{1}{13.6} \times 12$ — we multiplied both sides by the reciprocal of 13.6*t* coefficient

$t = \frac{12}{13.6} = 0.88.$

Therefore, the runners will meet after 0.88 hour, or 52.8 minutes.

5.12. Example: Age Problem

Peter is 11 years older than Michel and David is 3 years younger than Peter. Seven years ago, the sum of their ages was 57. How old are they now?

Solution

Let x represents Michel's age. Then, Peter's age is:

Peter age = 11 + Michel's age

Peter age = $11 + x$ — we substituted Michel's age = x into the equation.

Since David is 3 years younger than Peter, then,

David's age = Peter's age – 3

David's age = $11 + x - 3$ — we substituted Peter's age = $11 + x$ into the equation.

Seven years ago, the sum of their ages was 57, so now sum of their ages is 57 +7 = 64. Hence,

Peter's age + Michel's age + David's age = 64

$11 + x + x + 11 + x - 3 = 64$ — we substituted Peter's age = $11 + x$, Michel's age = x, and David's age = $11 + x - 3$ into the equation

$3x + 19 = 64$

$3x + 19 - 19 = 64 - 19$ — we added – 19 to both sides

$3x = 45$

$\frac{1}{3} \times 3x = \frac{1}{3} \times 45$ — we multiplied both sides by reciprocal of the coefficient of the variable

$x = 15.$

Therefore, Michel age is 15. To find the other ages we have.

Peter's age = 11 + Michel's age

Peter's age = 11 + 15 — we substituted Michel's age = 15 into the equation

Peter's age = 26.

David's age = Peter's age – 3

David's age = 26 – 3   we substituted Peter's age = 26 into the equation

David's age = 23.

Check: To verify the answers, we need to add up their ages to get 64 years (seven year ago their ages was 57, hence 57 + 7 = 64 years). So,

$$\text{Peter's age} + \text{Michel's age} + \text{David's age} \overset{?}{\Leftrightarrow} 64$$

$$26 + 15 + 23 \overset{?}{\Leftrightarrow} 64$$

$$64 = 64$$ true.

5.13. Example: Variation

The speed (or rate) Susan travels from her house to work is inversely proportional to time it takes to get there. If she travels 46 miles per hour, it will take her 2.1 hours to get to work. How long will it take her if she travels 69 miles per hour?

Solution

First, we need to write the equation for the case. The inversely proportional indicates multiplying rate and time to obtain distance. Hence, the equation is $rt = k$, where r is the rate, t is the time, and k is the distance from her house to her work.

Using this equation, we have:

$$rt = k$$

$$(46)\,(2.1) = k$$ we substituted $r = 46$ $r = 2.1$ into the equation

$$96.6 = k.$$

Again, we can use the equation with the rate she might use and the same distance she travels from her house to her work, to get the required time. So,

$$rt = k$$

$$(69)\,(t) = k$$ we substituted $r = 69$ into the equation

$$69t = k.$$

Considering the same distance in the above both equations, we get:

$$69t = 96.6$$

$$\frac{1}{69} \times 69t = \frac{1}{69} \times 96.6$$ we multiplied both sides by the reciprocal of the variable

$$t = \frac{96.6}{69} = 1.4.$$

Therefore, it takes 1.4 hours when she travels 69 miles per hour.

5.14. Example: Percentage and Rates

Antony works at a used car dealership. His monthly pay will be either a fixed amount, \$2,846, or \$800 plus 1.6% of his total sales. How much should his total sales be for his variable pay option to exceed the fixed amount of \$2,846?

Solution
First, we need to write the equation for the case. Let us define s as the total soles amount. Then Antony variable payment based on commission 1/6% will be \$800 plus 1.6% of total amount of sales. That is:

Variable pay =\$800 + 0.016s

To get his variable pay option to exceed the fixed amount of \$2,846, we will have:

Variable pay > \$2,846

$\$800 + 0.016s > \$2,846$ — we substituted variable pay = \$800 + 0.016s into the inequality

$\$800 + 0.016s - \$800 > \$2,846 - \$800$ — we add – \$800 to both sides

$0.016s > \$2,046$

$\frac{1}{0.016} \times 0.016s > \frac{1}{0.016} \times \$2,046$ — we multiplied both sides by the reciprocal of the variable

$s > \frac{\$2,046}{0.016}$

$s > \$127,875.$

Therefore, Antony's total sales must exceed \$127,875.

5.15. Example: Mixtures Problem

A mixture of 14 ounces of vinegar and oil contains 35 percent vinegar. How many ounces of oil must be added to the mixture to produce a new mixture that is only 20 percent vinegar?

Solution
First, we need to write the equation for the mixture problem. Let x represent the number of ounces of oil to be added. Then the total number of ounces of the new mixture will be $14 + x$. Since we only add vinegar to the new mixture, the weight of vinegar is the same in both mixtures. The weight of vinegar in the first mixture is:

Weight of vinegar in the first mixture = 0.35×14.

The weight of the new mixture is $14 + x$.

To have 20% vinegar in the second mixture its weight is determined as:

Weight of vinegar in the second mixture = $0.20 \times (14 + x)$.

Just oil is added to the new mixture, so the weight of vinegar in both mixtures are equal. So,

Weight of vinegar in the first mixture = Weight of vinegar in the second mixture

$0.35 \times 14 = 0.20 \times (14 + x)$ — we substituted vinegar weights into the equation

$4.9 = 2.8 + 0.20x$ — we applied the distributive rule

$-2.8 + 4.9 = -2.8 + 2.8 + 0.20x$ — we added –2.8 to both sides

$2.1 = 0.20x$

$\frac{1}{0.20} \times 2.1 = \frac{1}{0.20} \times 0.20x$ — we multiplied both sides by the reciprocal of the variable

$10.5 = x.$

Therefore, we need to add 10.5 ounces of vinegar.

Check:
In the first mixture, the weights of vinegar and oil are $0.35 \times 14 = 4.9$, and $0.65 \times 14 = 9.1$, respectively.

In the second mixture, the weights of vinegar and oil are $0.20 \times (14 + 10.5) = 4.9$, and $0.80 \times (14+10.5) = 19.6$, respectively.

Weight of oil in the first mixture = 9.1.

Weight of oil in the second mixture = 19.6.

Weight of oil in the second mixture = 9.1 +10.5 (our solution) = 19.6 true.

5.16. Example: Linear Depreciation Problem

Depreciation is the decrease in value of an asset. Straight line depreciation method charges cost evenly throughout the useful life of a fixed asset. A fixed asset having a useful life of 5 years is purchased on 1 January 2020. The cost of the asset is \$10,000 whereas its residual value is expected to be \$1,000.

Calculate the depreciation expense for the financial years ending June 30-2020 and the total depreciation expense at 30th of June 2021.

Solution

The formula for depreciation expense per annum, is $d = \frac{c-r}{n}$, where d is the depreciation expense per annum, c is the cost of the asset, r is the residual value of the asset and n is the number of years used. So,

$$d = \frac{c-r}{n}$$

$$d = \frac{10{,}000 - 1{,}000}{5}$$ we plugged $c = 10{,}000$, and $r = 1.000$ into the equation

$$d = \frac{9{,}000}{5} = 1{,}800.$$

Therefore, the depreciation expense for the end of December 2020 is \$1,800. To find depreciation expense for the years ending June-30-2020, we need to multiply the resulted expense by $\frac{6 \text{ months}}{12 \text{ months}} = \frac{1}{2}$. So,

depreciation expense for the years ending June-30-2020 $= \frac{1}{2} \times \$1{,}800 = \$900.$

Similarly, to find the total depreciation expense for the years ending 30 June 2021, we need to multiply the resulted expense by $\frac{18 \text{ months}}{12 \text{ months}} = \frac{3}{2}$. So,

Total depreciation expense for the years ending at 30th of June 2021 $= \frac{3}{2} \times \$1{,}800 = \$2{,}700.$

5.17. Example: Voting Problem

In a small town the governor election contested by two candidates A and B. Candidate A got 40% of total votes and still lost to candidate B by 30,000 votes. If there is no scrubbed vote, find the total number of votes cast.

Solution

Let V be the total votes cast. Since candidate A got 40%, we conclude candidate B got 60%. Then we will have:

The number of votes casted for candidate A = $V \times 40\% = 0.40V$

The number of votes cast for candidate B = $V \times 60\% = 0.60V$.

Since candidate A lost by 30,000 votes, we conclude the number of votes cast for candidate B is 30,000 more the number of votes casted for candidate A. So,

Number of votes casted for candidate A + 30,000 = Number of votes casted for candidate B

$0.40V + 30{,}000 = 0.60Vt$

$0.40V + 30{,}000 - 0.40V = 0.60V - 0.40V$ to isolate the variable, we added $-\,0.40V$ to both sides

$30{,}000 = 0.60V - 0.40V$

$30{,}000 = 0.20V$

$\frac{1}{0.20} \times 30{,}000 = \frac{1}{0.20} \times 0.20V$ we multiplied both sides by the reciprocal of the variable

$\frac{30{,}000}{0.20} = \frac{0.20}{0.20} \times V$

$V = \frac{30{,}000}{0.20} = 150{,}000.$

Therefore, the total number of cast votes is 150,000.

5.18. Exercises

1. A rental company charges a flat fee of \$28 and an additional \$0.26 per mile to rent a moving pickup truck. Write a linear equation to approximate the cost y (in dollars) in terms of x.

2. Trains A and B are 298 miles away from each other. Train A is traveling towards B at 58 miles per hour and train B is traveling towards A at 66 miles per hour.
   a) At what time will the two trains meet?
   b) At this time, how far did train A travel?
   c) At this time, how far did train B travel?

3. Two cars are 188 miles away from each other. Car A is traveling towards car B at 64 miles per hour and car B is traveling towards car A at 58 miles per hour.
   a) At what time will the two cars meet?
   b) At this time, how far did car A travel?
   c) At this time, how far did car B travel?
   d) If car A travels with the speed of 70 miles per hour and car B travel with the speed of 69 miles per hour, what time will the two cars meet?

4. Suppose a company sells 68,000 units of a product per month at a price of \$24 each. The variable cost associated with the product is \$10 per unit and the fixed cost is \$28,000 per month. Find the break-even point for the given company.

5. Coby has scored 89, 78, 94, 88, and 92 on his first five tests in geometry. He is hoping to get an A in his college geometry class. There is only one exam left, and the maximum points could be 100. Find if it is possible for Coby to end the course with an A.

6. Cell phone company A offers a package which includes a monthly service fee of \$36 plus a charge per minute of \$0.04 talk-time. Cell phone company B offers a package which charges a monthly service fee of \$28 plus a charge per minute of \$0.05 talk-time. For 1280 minutes talk time, which company offers a better deal?

7. One number exceeds another number by 29 and their sum is 73. Find the two numbers.

8. The sum of five consecutive odd integers is 135. What are these numbers?

9. The sum of three consecutive even integers is 78. What are these numbers?

10. The sum of six consecutive integers is 96. What are these numbers?

11. The speed Jordan travels from his work to the gym is inversely proportional to time it takes to get there. If he travels 56 miles per hour, it will take him 0.2 hours to get to the gym. How long will it take him if he travels 68 miles per hour?

12. Cell-Phone USA company offers a package by which they charge a monthly service fee of \$46 plus \$0.03/minutes talk-time. Another company American Cell company offers a package by which it charges a monthly service fee of \$32 plus \$0.04/ minutes talk-time.
    a) Write a linear equation that models the packages offered by both companies.

b) For the average number of minutes of 1,200 used each month, find which company offers a better plan.

c) How many minutes of talk-time would yield equal monthly statements from both companies?

13. Fred works in the electronic department of a wholesale corporation. His monthly pay will be either a fixed amount, \$2,500, or \$980 plus 2.6% of his total sales. How much should his total sales be for his variable pay option to exceed the fixed amount of \$2,500?

14. The cost of living for a family of 5 is given by $52{,}000x - 3{,}000$, where x represents the number of years. Determine after many years the cost of living would exceed \$294,000.

15. Resistors are said to be connected in series when they are daisy chained together in a single line resulting in a common current flowing through them. The total resistance in the circuit with resistors connected in series is equal to the sum of the individual resistances. Given three resistors in series as shown below, the resistance of the second resistor is 2 ohms less than the first resistor and the resistance of the third resistor is 3 ohms more than the second resistor. The equivalent total resistances have been measured at 8 Ohms. Find the resistance of the three resistors.

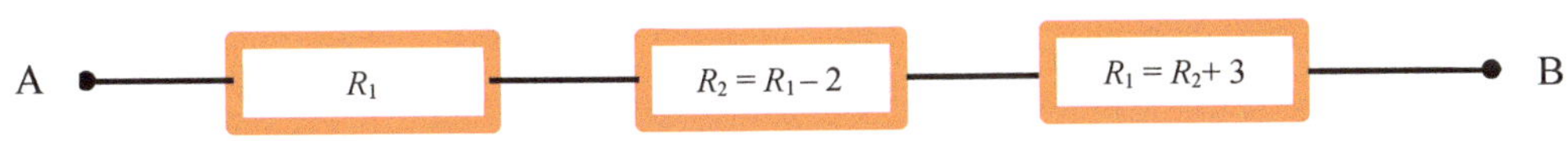

16. Eric and Eddy are planning to paint the inside of a church together. Eddy has less experience, and it would take him 2 times as long as it would take Eric to paint the entire church. Working together, they can complete the job in 36 hours. How long would it take each of them, working alone, to complete the job?

17. Merry takes 2 hours for transplanting 60 tree seedlings. Souzan takes 3 hours for transplanting 75 tree seedlings. Working together, how long would it take them for transplanting 180 tree seedlings?

18. The width of the rectangle is 22 inches. What must the length be if the perimeter is at least 104 inches?

    Hint: To find the perimeter of a rectangle, multiply both the height and width by two and add the results.

19. A rectangular garden has one side with a length in feet of $x + 8$ and another side with a length in feet of $3x - 1$. If the perimeter of the garden is 174, find the length of each side of the garden.

    Hint: To find the perimeter of a rectangle, multiply both the sides by two and add the results.

20. A mixture of $60ml$ after shave lotion of alcohol and water contains 32 percent alcohol. How many milliliters of water must be added to the mixture to produce a new mixture which contains only 28 percent alcohol?

21. Jacob is 5 years younger than Harold and Max is 3 years older than Jacob. Five years ago, the sum of their ages was 47. How old are they now?

22. It costs USA Electronics company $4.80 per unit to produce an electronic part used in a brand of laptop computers. The monthly operating expenses of the company are $2,060 for utilities and overhead and $4,680 for salaries.
    a) What are the company's monthly expenses for producing 12,000 units per month?
    b) What is the company's monthly revenue for selling 11,580 units per month, where the unit selling price of $6.99?

23. A company just purchased a car for admin staff that cost $44,000. The car is estimated to be used for four years and the residual of the car at the year fourth would be $8,000. What is the depreciation expense of the car if we use the straight-line depreciation method?

24. A business purchased important machinery for $250,000. It is estimated that this machine will be useful for 10 years, and that at the end of those 10 years its scrap value will be about $50,000. Using the straight-line depreciation method, find the total depreciation expenses for the first four years.

25. In a city the mayor election was contested by two candidates, Mr. Stewed and Mr. Douglas. Mr. Stewed got 45% of the total votes and still lost to Mr. Douglas by 960,000 votes. If there is no scrubbed vote, find the total number of votes cast.

6. Multiple-Choice Questions

1. Which of the following is equivalent to $\frac{b^3}{9a^2(b^4-1)} \div \frac{b^2}{27a^3(b^4-1)^2}$.

A. $\frac{3b(b^4-1)}{a}$

B. $\frac{3ab}{(b^4-1)}$

C. $\frac{3a(b^4-1)}{b}$

D. $3ab(b^4-1)$

2. Solve the linear inequality $-8x + 22 \leq 6$.

A. $x \leq 8$

B. $x \leq 2$

C. $x \geq -8$

D. $x \geq 2$

3. Solve the linear equation $10x + 7 = 2 + 4x$.

A. $\frac{-6}{5}$

B. $\frac{6}{5}$

C. $\frac{-5}{6}$

D. $\frac{14}{9}$

4. What is the value of x when $2x + 3 = 3x - 4$?

A. -7

B. $\frac{-1}{5}$

C. $\frac{1}{5}$

D. 7

5. In the equation below, what is the value of d?

$$\frac{2d^2 - d - 10}{d^2 + 7d + 10} = \frac{d^2 - 4d + 3}{d^2 + 2d + 15}.$$

A. -4

B. 2

C. 4

D. 6

6. Which of the following is equivalent to $\frac{2m+6}{4} \times \frac{6m-36}{3m+9}$?

A. $\frac{6m-36}{3m+9}$

B. $\frac{6m-36}{3m+9}$

C. $\frac{6m-36}{3m+9}$

D. $m-6$

7. Mary placed an online order for woman blouses that cost \$31.99 per blouse. A tax of 8% is added to the cost of the blouses, before a flat, untaxed shipping rate of \$9.99 is charged. Which of the following represents Mary's total cost for k blouses, in dollars?

A. $0.08(31.99k + 9.99)$

B. $1.08(31.99 + 9.99)k$

C. $1.08(31.99k) + 9.99$

D. $1.08(31.99 + k) + 9.99$

8. At early stages of growth of a certain plant, its height increases at a linear rate. After seven weeks, the plant is 28 inches tall. Which of the following functions best models the relationship between $h(w)$, the height, in inches, of the plant, and w, the number of weeks that the plant has been growing?

A. $h(w) = 7w$

B. $h(w) = 4w$

C. $h(w) = 28w$

D. $h(w) = 28 + w$

9. Which of the following is equivalent to $(9x^2 + 10x + 5xy) + (4x^2 - 8x + 3xy - 7)$?

A. $13x^2 + 2x + 8xy - 7$

B. $13x^2 - 8x + 8xy - 7$

C. $9x^2 - 8x + 5xy - 7$

D. $9x^2 + 10x + 5xy - 7$

10. Which of the following is equivalent to $\frac{1}{3}a + 7b + 6 = \frac{5}{6}a - 9b + 8$?

A. $-\frac{1}{2}a + 16b - 2$

B. $-\frac{4}{3}a - 2b - 2$

C. $\frac{5}{3}a + 16b + 2$

D. $\frac{1}{3}a - 2b + 2$

11. In the equation below, what is the value of x?

$$\frac{3}{5}x + 6 = \frac{9}{10}x + 9$$

A. -15

B. 6

C. -10
D. 2

12. Dogs need 0.25 to 0.51 liters of water each day for every 4.50 kilo grams (kg) of their weight. Helen has a 2.90kg Pomeranian mix, and Monica has a 19.60kg bulldog. Which of the following ranges represents the approximate total number of liters of water, w, that Helen and Monica need in a week?

A. $2.25 \leq w \leq 4.5$
B. $2.05 \leq w \leq 4.0$
C. $1.75 \leq w \leq 3.5$
D. $1.25 \leq w \leq 2.5$

13. In the equation below, what is the value of x?
$\sqrt[3]{7x-8} - \sqrt[3]{4x+10} = 0.$

A. -6
B. 4
C. 6
D. -4

14. What is the value of the expression below at $x = -1$.
$\sqrt[3]{8(2x^2-1)} - 4\sqrt[3]{3x^3+30} + \frac{1}{\sqrt{4x+8}}.$

A. -10.5
B. $2.5 - 4\sqrt[3]{35}$
C. $2.5\sqrt[3]{35}$
D. -9.5

15. What is the solution of $\sqrt[3]{20-4x} \leq 2$?

A. $x \leq -3$
B. $x \geq 3$
C. $x \leq -\frac{9}{2}$
D. $x \geq \frac{9}{2}$

16. If two times a number is equal to that number minus 4, what is the number?

A. -2
B. -3
C. -6
D. -4

17. If $-\frac{20}{7} < -5x + 15 < -\frac{17}{8}$, what is the greatest possible integer value of $15x - 45$?

A. 6
B. 7
C. 8
D. 9

18. If $x + y = 0$, which of the following must be equivalent to $x - y$?

A. $-2y$
B. $\frac{x}{y}$
C. x
D. x^2

19. The number of specified liquid samples, s cups, that George needs for an experiment must be greater than 5 but less than or equal to 14. Which of the following represents an acceptable number of liquid samples for George's experiment?

A. $5 < s \leq 14$
B. $5 < s < 14$
C. $5 \leq s \leq 14$
D. $5 \leq s < 14$

20. If $2\sqrt{5x} - 1 \leq 4$, which of the following is the solution for x?

A. $-\infty < x \leq 1$
B. $1 \leq x < \infty$
C. $-1 < x < \infty$
D. $-1 \leq x \leq 1$

21. If $7u = 5u + 2$, what is the value of $2u$?

A. 1
B. 2
C. 3
D. 4

22. Janet enrolled in a retirement benefit plan that charges a monthly fee of \$69. A one-time enrollment fee of \$35 was charged when she joined. Which of the following represents the total amount of fees that Janet has paid to the retirement benefit plan organizers after m months, in dollars?

A. $(35 + 69)m$
B. $35m + 69$
C. $69m + 35$
D. $69m - 35$

23. Officer Henry drives his squad car 1 kilo meter (*km.*) per minute while patrolling local highways during his shift. If he has driven 540*km.* by the end of his shift, how many total hours did he drive his car at the above rate?

A. 8
B. 9
C. 10
D. 11

24. In the inequality $16 \leq -3x + 5$, what is the appropriate order of steps needed to solve the inequality for x?

A. Add 5 to both sides, divide both sides by -3, and flip the inequality sign to $\geq$.
B. Subtract 5 from both sides, divide both sides by -3, and flip the inequality sign to $\geq$.
C. Add 5 to both sides, divide both sides by 3, and keep the original inequality sign.
D. Subtract 5 from both sides, divide both sides by 3, and keep the original inequality sign.

25. If $b = \frac{4a^{\frac{1}{2}}}{16}$ and b is a nonzero integer, which of the following is equivalent to b?

A. $\frac{1}{4}a$
B. $\frac{1}{4}\sqrt{a}$
C. $\frac{1}{4}a^2$
D. $\frac{1}{4}a\sqrt{2a}$

26. Jolie needs to receive completed surveys from at least 4,200 potential voters in her city. She has noticed that for every 4 surveys she sends out, only 1 survey is completed. Last week she received 1,420 completed surveys, and this week she received 1,150 completed surveys. Jolie plans to send out s additional surveys.

Which of the following inequalities shows all possible values for s that would ensure that Jolie receives 4,200 completed surveys?

A. $s \geq 2{,}570$.
B. $s \geq 1{,}630$.
C. $s \geq 8{,}150$.
D. $s \geq 4{,}200$.

27. Which of the following is equivalent to the expression $\frac{x^3 - 2x^2}{x^5 - 2x^4}$?

A. x^{-2}

B. $\frac{1}{x}$

C. $\frac{1}{x^2 - 2x}$

D. $\frac{x-2}{x^2 - 2x}$

28. For which of the following values of a does $\sqrt[4]{16a^3b^{\frac{9}{a}}} = (2)\left(3^{\frac{3}{4}}\right)\left(b^{\frac{3}{4}}\right)$?

A. 2
B. 3
C. 4
D. 6

29. If $|2x - 5| = 9$, what are the values of x?

A. -2 and 7
B. 2 and -7
C. -2 and -7
D. 2 and 7

30. The price of a specific pair of shoes that cost \$76 in 2020 always increases by \$3 per year. The current price in dollars, P, of the pair of shoes can be represented by the equation $P = 3t + 43$, where t is the number of years since the pair of shoes was first came to markets. Which of the following best explains the meaning of the number 43 in the equation?

A. It is the price of the item in 2007.
B. It is the price of the item in 2008.
C. It is the price of the item in 2009.
D. It is the price of the item in 2010.

31. If $\frac{5}{x} - \frac{1}{3} = \frac{1}{x}$, what is the value of $\frac{1}{3}x$?

A. $\frac{1}{3}$.
B. 2.
C. $\frac{5}{3}$.
D. 4.

32. If $2|3x + 9| < 36$, what is the solution?

A. $-3 < x < 9$.
B. $-9 < x < -3$.

C. $-9 < x < 3$.
D. $-3 < x < 3$.

33. If $\frac{x+1}{x+3} \leq 2$, what is the solution set?

A. $(-\infty, -5) \cup [3, +\infty)$.
B. $(-\infty, 5] \cup [-3, +\infty)$.
C. $(-\infty, 5] \cup (3, +\infty)$.
D. $(-\infty, -5] \cup (-3, +\infty)$.

34. David is ordering sugar, which is only sold in 25-pound bags. He currently has 60 pounds of sugar, and he needs to have a minimum of 190 pounds. Which of the following inequalities shows all possible values for the number of bags, *b*, that David needs to order to meet his minimum requirement?

A. 4.
B. 5.
C. 6.
D. 7.

35. A website hopes to sign up 150,000 subscribers. So far, the website has signed up an average of 450 subscribers per day for *d* days. Which of the following represents the number of additional subscribers, *W*, the website must sign up to reach its goal?

A. $W = 450d$.
B. $W = 149{,}550 - 450d$.
C. $W = 150{,}000 - 450d$.
D. $W = 150{,}000 + 450d$.

36. If $a = \frac{2}{3}(k + t + kt)$ and $b = -\frac{3}{4}k(t + 3) + 6t$, what is $3a - 4b$ in terms of k and t?

A. $4k - 23t + 4kt$.
B. $10k - 23t + 4kt$.
C. $8k + 25t + 2kt$.
D. $-8k - 24t + 2kt$.

37. Ohm's Law states, $V = IR$, where V is the voltage measured across the conductor in units of volts, I is the current through the conductor in units of amperes, and R is the resistance of the conductor in units of ohms. The power P in watts can be related to I and R by the equation $I = \sqrt{\frac{P}{R}}$. Which of the following gives P in terms of V and R?

A. $P = \frac{V}{R}$.
B. $P = \frac{V^2}{R}$.
C. $P = V^2R$.

D. $P = \frac{R}{V^2}$.

38. To receive an *B* in her mathematic class, Stephany needs to earn an average score from 80 to 89, inclusive. Her grade is based only on 3 tests, with equal score weight. The highest possible score on each of these tests is 100 points. She scored 78 on her first test and 97 on her second test. If *x* represents her score on the third test, which of the inequalities below shows all values of *x* that would earn Stephany an *A* in her mathematic class?

A. $65 \leq x \leq 92$.
B. $80 \leq x \leq 89$.
C. $80 \leq x \leq 92$.
D. $65 \leq x \leq 89$.

39. Which of the following expressions is equivalent to $(5k)^{\frac{1}{4}}$?

A. $\frac{4}{\sqrt{5k}}$.
B. $\sqrt[4]{5k}$.
C. $\frac{\sqrt{5k}}{4}$.
D. $\frac{4}{\sqrt[4]{5k}}$.

40. If the expression $\frac{w - 12}{\sqrt{8}}$ is most nearly equal to $\frac{w\sqrt{2}}{4} - K$, then what is the value of *K*?

A. 2.
B. -3.
C. $\sqrt{3}$.
D. $3\sqrt{2}$.

41. Water is being drained from a water tank at a constant linear rate. Three hours after draining of the tank began, the volume of water in the tank was 805 gallons, and six hours after draining of the tank began, the volume was 610 gallons.

Which of the following functions best models *v*(*t*), the volume of water in the tank, in gallons, *t* hours after draining of the tank began?

A. $v(t) = 740 - t$.
B. $v(t) = 1000 - 65t$.
C. $v(t) = 1000 - 545t$.
D. $v(t) = 740 - 65t$.

42. A chemist prepares a mixture of citric acid with concentration, by volume, equal to *V*. It is prepared by mixing a volume of citric acid given by *CA* with a volume of water given by *W*. The expression $V = \frac{CA}{CA + W}$ represents the mixture described. What physical quantity does the term *CA* + *W* represent in the equation above?

A. The volume of the citric acid in the mixture.
B. The concentration of the citric acid in the mixture.
C. The volume of the mixture.
D. The volume of the water in the mixture.

43. In expression $\left(\frac{8}{2k + 5}\right)(2k + 5) - \left(3 - \frac{4k}{2k + 5}\right)(2k + 5)$, what is the value of k?

A. $-\frac{5}{2}$.
B. $-\frac{5}{8}$.
C. $-\frac{8}{5}$.
D. $-\frac{7}{2}$.

44. What is the result of multiplying $2a^2 - 3a - 4$ by $5a - 3$?

A. $10a^3 - 11a^2 - 14a + 8$.
B. $10a^3 - 19a^2 - 14a + 8$.
C. $10a^3 - 19a^2 - 26a + 8$.
D. $-11a^2 - 26a + 8$.

45. An air pump at a gas station dispenses 82 pounds of air for \$0.20. Which of the following expressions gives the number of pounds of air dispensed, P, for the d dollars?

A. $P = d + 410$.
B. $P = 410d$.
C. $P = 82d$.
D. $P = d + 82$.

46. A developer is creating a plan for a 48-acre park that includes a 7-acre lake that cannot be developed. If 4 to 6 acres, inclusive, must be reserved for basketball fields, which of the following inequalities shows all possible values for p, the amount of land that within the park that is available for development?

A. $28 \leq p \leq 38$.
B. $34 \leq p \leq 36$.
C. $35 \leq p \leq 37$.
D. $31 \leq p \leq 33$.

47. Which of the following expressions is equivalent to $\frac{5x}{3x - 4} - \frac{x}{6x - 8} = \frac{24}{12x - 16}$?

A. $\frac{22x - 24}{4(3x - 4)}$.
B. $\frac{3}{2}$
C. $\frac{22x - 24}{(6x - 8)}$.

D. $\frac{16x-24}{4(3x-4)}$.

48. If $a = b^{-\frac{1}{4}}$, where $a > 0$ and $b > 0$, which of the following equations gives b in terms of a?

A. $b = \frac{1}{\sqrt[4]{a}}$.
B. $b = -\sqrt[4]{a}$.
C. $b = -a^4$
D. $b = \frac{1}{a^4}$.

49. If $\frac{1}{5}k - \frac{7}{10}h + \frac{1}{2}l = 2$, what is the value of $2k - 7h + 5l$?

A. 4.
B. 5.
C. 10.
D. 20.

50. An energy drink manufacturer packs his energy drink cans in standard 16-hole cartons and then packs the cartons in large shipping boxes. The number of boxes needed, b, to transport c cartons of cans can be found using the equation $b = \frac{c}{36}$. If the carton-packing machine can pack a maximum of 7,200 cartons per day, and it does not pack partial boxes, what is the range of the b in this framework?

A. $0 < b \leq 7{,}200$, where b is an integer number.
B. $0 < b \leq 36$, where b is an integer number.
C. $0 < b \leq 200$, where b is an integer number.
D. $0 < b \leq 450$, where b is an integer number.

51. Suppose a company is expecting to sell 13,000 units at a price of \$15 each. The variable cost associated with the product is \$10 per unit and the fixed cost is \$14,000 per year. The break-even point for the given company is.

A. 1,400.
B. 2,000.
C. 934
D. 2,800.

52. Jack is 4 years younger than Joe and the age of David is 2 times the age of Jack. Their total age is 52 years. What is the age of Jack?

A. 16.
B. 12.
C. 24
D. 8.

53. What is the area of a triangle with base $b = 3$ cm and height $h = 4$ cm? Hint: To find the area of a triangle, multiply its base by one half of its height.

A. 12*cm*
B. 6*cm*
C. 7*cm*
D. 24*cm*

Questions 54 - 56 refer to the problem below.

Consider a company producing a product. assembled by 2 units of component *A*, 3 units of component *B* and one unit of component *C*. Components, *A, B* and *C* require certain amount of raw material, machine hours and man-hours to make. The required and available amounts of raw materials, machine hours and man-hours are presented in table below.

	Component *A*	Component *B*	Component *C*	Total available/week
Raw material	12 lbs.	18 lbs.	9 lbs.	4,856 lbs.
Machine-hour	6 hours	4 hours	7 hours	992 hours
Man-hour	3 hours	5 hours	4 hours	886 hours

The production manager of the company decided that at least 141 units of the product to be produced.

Considering
x_1 = The number of component *A* to be produced.
x_2 = The number of component *B* to be produced.
x_3 = The number of component *C* to be produced.

54. The inequality concerning the machine-hour is.

A. $12x_1 + 18x_2 + 9x_3 > 4{,}856.$
B. $12x_1 + 18x_2 + 9x_3 < 4{,}856.$
C. $12x_1 + 18x_2 + 9x_3 \geq 4{,}856.$
D. $12x_1 + 18x_2 + 9x_3 \leq 4{,}856.$

55. The inequality concerning the raw material is.

A. $3x_1 + 5x_2 + 4x_3 < 886.$
B. $3x_1 + 5x_2 + 4x_3 > 886.$
C. $6x_1 + 4x_2 + 7x_3 \leq 992.$
D. $6x_1 + 4x_2 + 7x_3 \geq 992.$

56. The inequality concerning the number of products to be produced is.

A. $2x_1 + 3x_2 + x_3 \geq 141.$
B. $2x_1 + 3x_2 + 2x_3 > 141.$

C. $3x1 + 2x2 + x3 > 141$.

D. $x1 + 3x2 + 2x3 \geq 141$.

57. Nadia sells laptop computers, and her salary is $26,640 per year plus a 9% commission on sales. The equation $S = 26640 + 0.09x$ represents her salary S in dollars in terms of her total sales x in dollars. How much money in sales does Nadia need to earn a salary that exceeds $54,000?

A. $x \geq \$304,000$.

B. $x > \$600,000$.

C. $x \geq \$296,000$.

D. $x > \$420,000$.

58. Two boats travel down a river, starting at 10:00*am*. One of the boats travels twice as fast as the other, after 4 hours, the boats are 4.8 kilometers (*km*) apart. Find the average rate of each boat.

A. The average rate for one of the boats is 1.2*km/hr* and the other is 2.4*km/hr*.

B. The average rate for one of the boats is 2.4*km/hr* and the other is 4.8*km/hr*.

C. The average rate for one of the boats is 2.2*km/hr* and the other is 4.4*km/hr*.

D. The average rate for one of the boats is 0.6*km/hr* and the other is 1.2*km/hr*.

59. A car dealer offers two options for a car loan. The first option is a 3-year car loan at 8.4% simple interest and the second is a 4-year car loan at 6.5% simple interest. For 12,000 car loan which option will demand less interest and by how much?

A. The second option demands less by $94.

B. The first option demands less by $102.8.

C. The first option demands less by $104.2

D. The second option demands less by $96.

60. If $|\frac{2}{3}x - \frac{3}{4}| \leq \frac{5}{6}$, what is the solution?

A. $-\frac{1}{3} \leq x \leq \frac{5}{6}$.

B. $-\frac{1}{12} \leq x \leq \frac{5}{12}$.

C. $-\frac{1}{8} \leq x \leq \frac{19}{8}$.

D. $-\frac{1}{3} < x < \frac{5}{6}$.

6.3. Key to the Multiple-Choice Questions

Question No.	Answer	Question No.	Answer	Question No.	Answer
1	D	21	B	41	B
2	D	22	C	42	C
3	C	23	B	43	D
4	D	24	B	44	B
5	C	25	B	45	B
6	D	26	C	46	C
7	C	27	A	47	B
8	B	28	B	48	D
9	A	29	A	49	D
10	A	30	C	50	C
11	C	31	D	51	D
12	D	32	C	52	B
13	C	33	D	53	B
14	D	34	C	54	D
15	B	35	C	55	C
16	D	36	B	56	A
17	D	37	B	57	A
18	A	38	A	58	A
19	A	39	B	59	D
20	A	40	D	60	C

Some for the glories of this world; and some,
sigh for the prophet's paradise to come;
ah, take the cash, and let the credit go,
nor heed the rumble of a distant drum!

Omar Khayyam
Persian Mathematician, Astronomer, Philosopher, and Poet.

Chapter 4

Polynomials

Contents

6.1. Key to the Multiple-Choice Questions

Introduction

Polynomial equations have been central to the development of mathematics over the centuries, significantly shaping its evolution. The quest to solve nth degree polynomial equations has ancient origins, with the Babylonians around 2000 BC pioneering methods for solving quadratic equations. Later, around 250 ADS, Diophantus made notable contributions to polynomial theory, particularly over rational numbers, and introduced innovative techniques for translating word problems into polynomial equations.

The exploration of quadratic equations in two variables, also known as conic sections, was significantly advanced geometrically by Apollonius. During Europe's "Dark Age," the Persian mathematician Omar Khayyam made notable contributions to the study of polynomials, particularly focusing on cubic equations.

The Renaissance marked a breakthrough with the general solution of cubic equations, achieved by Italian mathematician Cardano in 1545. Polynomial equations in multiple variables were systematically explored thereafter.

The emergence of modern methods for solving equations, which included the introduction of equal signs and modern notation, occurred during the 16th century. René Descartes played a pivotal role in this development by introducing the graphing of polynomial equations in 1637 and standardizing the use of letters for constants and variables in polynomial formulas.

Polynomials have wide-ranging applications across diverse fields such as finance, electronics, physics, and engineering. They are utilized in financial analysis for present value assessments, in electronics for power loss calculations, and in physics for modeling motion and gravitation.

In this chapter, we delve deeper into polynomial concepts, exploring solutions for higher-degree polynomials and methods for sketching their graphs. Our aim is to illustrate how mathematics, particularly polynomial theory, finds practical utility in various real-world applications.

1. Polynomials

Polynomial equations and inequalities play vital roles in real-world scenarios across multiple disciplines. They are indispensable in calculating the path of a projectile launched in a gravitational environment, adhering to well-defined physical laws. Furthermore, in economics, polynomial equations facilitate cost analysis, empowering businesses to make strategic decisions. Engineers rely on polynomial equations for designing roller coaster tracks, ensuring both excitement and safety for riders. Additionally, these equations are pivotal in evaluating the curvature of bridges, guiding structural assessments and design improvements. These examples underscore the versatility and practical importance of polynomial equations in scientific and engineering domains.

Standard form of a polynomial expression - A polynomial is an algebraic expression. The standard form of a polynomial expression is an expression in the form of

$$a_n x^n + a_{n-1} x^{n-1} + \cdots + a_i x^i + \cdots + a_2 x^2 + a_1 x + a_0,$$

where, n is a non-negative integer, each a_i, multiplied by a variable is a coefficient and can be any real number other than zero, a_0 that is not multiplied by a variable is called a constant, and each term $a_i x^i$ is a term of a polynomial expression. The highest power of the variable in a polynomial is called the degree of a polynomial.

Another way of presenting polynomial expressions in a more condensed form is the use of summation notation. A condensed form of polynomial expressions is:

$$\sum_{k=0}^{n} a_k x^k = a_0 + a_1 x + a_2 x^2 + \cdots + a_n x^n.$$

Note that $a_0 x^0 = a_0$, since $x^0 = 1$.

Definition - Summation Notation

Given a sequence $a_0 + a_1 + a_2 + \cdots + a_n$, the summation from 0 to n of the sequence is written as:

$$\sum_{k=0}^{n} a_k = a_0 + a_1 + a_2 + \cdots + a_n,$$

where k is called the summation index and n is called the summation upper limit.

1.1. Polynomial Classification

There are seven main types of Polynomial expressions. These are:

1. Monomial polynomial.
2. Binomial polynomial.
3. Trinomial polynomial.
4. Linear polynomial.
5. Quadratic polynomial.
6. Cubic polynomial.
7. Homogeneous polynomial of degree n.

Let us have a look at them one by one:

Monomial polynomial – A **monomial polynomial** is an algebraic expression with a single term. monomial expressions such as: $8x$, $3x^3$, $5x^2y^4z$, etc. are some examples of monomial polynomials.

Binomial polynomial – A **binomial polynomial** is an algebraic expression with two terms. Binomial expression*s* such as: $2x+5$, $7x^2 - y$, $9x^3y - 12z$, etc. are some examples of binomial polynomials.

Trinomial polynomial–A **trinomial polynomial** is an algebraic expression with three terms. *Trinomial expressions* such as: $15x^2y - 2yz + 7xz^2$, $7x^4yz^2 - 24xz^3 + 5yz$, $43x^4y^4z - xy + 8$, etc. are some examples of trinomial polynomials.

Linear Polynomial–A polynomial of degree one is called a linear polynomial. In other words, in a **linear polynomial**, the highest power of the variable is one. Linear expressions such as: $9x - 5$, $18y + 45$, $7u - 25$ are examples of linear polynomials.

Quadratic Polynomial–A polynomial of degree two is called a **quadratic polynomial**. In other words, the highest power of the variable is two. Quadratic expressions such as: $6x^2 - 22$, $13u^2 + 8u - 15$, $9y^2 - 11y + 86$, etc. are examples of quadratic polynomials.

Cubic Polynomial–A polynomial of degree three is called a **cubic polynomial**. This means, the highest exponent of the variable is three. cubic expressions such as: $17x^3 - 34x^2 + 12x + 19$, $5u^3 + 22u - 45$, $13y^3 - 14y^2 + 26$, etc. are examples of cubic polynomials.

Homogeneous polynomial of degree *n* - A polynomial of degree n is called a **homogeneous polynomial**. This means that the highest exponent of the variable is n. Homogeneous expressions such as: $17x^6 - 34x^4 + 34x^3 - 42x + 9$, $3u^5 + 54u^4 - 8u^3 + 16u$, $53y^8 - 7y^5 + 14y^3 - 21y^2 + 35y + 6$, etc. are examples of homogeneous polynomials.

1.2. Polynomial Equations

A **polynomial equation** is an expression consisting of variables (also called indeterminates) and coefficients, that involves only the operations of addition, subtraction, multiplication, and non-negative integer exponents of variables. It consists of more than two algebraic terms, especially the sum of several terms that contain different powers of the same variable(s). Examples of polynomial equations are: $3x + 2 = 0$, $5x^2 + 7x - 14 = 0$, $9x^5 + 4x^3 - 8x - 11 = 0$, etc.

1.2.1. Standard Form of Polynomial Equations

Let n be a non-negative integer. The standard form of a polynomial equation is an expression in the form of:

$$a_n x^n + a_{n-1} x^{n-1} + \cdots + a_i x^i + \cdots + a_2 x^2 + a_1 x + a_0 = 0.$$

This is called the general form of a polynomial equation. Each a_i is a coefficient and can be any real number other than zero. Each term $a_i x^i$ is a term of a polynomial Equation.

Terminology of Polynomial Equations - We often rearrange polynomials so that the powers are descending. When a polynomial is written in this way, we say that it is in general form:

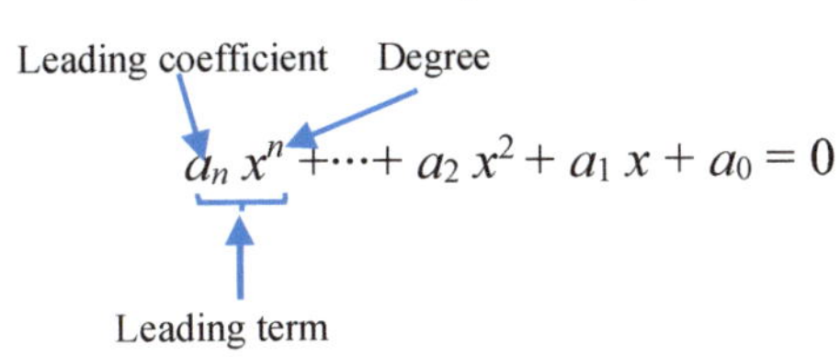

Degree of a Polynomial - Degree of a polynomial is the highest degree of the variable in a polynomial expression. Polynomials of degree one, two or three are respectively **linear polynomials, quadratic polynomials,** and **cubic polynomials**. For higher degrees, the specific names are not commonly used, although quartic polynomial (for degree four) and quintic polynomial (for degree five) are sometimes used.

1.3. Polynomials Arithmetic

Polynomial's arithmetic consists of addition, multiplication, and division. While addition and multiplication of a polynomial by another polynomial will result another polynomial, the division of one polynomial by another is not typically a polynomial. Here we look at two arithmetic operations, addition, multiplication, and division.

Addition - Polynomials can be added using the associative law of addition, grouping all their terms together into a single sum, possibly followed by reordering, using the commutative law, and combining of like terms.

Example
Given polynomials, $P = 9x^5 + 6x^4 - 3xy + 5y^3 - 7y^2 + 14$, and $Q = 2x^5 - 3x^4 - 12x^2 + 8y^3 - 16y^2 + 2xy + 8$, find $P + Q$.

Solution
Using associative law of addition and combining of like terms, we will have:

$$P = 9x^5 + 6x^4 - 3xy + 5y^3 - 7y^2 + 14$$

$$Q = 2x^5 - 3x^4 - 12x^2 + 8y^3 - 16y^2 + 2xy + 8$$

$$P + Q = (9x^5 + 6x^4 - 3xy + 5y^3 - 7y^2 + 14) + (2x^5 - 3x^4 - 12x^2 + 8y^3 - 16y^2 + 2xy + 8)$$

$$P + Q = (9x^5 + 2x^5) + [6x^4 + (-3x^4)] + (-12x^2) + (-3xy + 2xy) + (5y^3 + 8y^3) + [-7y^2 + (-16y^2)] + (14 + 8)$$

$$P + Q = (11x^5) + (3x^4) + (-12x^2) + (-xy) + (13y^3) + (-23y^2) + (22)$$

$$P + Q = 11x^5 + 3x^4 - 12x^2 - xy + 13y^3 - 23y^2 + 22.$$

Multiplication - Polynomials can also be multiplied. To expand the product of two polynomials into a sum of terms, the distributive law is repeatedly applied, which results in each term of one polynomial being multiplied by every term of the other.

Example
Given polynomials, $P = 2x^4 - 4xy - 6y^2 + 5y + 3$, and $Q = 3x^4 - 7xy - 11y^2 + - 3y - 2$, find PQ.

Solution
Using the distributive law, we will have:

$$P = 2x^4 - 4xy - 6y^2 + 5y + 3$$

$$Q = 3x^4 - 7xy - 11y^2 + - 3y - 2$$

$$PQ = (2x^4 - 4xy - 6y^2 + 5y + 3)(3x^4 - 7xy - 11y^2 + - 3y - 2)$$

$$PQ = (2x^4 \times 3x^4) + [(-4xy)\times(-7xy)] + [(-6y^2)\times(-11y^2)] + (5y\times(-3y)) + (3 \times (-2))$$

$$PQ = (6x^8) + (28x^2y^2) + (66y^4) + (-15y^2) + (-6)$$

$$PQ = 6x^8 + 28x^2y^2 + 66y^4 - 15y^2 - 6.$$

Division - Given a polynomial $P(x)$ with degree at least 1 and any number k there is another polynomial $Q(x)$, called the quotient, with degree one less than the degree of $P(x)$ and a number R, called the remainder, such that:

$$P(x) = (x - k)Q(x) + R.$$

Polynomials can be divided using the simple methods explained below.

1. Divide the first term of the numerator by the first term of the denominator and put that in the answer.
2. Multiply the denominator by that answer, put that below the numerator.
3. Subtract to create a new polynomial.

Example

Divide the polynomial in one variable, $6x^3 - 3x^2 + 5$ by $x - 2$.

Solution

Let us first get the problem set up.

$$x - 2\ \overline{)\,6x^3 - 3x^2 + 5}\,.$$

Keep in mind that we need to have the terms written down with the exponents in decreasing order and to make sure we do not make any mistakes we add in any missing terms with a zero coefficient.

Now, we divide the first term of the numerator by the first term of the denominator and put that in the answer. In this case, dividing $6x^3$ by x, we get $6x^2$as the first term of the resulted polynomial (our division result). So,

$$\begin{array}{r} 6x^2 \\ x - 2\ \overline{)\,6x^3 - 3x^2 + 5}\,. \end{array}$$

Now, we multiply $x - 2$ by $6x^2$ and subtract the results from the first polynomial. So,

$$\begin{array}{rl} 6x^2 & \\ x - 2\ \overline{)\,6x^3 - 3x^2 + 5} & \\ -(6x^3 - 12x^2) & \text{We multiplied } x - 2 \text{ by } 6x^2. \\ \hline 9x^2 + 5 & \text{We subtracted the result of the above multiplication from the first polynomial.} \end{array}$$

The new polynomial is called the remainder. We continue the above procedure until the degree of the remainder is less than the degree of the divisor, which is $x - 2$ in this case. That is, we need to continue until the degree of the remainder is less than 1.

Keep in mind that the degree of a polynomial is the highest exponent in the polynomial, 3 in this case. Also, recall that a constant is thought of as a polynomial of degree zero.

Therefore, we will need to continue until we get a constant in this case.

Here is the rest of the work for this example.

$$
\begin{array}{r|l}
 & 9x \\
\hline
x-2 & 9x^2+5 \\
 & -(9x^2-18x) \\
\hline
 & 18x+5
\end{array}
$$

We multiplied $x-2$ by $9x$.

We subtracted the result of the above multiplication from the reminder polynomial.

Proceeding with the same procedure, we get:

$$
\begin{array}{r|l}
 & 18 \\
\hline
x-2 & 18x+5 \\
 & -(18x-36) \\
\hline
 & 41
\end{array}
$$

We multiplied $x-2$ by 18.

We subtracted the result of the above multiplication from the next reminder polynomial.

Therefore, the answer is $6x^3 - 3x^2 + 5 = (x-2)(6x^2 + 9x + 18) + 41$.

Check: to verify our answer, we need to multiply $(x-2)$ by $(6x^2 + 9x + 18) + 41$. If we will get the first polynomial our answer would be correct. So,

$(x-2)(6x^2 + 9x + 18) + 41$

$= 6x^3 + 9x^2 + 18x - 12x^2 - 18x - 36 + 41$

$= 6x^3 - 3x^2 + 5$ true, it is the same as the original polynomial.

Example
Divide $x^2 + 2x^2y - 2xy + 2xy^2 - 3y^2$ by $x + y$.

Solution
Let us first get the problem set up.

$$x+y\ \overline{)\,x^2 + 2x^2y - 2xy + 2xy^2 - 3y^2}$$

Now we divide the first term of the numerator by the first term of the denominator and put that in the answer. In this case, dividing x^2 by x, we get x as the first term of the resulted polynomial (our division result). So,

$$
\begin{array}{r|l}
 & x \\
\hline
x+y & x^2 + 2x^2y - 2xy + 2xy^2 - 3y^2 .
\end{array}
$$

Now, we multiply $x + y$ by x and subtract the results from the first polynomial. So,

$$
\begin{array}{r|l}
 & x \\
\hline
x+y & x^2 + 2x^2y - 2xy + 2xy^2 - 3y^2 \\
 & -(x^2 + xy) \\
\hline
 & 2x^2y - 3xy + 2xy^2 - 3y^2
\end{array}
$$

we multiplied $x + y$ by x

we subtracted the result of the above multiplication from the first polynomial.

Proceeding with a similar procedure, we get:

$$\begin{array}{r l}
 & 2xy \\
x+y \,\big)\! & \overline{2x^2y - 3xy + 2xy^2 - 3y^2} \\
 & -(2x^2y + 2xy^2) \\
\hline
 & -3xy - 3y^2.
\end{array}$$

we multiplied $x + y$ by $2xy$

we subtracted the result of the above multiplication from the reminder polynomial

Proceeding again, we will have:

$$\begin{array}{r l}
 & -3y \\
x+y \,\big)\! & \overline{-3xy - 3y^2} \\
 & -(-3xy - 3y^2) \\
\hline
 & 0
\end{array}$$

we multiplied $x + y$ by $2xy$

we subtracted the result of the above multiplication from the reminder polynomial.

Therefore, the answer is $x^2 + 2x^2y - 2xy + 2xy^2 - 3y^2 = (x + y)(x + 2xy - 3y)$.

Check: To verify our answer, we need to multiply $(x + y)$ by $(x + 2xy - 3y)$. If we get the first polynomial our answer would be correct. So,

$(x + y)$ by $(x + 2xy - 3y)$

$= x^2 + 2x^2y - 3xy + xy + 2xy^2 - 3y^2$

$= x^2 + 2x^2y - 2xy + 2xy^2 - 3y^2$ True, it is the same as the original polynomial.

1.4. Expanding Binomial Expressions

A binomial is an algebraic expression with two terms. For example, $a + b$, $x - y$, $k + t$, etc. are binomials.

Let us learn the binomial expansion formula along with a few solved examples.

1.4.1. Perfect Square Trinomials

When a binomial is squared, the result is the first term squared added to double the product of both terms and the last term, the constant, is squared.

$$(x + a)^2 = (x + a)\,(x + a) = x^2 + ax + ax + a^2 = x^2 + 2ax + a^2.$$

Binomial theorem
Given expression $(x + a)^2$, its expanded expansion is:

$$(x + a)^2 = x^2 + 2ax + a^2.$$

A quick look at the diagram below makes the relationship clearer. Let as consider:

$$(a + b)^2 = (a + b)\,(a + b) = a^2 + ab + ab + b^2.$$

Each term of the expression $a^2 + ax + ax + b^2$ gives the area of one of the four pieces that form the square. The areas of the pieces add up to the area of the original square, that is $(x + a)^2$.

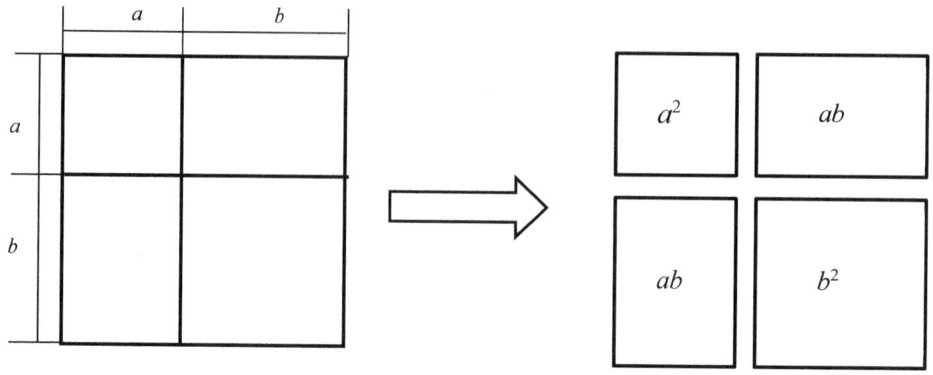

A geometric interpretation of binomial theorem

Example
Expand the squared binomial $(3x + 8)^2$.

Solution
We know when a binomial is squared, the result is the first term squared added to double the product of both terms and the constant, is squared. Then we have:

$(3x +8)^2$

$= (3x)^2 + 2 \times 3x \times 8 + 8^2$

$= 9x^2 + 48x + 64.$

The perfect square formula takes another form as below.

$(x - a)^2 = (x - a)\,(x - a) = x^2 - ax - ax + a^2 = x^2 - 2ax + a^2.$

> **Perfect Square Trinomial**
> Given expression $(x - a)^2$, its expanded expansion is:
> $(x - a)^2 = x^2 - 2ax + a^2.$

Example
Expand the squared binomial $(2u - 5)^2$.

Solution
We know when a binomial is squared, the result is the first term squared subtracted from double the product of both terms and the constant, is squared. Then we have:

$(2u - 5)^2$

$= (2u)^2 - 2\times 2u \times 5 + 5^2$

$= 4u^2 - 20u + 25.$

1.4.2. Difference of Squares

For a binomial which is multiplied by another binomial with the same terms separated by the opposite sign, the result is the square of the first term minus the square of the last term.

$(a + b)\,(a - b) = a^2 - b^2.$

Example

Expand the squared binomial $16x^2 - 81$.

Solution
We learned that when a binomial is multiplied by another binomial with similar terms separated by the opposite sign, the result is the square of the first term minus the square of the last term. Then we have:

$16x^2 - 81$

$= (4x)^2 - 9^2$

$= (4x + 9)(4x - 9).$

> **Difference of Squares**
> Given expression $a^2 - b^2$, its expanded expansion is:
>
> $a^2 - b^2 = (a + b)(a - b).$

1.4.3. Sum of Squares

For a binomial $a^2 + b^2$ no general formula exists.

1.4.4. Multiplication of Two Binomials

For a binomial which is multiplied by another binomial with the different terms separated by the same sign (+ sign), the result is the square of the first term plus the sum of two different terms times the first term plus the multiplication of the two different terms. That is:

$(x + a)(x + b) = x^2 + (a + b)x + ab.$

Also, for a binomial which is multiplied by another binomial with the different terms separated by the same sign (− sign), the result is the square of the first term minus the sum of two different terms times the first term plus the multiplication of the two different terms. That is:

$(x - a)(x - b) = x^2 - (a + b)x + ab.$

> **Multiplication of Two Binomials**
> - Given expression $(x + a)(x + b)$, its expanded expansion is:
>
> $(x + a)(x + b) = x^2 + (a + b)x + ab.$
>
> - Given expression $(x - a)(x - b)$, its expanded expansion is:
>
> $(x - a)(x - b) = x^2 - (a + b)x + ab.$

Example
Expand the multiplication of two binomials $(x + 10)(x + 3)$.

Solution
We learned that when a binomial is multiplied by another binomial with different terms separated by the same sign, the result is the square of the first term plus the sum of two different terms times the first term plus the multiplication of the two different terms.

$(x + 10)(x + 3)$

$= x^2 + (10 + 3)x + (10 \times 3)$

$= x^2 + 13x + 30.$

Example

Expand the multiplication of two binomial $(x - 6)(x - 5)$.

Solution

We learned that when a binomial is multiplied by another binomial with different terms separated by the same sign ($-$ sign), the result is the square of the first term minus the sum of two different terms times the first term plus the multiplication of the two different terms.

$$(x - 6)(x - 5)$$
$$= x^2 - (6 + 5)x + (6 \times 5)$$
$$= x^2 - 15x + 30.$$

1.4.5. Binomial to the 3rd Power

For the binomial of third power, $(x + a)^3$ we will have.

$$(x + a)^3 = (x + a)(x + a)(x + a)$$
$$(x + a)^3 = (x + a)(x^2 + 2xa + a^2)$$
$$(x + a)^3 = x^3 + 2ax^2 + a^2x + ax^2 + 2a^2x + a^3$$
$$(x + a)^3 = x^3 + 3ax^2 + 3a^2x + a^3.$$

> **Binomial to the 3rd Power**
> Given expression $(x + a)^3$, its expanded expansion is:
> $$(x + a)^3 = x^3 + 3ax^2 + 3a^2x + a^3.$$

Also, For the binomial of third power, $(x - a)^3$ we will have.

$$(x - a)^3 = (x - a)(x - a)(x - a)$$
$$(x - a)^3 = (x - a)(x^2 - 2xa + a^2)$$
$$(x - a)^3 = x^3 - 2ax^2 + a^2x - ax^2 + 2a^2x - a^3$$
$$(x - a)^3 = x^3 - 3ax^2 + 3a^2x - a^3.$$

> **Binomial to the 3rd Power**
> Given expression $(x - a)^3$, its expanded expansion is:
> $$(x - a)^3 = x^3 - 3ax^2 + 3a^2x - a^3.$$

Example

Expand the third power binomial $(2x + 5)^3$.

Solution

$$(2x + 5)^3$$
$$(2x + 5)^3 = (2x)^3 + 3(5)(2x)^2 + 3(5)^2(2x) + (5)^3$$
$$(2x + 5)^3 = 8x^3 + 3(5)(4x^2) + 3(25)(2x) + (125)$$
$$(2x + 5)^3 = 8x^3 + 60x^2 + 150x + 125.$$

Example
Expand the third power binomial $(a - 2)^3$.

Solution

$$(a - 2)^3 = a^3 - 3(2)a^2 + 3(2)^2(a) - 2^3$$

$$(a - 2)^3 = a^3 - 6a^2 + 12a - 8.$$

1.4.6. Difference of Cubes

For binomial of difference cubes, $a^3 - b^3$ we will have.

$$a^3 - b^3 = (a - b)(a^2 + ab + b^2).$$

> **Difference of Cubes**
> Given expression $a^3 - b^3$, its expanded expansion is:
> $$a^3 - b^3 = (a - b)(a^2 + ab + b^2).$$

Example
Expand the difference of cubes $27u^3 - 64$.

Solution

$$27u^3 - 64 = (3u - 4)(9u^2 + 3u(4) + 16)$$

$$27u^3 - 64 = (3u - 4)(9u^2 + 12u + 16).$$

1.4.7. Sum of Cube

For binomial of sum cubes, $a^3 + b^3$ we will have.

$$a^3 + b^3 = (a + b)(a^2 - ab + b^2).$$

> **Sum of Cubes**
> Given expression $a^3 + b^3$, its expanded expansion is:
> $$a^3 + b^3 = (a + b)(a^2 - ab + b^2).$$

Example
Expand the sum of cubes $216w^3 + 125$.

Solution

$$216w^3 + 125 = (3u - 4)(9u^2 + 3u(4) + 16)$$

$$27u^3 - 64 = (3u - 4)(9u^2 + 12u + 16).$$

1.4.8. Binomial to the nth Power

For nonzero real numbers a and b,

$$(a + b)^n = \sum_{k=0}^{n} \binom{n}{k} a^{n-k} b^k$$

$$= \binom{n}{0} a^{n-0} b^0 + \binom{n}{1} a^{n-1} b^{k1} + \binom{n}{2} a^{n-2} b^2 + \cdots + \binom{n}{n} a^{n-n} b^n,$$

where $\binom{n}{k} = \frac{n!}{k!(n-k)!}$

$$\binom{n}{0}a^{n-0}b^0 + \binom{n}{1}a^{n-1}b^{k1} + \binom{n}{2}a^{n-2}b^2 + \cdots + \binom{n}{n}a^{n-n}b^n$$

$$= \left(\frac{n!}{0!(n-0)!}\right)a^{n-0}b^0 + \left(\frac{n!}{1!(n-1)!}\right)a^{n-1}b^1 + \left(\frac{n!}{2!(n-2)!}\right)a^{n-2}b^2 + \cdots + \left(\frac{n!}{n!(n-n)!}\right)a^{n-n}b^n,$$

where, $0! = 1$ and $n! = n(n-1)!$ for $n \geq 1$. Or $n! = n \times (n - 1) \times (n - 2) \times \cdots \times 2 \times 1$. with $0! = 1$ as our 'base case. And $1! = 1$, $2! = 2 \times 1! = 2 \times 1 = 2$, $3! = 3 \times 2! = 3 \times 2 \times 1 = 6$ and $4! = 4 \times 3! = 4 \times 3 \times 2 \times 1 = 24$, etc. then:

$$(a+b)^n = a^n + na^{n-1}b + \frac{n(n-1)}{2}a^{n-2}b^2 + \frac{n(n-1)(n-2)}{6}a^{n-3}b^3 + \cdots + b^n.$$

Theorem: Binomial Coefficients: Given two whole numbers n and k with $n \geq k$, the binomial coefficient $\binom{n}{k}$, which is read, n choose k, is the whole number given by

$$\binom{n}{k} = \frac{n!}{k!(n-k)!}.$$

Example
Expand the polynomial $(a + b)^4$.

Solution
Using the binomial to the nth power formula, we have:

$$(a + b)^4 = a^4 + 4a^3b + 6a^2b^2 + 4ab^3 + b^4.$$

Definition: Factorials:
For a whole number n, n factorial, denoted $n!$, is the term f_n of the sequence

$$f_0 = 1, f_n = n \times f_{n-1}, n \geq 1.$$

1.5. Evaluation of a Polynomial

The evaluation of a polynomial consists of substituting a numerical value to each variable and carrying out the indicated multiplications and additions.

Example
Evaluate the polynomial of one variable, $6x^4 + 2x^3 - 5x + 7$, for $x = -3$.

Solution
To evaluate the given polynomial, we simply substitute the given value of $x = -3$ into the equation. So,

$$6x^4 + 2x^3 - 5x + 7$$

$$6x^4 + 2x^3 - 5x + 7\,|_{x=-3} = 6(-3)^4 + 2(-3)^3 - 5(-3) + 7$$ we substituted $x = -3$ into the equation

$$6x^4 + 2x^3 - 5x + 7\,|_{x=-3} = 6(81) + 2(-27) - 5(-3) + 7$$

$$6x^4 + 2x^3 - 5x + 7\,|_{x=-3} = 486 - 54 + 15 + 7$$

$$6x^4 + 2x^3 - 5x + 7\,|_{x=-3} = 493.$$

Therefore, the value of the polynomial, $6x^4 + 2x^3 - 5x + 7$, at $x = -3$ is 493.

Example

Evaluate the polynomial, $5x^3 - y^2z + 2x^2 - 4xyz^3 + 2xy + 3yz + 1$, for $x = -1$, $y = 1$, and $z = -3$.

Solution

To evaluate the given polynomial, we simply substitute the given values of $x = -1$, $y = 1$, and $z = -3$ into the equation. So,

$$5x^3 - y^2z + 2x^2 - 4xyz^3 + 2xy + 3yz + 1$$

$$5x^3 - y^2z + 2x^2 - 4xyz^3 + 2xy + 3yz + 1 \,|_{x=-1,\, y=1\, z=-3}$$

$$= 5(-1)^3 - (1)^2(-3) + 2(-1)^2 - 4(-1)(1)(-3)^3 + 2(-1)(1) + 3(1)(-3) + 1$$

we substituted $x = -1$, $y = 1$, and $z = -3$ into the equation

$$= 5(-1) - (1)(-3) + 2(1) - 4(-1)(1)(-27) + 2(-1)(1) + 3(1)(-3) + 1$$

$$= -5 + 3 + 2 - 108 - 2 - 9 + 1$$

$$= -5 + 6 - 108 - 2 - 9$$

$$= 6 - 124$$

$$= -118.$$

Therefore, the value of the polynomial, $5x^3 - y^2z + 2x^2 - 4xyz^3 + 2xy + 3yz + 1$ is -118.

1.6. Exercises

1. Given polynomials, $P = 11x^6 + 9x^5 - 3xy + 12y^4 - 6y^2 + 10$, and $Q = -2x^6 - 7x^5 + x^4 + 13y^4 - 19y^2 + 6xy - 22$, find $2P + Q$.

2. Given polynomials, $P = 8x^4 + 14x^3 - 5xy + 4y^5 - 7y^3 - 17$, and $Q = -4x^4 - 2x^3 + x^2 + 7y^5 - 12y^3 + 9xy - 33$, find $P - 3Q$.

3. Given polynomials, $P = 2u^3 - 4uv - 4v + 2$, and $Q = -u^2 + uv - 2v$, find $3P \times (-2Q)$.

4. Given polynomials, $P = 8a^2 + 3b^2$, $Q = -2a^2 + 3ab - b^2 + 1$ and $R = -4a^2 + 5ab - 3b^2$, find
   a) $5P + 6Q$.
   b) $P + 3R$.
   c) $P + 2R + Q$.

5. Given polynomials, $P = 2b + 5a$, $Q = -3b - a$ and $R = -3a + b$, find
   a) $4P - 3Q - 2R$.
   b) $P \times 2Q$.
   c) $(-2P) \times R$.
   d) $2R \times (-2Q)$.

6. Divide the polynomial in one variable,
   a) $P = 12x^3 - 5x^2 - 10x - 24$ by $Q = 3x - 4$.
   b) $R = 8k^4 - 5k^2 + 7k + 12$ by $S = k - 5$.

7. Divide the polynomial, $P = 2u^3 + 3u^2v + 6uv + 9v^2$ by $Q = u^2 + 3v$.

8. The formula for compound interest is $F = P\left(1 + \frac{r}{n}\right)^{nt}$, where P is the initial principal balance, r is the interest rate, n is the number of times interest is compounded per period and t is the number of time periods. If an amount of $10,000 is deposited into a savings account for 5 years, at an annual interest rate of 9%, compounded monthly or at an annual interest rate of 9.2%, compounded semiannually, by which option the earned interest is higher.

   Hint: Let F_1 denote the principal balance with the rate of 9%, compounded monthly and F_2 denote the principal balance with the rate of 9.2%, compounded semiannually, then find the difference between F_1 and F_2.

9. Expand each squared binomial:
   a) $(12 + 5a^2)^2$.
   b) $(a^2b + 5)^2$.
   c) $(3x + 5x^2y)^2$.

10. Expand each squared binomial:
    a) $(8 - 7u)^2$.

b) $(k^2h - 5)^2$.

c) $[tu - (5t + 1)]^2$.

11. Expand the squared binomial $4a^2 - 49$.

12. Expand each multiplication of two binomials:
   a) $(a + 5)(a + 7)$.
   b) $(x + y)(x + 2y)$.
   c) $(2u + 3t)(1 + 3t)$.

13. Expand each multiplication of two binomials:
   d) $(2x - 9)(2x - 3)$.
   e) $(5a - 6)(5a - 2)$.
   f) $(8 - 3t)(4 - 3t)$.

14. Expand the third power binomial.
   a) $(2x + 5)^3$.
   b) $(a + 3b)^3$.

15. Expand the third power binomial.
   a) $(2a - 5)^3$.
   b) $(3k - 2t)^3$.

16. Expand the difference of cubes.
   a) $8x^3 - 125$.
   b) $64a^3 - 343$.
   c) $216u^3 - 8v^3$.

17. Expand the sum of cubes.
   a) $729w^3 + 125v^3$.
   b) $512x^3 + 343y^3$.
   c) $64k^3 + 216t^3$.

18. Expand the polynomial $(2x + 3)^4$.

19. Evaluate the polynomial of one variable:

   a) $9x^2 - 8x + 15$, for $x = -3$.
   b) $3x^5 + 2x^3 - 5x^2 - 6x + 10$, for $x = -2$.

20. Evaluate the polynomial,
   a) $11a^3c^2 - 2b^3c + 8ac^2 - 7abc + 14ab + 6abc$, for $a = 4$, $b = -2$, and $c = -1$
   b) $3k^4l - 2h^4l + 5hl^2 - 4khl + 2kl + 7kl + 6l + 1$, for $k = 1$, $h = -1$, and $l = -2$.

2. Factoring Polynomials

Factoring polynomials is the inverse operation of multiplying them together. The process of factoring polynomials offers numerous advantages. Factored polynomials aid in solving equations, understanding graph behaviors, manipulating fractions, and more. Given that many algebraic concepts rely on the ability to factor polynomials, possessing strong factoring skills is especially crucial.

> **Factoring**
> The opposite of multiplying polynomials together is **factoring** polynomials. That is, you are given the product, then asked to find the original multiplication problem.

Given a polynomial, $\sum_{k=0}^{n} a_k x^k = a_0 + a_1 x + a_2 x^2 + \cdots + a_n x^n$, there are four methods for factoring polynomials, factoring by the greatest common factor, factoring by the leading coefficient, factoring by the grouping, and factoring by the substitution. Before presenting these methods, we must first define the greatest common factor of a polynomial.

2.1. Greatest Common Factor

The first method for factoring polynomials will be factoring out the greatest common factor. We presented the GCF of two or more numbers previously. We have learned that the greatest common factor of two numbers such as a and b is the largest positive number that divides evenly (no remainder) both a and b.

> **Greatest Common Factor of Two Numbers.**
> The greatest common factor of a and b is the largest positive number that divides evenly (no remainder) both a and b.

Similarly, the GCF of polynomials is the largest polynomial that divides evenly into the polynomials.

> **Greatest Common Factor (GCF) of Polynomials**
> The greatest common factor (GCF) of polynomials is the largest polynomial that divides evenly into the polynomials.

To use this method, we look for all the terms and determine if there is a factor that is in common to all the terms. If we find it, we will factor it out of the polynomial. It is interesting to realize that in the process of factoring we are using the distributive law in reverse.

> **First Rule of Factoring**
> If the terms of the given polynomial have a greatest common factor (GCF), then factor out the GCF.

2.1.1. Finding the GCF in a Polynomial

Regardless of how many terms a polynomial has, we always want to check for GCF first. The steps to find the GCF of a polynomial are as follows.

1. Break down every coefficient into prime factors to find the GCF of the coefficients.

2. Break down every term into prime factors to find the GCF of the variables.
3. Combine to find the GCF of the polynomial expression.
4. Determine what the GCF needs to be multiplied by obtaining each term in the polynomial expression.
5. Factor the GCF out from every term in front of parentheses and group the remnants inside the parentheses.

> **Factor the GCF From a Polynomial**
>
> 1. Find the GCF of all the terms of the polynomial.
> 2. Rewrite each term as a product using the GCF.
> 3. Use the "reverse" distributive property to factor the expression.
> 4. Check by multiplying the factors.

Example
Find the GCF of expression $6x^4y^2 + 45x^3y^3$.

Solution
Following the steps of finding the GCF of $6x^4y^2 + 45x^3y^3$, we first break down every coefficient into prime factors.

Breaking 6, we get, 2×3

Breaking 45, we get, $3 \times 3 \times 5$.

We found GCF of coefficient 3.

Breaking x^4y^2, we get, $x \times x \times x \times x \times y \times y$

Breaking x^3y^3, we get $x \times x \times x \times y \times y \times y$.

We found GCF of variables $x \times x \times x \times y \times y$.

Combining the above GCFs, we get $3 \times x \times x \times x \times y \times y = 3x^3y^2$.

Looking for GCF in the above three terms, we get, $3x^3y^2$ as the GCF of the expression to be factored out.

$6x^4y^2 + 45x^3y^3$

$3x^3y^2(2x + 15y)$.

Therefore, the GCF of expression the $6x^4y^2 + 45x^3y^3$ is, $3x^3y^2(2x + 15y)$.

Example
Find the GCF of expression $6x^4y^2 + 45x^3y^3 + 21x^2y$.

Solution
Following the steps of finding the GCF of $6x^4y^2 + 45\,x^3y^3 + 21x^2y$, we first break down every term into prime factors.

Breaking $6x^4y^2$, we get, $2 \times 3 \times x \times x \times x \times x \times y \times y$

Breaking $45x^3y^3$, we get, $3 \times 3 \times 5 \times x \times x \times x \times y \times y \times y$

Breaking $21x^2y$, we get, $3 \times 7 \times x \times x \times y$.

Looking for GCF in the above three terms, we get, $3x^2y$ as the GCF of the expression.

$6x^4y^2 + 45x^3y^3 + 21x^2y$

$3x^2y(2x^2y + 15xy^2 + 7)$.

Therefore, the GCF of expression $6x^4y^2 + 45x^3y^3 + 21x^2y$ is, $3x^2y(2x^2y + 15xy^2 + 7)$.

Example
Given the binomial $4x^3 + 12x^2$, factor by the GCF.

Solution
We first break every term into prime factors.

Breaking $4x^3$, we obtain the GCF, $2 \times 2 \times x \times x \times x$

Breaking $12x^2$, we obtain the GCF, $2 \times 2 \times 3 \times x \times x$.

The GCF of expression is $2 \times 2 \times x \times x = 4x^2$.

$4x^3 + 12x^2$

$4x^2(x + 3)$.

Therefore, the binomial $4x^3 + 12x^2$ can be factored as, $4x^2(x + 3)$.

Example
Given the trinomial $24a^3b^2 + 48ab^2 - 150a^2b^3$, factor by the GCF.

Solution
We first break every term into prime factors.

Breaking $24a^3b^2$, we obtain the GCF, $2 \times 2 \times 2 \times 3 \times a \times a \times a \times b \times b$

Breaking $48ab^2$, we obtain the GCF, $2 \times 2 \times 2 \times 2 \times 3 \times a \times b \times b$

Breaking $150a^2b^3$, we obtain the GCF, $2 \times 3 \times 5 \times 5 \times a \times a \times b \times b \times b$.

The GCF of expression is $2 \times 3 \times a \times b \times b = 6ab^2$.

$24a^3b^2 + 48ab^2 - 150a^2b^3$

$6ab^2(4a^2 + 8 - 25ab)$.

Therefore, the trinomial $24a^3b^2 + 48ab^2 - 150a^2b^3$ can be factored as, $6ab^2(4a^2 + 8 - 25ab)$.

2.2. Factoring a Polynomial with Leading Coefficient 1

In the previous section, we presented polynomials factoring by GCF. While we may always begin by looking for a GCF, pulling out the GCF is not the only way that polynomial expressions can be factored. Another way for factoring a polynomial is factoring by its leading coefficient.

Remember that the term with the highest power of is called the leading term and its coefficient is called the leading coefficient. If the leading coefficient is 1, then the polynomial is called monic. The index of the leading term is called the degree of the polynomial. The term independent of is called the constant term. Thus, $x^6 - 2x^5 + 8x + 26$ is a monic polynomial of degree 6 with constant term 26.

The polynomial of the form $x^2 + bx + c$, with the leading coefficient of 1, can be factored by finding two numbers p and q, with a product of c and a sum of b. Hence, the polynomial of $x^2 + bx + c$ can be written in factored form as $(x + p)(x + q)$ where $pq = c$ and $p + q = b$.

As an example, $x^2 + 10x + 7$ with the leading coefficient of 1, can be written as the product of $(x + 2)$ and $(x + 5)$.

Factoring a Trinomial with Leading Coefficient 1
A trinomial of the form $x^2 + bx + c$ can be written in factored form as $(x + p)(x + q)$ where $pq = c$ and $p + q = b$.

The steps to factor a trinomial with leading coefficient 1, $x^2 + bx + c$, are:

1. List factors of c.
2. Find p and q, a pair of factors of c with a sum of b.
3. Write the factored expression $(x + p)(x + q)$.

Example
Factor a polynomial in the form $x^2 + 7x - 18$.

Solution
We list factors of − 18 until we get the sum of factors equal to 7. So,

Factors of − 18	Sum of factors
1, − 18	− 17
− 1, 18	17
2, − 9	− 7
− 2, 9	**7**
3, − 6	− 3
− 3, 6	3

As we realize, there are two numbers, − 2, and 9 which their product is − 18 and their sum 7. So,

$$x^2 + 7x - 18$$

$$x^2 + 7x - 18 = (x - 2)(x + 9).$$

Therefore, the trinomial of $x^2 + 7x - 18$ is factored as $(x - 2)(x + 9)$.

Example
Factor a polynomial in the form $x^2 + 13x + 36$.

Solution
We list factor of 36. So,

Factors of 36	Sum of factors
1, 36	37
2, 18	20
3, 12	15
4, 9	**13**
6, 6	12

We realize that there are two numbers, 4, and 9 where their product is 36 and their sum 13. So,

$x^2 + 13x + 36$

$x^2 + 13x + 36 = (x + 4)(x + 9).$

Therefore, the polynomial of $x^2 + 13x + 36$ is factored as $(x + 4)(x + 9)$.

Check: We need to multiply $(x + 4)$ by $(x + 9)$to see if we get the original polynomial,

$(x + 4)(x + 9) = x^2 + 9x + 4x - 36$

$(x + 4)(x + 9) = 3x^2 + 13x - 12$ true, our answer is correct.

2.3. Factoring a Trinomial with Leading Coefficient Not 1

Trinomials with leading coefficient other than 1 are slightly more complicated to factor. In this section we continue to factor trinomials of the form $ax^2 + bx + c$ when $a \neq 1$. In the last section, all our examples had $a = 1$, and we were able to factor by finding two numbers p and q, with a product of c and a sum of b. However, in this section, $a \neq 1$, and we will soon see that we will not be able to easily use the above procedure.

In case of $a \neq 1$, we need to find two numbers p and q, with a product of ac and a sum of b. Note that finding p and q, will not imply that the polynomial $ax^2 + bx + c$ can be factored as $(x + p)(x + q)$, because the product of the terms in the "first" position does not equal ax^2. So, we need to do some extra operations. However, it will be easier to explain the extra operations by presenting an example.

Example
Factor $3x^2 + 16x - 12$.

Solution
For finding factor of $3x^2 + 16x - 12$, we need to proceed as follows.

1. We compare $3x^2 + 16x - 12$ with $ax^2 + bx + c$. It is recognized that $a = 3$, $b = 16$, and $c = -12$.
2. We calculate ac. Given $a = 3$ and $c = -12$, hence $ac = -36$.
3. We list all integer pairs whose product is $ac = -36$.
4. Designate the ordered pair whose sum is $b = 16$. We have already bolded them in the above table. They are -2 and 18 (see table below).
5. We note that, with the numbers -2 and 18 we cannot factor $3x^2 + 16x - 12$. That is the product.

$(x - 2)(x + 18)$ is not equal to $3x^2 + 16x - 12$. Note that, the product of the terms in the "first" position does not equal $3x^2$. The extra operations we need to perform is to break up the middle term of $3x^2 + 16x - 12$ into a sum of like terms using our pair of integers -2 and 18. So,

$$3x^2 + 16x - 12 = 3x^2 + 18x - 2x - 12.$$

Now we factor by grouping. Factor $3x$ out of the first two terms and -2 out of the second two terms.

$$3x^2 + 16x - 12 = 3x(x + 6) - 2(x + 6).$$

Now we can factor out $(x + 6)$.

$$3x^2 + 16x - 12 = (x + 6)(3x - 2).$$

6. Next, we check our answer. We need to multiply $(x + 6)$ by $(3x - 2)$ to see if we get the original polynomial. So,

$$(x + 6)(3x - 2) = 3x^2 - 2x + 18x - 12$$

$$(x + 6)(3x - 2) = 3x^2 + 16x - 12$$ true, our answer is correct.

Factors of −36	Sum of factors
1, −36	−35
−1, 36	35
2, −18	−16
−2, 18	**16**
3, −12	−9
−3, 12	9
4, −9	−5
−4, 9	5
6, −6	0
−6, 6	0

2.4. Factoring Polynomials by Grouping

The third method of factoring polynomial expressions, involving four or more terms, is **factoring polynomials by grouping.** Occasionally there is no common factor of all the terms of a polynomial. Considering a polynomial expression involving four terms, we need to separate the polynomial into two parts with two terms in each part. Then we look for the GCF in each part. If the polynomial can be factored, you will find a common factor emerges from both parts.

Note that there are some polynomials that cannot be factored. Just like some numbers are prime, some polynomials can also be prime.

The steps to factor a polynomial by grouping are given below.

Step 1. Group terms with common factors.

Step 2. Factor out the common factors in each group.

Step 3. Factor the common factor from the polynomial expression.

Step 4. Check your answer by multiplying the factors.

Example
Given a polynomial $10x^3 + 30x^2 - 7x - 21$, factor by grouping.

Solution
We "group" the first and second terms, hence we can factor the $10x^2$ out of both terms. Likewise, we "group" the third and fourth terms, and we can factor -7 out of both terms.

$$10x^3 + 30x^2 - 7x - 21 = 10x^2(x + 3) - 7(x + 3)$$

Next, we can factor $x + 3$ out of both terms.

$= (x + 3)(10x^2 - 7)$ this our answer.

Check: We need to multiply $(x + 3)$ by $(10x^2 - 7)$ to see if we get the original polynomial,

$$(x + 3)(10x^2 - 7) = 10x^3 - 7x + 30x^2 - 21$$

$(x + 3)(10x^2 - 7) = 10x^3 + 30x^2 - 7x - 21$ true, our answer is correct.

Example
Given a polynomial $9x^5 - 18x^3 + 12x^2 - 24$, factor by grouping.

Solution
We "group" the first and second terms, hence we can factor the $9x^3$ out of both terms. Likewise, we "group" the third and fourth terms, and we can factor 12 out of both terms.

$$9x^5 - 18x^3 + 12x^2 - 24 = 9x^3(x^2 - 2) + 12(x^2 - 2).$$

Next, we can factor $x^2 - 2$ out of both terms.

$= (x^2 - 2)(9x^3 + 12)$ this our answer.

Check: We need to multiply $(x^2 - 2)$ by $(9x^3 + 12)$ to see if we get the original polynomial,

$$(x^2 - 2)(9x^3 + 12) = 9x^5 + 12x^2 - 18x^3 - 24$$

$(x^2 - 2)(9x^3 + 12) = 9x^5 - 18x^3 + 12x^2 - 24$ true, our answer is correct.

We next present an example to demonstrate the combined factoring methods.

Example
Factor: $60x^3 - 3x^2 - 36x$.

Solution
We need to find the GCF of the polynomial.

Breaking $60x^3$, we obtain the GCF, $2 \times 2 \times 3 \times 5 \times x \times x \times x$.

Breaking $3x^2$, we obtain the GCF, $3 \times x \times x$.

Breaking $36x$, we obtain the GCF, $2 \times 2 \times 3 \times 3 \times x$.

The GCF of $60x^3$, $3x^2$, $36x$ is $3 \times x = 3x$.

Factoring the GCF, we get:

$$60x^3 - 3x^2 - 36x$$

$$3x(20x^2 - x - 12).$$

We now have a trinomial with a leading coefficient, $a \neq 1$. Next, we need to compare $20x^2 - x - 12$ with $ax^2 + bx + c$. We get $a = 20$, $b = -1$, and $c = -12$, to find p and q. Listing the integer pairs whose product is $ac = -240$, and an integer pair whose sum is $b = -1$.

Factors of −240	Sum of factors
1, −240	−239
−1, 240	239
2, −120	−118
−2, 120	118
3, −80	−77
−3, 80	77
4, −60	−56
−4, 60	56
5, −48	−43
−5, 48	43
6, −40	−34
−6, 40	34
8, −30	−22
−8, 30	22
10, −24	−14
−10, 24	14
12, −20	−8
−12, 20	8
15, −16	**−1**
−15, 16	1

With the numbers 15 and − 16 we cannot factor $20x^2 - x - 12$. This is because $(x + 15)(x - 16) \neq 20x^2 - x - 12$ since the product of the terms in the "first" position does not equal $20x^2$. The extra operations we need to perform is to break up the middle term of $20x^2 - x - 12$ into a sum of like terms using our pair of integers 15 and − 16. So,

$3x(20x^2 - x - 12)$

$= 3x(20x^2 - 16x + 15x - 12)$

$= 3x[(4x(5x - 4) + 3(5x - 4)]$ we factored by grouping

$= 3x[(5x - 4)(4x + 3)]$ we factored out $(5x - 4)$.

Therefore, $60x^3 - 3x^2 - 36x = 3x(5x - 4)(4x + 3)$.

Check: We need to multiply $3x(5x - 4)$ by $(4x + 3)$ to see if we get the original polynomial,

$3x(5x - 4)(4x + 3) = (15x^2 - 12x)(4x + 3)$

$3x(5x - 4)(4x + 3) = 60x^3 + 45x^2 - 48x^2 - 36x$

$3x(5x - 4)(4x + 3) = 60x^3 - 3x^2 - 36x$ true, our answer is correct.

2.5. Factoring Polynomials by Substitution

In certain scenarios, we can factor a complex polynomial by employing a straightforward substitution technique. Typically, factoring polynomials involves decomposing a polynomial into its constituent parts. When dealing with especially intricate polynomials, it can be advantageous to substitute a simpler term and then proceed with factoring. Substitution allows for the breakdown of complex expressions into more manageable components, even when dealing with higher-degree polynomials.

Additionally, multi-variable polynomials can be simplified by substituting them with a single variable.

Example
Given a polynomial $9x^4 - 18x^2 + 5$, factor using substitution.

Solution
We note that the degree of this polynomial is double that of the degree of the trinomial we have looked at (degree of 2). To match this polynomial as the pattern of usual trinomial, let $y = x^2$.

Using the rules of exponents, we can replace each x^2 with a y.

$9x^4 - 18x^2 + 5$

$9(\mathbf{x^2})^2 - 18x^2 + 5$ we used the power of power rule of exponents

$9(\mathbf{y})^2 - 18y + 5 = 9y^2 - 18y + 5$ we substituted $y = x^2$ into the equation.

Using the factoring a trinomial with leading coefficient $a \neq 1$, we need to find two numbers p and q, with a product of ac and a sum of b, where $a = 9$, $c = 5$, and $b = -18$. Hence, we list all integer pairs whose product is $ac = 45$ and whose sum is $b = -18$.

Factors of 45	Sum of factors
1, 45	46
−1, −45	−46
−3, −15	**−18**
3, 15	18
−5, −9	−14
5, 9	14

With the numbers −3 and −15 we cannot factor $9y^2 - 18y + 5$. Because $(x - 3)(x - 15) \neq 9y^2 - 18y + 5$, because the product of the terms in the "first" position does not equal $9y^2$. Hence, we need to break up the middle term of $9y^2 - 18y + 5$ into a sum of like terms using our pair of integers − 3 and − 15. So,

$9y^2 - 18y + 5$	
$= 9y^2 - 3y - 15y + 5$	we broke $-18y$ down into a sum of like terms using our pair of integers -3 and -15
$= 3y(3y - 1) - 5(y - 1)$	we factored out $3y$ from the first and second terms, and we factored out -5 from the third and fourth terms
$= (3y - 1)(3y - 5)$	we factored out $(3y - 1)$.

We have not found the answer yet. We need to obtain our answer in terms of x. We will substitute an x^2 for y in our factored form. So,

$$9y^2 - 18y + 5 = (3y - 1)(3y - 5)$$

$$9(x^2)^2 - 18x^2 + 5 = (3x^2 - 1)(3x^2 - 5).$$

Therefore, the polynomial of $9x^4 - 18x^2 + 5$ is factored as $(3x^2 - 1)(3x^2 - 5)$.

Check: We need to multiply $(3x^2 - 1)$ by $(x^2 - 5)$ to see if we get the original polynomial,

$$(3x^2 - 1)(3x^2 - 5) = 9x^4 - 15x^2 - 3x^2 + 5$$

$$(x^2 - 3)(x^2 - 5) = 9x^4 - 18x^2 + 5$$ true, our answer is correct.

Example

Given a polynomial $16(2x + 5)^2 - 19(2x + 5) + 3$, factor using substitution.

Solution

In this case, the binomial $(2x + 5)$ has an exponent of two and an exponent of one. So, we can use a simple substitution. Let z be equal to the binomial $(2x + 5)$.

$$z = (2x + 5).$$

Replacing $(2x + 5)$ in the polynomial $16(2x + 5)^2 - 19(2x + 5) + 3$ with z, we have:

$$16(2x + 5)^2 - 19(2x + 5) + 3$$

$$16(z)^2 - 19(z) + 3 = 16z^2 - 19z + 3.$$

Using the factoring a trinomial with leading coefficient $a \neq 1$, we need to find two numbers p and q, with a product of ac and a sum of b, where $a = 16$, $c = 3$, and $b = -19$. Hence, we list all integer pairs whose product is $ac = 48$ and whose sum is $b = -19$.

Factors of 48	Sum of factors
1, 48	49
−1, −48	−49
2, 24	26
−2, −24	−26
3, 16	19
−3, −16	**−19**

4, 12	16
−4, −12	−16
6, 8	14
−6, −8	−14

As the previous example, with the numbers − 3 and − 16 we cannot factor $16z^2 - 19z + 3$. Hence, we again need to break up the middle term of $16z^2 - 19z + 3$ into a sum of like terms using our pair of integers − 3 and − 16. So,

$16z^2 - 19z + 3$

$= 16z^2 - 16z - 3z + 3$ — we broke $- 19z$ into a sum of like terms using our pair of integers − 3 and − 16

$= 16z(z - 1) - 3(z - 1)$ — we factored out $16z$ from the first and second terms, and we factored out − 3 from the third and fourth terms

$= (z - 1)(16z - 3)$ — we factored out $(z - 1)$.

We have not found the answer yet. We need to obtain our answer in terms of $(2x + 5)$. We will substitute the binomial $(2x + 5)$ for z in our factored form. So,

$$16z^2 - 19z + 3 = (z - 1)(16z - 3)$$

$$16(2x + 5)^2 - 19(2x + 5) + 3 = [(2x + 5) - 1][16(2x + 5) - 3]$$

$$16(2x + 5)^2 - 19(2x + 5) + 3 = (2x + 4)[(32x + 80) - 3]$$

$$16(2x + 5)^2 - 19(2x + 5) + 3 = (2x + 4)(32x + 77).$$

Therefore, the polynomial of $16(2x + 5)^2 - 19(2x + 5) + 3$ is factored as $(2x + 4)(32x + 77)$.

Check: We need to multiply $(2x + 4)$ by $(32x + 77)$ to see if we get the original polynomial,

$$16(2x + 5)^2 - 19(2x + 5) + 3 \overset{?}{\Leftrightarrow} (2x + 4)(32x + 77)$$

$$16(2x + 5)^2 - 19(2x + 5) + 3 \overset{?}{\Leftrightarrow} 64x^2 + 154x + 128x + 308$$

$$16(2x + 5)^2 - 19(2x + 5) + 3 \overset{?}{\Leftrightarrow} 64x^2 + 282x + 308.$$

Since the expression on the right side of the equation is in the form of a simplified version, we need to simplify the right side, so,

$$16(2x + 5)^2 - 19(2x + 5) + 3$$

$$16(2x + 5)^2 - 19(2x + 5) + 3 = 16(4x^2 + 20x + 25) - 38x - 95 + 3$$

$$16(2x + 5)^2 - 19(2x + 5) + 3 = 64x^2 + 320x + 400 - 38x - 95 + 3$$

$16(2x + 5)^2 - 19(2x + 5) + 3 = 64x^2 + 282x + 308$ — true, our answer is correct.

2.6. Exercises

1. Find the GCF of expressions, $24u^5t^2v$, $48u^3t^5v^2$, $36u^3t^6v$, $32u^6t^2v^3$, $18u^4t^3v$.
2. Find the GCF of expression $64x^5y^3 + 108\,x^4y^6 - 36\,x^3y^5 + 132x^2y^2$.
3. Find the GCF of expression $\frac{15}{36}x^3y^2z^3 - \frac{55}{126}x^4y^2z^2 - \frac{75}{108}x^3yz^3 + \frac{45}{72}x^3y^2z^4 + \frac{125}{216}x^2y^2z^2$.
4. Given the trinomial $16t^3 + 56t^4 + 64t^7$, factor by the GCF.
5. Given the polynomial $49t^3u^4 + 63t^4u^5 + 77t^7u^8$, factor by the GCF.
6. Given the polynomial $125a^4b^6c^7 - 75a^9b^3c^5 - 85a^3b^5c^{12} + 50a^6b^7c^{11}$, factor by the GCF.
7. Factor a trinomial in the form $x^2 - 12x - 28$.
8. Factor a trinomial in the form $x^2 + \frac{7}{95}x - \frac{18}{95}$.
9. Given a trinomial in the form $8u^2 - 18u - 11$, factor it.
10. Factor a trinomial in the form $\frac{1}{9}x^2 - \frac{1}{2}x - \frac{5}{18}$.
11. Given a polynomial $12h^4 + 9h^3 - 28h - 21$, factor by grouping.
12. Given a polynomial $18u^5t^3 + 12u^2t^2 + 18ut + 27u^4t^2$, factor by grouping.
13. Given a polynomial $\frac{10}{21}a^4b^2 - \frac{14}{21}a^3b^3 - \frac{7}{3}a^2b^2 + \frac{5}{3}a^3b$, factor by grouping.
14. Given a polynomial $3y^2(4x - 5) + 5y(12x - 15) - (24x - 30)$, factor by grouping.
15. Factor trinomial $120x^4 - 6x^3 - 72x^2$ using combined factoring methods.
16. Factor the polynomial $30y^4z^2(3x - 7) + 45y^3z(9x - 21) - 24y^2z^5(15x - 35)$, using combined factoring methods.
17. Given a polynomial $18t^4 - 33t^2 + 9$, factor using substitution.
18. Given a polynomial $-27x^6 - 54x^3 - 15$, factor using substitution.
19. Given a polynomial $12(3x + 2)^2 - 16(3x + 2) + 5$, factor using substitution.
20. Given a polynomial $8(\frac{1}{2x-1})^2 \quad 15(\frac{1}{2x-1}) + 7$, factor using substitution.

3. Quadratic Equations and Inequalities

A quadratic equation is a second-degree polynomial equation that can be rearranged in standard form as:

$$ax^2 + bx + c = 0,$$

where x represents an unknown value, and a, b, and c represent known numbers, where $a \neq 0$.

A quadratic inequality is an equation of second degree that uses an inequality sign instead of an equal sign. That is: "<", "≤", ">", or "≥". We will discuss them later in more detail.

3.1. Quadratic Expression

As we discussed earlier in this chapter, a polynomial expression that consists only of a non-zero constant, is called a constant polynomial, and has degree 0. A polynomial expression with degree one is a linear polynomial. A polynomial with degree two is a quadratic polynomial or **quadratic expression**.

When we are faced with an equation containing polynomials of degree greater than or equal to two, we can usually solve them by factoring. As we have already learned, a polynomial of degree n is an expression of the type $a_n x^n + a_{n-1} x^{n-1} + \cdots + a_1 x + a_0$. Setting the polynomial equal to zero gives a polynomial equation. We also know that the total number of solutions to a polynomial equation is equal to the highest exponent n. We have already explored solving linear polynomial equations. Hear we look at solving polynomial equations and inequalities with degree two, or more precisely, quadratic equations and inequalities. We also learned that to obtain the zeros of a polynomial equation, $p(x)$, we need to have it as $p(x) = 0$.

To explore methods for solving polynomial equations and polynomial inequality, let us start with polynomial equations of degree two, called quadratic equations.

3.2. Quadratic Equations

A quadratic equation is a polynomial equation with degree of two. The standard form of a quadratic equation is in the form of $ax^2 + bx + c = 0$, where x is the variable and $a \neq 0$, b, and c are constants. Quadratic equations differ from linear equations by involving a quadratic term with the variable raised to the second power of the form ax^2. There are several methods we can use to solve a quadratic equation. In this section we look for four methods, factoring, using the square roots, completing the square, and quadratic formula. The latter method is useful in case a quadratic equation cannot be factored.

Also, the square root is a method which can be more easily used when there is no linear term, bx in the quadratic equations. In case when the linear term exists, the solution using the square root becomes more complicated. We will explain this later when we present the square root method. Let us first look at the factoring method.

3.2.1. Solving Quadratic Equations, Using Factoring

A quadratic equation can be factored into an equivalent equation.

$$ax^2 + bx + c = a(x\ \ r)(x\ \ s) = 0,$$

where r and s are the solutions for x. Completing the square on a quadratic equation in standard form results in the quadratic equation, which expresses the solutions in terms of a, b, and c. Solutions to problems that can be expressed in terms of quadratic equations were known as early as 2000 BC.

We have previously studied factoring a trinomial with leading coefficient 1, as well as with leading coefficient not 1. We also discovered that the case when leading coefficient is not 1 is an extended version of the first case, leading coefficient of 1. Applying the factoring method here, when the leading coefficient is not one, we factor a quadratic equation using the method called grouping, which requires four terms. With the equation in standard form, let us review the grouping procedures:

1. Given a quadratic equation in standard form, $ax^2 + bx + c = 0$, list factors of ac.
2. Find two numbers, p and q, a pair of factors of ac with sum of b.
3. Rewrite the original equation replacing the bx term with two terms using the numbers found in step 2 as coefficients of x to get $ax^2 + px + qx + c = 0$.
4. Factor the first two terms and then factor in the last two terms. The expressions in parentheses must appear as exactly two same expressions to be used for grouping.
5. factor out the expression in parentheses.
6. Set each of the two expressions equal to zero and solve each for the variable.

Example
Use factoring method to solve $2x^2 + 7x - 9 = 0$.

Solution
Comparing with $ax^2 + px + qx + c = 0$, we have an equation with $a = 2$, $b = 7$, and $c = -9$. First, we determine $ac = -18$. Next, we need to find two numbers with a product of -18 and a sum of 7. In the following table, we list all factors of $ac = -18$ until a pair of the anticipated sum, $b = 7$, is obtained.

Factors of − 18	Sum of factors
1, -18	-17
-1, 18	17
2, -9	-7
-2, 9	**7**
3, -6	-3
-3, 6	3

So, $p = -2$ and $q = 9$.

We now, rewrite the original expression as $ax^2 + px + qx + c = 0$:

$2x^2 + 7x - 9 = 0$

$2x^2 - 2x + 9x - 9 = 0$

$2x(x - 1) + 9(x - 1) = 0$ we factored out thc GCF of each term

$(x - 1)(2x + 9) = 0$ again we factored out the GCF of the expression

$(x - 1)(2x + 9) = 0.$

Next, using zero-factor principle, we set each of the two expressions $(x - 1)$ and $(2x + 9)$ equal to zero and solve each for x.

$x - 1 = 0$

$x - 1 + 1 = 0 + 1$ we added 1 to both sides

$x = 1.$

$2x + 9 = 0$

$2x + 9 - 9 = 0 - 9$ we added – 9 to both sides

$2x = - 9$

$\frac{1}{2} \times 2x = \frac{1}{2} \times (-9)$ to make the coefficient of the variable equal to 1, we multiplied both sides by the reciprocal of its coefficient

$x = -\frac{9}{2} =.- 4.5.$

Therefore, the solutions of $2x^2 + 7x - 9 = 0$ are, $x = 1$ and $x = -\frac{9}{2} =.- 4.5.$

3.2.2. Solving Quadratic Equations, Square Root

When there is no linear term, bx, in quadratic equations, a straightforward method of solving this type of equation is the use of the **square root** method. To perform this, we must isolate the x^2 term, often by manipulation, and take the square root of the number on the other side of the equals' sign.

As an example, consider a quadratic equation with an $x^2 - k = 0$. Note that there is no x term. To solve such a quadratic equation, we must first isolate the x^2 term on one side of the equal sign. Next, we take the square root of both sides of the equation, putting a."±" sign before the expression k, which is on the side opposite the equal's sign. We simply obtain $x = \pm\sqrt{k}$, where k is a nonzero real number.

Example
Solve the quadratic equation $x^2 - 9 = 0$ using the square root.

Solution
We must first isolate the x^2 term on one side of the equal sign and –9 on the other side of equal sign. Next, we take the square root of both sides of the equation, and we put a ± sign before the square root of 9.

$x^2 - 9 = 0$

$x^2 = 9$ we isolated the x^2 term

$x = \pm\sqrt{9}$

$x = \pm 3$ we took the square root of both sides of the equation to get the solution.

Therefore, the solutions of $x^2 - 9 = 0$ are, $x = 3$ and $x = - 3$.

Example
Solve the quadratic equation $6x^2 - 74 = 0$ using the square root.

Solution
Again, we must first isolate the $6x^2$ term on one side of the equal sign and 74 on the other side of equal sign.

$6x^2 - 74 = 0$

$6x^2 = 74$ — we isolated the $6x^2$ term

$\frac{1}{6} \times 6x^2 = \frac{1}{6} \times 74$ — to make the coefficient of the variable equal to 1, we multiplied both sides by the reciprocal of its coefficient

$x^2 = \frac{74}{6}$

$x^2 = \frac{37}{3}$

$x = \pm\sqrt{\frac{37}{3}}$ — we took the square root of both sides of the equation to get the solution.

Therefore, the solutions of $6x^2 - 74 = 0$ are, $x = \sqrt{\frac{37}{3}}$ and $x = -\sqrt{\frac{37}{3}}$.

Example
Solve the quadratic equation $9(x - 7)^2 - 36 = 0$ using the square root.

Solution
In this example, we must isolate the $(x - 7)^2$ term on one side of the equal sign. To do so we first take -36 on the other side of equal sign, and we then multiply both sides of equation by the reciprocal of the coefficient $99(x - 7)^2$. The next step is to take the square root of both sides of the equation to get $(x - 7)$. Finally, we must isolate x on one side of the equation and constant term on the other side of the equation.

$9(x - 7)^2 - 36 = 0$

$9(x - 7)^2 = 36$

$\frac{1}{9}\times 9(x - 7)^2 = \frac{1}{9}\times 36$ — we multiplied both sides by the reciprocal of its coefficient

of $9(x - 7)^2$.

$(x - 7)^2 = \frac{36}{9}$

$(x - 7)^2 = 4$

$x - 7 = \pm 2$ — we took the square root of both sides.

$x = 7 \pm 2$ — we isolated the x term

$x = 7 \pm 2$

$x = 7 + 2$ and $x = 7 - 2$ — due to ± 2 we have two distinct solutions

$x = 9$, and $x = 5$.

Therefore, the final solutions of $9(x - 7)^2 - 36 = 0$ are $x = 9$ and $x = 5$.

Check: To verify our answer, we need to plug each solution into the equation. So,

For $x = 9$ we will have,

$9(x - 7)^2 - 36 = 0$

$9(9-7)^2 - 36 \overset{?}{\Leftrightarrow} 0$ we plugged $x = -5$ into the right side

$9(2)^2 - 36 \overset{?}{\Leftrightarrow} 0$

$9(4) - 36 \overset{?}{\Leftrightarrow} 0$

$36 - 36 \overset{?}{\Leftrightarrow} 0$

$0 = 0$ true, one of the solutions is a correct answer.

For $x = 5$ we will have:

$9(x-7)^2 - 36 = 0$

$9(5-7)^2 - 36 = 0$

$9(2)^2 - 36 \overset{?}{\Leftrightarrow} 0$

$9(4) - 36 \overset{?}{\Leftrightarrow} 0$

$36 - 36 \overset{?}{\Leftrightarrow} 0$

$0 = 0$ true, the other solution is a correct answer.

For quadratic equations $ax^2 + bx + c = 0$, in which bx term is existed, we need to find the third term, c, by the formula $c = (\frac{1}{2} \times b)^2$, that complete the square and then we factor the perfect square. We will illustrate this procedure by some examples.

Example
Given the quadratic expression $x^2 + 12x + c$, rewrite it as the complete square.

Solution
Using the formula $c = (\frac{1}{2} \times b)^2$, we will have:

$c = (\frac{1}{2} \times b)^2$

$c = (\frac{1}{2} \times 12)^2 = (6)^2 = 36$ we obtained he third term to complete the square is 36.

We now can plug $c = 36$ into the expression. So,

$x^2 + 12x + c$

$x^2 + 12x + 36$ we plugged $c = 36$ into the expression

$(x^2 + 6)^2$.

Therefore, quadratic expression $x^2 + 12x + c$ can be written as the complete square, $(x^2 + 6)^2$.

Example
Solve the quadratic equation $x^2 - 4x - 12 = 0$ using the square root.

Solution

Let us separate constants from variables by adding 12 to both sides.

$x^2 - 4x - 12 = 0$

$x^2 - 4x - 12 + 12 = 0 + 12$ we added 12 to both sides

$x^2 - 4x = 12.$

We need to find a number to complete the square. So,

$(\frac{1}{2} \times 4)^2 = 2^2 = 4.$

Now that we have obtained 4, we need to add 4 to both sides. So,

$x^2 - 4x = 12$

$x^2 - 4x + 4 = 12 + 4$ we added 4 to both sides

$x^2 - 4x + 4 = 16$

$(x - 2)^2 = 16$ we applied binomial, perfect square rule

$(x - 2) = \pm\sqrt{16} = \pm 4$ we took the square root of both sides.

The right-hand side of the equation has two values, + 4 and – 4. For getting the answer, we looked at them one at a time. So,

$x - 2 = 4$ we considered the positive value of the right-hand side

$x - 2 + 2 = 4 + 2$ we added 2 to both sides

$x = 6.$

Now, we need to consider the negative value of the right-hand side. So,

$x - 2 = -4$ we considered the negative value of the right-hand side

$x - 2 + 2 = -4 + 2$ we added 2 to both sides

$x = -2.$

Therefore, the solutions of $x^2 - 4x - 12 = 0$ are $x = 6$ and $x = -2$.

Check: To verify our answer, we need to plug each solution into the equation. So,

$x^2 - 4x - 12 = 0$

$(6)^2 - 4(6) - 12 \overset{?}{\Leftrightarrow} 0$

$36 - 24 - 12 \overset{?}{\Leftrightarrow} 0$

$0 = 0$ true, one of the solutions is a correct answer

$x^2 - 4x - 12 = 0$

$(-2)^2 - 4(-2) - 12 \overset{?}{\Leftrightarrow} 0$

$4 + 8 - 12 \overset{?}{\Leftrightarrow} 0$

$12 - 12 \overset{?}{\Leftrightarrow} 0$

$0 = 0$ True, the other solution is a correct answer.

Example

Solve the quadratic equation $4x^2 + 8x - 96 = 0$ using the square root.

Solution

We note that in this equation the leading coefficient is not 1. Hence, we need to force this coefficient to become 1. So,

$4x^2 + 8x - 96 = 0$

$4x^2 + 8x - 96 + 96 = 0 + 96$ we added 96 to both sides

$4x^2 + 8x = - 96$

$\frac{1}{4} \times (4x^2 + 8x) = \frac{1}{4} \times (96)$ we multiplied both sides by the reciprocal of the coefficient of $4x^2$

$x^2 + 2x = 24.$

We need to find a number to complete the square. So,

$(\frac{1}{2} \times 2)^2 = 1^2 = 1.$

Now that we have obtained 1, we need to add 1 to both sides. So,

$x^2 + 2x = 24$

$x^2 + 2x + 1 = 24 + 1$ we added 1 to both sides

$(x + 1)^2 = 25$ we applied binomial, perfect square rule

$(x + 1) = \pm\sqrt{25}$ we toke the square root of both sides

$(x + 1) = \pm 5.$

The right-hand side of the equation has two values, + 5 and – 5. For getting the answer, we looked at them one at time. So,

$(x + 1) = 5$ we considered the positive value of the right-hand side

$x + 1 - 1 = 5 - 1$ we added – 1 to both sides

$x = 4.$

Now we need to consider the negative value of the right-hand side. So,

$(x + 1) = - 5$ we considered the negative value of the right-hand side

$X + 1 - 1 = - 5 - 1$ we added – 1 to both sides

$x = - 6.$

Therefore, the solutions of $4x^2 + 8x - 96 = 0$ are $x = 4$ and $x = - 6$.

Check: To verify our answer, we need to plug each solution separately into the equation. So,

For $x = 4$ we will have:

$4x^2 + 8x - 96 = 0$

$4(4)^2 + 8(4) - 96 \overset{?}{\Leftrightarrow} 0$

$4(16) + 32 - 96 \overset{?}{\Leftrightarrow} 0$

$64 + 32 - 96 \overset{?}{\Leftrightarrow} 0$

$96 - 96 \overset{?}{\Leftrightarrow} 0$

$0 = 0$ true, one of the solutions is a correct answer.

For $x = -6$ we will have:

$4x^2 + 8x - 96 = 0$

$4(-6)^2 + 8(-6) - 96 \overset{?}{\Leftrightarrow} 0$

$4(36) - 48 - 96 \overset{?}{\Leftrightarrow} 0$

$144 - 144 \overset{?}{\Leftrightarrow} 0$

$0 = 0$ true, the other solution is a correct answer.

3.2.3. Solving Quadratic Equations, Using Completing the Square

When the quadratic equation does not factor, we cannot solve the equation by factoring. Also, when the quadratic equation is not ready to square root, we cannot solve the equation by the square root. But there is a way for us to manipulate the quadratic to put it into that ready-for-square-rooting form, so we can solve. This is called completing the square.

For solving quadratic equations, $ax^2 + bx + c = 0$, using completing square we need to pursue steps below

Step 1. Transform the equation so that the constant term, c, is alone on the right side.

Step 2. If a, the leading coefficient of quadratic expression, the coefficient of the x^2 term, is not equal to one, divide both sides by a.

Step 3. Add the square of half the coefficient of the x-term, $\left(\frac{b}{2a}\right)^2$.

Step 4. Factor the left side as the square of a binomial.

Step 5. Take the square root of both sides. Note that $(x + q)^2 = r$ is equivalent to $x + q = \pm\sqrt{r}$.

Step 6. Solve for x.

Step 7. Check to verify the solutions of x.

Example

Solve $x^2 - 8x - 9 = 0$ using completing the square.

Solution

We need to transform the equation so that the constant term, c, is alone on the right side. So,

$x^2 - 8x - 9 = 0$

$x^2 - 8x = 9.$

We realize that the leading coefficient, the coefficient of x^2, is one. Hence, we need to look at the coefficient of the x-term, which is – 8 in this case. We take half of this number (including the sign), which gives me – 4. Then we square this value to get + 16, and we add this squared value to both sides of the equation.

This process creates a quadratic expression that is a perfect square on the left-hand side of the equation. So,

$x^2 - 8x = 9$

$x^2 - 8x + 16 = 9 + 16$ — we added 16 to both sides

$x^2 - 8x + 16 = 25$

$(x - 4)^2 = 25$ — we factored the left side as the square of a binomial

$x - 4 = \pm\sqrt{25}$ — we took the square root of both sides

$x - 4 = \pm 5$

$x - 4 + 4 = \pm 5 + 4$ — we added 4 to both sides

$x = \pm 5 + 4$

$x = 5 + 4 = 9.$

$x = -5 + 4 = -1.$

Therefore, the solutions of $x^2 - 8x - 9 = 0$ are $x = 9$ and $x = -1$.

Check: To verify our answer, we need to plug each solution separately into the equation. So,

For $x = 9$ we will have:

$x^2 - 8x - 9 = 0.$

$9^2 - 8(9) - 9 \overset{?}{\Leftrightarrow} 0$

$81 - 72 - 9 \overset{?}{\Leftrightarrow} 0$

$81 - 81 \overset{?}{\Leftrightarrow} 0$

$0 = 0$ — true, $x = 9$ is a correct answer.

For $x = -1$ we will have.

$x^2 - 8x - 9 = 0$

$(-1)^2 - 8 \times (-1) - 9 \overset{?}{\Leftrightarrow} 0$

$1 + 8 - 9 \overset{?}{\Leftrightarrow} 0$

$9 - 9 \overset{?}{\Leftrightarrow} 0$

$0 = 0$ — true, $x = -1$ is a correct answer.

Example

Solve $9x^2 + 3x - 2 = 0$ using completing the square.

Solution

We need to transform the equation so that the constant term is alone on the right side. So,

$$9x^2 + 3x - 2 = 0$$

$$9x^2 + 3x = 2.$$

Since the coefficient of the x^2 term is not equal to one, we need to divide both sides by the leading coefficient which is 9.

$$9x^2 + 3x = 2$$

$$\frac{1}{9}(9x^2 + 3x) = \frac{1}{9} \times 2$$ we multiplied both sides by the reciprocal of x^2

$$\frac{1}{9} \times 9x^2 + \frac{1}{9} \times 3x = \frac{1}{9} \times 2$$

$$x^2 + \frac{1}{3}x = \frac{2}{9}.$$

Now, we need to add the square of half the coefficient of the x-term, which is $\frac{1}{3}$. The square of half of $\frac{1}{3}$ is $(\frac{1}{2} \times \frac{1}{3})^2 = (\frac{1}{6})^2 = \frac{1}{36}$. So,

$$x^2 + \frac{1}{3}x = \frac{2}{9}$$

$$x^2 + \frac{1}{3}x + \frac{1}{36} = \frac{2}{9} + \frac{1}{36}$$ we added $\frac{1}{36}$ to both sides

$$x^2 + \frac{1}{3}x + \frac{1}{36} = \frac{2}{9} \times \frac{4}{4} + \frac{1}{36}$$ we multiplied $\frac{4}{4}$ to the first fraction of the right-side

$$x^2 + \frac{1}{3}x + \frac{1}{36} = \frac{8}{36} + \frac{1}{36}$$

$$x^2 + \frac{1}{3}x + \frac{1}{36} = \frac{8+1}{36}$$

$$x^2 + \frac{1}{3}x + \frac{1}{36} = \frac{9}{36}$$

$$x^2 + \frac{1}{3}x + \frac{1}{36} = \frac{1}{4}$$

$$(x + \frac{1}{6})^2 = \frac{1}{4}$$ we factored the left side as the square of a binomial

$$x + \frac{1}{6} = \pm\sqrt{\frac{1}{4}}$$ we took the square root of both sides

$$x + \frac{1}{6} = \pm\frac{1}{2}$$

$$x + \frac{1}{6} - \frac{1}{6} = \pm\frac{1}{2} - \frac{1}{6}$$ we added $-\frac{1}{6}$ to both sides

$$x = \pm\frac{1}{2} - \frac{1}{6}$$

$$x = \pm\left(\frac{1}{2} \times \frac{3}{3}\right) - \frac{1}{6}$$ we multiplied $\frac{3}{3}$ to the first fraction of the right-side

$$x = \pm \frac{3}{6} - \frac{1}{6}$$

$$x = \frac{3-1}{6} = \frac{2}{6} = \frac{1}{3} \quad \text{and} \quad x = \frac{-3-1}{6} = \frac{-4}{6} = -\frac{2}{3}.$$

Therefore, the solutions of $9x^2 + 3x - 2 = 0$ are $x = \frac{1}{3}$ and $x = -\frac{2}{3}$.

Check: To verify our answer, we need to plug each solution separately into the equation. So,

For $x = \frac{1}{3}$ we will have:

$$9x^2 + 3x - 2 = 0$$

$$9(\frac{1}{3})^2 + 3 \times \frac{1}{3} - 2 \overset{?}{\Leftrightarrow} 0$$

$$9\times \frac{1}{9} + 1 - 2 \overset{?}{\Leftrightarrow} 0$$

$$1 + 1 - 2 \overset{?}{\Leftrightarrow} 0$$

$$2 - 2 \overset{?}{\Leftrightarrow} 0$$

$0 = 0$ true, $x = \frac{1}{3}$ is a correct answer.

For $x = -\frac{2}{3}$ we will have:

$$9x^2 + 3x - 2 = 0$$

$$9(-\frac{2}{3})^2 + 3(-\frac{2}{3}) - 2 \overset{?}{\Leftrightarrow} 0$$

$$9 \times \frac{4}{9} - 2 - 2 \overset{?}{\Leftrightarrow} 0$$

$$4 - 2 - 2 \overset{?}{\Leftrightarrow} 0$$

$$4 - 4 \overset{?}{\Leftrightarrow} 0$$

$0 = 0$ true, $x = -\frac{2}{3}$ is a correct answer.

3.2.4. Solving Quadratic Equations, Using Quadratic Formula

Given a quadratic equation in standard form, $ax^2 + bx + c = 0$, it can be solved using the **quadratic formula.** The quadratic formula of this quadratic equation is.

$$x = \frac{-b \pm \sqrt{b^2 - 4ac}}{2a},$$

where a, b, and c are real numbers and $a \neq 0$.

Discriminant of quadratic equation - For $ax^2 + bx + c = 0$, where a, b, and c are real numbers and $a \neq 0$, the discriminant, usually denoted as delta, Δ, is the expression under the radical in the quadratic formula, $\Delta = b^2 - 4ac$. It tells us whether the solutions are real numbers or complex numbers and how many solutions of each type to expect.

Discriminant of a Quadratic Equation
The discriminant of a quadratic equation, usually denoted as delta, Δ, is:

$$\Delta = b^2 - 4ac.$$

Discriminant Trichotomy - Let a, b and c be real numbers with $a \neq 0$.

- If $\Delta = b^2 - 4ac < 0$, the equation $ax^2 + bx + c = 0$ has no real solution.
- If $\Delta = b^2 - 4ac = 0$, the equation $ax^2 + bx + c = 0$ has exactly one real solution.
- If $\Delta = b^2 - 4ac > 0$, the equation $ax^2 + bx + c = 0$ has exactly two real solutions.

Example
Solve the quadratic equation $2x^2 + 8x + 12 = 0$ using the quadratic formula.

Solution
We can use the quadratic formula, $x = \frac{-b \pm \sqrt{b^2 - 4ac}}{2a}$ to solve the quadratic equation, where $a = 2$, $b = 8$, and $c = 12$

$$2x^2 + 8x - 12 = 0$$

$$x = \frac{-8 \pm \sqrt{8^2 - 4 \times 2 \times 12}}{2 \times 2}$$ we substituted $a = 2$, $b = 8$, and $c = 12$ into the equation

$$x = \frac{-8 \pm \sqrt{64 - 96}}{4}$$

$$x = \frac{-8 \pm \sqrt{-32}}{4}.$$

As it is seen, the discriminant is a negative number. Therefore, the quadratic equation $2x^2 + 8x - 12 = 0$, has no real solution.

Example
Solve the quadratic equation $4x^2 + 8x + 4 = 0$ by the quadratic formula.

Solution
We can use the quadratic formula, $x = \frac{-b \pm \sqrt{b^2 - 4ac}}{2a}$ to solve the quadratic equation, where $a = 4$, $b = 8$, and $c = 4$.

$$4x^2 + 8x + 4 = 0$$

$$x = \frac{-8 \pm \sqrt{8^2 - 4 \times 4 \times 4}}{2 \times 4}$$ we substituted $a = 4$, $b = 8$, and $c = 4$ into the equation

$$x = \frac{-8 \pm \sqrt{64 - 64}}{8}.$$

As it is seen, the discriminant is equal to zero, so we have a unique solution $x = -1$.

$$x = \frac{-8 \pm \sqrt{0}}{8} = \frac{-8}{8} = -1.$$

Therefore, the quadratic equation $4x^2 + 8x + 4 = 0$, has a unique solution $x = -1$.

Check: To verify our answer, we need to plug the solution into the equation. So,

$$4x^2 + 8x + 4 = 0$$

$4(-1)^2 + 8(-1) + 4 \overset{?}{\Leftrightarrow} 0$ we plugged $x = -1$ into the equation

$4(1) + 8(-1) + 4 \overset{?}{\Leftrightarrow} 0$

$4 - 8 + 4 \overset{?}{\Leftrightarrow} 0$

$8 - 8 \overset{?}{\Leftrightarrow} 0$

$0 = 0$ true, the solution $x = -1$ is a correct answer

Example
Solve the quadratic equation $3x^2 + 9x + 6 = 0$ using the quadratic formula.

Solution
We can use the quadratic formula, $x = \frac{-b \pm \sqrt{b^2 - 4ac}}{2a}$ to solve the quadratic equation, where $a = 3$, $b = 9$, and $c = 6$.

$3x^2 + 9x + 6 = 0$

$x = \frac{-9 \pm \sqrt{9^2 - 4 \times 3 \times 6}}{2 \times 3}$ we plugged $a = 3$, $b = 9$, and $c = 6$ into the equation

$x = \frac{-9 \pm \sqrt{81 - 72}}{6}$

$x = \frac{-9 \pm \sqrt{9}}{6} = \frac{-9 \pm 3}{6}.$

As it is seen, the discriminant is greater than zero, so we have two solutions.

$x = \frac{-9 \pm 3}{6}$

$x = \frac{-9 + 3}{6}$ we used the plus sign

$x = \frac{-6}{6} = -1$

$x = \frac{-9 - 3}{6}$ we now used the minus sign

$x = \frac{-12}{6} = -2.$

Therefore, the quadratic equation $3x^2 + 9x + 6 = 0$, has two solutions, $x = -1$ and $x = -2$.

Check: To verify our answer, we need to plug each solution into the equation. So,

For $x = -1$:

$3x^2 + 9x + 6 = 0$

$3(-1)^2 + 9(-1) + 6 \overset{?}{\Leftrightarrow} 0$ we plugged $x = -1$ into the equation

$3(1) + 9(-1) + 6 \overset{?}{\Leftrightarrow} 0$

$3 - 9 + 6 \overset{?}{\Leftrightarrow} 0$

$$9 - 9 \overset{?}{\Leftrightarrow} 0$$

$0 = 0$ true, the solution $x = -1$ is a correct answer.

For $x = -2$:

$$3x^2 + 9x + 6 = 0$$

$3(-2)^2 + 9(-2) + 6 \overset{?}{\Leftrightarrow} 0$ we plugged $x = -2$ into the equation

$$3(4) + 9(-2) + 6 \overset{?}{\Leftrightarrow} 0$$

$$12 - 18 + 6 \overset{?}{\Leftrightarrow} 0$$

$$18 - 18 \overset{?}{\Leftrightarrow} 0$$

$0 = 0$ true, the solution $x = -2$ is a correct answer.

3.3. Graph of Quadratic Function

The graph of a quadratic function is called a **parabola** and has a curved shape. One of the main points of a parabola is its vertex. It is the highest, maximum, or the lowest, minimum, point on its graph. The function $f(x) = ax^2 + bx + c$ is a quadratic function. The location and size of the parabola, and how it opens, depend on the values of a, b, and c. As shown in figure below, if $a > 0$, the parabola has a minimum point and opens upward. If $a < 0$, the parabola has a maximum point and opens downward. The extreme point of the parabola, whether minimum or maximum, corresponds to its vertex.

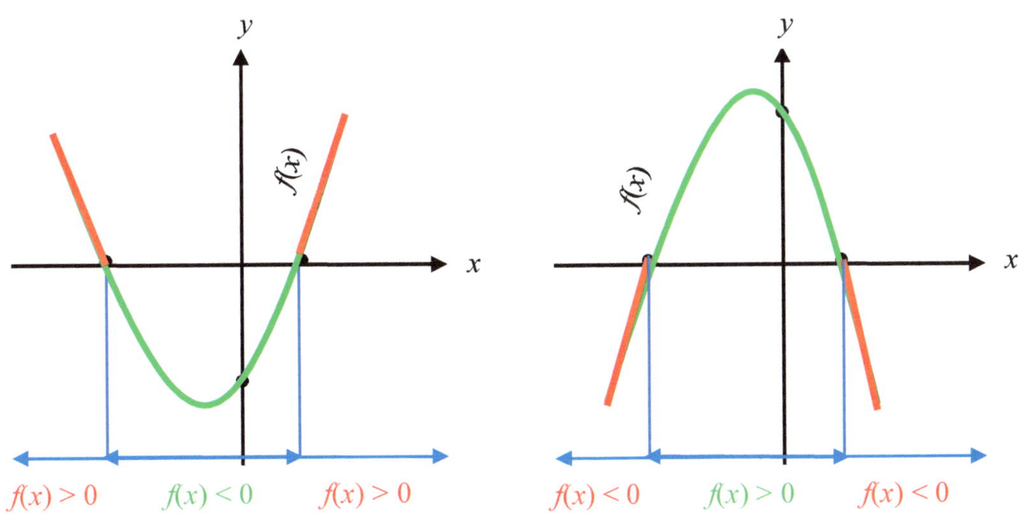

$f(x) = ax^2 + bx + c$, where $a > 0$

$y = ax^2 + bx + c$, where $a > 0$

$f(x) = ax^2 + bx + c$, where $a < 0$

$y = ax^2 + bx + c$, where $a < 0$.

As illustrated in figure below, the x-coordinate of the vertex will be located at $x = -\frac{b}{2a}$, and the y-coordinate of the vertex may be found by substituting this x-value into the function, $y = -\frac{b^2 - 4a}{4a} =$

$\frac{-b^2+4ac}{4a}$. The y-intercept is located at the point $(0, c)$. The dashed line presents the line of symmetry by the equation of $x = -\frac{b}{2a}$.

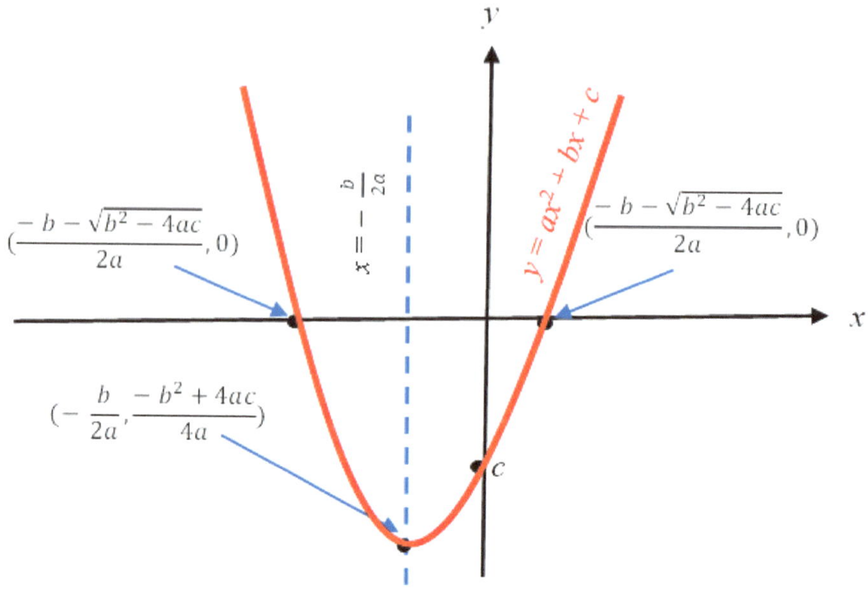

Discriminant Trichotomy - Let a, b and c be real numbers with $a \neq 0$.

- If $\Delta = b^2 - 4ac < 0$, the equation $ax^2 + bx + c = 0$ has no real solution.
- If $\Delta = b^2 - 4ac = 0$, the equation $ax^2 + bx + c = 0$ has exactly one real solution.
- If $\Delta = b^2 - 4ac > 0$, the equation $ax^2 + bx + c = 0$ has exactly two real solutions.

For a quadratic equation of the form $ax^2 + bx + c = 0$, $a \neq 0$.

Discriminant	Number/Type of solution	Typical graph
$b^2 + 4ac > 0$	2 real solutions two x-intercept on graph	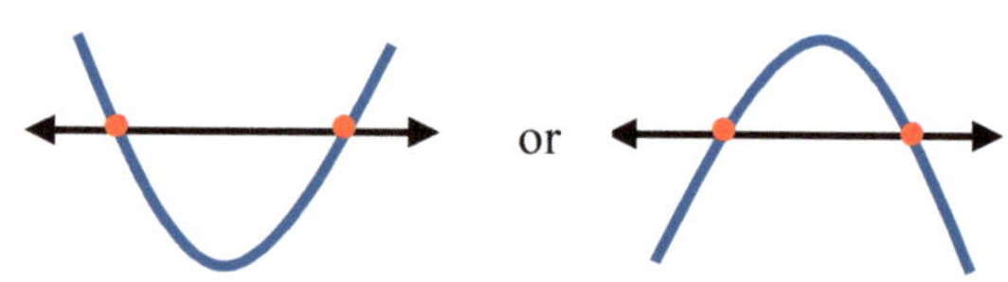
$b^2 + 4ac = 0$	1 real solutions one x-intercept on graph	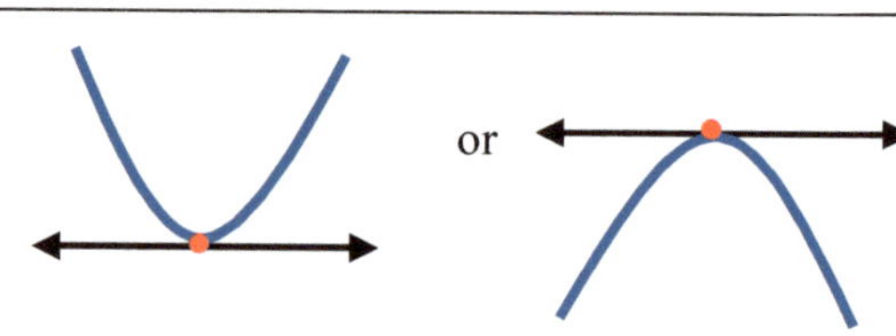
$b^2 + 4ac < 0$	2 complex solutions no x-intercept on graph	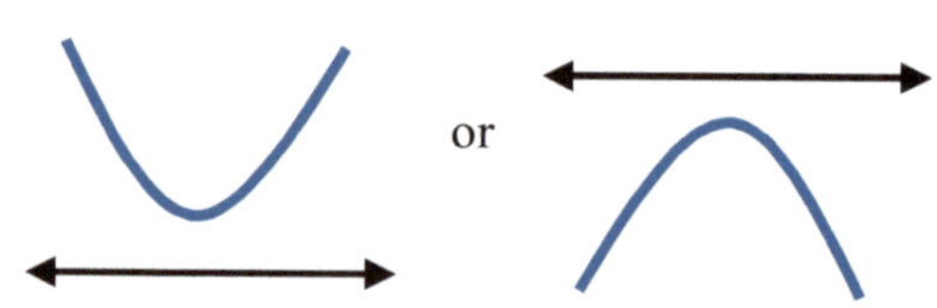

Example

Given the quadratic function $f(x) = 2x^2 - 4x + 1$, determine:

a) It an open upward or an open downward?

b) Sketch the graph of the function.

c) Find minimum/maximum point of the function.

Solution

a) We have learned that given $f(x) = ax^2 + bx + c$, where $a > 0$, it is an open upward function. In this case $a = 2 > 0$. Hence, the function is open upward.

b) To sketch the graph of the function, we need to get the coordinate of its vertex and the zeros of the function. Give a quadratic function $f(x) = ax^2 + bx + c$, the x-coordinate of the vertex is $x = -\frac{b}{2a}$ and the y-coordinate of the vertex is $y = -\frac{b^2 - 4ac}{4a}$. In this case, $a = 2$, $b = -4$, and $c = 1$. So,

$$x = -\frac{b}{2a} = -\frac{-4}{2(2)} = -\frac{-4}{4} = 1$$

$$y = -\frac{b^2 - 4ac}{4a} = -\frac{(-4)^2 - 4(2)(1)}{4(2)} = -\frac{16-8}{8} = -\frac{8}{8} = -1.$$

To find the zeros, we let $f(x) = 2x^2 - 4x + 1 = 0$. So,

$$x = \frac{-b \pm \sqrt{b^2 - 4ac}}{2a}$$

$$x = \frac{-(-4) \pm \sqrt{(-4)^2 - 4 \times 2 \times 1}}{2(2)}$$

$$x = \frac{4 \pm \sqrt{(16) - 8}}{4}$$

$$x = \frac{4 \pm \sqrt{8}}{4}$$

$$x = \frac{4 + \sqrt{8}}{4} = 1.71. \quad \text{or} \quad x = \frac{4 - \sqrt{8}}{4} = 0.29.$$

The graph of the function is.

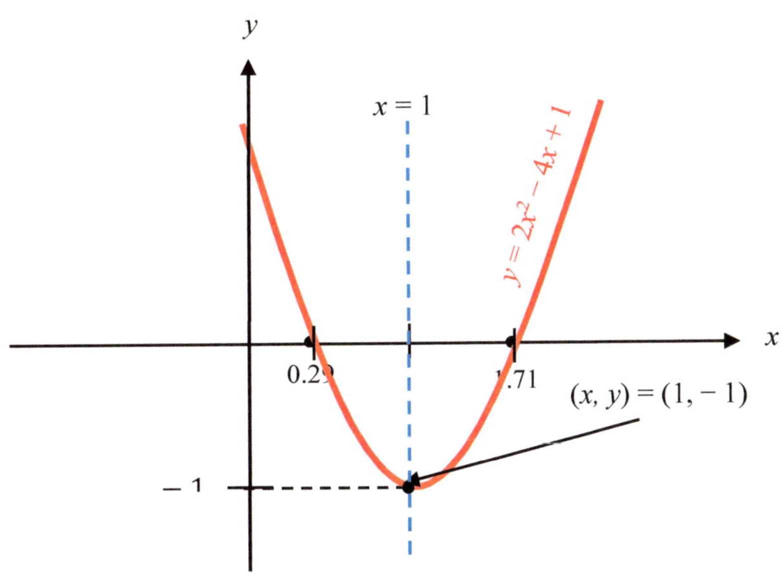

c) The function is open upward. Hence it has a minimum. We know that the minimum of the function is at its vertex. So, the minimum is lactated at $(x, y) = (1, -1)$.

Vertex of a Quadratic Function
This is what the x-coordinate and the y-coordinate of the vertex of a quadratic function tells us.

- The x-coordinate of the vertex determines:
  - if the parabola opens up, the graph is decreasing from $(-\infty, x)$ and increasing from (x, ∞).
  - if the parabola opens down, the graph is increasing from $(-\infty, x)$ and decreasing from (x, ∞).
  - the line of symmetry is the vertical line that cuts the parabola in half.
  - the equation of the line of symmetry is $x =$
- The y-coordinate of the vertex determines:
  - if a parabola open up, the vertex is a minimum and the y-coordinate is smallest function value.
  - if a parabola opens down, the vertex is a maximum and the y-coordinate is largest function value.
  - if the vertex is the minimum function value, the range is the y-coordinate to infinity $[y, \infty)$ because the parabola open up.
  - if the vertex is the maximum function value, the range is negative infinity to the y-coordinate $(-\infty, y]$.

Example
Given $y = 6(x - 1)^2 + 5$, find the x-coordinate and the y-coordinate of the vertex and write the equation of the line of symmetry.

Solution
We have learned that, give a quadratic function $f(x) = ax^2 + bx + c$, the x-coordinate of the vertex is $x = -\frac{b}{2a}$ and the y-coordinate of the vertex is $y = -\frac{b^2 - 4ac}{4a}$. In this case, we need to rewrite $y = 6(x - 1)^2 + 5$ as a form of $y = ax^2 + bx + c$. So,

$$y = 6(x - 1)^2 + 5$$

$$y = 6(x^2 - 2x + 1) + 5$$

$$y = 6x^2 - 12x + 6 + 5$$

$$y = 6x^2 - 12x + 11.$$

Comparing $y = 6x^2 - 12x + 11$ with $y = ax^2 + bx + c$, $a = 6$, $b = -12$, and $c = 11$. The coordinates of the vertex, (x_V, y_V), are:

$$x_V = -\frac{b}{2a}$$

$$x_V = -\frac{-12}{2(6)}$$

$$x_V = -\frac{-12}{12} = 1.$$

$$y_V = -\frac{b^2 - 4ac}{4a}$$

$$y_V = -\frac{(-12)^2 - 4 \times 6 \times 11}{4 \times 6}$$

$$y_V = -\frac{144 - 264}{24}$$

$$y_V = -\frac{-120}{24} = -(-5) = 5.$$

The equation for the line symmetry is $x = -\frac{b}{2a} = -\frac{-12}{2(6)} = 1.$

Forms of a Quadratic Equation - A quadratic equation can be expressed in three forms: the standard form, the factored form, and the vertex form. These are the three forms.

$y = ax^2 + bx + c$ **Standard form.**

$y = a(x - r_1)(x - r_2)$ **Factored form:** where, r_1 and r_2 are the roots of the equation.

$y = a(x - h)^2 + k$ **Vertex form:** where, h and k are the x and y coordinates of the vertex, respectively.

Example

Find the coordinates of the vertex in each of the quadratic equations,

a) $y = 5(x - 2)^2 + 6.$

b) $y = 7(x + 4)^2 - 9.$

Solution

a) Comparing the equation $y = 5(x - 2)^2 + 6$, with the vertex form $y = a(x - h)^2 + k$, we find $h = 2$ and $k = 6$. Therefore, the x-coordinate of the vertex is 2, and the y-coordinate of the vertex is 6.

b) Comparing the equation $y = 7(x + 4)^2 - 9$, with the vertex form $y = a(x - h)^2 + k$, we find $h = -4$ and $k = -9$. Thus, the x-coordinate of the vertex is -4, and the y-coordinate of the vertex is -9.

Example

Rewrite the quadratic equation $y = 3x^2 + 3x - 90$ as the factored form.

Solution

To rewrite the standard form of the quadratic equation as the factored form, we need to find the roots of the equation. So,

$$3x^2 + 3x - 90 = 0$$

$$x = \frac{-b \pm \sqrt{b^2 - 4a}}{2a}$$

$$x = \frac{-3 \pm \sqrt{(3)^2 - 4 \times 3 \times (-90)}}{2(3)}$$

$$x = \frac{-3 \pm \sqrt{(9) + 1080}}{6}$$

$$x = \frac{-3 \pm 33}{6}$$

$$x = \frac{-3 + 33}{6} = \frac{30}{6} = 5. \quad \text{and} \quad x = \frac{-3 - 33}{6} = \frac{-36}{6} = -6.$$

Now, with the roots of 5 and -6, we can rewrite the standard form of the quadratic equation as the factored form. So,

$y = 3x^2 + 3x - 90$

$y = (x - 5)(x + 6)$.

Therefore, the factored form of the quadratic equation is $y = (x - 5)(x + 6)$.

Example

Given the standard form of a quadratic equation $y = 5x^2 - 4x - 12$, with points P and Q on the graph of this quadratic equation, and points P and Q are symmetric with respect to its line of symmetry. Find.

a) The equation for the line of symmetry of its graph.

b) If point $P = (3, 21)$ find the coordinates of Q.

Solution

a) Comparing $y = 5x^2 - 4x - 12$ with $y = ax^2 + bx + c$, $a = 5$, $b = -4$, and $c = -12$. We learned that the line of symmetry for the graph is $x = \frac{-b}{2a}$. So,

$x = \frac{-b}{2a}$

$x = \frac{-(-4)}{2(5)} = \frac{4}{10} = 0.4$ we substituted $a = 5$ and $b = -4$ into the equation

$x = 0.4$.

Therefore, the line of symmetry of $y = 5x^2 - 4x - 12$ is $x = 0.4$.

b) Given that points P and Q are symmetric with respect to the line of symmetry, the y-coordinate of Q is the same as the y-coordinate of P, 21. We just need to determine the x-coordinate of Q. Knowing points P and Q are symmetric with respect to the line of symmetry means that distance of x_P from the line of symmetry is equal to the distance of x_Q from the line of symmetry.

We let $d(x_P)$ and $d(x_Q)$ denote as the distance of x-coordinate of P and Q from the line of symmetry respectively. Also, we let x_L represent the x-coordinate of the line of symmetry, then we will have:

$d(x_P) = x_P - x_L$

$d(x_P) = 3 - 0.4 = 2.6$.   we substituted $x_P = 3$ and $x_L = 0.4$ into the equation

$d(x_Q) = x_L - x_Q$

$d(x_Q) = 0.4 - x_Q$ we substituted $x_L = 0.4$ into the equation.

Now, letting $d(x_Q) = d(x_P)$, we get

$d(x_Q) = d(x_P)$

$0.4 - x_Q = 2.6$ we substituted $d(x_Q) = 0.4 - x_Q$ and $d(x_P) = 2.6$ into the equation

$0.4 - x_Q - 0.4 = 2.6 - 0.4$ we added -0.4 to both sides

$-x_Q = 2.2$

$(-1)(-x_Q) = (-1)(2.2)$ we multiplied both sides by (-1)

$x_Q = -2.2$.

Therefore, the coordinates of Q are $x_Q = -2.2$ and $y_Q = 21$. Or $Q(-2.2, 21)$.

Quadratic Inequalities

Quadratic inequalities can be of the following forms:

$$ax^2 + bx + c > 0$$

$$ax^2 + bx + c \geq 0$$

$$ax^2 + bx + c < 0$$

$$ax^2 + bx + c \leq 0.$$

3.4. Solving Quadratic Inequalities

A quadratic inequality can be solved graphically using a graph or algebraically using a table of signs to find out where the equation has a positive value and a negative value. To solve a quadratic inequality graphically, we must find out which part of the graph of a quadratic function lies above or below the x-axis. To solve a quadratic inequality algebraically, we need to use a table of signs by which we can find out where the equation has a positive value and a negative value.

The steps below present the process for solving quadratic inequalities graphically as well as algebraically.

Step 1. Factorize the quadratic inequality.

Step 2. Convert the inequality to an equality and determine the zeros of the equality.

Step 3. Create a table sign and a number line to determine where each factor of the inequality has positive value and negative value on the number line. The sign within each interval on the number line is determined by using test values.

Step 4. Considering the quadratic inequality as $y = ax^2 + bx + c$, sketch its graph.

Step 5. Find the x-intercept, the final answers of x, and determine in which intervals the y has positive and negative values.

Step 6. Represent the final answers on a number line.

Example

Solve $x^2 - 4x + 3 \geq 0$.

Solution

The inequality is factorized as:

$$(x - 3)(x - 1) \geq 0.$$

We convert $x^2 - 4x + 3 = (x - 3)(x - 1) \geq 0$ to $x^2 - 4x + 3 = (x - 3)(x - 1) = 0$, to find its zeros. So,

$$(x - 3)(x - 1) = 0$$

$(x - 3) = 0$ and $(x - 1) = 0$

$x = 3$ and $x = 1$.

The table sign is:

x values	$x < 1$	$x = 1$	$1 < x < 3$	$x = 3$	$x > 3$
$x - 3$	–	–	–	0	+
$x - 1$	–	0	+	+	+
$y = (x - 3)(x - 1)$	+	0	–	0	+

The graph of $y = x^2 - 4x + 3$ is.

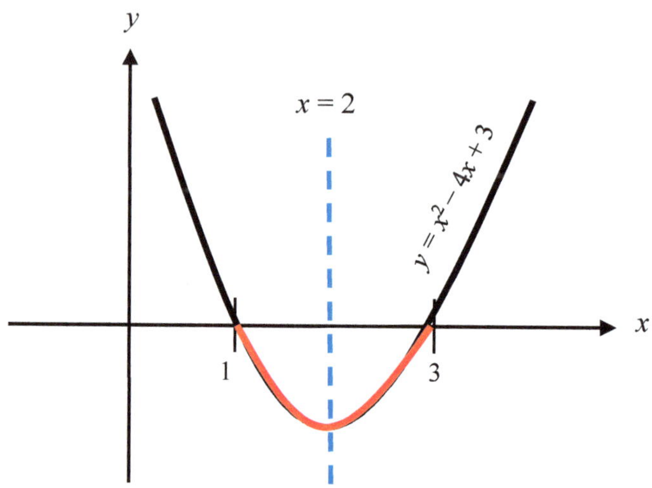

Referring to the graph, it is seen for $x < 1$, the value of inequality is positive, $y > 0$, for $1 < x < 3$, the value of inequality is negative, $y < 0$, and for $x > 3$, the value of inequality is positive $y > 0$.

Therefore, the solution of $x^2 - 4x + 3 \geq 0$ is $x < 1$ and $x > 3$.

To represent the final answers on a number line we will have:

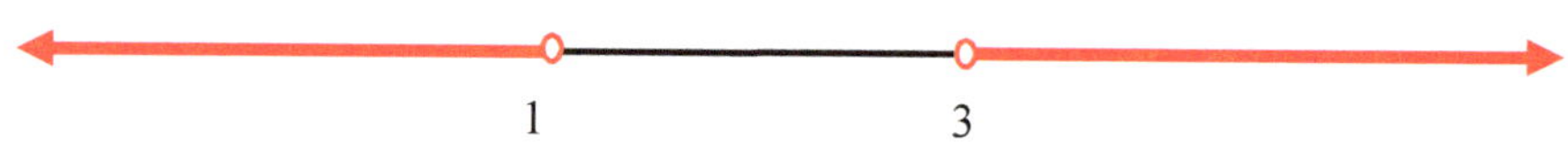

Example
Solve $9x^2 - 6x + 1 \leq 0$.

Solution
The inequality is factorized as.

$$(3x - 1)(3x - 1) \leq 0$$

$$(3x - 1)^2 \leq 0$$

We convert $9x^2 - 6x + 1 = (3x - 1)^2 \leq 0$ to $9x^2 - 6x + 1 = (3x - 1)^2 = 0$, for finding its zeros. So,

$$(3x - 1)^2 = 0$$

$$3x - 1 = 0$$

$$x = \frac{1}{3}.$$

The table sign is:

x values	$x < \frac{1}{3}$	$x = \frac{1}{3}$	$x > \frac{1}{3}$
$3x - 1$	+	0	+
$y = (3x - 1)^2$	+	0	+

The graph of $y = 9x^2 - 6x + 1$ is.

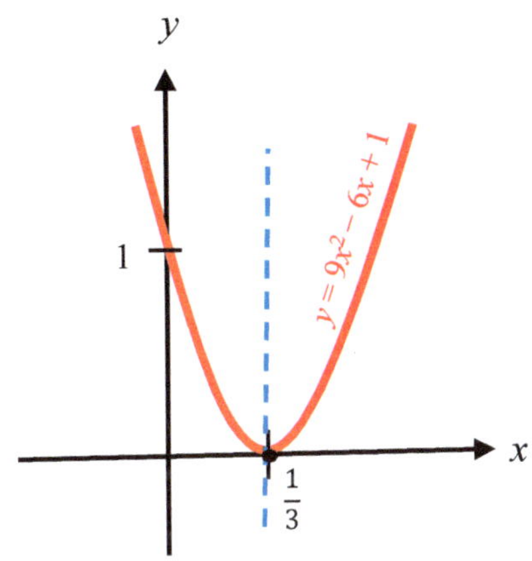

Referring to the graph, it is seen for $x < \frac{1}{3}$, and for $x > 3$, the value of inequality is positive $y > 0$. Therefore, the solution of $x^2 - 4x + 3 \leq 0$ is only $x = \frac{1}{3}$.

To represent the final answers on a number line we will have:

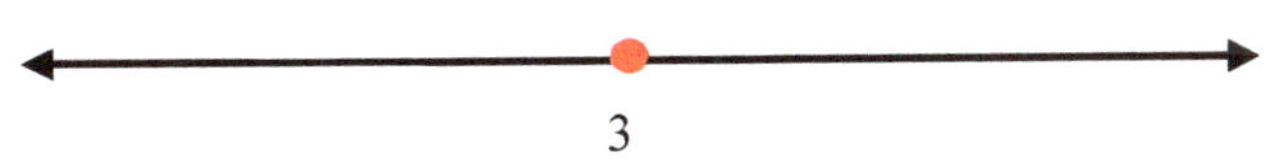

Up to this point we were looking at the quadratic inequality that could be factored. For solving those which are not factorable, we simply use the quadratic formula. To illustrate these cases, let us present an example.

Example
Solve $3x^2 - 4x - 7 \leq 0$.

Solution
As it is realized, the inequality can be easily factored. Hence, we convert it to equality and solve it using the quadratic formula. So,

$$3x^2 - 4x - 7 = 0$$

We know, for a quadratic equation in form of $ax^2 + bx + c = 0$, the solution can be found using the quadratic formula $x = \frac{-b \pm \sqrt{b^2 - 4ac}}{2a}$. For $3x^2 - 4x - 7 = 0$, $a = 3$, $b = -4$, and $c = -7$, So,

$$x = \frac{-(-4) \pm \sqrt{(-4)^2 - 4 \times 3 \times (-7)}}{2 \times 3}$$

$$x = \frac{4 \pm \sqrt{16 + 84}}{6}$$

$$x = \frac{4 \pm \sqrt{100}}{6}$$

$$x = \frac{4 \pm 10}{6}$$

$$x = \frac{4+10}{6} = \frac{14}{6} = \frac{7}{3} = 2.33 \quad \text{or} \quad x = \frac{4-10}{6} = \frac{-6}{6} = -1.$$

Therefore, we can write the correct answer to one decimal place:

$$3x^2 - 4x - 7 \le 0$$

$$(x + 1)(x - 2.3) \le 0$$

The table sign is:

x values	$x < -1$	$x = -1$	$-1 < x < 2.3$	$x = 2.3$	$x > 2.3$
$x + 1$	+	0	−	−	−
$x - 2.3$	+	+	+	0	−
$y = (x + 1)(x - 2.3)$	+	0	−	0	+

The graph of $y = 3x^2 - 4x - 7$ is.

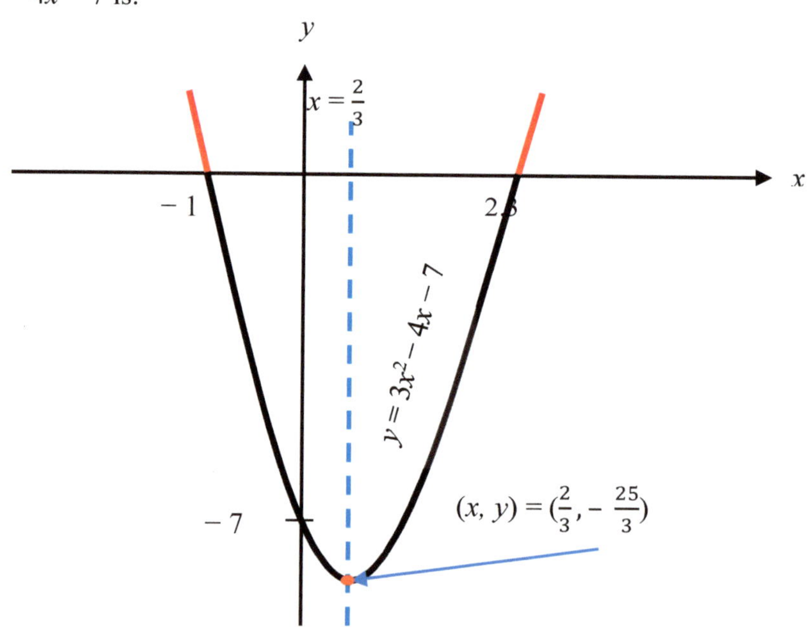

Referring to the graph, it is seen for $x < -1$, the value of inequality is positive $y > 0$, for $-1 < x < 2.3$ the value of inequality is negative $y < 0$, and for $x > 2.3$, the value of inequality is positive $y > 0$.

Therefore, the solution of $3x^2 - 4x - 7 \le 0$ is $-1 < x < 2.3$.

To represent the final answers on a number line we have:

Key Concepts

- How to Solve a Quadratic Inequality Graphically

  1. Write the quadratic inequality in standard form.
  2. Graph the function $f(x)=ax2+bx+c$ using properties or transformations.
  3. Determine the solution from the graph.

- How to Solve a Quadratic Inequality Algebraically

  1. Write the quadratic inequality in standard form.
  2. Determine the critical points, the solutions to the related quadratic equation.
  3. Use the critical points to divide the number line into intervals.
  4. Above the number line show the sign of each quadratic expression using test points from each interval substituted into the original inequality.
  5. Determine the intervals where the inequality is correct. Write the solution in set interval notation/number line.

3.5. Exercises

1. Use the factoring method to solve each of the following quadratic equation.
   a) $x^2 + 7x + 12 = 0$.
   b) $x^2 - 5x - 24 = 0$.
   c) $5x^2 - 18x - 8 = 0$.
   d) $3x^2 + 8x = 16$.

2. Solve each of the quadratic equations using the square root.
   a) $x^2 - 16 = 0$.
   b) $x^2 = 64$.
   c) $(5 + h)^2 - 36 = 0$.
   d) $(5 + k)^2 - 81k^2 = 0$.
   e) $16(x - 7)^2 - 225 = 0$.
   f) $(x + 1)^2 = 36$.
   g) $(3x - 6)^2 = 121$.

3. Solve each of the quadratic equations.
   a) $u^2 - 6u + 9 = 0$.
   b) $3l^2 + 12l - 25 = 0$.

4. Solve each of the quadratic equations.
   a) $x^2 - 8x + 16 = 0$.
   b) $x^2 - 6x + 9 = 0$.
   c) $x^2 + 2x = -1$.
   d) $5x^2 + 3x - 14 = 0$.

5. For the following exercises, determine the discriminant, and then state how many solutions there are and the nature of the solutions. Do not solve.
   a) $x^2 + 3x - 6 = 0$.
   b) $4x^2 - 2x - 4 = 0$.
   c) $x^2 - 12x - 2 = 0$.
   d) $2x^2 + 5x - 3 = 0$.
   e) $6x^2 - 2x + 7 = 0$.
   f) $x^2 - 4x + 4 = 0$.

6. Solve each of the quadratic equations using the quadratic formula.
   a) $x^2 - 2x - 1 = 0$.
   b) $2t^2 + 4t - 3 = 0$.
   c) $3w^2 - w - 8 = 0$.

d) $6k^2 - 2k = 4$.

e) $5x^2 = -4x + 3$.

f) $2x^2 + x - 7 = 0$.

g) $7u^2 + 9u + 24 = 0$.

h) $3y^2 - 12y + 22 = 0$.

i) $h^2 + 8h = -16$.

7. Solve each of the quadratic equations.
   a) $3(x + 2)^2 - 10(x + 2) + 8 = 0$.

   b) $7(\frac{4}{5}s + 6)^2 - 11(\frac{4}{5}s + 6) + 4 = 0$.

   c) $5(\frac{4h-3}{7})^2 - 16(\frac{4h-3}{7}) + 3 = 0$.

8. Given $y = 9(x - 5)^2 + 12$, find the x-coordinate and the y-coordinate of the vertex and write the equation of the line of symmetry.

9. Given $y = 8(x + 13)^2 - 16$, find the x-coordinate and the y-coordinate of the vertex and write the equation of the line of symmetry.

10. Given the quadratic function $f(x) = -4x^2 - 8x + 7$, determine:
    a) If it is an open upward or an open downward?

    b) Sketch the graph of the function.

    c) Find the minimum/maximum point of the function.

11. Given the quadratic function $f(x) = x^2 - 10x + 16$, determine:
    a) If it is an open upward or an open downward?

    b) Sketch the graph of the function.

    c) Find the minimum/maximum point of the function.

12. Rewrite the standard quadratic equation $y = 5x^2 + 4x - 12$ as the factored form.

13. Rewrite the factored quadratic equation $y = 7(k - 3)(k + 8)$ as the standard form.

14. Rewrite the factored quadratic equation $y = 2(x + 6)(x - 4)$ as the vertex form.

15. Rewrite the standard quadratic equation $y = 12t^2 + 6t - 5$ as the vertex form.

16. Rewrite the vertex quadratic equation $y = 6(x + 4)^2 - 13$ as the standard form.

17. Rewrite the vertex quadratic equation $y = 9(x - 2)^2 + 8$ as the factored form.

18. Given the standard form of a quadratic equation $y = x^2 + 7x - 18$, with points P and Q on the graph of this quadratic equation, and points P and Q are symmetric with respect to its line of symmetry. Find:
    a) The equation for the line of symmetry of its graph.

b) If point $P = (3, 12)$ find the coordinates of Q.

c) If point $P = (2, 0)$ find the coordinates of Q.

19. Solve each quadratic inequality.

a) $x^2 - 10x \geq -16$.

b) $y^2 + 12y \leq 45$.

c) $5z^2 - 5z > 10$.

20. Solve each quadratic inequality.

a) $6(x + 7)^2 - 14 \leq 0$.

b) $-8(x - 3)^2 - 12 \leq 0$.

c) $-2(y - 1)(y - 9) \geq 0$.

4. Polynomial Equations and Inequalities of Degree $n > 2$

Up to now, we have looked at polynomial equations and inequalities with degree of less than or equal to two. Here we will consider polynomial equations and inequalities with degree greater than two, $n > 2$. First, we look at the polynomial equations.

4.1. Polynomial Equation of Degree $n > 2$

Looking at polynomial equations in one variable, the variable of polynomial equations is called unknown, and the solution(s) is(are) the possible value(s) of the unknown for which the equality is true. Consider a polynomial, P, given below:

$$P = a_n x^n + a_{n-1} x^{n-1} + \cdots + a_i x^i + \cdots + a_2x^2 + a_1x + a_0.$$

Then letting $P = 0$ is the polynomial equation and can be written as.

$$a_n x^n + a_{n-1} x^{n-1} + \cdots + a_i x^i + \cdots + a_2 x^2 + a_1 x + a_0 = 0.$$

A number a is deemed a root of a polynomial P if and only if the linear polynomial $x - a$ divides P.

In other words, if there exists another polynomial Q such that $P =(x - a)Q$, then a qualifies as a root of the polynomial equation.

4.1.1. Solving Polynomial Equations, $n > 2$

Consider a polynomial equation represented by $P = 0$, where P is expressed as $a_n x^n + a_{n-1} x^{n-1} + \cdots + a_1x + a_0 = 0$. The solutions of this equation are known as the **roots** of the polynomial equation.

A number a is a root of a polynomial P if and only if the linear polynomial $x - a$ divides P. This means if there is another polynomial Q such that $P = (x - a)\,Q$, then a is a root of the polynomial equation.

It's possible for $x - a$ to divide P more than once: if $(x - a)^2$ divides P, then a is termed a multiple root of P; otherwise, a is labeled as a simple root of P. When P is a nonzero polynomial, there exists a highest power k such that $(x - a)^k$ divides P; this is referred to as the multiplicity of the root a in P. However, in the case of the zero polynomial, the corresponding polynomial equation is trivial. Thus, this case is typically excluded when considering roots, as every number is a root of the zero polynomial, with an undefined multiplicity. With this exception made, the total count of roots of P.

Solving Polynomial Equations

A polynomial of degree n is an expression of the type,

$$a_n x^n + a_{n-1} x^{n-1} + \cdots + a_i x^i + \cdots + a_2 x^2 + a_1 x + a_0,$$

where n is a positive integer and $a_n, a_{n-1}, \cdots, a_0$ are real numbers and $a_n \neq 0$.

Setting the polynomial equal to zero gives a polynomial equation. The total number of solutions (real and complex) to a polynomial equation is equal to the highest exponent n.

The three most popular methods for solving a polynomial equation are solving by factoring, solving by substitution factoring, and solving by grouping.

Let us first look at the procedure for solving a polynomial with degree of greater than two, $n > 2$, using factoring method.

4.1.2. Solving Polynomial Equations by Factoring, $n > 2$

Solving polynomial equations by factoring requires performing steps below.

Step 1. Write the equation in the standard form, preferably with its terms are in descending order.

Step 2. Use one of the factoring methods to factor the equation.

Step 3. Use the "zero-factor property" or "zero-factor principle" and set each factor containing a variable equal to zero (denoted as the factored equation).

Step 4. Solve each factored equation.

Step 5. Check the solution by substituting each solution of the factored equation separately into the original equation.

Example
Solve $3x^3 + 10x^2 - 8x = 0$.

Solution
We note that the equation is in the standard form, with the terms written in descending order. Hence, we need to apply factoring to factor the equation. So,

$3x^3 + 10x^2 - 8x = 0$

$x(3x^2 + 10x - 8) = 0$ we factored out x.

Now, considering the expression $3x^2 + 10x - 8$, we use the factoring polynomial method with leading coefficient, $a \neq 1$. So,

$3x^2 + 10x - 8$

$3x^2 + 10x - 8 = (x + 4)(3x - 2)$ we applied the factoring polynomial method.

Therefore, the equation $3x^3 + 10x^2 - 8x = 0$, is rewritten using the results of the above factorings. So,

$3x^3 + 10x^2 - 8x = 0$

$x(x + 4)(3x - 2) = 0$.

Using the zero-factor property, we set each factor equal to zero to get the factored equations. So,

$x(x + 4)(3x - 2) = 0$.

$x = 0$ or $(x + 4) = 0$ or $(3x - 2) = 0$

$x = 0$ or $x = -4$ or $3x = 2$

$x = 0$ or $x = -4$ or $x = \frac{2}{3}$.

Therefore, the solutions of the equation $3x^3 + 10x^2 - 8x = 0$, are $x = 0$, $x = -4$ and $x = \frac{2}{3}$.

Check: To verify our answer, we need to substitute each solution of the factored equation separately into the original equation to see if the left-hand side of the equation is also 0. So,

$3x^3 + 10x^2 - 8x = 0$.

For $x = 0$, we will have:

$3(0)^3 + 10(0)^2 - 8(0) \overset{?}{\Leftrightarrow} 0$ we substituted $x = 0$ into the equation

$0 + 0 - 0 \overset{?}{\Leftrightarrow} 0$

$0 = 0$ true $x = 0$ is a correct answer.

For $x = -4$, we will have:

$3(-4)^3 + 10(-4)^2 - 8(-4) \overset{?}{\Leftrightarrow} 0$ we substituted $x = -4$ into the equation

$3(-64) + 10(16) - 8(-4) \overset{?}{\Leftrightarrow} 0$

$-192 + 160 + 32 \overset{?}{\Leftrightarrow} 0$

$0 = 0$ true $x = -4$ is a correct answer.

For $x = \frac{2}{3}$, we will have:

$3(\frac{2}{3})^3 + 10(\frac{2}{3})^2 - 8(\frac{2}{3}) \overset{?}{\Leftrightarrow} 0$ we substituted $x = \frac{2}{3}$ into the equation

$3(\frac{8}{27}) + 10(\frac{4}{9}) - 8(\frac{2}{3}) \overset{?}{\Leftrightarrow} 0$

$\frac{24}{27} + \frac{40}{9} - \frac{16}{3} \overset{?}{\Leftrightarrow} 0$

$\frac{24}{27} + (\frac{40}{9} \times \frac{3}{3}) - (\frac{16}{3} \times \frac{9}{9}) \overset{?}{\Leftrightarrow} 0$ for having a common denominator, we multiplied the second and the third terms of the left side by $\frac{3}{3}$ and $\frac{9}{9}$ respectively

$\frac{24}{27} + \frac{120}{27} - \frac{144}{27} \overset{?}{\Leftrightarrow} 0$

$\frac{24 + 120 - 144}{27} \overset{?}{\Leftrightarrow} 0$

$\frac{0}{27} \overset{?}{\Leftrightarrow} 0$

$0 = 0$ true $x = \frac{2}{3}$ is a correct answer.

Example

Solve $6x^5 + 7x^4 - 26x^3 - 17x^2 + 30x = 0$.

Solution

We note that the equation is in the standard form, with the terms written in descending order. Hence, we need to apply factoring to factor the equation. So,

$6x^5 + 7x^4 - 26x^3 - 17x^2 + 30x = 0$

$x(6x^4 + 7x^3 - 26x^2 - 17x + 30) = 0$ we factored out x.

Now, considering the expression $6x^4 + 7x^3 - 26x^2 - 17x + 30$, we use the factoring polynomial method to factor the expression. So,

$6x^4 + 7x^3 - 26x^2 - 17x + 30$

$= (x - 1)(x + 2)(2x - 3)(3x + 5)$ we applied the factoring polynomial method.

Therefore, the equation $6x^5 + 7x^4 - 26x^3 - 17x^2 + 30x = 0$, is rewritten using the results of the above factorings. So,

$$6x^5 + 7x^4 - 26x^3 - 17x^2 + 30x = 0$$

$$x(x - 1)(x + 2)(2x - 3)(3x + 5) = 0.$$

Using the zero-factor property, we set each factor equal to zero to get the factored equations. So,

$$x(x - 1)(x + 2)(2x - 3)(3x + 5) = 0.$$

$x = 0$ or $(x - 1) = 0$ or $(x + 2) = 0$ or $(2x - 3) = 0$ or $(3x + 5) = 0$

$x = 0$ or $x = 1$ or $x = -2$ or $x = \frac{3}{2}$ or $x = -\frac{5}{3}$.

Therefore, the solutions of the equation $6x^5 + 7x^4 - 26x^3 - 17x^2 + 30x = 0$, are $x = 0$, $x = 1$, $x = -2$, $x = \frac{3}{2}$, and $x = -\frac{5}{3}$.

Check: To verify our answer, we need to substitute each solution of the factored equation separately into the original equation to see if the left-hand side of the equation is also 0. So,

$$6x^5 + 7x^4 - 26x^3 - 17x^2 + 30x = 0.$$

For $x = 0$, we will have:

$6(0)^5 + 7(0)^4 - 26(0)^3 - 17(0)^2 + 30(0) \overset{?}{\Leftrightarrow} 0$ we substituted $x = 0$ into the equation.

$0 + 0 - 0 - 0 + 0 \overset{?}{\Leftrightarrow} 0$

$0 = 0$ true $x = 0$ is a correct answer.

For $x = 1$, we will have:

$6(1)^5 + 7(1)^4 - 26(1)^3 - 17(1)^2 + 30(1) \overset{?}{\Leftrightarrow} 0$ we substituted $x = 1$ into the equation

$6 + 7 - 26 - 17 + 30 \overset{?}{\Leftrightarrow} 0$

$43 - 43 \overset{?}{\Leftrightarrow} 0$

$0 = 0$ true $x = 1$ is a correct answer.

For $x = -2$, we will have:

$6(-2)^5 + 7(-2)^4 - 26(-2)^3 - 17(-2)^2 + 30(-2) \overset{?}{\Leftrightarrow} 0$ we substituted $x = -2$ into the equation.

$6(-32) + 7(16) - 26(-8) - 17(4) + 30(-2) \overset{?}{\Leftrightarrow} 0$

$-192 + 112 + 208 - 68 - 60 \overset{?}{\Leftrightarrow} 0$

$-320 + 320 \overset{?}{\Leftrightarrow} 0$

$0 = 0$ true $x = -2$ is a correct answer.

For $x = \frac{3}{2}$, we will have:

$$6(\tfrac{3}{2})^5 + 7(\tfrac{3}{2})^4 - 26(\tfrac{3}{2})^3 - 17(\tfrac{3}{2})^2 + 30(\tfrac{3}{2}) \overset{?}{\Leftrightarrow} 0$$

we substituted $x = \frac{3}{2}$ into the equation

$$6(\tfrac{243}{32}) + 7(\tfrac{81}{16}) - 26(\tfrac{27}{8}) - 17(\tfrac{9}{4}) + 30(\tfrac{3}{2}) \overset{?}{\Leftrightarrow} 0$$

$$\frac{1458}{32} + \frac{567}{16} - \frac{702}{8} - \frac{153}{4} + \frac{90}{2} \overset{?}{\Leftrightarrow} 0$$

$$\frac{1458}{32} + (\frac{567}{16} \times \frac{2}{2}) - (\frac{702}{8} \times \frac{4}{4}) - (\frac{153}{4} \times \frac{8}{8}) + (\frac{90}{2} \times \frac{16}{16}) \overset{?}{\Leftrightarrow} 0$$

for having a common denominator, we multiplied the second the third, the fourth, and fifth terms of the left side by $\frac{2}{2}$, $\frac{4}{4}$, $\frac{8}{8}$, and $\frac{16}{16}$ respectively

$$\frac{1458}{32} + \frac{1134}{32} - \frac{2808}{32} - \frac{1224}{32} + \frac{1440}{32} \overset{?}{\Leftrightarrow} 0$$

$$\frac{1458 + 1134 - 2808 - 1224 + 1440}{32} \overset{?}{\Leftrightarrow} 0$$

$$\frac{4032 - 4032}{3125} \overset{?}{\Leftrightarrow} 0$$

$$0 = 0$$

true $x = \frac{3}{2}$ is a correct answer.

For $x = -\frac{5}{3}$, we will have:

$$6(-\tfrac{5}{3})^5 + 7(-\tfrac{5}{3})^4 - 26(-\tfrac{5}{3})^3 - 17(-\tfrac{5}{3})^2 + 30(-\tfrac{5}{3}) \overset{?}{\Leftrightarrow} 0$$

we substituted $x = -\frac{5}{3}$ into the equation.

$$6(-\tfrac{3125}{243}) + 7(\tfrac{625}{81}) - 26(-\tfrac{125}{27}) - 17(\tfrac{25}{9}) + 30(-\tfrac{5}{3}) \overset{?}{\Leftrightarrow} 0$$

$$-\frac{18750}{243} + \frac{4375}{81} + \frac{3250}{27} - \frac{425}{9} - \frac{150}{3} \overset{?}{\Leftrightarrow} 0$$

$$-\frac{18750}{243} + (\frac{4375}{81} \times \frac{3}{3}) + (\frac{3250}{27} \times \frac{9}{9}) - (\frac{425}{9} \times \frac{27}{27}) - (\frac{150}{3} \times \frac{81}{81}) \overset{?}{\Leftrightarrow} 0$$

for having a common denominator, we multiplied the second the third, the fourth, and fifth terms of the left side by $\frac{3}{3}$, $\frac{9}{9}$, $\frac{27}{27}$, and $\frac{81}{81}$ respectively

$$-\frac{18750}{243} + \frac{13125}{243} + \frac{29250}{243} - \frac{11475}{243} - \frac{12150}{243} \overset{?}{\Leftrightarrow} 0$$

$$\frac{-18750 + 13125 + 29250 - 11475 - 12150}{243} \overset{?}{\Leftrightarrow} 0$$

$$\frac{42375 - 42375}{243} \overset{?}{\Leftrightarrow} 0$$

$$0 = 0$$

true $x = -\frac{5}{3}$ is a correct answer.

4.1.3. Solving Polynomial Equations by Substitution, $n > 2$

We can solve polynomial equations by simplifying the equation using a substitution. When polynomials are particularly complicated, it is easiest to substitute a simple term and factor down.

Example

Solve the polynomial equation by substitution factoring $5x^4 - 14x^2 + 8 = 0$.

Solution

Using the substitution method, we let $y = x^2$. We then have.

$5x^4 - 14x^2 + 8 = 0$

$5x^{2^2} - 14x^2 + 8 = 0$

$5y^2 - 14y + 8 = 0$ — we substituted $y = x^2$ into the equation.

We now have a trinomial with leading coefficient not 1. The procedure to solve this trinomial is to compare it with the standard form of a trinomial equation, $ax^2 + bx + c = 0$. By this comparison we get $a = 5$, $b = -14$,.and $c = 8$. Then, we get $ac = 40$ and $b = -14$. This implies that we need to look for two numbers p and q which their product is 40 and their sum is – 14. Hence, we list all factors of $ac = 40$ until a pair of the anticipated sum $b = -14$ is attained.

Factors of ac =40	Sum of factors
1, 40	41
–1, – 40	– 41
2, 20	22
– 2, – 20	– 22
4, 10	14
– 4, – 10	**– 14**
5, 8	13
– 5, – 8	– 13

So, $p = -10$ and $q = -4$. Rewrite the original expression as $ax^2 + px + qx + c$.

$5y^2 - 14y + 8 = 0.$

$5y^2 - 10y - 4y + 8 = 0$

$5y(y - 2) - 4(y - 2) = 0$ — we factored out the GCF of each part

$(y - 2)(5y - 4) = 0$ — again we factored out the GCF of the expression.

Using the zero-factor principle we will have:

$y - 2 = 0$

$y - 2 + 2 = 0 + 2$ — we added 2 to both sides

$y = 2$ — one of the answers

$5y - 4 = 0$

$5y - 4 + 4 = 0 + 4$ we added 4 to both sides

$5y = 4$

$\frac{1}{5} \times 5y = \frac{1}{5} \times 4$ we multiplied both sides by the reciprocal of the variable

$y = \frac{1}{5} \times 4$

$y = \frac{4}{5}$ the other answer.

We now need to find the solutions in in term of x. So,

For $y = 2$:

$y = x^2$

$2 = x^2$ we substituted $y = 2$ into the equation

$x = \pm\sqrt{2}$ we took square root of both sides.

For $y = \frac{4}{5}$:

$y = x^2$

$x^2 = \frac{4}{5}$ we substituted $y = \frac{4}{5}$ into the equation

$x = \pm\sqrt{\frac{4}{5}}$ we took square root of both sides

$x = \pm\sqrt{\frac{2^2}{5}}$

$x = \pm 2\sqrt{\frac{1}{5}}$ we factored the perfect squares from radicands

$x = \pm\frac{2}{\sqrt{5}}$

$x = \pm\frac{2}{\sqrt{5}} \times \frac{\sqrt{5}}{\sqrt{5}}$ we multiplied the left-side by $\frac{\sqrt{5}}{\sqrt{5}}$

$x = \pm\frac{2\sqrt{5}}{\sqrt{5^2}}$

$x = \pm\frac{2\sqrt{5}}{5}$ we factored the perfect squares from radicands.

Therefore, the solutions of $5x^4 - 14x^2 + 8 = 0$, are $x = \sqrt{2}$, $x = -\sqrt{2}$, $x = \frac{2\sqrt{5}}{5}$, and $x = -\frac{2\sqrt{5}}{5}$.

Check: To verify our answer, we need to plug each solution into the equation. So,

For $x = \sqrt{2}$:

$5x^4 - 14x^2 + 8 = 0$

$5(\sqrt{2})^4 - 14(\sqrt{2})^2 + 8 \overset{?}{\Leftrightarrow} 0$ we plugged $x = \sqrt{2}$ into the equation

$5(\sqrt{2^4}) - 14(\sqrt{2^2}) + 8 \overset{?}{\Leftrightarrow} 0$

$5(2^2) - 14(2) + 8 \overset{?}{\Leftrightarrow} 0$ — we factored the perfect squares from radicands

$5(4) - 14(2) + 8 \overset{?}{\Leftrightarrow} 0$

$20 - 28 + 8 \overset{?}{\Leftrightarrow} 0$

$28 - 28 \overset{?}{\Leftrightarrow} 0$

$0 = 0$ — true, the solution $x = \sqrt{2}$ is a correct answer.

For $x = -\sqrt{2}$:

$5x^4 - 14x^2 + 8 = 0$

$5(-\sqrt{2})^4 - 14(-\sqrt{2})^2 + 8 \overset{?}{\Leftrightarrow} 0$ — we plugged $x = -\sqrt{2}$ into the equation

$5(\sqrt{2^4}) - 14(\sqrt{2^2}) + 8 \overset{?}{\Leftrightarrow} 0$

$5(2^2) - 14(2) + 8 \overset{?}{\Leftrightarrow} 0$ — we factored the perfect squares from radicands

$5(4) - 14(2) + 8 \overset{?}{\Leftrightarrow} 0$

$20 - 28 + 8 \overset{?}{\Leftrightarrow} 0$

$28 - 28 \overset{?}{\Leftrightarrow} 0$

$0 = 0$ — true, the solution $x = -\sqrt{2}$ is a correct answer.

For $x = \frac{2\sqrt{5}}{5}$:

$5x^4 - 14x^2 + 8 = 0$

$5(\frac{2\sqrt{5}}{5})^4 - 14(\frac{2\sqrt{5}}{5})^2 + 8 \overset{?}{\Leftrightarrow} 0$ — we plugged $x = \frac{2\sqrt{5}}{5}$ into the equation

$5(\frac{2^4\sqrt{5^4}}{5^4}) - 14(\frac{2^2\sqrt{5^2}}{5^2}) + 8 \overset{?}{\Leftrightarrow} 0$

$5(\frac{16(5^2)}{5^{2^2}}) - 14(\frac{4(5)}{5^2}) + 8 \overset{?}{\Leftrightarrow} 0$

$5(\frac{16}{5^2}) - 14(\frac{20}{5^2}) + 8 \overset{?}{\Leftrightarrow} 0$

$\frac{80}{25} - \frac{280}{25} + 8 \overset{?}{\Leftrightarrow} 0$

$\frac{80}{25} - \frac{280}{25} + 8 \times \frac{25}{25} \overset{?}{\Leftrightarrow} 0$ — to get a common denominator, we multiplied the last term by $\frac{25}{25}$

$\frac{80}{25} - \frac{280}{25} + \frac{200}{25} \overset{?}{\Leftrightarrow} 0$

$\frac{80-280+200}{25} \overset{?}{\Leftrightarrow} 0$

$\frac{280-280}{25} \overset{?}{\Leftrightarrow} 0$

$\frac{0}{25} \overset{?}{\Leftrightarrow} 0$

$0 = 0$ true, the solution $x = \frac{2\sqrt{5}}{5}$ is a correct answer.

For $x = -\frac{2\sqrt{5}}{5}$:

$5x^4 - 14x^2 + 8 = 0$

$5(-\frac{2\sqrt{5}}{5})^4 - 14(-\frac{2\sqrt{5}}{5})^2 + 8 \overset{?}{\Leftrightarrow} 0$ we plugged $x = -\frac{2\sqrt{5}}{5}$ into the equation

$5(\frac{2^4\sqrt{5^4}}{5^4}) - 14(\frac{2^2\sqrt{5^2}}{5^2}) + 8 \overset{?}{\Leftrightarrow} 0$

$5(\frac{16(5^2)}{5^{2^2}}) - 14(\frac{4(5)}{5^2}) + 8 \overset{?}{\Leftrightarrow} 0$

$5(\frac{16}{5^2}) - 14(\frac{20}{5^2}) + 8 \overset{?}{\Leftrightarrow} 0$

$\frac{80}{25} - \frac{280}{25} + 8 \overset{?}{\Leftrightarrow} 0$

$\frac{80}{25} - \frac{280}{25} + 8 \times \frac{25}{25} \overset{?}{\Leftrightarrow} 0$ to get a common denominator, we multiplied the last term by $\frac{25}{25}$

$\frac{80}{25} - \frac{280}{25} + \frac{200}{25} \overset{?}{\Leftrightarrow} 0$

$\frac{80-280+200}{25} \overset{?}{\Leftrightarrow} 0$

$\frac{280-280}{25} \overset{?}{\Leftrightarrow} 0$

$\frac{0}{25} \overset{?}{\Leftrightarrow} 0$

$0 = 0$ true, the solution $x = -\frac{2\sqrt{5}}{5}$ is a correct answer.

4.1.4. Solving Polynomial Equations by Factor Grouping, $n > 2$

When an expression has an even number of terms and there are no common factors for all the terms, we may group the terms into pairs and find the common factor for each pair.

Example

Solve polynomial equations by factor grouping $5x^5 - 10x^4 - 20x^3 + 40x^2 + 15x - 30 = 0$.

Solution

This polynomial consists of 6 terms, which we can solve by grouping. We need to group the first two terms, the second two terms, and the third two terms because by doing this we get factors in the parentheses are identical. So,

$$5x^5 - 10x^4 - 20x^3 + 40x^2 + 15x - 30 = 0$$

$$5x^4(x - 2) - 20x^2(x - 2) + 15(x - 2) = 0.$$

We can now factor out $(x - 2)$ from the three terms. So,

$$(x - 2)(5x^4 - 20x^2 + 15) = 0.$$

We need further to perform some factoring methods for factoring $5x^4 - 20x^2 + 15$. We first use the substitution method by letting $y = x^2$. So,

$$5x^4 - 20x^2 + 15$$

$$5y^2 - 20y + 15.$$

For factoring $5y^2 - 20y + 15$ by the leading coefficient is not 1, we compare $5x^4 - 20x^2 +15$ with $ax^2 + bx + c$, we get $a = 5$, $b = - 20$, and $c = 15$. First, we determine $ac = 75$. Next, we need to find two numbers, p and q, with a product of 75 and a sum of $- 20$.

In the following table, we list all factors until a pair of the anticipated sum is attained:

Factors of 75	Sum of factors
1, 75	76
−1, − 75	− 76
3, 25	28
− 3, − 25	− 28
5, 15	20
− 5, − 15	**− 20**

So, $p =$ 5 and $q = - 15$.

Rewrite the original expression as $ax^2 + px + qx + c$.

$$5y^2 - 20y + 15$$

$$5y^2 - 5y - 15y + 15$$

$5y(y - 1) - 15(y - 1)$ we factored out the GCF of each part

$(y - 1)(5y - 15)$ again we factored out the GCF of the expression.

We need to rewrite the products in term of x. So,

$$(y - 1)(5y - 15)$$

$(x^2 - 1)(5x^2 - 15)$ we substituted $y = x^2$.

The original equation $5x^5 - 10x^4 - 20x^3 + 40x^2 + 15x - 30 = 0$ can be written as.

$$5x^5 - 10x^4 - 20x^3 + 40x^2 + 15x - 30 = 0$$

$$(x - 2)(x^2 - 1)(5x^2 - 15) = 0.$$

Note that $x^2 - 1$ is the difference of squares and can be written as $(x - 1)(x + 2)$. So, the original equation $5x^5 - 10x^4 - 20x^3 + 40x^2 + 15x - 30 = 0$ can be factored as.

$$5x^5 - 10x^4 - 20x^3 + 40x^2 + 15x - 30 = 0$$

$$(x - 2)\,(x - 1)(x + 1)(5x^2 - 15) = 0.$$

Using the zero-factor property we set each factor equal to zero. So,

$$(x - 2)\,(x - 1)(x + 1)(5x^2 - 15) = 0$$

$(x - 2) = 0$ or $(x - 1) = 0$ or $(x + 1) = 0$ or $(5x^2 - 15) = 0$

$x = 2$ or $x = 1$ or $x = -1$ or $x = \pm\sqrt{3}$.

Therefore, the solutions for the polynomial equation, $5x^5 - 10x^4 - 20x^3 + 40x^2 + 15x - 30 = 0$, are $x = 2$, $x = 1$, $x = -1$, $x = \sqrt{3}$ and $x = -\sqrt{3}$.

Check: To verify our answer, we need to substitute each solution of the factored equation separately into the original equation to see if the left-hand side of the equation is also 0. So,

For $x = 2$, we will have:

$$5x^5 - 10x^4 - 20x^3 + 40x^2 + 15x - 30 = 0$$

$5(2)^5 - 10(2)^4 - 20(2)^3 + 40(2)^2 + 15(2) - 30 \overset{?}{\Leftrightarrow} 0$ we substituted $x = 2$ into the equation

$$5(32) - 10(16) - 20(8) + 40(4) + 15(2) - 30 \overset{?}{\Leftrightarrow} 0$$

$$160 - 160 - 160 + 160 + 30 - 30 \overset{?}{\Leftrightarrow} 0$$

$$0 + 0 + 0 \overset{?}{\Leftrightarrow} 0$$

$0 = 0$ true $x = 2$ is a correct answer.

For $x = 1$, we will have:

$$5x^5 - 10x^4 - 20x^3 + 40x^2 + 15x - 30 \overset{?}{\Leftrightarrow} 0$$

$5(1)^5 - 10(1)^4 - 20(1)^3 + 40(1)^2 + 15(1) - 30 \overset{?}{\Leftrightarrow} 0$ we substituted $x = 1$ into the equation

$$5(1) - 10(1) - 20(1) + 40(1) + 15(1) - 30 \overset{?}{\Leftrightarrow} 0$$

$$5 - 10 - 20 + 40 + 15 - 30 \overset{?}{\Leftrightarrow} 0$$

$$60 - 60 \overset{?}{\Leftrightarrow} 0$$

$0 = 0$ true $x = 1$ is a correct answer.

For $x = -1$, we will have:

$$5x^5 - 10x^4 - 20x^3 + 40x^2 + 15x - 30 \overset{?}{\Leftrightarrow} 0$$

$5(-1)^5 - 10(-1)^4 - 20(-1)^3 + 40(-1)^2 + 15(-1) - 30 \overset{?}{\Leftrightarrow} 0$ we substituted $x = -1$ into the equation

$5(-1) - 10(1) - 20(-1) + 40(1) + 15(-1) - 30 \overset{?}{\Leftrightarrow} 0$

$-5 - 10 + 20 + 40 - 15 - 30 \overset{?}{\Leftrightarrow} 0$

$-60 + 60 \overset{?}{\Leftrightarrow} 0$

$0 = 0$ true $x = 1$ is a correct answer.

For $x = \sqrt{3}$, we will have:

$5x^5 - 10x^4 - 20x^3 + 40x^2 + 15x - 30 \overset{?}{\Leftrightarrow} 0$

$5(\sqrt{3})^5 - 10(\sqrt{3})^4 - 20(\sqrt{3})^3 + 40(\sqrt{3})^2 + 15(\sqrt{3}) - 30 \overset{?}{\Leftrightarrow} 0$ we substituted $x = \sqrt{3}$ into the equation

$5(9\sqrt{3}) - 10(9) - 20(3\sqrt{3}) + 40(3) + 15(\sqrt{3}) - 30 \overset{?}{\Leftrightarrow} 0$

$45\sqrt{3} - 90 - 60\sqrt{3} + 120 + 15\sqrt{3} - 30 \overset{?}{\Leftrightarrow} 0$

$60\sqrt{3} - 60\sqrt{3} - 120 + 120 \overset{?}{\Leftrightarrow} 0$

$0 + 0 \overset{?}{\Leftrightarrow} 0$

$0 = 0$ true $x = \sqrt{3}$ is a correct answer.

For $x = -\sqrt{3}$, we will have:

$5x^5 - 10x^4 - 20x^3 + 40x^2 + 15x - 30 \overset{?}{\Leftrightarrow} 0$

$5(-\sqrt{3})^5 - 10(-\sqrt{3})^4 - 20(-\sqrt{3})^3 + 40(-\sqrt{3})^2 + 15(-\sqrt{3}) - 30 \overset{?}{\Leftrightarrow} 0$ we substituted $x = -\sqrt{3}$ into the equation

$5(-9\sqrt{3}) - 10(9) - 20(-3\sqrt{3}) + 40(3) + 15(-\sqrt{3}) - 30 \overset{?}{\Leftrightarrow} 0$

$-45\sqrt{3} - 90 + 60\sqrt{3} + 120 - 15\sqrt{3} - 30 \overset{?}{\Leftrightarrow} 0$

$-60\sqrt{3} - 60\sqrt{3} - 120 + 120 \overset{?}{\Leftrightarrow} 0$

$0 + 0 \overset{?}{\Leftrightarrow} 0$

$0 = 0$ true $x = -\sqrt{3}$ is a correct answer.

4.2. Polynomial Inequalities of Degree $n > 2$

Let's now explore solving inequalities of degree higher than two, $n > 2$. Polynomial inequality differs from a polynomial equation in that the equality sign is replaced by one of the inequality symbols. To solve such an inequality, the first step involves finding the roots of the polynomial equation.

Like linear and quadratic inequalities, the objective is to identify all values of the variable that satisfy the inequality. The roots of the polynomial equation divide the x-axis into intervals where the polynomial's sign alternates between positive and negative. Within each interval, the sign of the polynomial can be determined using test values.

Finally, we determine the intervals that satisfy the inequality and express the solution set using interval notation.

The steps below present the process of solving polynomial inequalities algebraically.

Step 1. Get zero on one side of the inequality and factorize the polynomial inequality.

Step 2. Convert the inequality to an equality and determine the zeros, the key or critical x values, of the equality.

Step 3. Establish a table sign and a number line to determine where each factor of the inequality has positive value and negative value on the number line. The sign within each interval on the number line is determined by using test values.

Step 4. Represent the final answers on a number line and write the solution set interval notation.

The following examples will more clearly illustrate the above steps.

Example

Solve $x^4 - 2x^3 - 24x^2 \leq 0$.

Solution

We have already got zero on one side so we can go straight to factoring.

$$x^4 - 2x^3 - 24x^2 \leq 0$$

$$x^2(x^2 - 2x - 24) \leq 0$$

$$x^2(x - 6)(x + 4) \leq 0.$$

We will now convert inequality to equality. So,

$$x^2(x - 6)(x + 4) \leq 0$$

$$x^2(x - 6)(x + 4) = 0.$$

Using the zero-factor principle we will have:

$$x^2 = 0$$

$$x - 6 = 0$$

$$x + 4 = 0.$$

For $x^2 = 0$, we get:

$$x^2 = 0$$

$$x = 0.$$

For $x - 6 = 0$, we get:

$$x - 6 = 0$$

$$x - 6 + 6 = 0 + 6 \qquad \text{we added 6 to both sides}$$

$$x = 6.$$

For $x + 4 = 0$, we get:

$x + 4 - 4 = 0 - 4$ we added – 4 to both sides

$x = -4.$

Therefore, the solutions of $x^2(x-6)(x+4) = 0$ are $x = 0$, $x = 6$, and $x = -4$.

Before we go to the next step, let us verify our solutions.

Check: To verify our answer, we need to plug each solution into the equation. So,

For $x = 0$:

$x^4 - 2x^3 - 24x^2 = 0$

$0^4 - 2\times 0^3 - 24 \times 0^2 \overset{?}{\Leftrightarrow} 0$ we plugged $x = 0$ into the equation

$0 - 0 - 0 \overset{?}{\Leftrightarrow} 0$

$0 = 0$ true, the solution $x = 0$ is a correct answer.

For $x = 6$:

$x^4 - 2x^3 - 24x^2 = 0$

$6^4 - 2 \times 6^3 - 24 \times 6^2 \overset{?}{\Leftrightarrow} 0$ we plugged $x = 6$ into the equation

$1.296 - 432 - 864 \overset{?}{\Leftrightarrow} 0$

$1.296 - 1{,}296 \overset{?}{\Leftrightarrow} 0$

$0 = 0$ true, the solution $x = 6$ is a correct answer.

For $x = -4$:

$x^4 - 2x^3 - 24x^2 = 0$

$(-4)^4 - 2 \times (-4)^3 - 24 \times (-4)^2 \overset{?}{\Leftrightarrow} 0$ we plugged $x = 6$ into the equation

$256 + 128 - 384 \overset{?}{\Leftrightarrow} 0$

$384 - 384 \overset{?}{\Leftrightarrow} 0$

$0 = 0$ true, the solution $x =-4$ is a correct answer.

The table sign is:

x values	$x < -4$	$x = -4$	$-4 < x < 0$	$x = 0$	$0 < x < 6$	$x = 6$	$x > 6$
x^2	+	+	+	0	+	+	+
$x + 4$	–	0	+	+	+	+	+
$x - 6$	–	–	–	–	–	0	+
$y = x^4 - 2x^3 - 24x^2$	+	0	–	0	–	0	+

Let us select the test points $x = -5, x = -1, x = 1$ and $x = 7$. Here is the number line for this problem.

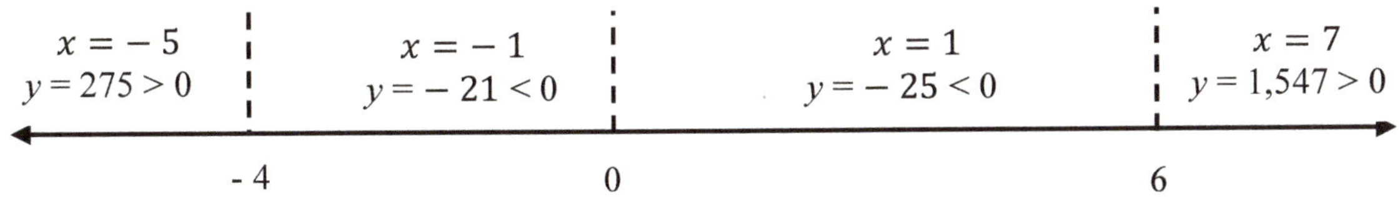

Based on the above information, the solution of $x^4 - 2x^3 - 24x^2 \leq 0$ is $-4 \leq x \leq 6$, or as solution set interval notation $[-4, 6]$. To represent the final answers on a number line we will have.

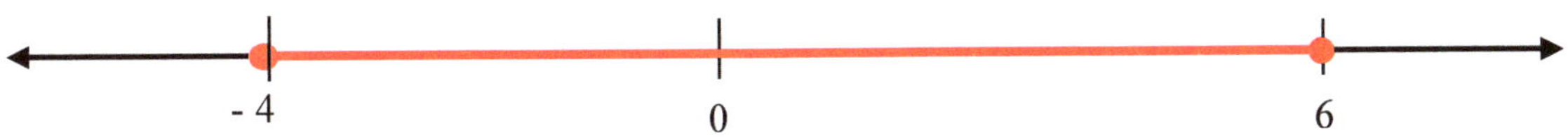

Example
Solve $x^4 + 3x^3 - 10x^2 \leq 0$.

Solution
It can be easily seen that we can factor x^2out. So,

$$x^4 + 3x^3 - 10x^2 \leq 0$$

$$x^2 (x^2 + 3x - 10) \leq 0.$$

The terms inside the parentheses can be rewritten as products of two linear terms. So,

$$x^2 (x + 5) (x - 2) \leq 0.$$

We will now need to find the zeroes of $x^2 (x + 5) (x - 2) = 0$. So,

$$x^2 (x + 5) (x - 2) = 0$$

$$x^2 = 0 \quad \text{or} \quad (x + 5) = 0 \quad \text{or} \quad (x - 2) = 0.$$

Therefore, this polynomial is zero at, $x = 0$, $x = -5$, and $x = 2$. Here is the table sign for this problem:

x values	$x < -5$	$x = -5$	$-5 < x < 0$	$x = 0$	$0 < x < 2$	$x = 2$	$x > 2$
x^2	+	+	+	0	+	+	+
$x + 5$	–	0	+		+	+	+
$x - 2$	–	–	–	–	–	0	+
$x^2 (x + 5) (x - 2)$	+	0	–	0	–	0	+

First, notice that unlike the previous examples the regions within zeroes do NOT alternate between positive and negative.

Now, for our solution we want regions where the polynomial will be negative. So, we can combine up the middle two regions and the three points into a single inequality in this case. The solution, in both inequality and interval notation form, is $-6 \leq x \leq 2$ or $[-6, 2]$.

Example
Solve $(x + 6)(x + 1)(x - 4) > 0$.

Solution
It is seen that polynomials are in the form of the product of three linear terms. Hence, to find its zeroes, we can easily let $(x + 6)(x + 1)(x - 4) = 0$. So,

$$(x + 6)(x + 1)(x - 4) = 0$$

$$x + 6 = 0 \qquad \text{or} \qquad x + 1 = 0 \qquad \text{or} \qquad x - 4 = 0.$$

Therefore, this polynomial is zero at, $x = -6$, $x = -1$, and $x = 4$. Here is the table sign for this problem:

x values	$x < -6$	$x = -6$	$-6 < x < -1$	$x = -1$	$-1 < x < 4$	$x = 4$	$x > 4$
$x + 6$	–	0	+	+	+	+	+
$x + 1$	–	–	–	0	+	+	+
$x - 4$	–	–	–	–	–	0	+
$(x + 6)(x + 1)(x - 4)$	–	0	+	0		0	+

Now, for our solution we want regions where the polynomial will be positive. The solution, in inequality and interval notation form, is:

$$-6 < x \leq -1 \text{ and } x > 4$$

$$(-6, -1) \text{ and } (4, \infty).$$

4.3. Solving Polynomial Equations Involving Rational Expressions

Solving a polynomial equation that contains rational expressions in nominator is like solving any equation that has fractions. We can solve these equations using the techniques for performing operations with rational expressions and for solving algebraic equations.

Solving a polynomial equation involving rational expressions in denominator, a method is to rewrite the rational expressions in terms of a common denominator. Then, again we solve these equations by performing operations with rational expressions and by performing algebraic operations for solving equations.

Let us present two examples of each polynomial equation, one with rational expressions in its nominator and one with rational expressions in its denominator.

Example
Solve the polynomial equation, $\frac{x^3 - 4x^2 + 2x}{2} + \frac{6x^3 + 4x^2 - 8x + 12}{4} = 3$.

Solution
We need to perform operations with rational expressions. So,

$$\frac{x^3 - 4x^2 + 2x}{2} + \frac{6x^3 + 4x^2 - 8x + 12}{4} = 3$$

$$\frac{x^3 - 4x^2 + 2x}{2} \times \frac{2}{2} + \frac{6x^3 + 4x^2 - 8x + 12}{4} = 3$$

to have a common denominator, we multiplied the first fraction by $\frac{2}{2}$

$$\frac{2(x^3 - 4x^2 + 2x)}{4} + \frac{6x^3 + 4x^2 - 8x + 12}{4} = 3$$

$$\frac{2x^3 - 8x^2 + 4x}{4} + \frac{6x^3 + 4x^2 - 8x + 12}{4} = 3$$

$$\frac{2x^3 - 8x^2 + 4x + 6x^3 + 4x^2 - 8x + 12}{4} = 3$$

$$\frac{8x^3 - 4x^2 - 4x}{4} + \frac{12}{4} = 3$$

$$\frac{8x^3 - 4x^2 - 4x}{4} + 3 = 3$$

$$\frac{8x^3 - 4x^2 - 4x}{4} + 3 - 3 = 3 - 3$$ we added − 3 to both sides

$$\frac{8x^3 - 4x^2 - 4x}{4} = 0$$

$$8x^3 - 4x^2 - 4x = 0.$$

We now need to solve the equation $8x^3 - 4x^2 - 4x = 0$.

$8x^3 - 4x^2 - 4x = 0$

$x(8x^2 - 4x - 4) = 0$ we factored x out.

Using the zero-factor principle, we will have:

$x(8x^2 - 4x - 4) = 0$

$x = 0$ or $8x^2 - 4x - 4 = 0$.

For solving $8x^2 - 4x - 4 = 0$, we use the factoring polynomial method with leading coefficient, $a \neq 1$. We compare $8x^2 - 4x - 4$ with $ax^2 + bx + c$, we get $a = 8$, $b = -4$, and $c = -4$. First, we determine $ac = 8(-4) = -32$. Next, we need to find two numbers, p and q, with a product of -32 and a sum of -4.

In the following table, we list all factors until a pair of the anticipated sum is attained:

Factors of $ac = -32$	Sum of factors
1, − 32	− 31
− 1, 32	31
2, − 16	− 14
−2, 16	14
4, − 8	**− 4**
− 4, 8	4

So, $p = 8$ and $q = 4$. Rewrite the original expression as $ax^2 + px + qx + c$.

$8x^2 - 4x - 4 = 0$

$8x^2 - 8x + 4x - 4 = 0$

$8x(x - 1) + 4(x - 1)$ — we factored out the GCF of each part.

$(x - 1)(8x + 4)$ — again we factored out the GCF of the expression.

Using the zero-factor property, we set each factor equal to zero. So,

$$(x - 1)(8x + 4) = 0.$$

For $(x - 1) = 0$:

$x - 1 = 0$

$x - 1 + 1 = 0 + 1$ — we added 1 to both sides

$x = 1.$

For $(8x + 4) = 0$:

$8x + 4 = 0$

$8x + 4 - 4 = 0 - 4$ — we added -4 to both sides

$8x = -4$

$\frac{1}{8} \times 8x = \frac{1}{8} \times (-4)$ — we multiplied both sides by the reciprocal of the variable

$x = \frac{-4}{8} = -\frac{1}{2}.$

Therefore, the solutions of the equation $\frac{x^3 - 4x^2 + 2x}{2} + \frac{6x^3 + 4x^2 - 8x + 12}{4} = 3$, are $x = 0$, $x = 1$ and $x = -\frac{1}{2}$.

Check: To verify our answer, we need to substitute each solution separately into the original equation to see if the left-hand side of the equation is also 0. So,

$$\frac{x^3 - 4x^2 + 2x}{2} + \frac{6x^3 + 4x^2 - 8x + 12}{4} = 3$$

$$\frac{x^3 - 4x^2 + 2x}{2} + \frac{6x^3 + 4x^2 - 8x + 12}{4} - 3 = 0.$$

For $x = 0$, we will have:

$$\frac{x^3 - 4x^2 + 2x}{2} + \frac{6x^3 + 4x^2 - 8x + 12}{4} - 3 = 0$$

$\frac{0^3 - 4 \times 0^2 + 2 \times 0}{2} + \frac{6 \times 0^3 + 4 \times 0^2 - 8 \times 0 + 12}{4} - 3 \overset{?}{\Leftrightarrow} 0$ — we substituted $x = 0$ into the equation

$0 + \frac{0 + 12}{4} - 3 \overset{?}{\Leftrightarrow} 0$

$0 + \frac{12}{4} - 3 \overset{?}{\Leftrightarrow} 0$

$0 + 3 - 3 \overset{?}{\Leftrightarrow} 0$

$0 = 0$ — true $x = 0$ is a correct answer.

For $x = 1$, we will have:

$$\frac{x^3 - 4x^2 + 2x}{2} + \frac{6x^3 + 4x^2 - 8x + 12}{4} - 3 = 0$$

$$\frac{1^3-(4\times 1^2)+(2\times 1)}{2}+\frac{(6\times 1^3)+(4\times 1^2)-(8\times 1)+12}{4}-3\overset{?}{\Leftrightarrow}0$$

we substituted $x = 1$ into the equation

$$\frac{1-4+2}{2}+\frac{6+4-8+12}{4}-3\overset{?}{\Leftrightarrow}0$$

$$\frac{-1}{2}+\frac{22-8}{4}-3\overset{?}{\Leftrightarrow}0$$

$$\frac{-1}{2}+\frac{14}{4}-3\overset{?}{\Leftrightarrow}0$$

$$\frac{-1}{2}+\frac{7}{2}-3\overset{?}{\Leftrightarrow}0$$

$$\frac{-1+7}{2}-3\overset{?}{\Leftrightarrow}0$$

$$\frac{6}{2}-3\overset{?}{\Leftrightarrow}0$$

$$3-3\overset{?}{\Leftrightarrow}0$$

$0 = 0$ true $x = 1$ is a correct answer.

For $x = -\frac{1}{2}$, , we will have:

$$\frac{x^3-4x^2+2x}{2}+\frac{6x^3+4x^2-8x+12}{4}-3=0$$

$$\frac{(-\frac{1}{2})^3-4\left(-\frac{1}{2}\right)^2+2(-\frac{1}{2})}{2}+\frac{6(-\frac{1}{2})^3+4(-\frac{1}{2})^2-8\left(-\frac{1}{2}\right)+12}{4}-3\overset{?}{\Leftrightarrow}0$$

we substituted $x = -\frac{1}{2}$ into the equation

$$\frac{-\frac{1}{8}-4\left(\frac{1}{4}\right)-\frac{2}{2}}{2}+\frac{-\frac{6}{8}+4\left(\frac{1}{4}\right)+\left(\frac{8}{2}\right)+12}{4}-3\overset{?}{\Leftrightarrow}0$$

$$\frac{-\frac{1}{8}-\frac{4}{4}-\frac{2}{2}}{2}+\frac{-\frac{6}{8}+\frac{4}{4}+\frac{8}{2}+12}{4}-3\overset{?}{\Leftrightarrow}0$$

$$\frac{\frac{-1-8-8}{8}}{2}+\frac{\frac{-6+8+32}{8}+12}{4}-3\overset{?}{\Leftrightarrow}0$$

$$\frac{\frac{-17}{8}}{2}+\frac{\frac{34}{8}+12}{4}-3\overset{?}{\Leftrightarrow}0$$

$$\frac{\frac{-17}{8}}{2}\times\frac{2}{2}+\frac{\frac{34}{8}+12}{4}-3\overset{?}{\Leftrightarrow}0$$

we multiplied the first fraction by $\frac{2}{2}$

$$\frac{\frac{-34}{8}}{4}+\frac{\frac{34}{8}+12}{4}-3\overset{?}{\Leftrightarrow}0$$

$$\frac{\frac{-34}{8}+\frac{34}{8}+12}{4}-3\overset{?}{\Leftrightarrow}0$$

$$\frac{12}{4}-3\overset{?}{\Leftrightarrow}0$$

$$3-3\overset{?}{\Leftrightarrow}0$$

$0 = 0$ true $x = -\frac{1}{2}$ is a correct answer.

Example

Solve $\frac{6}{x+2} + \frac{4}{x-2} + \frac{14}{x^2-4} = 0.$

Solution

First, we need to determine the excluded values of x, those which result in a 0 denominator. So,

For $x + 2 = 0$:

$x + 2 - 2 = 0 - 2$ we added -2 to both sides.

$x = -2.$

For $x - 2 = 0$:

$x - 2 = 0$

$x - 2 + 2 = 0 + 2$ we added 2 to both sides

$x = 2.$

For $x^2 - 4 = 0$:

$x^2 - 4 = 0$

$x^2 - 4 + 4 = 0 + 4$ we added 4 to both sides

$x^2 = 4.$

$x = \pm\sqrt{4} = \pm 2.$

Therefore, the values of $x = 2$ and $x = -2$ are excluded values.

Now, to perform algebraic operations on the equation $\frac{6}{x+2} + \frac{4}{x-2} + \frac{14}{x^2-4} = 0$. We need to find the common denominator of $x + 2$, $x - 2$, and $x^2 - 4$. Base on the difference of squares rule, the expression $x^2 - 4$ can be written as $(x + 2)(x - 2)$. Hence, the common denominator of $x + 2$, $x - 2$, and $x^2 - 4$ which is divisible to all of them is $(x + 2)(x - 2)$. So,

$\frac{6}{x+2} + \frac{4}{x-2} + \frac{14}{x^2-4} = 0$

$\frac{x-2}{x-2} \times \frac{6}{x+2} + \frac{x+2}{x+2} \times \frac{4}{x-2} + \frac{14}{x^2-4} = 0$ we multiplied the first and the second term by $\frac{x-2}{x-2}$ and $\frac{x+2}{x+2}$ respectively, to make a common denominator

$\frac{6(x-2) + 4(x+2) + 14}{(x-2)(x+2)} = 0$

$(x-2)(x+2) \times \frac{6(x-2) + 4(x+2) + 14}{(x-2)(x+2)} = (x-2)(x+2) \times 0$ we multiplied both sides by $(x-2)(x+2)$

$6(x-2) + 4(x+2) + 14 = 0$

$6x - 12 + 4x + 8 + 14 = 0$

$10x + 10 = 0$

$10x + 10 - 10 = 0 - 10$ we added -10 to both sides

$10x = -10$

$\frac{1}{10} \times 10x = \frac{1}{10} \times (-10)$ we multiplied both sides by the reciprocal of the variable

$x = \frac{-10}{10} = -1.$

Check: To verify our answer, we need to substitute the solution into the original equation to see if the left-hand side of the equation is also 0. So,

$$\frac{6}{x+2} + \frac{4}{x-2} + \frac{14}{x^2-4} = 0$$

$\frac{6}{(-1)+2} + \frac{4}{(-1)-2} + \frac{14}{(-1)^2-4} \overset{?}{\Leftrightarrow} 0$ we substituted $x = -1$ into the equation

$$\frac{6}{1} + \frac{4}{-3} + \frac{14}{(1)-4} \overset{?}{\Leftrightarrow} 0$$

$$\frac{6}{1} + \frac{4}{-3} + \frac{14}{-3} \overset{?}{\Leftrightarrow} 0$$

$\frac{6}{1} \times \frac{-3}{-3} + \frac{4}{-3} + \frac{14}{-3} \overset{?}{\Leftrightarrow} 0$ we multiplied both side by $\frac{-3}{-3}$ to make a common denominator

$$\frac{-18}{-3} + \frac{4}{-3} + \frac{14}{-3} \overset{?}{\Leftrightarrow} 0$$

$$\frac{-18+4+14}{-3} \overset{?}{\Leftrightarrow} 0$$

$$\frac{0}{-3} \overset{?}{\Leftrightarrow} 0$$

$0 = 0$ true, $x = -1$ is correct answer.

4.4. Solving Polynomial Equations Involving Rational Exponents

As we mentioned earlier, trinomials are polynomials with exactly three terms and are usually polynomials of degree two. Fractional exponents make polynomials hard to factor and hence we need to make them integer. To do so, a typical is substitution. The reason polynomials are factored is that the factors are much easier to solve than the polynomial, and the roots of the factors are the same as the roots of the polynomial. To describe the substitution, we present an example below.

Example

Solve the trinomial $x^{\frac{1}{2}} = 3x^{\frac{1}{4}} - 2$.

Solution

Let us rewrite the trinomial $x^{\frac{1}{2}} = 3x^{\frac{1}{4}} - 2$ by applying following operations.

$$x^{\frac{1}{2}} = 3x^{\frac{1}{4}} - 2$$

$$x^{\frac{1}{2}} - 3x^{\frac{1}{4}} + 2 = 0$$

$$(x^{\frac{1}{4}})^2 - 3x^{\frac{1}{4}} + 2 = 0.$$

Now, to solve the trinomial $x^{\frac{1}{2}} = 3x^{\frac{1}{4}} - 2$, let us first make the substitution $y = x^{\frac{1}{4}}$. So,

$$(x^{\frac{1}{4}})^2 - 3x^{\frac{1}{4}} + 2 = 0$$

$(y)^2 - 3y + 2 = 0$ we substituted $x^{\frac{1}{4}} = y$ into the equation.

The resulted trinomial $y^2 - 3y + 2 = 0$, can be solved by factoring. So,

$y^2 - 3y + 2 = 0$

$(y - 1)(y - 2) = 0.$

Then $(y - 1) = 0$ and $(y - 2) = 0$,

$(y - 1) = 0$

$y = 1.$

$(y - 2) = 0$

$y = 2.$

Now, to obtain the solution of the trinomial $x^{\frac{1}{2}} = 3x^{\frac{1}{4}} - 2$, we put the values of y into the $y = x^{\frac{1}{4}}$ equation.

$y = x^{\frac{1}{4}}$

$1 = x^{\frac{1}{4}}$ we substituted $y = 1$ into the equation

$(1)^4 = (x^{\frac{1}{4}})^4$ we raised both sides to the power of 4

$x = 1.$

Similarly,

$y = x^{\frac{1}{4}}$

$2 = x^{\frac{1}{4}}$ we substituted $y = 2$ into the equation

$(2)^4 = (x^{\frac{1}{4}})^4$ we raised both sides to the power of 4

$x = 16.$

Therefore, the roots of the trinomial $x^{\frac{1}{2}} = 3x^{\frac{1}{4}} - 2$, are $x = 1$ and $x = 16$.

Check: To verify our answer, we need to substitute each solution separately into the original equation to see if the left-hand side of the equation is also 0. So,

For $x = 1$, we will have:

$x^{\frac{1}{2}} = 3x^{\frac{1}{4}} - 2$

$x^{\frac{1}{2}} - 3x^{\frac{1}{4}} + 2 = 0$

$(1)^{\frac{1}{2}} - 3(1)^{\frac{1}{4}} + 2 \overset{?}{\Leftrightarrow} 0$ we substituted $x = 1$ into the equation.

We know that $1^a = 1$, for all values of a. So,

$1 - 3 + 2 \overset{?}{\Leftrightarrow} 0$

$3 - 3 \overset{?}{\Leftrightarrow} 0$

$0 = 0$ true $x = 1$ is a correct answer.

For $x = 16$, we will have:

$x^{\frac{1}{2}} - 3x^{\frac{1}{4}} + 2 = 0$

$(16)^{\frac{1}{2}} - 3(16)^{\frac{1}{4}} + 2 \overset{?}{\Leftrightarrow} 0$ we substituted $x = 16$ into the equation.

We have learned that $x^{\frac{m}{n}} = \sqrt[n]{x^m} = \left(\sqrt[n]{x}\right)^m$. So,

$\sqrt{16} - 3(\sqrt[4]{16}) + 2 \overset{?}{\Leftrightarrow} 0$

$\sqrt{4^2} - 3(\sqrt[4]{2^4}) + 2 \overset{?}{\Leftrightarrow} 0$

$4 - 3(2) + 2 \overset{?}{\Leftrightarrow} 0$ we factored perfect square

$4 - 6 + 2 \overset{?}{\Leftrightarrow} 0$

$6 - 6 \overset{?}{\Leftrightarrow} 0$

$0 = 0$ true $x = 16$ is a correct answer.

Example

Solve $3^{\frac{378}{x^3+1}} = 27$.

Solution

We can rewrite the right-hand side as an exponent number. So,

$3^{\frac{378}{x^3+1}} = 27$

$3^{\frac{378}{x^3+1}} = 3^3$ we rewrite the right-hand side as an exponent number.

We have a common base. Hence, we equate the exponents. So,

$3^{\frac{378}{x^3+1}} = 3^3$

$\frac{378}{x^3+1} = 3$

$(x^3 + 1) \times \frac{378}{x^3+1} = (x^3 + 1) \times 3$ we multiplied both sides by $(x^3 + 1)$

$378 = (x^3 + 1) \times 3$

$378 = 3x^3 + 3$ we applied the distributive property

$378 - 3 = 3x^3 + 3 - 3$ we added -3 to both sides

$375 = 3x^3$

$\frac{1}{3} \times 375 = \frac{1}{3} \times 3x^3$ we multiplied both sides by the reciprocal of the variable

$125 = x^3$

$5^3 = x^3$ we rewrite the left-hand side as an exponent number

$x = 5.$

Therefore, the solution of $3^{\frac{378}{x^3+1}} = 27$ is $x = 5$.

Check: To verify our answer, we need to substitute the solution into the original equation to see if the left-hand side of the equation is the same as the right-hand side. So,

$3^{\frac{378}{x^3+1}} = 27$

$3^{\frac{378}{x^3+1}} \overset{?}{\Leftrightarrow} 27$ we substituted $x = 5$ into the equation

$3^{\frac{378}{125+1}} \overset{?}{\Leftrightarrow} 27$

$3^{\frac{378}{126}} \overset{?}{\Leftrightarrow} 27$

$3^3 \overset{?}{\Leftrightarrow} 27$

$27 = 27$ true.

Therefore, the answer $x = 5$ is the correct one.

4.5. Solving Polynomial Equations Involving Radicals

For solving polynomial equations involving radicals, we isolate the radical, and then raise both sides of the equation to the power of radical index. We must always check our answers in the original equation because raising both sides of an equation to the power of index sometimes generates an equation that has roots that are not roots of the original equation.

Example
Solve the polynomial equation $\sqrt[3]{2x-7} - 2\sqrt[3]{x^2-1} = 0$.

Solution
We first rewrite the equation to isolate the radicals. So,

$\sqrt[3]{2x-7} - 2\sqrt[3]{x^2-1} = 0$

$\sqrt[3]{2x-7} - 2\sqrt[3]{x^2-1} + 2\sqrt[3]{x^2-1} = 0 + 2\sqrt[3]{x^2-1}$

$\sqrt[3]{2x-7} = 2\sqrt[3]{x^2-1}.$

Next, we raise both sides to the power of the radical index. So,

$\sqrt[3]{2x-7} = 2\sqrt[3]{x^2-1}$

$\left(\sqrt[3]{2x-7}\right)^3 = \left(2\sqrt[3]{x^2-1}\right)^3$ we raised both sides to the power of the radical index, 3

$2x - 7 = 8(x^2 - 1)$

$2x - 7 = 8x^2 - 8$

$2x - 7 + 7 = 8x^2 - 8 + 7$ we added 7 to both sides

$2x = 8x^2 - 1$

$2x - 2x = 8x^2 - 1 - 2x$ we added $-2x$ to both sides

$8x^2 - 2x - 1 = 0.$

To solve $8x^2 - 2x - 1 = 0$ using factoring by the leading coefficient is not 1, we need to compare it with $ax^2 + bx + c$. We get $a = 8$, $b = -2$, and $c = -1$. We found $ac = -8$. Next, we need to find two numbers, p and q, with a product of -8 and a sum of -2.

In the following table, we list all factors until a pair of the anticipated sum is attained.

Factors of $ac = -8$	Sum of factors
1, − 8	− 7
− 1, 8	7
2, − 4	**− 2**
− 2, 4	2

So, $p =$ 4 and $q = 2$. Rewrite the original expression as $ax2 + px + qx + c = 0$.

$8x^2 - 2x - 1 = 0$

$8x^2 - 4x + 2x - 1 = 0$

$4x(2x - 1) + 2x - 1 = 0$ we factored out the GCF of each part

$(2x - 1)(4x + 1) = 0$ again we factored out the GCF of the expression.

Using the zero-factor property we set each factor equal to zero. So,

$(2x - 1)(4x + 1) = 0$

For $(2x - 1) = 0$, we will have:

$(2x - 1) = 0$

$2x - 1 + 1 = 0 + 1$ we added 1 to both sides

$2x = 1$

$\frac{1}{2} \times 2x = \frac{1}{2} \times 1$ we multiplied both sides by the reciprocal of the variable

$x = \frac{1}{2}.$

For $(4x + 1) = 0$, we will have:

$(4x + 1) = 0$

$4x + 1 - 1 = 0 - 1$ we added -1 to both sides

$4x = -1$

$\frac{1}{4} \times 4x = \frac{1}{4} \times (-1)$ we multiplied both sides by the reciprocal of the variable

$x = -\frac{1}{4}.$

Therefore, the solutions for the polynomial equation, $8x^2 - 2x - 1 = 0$, are $x = \frac{1}{2}$, and $x = -\frac{1}{4}$.

Check: To verify our answer, we need to substitute each solution separately into the equation to see if the left-hand side of the equation is also 0. So,

For $x = \frac{1}{2}$, we will have:

$8x^2 - 2x - 1 = 0$

$8(\frac{1}{2})^2 - 2(\frac{1}{2}) - 1 \overset{?}{\Leftrightarrow} 0$ we substituted $x = \frac{1}{2}$ into the equation

$8(\frac{1}{4}) - 2(\frac{1}{2}) - 1 \overset{?}{\Leftrightarrow} 0$

$\frac{8}{4} - \frac{2}{2} - 1 \overset{?}{\Leftrightarrow} 0$

$2 - 1 - 1 \overset{?}{\Leftrightarrow} 0$

$2 - 2 \overset{?}{\Leftrightarrow} 0$

$0 = 0$ true $x = \frac{1}{2}$ is a correct answer.

For $x = -\frac{1}{4}$, we will have:

$8x^2 - 2x - 1 = 0$

$8(-\frac{1}{4})^2 - 2(-\frac{1}{4}) - 1 \overset{?}{\Leftrightarrow} 0$ we substituted $x = -\frac{1}{4}$ into the equation

$8(\frac{1}{16}) - 2(-\frac{1}{4}) - 1 \overset{?}{\Leftrightarrow} 0$

$\frac{8}{16} + \frac{2}{4} - 1 \overset{?}{\Leftrightarrow} 0$

$\frac{1}{2} + \frac{1}{2} - 1 \overset{?}{\Leftrightarrow} 0$

$1 - 1 \overset{?}{\Leftrightarrow} 0$

$0 = 0$ true $x = -\frac{1}{4}$ is a correct answer.

4.6. Solving Polynomial Inequalities Involving Exponential Expressions

To solve a polynomial inequality, we first need to find the zeros of the polynomial equation consisting exponential expression. To do so, we next need to isolate the expression consisting exponential expression. In finding the final solution. We may face two distinct cases.

- If it is convenient, express both sides with a common base and equate the exponents.
- Otherwise, take the logarithm or natural log of both sides of the equation and use the power rule.

After verifying the solutions, we then establish a table sign and a number line to determine where the inequality has positive value and negative negetive value on the number line. The sign within each interval on the number line is determined by using test values.

Finally, we represent the final answers on a number line and additionally we can write the solution set interval notation.

Example

Solve $3^{x^2 - 3x} - 81 \geq 0$.

Solution

We have already got zero on one side. Hence, we need to convert inequality to equality. So,

$$3^{x^2 - 3x} - 81 \geq 0.$$

$$3^{x^2 - 3x} - 81 = 0.$$

Next, we can go to isolating the expression consisting exponential function.

$$3^{x^2 - 3x} - 81 = 0.$$

$$3^{x^2 - 3x} = 81$$

$$3^{x^2 - 3x} = 3^4.$$

We have a common base. Hence, we equate the exponents.

$$x^2 - 3x = 4$$
$$x^2 - 3x - 4 = 0.$$

The above trinomial can be factorized as below:

$$(x - 4)(x + 1) = 0.$$

Using the zero-factor principle we will have:

$$x - 4 = 0$$

$$x + 1 = 0.$$

For $x - 4 = 0$, we get:

$$x - 4 = 0$$

$$x - 4 + 4 = 0 + 4 \qquad \text{we added 4 to both sides}$$

$$x = 4.$$

For $x + 1 = 0$, we get:

$$x + 1 = 0$$

$$x + 1 - 1 = 0 - 1 \qquad \text{we added } -1 \text{ to both sides}$$

$$x = -1.$$

Therefore, the solutions of $3^{x^2 - 3x} - 81 = 0$ are $x = 4$, and $x = -1$.

Before we go to the next step, let us verify our solutions.

Check: To verify our answer, we need to plug each solution into the equation. So,

For $x = 4$:

$$3^{x^2 - 3x} - 81 = 0$$

$$3^{4^2 - 3(4)} - 81 \overset{?}{\Leftrightarrow} 0. \qquad \text{we plugged } x = 4 \text{ into the equation}$$

$3^{16\,-\,12} - 81 \overset{?}{\Leftrightarrow} 0$

$3^4 - 81 \overset{?}{\Leftrightarrow} 0$

$81 - 81 \overset{?}{\Leftrightarrow} 0$

$0 = 0$ true, the solution $x = 4$ is a correct answer.

For $x = -1$:

$3^{x^2 - 3x} - 81 = 0$

$3^{(-1)^2 - 3(-1)} - 81 \overset{?}{\Leftrightarrow} 0.$ we plugged $x = 4$ into the equation

$3^{1+3} - 81 \overset{?}{\Leftrightarrow} 0$

$3^4 - 81 \overset{?}{\Leftrightarrow} 0$

$81 - 81 \overset{?}{\Leftrightarrow} 0$

$0 = 0$ true, the solution $x = 4$ is a correct answer.

Now that we verified the solution of $3^{x^2 - 3x} - 81 = 0$, we can build the table sign. So,

x values	$x < -1$	$x = -1$	$-1 < x < 4$	$x = 4$	$x > 4$
$x + 1$	$-$	0	$+$	$+$	$+$
$x - 4$	$-$	$-$	$-$	0	$+$
$y = 3^{x^2 - 3x} - 81$	$+$	0	$-$	0	$+$

Let us select the test points $x = -2, x = 0$, and $x = 5$, here is the number line for this problem:

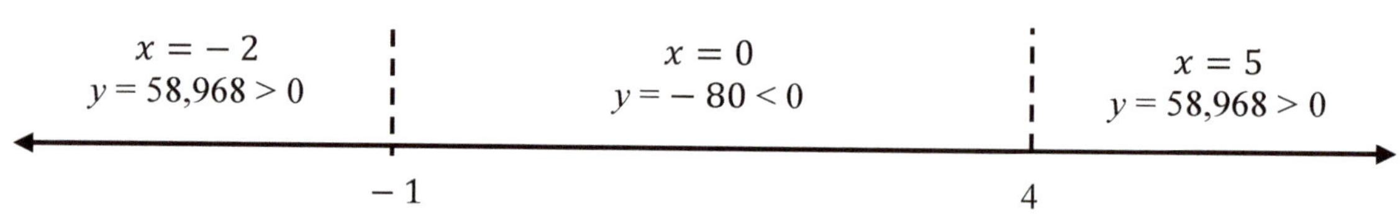

Based on the above number line the solution of $x^4 - 2x^3 - 24x^2 \leq 0$ is $-4 \leq x \leq 6$, or as solution set interval notation $[-4, 6]$. To represent the final answers on a number line we will have:

Example

Solve $log2^{(2x^2 - 10)lo} \quad - 2log5^{log2} \leq 0.$

Solution

First, we need to find the zeros of $log2^{(2x^2 - 10)log5} + 2log5^{log2} \leq 0$. So,

$log2^{(2x^2 - 10)log5} + 2log5^{log2} \leq 0$

$$log2^{(2x^2 - 10)log5} + 2log5^{log2} = 0.$$

Now, we need to isolate the variable in one side. So,

$$log2^{(2x^2 - 10)log5} + 2log5^{log2} = 0$$

$$log2^{(2x^2 - 10)log5} = -2logg5^{log2}.$$

Using the logarithm property stating $log_b a^x = x\ log_b a$, we have:

$$log2^{(2x^2 - 10)log5} = -2log5^{log2}$$

$$[(2x^2 - 10)log5]log2 = -2log2 \times log5$$

$$\frac{1}{log2\ log5} 2(x^2 - 10)log5log2 = \frac{1}{log2\ log5}(-2)log2log5$$ we multiplied both sides by $\frac{1}{log2\ log5}$

$$2x^2 - 10 = -2$$

$$2x^2 - 10 + 10 = -2 + 10$$ we added 10 to both sides

$$2x^2 = 8$$

$$\frac{1}{2} \times 2x^2 = \frac{1}{2} \times 8$$ we multiplied both sides by the reciprocal of the variable

$$x^2 - 4 = 4 - 4$$

$$x^2 - 4 = 0$$

$$(x - 2)(x + 2) = 0$$ we applied difference of squares rule.

Using the zero-factor principle, we will have:

$$(x - 2)(x + 2) = 0$$

$$(x - 2) = 0$$

$$(x + 2) = 0.$$

For $x - 2 = 0$:

$$x - 2 = 0$$

$$x - 2 + 2 = 0 + 2$$ we added 2 to both sides

$$x = 2.$$

For $(x + 2) = 0$:

$$x + 2 = 0$$

$$x + 2 - 2 = 0 - 2$$ we added -2 to both sides

$$x = -2.$$

Therefore, the solutions of $log2^{(2x^2 - 10)log5} + 2log5^{log2} = 0$ are $x = 2$, and $x = -2$.

Check: To verify our answer, we need to plug each solution into the equation. So,

For $x = 2$:

$$log2^{(2x^2 - 10)log5} + 2log5^{log2} = 0$$

$$log2^{(2 \times 2^2 - 10)log5} + 2log5^{log2} \overset{?}{\Leftrightarrow} 0$$ we have plugged $x = 2$ into the equation

$$log2^{(8 - 10)log5} + 2log5^{lo} \overset{?}{\Leftrightarrow} 0$$

$$log2^{-2log5} + 2log5^{log2} \overset{?}{\Leftrightarrow} 0$$

$$-2\log5\log2 + 2\log2\log5 \overset{?}{\Leftrightarrow} 0$$ we employed property of logarithmic functions, $\log_b a^x = x \log_b a$

$$0 = 0$$ True, the solution $x = 2$ is a correct answer.

For $x = -2$:

$$log2^{(2x^2 - 10)log5} + 2log5^{log2} = 0$$

$$log2^{(2(-2)^2 - 10)log5} + 2log5^{log2} \overset{?}{\Leftrightarrow} 0$$ we plugged $x = -2$ into the equation

$$log2^{(2 \times 4 - 10)log5} + 2log5^{log2} \overset{?}{\Leftrightarrow} 0$$

$$log2^{(8 - 10)log5} + 2log5^{log2} \overset{?}{\Leftrightarrow} 0$$

$$log2^{-2\log5} + 2log5^{log2} \overset{?}{\Leftrightarrow} 0$$

$$-2log5\, log2 + 2log2\, log5 \overset{?}{\Leftrightarrow} 0$$ we employed property of logarithmic functions, $\log_b a^x = x\, log_b a$

$$0 = 0$$ true, the solution $x = -2$ is a correct answer.

Now that we verified the solution , we can build the table sign. So,

x values	$x < -2$	$x = -2$	$-2 < x < 2$	$x = 2$	$x > 2$
$x - 2$	$-$	$-$	$-$	0	$+$
$x + 2$	$-$	0	$+$	$+$	$+$
$y = log2^{(2x^2 - 10)log5} + 2log5^{log2}$	$+$	0	$-$	0	$+$

Let us select the test points $x = -3, x = 0$, and $x = 3$, here is the number line for this problem:

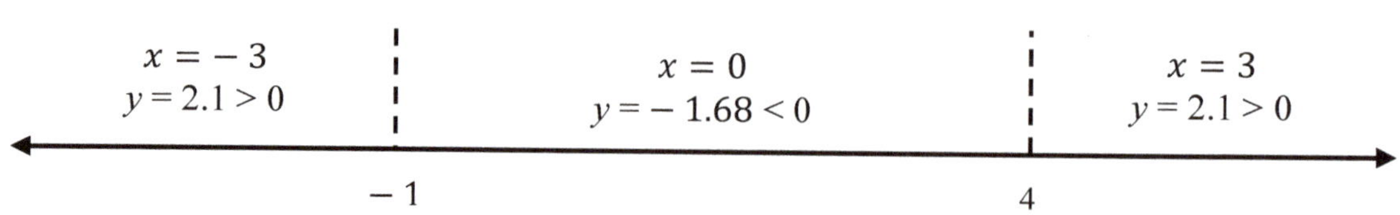

Based on the above number line the solution of $log2^{(2x^2-10)log5} + 2log5^{log2} \leq 0$ is $-2 \leq x \leq 2$, or as solution set interval notation $[-2, 2]$. To represent the final answers on a number line we will have:

4.7. Solving Polynomial Inequalities Involving Radicals

When dealing with polynomial inequalities involving radicals, the first step is to change the inequality sign to an equality sign, thus converting it into polynomial equality. To solve this type of equality, we isolate the radical and then raise both sides of the equation to the index value of the radical. However, it's crucial to check the obtained solutions in the original equation because raising both sides of an equation may introduce extraneous solutions.

Next, we create a sign table and a number line to determine where the inequality is positive and negative on the number line. The sign within each interval on the number line is determined by using test values. Subsequently, the final solutions are represented on a number line, and the solution set can be expressed in interval notation.

In summary, the process involves converting the inequality into an equality, solving the resulting polynomial equation, verifying solutions in the original inequality, constructing a sign table and number line, determining the signs within intervals using test values, and finally representing the solutions on a number line and in interval notation.

Example

Solve $\sqrt[3]{x^3 - x^2 - 6x} + x \leq 0$.

Solution

We convert inequality to an equality. So,

$$\sqrt[3]{x^3 - x^2 - 6x} + x \leq 0$$

$$\sqrt[3]{x^3 - x^2 - 6x} + x = 0.$$

We now isolate the radical in one side of the equation. So,

$$\sqrt[3]{x^3 - x^2 - 6x} + x = 0$$

$$\sqrt[3]{x^3 - x^2 - 6x} = -x.$$

Next, we raise both sides to the power of the index value. So,

$$\sqrt[3]{x^3 - x^2 - 6x} = -x$$

$$\left(\sqrt[3]{x^3 - x^2 - 6x}\right)^3 = (-x)^3 \qquad \text{we raised both sides to 3}$$

$$x^3 - x^2 - 6x = -x^3$$

$$x^3 - x^2 - 6x + x^3 = -x^3 + x^3 \qquad \text{we added } x^3 \text{ to both sides}$$

$$2x^3 - x^2 - 6x = 0$$

$$x(2x^2 - x - 6) = 0 \qquad \text{we factored out } x.$$

Using the zero-factor principle we will have.

$$x(2x^2 - x - 6) = 0$$

$$x = 0$$

$$2x^2 - x - 6 = 0.$$

For $2x^2 - x - 6 = 0$, we have a quadratic equation with leading coefficient $a \neq 1$. Comparing this equation with $ax^2 + bx + c = 0$, we have $a = 2$, $b = -1$, $c = -6$ and $ac = -12$.

We need to list factor of $ac = -12$ to get two numbers with sum of -1. So,

Factor of $ac = -12$	Sum of factors
$1, -12$	-11
$-1, 12$	11
$2, -6$	-4
$-2, 6$	4
$3, -4$	**-1**
$-3, 4$	1

So, $p = -4$ and $q = 3$. We now, rewrite the original expression as $ax^2 + px + qx + c = 0$

$2x^2 - x - 6 = 0$

$2x^2 - 4x + 3x - 6 = 0$

$2x(x - 3) + 3(x - 2) = 0$ — we factored out the GCF of each term

$(x - 2)(2x + 3) = 0$ — again we factored out the GCF of the expression

$(x - 2)(2x + 3) = 0.$

Next, using zero-factor principle, we set each of the two expressions $(x - 2)$ and $(2x + 3)$ equal to zero and solve each for x.

$x - 2 = 0$

$x - 2 + 2 = 0 + 2$ — we added 2 to both sides

$x = 2.$

$2x + 3 = 0$

$2x + 3 - 3 = 0 - 3$ — we added -3 to both sides

$2x = -3$

$\frac{1}{2} \times 2x = \frac{1}{2} \times (-3)$ — we multiplied both sides of equation by the reciprocal of $2x$ coefficient

$x = -\frac{3}{2}$

$x = 1.5.$

Therefore, the solutions of $\sqrt[3]{x^3 - x^2 - 6x} = -x$ are, $x = 0$, $x = 2$ and $x = -\frac{3}{2} =.- 1.5.$

Check: To verify our answer, we need to plug each solution separately into the equation. So,

For $x = 0$:

$$\sqrt[3]{x^3 - x^2 - 6x} + x = 0$$

$\sqrt[3]{(0)^3-(0)^2-6(0)}+(0)\overset{?}{\Leftrightarrow}0$ we plugged $x=0$ into the equation

$0=0$ true, the solution $x=0$ is a correct answer.

For $x=2$:

$\sqrt[3]{x^3-x^2-6x}+x=0$

$\sqrt[3]{(2)^3-(2)^2-6(2)}+(2)\overset{?}{\Leftrightarrow}0$ we plugged $x=2$ into the equation

$\sqrt[3]{(8)-(4)-12}+(2)\overset{?}{\Leftrightarrow}0$

$\sqrt[3]{-8}+2\overset{?}{\Leftrightarrow}0$

$-2+2\overset{?}{\Leftrightarrow}0$

$0=0$ true, the solution $x=2$ is a correct answer.

For $x=-\frac{3}{2}$:

$\sqrt[3]{x^3-x^2-6x}+x=0$

$\sqrt[3]{(-\frac{3}{2})^3-\left(-\frac{3}{2}\right)^2-6(-\frac{3}{2})}+(-\frac{3}{2})\overset{?}{\Leftrightarrow}0$ we plugged $x=-\frac{3}{2}$ into the equation

$\sqrt[3]{\left(-\frac{27}{8}\right)-\left(\frac{9}{4}\right)+\frac{18}{2}}+\left(-\frac{3}{2}\right)\overset{?}{\Leftrightarrow}0$

$\sqrt[3]{\left(-\frac{27}{8}\right)-\left(\frac{9}{4}\right)\left(\frac{2}{2}\right)+(\frac{18}{2})(\frac{4}{4})}+\left(-\frac{3}{2}\right)\overset{?}{\Leftrightarrow}0$ for having a common denominator, we multiplied the second and the third fraction by $\frac{2}{2}$ and $\frac{4}{4}$ respectively

$\sqrt[3]{\left(-\frac{27}{8}\right)-\left(\frac{18}{8}\right)+(\frac{72}{8})}+\left(-\frac{3}{2}\right)\overset{?}{\Leftrightarrow}0$

$\sqrt[3]{\frac{-27-18+72}{8}}+\left(-\frac{3}{2}\right)\overset{?}{\Leftrightarrow}0$

$\sqrt[3]{\frac{27}{8}}+\left(-\frac{3}{2}\right)\overset{?}{\Leftrightarrow}0$

$\sqrt[3]{\frac{3^3}{2^3}}+\left(-\frac{3}{2}\right)\overset{?}{\Leftrightarrow}0$ we factored the third power from radicand

$\frac{3}{2}-\frac{3}{2}\overset{?}{\Leftrightarrow}0$

$0=0$ true, the solution $x==-\frac{3}{2}$ is a correct answer.

Now that we verified the solution, we can build the table sign. So,

x values	$x<-\frac{3}{2}$	$x=-\frac{3}{2}$	$-\frac{3}{2}<x<0$	$x=0$	$0<x<2$	$x=2$	$x>2$
x	$-$	$-$	$-$	0	$+$	$+$	$+$
$x-2$	$-$	$-$	$-$	$-$	$-$	0	$+$
$2x+3$	$-$	0	$+$	$+$	$+$	$+$	$+$
$y=\sqrt[3]{x^3-x^2-6x}+x$	$-$	0	$+$	0	$-$	0	$+$

Let us select the test points $x=-2, x=-1, x=1$, and $x=3$. Here is the number line for this problem:

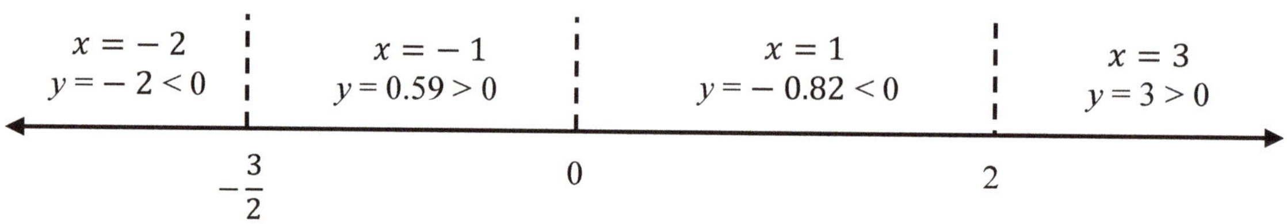

Based on the above number line the solution of $\sqrt[3]{x^3-x^2-6x}+x \le 0$ is $x \le -\frac{3}{2}$ and $0 \le x \le 2$ or as solution set interval notation $(-\infty, -\frac{3}{2}] \cup [0, 2]$.

To represent the final answers on a number line, we will have:

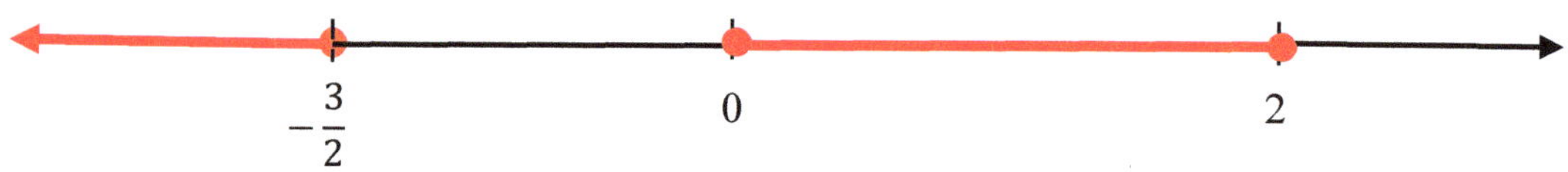

4.8. Exercises

1. Solve each polynomial.
   a) $(8x + 4)(12x - 5)(3x + 6) = 0.$
   b) $(x - 7)(2x + 3)(5x - 15)(8x - 24) = 0.$
   c) $\frac{1}{x^2 + 8} \div \frac{3x}{x^4 - 64} = 0.$

2. Solve each polynomial.
   a) $x^3 - 12x^2 - 24x = 0.$
   b) $4x^3 + 12x^2 - 8x = 0.$
   c) $6x^3 + 15x^2 - 81x = 0.$
   d) $3x^3 + 10x^2 + 8x = 0.$

3. Let the polynomial expression $x^3 - 17x^2 + 54x - 8$ be denoted by $P(x)$.
   a) Verify that $x = 4$ is a solution of the equation, $P(x = 4) = 0$.
   b) Find the other solutions.

4. Solve each polynomial equation by substitution.
   a) $x^4 - 11x^2 + 24 = 0.$
   b) $3(x + 5)^4 - 12(x + 5)^2 - 15 = 0.$
   c) $4(2x + 6)^6 - 17(2x + 6)^4 - 15(2x + 6)^2 = 0$

5. Solve each polynomial equation by factor grouping.
   a) $6x^3 - 12x^2 - 9x + 18 = 0.$
   b) $15x^3 - 5x^2 - 27x = -9.$
   c) $7x^5 - 21x^4 - 8x^3 + 24x^2 - 11x + 33 = 0.$
   d) $9x^5 - 3x^4 - 12x^3 + 4x^2 + 21x = 7.$

6. Consider the polynomial equation $P(x) = 10x^3 - 47x^2 + 42x$. By evaluating $P(3)$ and $P(4)$ show that at least one root of $P(x) = 0$ lies between $x = 3$ and $x = 4$.

7. Solve each polynomial equation.
   a) $\frac{2x^4 - 3x^2 + 6x}{3} + \frac{7x^4 + 3x^2 - 12x + 24}{6} = 36.$
   b) $\frac{6x^2 + 2}{5} + \frac{5x^2 - 3x - 9}{15} + \frac{4x^2 - 15}{10} = 18.$
   c) $\frac{6x^2}{5} - \frac{-3x - 9}{15} + \frac{4x^3 - 3x^2 + 9}{10} - \frac{24}{5} = \frac{2x^3 - 2x}{5}.$

8. Solve each rational polynomial equation.
   a) $\frac{3}{5x + 20} + \frac{5}{x - 4} + \frac{9}{2(x^2 - 16)} = 0.$
   b) $\frac{1}{x^2 + 9} = -\frac{6}{2(x^2 + 9)} - \frac{5}{x^4 - 81}.$

c) $\frac{4}{x^2 - 25} - \frac{4}{(x-5)^2} = \frac{3x}{(x-5)^3}$.

9. Solve each polynomial equation involving rational expressions.
   a) $3x^{\frac{1}{2}} - 5x^{\frac{1}{4}} - 8 = 0.$
   b) $x^{\frac{1}{3}} - 8x^{\frac{1}{6}} - 48 = 0.$
   c) $x^{\frac{1}{4}} - 9x^{\frac{1}{8}} = 36.$

10. Solve each polynomial equation involving rational exponents.
    a) $\frac{2}{x^2 - 2x - 8} = 4.$
    b) $\frac{56}{9x^3 + 1} - 81 = 0.$
    c) $\frac{112}{52\ \ ^4 + 5} - 25 = 0.$
    d) $\frac{2x^2}{7x^4 - 3} - 49 = 0.$

11. Solve each polynomial equation involving radicals.
    a. $\sqrt{x^2 - 4} - 2\sqrt{3x - 8} = 0.$
    b. $3\sqrt[3]{2x + 9} - 2\sqrt[3]{x^2 - 1} = 0.$
    c. $2\sqrt[4]{x + 11} - \sqrt[4]{x^2 - 8} = 0.$
    d. $\sqrt[3]{\frac{4}{5x + 10}} - 6\sqrt[3]{\frac{3x - 1}{x^2 - 4}} = 0.$

12. Solve each polynomial equation involving logarithms and natural logarithms.
    a) $log3log8^{(x^2 - 4x - 10)} - log3^{2log8} = 0.$
    b) $ln8^{(2x^2 - 8)log5} - ln8^{10lo} \ \ = 0.$
    c) $ln4^{(4x^3 - x^2 - 5x + 13)} - ln4^{13} = 0.$

13. Solve each polynomial inequality. Present your solutions as the set interval notation and on a number line.
    a) $(4x + 12)(x - 5) < 0.$
    b) $(5x - 20)(2x - 6)(5x + 15) > 0.$
    c) $(x + 2)(2x - 4)(7x + 21) \geq 0.$
    d) $(6x + 18)(2x - 1)(25x + 5) \leq 0.$

14. Solve each polynomial inequality. Present your solutions as the set interval notation and on a number line.
    a) $x^4 - 81x^2 > 0.$
    b) $x^2 - 5x \geq -6.$
    c) $x^3 - 2x^2 - 24x \geq 0.$

d) $3x^3 - 9x^2 - 12x > 0.$

e) $2x^4 - 3x^3 -.8x^2 + 12x < 0.$

15. Solve each polynomial inequality. Present your solutions as the set interval notation and on a number line.

a) $x^2 - 81 \leq 0.$

b) $x^3 - 27 > 0.$

c) $5^{x^2 - 2} - 25 \geq 0.$

d) $3^{(x^3 - 4x^2 - 12x + 3)} - 27 \geq 0.$

16. Solve each polynomial inequality. Present your solutions as the set interval notation and on a number line.

a) $log2^{(3x^2 - 3)} - log2^9 \leq 0.$

b) $log\left(\frac{2x^2 + 16}{3x^2 - 6x}\right) > 0.$

c) $log_3\left(\frac{5x^3 - 3x^2 + 2x - 3}{4x^3 - 3}\right) \geq 0.$

d) $log7^{(3x^2 - 9)log8} \leq log8^{6xlog7}.$

17. Solve each polynomial inequality. Present your solutions as the set interval notation.

a) $\sqrt[3]{9x^3 - 3x^2 - 8x} - 2x \leq 0.$

b) $\sqrt[5]{5x^3 - 6x^2 - 12x} > \sqrt[5]{4x^3 - 4x^2}$

c) $\sqrt[3]{x^2 - 10x + 16} + 2 \leq 0.$

d) $\sqrt[3]{4x^3 - 6x^2 - 17x} + 2x \geq 0.$

18. Solve each polynomial inequality. Present your solutions on a number line.

a) $\sqrt[7]{9x^3 - 5x^2 - 12x} \div \frac{\sqrt[7]{8x^3 - x^2}}{2} - 2 \leq 0.$

b) $\sqrt[3]{4x^3 - 12x^2 - 9x} + 3 \leq 0.$

c) $\sqrt[4]{4x^3 - 7x^2 - 14x} \geq \sqrt[4]{x}.$

d) $\sqrt[3]{x^2 - 17} + 2 \geq 0.$

5. Real - World Applications of Polynomials

Polynomials play a vital role in addressing real-world challenges across diverse domains. Their ability to describe a wide array of curves makes them invaluable for graphing purposes. Additionally, polynomials find application in modeling various scenarios, such as predicting stock market fluctuations, analyzing market trends in response to price changes, or computing the height above ground of a projectile fired into the air at any given time.

In physics, polynomials are instrumental in accurately portraying the trajectory of projectiles, facilitating analysis and prediction of their paths. By using polynomials, scientists and engineers can model and simulate projectile motion with precision. Furthermore, in real-world scenarios, polynomials find application in financial analysis, enabling the examination of costs, revenues, and profits for a firm.

This section explores real-world applications of polynomials through practical examples spanning different contexts. These examples showcase the versatility and effectiveness of polynomials in tackling complex issues and solving practical problems encountered in numerous fields.

5.1. Example: Architectural Design

An architect is designing a triangular window in front of a mountain cabin. Due to exceptionally wintry weather, the window can only have an area of 180 square inches and the architect wants the base to be 6 inches less than twice the height. Find the base and height of the window.

Hint: For a triangle with base, b, and height, h, the area, A, is given by the formula $A = \frac{1}{2} bh$.

Solution

Drawing a picture would help solve this case. So,

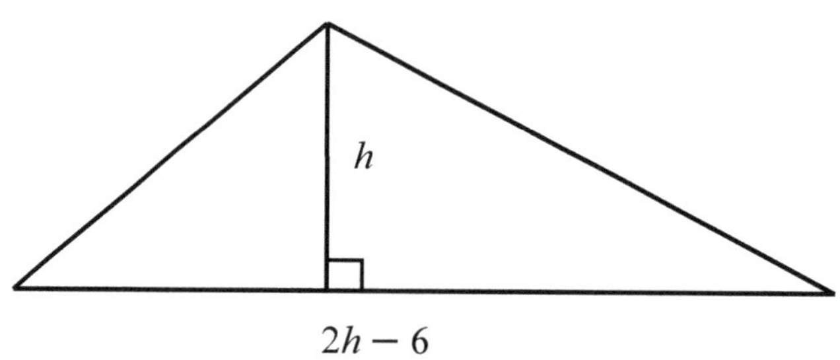

The area, A, of a triangle with base b and height h, is determined by $A = \frac{1}{2} bh$. So,

$A = \frac{1}{2} bh$

$180 = \frac{1}{2}(2h - 6)\, h$ — we replaced respectively the area and the base by 180 and $2h - 6$ into the equation

$180 = (h - 3)\, h$

$180 = h^2 - 3h$ — we applied the distributive property

$180 - 180 = h^2 - 3h - 180$ — we added -180 to both sides

$h^2 - 3h - 180 = 0.$

Using the factoring a trinomial with leading coefficient 1, we have learned that a trinomial in the form of $x^2 + bx + c$ can be written in factored form as $(x + p)(x + q)$ where $pq = c$ and $p + q = b$. Comparing our trinomial $h^2 - 3h - 180$ with $x^2 + bx + c$, we will have $c = -180$ and $b = -3$. Hence, by factoring -180 and looking for those with sum -3, we get $p = -15$ and $q = 12$. So,

$$h^2 - 3h - 180 = 0$$

$$(h - 15)(h + 12) = 0.$$

Using zero-factor principle, we will have:

$$(h - 15) = 0$$

$$(h + 12) = 0.$$

For $(h - 15) = 0$, we have:

$$h - 15 = 0$$

$$h - 15 + 15 = 0 + 15 \qquad \text{we added 15 to both sides}$$

$$h = 15.$$

For $(h + 12) = 0$, we have:

$$h + 12 = 0$$

$$h + 12 - 12 = 0 - 12 \qquad \text{we added } -12 \text{ to both sides}$$

$$h = -12.$$

Therefore, the solutions of $h^2 - 3h - 180 = 0$ are $h = 15in.$ and $h = -12in.$ However, h is the height of the window, a value of $h = -12in.$ does not make sense. Hence, the only acceptable solution is $h = 15in.$

Check: To verify our answer, we need to plug the accepted solution into the equation. So,

$$h^2 - 3h - 180 = 0$$

$$15^2 - 3\times15 - 180 \overset{?}{\Leftrightarrow} 0 \qquad \text{we plugged } x = 15 \text{ into the equation}$$

$$225 - 45 - 180 \overset{?}{\Leftrightarrow} 0$$

$$225 - 225 \overset{?}{\Leftrightarrow} 0$$

$$0 = 0 \qquad \text{true, the solution } x = 15in. \text{ is a correct answer.}$$

5.2. Example: Landscaping

Steven wants to put 198 square feet of artificial lawn in his back yard. This is the maximum area of artificial lawn allowed by his homeowner's association. He wants to have a rectangular area of lawn with a length 4 feet less than twice the width. Find the length and width.

Hint: For a rectangle with length, L, and width, W, the area, A, is given by the formula $A = LW$.

Solution

Drawing a picture would help solve this case. So,

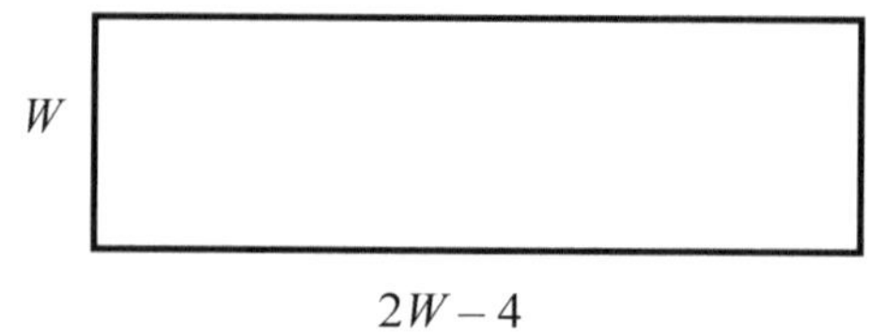

It is given that for a rectangle yard with length, L, and width, W, the area, A, is determined by the formula $A = LW$. So,

$A = LW$

$198 = (2W - 4)W$ we replaced respectively the area and the width by 198 and $2W - 4$ into the equation

$198 = 2W^2 - 4W$

$2W^2 - 4W - 198 = 198 - 198$ we added -198 to both sides

$2W^2 - 4W - 198 = 0.$

Comparing $2W^2 - 4W - 198 = 0$ with $ax^2 + bx + c = 0$, we find $a = 2$, $b = -4$, and $c = -198$. Hence using the quadratic formula, we will have.

$$W = \frac{-b \pm \sqrt{b^2 - 4ac}}{2a}$$

$$W = \frac{-(-4) \pm \sqrt{(-4)^2 - 4(2)(-198)}}{2(2}$$ we plugged $a = 2$, $b = -4$, and $c = -198$ into the equation

$$W = \frac{4 \pm \sqrt{16\ \ (-1584)}}{2(2)}$$

$$W = \frac{4 \pm \sqrt{16 + 1584}}{4}$$

$$W = \frac{4 \pm \sqrt{1600}}{4}$$

$$W = \frac{4 \pm \sqrt{40^2}}{4}$$ we factored the squared from the radical

$$W = \frac{4 \pm 40}{4}$$

$$W = \frac{4 + 40}{4} = \frac{44}{4} = 11 \quad \text{or} \quad W = \frac{4 - 40}{4} = \frac{-36}{4} = -9.$$

Therefore, the solutions of $2W^2 - 4W - 198 = 0$ are $W = 11ft.$ and $W = -9ft.$ However, W is the width of the lawn yard, a value of $W = -9ft.$ does not make sense. Hence the only acceptable solution is $W = 11ft.$ Given the length, L, is $4\,ft.$ less than twice the width, the length is $2W - 4$, the measure of the length is

$L = 2W - 4$

$L = 2\times 11 - 4$ we plugged $W = 11$ into the equation

$L = 22 - 4$

$L = 18.$

Check: To verify our answer, we need to plug the accepted solution into the equation. So,

$2W^2 - 4W - 198 = 0$

$2(11)^2 - 4\times 11 - 198 \overset{?}{\Leftrightarrow} 0$ we plugged $W = 11$ into the equation

$2(121) - 44 - 198 \overset{?}{\Leftrightarrow} 0$

$242 - 44 - 198 \overset{?}{\Leftrightarrow} 0$

$242 - 242 \overset{?}{\Leftrightarrow} 0$

$0 = 0$ true, the solution $W = 11$ is a correct answer.

Now that we verified the width = 11*ft.*, the length is also verified to be 18*ft.*

5.3. Example: Projectile Motion

A firework is shot upwards with an initial velocity of 140 feet per second. How many seconds will it take to reach a height of 280 feet? Round to the nearest tenth of a second.

Hint: The height, h, of an object shot upwards into the air is determined by the formula $h = -16t^2 + v_0t$, where h is the height in feet, v_0 is the initial velocity in feet per second, and t is the time in second.

Solution

To determine the time, t, we are given the formula $h = -16t^2 + v_0t$, so,

$h = -16t^2 + v_0t$

$280 = -16t^2 + 140t$ we replaced h and v_0, by 280*ft.* and 140*ft/s* respectively

$280 - 280 = -16t^2 + 140t - 280$ we added – 280 to both sides

$0 = -16t^2 + 140t - 280$

$16t^2 - 140t + 280 = 0$ we multiplied both sides by – 1.

To solve the above quadratic equation, we compare it with $ax^2 + bx + c = 0$, we find $a = 16$, $b = -140$, and $c = 280$. Hence using the quadratic formula, we will have:

$t = \frac{-b \pm \sqrt{b^2 - 4a}}{2a}$

$t = \frac{-(-140) \pm \sqrt{(-140)^2 - 4(16)(280)}}{2(16)}$ we plugged $a = 16$, $b = -140$, and $c = 280$ into the equation

$t = \frac{140 \pm \sqrt{19{,}600 - 17{,}920}}{32}$

$t = \frac{140 \pm \sqrt{1{,}680}}{-32}$

$t = \frac{140 \pm 41}{32}$

$t = \frac{140 + 41}{32} = \frac{181}{32} = 5.7$ or $t = \frac{140 - 41}{32} = \frac{99}{32} = 3.1.$

Therefore, the solutions of $-16t^2 + 140t - 280 = 0$ are $t = 5.7s.$ and $t = 3.1s.$

Check: To verify our answer, we need to plug each solution into the equation. So,

For t = 3.1s. we will have:

$$-16t^2 + 140t - 280 = 0$$

$$-16(3.1)^2 + 140(3.1) - 280 \overset{?}{\Leftrightarrow} 0$$ we plugged $t = 3.1$ into the equation

$$-16(9.61) + 434 - 280 \overset{?}{\Leftrightarrow} 0$$

$$-154 + 434 - 280 \overset{?}{\Leftrightarrow} 0$$

$$-434 + 434 \overset{?}{\Leftrightarrow} 0$$

$$0 = 0$$ true, the solution $W = t = 3.1s$. is a correct answer.

For t = 5.7s. we will have:

$$-16t^2 + 140t - 280 = 0$$

$$-16(5.7)^2 + 140(5.7) - 280 \overset{?}{\Leftrightarrow} 0$$ we plugged $t = 5.7$ into the equation

$$-16(32.4) + 798 - 280 \overset{?}{\Leftrightarrow} 0$$

$$-518 + 798 - 280 \overset{?}{\Leftrightarrow} 0$$

$$-798 + 798 \overset{?}{\Leftrightarrow} 0$$

$$0 = 0$$ true, the solution $W = t = 5.7s$. is a correct answer.

Now that we verified both answers, we can conclude that the time will it take to reach a height of 280 feet is t = 3.1s. and time the way down at t = 5.7s.

5.4. Example: Cost, Revenue, and Profit

Profit is typically defined in business as the difference between the amount of revenue and the amount of costs. For example, let us assume that the cost to a manufacturer to produce a certain number of goods is C and the revenue generated by selling those goods is R. The profit, P, can then be defined as $P = R - C$. Let us present two related examples.

As an example, consider a soft-drink manufacturer can produce 4,000 cases of soda in a month at a total cost of \$26,000, and 6,000 cases of soda at a total cost of \$36,000. The revenue for selling x box of soda per month is given as $R(x) = 0.01x^2 - 45x$. Find the manufacturer's monthly fixed costs, marginal cost per case of soda, and the monthly profit for selling 6,000 boxes.

Solution

We first need to find the cost function $C(x) = mx + b$, where m is the marginal cost, and b is the fixed costs for producing x boxes of soft drink per month. this case. To find the cost function, we need to find both m and b. Since this is a linear function, we can find using the formula for finding the slope of a line between two points, where the points with the number of boxes and their associated cost are (x_1, C_1) and (x_2, C_2).

The slope formula is:

$$m = \frac{C_2 - C_1}{x_2 - x_1} = \frac{36{,}000 - 2\ {,}000}{6{,}000 - 4{,}000} = \frac{10{,}000}{2{,}000} = 5.$$

So, the marginal cost is $5. This means it costs $5 per case to manufacture the soft drinks. Now that we have *m,* we can find *b*. We know that $C(x) = 5x + b$, so we can use what we were given to find *b*. We can choose either ordered pair to do this, either (4,000, 26,000) or (6,000, 36,000). Choosing the first ordered pair we have.

$C(x) = mx + b$

$26{,}000 = 5(4{,}000) + b$ — we plugged respectively, 26,000, 5, and 4000 as the production cost, marginal cost, and the number boxes into the equation

$26,\ 000 = 20{,}000 + b$

$26,\ 000 - 26{,}000 = 20{,}000 + b - 26{,}000$ — we added – 26,000 to both sides

$0 = b - 6{,}000$

$0 + 6{,}000 = b - 6{,}000 + 6{,}000$ — we added 6,000 to both sides

$b = 6{,}000.$

Therefore, the fixed costs (rent, electricity, and other bills) are $b = 6{,}000$ and the cost function is $C(x) = 5x + 6{,}000$.

We now can find the profit of selling 6,000 boxes. We have learned that profit selling x number of goods, $P(x)$, is typically defined as the difference between the amount of revenue $R(x)$ and the amount of costs $C(x)$, or $P(x) = R(x) - C(x)$. So,

$P(x) = R(x) - C(x)$

$P(x) = 0.01x^2 - 45x - (5x + 6{,}000)$ — we replaced $R(x)$ and $C(x)$ in the equation

$P(x) = 0.01x^2 - 45x - 5x - 6{,}000$

$P(x) = 0.01x^2 - 50x - 6{,}000$

$P(x) = 0.01(6{,}000)^2 - 50(6{,}000) - 6{,}000$ — we plugged $x = 6{,}000$ into the equation

$P(x) = 0.01(36{,}000{,}000) - 300{,}000 - 6{,}000$

$P(x) = 360{,}000 - 306{,}000$

$P(x) = 54{,}000.$

Therefore, the amount of profit for selling 6,000 boxes is $P(x) = \$54{,}000$.

5.5. Example: Cost, Revenue, and Maximum Profit

Duke's Appliances Company manufactures and sells countertop microwave ovens. Each oven costs $200 to make, and the company's fixed costs are $20,000. In addition, Duke knows that the price of each oven is expected by the function $p(x) = 2{,}000 - 2x$, where $p(x)$ is the price as a function of units, *x,* sold. Find:

a) The company's revenue function, $R(x)$.

b) The company's cost function, $C(x)$.

c) The company's profit function, $P(x)$.

d) The two points, (x1, y1) and (x2, y2) at which the company breaks even.

e) The output level that maximizes the company's profit, and the maximum profit.

Solution

a) Revenue is equal to the number of units sold times the price per unit. To obtain the revenue function, we need to multiply the output level by the price function.

$$R(x) = xp(x)$$

$$R(x) = x(2{,}000 - 2x)$$

$$R(x) = 2{,}000x - 2x^2.$$

b) A business' costs include the fixed cost of \$20,000 as well as the variable cost of \$300 per oven. To obtain the cost function, we need to add fixed cost and variable cost together.

$$C(x) = (\text{fixed costs}) + x(\text{variable costs})$$

$$C(x) = 20{,}000 + x(200)$$

$$C(x) = 20{,}000 + 200x.$$

c) The profit that this business makes is equal to its revenue minus its costs. To obtain the profit function, we need to subtract costs from revenue.

$$P(x) = R(x) - C(x)$$

$$P(x) = (20{,}00x - 2x^2) - (20{,}000 + 200x)$$

$$P(x) = 2{,}000x - 2x^2 - 20{,}000 - 200x$$

$$P(x) = -2x^2 + 1{,}800x - 20{,}000.$$

d) In the previous chapter of this text, we learned that a company's break-even points occur where the revenue function and the cost function have the same value. This also implies that the profit function equals zero at break-even points. These points are found most easily by graphing two functions.

Earlier, we learned the graph of a quadratic function, $f(x) = ax^2 + bx + c$ is a parabola in which the x-coordinate of its vertex is located at $x = -\frac{b}{2a}$, and the y-coordinate of the vertex is located at $y = \frac{-b^2+4ac}{4a}$. Also, the y-intercept is located at the point $(0, c)$, the line of symmetry has the equation of $x = -\frac{b}{2a}$, and if a is negative the graph is open downward.

Comparing $R(x) = -2x^2 + 2{,}000x$ with $f(x) = ax^2 + bx + c$, we find $a = -2$, $b = 2{,}000$, and $c = 0$. Hence, the vertex x-coordinate, y-coordinate, and the equation of the line of symmetry can be obtained as:

$$x\text{-coordinate} = -\frac{b}{2a}$$

x-coordinate $= -\frac{2{,}000}{2(-2)}$ we plugged $a = -2$ and $b = 2{,}000$ into the equation

x-coordinate $= -\frac{2{,}000}{-4}$

x-coordinate $= \frac{2{,}000}{4} = 500$

y-coordinate $= \frac{-b^2+4ac}{4a}$

y-coordinate $= \frac{-2{,}000^2+4(-2)(0)}{4(-2)}$ we plugged $a = -2$, $b = 2{,}000$ and $c = 0$ into the equation

y-coordinate $= \frac{-4{,}000{,}000}{-8} = 500{,}000$

the line of symmetry has the equation $= -\frac{b}{2a}$

the line of symmetry has the equation, $x = -\frac{2{,}000}{2(-2)}$ we plugged $a = -2$ and $b = 2{,}000$ into the equation

the line of symmetry has the equation, $x = 500$.

To graph we just need to find the zeroes of $R(x) = -2x^2 + 2{,}000x$. So,

$-2x^2 + 2{,}000x = 0$

$x(-2x + 2{,}000) = 0$ we factored out x.

Using the zero-factor principle, we will have.

$x = 0$, and $-2x + 2{,}000 = 0$.

For $-2x + 2{,}000 = 0$, we have:

$-2x + 2{,}000 = 0$

$-2x + 2{,}000 - 2{,}000 = 0 - 2{,}000$ we added $-2{,}000$ to both sides

$-2x = -2{,}000$

$-\frac{1}{2}(-2x) = -\frac{1}{2}(-2{,}000)$ we multiplied both sides by the reciprocal of the variable

$x = 1{,}000$.

Therefore, the zeroes of the $R(x) = -2x^2 + 2{,}000x$ are $x = 0$ and $x = 1{,}000$.

Having the above information, the graphs of $R(x) = -2x^2 + 2{,}000x$ and $C(x) = 20{,}000 + 200x$, is depicted as follows. The graph shows two points where the linear cost function and the parabolic revenue function intersect.

We let $R(x) = C(x)$ to find the break-even points coordinates algebraically and round to the nearest integer as needed. So,

$R(x) = C(x)$

$-2x^2 + 2{,}000x = 20{,}000 + 200x$ we replaced $R(x)$ and $C(x)$ into the equation.

$-2x^2 + 2{,}000x - 20{,}000 - 200x = 0$ we moved all terms to the left-side of equation

$-2x^2 + 1{,}800x - 20{,}000 = 0.$

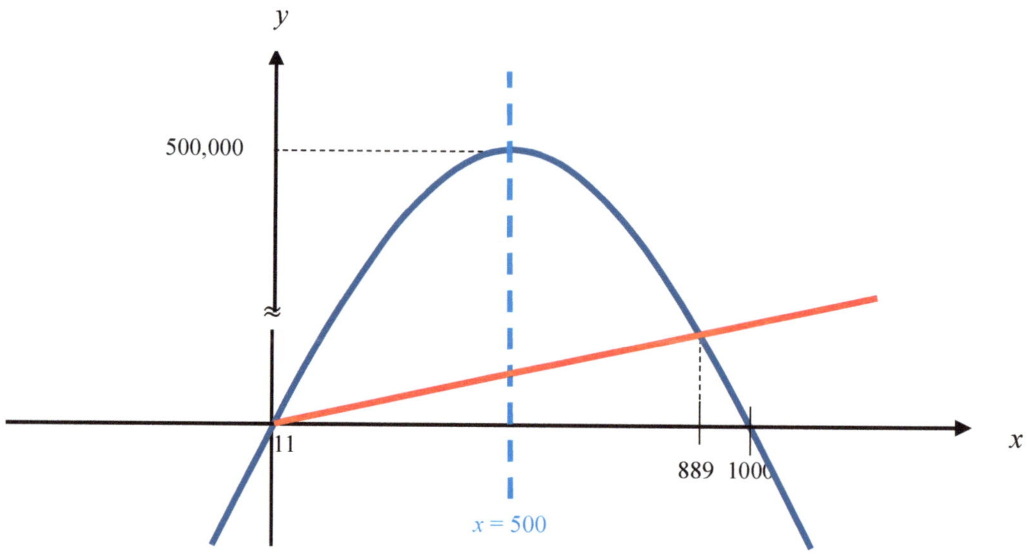

Comparing $-2x^2 + 1{,}800x - 20{,}000 = 0$ with $ax^2 + bx + c = 0$, we find $a = -2$, $b = 1{,}800$, and $c = -20{,}000$. Using the quadratic formula, we will have:

$$x = \frac{-b \pm \sqrt{b^2 - 4ac}}{2a}$$

$$x = \frac{-1{,}800 \pm \sqrt{1800^2 - 4(-2)(-20{,}000)}}{2(-2)}$$ we plugged $a = -2$, $b = 1{,}800$, and $c = -20{,}000$ into the equation

$$x = \frac{-1{,}800 \pm \sqrt{3{,}240{,}000 - 4(40{,}000)}}{-4}$$

$$x = \frac{-1{,}800 \pm \sqrt{3{,}240{,}000 - 160{,}000}}{-4}$$

$$x = \frac{-1{,}800 \pm \sqrt{3{,}080{,}000}}{-4}$$

$$x = \frac{-1{,}800 \pm 1{,}754.99}{-4}$$

$$x = \frac{-1{,}800 + 1{,}754.99}{-4} = \frac{-45.01}{-4} = 11.25.$$

$$x = \frac{-1{,}800 - 1{,}754.99}{-4} = \frac{-3{,}554.99}{-4} = 888.75.$$

Therefore, the x-coordinates of the break-even points are $x = 11.25$ and $x = 888.75$. Rounding these values, we have $x = 11$and $x = 889$. To find their associated y-coordinates, we plugged them separately into the either $y = -2x^2 + 2{,}000x$ or $y = 20{,}000 + 200x$. Using the later equation, which is easier to use, we will have.

$y = 20{,}000 + 200x$

$y = 20{,}000 + 200(11.25)$ we plugged $x = 11.25$ into the equation

$y = 20{,}000 + 2{,}250$

$y = 22,250.$

$y = 20,000 + 200(888.75)$ we plugged $x = 11.25$ into the equation

$y = 20,000 + 177,750$

$y = 197,750.$

Therefore, the break-even points' coordinates are, (11, 22,250) and (889, 197,750).

e) To find the output level that maximizes the company's profit, we look at the quadratic function of profit, $P(x) = -2x^2 + 1,800x - 20,000$. Comparing this function with the standard form of a quadratic function, $f(x) = ax^2 + bx + c = 0$, we find $a = -2$, $b = 1,800$, and $c = -20,000$. Since a is negative, we know that the curve of the profit function $P(x)$ is a downward open curve, and its maximum is located at its vertex. We also know that the x-coordinate of the vertex is $x = -\frac{b}{2a}$, and the y-coordinate of the vertex is $y = \frac{-b^2 + 4ac}{4a}$. So,

$$x = -\frac{b}{2a}$$

$$x = -\frac{1,800}{2(-2)} = -\frac{1,800}{-4} = 450$$ we plugged $a = -2$ and $b = 1,800$ into the equation

$$y = \frac{-b^2 + 4ac}{4a}$$

$$y = \frac{-180\ ^2 + 4(-2)(-20,000)}{4(-2)}$$ we plugged $a = -2$, $b = 1,800$, and $c = -20,000$ into the equation

$$y = \frac{-3,240,000 + 160,000}{-8}$$

$$y = \frac{-3,240,000 + 160,000}{-8}$$

$$y = \frac{-3,080,000}{-8} = 385,000.$$

Therefore, the number of microwave ovens sold for maximizing the company's profit is and the maximum profit for this sell's volume is $.

5.6. Example: Age Problem

Helen is 3 years older than her brother Peter. If the product of their ages is 2 more than four times the sum of their ages, find the ages of Helen and Peter.

Solution

We first need to assign a variable. Let us denote x = Helen's age. This means, since Peter is 3 years younger than Helen, $x - 3$ = Peter's age. Now we just need to make an equation. Since their ages is 2 less than four times the sum of their ages, we will make the below equation.

Helen's age $= x$

Peter's age $= x - 3$

Product of their ages $= x(x - 3)$

Sum of their ages = $x + (x - 3)$

Four times of the sum of their ages = $4[x + (x - 3)]$

Product of their ages = Four times of the sum of their ages + 2

$x(x - 3) = 4[x + (x - 3)] + 2$ — we replaced the product of their ages and four times of the sum of their ages into the equation

$x^2 - 3x = 4x + 4x - 12 + 2$ — we applied the distributive property

$x^2 - 3x = 8x - 10$

$x^2 - 3x - 8x + 10 = 0$ — we moved all terms to the right-side

$x^2 - 11x + 10 = 0$

$(x - 1)(x - 10) = 0$ — we wrote the quadratic equation in a factored form.

Using the zero-factor principle, we will have:

$(x - 1) = 0$ and $(x - 10) = 0$

For $(x - 1) = 0$

$x - 1 = 0$

$x - 1 + 1 = 0 + 1$ — we added 1 to both sides

$x = 1.$

For $(x - 10) = 0$

$x - 10 = 0$

$x - 10 + 10 = 0 + 10$ — we added 10 to both sides

$x = 10.$

Therefore, Helen's age is $x = 1$ or $x = 10$. However, since Peter is 3 years younger than Helen, Helen age cannot be 1. So, Helen must be 10 years old, and Peter must be $10 - 3 = 7$ years old.

5.7. Example: Geometry, Rectangular Area

Given a rectangular swimming pool with is the length of 4 yards less than three times its width and an area of 84 square yards, find the dimensions of the pool.

Hint: To find the area of a rectangle, multiply the length by the width. The formula is: $A = l \times w$, where A is the area, l is the length, and w is the width.

Solution

To solve this case, we first need to define a variable to make an equation. Let us denote:

l = the length of the swimming pool

w = the width of the swimming pool

l = the area of the swimming pool

We have a geometric figure problem which is a common challenge in the real-world. Hence, as we always do in such a situation, we draw a picture to help us with solving this case. Our rectangle must look like this.

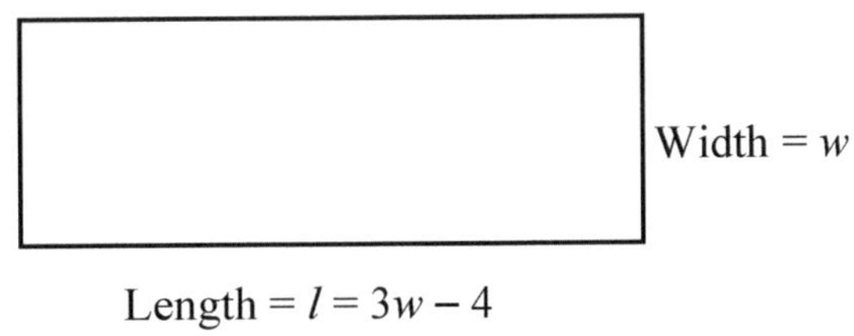

Since the length of the swimming pool is 5 yards less than three times of the width, the equation presenting the area can be written as:

$A = l\,w$ and $l = 3w - 4$

$A = l\,w = (3w - 4)\,w$	we replace $l = 3w - 4$ into the equation.
$A = 3w^2 - 4w$	we applied the distribution property
$84 = 3w^2 - 4w$	we plugged $A = 84$ into the equation
$84 - 84 = 3w^2 - 4w - 84$	we added -84 to both sides
$3w^2 - 4w - 84 = 0.$	

We have a quadratic equation with a leading coefficient not 1. The procedure to solve this quadratic equation is to compare it with the standard form of a quadratic equation, $ax^2 + bx + c = 0$. By this comparison we get $a = 3$, $b = -1$,.and $c = -84$. Then, we get $ac = -252$.

Hence, we need to look for two numbers p and q which their product is -420 and their sum is -1. We need to list all factors of $ac = -252$ until a pair of the anticipated sum $b = -1$ is attained.

Factors of $ac = -252$	Sum of factors
$1, -252$	-251
$-1, 252$	251
$2, -126$	-124
$-2, 126$	124
$3, -84$	-81
$-3, 84$	81
$4, -63$	-59
$-4, 63$	59
$6, -42$	-36
$-6, 42$	36
$7, -36$	-29
$-7, 36$	29
$9, -28$	-19

$-9, 28$	19
$12, -21$	-9
$-12, 21$	9
$14, -18$	-4
$-14, 18$	4

So, $p = -18$ and $q = 14$.

Rewrite the original expression as $ax2 + px + qx + c$.

$3w^2 - 4w - 84 = 0$

$3w^2 - 18w + 14w - 84 = 0$

$3w(w - 6) + 14(w - 6) = 0$ we factored out the GCF of each part

$(w - 6)(3w + 14) = 0$ again we factored out the GCF of the expression.

Using the zero-factor principle we will have.

$w - 6 = 0$

$w - 6 + 6 = 0 + 6$ we added 6 to both sides

$w = 6$ One of the answers

$3w + 14 = 0$

$3w + 14 - 14 = 0 - 14$ we added -14 to both sides

$3w = -14$

$\frac{1}{3} \times 3w = \frac{1}{3} \times (-14)$ we multiplied both sides by the reciprocal of the variable

$w = \frac{-14}{3}$ the other answer.

Therefore, the solutions of $3w^2 - 4w - 84 = 0$, are $w = 6$ and $w = -\frac{14}{3}$.

Since w represents the width of a rectangle, it cannot be negative. So, $w = 6$ yds. We now need to find l. Recall, $l = 3w - 4$. So, plugging 6 in for w we get

$l = 3w - 4$

$l = 3(6) - 4$ we plugged $w = 6$ into the equation

$l = 18 - 4$

$l = 14$.

Therefore, the length of the swimming pool is $l = 14$ yds., and its width is $w = 6$ yds.

5.8. Example: Projectile Motion, Polynomial Inequality

Janet drops a ball off the top of an 8-foot pool slide. The height h in feet of the ball to hit the water is given by $h(t) = -16t^2 + 5$, where t is the time in seconds. Her friend Joe desires to catch the ball between 1 foot and 4 feet off the top of the water. Between what two times should Joe try to catch the ball?

Solution

We are looking for the time, t, the ball should be caught. Hence, we need to solve $h(t)$ using an inequality for the time between 1 and 4 seconds. Considering word between 1 and 4 seconds does not mean an inclusive interval, we use $<$ sign instead of $\leq$ sign. So, the inequality we need to solve is.

$$1 < -16t^2 + 5 < 4.$$

The above inequality is a compound inequality. Hence, we need first to rewrite the compounded inequality $1 < -16t^2 + 5 < 4$ as two inequalities, $1 < -16t^2 + 5$ and $-16t^2 + 5 < 4$. Considering the first one we have:

For $1 < -16t^2 + 8$, we have.

$$1 < -16t^2 + 5$$

$$1 - 5 < -16t^2 + 5 - 5$$

to isolate the variable, we added – 5 to both sides

$$-4 < -16t^2$$

$$-\frac{1}{16}(-4) > -\frac{1}{16}(-16t^2)$$

we multiplied both side by the reciprocal of the variable, since the reciprocal is negative the sign has been changed

$$\frac{4}{16} > t^2$$

$$\sqrt{\frac{4}{16}} > t$$

we took the square root. we don't need to worry about the negative values, since t can't be negative

$$\sqrt{\frac{2^2}{4^2}} > t$$

$$t < \frac{2}{4}$$

we took out the squared from radicand

$$t < \frac{1}{2}.$$

For $-16t^2 + 5 < 4$, we have:

$$-16t^2 + 5 < 4$$

$$-16t^2 + 5 - 5 < 4 - 5$$

to isolate the variable, we added – 5 to both sides

$$-16t^2 < -1$$

$$-\frac{1}{16}(-16t^2) > -\frac{1}{16}(-1)$$

we multiplied both side by the reciprocal of the variable, since the reciprocal is negative the sign has been changed

$$t^2 > \frac{1}{16}$$

$$t > \sqrt{\frac{1}{16}}$$

we took the square root. we do not need to worry about the negative values, since t cannot be negative

$t > \sqrt{\frac{1}{4^2}}$

$t > \frac{1}{4}$ we took out the squared from radicand.

Therefore, the two times should Joe try to catch the ball are $t < \frac{1}{2}$ and $t > \frac{1}{4}$ seconds. Hence, $\frac{1}{4} < t < \frac{1}{2}$.

Based on the above information, the solution of $1 < -16t^2 + 5 < 4$ as a set interval notation is $(\frac{1}{4}, \frac{1}{2})$. To represent the final answers on a number line we will have:

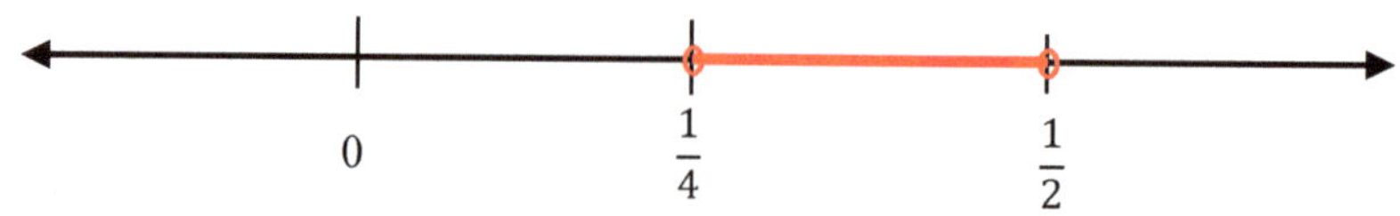

It would be interesting to look at the graphical solution of the given compound inequality. The graph of $-16t^2 + 5$ is a parabola. If we graph this parabola and the two graphs of line $t = 1$, and line $t = 4$, the intersections of lines and parabola would obtain the interval we are looking for.

Figure below illustrates the graph of $y = -16t^2 + 5$ and two lines, $t = 1$, and line $t = 4$. In this figure we see that line $t = 1$ intersects the parabola at two points of $t = \frac{1}{2}$ and $t = -\frac{1}{2}$. Apparently, $t = -\frac{1}{2}$ cannot be an acceptable answer, since we cannot have a negative value for the time Joe is willing to catch the ball.

Likewise, line $t = 4$ intersects the parabola at two points of $t = \frac{1}{4}$ and $t = -\frac{1}{4}$, in which the negative time cannot be a correct one. Due to the inequality of $1 < -16t^2 + 5 < 4$, the only part of parabola which is greater than one and less than four is the solution of the compound inequality. To better reveal this part, we exposed it with a dark red color.

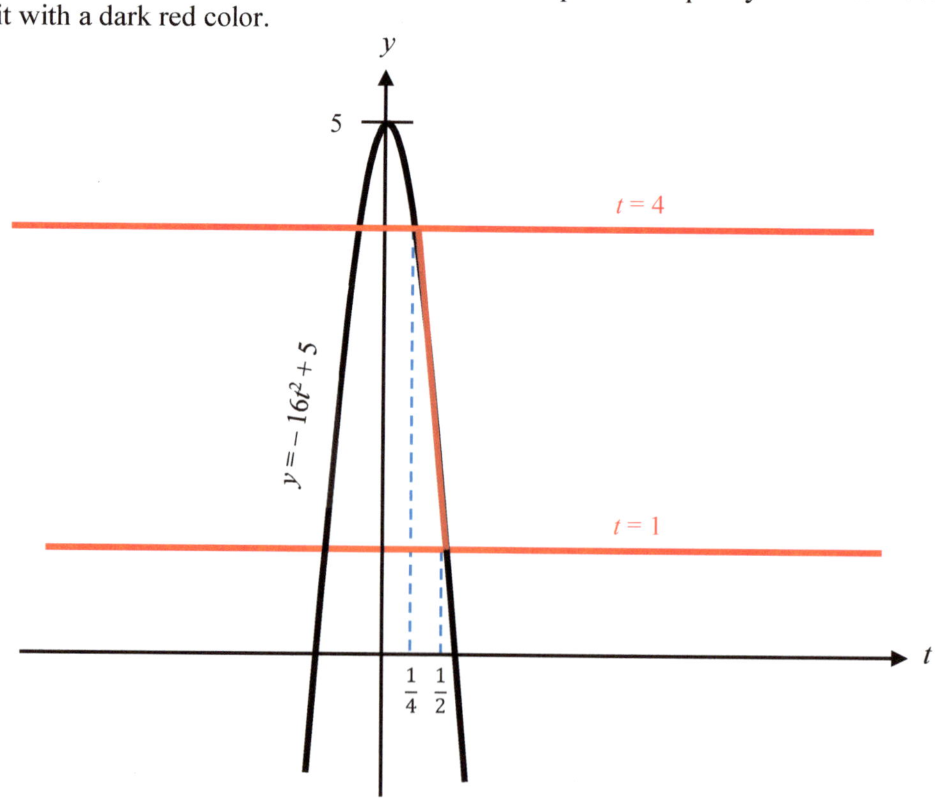

5.9. Example: Volume Analysis, Cylinder

Given a cylindrical barrel wine with a height of 12 inches and the base with the external diameter of 12 inches and the internal diameter of 11 inches, find the volume of the barrel wine.

Hint: The formula for the volume V of a cylinder is found by multiplying the area of its base by the height and is defined as: $V=\pi r^2 h = \frac{\pi d^2 h}{4}$. , where r is the base radius, d is the base dimeter, h is the height of the cylinder and in decimal form, the value of π is approximately 3.14.

Solution

To find the volume of the barrel wine, we simply need to use its formula, $V=\frac{\pi d^2 h}{4}$, and we plugged in the given parameters. Note that for the diameter of the base, we need to use the internal diameter. Hence, we have a cylinder with a base diameter of 11 inches, and height of 12 inches. so,

$$V=\frac{\pi d^2 h}{4}$$

$$V=\frac{3.14(11^2)(12)}{4}$$

we plugged $\pi = 3.14$, the base diameter and the height into the equation

$$V=\frac{3.14(121)(12)}{4}$$

$$V=\frac{4{,}559.28}{4}$$

$$V= 1{,}139.82.$$

Therefore, the volume of the barrel wine is $V= 1{,}139.82$ cubic inches.

5.10. Example: Volume Analysis, Cuboid

Given a rectangular sheet of cardboard 14 inches by 24 inches. A small square of the same size is cut from each corner and each side is folded up along cuts to form a lidless box.

a) What is the maximum volume, $V(x)$, of the box?
b) What would be the box volume if the size of the square for producing the box is equal to 2 inches?

Hint: The volume of the cuboid (rectangular box) is $V=lwh$, where l is the length, w is the width and h is the height.

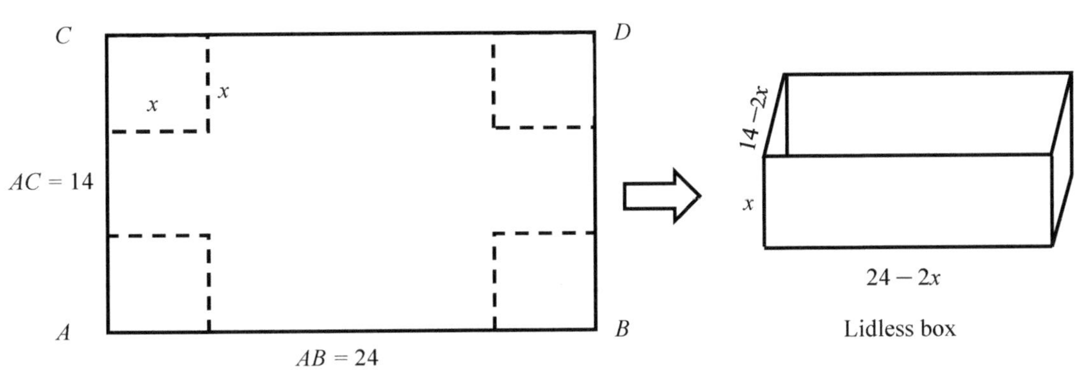

Cardboard sheet

Lidless box

Solution

a) As it can be seen from the above figure, the dimensions of the box are $(24 - 2x)$ inches long by $(14 - 2x)$ inches wide by x inches high. Give the formula for the volume of the rectangular box, we will have.

$V(x) = (length)(width)(height)$

$V(x) = (24 - 2x)\,(14 - 2x)\,x$ — we replace length, width, and height into the equation.

To find the maximum volume, max. $V(x)$, of the box we need to use the graphical approach. To do so, we know that the polynomial $V(x)$ has three zeros (since its degree is three) and has two turning points.

Hence, using this information, we need to find zeros and plug some arbitrary point into the $V(x)$, and then schematically sketch the graph. To find zeros, we let $V(x) = (24 - 2x)\,(14 - 2x)\,x = 0$ and then we use the zero-factor principle. So,

$(24 - 2x)\,(14 - 2x)x = 0$

$24 - 2x = 0, \quad 14 - 2x = 0 \quad \text{and} \quad x = 0.$

For $24 - 2x = 0$ we will have:

$24 - 2x = 0$

$24 - 2x - 24 = 0 - 24$ — we added $- 24$ to both sides

$- 2x = - 24$

$-\frac{1}{2}(- 2x) = -\frac{1}{2}(- 24)$ — we multiplied both sides by the reciprocal of the variable

$x = \frac{24}{2}$

$x = 12.$

For $14 - 2x = 0$ we will have:

$14 - 2x = 0$

$14 - 2x - 14 = 0 - 14$ — we added $- 14$ to both sides

$- 2x = - 14$

$-\frac{1}{2}(- 2x) = -\frac{1}{2}(- 14)$ — we multiplied both sides by the reciprocal of the variable

$x = \frac{14}{2}$

$x = 7.$

Therefore, the zeros of the polynomial $V(x) = (24 - 2x)(14 - 2x)x$, are $x = 0$, $x = 12$, and $x = 7$. If we plug an arbitrary point $x = 1$ into $V(x)$ we get $V(1) = [24 - 2(1)][14 - 2(1)](1) = 264$. Since, $V(x) = 0$ at $x = 0$ and $V(x) = 264$ at $x = 1$, we figure out that the curve of $V(x)$ increases in the range of $0 < x < 1$. Hence, here is the schematic graph of the polynomial $V(x) = (24 - 2x)(14 - 2x)x$.

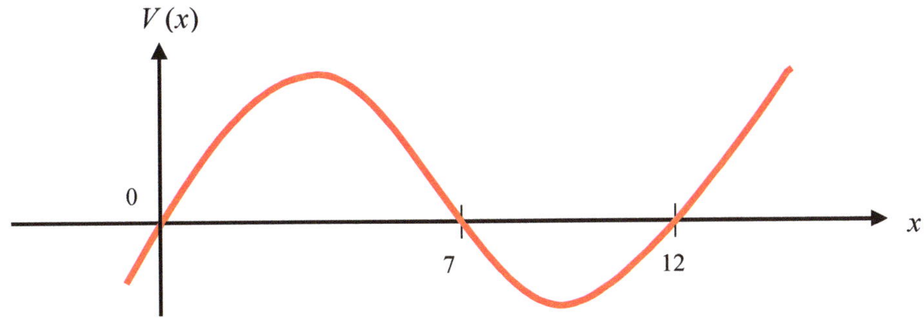

To find the maximum volume, $V(x)$, of the box, we refer to the above graph and we will spot that the maximum is somewhere around $x = 3$. However, we need to find a closer estimate which can be done by magnifying the peak of the graph. Hence, we list some points around $x = 3$ and we get their associated $V(x)$. Let us start at $x = 2.7$ and increase it sequentially by one tenth.

For $x = 2.7$, $V(x) = 431.89$

For $x = 2.8$, $V(x) = 432.77$

For $x = 2.9$, $V(x) = 432.80$

For $x = 3.0$, $V(x) = 432.00$.

Since at a point after 2.9 the value of $V(x)$ declines, the maximum must be between $x = 2.8$ and $x = 3.0$. Let us choose the mid points of "$x = 2.8$ and $x = 2.9$", and "$x = 2.9$ and $x = 3.0$".

For $x = 2.85$, $V(x) = 432.89$

For $x = 2.95$, $V(x) = 432.50$.

Therefore, the maximum volume, $V(x)$, of the box is $V(x) = 432.89$, where $x = 2.85$.

b) To find the box volume for each of four cutting squares with the size of 2 inches, we simply need to plug $x = 2$ into the polynomial equation $V(x) = (24 - 2x)(14 - 2x)x$. So,

$V(x) = (24 - 2x)(14 - 2x)x$

$V(x) = [24 - 2(2)][(14 - 2(2)](2)$ we plugged $x = 2$ into the equation

$V(x) = (24 - 4)(14 - 4)(2)$

$V(x) = (20)(10)(2)$

$V(x) = 400$.

Therefore, for the size of each cutting square equal to 2 inches, they will produce a box with the volume of $V(x) = 400$ cubic inches.

5.11. Example: Machine Production Rate

The weekly sociopolitical magazine needs to publicize the Trump speech on issues such as the climate crisis, nuclear weapons and the future of America and the world. The editor wants the magazine to be printed as soon as possible. He has asked the printer department to run an extra printing press to get the printing done more quickly. Press machine 1 takes 6 hours more than Press machine 2 to accomplish the job. When both machines are running, they can print the job in 4 hours. How many hours it would take for each press machine to print the job alone?

Solution

We need to find the time that would take each machine separately to complete the job. To make a mathematical model for this case, we first need to define a variable. Let us denote x as the number of hours for machine 2 to complete the job. So,

$x =$ the number of hours to complete the job by machine 2

$x + 6 =$ the number of hours to complete the job by machine 1

$6 =$ the number of hours to complete the job by machine 1 and machine 2, working simultaneously.

Using the above information, we will get:

$p_1 =$ Portion of job completed per hour by machine 1 $= \frac{1}{x}$

$p_2 =$ Portion of job completed per hour by machine 2 $= \frac{1}{x+6}$

$p_3 =$ Portion of job completed per hour by two machines $= \frac{1}{4}$.

We know that the portion of a job completed per hour by two machines is equal to the portion of job completed per hour by machine 1 plus the portion of job completed per hour by machine 2. So,

$p_3 = p_1 + p_2$

$\frac{1}{4} = \frac{1}{x} + \frac{1}{x+6}$ we replaced $p_3 = \frac{1}{4}$, $p_1 = \frac{1}{x}$, and $p_2 = \frac{1}{x+6}$ in the equation.

We now need to solve the above equation for x. so,

$\frac{1}{4} = \frac{1}{x} + \frac{1}{x+6}$

$\frac{1}{4} - \frac{1}{4} = \frac{1}{x} + \frac{1}{x+6} - \frac{1}{4}$ we added $-\frac{1}{4}$ to both sides

$0 = \frac{1}{x} + \frac{1}{x+6} - \frac{1}{4}$.

$4x(x+6)(0) = 6x(x+6)(\frac{1}{x} + \frac{1}{x+6} - \frac{1}{6})$ we multiplied both sides of equation by the LCD, $4x(x+6)$

$4x(x+6)(\frac{1}{x} + \frac{1}{x+6} - \frac{1}{4}) = 0$

$\frac{4x(x+6)}{x} + \frac{4x(x+6)}{x+6} - \frac{4x(x+6)}{4} = 0$ we applied the distribution property

$4(x+6)\frac{x}{x} + (4x)\frac{(x+6)}{x+6} - x(x+6)\frac{4}{4} = 0$

$4(x+6) + (4x) - x(x+6) = 0$

$4x + 24 + 4x - x^2 - 6x = 0$ again, we applied the distribution property

$2x + 24 - x^2 = 0$

$-x^2 + 2x + 24 = 0$ — we rewrote the equation with the exponents in decreasing order

$-1(-x^2 + 2x + 24) = (-1)(0)$ — to get positive leading coefficient, we multiplied both sides by -1

$x^2 - 2x - 24 = 0.$

We now have a polynomial in the form $x^2 + bx + c$, with the leading coefficient of 1, in which $b = -2$ and $c = -24$. Hence, we can factor it by finding two numbers p and q, with a product of $c = -24$ and a sum of $b = -2$. To do so, we list factors of -24 until we get the sum of factors equal to -2. So,

Factors of − 24	Sum of factors
1, −24	−23
−1, 24	23
2, −12	−10
−2, 12	10
3, −8	−5
−3, 8	5
4, −6	**−2**
−4, 6	2

As we realize, there are two numbers, 4, and − 6 which their product is −24 and their sum − 2. So,

$x^2 - 2x - 24 = 0$

$(x - 6)(x + 4) = 0.$

Using the zero-factor principle, we will have.

$x - 6 = 0$ or $x + 4 = 0.$

For $x - 6 = 0$, we will have:

$x - 6 = 0$

$x - 6 + 6 = 0 + 6$ — we added 6 to both sides

$x = 6.$

For $x + 4 = 0$, we will have:

$x + 4 = 0$

$x + 4 - 4 = 0 - 4$ — we added -4 to both sides

$x = -4.$

Therefore, the solutions of $\frac{1}{4} = \frac{1}{x} + \frac{1}{x+6}$, are $x = 6$ and $x = -4$. However, since x represents the number of hours, it cannot have a negative value. Hence the only acceptable solution is $x = 6$.

Check: To verify our answer, we need to plug the solution into the equation. So,

$\frac{1}{4} = \frac{1}{x} + \frac{1}{x+6} = 0$

$\frac{1}{4} \overset{?}{\Leftrightarrow} \frac{1}{6} + \frac{1}{6+6}0$ we plugged $x = 6$ into the equation

$\frac{1}{4} \overset{?}{\Leftrightarrow} \frac{1}{6} + \frac{1}{12}$

$12(\frac{1}{4}) \overset{?}{\Leftrightarrow} 12(\frac{1}{6} + \frac{1}{12})$ we multiplied both sides by LCD, $\frac{1}{12}$

$\frac{12}{4} \overset{?}{\Leftrightarrow} \frac{12}{6} + \frac{12}{12}$

$3 \overset{?}{\Leftrightarrow} 2 + 1$

$3 = 3$ true, the solutions $x = 6$ is a correct answer.

Therefore, the number of hours to complete the job by machine 2 is $x = 6$. To find the number of hours it would take for machine 1 to print the job, we need to add 6 to the number of hours to complete the job by machine 2, $x = 6$. Hence, the number of hours takes for machine 1 to complete the job is $x + 6 = 6 + 6 = 12$ hours.

5.12. Example: Construction of a Christmas Tree

Charli is setting up a Christmas tree light display. He wants to make a Christmas tree in the shape of two right triangles, as shown below, and has two 10-foot strings of lights to use for each side. He will attach the lights to the top of a pole and to two stakes. He wants the height of the pole to be three times of the distance from the base of the pole to each stake plus 1-foot underneath of the tree to the ground. Find the height of the pole. Round the answer to the nearest tenth.

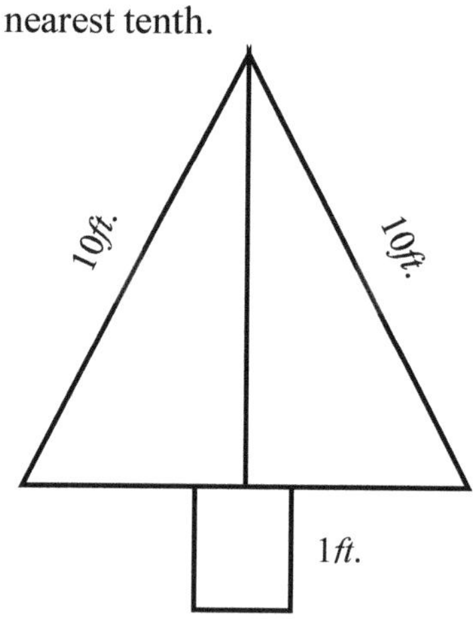

Hint: By Pythagorean theorem, in any right triangle, $a^2 + b^2 = c^2$, where a and b are the lengths of the legs, and c is the length of the hypotenuse.

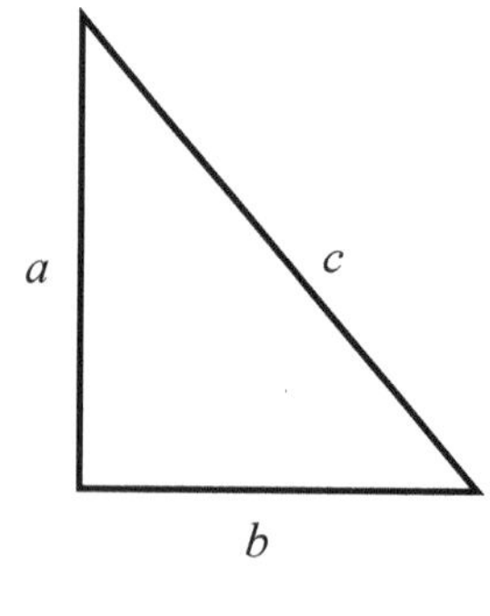

Solution

To solve this case let us draw the shape of one of the right triangles, triangle *ABC*, in which we denote *d* as the distance of the pole from ending point of the stake, as shown below.

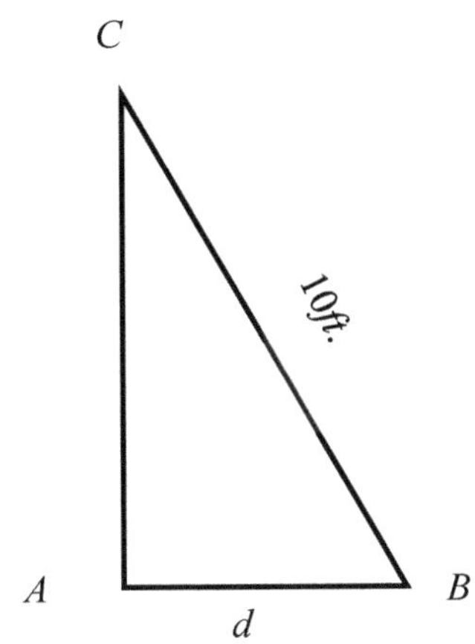

By the given information $AC = 3d$, using the Pythagorean theorem, we will have

$AB^2 + AC^2 = BC^2$

$d^2 + (3d)^2 = 10^2$ — we plugged $AB = d$, $AC = 3d$, $BC = 10$ into the equation

$d^2 + 9d^2 = 10^2$

$10d^2 = 100$

$\frac{1}{10}(10d^2) = \frac{1}{10} \times 100$ — we multiplied both sides by the reciprocal of the variable

$d^2 = 10$

$d = \sqrt{10}$ — we took the square root. we do not need to worry about the negative values, since *d* cannot be negative

$d = 3.1623 \cong 3.2$

$AC = 3d$

$AC = 3(3.1623)$ — we plugged $d = 3.1623$ into the equation

$AC = 9.4869$

$AC \cong 9.5.$

Therefore, the height of pole to the stake is $AC = 9.5$. Since the height of pole to the ground is 1-foot longer than its height to the stake, we need to add 1 *AC*. So,

The height of the pole = $AC + 1$

The height of the pole = $9.5 + 1$ — we plugged $AC = 9.5$ into the equation

The height of the pole = 10.5.

Therefore, the height of the pole is 10.5 feet.

5.13. Example: Resistors Resistance Analysis

Resistors are in parallel if their terminals are connected to the same two nodes. The equivalent overall resistance is smaller than the smallest parallel resistor. Given three resistors in parallel, equivalent overall resistance has been measured at 12 Ohms. Two of the resistors are known to be 6 ohms more than the first one, as shown below.

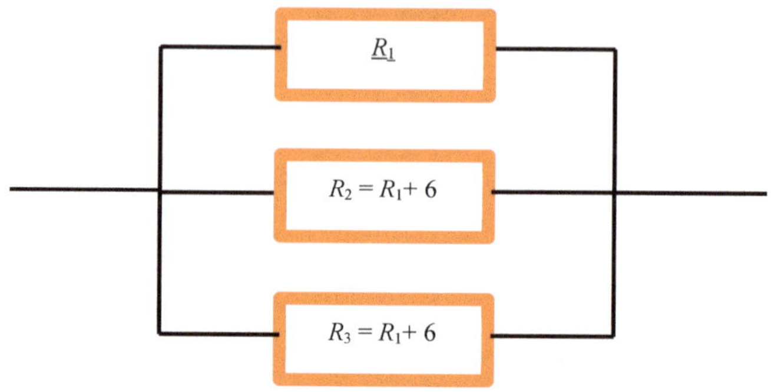

What are the values of the three resistors?

Hint: Given *n* resistors, each with the resistance value of R_j, $j = 1, 2, 3, \cdots, n$, the equivalent or overall resistance *R*, is determined by the equation $\frac{1}{R} = \frac{1}{R_1} + \frac{1}{R_2} + \frac{1}{R_3} + \cdots + \frac{1}{n}$.

Solution

To find the values for each of the individual resistors, we need to use the above equation. So,

$$\frac{1}{R} = \frac{1}{R_1} + \frac{1}{R_2} + \frac{1}{R_3}$$

$$\frac{1}{12} = \frac{1}{R_1} + \frac{1}{R_1+6} + \frac{1}{R_1+6}$$

we plugged $R = 12$, $R_2 = R_1 + 6$, and $R_3 = R_1 + 6$ into the equation

$$\frac{1}{12} = \frac{1}{R_1} + \frac{2}{R_1+6}$$

$$12R_1(R_1+6)\left(\frac{1}{12}\right) = 12R_1(R_1+6)\left(\frac{1}{R_1} + \frac{2}{R_1+6}\right)$$

to make a common denominator, we multiplied both sides by the LCD, $6R_1(R_1+2)$

$$R_1(R_1+6) = 12(R_1+6) + 24R_1$$

$$R_1^2 + 6R_1 = 12R_1 + 72 + 24R_1$$

$$R_1^2 + 6R_1 = 36R_1 + 72$$

$$R_1^2 + 6R_1 - 36R_1 - 72 = 0$$

we moved all the terms to the left-side of the equation

$$R_1^2 - 30R_1 - 72 = 0.$$

To solve the quadratic equation $R_1^2 - 30R_1 - 72 = 0$, we need to use the quadratic formula, $R_1 = \frac{-b \pm \sqrt{b^2 - 4ac}}{2a}$. Comparing $R_1^2 - 30R_1 - 72 = 0$ with $ax^2 + bx + c = 0$, we find $a = 1$, $b = -30$ and $c = -72$. So,

$$R_1 = \frac{-b \pm \sqrt{b^2 - 4ac}}{2a}$$

$$R_1 = \frac{-(-30) \pm \sqrt{(-30)^2 - 4(1)(-72)}}{2(1)}$$ we plugged $a = 1$, $b = -30$ and $c = -72$ into the equation

$$R_1 = \frac{30 \pm \sqrt{900 + 288}}{2}$$

$$R_1 = \frac{30 \pm \sqrt{1{,}188}}{2}$$

$$R_1 = \frac{30 \pm 34.5}{2}$$

$$R_1 = \frac{30 + 34.47}{2} \quad \text{and} \quad R_1 = \frac{30 - 34.47}{2}$$

$$R_1 = \frac{64.47}{2} \quad \text{and} \quad R_1 = \frac{-4.47}{2}$$

$$R_1 = 32.235 \quad \text{and} \quad R_1 = -2.235.$$

R_1 cannot be negative, so $R_1 = 32.2$ Ohms is the answer.

Check: To verify our answer, we need to plug the solution into the equation. So,

$$R_1^2 - 30R_1 - 72 = 0$$

$$(32.235)^2 - 30(32.235) - 72 \overset{?}{\Leftrightarrow} 0$$ we plugged $R_1 = 32.235$ into the equation

$$1{,}039.1 - 967.1 - 72 \overset{?}{\Leftrightarrow} 0$$

$$1{,}039.1 - 1{,}039.1 \overset{?}{\Leftrightarrow} 0$$

$$0 = 0$$ true, the solution $R_1 = 32.235$ohms is a correct answer.

5.14. Example: Speed, Time, and Distance

Erin and John are planning to participate in a river cruise. There is a 3-hour river cruise that goes 18 km upstream and then back again. The river has a current of $3km/h$. Find,

a) the boat's speed.

b) the time of the upstream journey and the downstream journey.

Hint: The formula for time, speed, and distance is $D = rt$, where D represents the distance, r denotes the rate or speed, and t is the time.

Solution

a) Let us denote v as the boat's speed relative to the water. Since the speed of water is $2km/h$, then the speed of the boat relative to the land, when it goes upstream, and it goes downstream are as below:

The speed of the boat relative to the land going upstream = $v - 3$

The speed of the boat relative to the land going downstream = $v + 3$.

The relation between time, speed, and distance is given as $D = rt$, by which, if we know D and r and need to find t, we will solve the equation for t and get the formula $t = \frac{D}{r}$. Uding this formula, we will have:

Time of going upstream $= \frac{D}{r} = \frac{18}{v-3}$ — we plugged $D = 18$ and $r = v - 3$ into the equation

Time of going downstream $= \frac{D}{r} = \frac{18}{v+3}$ — we plugged $D = 18$ and $r = v + 3$ into the equation.

We are given that the total time of the cruise journey is 3 hours. So,
total time = time upstream + time downstream = 3 hours.

The table below summarizes the information and our findings for this case.

Type	Speed	Time	Distance
Upstream	$v - 3$	$\frac{18}{v-3}$	18
Downstream	$v + 3$	$\frac{18}{v+3}$	18

We now use the total time formula to find the boat's speed.
total time = time upstream + time downstream

$3 = \frac{18}{v-3} + \frac{18}{v+3}$ — we plugged the total time = 3, time upstream = $\frac{18}{v-3}$ and time downstream = $\frac{18}{v+3}$ into the equation.

To get rid of the fractions, we need to multiply both sides by $(v - 3)(v + 3)$. So,

$3 = \frac{18}{v-3} + \frac{18}{v+3}$

$3(v-3)(v+3) = (v-3)(v+3)(\frac{18}{v-3} + \frac{18}{v+3})$ — we multiply both sides by $(v - 3)(v + 3)$

$3(v^2 - 9) = (v-3)(v+3)(\frac{18}{v-3} + \frac{18}{v+3})$ — we used difference of squares rule

$3(v^2 - 9) = \frac{18(v-3)(v+3)}{v-3} + \frac{18(v-3)(v+3)}{v+3}$ — we used the distributive property

$3v^2 - 27 = 18(v + 3) + 18(v - 3)$

$\frac{1}{3}(3v^2 - 27) = \frac{1}{3}[18(v + 3) + 18(v - 3)]$ — we multiplied both sides by the reciprocal of $3v^2$

$v^2 - 9 = 6(v + 3) + 6(v - 3)$ — we used the distributive property

$v^2 - 9 = 6v + 18 + 6v - 18$

$v^2 - 9 = 12v$

$v^2 - 12v - 9 = 0.$

We now have a quadratic equation $v^2 - 12v - 9 = 0$. Comparing it with $ax^2 + bx + c = 0$, we have $a = 1$, $b = -12$, and $c = -9$. Using the quadratic formula $v = \frac{-b \pm \sqrt{b^2 - 4ac}}{2a}$ we will have.

$v = \frac{-b \pm \sqrt{b^2 - 4ac}}{2a}$

$v = \frac{-(-12) \pm \sqrt{(-12)^2 - 4(1)(-9)}}{2(1)}$ we plugged $a = 1$, $b = -12$, and $c = -9$ into the equation

$v = \frac{12 \pm \sqrt{144 + 36}}{2}$

$v = \frac{12 \pm \sqrt{180}}{2}$

$v = \frac{12 \pm 13.42}{2}$ we took the square root of $\sqrt{180}$

$v = \frac{12 + 13.42}{2}$ and $v = \frac{12 - 13.42}{2}$

$v = \frac{25.42}{2}$ and $v = \frac{-1.42}{2}$

$v = 12.71$ and $v = -0.71.$

Therefore, the solutions of $v^2 - 12v - 9 = 0$, are $v = 12.71$ and $v = -0.71$ (rounded to 2 decimal places). However, $v = -0.71$ makes no sense for the boat's speed, but $v = 12.71$ makes perfect sense.

Check: To verify our answer, we need to plug the solution into the equation. So,

$v^2 - 12v - 9 = 0$

$(12.71)^2 - 12(12.71) - 9 \overset{?}{\Leftrightarrow} 0$ we plugged $v = 12.71$ into the equation

$161.5 - 152.5 - 9 \overset{?}{\Leftrightarrow} 0$

$161.5 - 161.5 \overset{?}{\Leftrightarrow} 0$

$0 = 0$ true, the solution $v = 12.71$ is a correct answer.

To find the time of the upstream journey and upstream journey, we use the time of going upstream $= \frac{18}{v-3}$. and the time of going downstream $= \frac{18}{v+3}$So,

The time of going upstream $= \frac{18}{v-3}$

The time of going upstream $= \frac{18}{12.71 - 3}$ we plugged $v = 12.71$ into the equation

The time of going upstream $= \frac{18}{9.71} = 1.85$

The time of going downstream $= \frac{18}{12.71 + 3}$ we plugged $v = 12.71$ into the equation

The time of going downstream = $\frac{18}{15.71} = 1.15$.

Note that the time of going upstream plus the time of going downstream must be 3. We see that:

The time of going upstream = 1.85

The time of going downstream = 1.15.

If we add up these two values, we get the total time of journey, 1.85 + 1.15 = 3 hours.

5.15. Example: Living Expenses

The cost of living for a family of 3 is given by $x^2 + 45{,}000x - 3{,}000$, where x represents the number of years. Determine after many years the cost of living would exceed $450,000.

Solution

To determine the number of years that the cost of living exceeding $450,000, we need to solve the inequality $x^2 + 45{,}000x - 3{,}000 >$ $450,000. For solving the inequality, we first need to convert it to an equality and then solve it for x. So,

$$x^2 + 45{,}000x - 3{,}000 > 450{,}000$$

$$x^2 + 45{,}000x - 3{,}000 = 450{,}000$$

$$x^2 + 45{,}000x - 3{,}000 - 450{,}000 = 450{,}000 - 450{,}000$$ to have all terms in the left-side, we added 450,000 to both sides

$$x^2 + 45{,}000x - 3{,}000 - 450{,}000 = 0$$

$$x^2 + 45{,}000x - 453{,}000 = 0.$$

We now need to solve the quadratic equation $x^2 + 45{,}000x - 453{,}000 = 0$. Comparing it with $ax^2 + bx + c = 0$, we have $a = 1$, $b = 45{,}000$, and $c = -453{,}000$. Using the quadratic formula $x = \frac{-b \pm \sqrt{b^2 - 4ac}}{2a}$ we will have.

$$x = \frac{-b \pm \sqrt{b^2 - 4ac}}{2a}$$

$$x = \frac{-(45{,}000) \pm \sqrt{(45{,}000)^2 - 4(1)(-453{,}000)}}{2(1)}$$ we plugged $a = 0.15$, $b = -45{,}000$, and $c = 453{,}000$ into the equation

$$x = \frac{-45{,}000 \pm \sqrt{2{,}025{,}000{,}000 + 1{,}812{,}000}}{2}$$

$$x = \frac{-45{,}000 \pm \sqrt{2{,}026{,}812{,}000}}{2}$$ we took the square root of $\sqrt{2{,}025{,}070{,}200}$

$$x = \frac{-45{,}000 \pm 45{,}020.1288}{2}$$

$$x = \frac{-45{,}000 + 45{,}020.1288}{2} \quad \text{and} \quad x = \frac{-45{,}000 - 45{,}020.1288}{2}$$

$$x = \frac{20.1288}{2} \quad \text{and} \quad x = \frac{-90{,}020.1288}{2}$$

$x = 10.0644$ or $x = -45{,}010.0644$.

Therefore, the solutions of $x^2 + 45{,}000x - 453{,}000 = 0$, are $x = 10.0644$ and $x = -45{,}010.0644$. However, the negative value of x makes no sense as the number of years, hence the only solution that can be acceptable is $x = 10.06$ (rounded to 2 decimal places).

Check: To verify our answer, we need to plug the solution into the equation. So,

$x^2 + 45{,}000x - 453{,}000 = 0$

$(10.0644)^2 + 45{,}000(10.0644) - 453{,}000 \overset{?}{\Leftrightarrow} 0$ we plugged $x = 10.0644$ into the equation

$101.29 + 452{,}898 - 453{,}000 \overset{?}{\Leftrightarrow} 0$

$452{,}999.29 - 453{,}000 \overset{?}{\Leftrightarrow} 0$

$-0.7 \cong 0$ True, the solution $x = 10.06$ (rounded to 2 decimal places) is a correct answer.

Therefore, after 10.06 years and one month the cost of living exceeds \$450,000.

5.16. Exercises

1. An arrow is shot from the top of a building with height of 12 feet into the air at an initial speed of 112 feet per second. Use the formula $h = -16t^2 + v_0t + 12$ to determine when the arrow will be 188 feet from the ground. Round you answer to the nearest tenth.

2. Trapeze Artist Clothing Manufactures high-priced jumpers for sale to college bookstores in runs of up to 180. The cost for a run of jumpers is $C(x) = 0.3x^2 + 12x + 1600$, $(0 \leq x \leq 180)$. Trapeze Artist Clothing sells the jumpers at $120 each.
   a. Find the revenue function.
   b. Find the profit function.
   c. How much should Trapeze Artist Clothing manufacture to make a profit?

3. Toys for Kids company manufactures and sells educational toys. For a particular toy, LEGO kit, its marketing department has estimated a price of $$p$ per unit. The monthly cost C and revenue R are given by the equations $C = 325 - p^2$ and $R = 52p - p^2$. Write the inequality that can be used to find the price per unit for which the company will lose money.

4. The distance of an object shot straight up from the ground with an initial velocity of 86 feet per second, at the end of t seconds is given by $d = 86t - 16t^2$. Find the time-interval for which the object is more than 102 feet above the ground.

5. Mama's Bakery offers decorated sheet cakes in the shape of a rectangular solid for children's birthday parties. The bakery wants the volume of a specific cake to be 1,000 cubic inches. They want the length of the cake to be twice the width of the cake and the height of the cake to be one-half of the width. Find the dimensions of the cake pan.

   Hint: the formula for calculating the volume, v, of a rectangular cub with the length l, the width w, and the height h is $v = l \times w \times h$.

6. Michel is 4 years older than his sister Erin. If the product of their ages eleven less than four times the sum of their ages, how old are Michel and Erin?

7. Sharon and Fatima are sisters. Sharon is the older one. The sum of the ages of Sharon and Fatima is 30. If the product of their ages is 225, how old are Sharon and Fatima?

8. In the figure below, we are given that the circle in middle will be extracted to form a ring with a diameter of $7in$. Calculate the area of shaded region and write a polynomial expression for the shaded region.

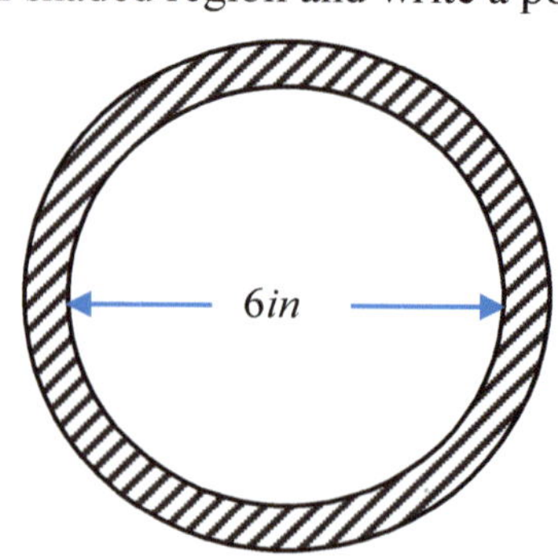

Hint: the formula for calculating the area, S, of a circle with dimeter d is $S = \frac{\pi d^2}{4}$.

9. Suppose we have a hypothetical laptop manufacturer whose cost to manufacture x number of laptops is $C = 800x + 640{,}000$, and the revenue generated from manufacturing x number of laptops is $R= -0.008x^2 + 2{,}000x$.
   a) Define a profit polynomial for the hypothetical laptop manufacturer.
   b) Find the profit for the laptop manufacturer, given the following numbers of laptops manufactured.

      For $x = 3{,}000$ laptops.

      For $x = 18{,}000$ laptops.

      For $x = 48{,}000$ laptops.

Hint: Given the cost to produce a certain number of items is C and the revenue generated by selling those items is R, the profit, P, can then be defined as $P = R - C$.

10. The length of a rectangular swimming pool is 3 feet more than its width. After a 2-foot cement border is placed around the pool, the area of the pool and its border is 280 square feet (see figure below). Find the original dimensions of the swimming pool.

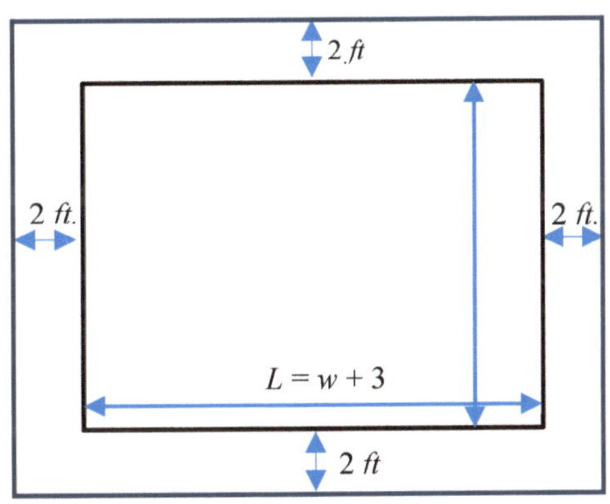

Hint: The area, A, a rectangle is the product of its length, l, and width, w, or $A = l \times w$.

11. A piece of sheet metal 32*cm* long and 26*cm* wide is used to make a box with no lid. Equal squares of side length x centimeters are cut from the corners and the sides are folded up.
    a) Write a polynomial function to represent the volume, V, of the box in terms of x.
    b) To the nearest cubic centimeters, what is the maximum volume of the box?
    c) What size of square should be cut out to create a box with this volume?
    d) To the nearest tenth of a centimeter, what are the dimensions of the box?

Hint: the volume of a box with the length l, the width w, and the height h is $V = l \times w \times h$.

12. The cost of living for a family of 4 is given by $0.11x^2 + 58{,}000x - 4{,}200$, where x represents the number of years. Determine after many years the cost of living would exceed \$620,000.

13. The diameter of the base of a cylinder is 2 meters. If its height is h, for what meter of h the volume of the cylinder is between $9m^3$ and $10m^3$?

Hint: The formula for the volume, V, of a cylinder is $V = \frac{\mu d^2 h}{4}$ where d is diameter of the base and h is the height of the cylinder.

14. Given a pyramid with a square base having each side equal to 3 inches and the height of 4 inches, find its volume.

Hint: The formula for the volume, V, of square pyramids is $V = \frac{1}{3}s^2h$, where s is the base length and h is the height.

15. A shipping container box in the shape of a rectangular solid has a length of 4 inches, a width of 3 inches and a height of 2 inches. How many cubes with each side of $\frac{1}{2}$ inch can be loaded in this container box?

Hint: The formula for volume of a cub is $V = a^3$, where V is the volume and a is the edge length.

16. The product of two consecutive even integers is 48. What are these numbers?

17. Two consecutive pages have a product that is 552. What are the page numbers?

18. Two consecutive positive odd numbers have a product that is 323. What are the numbers?

19. The number of possible homemade food tasters, N, within a group of n people is given by the formula $N = \frac{1}{2}(n^2 - 49n)$. How many people would there have to be in a group to have 25 tasters?

20. Consider a cylindrical barrel with a height of $15in$, a base with an external diameter of $13in$, and an internal base having a diameter $2in$ less than the external diameter. Find the volume of the barrel.

Hint: The formula for the volume V of a cylinder is: $V = \frac{\pi d^2 h}{4}$., where d is the base dimeter, and h is the height of the cylinder.

21. A farmer needs to enclose three sides of a field with a fence. In front of the fourth side along the length is a pond and no fencing is required. The farmer has an $80ft$ fence and wants a rectangular field with an area of $350ft^2$. What should the dimensions of the field be?

22. Consider a shipping container in the shape of a rectangular solid with a volume of $84m^3$, the length of $4m$, the width of $1m$ meter less than its length, and the height of one meter greater than twice of its width. What are the dimensions of the container.

23. A business owner has estimated the cost, C, of a specific product as $C = 320{,}000 + 450p$ and its revenue, R, as $R = 0.006p^2 + 900p$, where p is the price of the product. Find p, the price of the product, that would make his revenue greater than his costs.

24. The formula for the number of games to be played in a football league where each team is to play each other twice is, $N = x^2 - x$, where x is the number of teams in the league and N is the number of

games to be played. If a league wants to limit the games to a total of 90 games, how many teams can be in the league?

25. Two technicians can do the weekly power station maintenance in 58 minutes working together. The less experienced technician takes 18 minutes more than the more experienced technician to finish the job by himself. How long does it take for each technician to do the weekly power station maintenance individually?

26. It takes five hours for two injection molding machines to manufacture 18,000 plastic storage containers. If machine #1 can do the job one hour less than machine #2 can do the job, how long does it take for each machine to manufacture 18,000 parts alone?

6. Multiple-Choice Questions

1. A cube with sides of 4 centimeters in length weighs 32 grams. What is its density?

   A. 0.5 grams/ cm^3
   B. 1 gram/ cm^3
   C. 2 grams/ cm^3
   D. 4 grams/ cm^3

2. Which of the following are roots of the equation $x^2 - 4x - 8 = 0$?

   A. $\frac{2 \pm 3\sqrt{2}}{2}$
   B. $\frac{4 \pm 3\sqrt{2}}{2}$
   C. $2 \pm 3\sqrt{2}$
   D. $2 \pm 2\sqrt{3}$

3. Which of the following is equivalent to the polynomial $x^5 - 4x^4 - 8x^2$?

   A. $x(x^4 - 4x^2 - 8x)$
   B. $x(x^4 - 4x^3 - 8)$
   C. $x^2(x^3 - 4x^2 - 8)$
   D. $x^2(x^3 - 4x^2 - 8x)$

4. What is the GCF of expressions, $24a^2b^2$, $12a^5b^4$, $36a^6b$, $20a^2b^4$, $18a^3b^3$.

   A. $6\, a^2b^2$
   B. $2\, a^2b$
   C. $2\, a^2b^2$
   D. $6\, a^2b$

5. Given the quadratic equation $y = 5(x + 2)^2 - 6$, find its vertex coordinates.

   A. $V(2, -6)$
   B. $V(-2, -6)$
   C. $V(-2, 6)$
   D. $V(2, 6)$

6. Given the quadratic equation $y = -3(x + 2)^2 + 4$, with points P and Q on the graph of this quadratic equation, and points P and Q are symmetric with respect to its line of symmetry. If point $P = (-1, 1)$, find the coordinates of Q.

   A. $Q(-1, 2)$
   B. $Q(-3, 1)$
   C. $Q(3, -1)$
   D. $Q(-1, 0)$

7. Given the quadratic equation $y = x^2 + 4x + 3$, what are x-intercepts?

A. $x = -1$, and $x = -3$
B. $x = -1$, and $x = 3$
C. $x = 1$, and $x = -3$
D. $x = -1$, and $x = -2$

8. Find the coordinates of the vertex, $V(x, y)$, in the quadratic equation $2x^2 - 6x + 7$.

A. $V(-\frac{3}{2}, -\frac{5}{2})$
B. $V(\frac{3}{2}, \frac{2}{5})$
C. $V(\frac{2}{3}, \frac{5}{2})$
D. $V(\frac{3}{2}, \frac{5}{2})$

9. If $x^2 - 8x = 48$ and $x < 0$, what is the value of $x + 10$?

A. 4.5.
B. 5.0
C. 5.5.
D. 6.0

10. Josef has two more than twice as many bubble gums as does Helen, and half as many bubble gums as does Eave. If Helen has 'a' number of bubble gums, then in terms of 'a', how many bubble gums do Josef, Helen and Eave have?

A. $5a + 5$
B. $6a + 7$
C. $7a + 6$
D. $8a + 5$

11. If $3x^2 + 7x + 1$ is multiplied by $5x + 2$, what is the coefficient of x in the resulting polynomial?

A. 15
B. 35
C. 41
D. 19

12. Which of the following are roots of the equation $2x^2 - 4x - 12 = 0$?

A. $1 \pm 2\sqrt{7}$
B. $1 \pm \sqrt{7}$
C. $\frac{4 \pm \sqrt{7}}{4}$
D. $\frac{4 \pm 2\sqrt{7}}{4}$

13. If $\frac{C+x}{x-3} = \frac{x+8}{3}$, which of the following could be an expression of C in terms of x?

A. $C = x^2 + 2x - 24$
B. $C = \frac{1}{3}(x + 6)(x - 4)$
C. $C = \frac{1}{3}(x + 8)(x - 3)$
D. $C = \frac{1}{3}x^2 + \frac{2}{3}x - \frac{8}{3}$

14. Which of the following quadratic equations has a vertex of $(5, -5)$?

A. $y = 7(x - 5)^2 - 5$
B. $y = 7(x + 5)^2 - 5$
C. $y = 7(x - 5)^2 + 5$
D. $y = 7(x + 5)^2 + 5$

15. Which of the following quadratic equations has a vertex of $(1, -19)$?

A. $y = 4x^2 + 8x - 15$
B. $y = 4x^2 - 8x - 15$
C. $y = 4x^2 - 8x + 15$
D. $y = 4x^2 + 8x + 15$

16. A computer simulation experiment is conducted to study a bullet fired straight up from the roof of the physic lab. The software program generates the equation $h = -15.2t^2 + 114.8t + 16$ to model the data, where h is the height in meters of the bullet t seconds after it was fired. What does the number 16 most likely represent in this equation?

A. The number of seconds the bullet was in the air.
B. The height of roof of the physic lab from which the bullet was fired.
C. The number of seconds that it took the bullet to reach its maximum height.
D. The maximum height of the bullet.

17. Emma, who works in the computer department of a wholesale store, earns a base pay of \$11.00 per hour plus a certain percent commission on the sales that she helps to broker in the store. Emma worked an average of 35 hours per week over the past two weeks and helped to broker sales of \$3,000.00 worth of laptops and tablets during that same two-week period. If Emma's earnings for the two-week period were \$890.00, what percent commission on sales does Emma earn?

A. 1%
B. 2%
C. 3%
D. 4%

18. The quadratic equation is defined by $y = 2x^2 - ax - 6$, where a is a constant. If one of the roots of y is 6, what is the value of the other root of y?

A. $-\frac{1}{2}$

B. 2

C. -2

D. $\frac{1}{2}$

19. Kirk needs to buy t Italian opera concerts tickets for his classmates. Each ticket costs \$110, and the number of tickets he needs to buy can be modeled by the expression $2t^2 - 6t = 8$ when $t > 0$. What is the total cost of Italian opera concerts tickets that Kirk purchased?

A. \$640
B. \$400
C. \$860
D. \$440

20. Given an oblique hexagonal prism as shown below with $h = 3$ inches, $a = 2$ inches, and $s = 5$ inches, what is its volume?

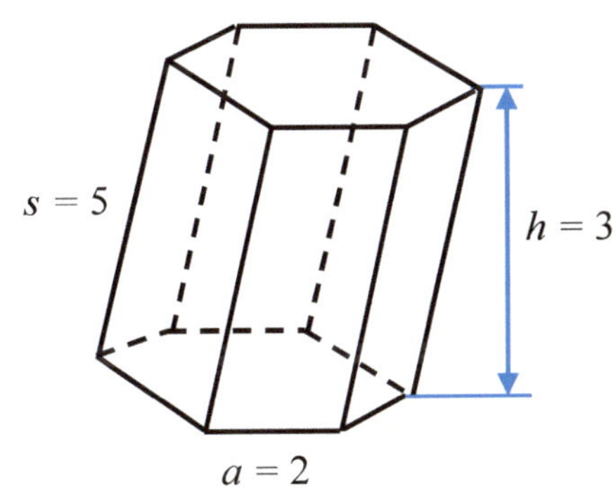

A. 252
B. 108
C. 78
D. 30

21. Given the polynomial equation $(x + 1)^3 = 216$, what is x?

A. 4
B. 5
C. 6
D. 7

22. Given the quadratic equation $y = ax^2 + 6x$, with the graph below, what is a?

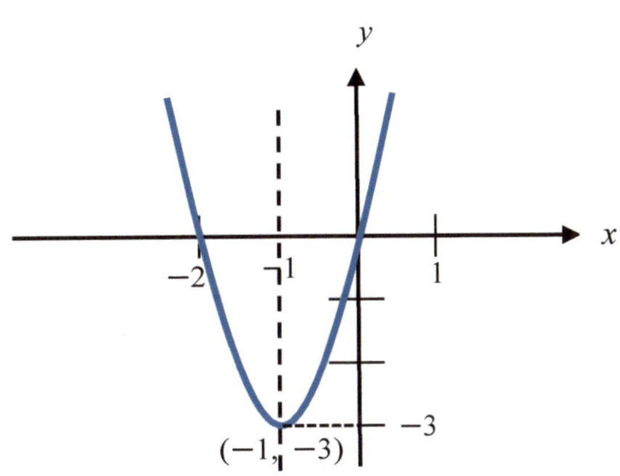

A. 3
B. −3
C. 6
D. − 6

23. The polynomial expression $0.08d^{\frac{3}{2}}$ describes the duration of a storm, in hours, whose diameter is d miles. Determine the duration of a storm with a diameter of 6 miles. Round to the nearest hundredth.

A. 1.18
B. 14.70
C. 0.26
D. 3.03

24. A cube with sides of 2 centimeters (cm.) in length weighs 4 grams. What is its density?

A. 0.5 grams/cm^3
B. 1 grams/cm^3
C. 1.5 grams/cm^3
D. 2 grams/cm^3

25. Given a quadratic equation $y = ax^2 + bx + c$, for $b^2 - 4ac = 0$, the typical graph is.

A. 

B. 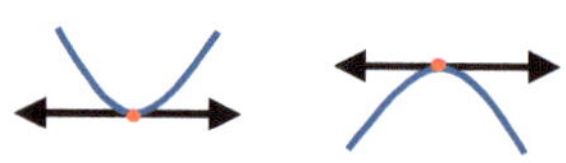

C. 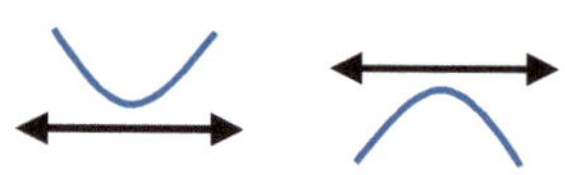

D. 2 x-intercept on graph

26. Given a quadratic equation $y = ax^2 + bx + c$, for $b^2 - 4ac > 0$, the typical graph is.

A. 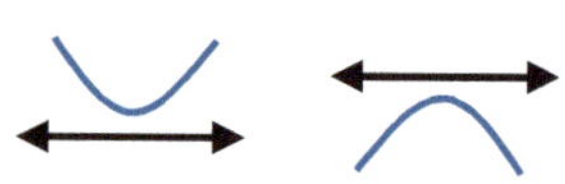

B. 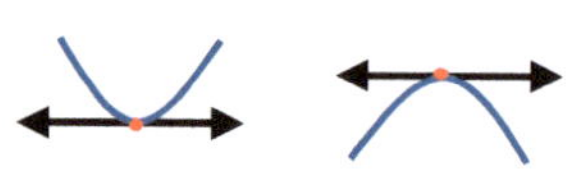

C. 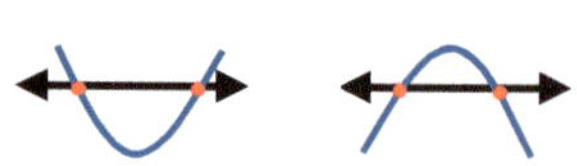

D. 1 x-intercept on graph

27. Given a quadratic equation $y = ax^2 + bx + c$, for $b^2 - 4ac < 0$, the typical graph is.

A. 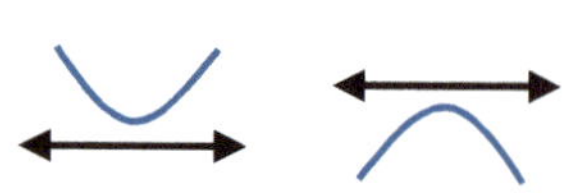

B. 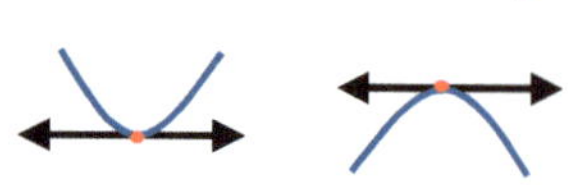

C. 

D. None of the above graph

Questions 28 to 30 relate to the problem below.

Problem: Home & Kitchen Appliances Inc. manufactures and sells multipurpose 12-cup food processor. Each food processor costs \$38 to make, and the company's fixed costs are \$32,000 per month. They know that the price of each food processor is expected by the function $p(x) = 86 - 1.5x$, where $p(x)$ is the price as a function of units, x, sold.

28. The company cost equation $C(x)$, is,

    A. $C(x) = 38x - 32{,}000$
    B. $C(x) = 86x + 32{,}000$
    C. $C(x) = 86x - 32{,}000$
    D. $C(x) = 38x + 32{,}000$

29. The company revenue equation $R(x)$, is,

    A. $R(x) = 38x - 1.5x^2$
    B. $R(x) = (86 - 38)x - 1.5\,x^2$
    C. $R(x) = 86x - 1.5x^2$
    D. $R(x) = (86 - 38)x - 1.5x$

30. The company profit equation $P(x)$, is,

    A. $P(x) = -\,1.5x^2 - 48x - 32{,}000$
    B. $P(x) = -\,1.5\,x^2 - 48x + 32{,}000$
    C. $P(x) = -\,1.5x^2 + 48x - 32{,}000$
    D. $P(x) = -\,1.5\,x^2 + 48x + 32{,}000$

31. The height, h, of an object shot upwards into the air is determined by the formula $h = -\,16t^2 + v_0t$, where h is the height in feet, v_0 is the initial velocity in feet per second, and t is the time in second. A bullet is shot from an experimental gun upwards with initial velocity 1260 feet per second. The time it takes to reach a height of 840 feet is determined by:

    A. $t = \dfrac{-1260 \pm \sqrt{(1260)^2 - 4(-16)(-840)}}{-32}$

    B. $t = \dfrac{-1260 \pm \sqrt{(1260)^2 - 4(-16)(840)}}{-32}$

    C. $t = \dfrac{-1260 \pm \sqrt{(1260)^2 - 4(-16)(-840)}}{-16}$

D. $t = \frac{-1260 \pm \sqrt{(1260)^2 - 4(-16)(840)}}{-16}$

32. Samuel is 4 years older than his sister Samantha. If the product of their ages is 4 more than three times the sum of their ages, find the ages of Samuel and Samantha.

A. Samantha is 3 years old, and Samuel is 7 years old
B. Samantha is 4 years old, and Samuel is 8 years old
C. Samantha is 5 years old, and Samuel is 9 years old
D. Samantha is 6 years old, and Samuel is 10 years old

33. Which of the following quadratic equation is equivalent to $(5x^2 + 9x - 6) - (3x^2 - 4x + 13)$?

A. $2x^2 + 5x - 7$
B. $5x^2 - 5x + 7$
C. $5x^2 - 5x - 7$
D. $2x^2 + 5x + 7$

34. A certain polynomial, P, has a degree of 2. Polynomial P has zeros of 2 and -3, and $a > 0$. Given this information, which of the following could be the graph of polynomial P?

A.

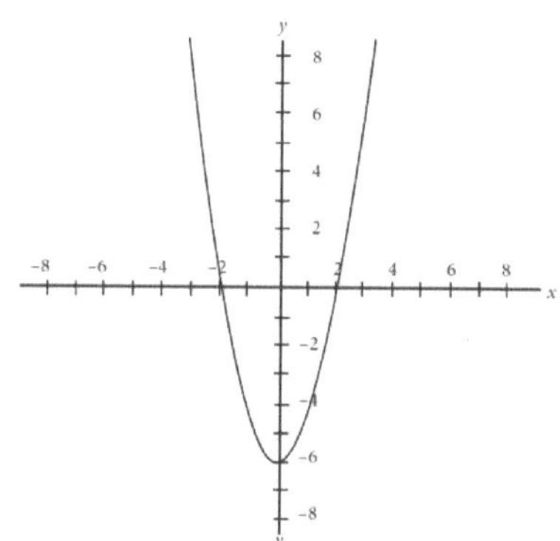

B.

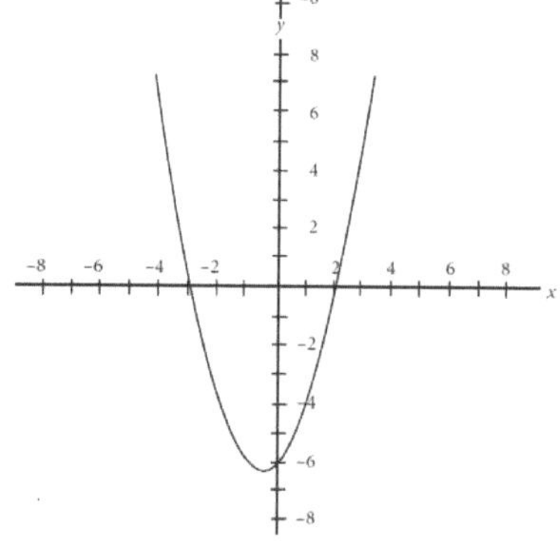

C. 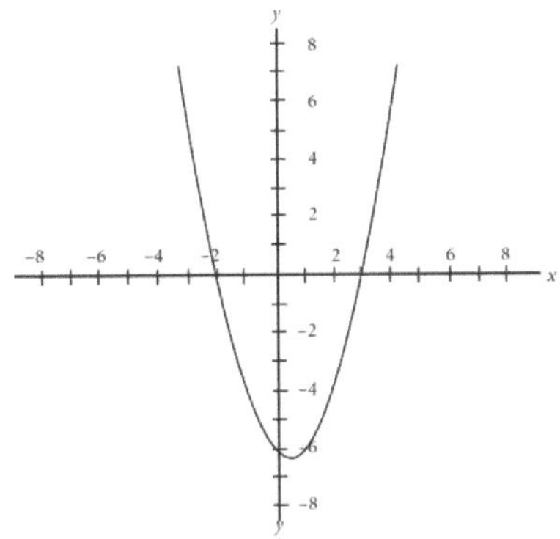

D. 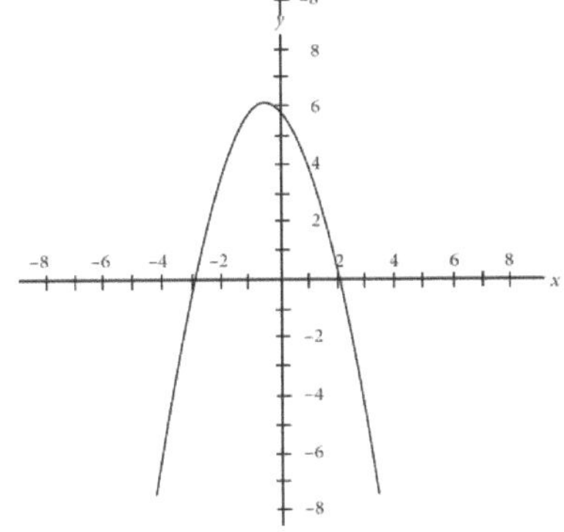

35. Given $10\sqrt{3x+1} + 13\sqrt{3x+1} - 12\sqrt{12x+4}$, it can be simplified as:

A. $-3\sqrt{3x+1}$
B. $3\sqrt{3x+1}$
C. $-25\sqrt{3x+1}$
D. $25\sqrt{3x+1}$

36. The figure below shows the graph in the *xy*-plane of a polynomial $P(x)$. How many distinct real roots does $P(x)$ have?

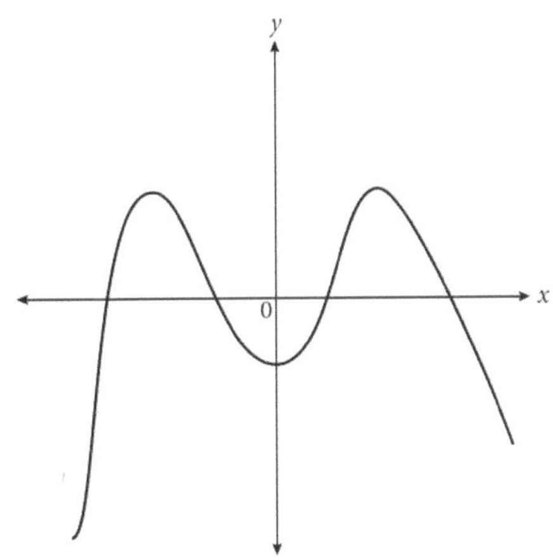

A. 2
B. 3
C. 4
D. 1

37. The cost in dollars, $C(x)$, of producing a custom-made baseball hat with a team logo is given by the formula $C(x) = 25x - \frac{1}{2}(x - 3)$, where x is the number of hats produced. When every hat produced is sold, the revenue from selling the customized custom-made baseball hat is given by $R(x) = 32x + 0.01x^2$. Which one of the following would be the formula for the profit $P(x)$ from producing and selling x baseball hats?

A. $P(x) = \frac{15}{2}x + 0.01x^2 - \frac{3}{2}$

B. $P(x) = \frac{15}{2}x + 0.01x^2 + \frac{3}{2}$

C. $P(x) = \frac{13}{2}x + 0.01x^2 - \frac{3}{2}$

D. $P(x) = \frac{13}{2}x + 0.01x^2 + \frac{3}{2}$

38. Which of the following equations best describes the graph below?

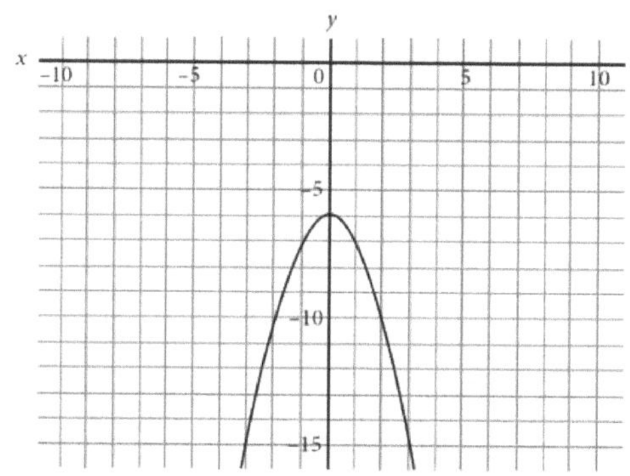

A. $y = -x^4 + 6$

B. $y = x^4 + 6$

C. $y = -x^2 + 6$

D. $y = -(x^2 + 6)$

39. Which of the following is the solution of $(x + 3)(5 - x) > 0$ in a set interval notation?

A. $[-3, 5)$

B. $(-3, 5)$

C. $(-3, 5]$

D. $[-3, 5]$

40. Given $2(x)^4 - 128 = 0$, what is x?

A. $x = 12$

B. $x = 8$

C. $x = 6$

D. $x = 4$

41. Which of the following is the solution of $(x + 2)(1 - x) > 0$ on a number line?

A. 

B. 

C. 

D. 

42. Newton's law of gravitation expresses the strength of the force F between two objects with masses m_1 and m_2 separated by a distance of d units and is defined as. $F = \frac{gm_1m_2}{d^2}$. Which of the following gives the value of Newton's gravitational constant g, in terms of F, m_1, m_2, and d?

A. $g = \frac{d^2F}{m_1m_2}$
B. $g = \frac{m_1m_2F}{d^2}$
C. $g = \frac{m_1m_2d^2}{F}$
D. $g = \frac{dF}{m_1m_2}$

43. If $16x^3 + 4x^2 - 6x = 0$, what are the possible values of x?

A. $x = 0$, $x = 0.5$, and $x = -0.75$
B. $x = 0$, $x = -0.5$, and $x = -0.75$
C. $x = 0$, $x = 0.5$, and $x = 0.75$
D. $x = 0$, $x = -0.5$, and $x = 0.75$

44. Given $(4ab^3 + 10a^3b^3) \div (2 + 5a^2)$, it can be simplified as:

A. $2ab^3$
B. $2ab^2$
C. $2a^2b^3$
D. $4ab^3$

45. Given $3\sqrt[3]{16a + 8b} - 4\sqrt[3]{2a + b}(4ab^3 + 10a^3b^3) \div (2 + 5a^2)$, it can be simplified as:

A. $2\sqrt[3]{2a + b}$
B. $-\sqrt[3]{2a + b}$
C. $-8\sqrt[3]{2a + b}$
D. $8\sqrt[3]{2a + b}$

46. When a virus breaks out, as soon as people are infected with a virus, it is possible they can transmit it to multiple new people. In a COVID-19 outbreak at an elementary school, the number

of infected people triples each day in the first school week of September. If 4 students were sick with a COVID-19 on Monday, which of the following equations best predicts the number of infected people. Consider $I(d)$ as the equation for predicting the number of infected people, d days after Monday?

A. $I(d) = 4(3^2d^2)$
B. $I(d) = 4(3d^3)$
C. $I(d) = 4(3d)$
D. $I(d) = 4(3^2d)$

47. Given the right triangle ABC shown in the figure below, with $AB = 8$ inches, $AC = 6$ inches, and $\angle CAB = 90°$, which inequality best represent the length of BC?

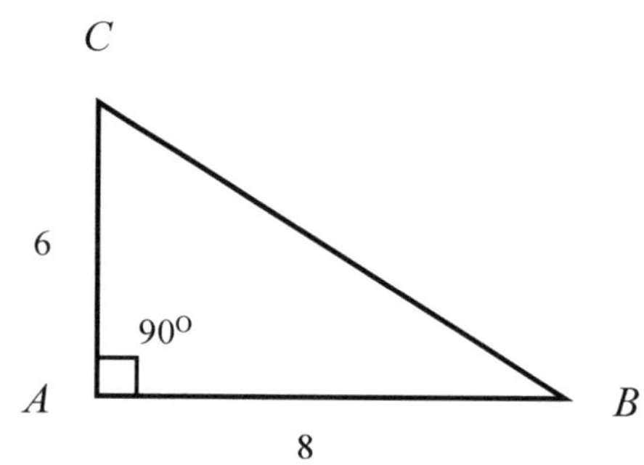

A. $BC > 9$
B. $BC \geq 10$
C. $BC < 9$
D. $BC \leq 10$

48. Assuming the size of pump required to pump out water from a tank depends on the height of the tank. If pumping out a tank that is h feet height requires a pump that can pump a minimum of $\frac{1}{6}h^2 + 8h - 2$ gallons of water per minute, pumping out a tank that is 12 feet height would require a pump that pumps a minimum of how many gallons of water per minute?

A. 98
B. 108
C. 128
D. 118

49. If $P(x) = 6x^3 + 3x - 7$, $R(x) = 2x^2 - 12$, and $Q(x) = 3x - 2$, then what is $P(x) - R(x) - Q(x)$?

A. $6x^3 - 2x^2 + 7$
B. $6x^3 - 2x^2 + x + 7$
C. $6x^3 - 2x^2 - x + 7$
D. $6x^3 - 2x^2 - x - 7$

50. Given the polynomial equation $P(x)$ has a simple zero at $x = 5$ and a double zero at $x = -\frac{1}{4}$, which of the following equation could be the factored form of $P(x)$?

A. $P(x) = (x - 5)(x + 4)^2$
B. $P(x) = (x - 5)(4x + 1)^2$
C. $P(x) = 2(x - 5)(x + 4)$
D. $P(x) = 2(x - 5)(4x + 1)$

51. If $10x^2 - \frac{3}{7}x + 13$ is multiplied by $14x + \frac{1}{5}$, what could be the coefficient of x^2?

A. 10
B. − 4
C. − 6
D. 6

52. Given $\sqrt[3]{2x+5} + 3 = 0$, what is x?

A. $x = 16$
B. $x = -16$
C. $x = 32$
D. $x = -32$

53. Given $\sqrt{x-13} - 5 \geq 0$ where $x \geq 13$, what is x?

A. $x \geq 38$
B. $13 \leq x \leq 22$
C. $-13 \leq x \leq 22$
D. $x \leq 38$

54. How many real values of x satisfy the quadratic equation $4x^2 - 12x + 9 = 0$?

A. 2
B. 0
C. 3
D. 1

55. A polynomial equation has three zeroes, x_1, x_2 and x_3, which their sum is $x_1 + x_2 + x_3 = -4$. Which of the factored form below presents the quadratic equation?

A. $x(x+4)(x-8)$
B. $x(x-6)(x+2)$
C. $x(x-3)(x+7)$
D. $x(x+1)(x-5)$

56. Which of the following quadratic equations has no solution?

A. $2(x+4)(x-6)$
B. $-2(x-6)^2 + 4 = 0$
C. $2(x-6)^2 + 4 = 0$
D. $-2(x+4)(x-6)$

57. The graph of the polynomial, $P(x)$ is sketched on the xy-plane, as shown in the figure below. If p, q, r, and t are distinct real numbers, which of the following could be $P(x)$?

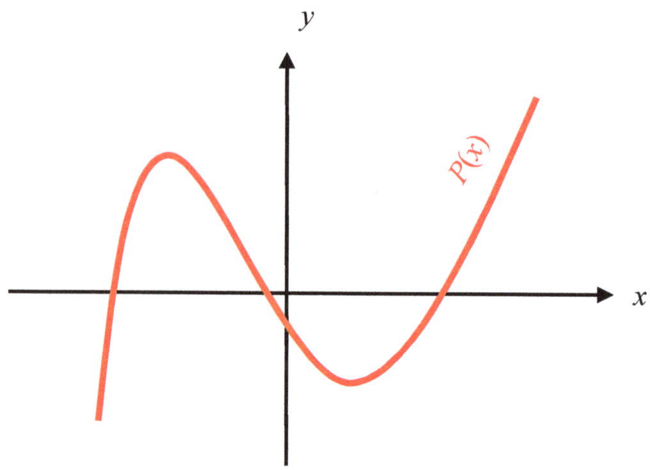

A. $P(x) = (x - p)^2$

B. $P(x) = (x - p)(x - q)$

C. $P(x) = (x - q)(x - r)(x - t)$

D. $P(x) = (x - p)(x - q))(x - r)(x - t)$

58. Which of the following equations could be the factored form of the polynomial $P(x)$, graphed in the figure below?

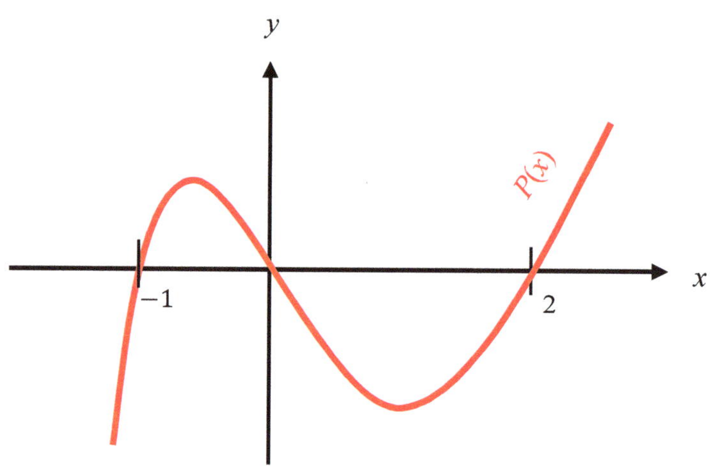

A. $P(x) = 2x(x - 1)(3x + 6)$

B. $P(x) = x(x + 1)(x - 2)$

C. $P(x) = x(x - 1)(x + 2)$

D. $P(x) = 2x(x + 1)(2x - 2)$

59. Which of the following polynomials is equivalent to the expression $\frac{32x^5 - 48x^4 + 16x^3 - 4x^2}{4x^2}$?

A. $P(x) = 8x^3 - 12x^2 - 1$

B. $P(x) = 8x^3 - 12x^2 + 4x + 1$

C. $P(x) = 8x^3 - 12x^2 + 4x - 1$

D. $P(x) = 8x^3 - 12x^2 + 4x$

60. The answer to the quadratic inequality $x^2 - 10 \geq 3x$, in a set interval notation is:

A. $(-\infty, -1)\cap(5, \infty)$.
B. $(-\infty, -1]\cup(5, \infty)$
C. $(-\infty, -2)\cap[5, \infty)$
D. $(-\infty, -2]\cup[5, \infty)$

6.1. Key to the Multiple-Choice Questions

Question No.	Answer
1	A
2	D
3	C
4	B
5	C
6	B
7	A
8	D
9	D
10	C
11	D
12	B
13	B
14	A
15	B
16	B
17	D
18	A
19	D
20	C

Question No.	Answer
21	B
22	A
23	A
24	A
25	B
26	C
27	A
28	D
29	C
30	C
31	A
32	B
33	D
34	B
35	A
36	C
37	A
38	D
39	B
40	D

Question No.	Answer
41	B
42	A
43	D
44	A
45	A
46	C
47	A
48	D
49	A
50	B
51	B
52	B
53	A
54	D
55	C
56	C
57	C
58	B
59	C
60	D

Test is the process a life without love is of no account.
Don't ask yourself what kind of love you should seek,
spiritual or material, divine or mundane, Eastern or Western.
Divisions only lead to more divisions.
Love has no labels, no definitions.
It is what it is, pure and simple.
Love is the water of life.
And a lover is a soul of fire!
The universe turns differently when fire loves water.

Rumi

Chapter 5

Functions

Contents

Introduction

The concept of functions originated to describe how one quantity changes concerning another, with its formal mathematical notion emerging during the 17th century alongside the development of calculus. In 1673, Gottfried Wilhelm Leibniz introduced the term "function" to describe a quantity tied to a curve, such as its slope at a specific point.

Functions underwent further development through infinitesimal calculus, which studies continuous change, analogous to how geometry examines shape and algebra deals with generalized arithmetic operations. By the late 19th century, functions were formally defined within the framework of set theory, greatly expanding their applicability.

Functions are extensively utilized in science and various mathematical disciplines, often being considered central objects of inquiry. Despite their widespread usage, functions may not always be explicitly recognized in everyday contexts. They play pivotal roles in financial planning, population growth analysis, production forecasting, and numerous other domains. By expressing relationships between quantities as functions, we can efficiently determine the value of one variable when another is known and visually represent these relationships to quickly comprehend their direction, magnitude, and rate of change across different values.

This chapter aims to facilitate the teaching and learning of mathematical functions. It will cover the formation of functions through ordered pairs, delve into arithmetic functions, discuss various concepts and methodologies associated with functions, introduce key functions, and explore their real-world applications. The discussion will commence with an examination of relations and functions.

1. Functions

In mathematics, **functions** represent a unique relationship between inputs and their corresponding outputs, where each input yields precisely one output, and this output can be directly linked back to its input. Within this framework, a function can be conceptualized as a process that assigns precisely one element from a designated set, known as the domain of the function, to each element of another set, termed the codomain of the function. This process essentially operates as an input-output rule: the function receives an input, which is an element from the domain, and generates an output, which is an element from the codomain. Traditionally in mathematics, the inputs and outputs primarily consist of real numbers.

Throughout this chapter, we will introduce the fundamental concepts of relations and functions, highlighting their distinctions. We will delve into various types of relations and functions and illustrate techniques for transforming a relation into a function.

1.1. Relations and Functions

Functions and Relations - Functions and relations are one the most important topics in Algebra. On most occasions, many people confuse the meaning of these two terms. Let us first look at some key concepts as used in functions and relations.

Set - A **set** is a group or collection of distinct elements or members. Sets are usually symbolized by uppercase, italicized, or bold face letters such as *A*, *B*, *R*, *S*, or *U*. Each object or number in a set is called a member or element of the set. Examples include the set of all English alphabet, the set of all students in a class, and the set of all integer numbers between 20 and 30.

An Ordered Pair - An **ordered pair** is two numbers for which the order in which they are given is important. Ordered pair numbers are represented within parentheses and separated by a comma.

For example, (9, 3) is an ordered-pair number whereby the numbers 9 and 3 are the first and the second element, respectively.

Domain - A **domain** is a set of all input or first values of a relation or a function.

Range - A **range** of a relation or a function is a collection of all output or second values.

Equal Sets - Two sets are equal if they contain the same members, regardless of the position of the members in the sets. For example, two sets, $A = \{2, 4, 6, 8\}$ and $B = \{4, 8, 2, 6\}$ are equal because they contain similar members.

A relation - A **relation** is any set of ordered-pair numbers. In other words, we can define a relation as a bunch ordered pairs.

Function - A **function** is a relation for which each value from the set the first components of the ordered pairs is associated with exactly one value from the set of second components of the ordered pair.

Diagrams below illustrate a relation representing a function and a relation not representing a function.

Relation is a **function**

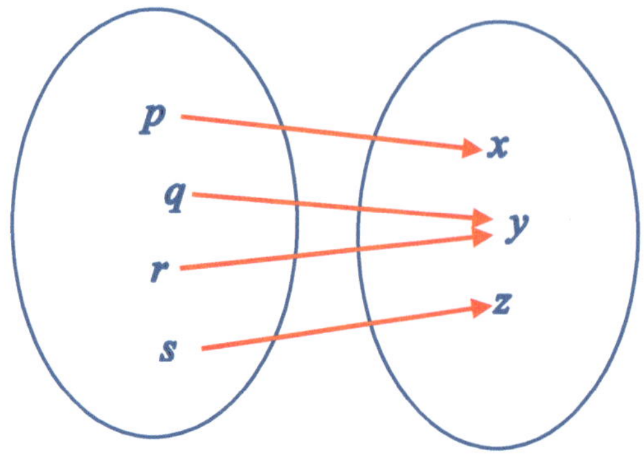

Relation is **NOT a function**

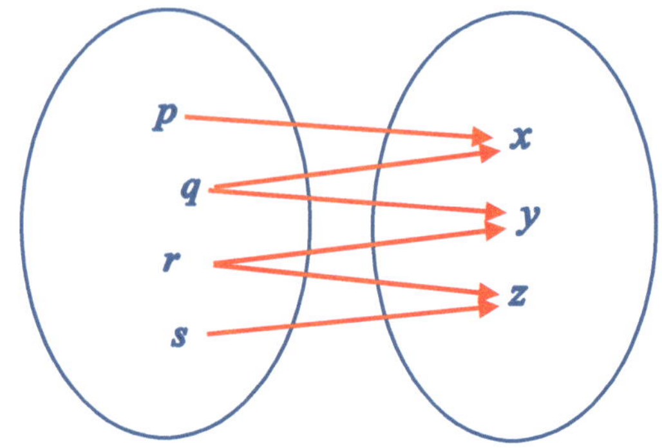

Function theory, introduced by Leonhard Euler in 1734, typically denotes functions with symbols, often single lowercase letters in italic font such as *f*, *g*, or *h*. A function represents a relation where each potential input value corresponds to exactly one output value. In simpler terms, a function is a rule or process that connects each element x from a set X (known as the domain) to a single element y from another set Y (known as the codomain).

The notation $y = f(x)$ defines a function named f. This expression is read as "y is a function of x." Here, the variable x denotes the input value or independent variable, while y (or $f(x)$) denotes the output value or dependent variable.

A function is uniquely defined by the set of all pairs $(x, f(x))$, which constitutes the graph of the function. In the context where both the domain and codomain are sets of real numbers, each pair can be interpreted as Cartesian coordinates representing a point on the plane. The collection of these points forms the graph of the function, a commonly used method to visually depict the function's behavior.

Function
A function is a relation in which each possible input value leads to exactly one output value. This is read as "the output is a function of the input".

The input values, independent variables, make up the domain, and the output values, dependent variables, make up the range.

1.2. Determining Whether a Relation Represents a Function

A relation consists of ordered pairs, where the first component of each pair forms the domain, and the second component forms the range. In the given example, the ordered pairs are (2, 1), (4, 2), (6, 3), and (8, 4). Here, the first numbers represent the first four even integers, while the second numbers are half of the first. Thus, the domain is {2, 4, 6, 8}, and the range is {1, 2, 3, 4}.

In a function, each element in the domain is paired with exactly one element in the range. As each element in the domain {2, 4, 6, 8} corresponds to precisely one element in the range {1, 2, 3, 4}, the relation is indeed a function.

Now let us consider another set of ordered pairs relating "vowel" and "consonant" to English letters: {(vowel, A), (consonant, B), (vowel, E), (consonant, C), (vowel, I), (consonant, D), (vowel, O), (consonant, F), (vowel, U), (consonant, G)}. In this case, the domain consists of "vowel" and "consonant", while the range consists of English letters.

However, it is observed that each element in the domain, {vowel, consonant}, is not paired with just one element in the range, {A, B, C, D, E, F, G, I, O, U}. For instance, "vowel" corresponds to five elements in the range {A, E, I, O, U}, and "consonant" corresponds to five elements in the range {B, C, D, F, G}. This discrepancy violates the definition of a function, indicating that this relation is not a function.

Note that each value in the domain is also known as an input value, or independent variable, and is often labeled with the lowercase letter x. Each value in the range is also known as an output value, or dependent variable, and is often labeled lowercase letter y.

Example
Is the relation expressed in the mapping diagram a function?

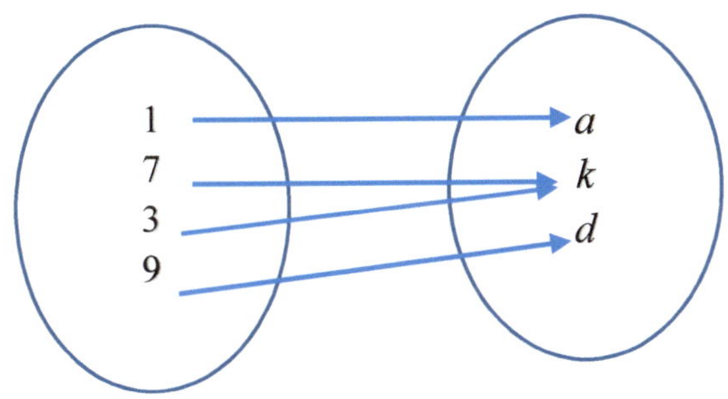

Solution
The domain is {1, 7, 3, 9} and the range is {*a*, *k*, *d*}. We understand that each element in the domain is mapped to exactly one element in the range. Hence, we conclude the relation is a function.

Example
Is the relation expressed in the mapping diagram a function?

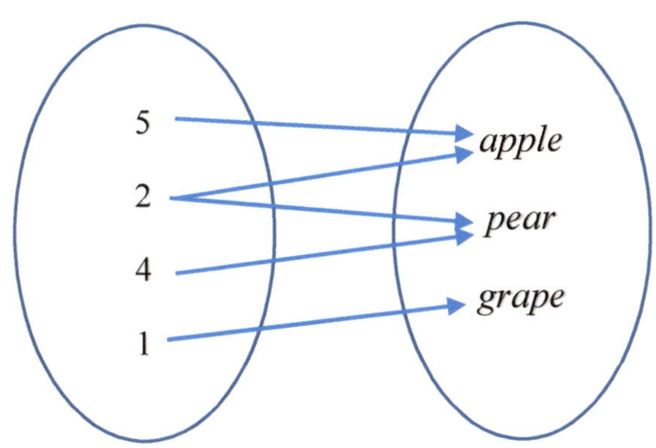

Solution
The domain is {5, 2, 4, 1} and the range is {*apple*, *pear*, *grape*}. We realize that each element in the domain is not mapped to exactly one element in the range, the element “2” in the range is mapped to apple and pear in the range. This violates the definition of a function, so this relation is not a function.

Example
Consider the relation pictured in the figure below. Is this relation a function?

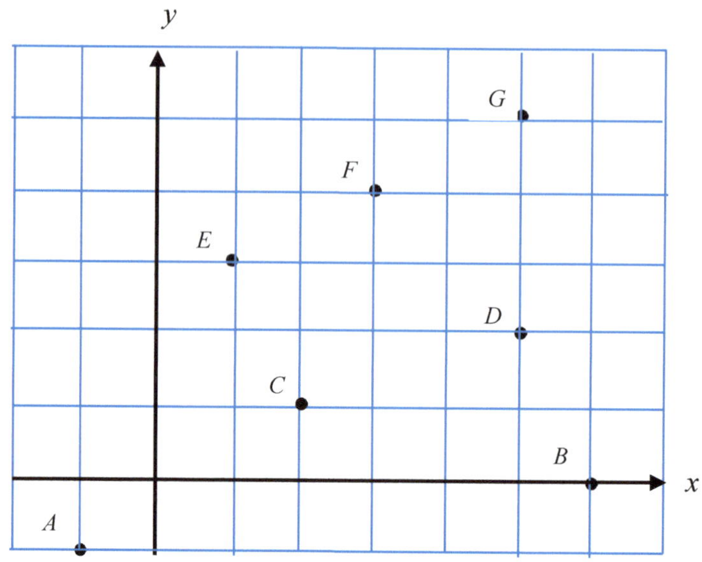

Solution

The graph in the figure consists of the points $A(-1, -1)$, $B(6, 0)$, $C(2, 1)$, $D(5, 2)$, $E(1, 3)$, $F(3, 4)$, and $G(5, 5)$. The domain is $DOM = \{-1, 1, 2, 3, 5, 6\}$ and the range is $RNG = \{-1, 0, 1, 2, 3, 4, 5\}$. The mapping diagram shown below will help us decide if the relation represented by the graph is a function or not. Put the domain on the left, the range on the right, then use arrows to indicate the ordered pairs. Let us call the relation $\mathbb{R}$, then we will have:

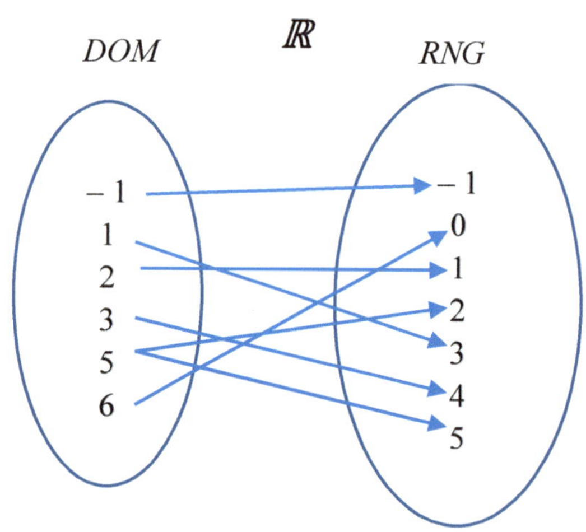

Note the domain object 5 is "sent to" or paired with two answers, 2 and 4. Therefore, the relation is not function.

1.3. Function Notation

Dependent and Independent Variables – Definition of a function states a function is a rule for taking an input and providing an output. A symbol that stands for an arbitrary input is called an **independent variable**, while a symbol that stands for an arbitrary output is called a **dependent variable**.

The most common symbol for the input is x, and the most common symbol for the output is y; the function itself is commonly written $y = f(x)$.

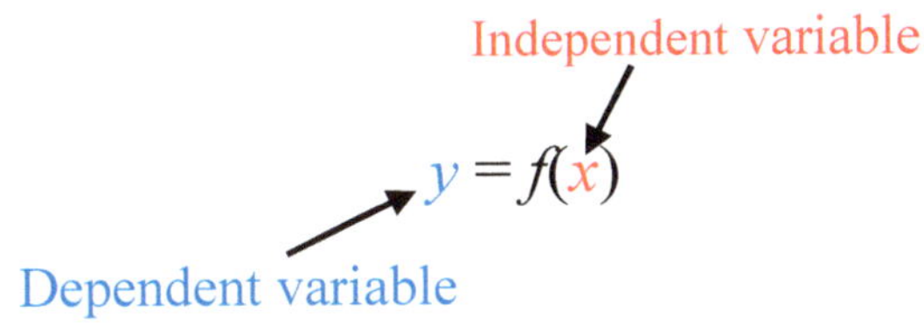

One-to-One Function - A **one-to-one function** is a function in which each output value corresponds to exactly one input value.

$f(x)$ is one-to one function

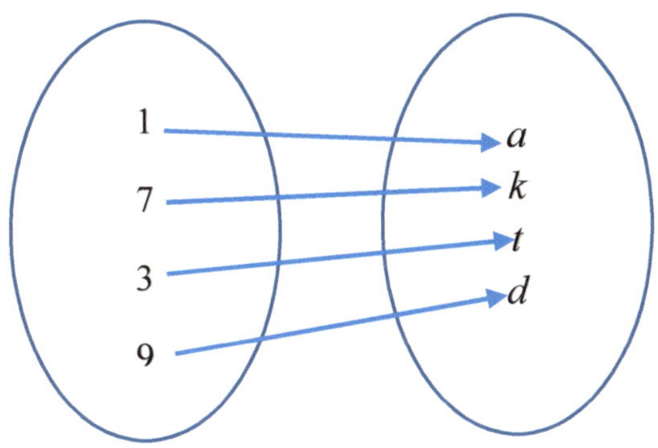

$g(x)$ is NOT one-to one function, $g(7) = k$ and $g(3) = k$

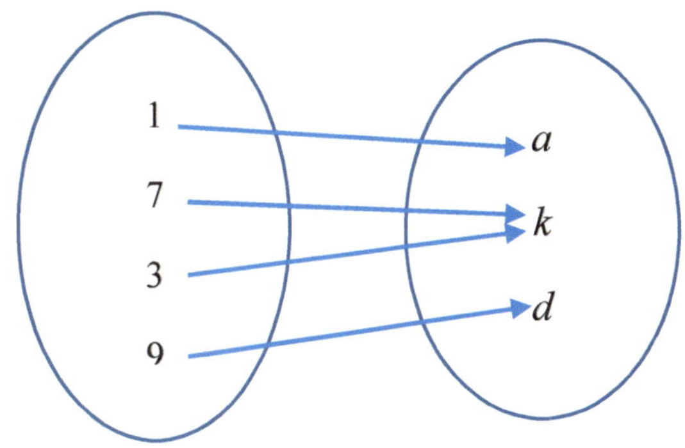

$f(x) = 3x - 3$ is one-to one function,

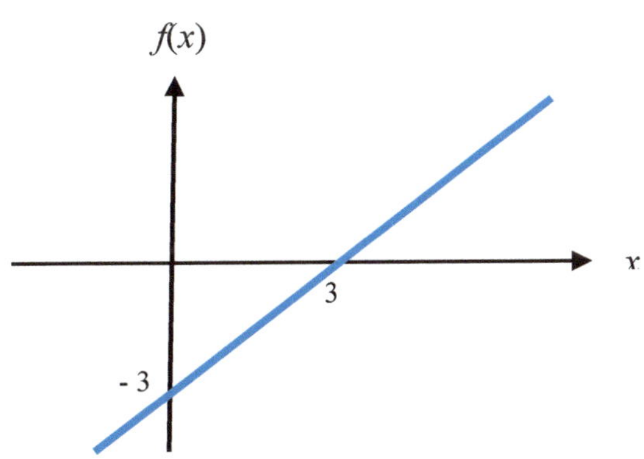

$g(x) = x^2$ is NOT one-to one function,

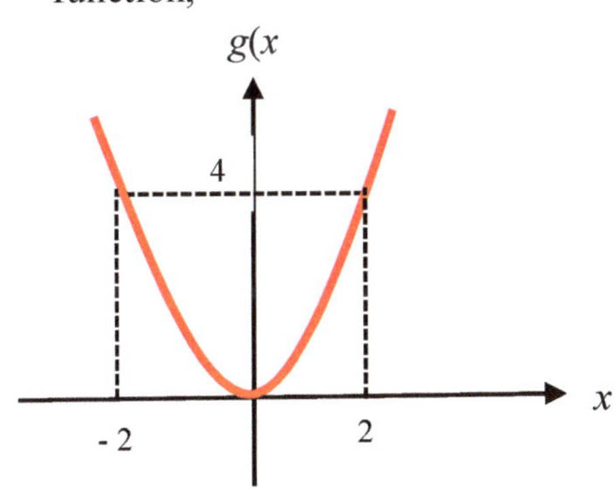

Range and Domain -The **domain** of a function is the complete set of possible values of the independent variable. The **range** of a function is the complete set of all possible resulting values of the dependent variable (y, usually), after we have substituted the domain.

Even Functions - A function ***f* is even** if the graph of f is symmetric with respect to the y-axis. Algebraically, f is **even** if and only if $f(-x) = f(x)$ for all x in the domain of f.

Odd Functions - A function ***f* is odd** if the graph of f is symmetric with respect to the origin. Algebraically, f is odd if and only if $f(-x) = -f(x)$ for all x in the domain of f.

Implicit Function - An **implicit function** is a function written in terms of both dependent and independent variables. In other words, when in a function the dependent variable is not explicitly isolated on either side of the equation then the function becomes an implicit function, like $2y - 5x^2 + 7x + 12 = 0$.

Explicit Function - An **explicit function** is one which is represented in terms of the independent variable, like $y = 13x^3 - 6x^2 + 2x + 8$. Note that y is the dependent variable and is given in terms of the independent variable x.

> **Function Notation**
>
> The notation $y = f(x)$ defines a function denoted f. This is read as "y is a function of x".
>
> The letter x represents the input value, or independent variable. The letter y, or $f(x)$, represents the output value, or dependent variable.

Example

State the domain and range of the following function.

$$(1, \tfrac{1}{2}), (2, 1), (3, \tfrac{3}{2}), (4, 2), (5, \tfrac{5}{2}), (6, 3), (7, \tfrac{7}{2}).$$

Solution

The first numbers in each pair are the first seven natural numbers. The second number in each pair is half of the first. The domain is $\{1, 2, 3, 4, 5. 6, 7\}$, and the range is $\{\frac{1}{2}, 1, \frac{3}{2}, 2, \frac{5}{2}, 3, \frac{7}{2}\}$.

Example

Is the function $f(x) = x^3 + 2x$ even, odd, or neither?

Solution

We can determine whether the function is even or odd by finding formulas for the reflections and determining if they return us to the original function. Let us begin with the rule for even functions.

$f(x) = x^3 + 2x$

$f(-x) = (-x)^3 + 2(-x)$ we plugged $x = -x$ into the equation

$f(-x) = -x^3 - 2x.$

This does not return us to the original function, so this function is not even. We can now test the rule for odd functions.

$f(x) = x^3 + 2x$

$(-1)(f(x) = (-1)(x^3 + 2x)$ we multiplied both sides by (-1)

$-f(x) = -x^3 - 2x$

$-f(-x) = -(-x)^3 - 2(-x)$ we plugged $x = -x$ into the equation

$-f(-x) = -(-x^3) + 2x$

$-f(-x) = x^3 + 2x$

$-f(-x) = f(x).$

Since $-f(-x) = f(x)$, the function is odd.

Example
Is the function $f(x) = x^2 - 13$ even, odd, or neither?

Solution
We can determine whether the function is even or odd by finding formulas for the reflections and determining if they return us to the original function. Let us begin with the rule for even functions.

$$f(x) = x^2 - 13$$

$$f(-x) = (-x)^2 - 13 \qquad \text{we plugged } x = -x \text{ into the equation.}$$

$$f(-x) = x^2 - 13$$

$$f(-x) = f(x).$$

Since $f(-x) = f(x)$, the function is even.

> **Functions, Rule of Three**
> A function consists of three parts:
>
> - A set of objects which is called the domain.
> - A second set of objects which is called the range.
> - A rule that describes how to assign each object in the domain to exactly one object in the range.

1.4. Representations of Functions: Tables, Equations, Graphs

A **function** is a rule that assigns a set of inputs to a set of outputs in such a way that each input has a unique output. That is, no input corresponds to more than one output. Because of this, the term 'is a function of' can be thought of as 'is determined by.' For example, if you were to go to the store with \$12.00 to buy some candy bars that were \$2.00 each, your total cost would be determined by how many candy bars you bought. Therefore, your total cost is a function of the number of candy bars you buy.

A function is a relationship between two variables, such that one variable, dependent variable, is determined by the other variable, independent variable. In our example, if we let x = the items we purchased and y = the amount of money we pay for purchasing items, then y is determined by x, and we say y is a function of x. In this case we have described this pricing example of a function in words. There are other ways to represent a function, as well. Functions can be represented in three more different ways:

- By a function table.
- Mathematically by an equation.
- Graphically.

1.4.1. Function Table

A **function table** is a table of ordered pairs that follows the relationship, or rule, of a function. Function tables can be vertical or horizontal. For example, consider a job where you get paid \$300 a day. If you only work a fraction of the day, you get that fraction of \$300. Thus, the total amount of money you make at the job is determined by the number of days you work. To create a function table, we must first figure out the rule that defines our function. We get \$300 for one-day worked, because $1 \times 300 = 300$. We get

$600 for two-day worked, because $2 \times 300 = 600$. And we get $450 for 1.5 days worked, because $1.5 \times 300 = 450$. So, we identify a pattern here. To find the total amount of money made for this job, we multiply the number of days we have worked by 300. Thus, letting x be the number of days worked, our rule is that we take a value of x and we multiply it by 300 to get y, where y is the total amount of money made.

A **function table** can be used to display this rule. In the first row of the **function table**, we list values of x, and in the second row of the table, we list the corresponding values of y according to the function rule.

x = # days worked	1	1.5	2	3	4	4.25	5	6
y = total money made	300	450	600	900	1200	1275	1500	1800

Function tables can be displayed using columns instead of rows. In this case, the first column displays x-values, and the second column displays y-values.

x = # days worked	y = total money made
1	300
1.5	450
2	600
3	900
4	1200
4.25	1275
5	1500
6	1800

Example
Given function table below, find the missing values of equivalent ratios:

5	25
7	x
9	45
y	55

Solution
Since the table gives values of equivalent ratios we will have:

$$\frac{x}{7} = \frac{25}{5}$$

$$7 \times \frac{x}{7} = 7 \times \frac{25}{5}$$ we multiplied both sides by 7

$$x = \frac{7 \times 25}{5}$$

$$x = \frac{175}{5}$$

$$x = 35.$$

Similarly, for getting the value of y we will have:

$\frac{y}{55} = \frac{9}{45}$

$55 \times \frac{y}{55} = 55 \times \frac{9}{45}$ we multiplied both sides by 55

$y = \frac{55 \times 9}{45}$

$y = \frac{5 \times 11 \times 9}{9 \times 5}$

$y = 11.$

Therefore, $x = 35$, and $y = 11$.

1.4.2. Function Equation

Another method to represent a function is using an equation. In this method, we basically just put our rule into equation form. For the above example, the rule is that we take the number of days worked, x, and multiply it by 300 to get the total amount of money made, y. Putting this in algebraic terms, we have those 300 times x is equal to y. In equation form, we have $y = 300\,x$.

As we obtain our equation that represents our function, we can use it to find y for different values of x by incorporating values of x into the equation. For example, if you wanted to know how much money you would make for a 15-day work, you would put $x = 15$ into the obtained equation.

$y = 300\,x$

$y = 300\,(15)$ we substituted $x = 15$ into the equation

$y = 4500.$

You see that if you worked 15 days, you would make $4,500.

Example
Determine which of the following equations are functions and which are not functions.

a) $y = 3x^2 + 5$
b) $y^2 = 3x + 4$
c) $y^2 = 2x^2 + 1$.

Solution

a) We need to show that no matter what x we plug into the equation and solve for y we will only get a single value of y.

Let us start by plugging in some values of x.

$y = 3x^2 + 5$

for $x = -1$ $y = 3(-1)^2 + 5 = 3(1) + 5 = 3 + 5 = 8$

for $x = 0$ $y = 3(0)^2 + 5 = 3(0) + 5 = 5$

for $x = 1$ $y = 3(1)^2 + 5 = 3(1) + 5 = 3 + 5 = 8$

for $x = 2$ $y = 3(2)^2 + 5 = 3(4) + 5 = 12 + 5 = 17.$

So, for each of the plugged values of x we have got a single value of y. Note that for $x = -1$ and for $x = 1$ we got the same y value, 8. Of course getting the same y value for different x's will not violate the function rule.

However, what we have done so far would not be sufficient to claim that this is a function. We need to show that this will work no matter which value of x we plug into the equation. Practically, we cannot plug all possible values of x into the equation.

Let us examine the functioning of the given relation. For each input value of x, it undergoes a series of operations: it is squared, then the squared value is multiplied by 3, and finally, 5 is added to the result. Notably, after each algebraic operation, only one specific value is produced. Hence, it appears logical that based on the sequence of operations involved in substituting x into the equation, we will consistently obtain a single corresponding value of y. Consequently, we can infer that this relation qualifies as a function.

b) Again, we need to plug in a couple of values of x and solve for y to see what happens.

$y^2 = 3x + 4$

for $x = -1$ $\quad y^2 = 3(\ 1) + 4 = -3 + 4 = 1 \quad \Rightarrow y = \pm 1$

for $x = 0$ $\quad y^2 = 3(0)^2 + 4 = 4 \quad \Rightarrow y = \pm 2$

for $x = 1$ $\quad y^2 = 3(1) + 4 = 3 + 4 = 7 \quad \Rightarrow y = \pm\sqrt{7}.$

As is seen, we get two different y values out of the x and so this relation is NOT a function. Note that in this case there was no need to go any farther than the first evaluation, since that gave us multiple values of y.

c) Since we have y^2 in the problem, as what we learned from part (b), we need to find if we can find a value of x that will give more than one value of y upon solving. So,

$y^2 = 2x^2 + 1$

for $x = -1$ $\quad y^2 = 2(-1)^2 + 1$

$y^2 = 2(1) + 1$

$y^2 = 3$

$y = \pm\sqrt{3}.$

Therefore, this equation is not a function.

Evaluating Function Equation - To evaluate a function, substitute the input (the given number or expression) for the function's variable (place holder, x). Replace the x with the number or expression.

Example

Given the function $f(x) = 12x - 8$, find $f(3)$.

Solution

We need to substitute 3 into the function in place of x. So,

$f(x) = 12x - 8$

$f(x) = 12(3) - 8$ $\quad$ we plugged $x = 3$ into the equation

$f(x) = 36 - 8$

$f(x) = 28.$

This answer can be presented as the ordered pair (3, 28).

Example
Given the function $g(k) = 5k^2 + 4k - 11$ find $g(-2)$.

Solution
We need to substitute – 2 into the function in place of k. So,

$g(x) = 5k^2 + 4k + 3$

$g(x) = 5(-2)^2 + 4(-2) - 11$ we plugged $k = -2$ into the equation

$g(x) = 5(4) + (-8) - 11$

$g(x) = 20 - 8 - 11$

$g(x) = 20 - 19 = 1.$

This answer can be presented as the ordered pair (– 2, 1).

Example
Find $h(4a)$ when $h(u) = 3u^3 - 3u$.

Solution
We need to substitute $4a$ into the function in place of u. So,

$h(u) = 3u^3 - 3u$

$h(4a) = 3(4a)^3 - 3(4a)$ we plugged $u = 4a$ into the equation

$h(4a) = 3(64a^3) - 12a$

$h(4a) = 192a^3 - 12a$

$h(4a) = 12a(16a^2 - 1).$

Example
Find $t(a-2)$ when $t(v) = 2v - 15$.

Solution
We need to substitute $(a - 2)$ into the function in place of v. So,

$t(v) = 2v - 15$

$t(a-2) = 2(a-2) - 15$ we plugged $v = a - 2$ into the equation

$t(a-2) = 2a - 4 - 15$

$t(a-2) = 2a - 19.$

> **Function Definition**
> A relation in which each x-coordinate is matched with only one y-coordinate is said to describe y as a function of x.

1.4.3. Function Graph

The last representation of a function is a graph. We have seen that a function can be represented by an equation, and because equations can be graphed, we can graph a function as well. To represent a function graphically, we find some ordered pairs that satisfy our function rule, plot them, and then connect them in a nice smooth curve. To do so, we have learned that a function is uniquely represented by the set of all pairs $(x, f(x))$, called the graph of the function. When the domain and the codomain are sets of real numbers, each such pair may be considered as the Cartesian coordinates of a point in the plane.

> **Function Graph**
> The graph of a function f is the set of ordered pairs (x, y), where $f(x) = y$. In the common case where x and $f(x)$ are real numbers, these pairs, $(x, f(x))$ are Cartesian coordinates of points in two-dimensional space.

To present a function on a graph, let us consider a simple example.

Example
Given the equation of line $y = f(x) = 3x$, represent it as a graph.

Solution
The graphical representation of this function will be as the figure below.

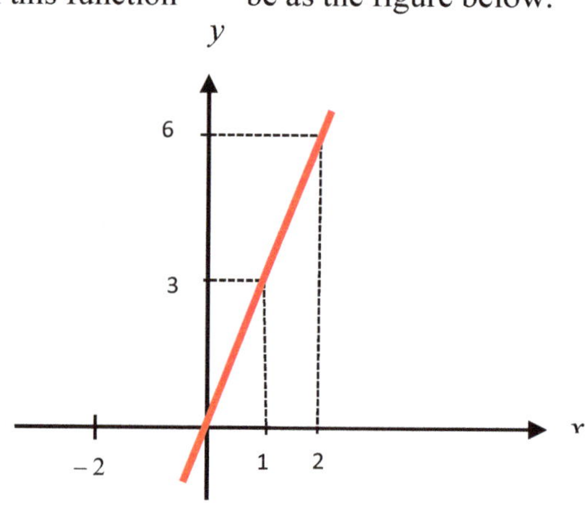

1.5. Function Arithmetic

We are used to adding subtracting, multiplying, and dividing real numbers. We do these operations every day in a variety of situations. We have also learned how to perform four basic operations on algebraic expressions. If you know how to perform the four basic operations on polynomials, then you can also add, subtract, multiply, and divide functions. The notation will look different at first but knowing a couple of steps can help you arrive at the correct answer.

Suppose f and g are functions and x is an element common to the domains of f and g.

- The sum of f and g, denoted $f + g$, is a function defined by the formula: $(f + g)(x) = f(x) + g(x)$.

- The difference of f and g, denoted $f - g$, is a function defined by the formula:

$(f - g)(x) = f(x) - g(x)$.

- The product of f and g, denoted fg, is a function defined by the formula: $(fg)(x) = f(x)g(x)$.

- The quotient of f and g denoted $\frac{f}{g}$, is a function defined by the formula: $\frac{f}{g}(x) = \frac{f(x)}{g(x)}$, provided $g(x) \neq 0$.

In other words, to add two functions, we add their outputs; to subtract two functions, we subtract their outputs, and so on.

Example

Let $f(x) = 3x - 6$ and $g(x) = \frac{3}{x^2} - 9$. Find and simplify expressions for.

a) $(f + g)(x)$.

b) $(g - f)(x)$.

c) $(fg)(x)$. $\frac{f}{g}(x)$.

d) $\frac{f}{g}(x)$.

Additionally, find the domain of each of these functions.

Solution

a) By the function arithmetic, $(f + g)(x)$ is defined to be $f(x) + g(x)$. So,

$$(f + g)(x) = f(x) + g(x)$$

$$(f + g)(x) = 3x - 6 + \frac{3}{x^2} - 9$$

$$(f + g)(x) = \frac{3x^3}{x^2} - \frac{6x^2}{x^2} + \frac{3}{x^2} - \frac{9x^2}{x^2}$$ we need to multiply the associated terms by $\frac{x^2}{x^2}$

$$(f + g)(x) = \frac{3x^3 - 6x^2 + 3 - 9x^2}{x^2}$$

$$(f + g)(x) = \frac{3x^3 - 15x^2 + 3}{x^2}$$

$$(f + g)(x) = 3x - 15 + \frac{3}{x^2}.$$

Except $x \neq 0$, everything else is fine. Hence, the domain is $(-\infty, 0) \cup (0, \infty)$

b) By the function arithmetic, $(f - g)(x)$ is defined to be $f(x) - g(x)$. So,

$$(f - g)(x) = f(x) - g(x)$$

$$(f - g)(x) = 3x - 6 - (\frac{3}{x^2} - 9)$$

$$(f - g)(x) = \frac{3x^3}{x^2} - \frac{6x^2}{x^2} - \frac{3}{x^2} + \frac{9x^2}{x^2}$$ we need to multiply the associated terms by $\frac{x^2}{x^2}$

$$(f-g)(x) = \frac{3x^3 - 6x^2 - 3 + 9x^2}{x^2}$$

$$(f-g)(x) = \frac{3x^3 + 3x^2 - 3}{x^2}$$

$$(f-g)(x) = 3x + 3 - \frac{3}{x^2}.$$

$$(f-g)(x) = 3(x + 1 - \frac{1}{x^2}).$$

Except $x \neq 0$, everything else is fine. Hence, the domain is $(-\infty, 0) \cup (0, \infty)$

c) By the function arithmetic, $(f\,g)(x)$ is defined to be $f(x)\,g(x)$. So,

$$(f\,g)(x) = f(x)\,g(x)$$

$$(f\,g)(x) = (3x - 6) \times (\frac{3}{x^2} - 9)$$

$$(f\,g)(x) = \frac{9x}{x^2} - 27x - \frac{18}{x^2} + 54$$ we applied the distributive property

$$(f\,g)(x) = \frac{9x}{x^2} - \frac{27x^3}{x^2} - \frac{18}{x^2} + \frac{54x^2}{x^2}$$ we need to multiply the associated terms by $\frac{x^2}{x^2}$

$$(f\,g)(x) = \frac{9x - 27x^3 - 18 + 54x^2}{x^2}$$

$$(f\,g)(x) = \frac{-27x^3 + 54x^2 + 9x - 18}{x^2}$$ we sorted the right-hand side nominator by the decreasing order of variables' exponent

$$(f\,g)(x) = \frac{-27x \times x^2 + 54 \times x^2 + 9x - 18}{x^2}$$

$$(f\,g)(x) = -27x + 54 + 9 \times \frac{x-2}{x^2}.$$

Except $x \neq 0$, everything else is fine. Hence, the domain is $(-\infty, 0) \cup (0, \infty)$

d) By the function arithmetic, $\frac{f}{g}(x)$ is defined to be $\frac{f(x)}{g(x)}$. So,

$$\frac{f}{g}(x) = \frac{f(x)}{g(x)}$$

$$\frac{f}{g}(x) = \frac{(3x-6)}{(\frac{3}{x^2} - 9)}$$

$$\frac{f}{g}(x) = \frac{(3x-6)}{(\frac{3-9x^2}{x^2})}$$

$$\frac{f}{g}(x) = (3x - 6) \times \frac{x^2}{3 - 9x^2}$$ we applied the inverse property

$$\frac{f}{g}(x) = \frac{3x^3 - 6x^2}{3 - 9x^2}$$ we need to multiply the associated terms by $\frac{x^2}{x^2}$

$$\frac{f}{g}(x) = \frac{3(x^3 - 2x^2)}{3(1 - 3x^2)}$$

$$\frac{f}{g}(x) = \frac{x^3 - 2x^2}{1 - 3x^2}.$$

To find the domain, we let

$$1 - 3x^2 = 0$$

$$3x^2 = 1$$

$$x = \pm\sqrt{3}.$$

So, except $x \neq \pm\sqrt{3}$, everything else is fine. Hence, the domain is $(-\infty, -\sqrt{3}) \cup (-\sqrt{3}, \sqrt{3}) \cup (\sqrt{3}, \infty)$.

Next, let us look at a example for the difference quotient of a function.

Example

Given three functions, $f(x) = 17x^3 + 24x^2$, $g(x) = 6x^2$, and $h(x) = 18x^4 - 5x^3$, find $\left(\frac{f+h}{g}\right)(x)$.

Solution

We learned, $(f + h)(x) = f(x) + h(x)$. So,

$$(f + h)(x) = f(x) + h(x)$$

$(f + h)(x) = 17x^3 + 24x^2 + 18x^4 - 5x^3$ — we replaced the $f(x)$ and the $h(x)$ with the equivalent polynomials

$(f + h)(x) = 18x^4 + 17x^3 - 5x^3 + 24x^2$ — we sorted the right-hand side by the decreasing order of variables' exponent

$$(f + h)(x) = 18x^4 + 12x^3 + 24x^2.$$

Now letting $t(x) = (f + h)(x)$. we learned $(\frac{t}{g})(x) = \frac{t(x)}{g(x)}$. So,

$$(\frac{t}{g})(x) = \frac{t(x)}{g(x)}$$

$(\frac{f+h}{g})(x) = \frac{(f+h)(x)}{g(x)}$ — we replace t with the equivalent $f + h$

$(\frac{f+h}{g})(x) = \frac{18x^4 + 12x^3 + 24\ \ ^2}{6x^2}$ — we replaced the $(f + h)(x)$ and the $g(x)$ with the equivalent polynomials

$$(\frac{f+h}{g})(x) = \frac{(18x^2 \times x^2)+(12x \times x^2) + 24 \times x^2}{6x^2}$$

$$(\frac{f+h}{g})(x) = \frac{18x^2 + 12x + 24}{6} \times \frac{x^2}{x^2}$$

$$(\frac{f+h}{g})(x) = \frac{18x^2 + 12x + 24}{6}$$

$$(\frac{f+h}{g})(x) = \frac{(6 \times 3x^2) + (6 \times 2x) + (6 \times 4)}{6}$$

$$(\frac{f+h}{g})(x) = \frac{6(3x^2 + 2x + 4)}{6}$$

$$(\frac{f+h}{g})(x) = 3x^2 + x + 4.$$

Therefore, the function $(\frac{f+h}{g})(x)$ is equal to the polynomial $3x^2 + x + 4$.

1.5.1. Functions, Difference Quotient

The **difference quotient** is usually the name for the expression $\frac{f(x+h)-f(x)}{h}$ which when taken to the limit as h approaches 0 gives the derivative of the function f. The difference quotient is a measure of the average rate of change of the function over an interval (in this case, an interval of length h).

> **Definition of Difference Quotient**
> Given a function f, the difference quotient of f is the expression:
> $$\frac{f(x+h)-f(x)}{h}.$$

By a slight change in notation, for an interval $[a, b]$, the difference quotient is.

$$\frac{f(b)-f(a)}{b-a}.$$

Which is called the mean (or average) value of the derivative of a function f over the interval $[a, b]$. Geometrically, this difference quotient measures the slope of the secant line passing through the points with coordinates $(a, f(a))$ and $(b, f(b))$.

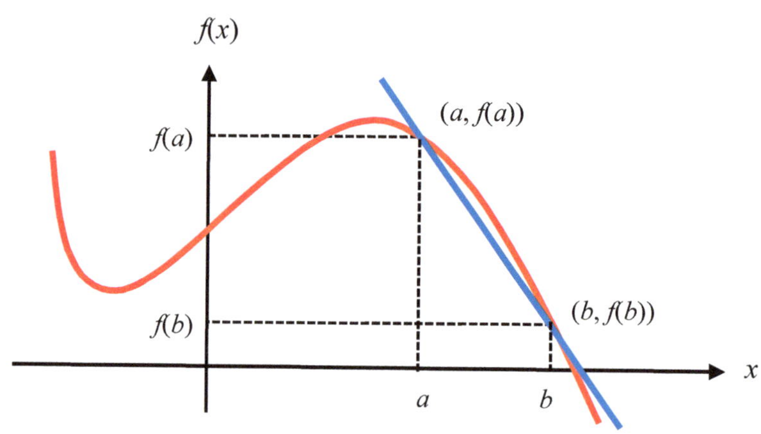

1.5.2. Functions, Simplifying of Difference Quotient

Simplifying a difference quotient involves rewriting it in a form where the 'h' in the definition of the difference quotient cancels out from the denominator. Once this cancellation occurs, we consider our task complete.

Example
Find and simplify the difference quotients for the following functions.

a) $f(x) = 2x^2 - 3x - 5$.

b) $g(x) = \frac{7}{4x+6}$.

c) $r(x) = \sqrt{3x}$.

Solution

a) To find $f(x + h)$, we substitute the expression $x + h$ for the function's variable, x. So,

$f(x) = 2x^2 - 3x - 5$

$f(x + h) = 2(x + h)^2 - 3(x + h) - 5$ — we substituted $x + h$ for x

$f(x + h) = 2(x^2 + 2hx + h^2) - 3x - 3h - 5$ — we applied binomial theorem expansion

$f(x + h) = 2x^2 + 4hx + 2h^2 - 3x - 3h - 5$

$f(x + h) = 2x^2 + 4hx + 2h^2 - 3x - 3h - 5.$

Now, the difference quotients will be:

$$\frac{f(x+h) - f(x)}{h}$$

$$\frac{f(x+h)-f(x)}{h} = \frac{2x^2 + 4hx + 2h^2 - 3x - 3h - 5 - (2x^2 - 3x - 5)}{h}$$

$$\frac{f(x+h) - f(x)}{h} = \frac{2x^2 + 4hx + 2h^2 - 3x - 3h - 5 - 2x^2 + 3x + 5}{h}$$

$$\frac{f(x+h) - f(x)}{h} = \frac{4hx + 2h^2 - 3h}{h}$$

$$\frac{f(x+h) - f(x)}{h} = \frac{h(4x + 2h - 3)}{h}$$

$$\frac{f(x+h) - f(x)}{h} = 4x + 2h - 3.$$

Since we have managed to cancel the original 'h' from the denominator, we are done.

b) Again, to find $g(x + h)$, we substitute the expression $x + h$ for the function's variable, x. So,

$gg(x) = \frac{7}{4x + 6}$

$g(x + h) = \frac{7}{4(x + h) + 6}$ — we substituted $x + h$ for x

$g(x + h) = \frac{7}{4x + 4h + 6}.$

Now, the difference quotients will be:

$$\frac{g(x+h) - g(x)}{h}$$

$$\frac{g(x+h) - g(x)}{h} = \frac{\frac{7}{4x + 4h + 6} - \frac{7}{4x + 6}}{h}$$

$$\frac{g(x+h) - g(x)}{h} = \frac{\frac{7}{4x + 4h + 6} - \frac{7}{4x + 6}}{h} \times \frac{(4x + 4h + 6)(4x + 6)}{(4x + 4h + 6)(4x + 6)}$$

$$\frac{g(x+h) - g(x)}{h} = \frac{7(4x + 6) - 7(4x + 4h + 6)}{h(4x + 4h + 6)(4x + 6)}$$

$$\frac{g(x+h) - g(x)}{h} = \frac{28x + 42 - 28x - 28h - 42}{h(4x + 4h + 6)(4x + 6)}$$

$$\frac{g(x+h)-g(x)}{h}=\frac{-28h}{h(4x+4h+6)(4x+6)}$$

$$\frac{g(x+h)-g(x)}{h}=-\frac{28}{(4x+4h+6)(4x+6)}.$$

Since we have managed to cancel the original 'h' from the denominator, we are done.

c) For $r(x)=\sqrt{3x}$ we need first to get $r(x+h)=\sqrt{3(x+h)}=\sqrt{3x+3h}$, then the difference quotient will be:

$$\frac{r(x+h)-r(x)}{h}$$

$$\frac{r(x+h)-r(x)}{h}=\frac{\sqrt{3x+3h}-\sqrt{3x}}{h}.$$

To cancel the 'h' from the denominator, we rationalize the numerator by multiplying by its conjugate. Note that a conjugate of an expression like $(a+b)$ is $(a-b)$. So,

$$\frac{r(x+h)-r(x)}{h}=\frac{\sqrt{3x+3h}-\sqrt{3x}}{h}$$

$$\frac{r(x+h)-r(x)}{h}=\frac{\sqrt{3x+3h}-\sqrt{3x}}{h}\times\frac{\sqrt{3x+3h}+\sqrt{3x}}{\sqrt{3x+3h}+\sqrt{3x}}$$

$$\frac{r(x+h)-r(x)}{h}=\frac{\sqrt{3x+3h}-\sqrt{3x}}{h}\times\frac{\sqrt{3x+3h}+\sqrt{3x}}{\sqrt{3x+3h}+\sqrt{3x}}$$

$$\frac{r(x+h)-r(x)}{h}=\frac{(\sqrt{3x+3h}-\sqrt{3x})(\sqrt{3x+3h}+\sqrt{3x})}{h(\sqrt{3x+3h}+\sqrt{3x})}$$

$$\frac{r(x+h)-r(x)}{h}=\frac{(\sqrt{3x+3h})^2-(\sqrt{3x})^2}{h(\sqrt{3x+3h}+\sqrt{3x})}$$ we applied the difference of squares rule

$$\frac{r(x+h)-r(x)}{h}=\frac{3x+3h-3x}{h(\sqrt{3x+3h}+\sqrt{3x})}$$

$$\frac{r(x+h)-r(x)}{h}=\frac{3h}{h(\sqrt{3x+3h}+\sqrt{3x})}$$

$$\frac{r(x+h)-r(x)}{h}=\frac{3}{\sqrt{3x+3h}+\sqrt{3x}}.$$

Since we have managed to cancel the original 'h' from the denominator, we are done.

1.6. Equality of Functions

Two functions are equal if they have the same domain and codomain, and their values are the same for all elements of the domain. In other words, two functions f and g are equal, if their domain and codomain sets are the same and their output values agree on the whole domain. More formally, $f=g$ if $f(x)=g(x)$ for all $x\in X$, where $f{:}X\to Y$ and $g{:}X\to Y$.

Equality of Functions
Let A and B be two sets and f: A→B and g: A→B be functions. We say that f and g are equal and write $f=g$ if $f(a)=g(a)$ for all $a\in A$.

Example

Determine whether $f(x)$ and $h(x)$ are identical functions.

$$f(x) = |x|$$

$$h(x) = \sqrt{x^2}.$$

Solution

Two function forms are equivalent as $f(x)$ is reduced to $h(x)$ on simplification. Now $f(x)$ is:

$$f(x) = |x|$$

$$f(x) = x \qquad \text{for } x \geq 0$$

$$f(x) = -x \qquad \text{for } x < 0$$

$$h(x) = \sqrt{x^2}$$

$$h(x) = \pm x.$$

Thus, domain of $f(x)$ is $\mathbb{R}$. On the other hand, domain of $h(x)$ is also $\mathbb{R}$. Clearly, the two given functions are equal.

1.7. Asymptotes of Functions

An **asymptote** of a curve is a line such that the distance between the curve and the line approaches zero as one or both of the x or y coordinates tends to infinity. In other words, an asymptote of a curve is a line which is tangent to the curve at a point at infinity.

1.7.1. Vertical Asymptotes

The constant line $x = a$ is a vertical asymptote of the graph of the function $y = f(x)$ if at least one of the following statements is true:

$$\lim_{x \to a^-} f(\mathrm{x}) = \pm\infty$$

$$\lim_{x \to a^+} f(\mathrm{x}) = \pm\infty,$$

where $\lim_{x \to a^-}$ is the limit as x approaches the value a from the left, and $\lim_{x \to a^+}$ is the limit as x approaches a from the right.

1.7.2. Horizontal Asymptotes

The constant line $y = b$ is a **horizontal asymptote** of the graph of the function $y = f(x)$ if:

$$\lim_{x \to -\infty} f(\mathrm{x}) = b$$

$$\lim_{x \to +\infty} f(\mathrm{x}) = b.$$

In the first case, $f(x)$ has $y = b$ as asymptote when x tends to and in the second case $f(x)$ has $y = b$ as an asymptote as x tends to .

1.7.3. Oblique Asymptotes

Oblique asymptote or slant asymptote is not parallel to the x-axis or y-axis. The line $y = mx + k$ $(m \neq 0)$ is an oblique asymptote of the function $f(x)$ if:

$$\lim_{x \to +\infty} [f(x) - (mx + k)] = 0 \text{ or } \lim_{x \to -\infty} [f(x) - (mx + k)] = 0.$$

In the first case the line $y = mx + k$ is an oblique of $f(x)$ when x tends line $y = mx + k$ is an oblique asymptote of $f(x)$ when x tends

Here are figures of the three asymptotes. The red line in each is the asymptote.

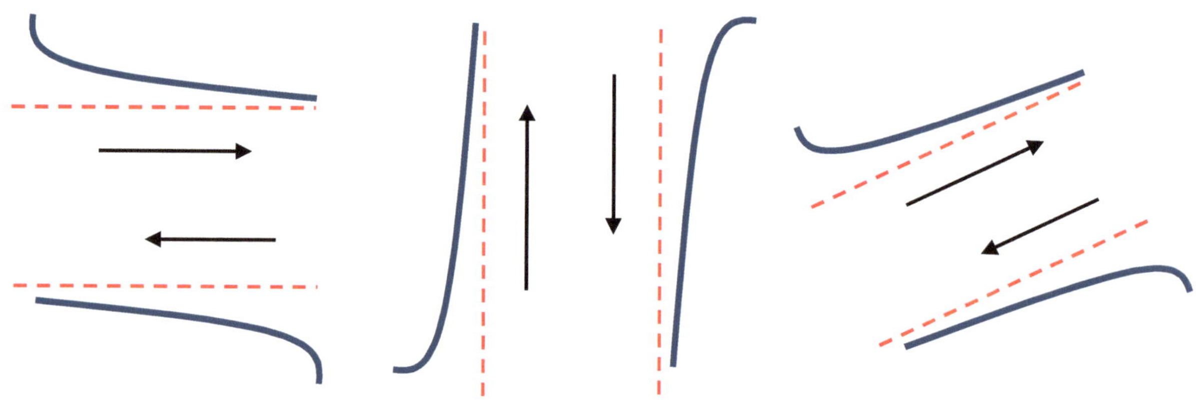

horizontal asymptote vertical asymptote oblique asymptote

Finding the asymptotes of a function requires a more detailed discussion. The discussion regarding how to obtain horizontal asymptotes and slant asymptotes will be presented in later sections of this chapter (Rational functions). We therefore only provided their definitions here, and we will present an example of each type.

Example

Find the asymptote of $\boldsymbol{f(x) = \frac{2x}{x-3}}$.

Solution

As x approaches the constant value 3 (from the left or right) then the curve goes towards plus infinity (or minus infinity).

$$\lim_{x \to 3^-} \frac{2x}{x-3} = -\infty$$

$$\lim_{x \to 3^+} \frac{2x}{x-3} = +\infty,$$

Therefore, the curve has a vertical asymptote $x = 3$.

Example

Find the horizontal asymptotes for $f(x) = \frac{4x + 3}{2x}$.

Solution:

Given, $f(x) = \frac{4x + 3}{2x}$, since the highest degree here in both numerator and denominator is 1, therefore, the horizontal asymptote is the coefficient of x.

Therefore, the horizontal asymptote is $y = \frac{4}{2} = 2$.

Example

Find the horizontal asymptotes for $f(x) = \frac{5x}{x^3 + 10}$.

Solution:

Given, $f(x) = \frac{5x}{x^3 + 10}$, since the degree of the numerator is less than the denominator, therefore, the horizontal asymptote is located at $y = 0$.

Example

Find the horizontal asymptotes for $f(x) = \frac{3x^2 + 7}{x + 8}$.

Solution:

Given, $f(x) = \frac{3x^2 + 7}{x + 8}$., since the degree of the numerator is greater than that of the denominator. Hence, there is no horizontal asymptote.

Example

Find the horizontal and the vertical asymptotes of the function $f(x) = \frac{2x + 5}{x + 3}$.

Solution:

Given, $f(x) = \frac{2x + 5}{x + 3}$, since $f(x)$ is not defined for $x = -3$, we find the one-sided limits for the given function at $x = -3$.

$$\lim_{x \to -3^{-0}} f(x) = \lim_{x \to -3^{+0}} \frac{2x + 5}{x + 3} = \frac{-1}{-0} = +\infty$$

$$\lim_{x \to -3^{+}} f(x) = \lim_{x \to -3^{+}} \frac{2x + 5}{x + 3} = \frac{-1}{0} = -\infty.$$

Therefore, the function $f(x)$ has a vertical asymptote at $x = -3$.

Now, let us find the horizontal asymptotes by taking $x \to \pm\infty$

$$\lim_{x \to \pm\infty} f(x) = \lim_{x \to \pm\infty} \frac{2x + 5}{x + 3} = \lim_{x \to \pm\infty} \frac{2 + \frac{5}{x}}{1 + \frac{3}{x}} = \frac{2}{1} = 2.$$

Therefore, the function $f(x)$ has a horizontal asymptote at $y = 2$.

Example

Find the asymptotes of the function $f(x) = \frac{2x^3-18}{x^2-9}$.

Solution:

The function has vertical asymptotes $x = \pm 3$.

After a long division, the function we find the oblique asymptote at $y = 2x$.

1.8. Exercises

1. Find if the relation expressed in the mapping diagram is a function?

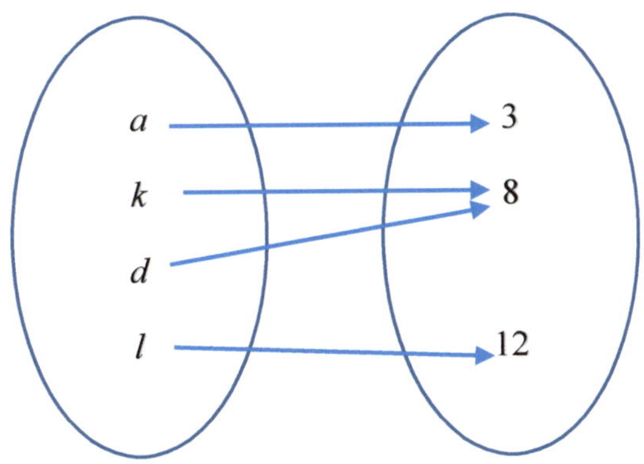

2. Given the relation pictured in the figure below. Find if this relation is a function or not?

•

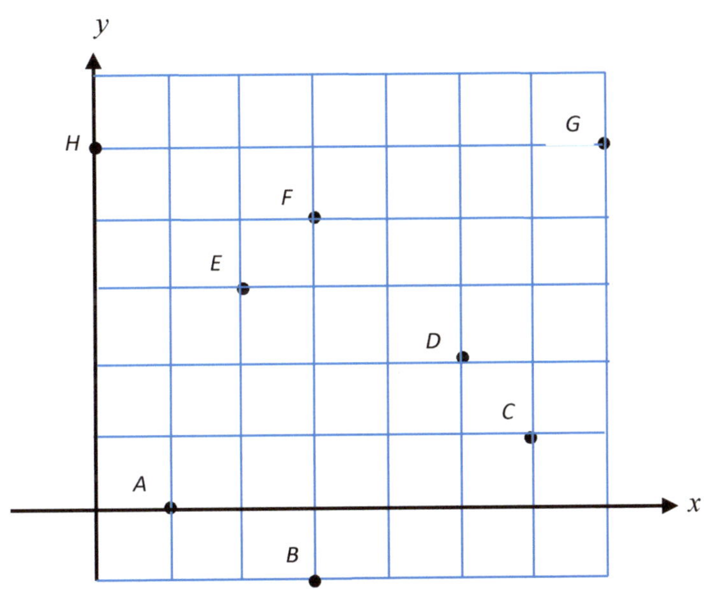

3. State the domain and range of the following function.
$(11, \frac{11}{2}), (12, 6), (13, \frac{13}{2}), (14, 7), (15, \frac{15}{2}), (16, 8).$

4 Find if each of the following functions is even, odd, or neither?

a) $f(x) = 5x^3 + 2x + 7.$

b) $g(v) = 2v^3 - 5v.$

c) $h(u) = \frac{u^3}{2u - u^3}.$

4. Given function table below, find the missing values:

3	9
4	12
5	15
6	x
7	21
8	y

5. Determine which of the following equations are functions and which are not functions.
   a) $4x^2 - 5y + 2 = 0$.
   b) $y^2 = \frac{3}{5}x + 1$.
6. Determine the domain of each of the following functions.
   a) $f(x) = \frac{2x+7}{4-x^2}$
   b) $g(x) = \sqrt{7-3x}$
   c) $h(x) = \frac{\sqrt{4x-12}}{16-9x^2}$
   d) $k(x) = -5x + 1$
   e) $f(t) = \frac{1}{t^2}$
   f) $f(t) = \frac{1}{t^2+1}$
   g) $f(x) = -\frac{7}{x^2-3x-4}$.

8 Given $f(x) = x^2 - 3x + 1$ and $g(x) = x - 3x^2$ evaluate each of the following.

   a) $(f+g)(x)$.
   b) $(f+g)(3)$.
   c) $(f-g)(x)$.
   d) $(f-g)(1)$.
   e) $(f\times g)(x)$.
   f) $(f\times g)(2)$.
   g) $(\frac{f}{g})(x)$.
   h) $(\frac{f}{g})(4)$.

9. Evaluate each function

   a) $f(x) = -6x^3 + 4x^2 - 5x + 1$ at $f(2)$.

b) $h(z) = \frac{3z + 9}{z^2 - 4z - 12}$ at $h(-4)$.

c) $t(x) = \frac{2x + 7}{4 - x^2}$ at $t(0)$.

d) $k(v) = \frac{\sqrt{8v + 1}}{-6 + 8v - 4v^2}$ at $k(1)$.

e) $g(x) = -\frac{15}{3x^2 - 4x - 5}$ at $g(0)$.

f) $l(z) = z^2 + 4z - 5$ at $l(a - b)$.

10. Given two functions, $f(x) = x^3 + 9x^2$ and $h(x) = -3x^3 + 11x^2$ find $3f - 4h$.

11. Given three functions, $f(x) = 5x^6 + 2x^4$, $g(x) = 3x^6 - 4x^4$ and $h(x) = -5x$, find $h(3f - 5g)(x)$.

12. Given three functions, $t(x) = 3x^5 - 6x^4 + 7x^2$, $u(x) = 6x^5 - 12x^4 + 4x^2$, $h(x) = -5x$, and $v(x) = 2 + (x + 1)^2$, find $v\left(\frac{2t - u}{h}\right)(x)$.

13. Find and simplify the difference quotients for the following functions.

a) $f(x) = x^2 - 10x + 7$.

b) $g(v) = \frac{12v^2}{3v + 6}$.

c) $h(u) = \sqrt{16(u - 5)}$.

14. Determine whether $f(x)$ and $g(x)$ are identical functions.

a) $f(x) = (x - 1)(x + 1)$ and $g(x) = x^2 - 1$.

b) $f(x) = 2x$ and $g(x) = x^2$.

c) $f(x) = (x - 3)(x + 1)$ and $g(x) = x^2 - 2x - 1$.

15. Find the asymptote of $f(x) = \frac{5x}{x - 9}$.

16. Find the horizontal asymptotes for $f(x) = \frac{12x + 7}{4x}$.

17. Find the horizontal asymptotes for $f(x) = \frac{3x^2}{2x^3 + 11}$.

18 Find the horizontal asymptotes for $f(x) = \frac{3x^2 + 7}{x + 8}$.

19. Find the horizontal and the vertical asymptotes of the function $f(x) = \frac{4x + 5}{x + 13}$.

20 Find the asymptotes of the function $f(x) = \frac{3x^3 - 12}{x^2 - 4}$.

2. Composition of Functions

Function composition is a process of performing a set of operations on an input to produce an output. Therefore, a **composition of functions** occurs when the output, or result, of one function becomes the input of another function. That is for functions represented by $f(x)$ or $g(x)$, the **composition** would be represented by $f(g(x))$ or $g(f(x))$.

In other words, given two functions f and g to produce a function h we can have either $h(x) = g(f(x))$ or $h'(x) = f(g(x))$. In input-output perception, for $h(x) = g(f(x))$, the function g is applied to the result of applying the function f to x. It implies that the functions f maps X to Y and g maps Y to Z are composed to yield a function that maps x in X to $g(f(x))$ in Z.

A composite function is generally a function that is written inside another function. Composition of a function is done by substituting one function into another function. That is, if we are given two functions, we can create another function by composing one function into the other, like $f[g(x)]$ or $(f \circ g)(x)$ is the composite function of $f(x)$ and $g(x)$.

The result of $f(\ )$ is sent through $g(\ )$

It is written: $(g \circ f)(x)$, which means $g\,[f(x)]$.

The result of $g(\ )$ is sent through $f(\ )$

It is written: $(f \circ g)(x)$, which means $f\,[g(x)]$.

> **Composition of Functions**
> When the output of one function is used as the input of another, we call the entire operation a composition of functions. For any input x and functions f and g, this action defines a composite function, which we write as $f \circ g$ such that:
>
> $$(f \circ g)(x) = f(g(x)).$$
>
> The domain of the composite function $f \circ g$ is all x such that x is in the domain of g and $g(x)$ is in the domain of f.
>
> It is important to realize that the product of functions fg is not the same as the function composition $(f \circ g)(x) = f(g(x))$, because, in general, $f(x)g(x) \neq f(g(x))$.

Example
Given $f = \{(1, 5), (2, 5), (3, 7), (4, 9)\}$, and $g = \{(5, 3), (7, 1), (9, 7)\}$, find the composite function of g and f, that is $g \circ f$.

Solution

The composite function of g and f is $g \circ f = \{(1, 3), (2, 3), (3, 1), (4, 7)\}$. This is based on the mapping that is shown in the figure below.

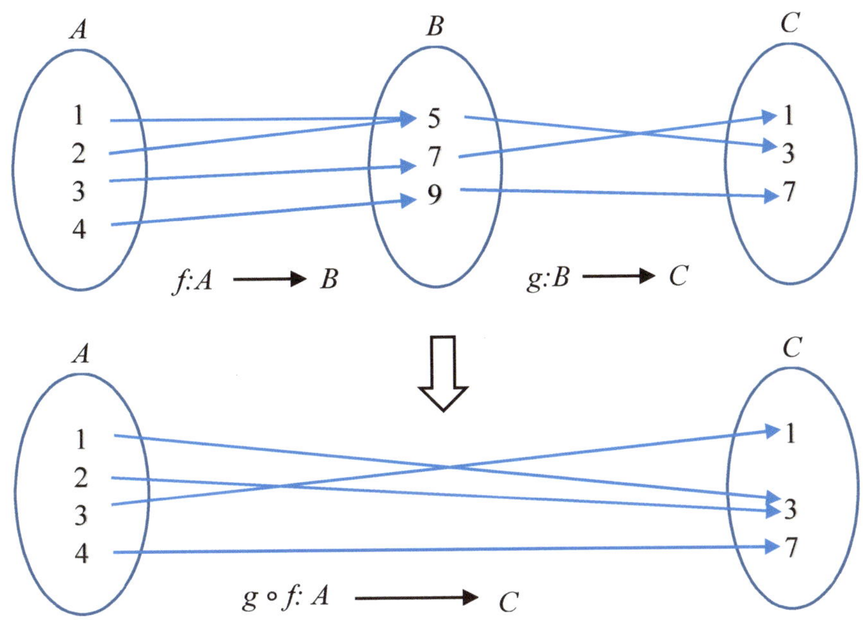

Example

Given $f(x) = 5x - 8$, and $g(x) = 4x^3$, find:

a) The composite function of $f[g(x)]$.

b) The composite function of $g[f(x)]$.

Solution

a) Note that "x" is just a placeholder. To avoid confusion let us just call it "input":

$$f(\text{input}) = 5(\text{input}) - 8$$

$$g(\text{input}) = 4(\text{input})^3.$$

Now for $f[g(x)]$, we will have:

$f(x) = 5x - 8$, and $g(x) = 4x^3$

$f[g(x)] = f[4x^3]$ we plugged $x = g(x)$ into $f(x)$.

$f[g(x)] = 5(4x^3) - 8$ we plugged $x = 4x^3$ into $f(x)$.

$f[g(x)] = 20x^3 - 8.$

b) Now for $g[f(x)]$, we will have:

$f(x) = 5x - 8$, and $g(x) = 4x^3$

$g[f(x)] = g[5x - 8]$ we plugged $x = f(x)$ into $g(x)$.

$g[f(x)] = 4(5x - 8)^3$ we plugged $x = 5x - 8$ into $g(x)$.

Example

Given $f(x) = 3x + 2$, $g(x) = 2x^2 - 1$, find:

a) $f(g(x))$.

b) $g(f(x))$.

Solution

a) To find $f(g(x))$, we have:

$$f(2x^2 - 1) = 3(2x^2 - 1) + 2$$

$$f(2x^2 - 1) = 6x^2 - 3 + 2$$

$$f(2x^2 - 1) = 6x^2 - 1.$$

b) Similarly, to get $g(f(x))$, we have:

$$g(f(x)) = g(3x + 2)$$

$$g(f(x)) = 2(3x + 2)^2 - 1$$

$$g(f(x)) = 2(9x^2 + 12x + 4) - 1$$

$$g(f(x)) = 18x^2 + 24x + 8 - 1$$

$$g(f(x)) = 18x^2 + 24x + 7.$$

Note that $f(g(x)) \neq g(f(x))$. This implies that the operation of function composition is not commutative.

2.1. Domain of a Composite Functions

The domain of a composite function $f(g(x))$ is the set of those inputs x in the domain of g for which $g(x)$ is in the domain of f.
To determine the domain of composition function f(g(x)),

1. Find the domain of g.
2. Find the domain of f.
3. Find those inputs x in the domain of g for which $g(x)$ is in the domain of f. That is, exclude those inputs x from the domain of g for which $g(x)$ is not in the domain of f. The resulting set is the domain of $f(g(x))$.

Example

Given $f(x) = \frac{8}{x-3}$ and $g(x) = \frac{3}{2x-3}$, find the domain of $f(g(x))$.

Solution

The domain of $g(x)$ consists of all real numbers except $x = \frac{3}{2}$, since that input value implies that we divide it by 0. Similarly, the domain of f consists of all real numbers except 3. So, we need to exclude from the domain of $g(x)$ that value of x for which $g(x) = 3$. So,

$$g(x) = \frac{3}{2x-3}$$

$3 = \frac{3}{2x-3}$

$6x - 9 = 3$

$6x = 12$ we isolated the variable in one side and constants in the other side of the equation

$\frac{1}{6} \times 6x = \frac{1}{6} \times 12$ we multiplied both side of the equation by the reciprocal of the variable coefficient

$x = \frac{12}{6} = 2.$

So, the domain of $f(g(x))$.is the set of all real numbers except $\frac{3}{2}$ and 2, $x \neq \frac{3}{2}$ and $x \neq 2$. The domain in interval notation is written as:

$$(-\infty, \tfrac{3}{2}) \cup \left(\tfrac{3}{2}, 2\right) \cup (2, \infty).$$

2.2. Transformation of Functions, Horizonal and Vertical Shift

A simple transformation comprises shifting the entire graph of a function up, down, right, or left. The simplest shift is a vertical shift, moving the graph up or down, because this transformation involves adding a positive or negative constant to the function.

Vertical Shift

Given a function $f(x)$ and a constant number k, a new function $g(x) = f(x) + k$ is a vertical shift of the function $f(x)$. Every output value changes by k units. If k is positive, the graph will shift up. If k is negative, the graph will shift down.

As an example, for a function $g(x) = f(x) + k$, the function $f(x)$ is shifted vertically k units up, and for a function $h(x) = f(x) - k$, the function $f(x)$ is shifted vertically k units down. See figure below:

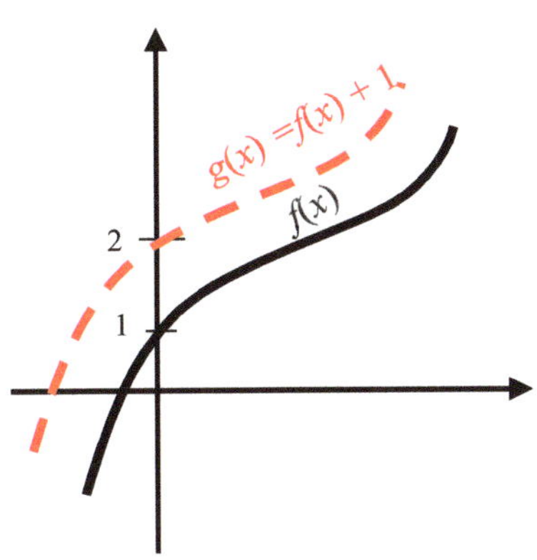

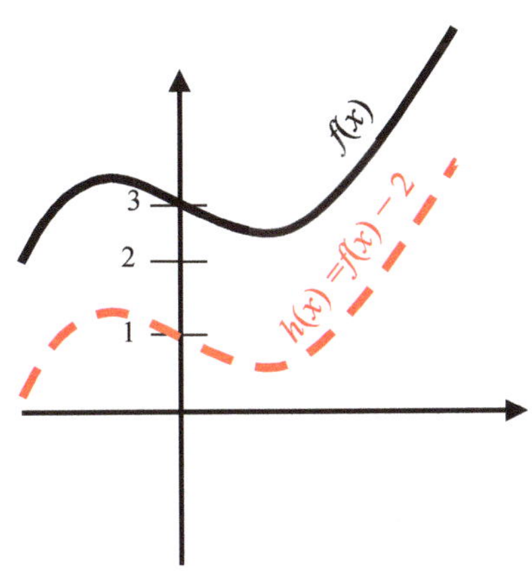

It can be realized that the vertical shift is a change to the output, or outside, of the function. A horizontal shift is a change to every value of input, on the inside of the function. A shift to the input results in a movement of the output of the function left or right in what is known as a horizontal shift.

For example, if $f(x) = x^2$, then $g(x) = (x - 5)^2$ is a new function. Each input is reduced by 5 prior to squaring the function. The result is that the graph is shifted 5 units to the right, because we would need to increase the prior input by 5 units to yield the same output value as given in f.

Horizonal Shift

Given a function $f(x)$ and a constant number k, a new function $g(x) = f(x - k)$ is a horizonal shift of the function f. For positive k, the graph will shift right. For negetive k, the graph will shift left.

Example

Given the function $f(x) = x^2$, graph $g(x) = x^2 - 3$, and $h(x) = x^2 + 1$.

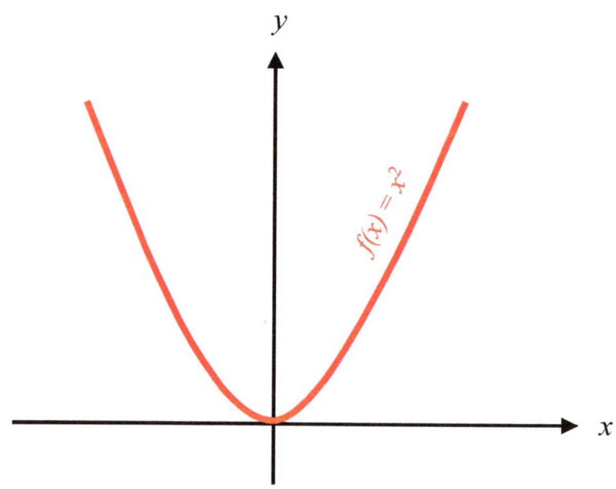

Solution

For graphing $g(x) = x^2 - 3$, we need to shift the function 3 units down and for graphing $h(x) = x^2 + 1$, we need to shift the function 1 units Up. So, here are the requested graphs.

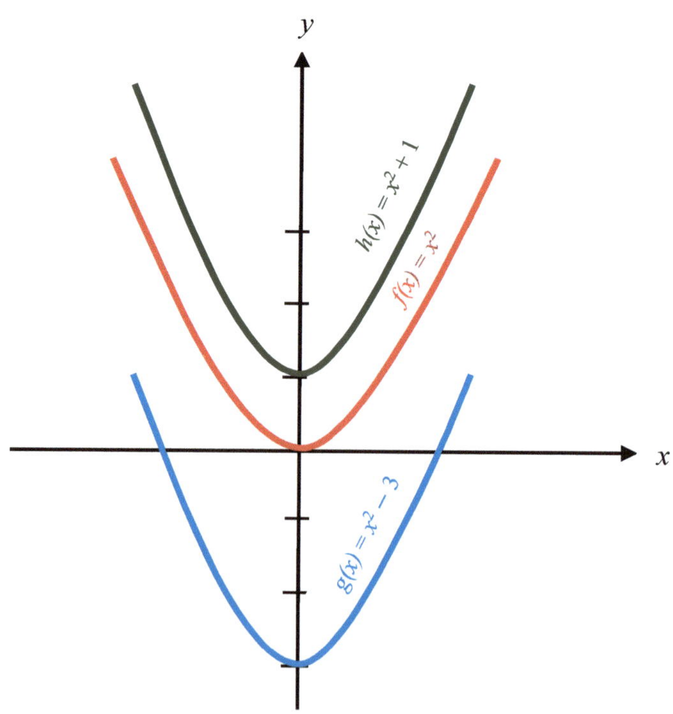

Example
Given the function $f(x) = 2|x|$, graph $g(x) = 2|x - 1|$, and $h(x) = 2|x + 2|$.

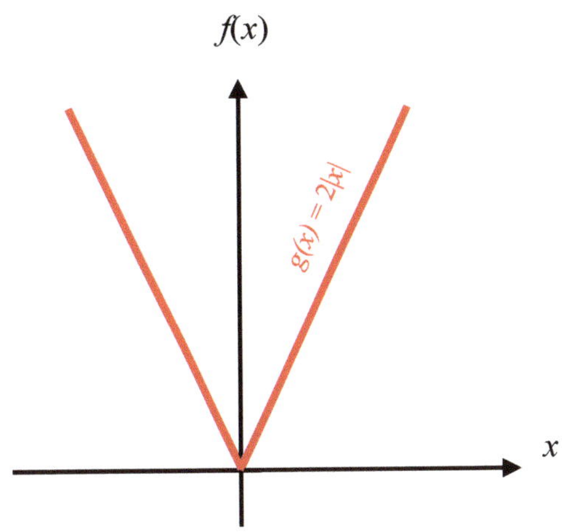

Solution
For graphing $g(x) = 2|x - 1|$, we need to shift the function 1 unit right and for graphing $h(x) = 2|x + 2|$, we need to shift the function 2 units left. So, here are the requested graphs.

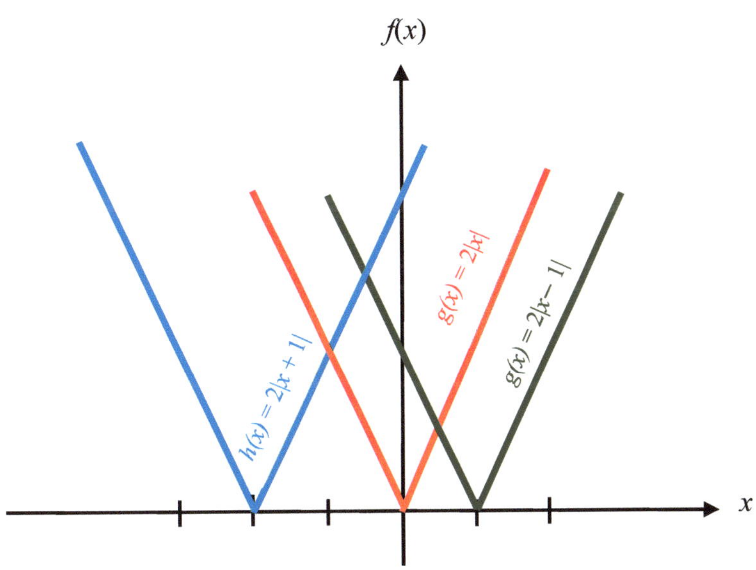

2.3. Combined Transformations, Vertical and Horizontal Shifts and Stretches

Previously we looked at two transformations, vertical shift, and horizontal shift. Now we want to explore the composition of them as well as functions stretch. When two or more transformations are combined to form a new transformation, the result is called a sequence of transformations, or a composition of transformations.

To sketch the graph of sequence of transformations for a vertical and a horizontal shift, the following steps can be performed.

Step 1. Identify the vertical and horizontal shifts from the formula.

Step 2. The vertical shift results from a constant added to the output. Move the graph up to a positive constant and down to a negative constant.

Step 3. The horizontal shift results from a constant added to the input. Move the graph left for a positive constant and right for a negative constant.

Step 4. Apply the shifts to the graph in either order.

When working with combined transformations, the order in which the transformations are applied often changes the outcome. For example, consider a parent function $f(x) = x^2$. Let us first apply a vertical stretch of 2 units followed by a vertical shift of 1 unit, $g(x) = 2x^2 + 1$.

As can be seen in the figures below these transformations will not produce the same graph when we first have a vertical shift of 1 unit followed by a vertical stretch of 2 units, $h(x) = 2x^2 + 2$.

vertical stretch of 2 followed by vertical shift of 1

vertical shift of 2 followed by vertical stretch of 2

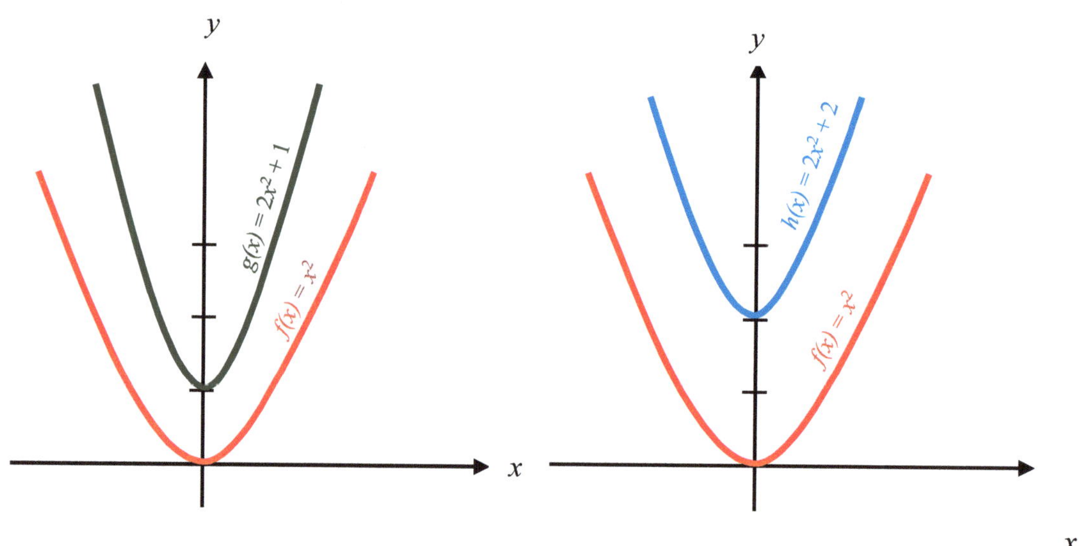

Functions, Combining Transformation

Given function $f(x)$, first a vertical stretch of by k and then a vertical shift by a will results the function $g(x) = kf(x) + a$.

Given function $f(x)$, first a horizontal shift left by b and then a horizontal stretch of by $\frac{1}{t}$ will results the function $h(x) = f(tx + b)$.

Given function $f(x)$, first a horizontally stretch by $\frac{1}{t}$ and then horizontal shift left of by c will results the function $l(x) = f(t(x + c))$.

Example

Given the function $f(x) = |x|$, graph the function by 2 units horizontal shift right, $g(x) = |x - 2|$, and then by 3 units vertical shift down, $h(x) = |x - 2| - 3$.

Solution
The function $f(x) = |x|$ is an absolute value function with a graph of V shape. Hence, here is the graph of $f(x) = |x|$.

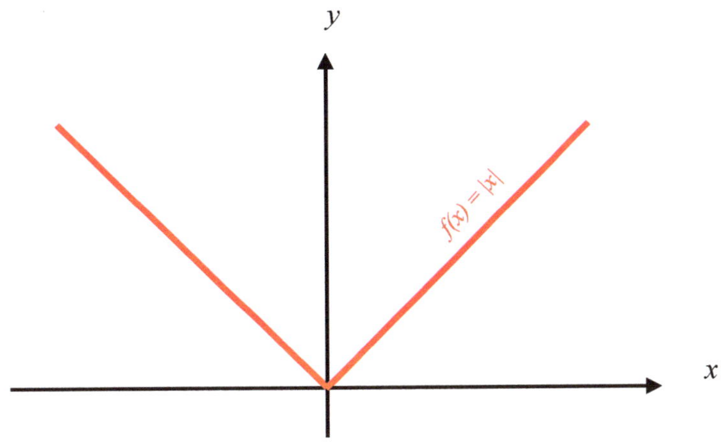

Now, the graphs for $g(x) = |x - 2|$ and $h(x) = |x - 2| - 3$ will be as below:

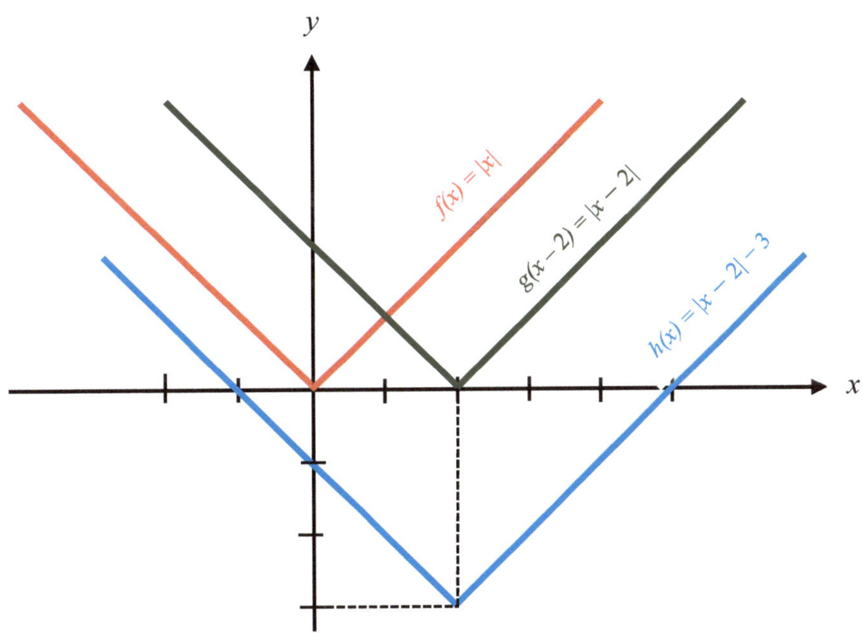

2.4. Inverse of Functions

An **inverse of a function** is a function that "reverses" another function: if the function f applied to an input x gives a result of y, then applying its inverse function g to y gives the result x, and vice versa, i.e., $f(x) = y$ if and only if $g(y) = x$. Given a function $f(x)$, we represent its inverse as $f^{-1}(x)$, read as "f inverse of x". The raised -1 is part of the notation. It is not an exponent; it does not imply a power of -1.

Given a function $f(x)$, we can verify whether some other function $g(x)$ is the inverse of $f(x)$ by checking whether either $g(f(x)) = x$ or $f(g(x)) = x$ is true.

An inverse function is a function that "reverses" another function: If the function f applied to an input x gives a result of y, then applying its inverse function g to y gives the result x, i.e., $g(y) = x$ if and only if $f(x) = y$. The inverse function of f is also denoted as f^{-1}.

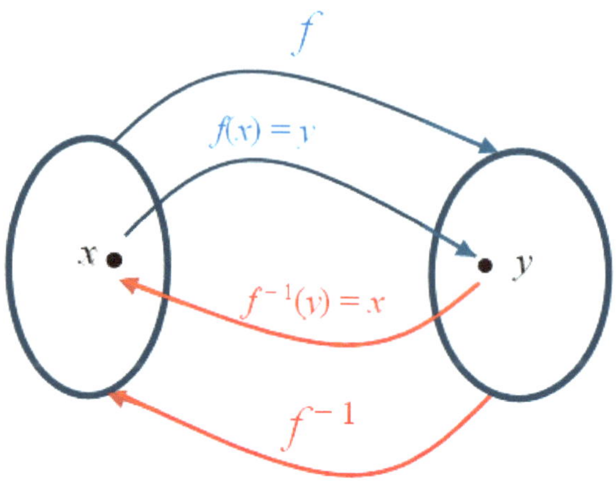

2.4.1. Finding the Inverse of a Function

The steps to find the inverse of a function $f(x)$ are as follows:

Step 1. Replace $f(x)$ with y.

Step 2. Replace every x with a y and replace every y with an x.

Step 3. Solve the equation from Step 2 for y.

Step 4. Replace y with $f^{-1}(x)$ and the inverse of the function is obtained.

Example

Find the inverse function of $f(x) = 4x - 7$.

Solution

Here we have the function $f(x) = 4x - 7$, written as a flow diagram:

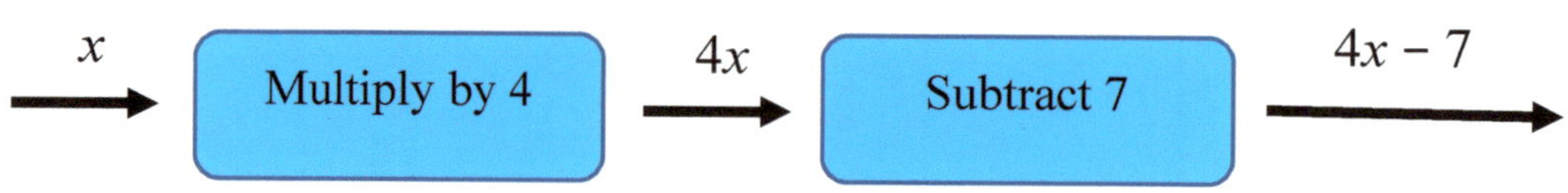

Inverse function goes the other way.

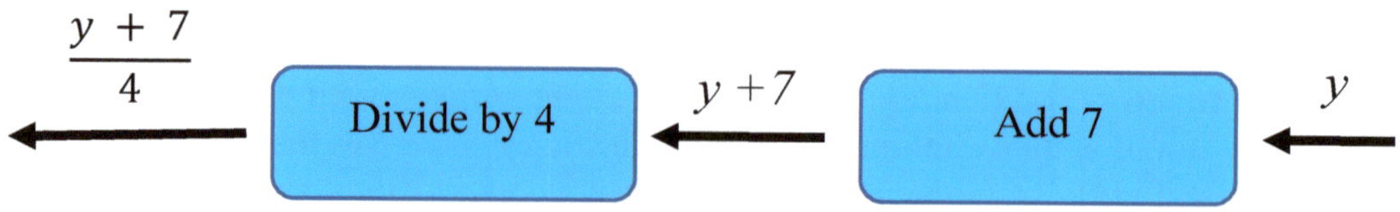

Hence, the inverse of the function $f(x) = 4x - 7$ is $f^{-1}(f(x)) = g(y) = \frac{y+7}{4}$.

Example
Find the inverse function of the function f.

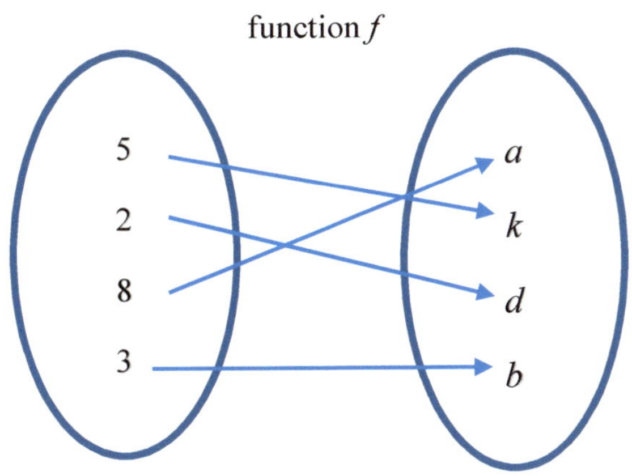

Solution
Here is the mapping diagram of f^{-1}.

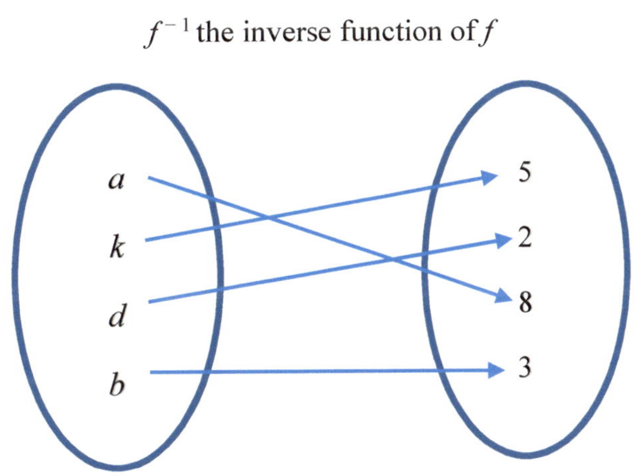

Example
Given $f(x) = \frac{4}{x} - 3$, $g(x) = \frac{4}{x+3}$, verify that $g = f^{-1}$.

Solution

$$g(f(x)) = \frac{4}{f(x) + 3}$$

$$g(f(x)) = \frac{4}{\frac{4}{x} - 3 + 3}$$

$$g(f(x)) = \frac{4}{\frac{4}{x}}$$

$$g(f(x)) = 4 \times \frac{x}{4}$$

$$g(f(x)) = x.$$

This verifies that $g = f^{-1}$. But we can also verify using the other formula. So,

$$f(g(x)) = \frac{4}{g(x)} - 3$$

$$f(g(x)) = \frac{4}{\frac{4}{x+3}} - 3$$

$$f(g(x)) = 4 \times \frac{x+3}{4} - 3$$

$$f(g(x)) = (x + 3) - 3$$

$$f(g(x)) = x.$$

Example

Find the inverse for the function $f(x) = \frac{3x+2}{x-1}$.

Solution:

First, replace $f(x)$ with y and the function becomes,

$$y = \frac{3x+2}{x-1}.$$

By replacing x with y we get,

$$x = \frac{3y+2}{y-1}.$$

Now, solve y in terms of x:

$$x\,(y - 1) = 3y + 2$$

$$xy - x = 3y + 2$$

$$xy - 3y = 2 + x$$

we isolated terms including y in one side add the other terms in the other side of the equation

$$y\,(x - 3) = 2 + x$$

$$y = \frac{x+2}{x-3}.$$

So, $y = f^{-1}(x) = \frac{x+2}{x-3}$.

We can verify the result by checking that $f(f^{-1}(x)) = x$. So,

$$f(f^{-1}(x)) = \frac{3f^{-1}(x) + 2}{f^{-1}(x) - 1}$$

$$f(f^{-1}(x)) = \frac{3\times\frac{x+2}{x-3} + 2}{\frac{x+2}{x-3} - 1}$$

$$f(f^{-1}(x)) = \frac{\frac{3x+6}{x-3} + 2}{\frac{x+2}{x-3} - 1}$$

$$f(f^{-1}(x))) = \frac{\frac{3x+6}{x-3} + 2\times\frac{x-3}{x-3}}{\frac{x+2}{x-3} - 1\times\frac{x-3}{x-3}}$$

$$f(f^{-1}(x)) = \frac{\frac{3x+6+2x-6}{x-3}}{\frac{x+2-x+3}{x-3}}$$

$$f(f^{-1}(x)) = \frac{\frac{5x}{\frac{x-3}{5}}}{x-3}$$

$$f(f^{-1}(x)) = \frac{5x}{x-3} \times \frac{x-3}{5}$$

$$f(f^{-1}(x)) = \frac{5x}{5} = x.$$

Therefore, the result is verified.

2.5. Finding Input and Output Values of a Function

In evaluating a function, it will always produce one result because each **input value of a function** corresponds to exactly one **output value**. When we know an output value and want to determine the input values that would produce that output value, we set the output equal to the function's formula and solve for the input. When we want to solve a function, we simply set the function equal to zero that would produce the input value(s). Solving can produce more than one solution because different input values can produce the same output value.

Example
Evaluate $f(x) = 3x^2 + 5x - 1$ at:

a) 7.

b) $t + 1$.

c) $f(t^2) - f(t)$.

Solution

a) We simply replace x with 7. So,

$$f(7) = 3(7)^2 + 5\,(7) - 1$$

$$f(7) = 3(49) + 5\,(7) - 1$$

$$f(7) = 147 + 35 - 1$$

$$f(7) = 181.$$

b) We simply replace x with $t + 1$. So,

$$f(t + 1) = 3(t + 1)^2 + 5(t + 1) - 1$$

$$f(t + 1) = 3(t^2 + 1 + 2t) + 5t + 5 - 1$$

$$f(t + 1) = 3t^2 + 3 + 6t + 5t + 5 - 1$$

$$f(t + 1) = 3t^2 + 11t + 7.$$

c) We first replace x with t^2 to get $f(t^2)$ and replace x with t to get $f(t)$. So,
$f(t^2) = 3(t^2)^2 + 5t^2 - 1 = 3t^4 + 5t^2 - 1$, and $f(t) = 3t^2 + 5t - 1$, then we have:

$$f(t^2) - f(t) = (3t^4 + 5t^2 - 1) - (3t^2 + 5t - 1)$$

$$f(t^2) - f(t) = 3t^4 + 5t^2 - 1 - 3t^2 - 5t + 1$$

$$f(t^2) - f(t) = 3t^4 + 2t^2 - 5t.$$

Example
Solve $f(x) = 16x^2 - 1$.

Solution
To solve a function, we simply set the function equal to zero to produce the input value(s), So,

$f(x) = 16x^2 - 1 = 0$

$16x^2 - 1 = 0$

$16x^2 = 1$

$\frac{1}{16} \times x^2 = \frac{1}{16} \times 1$ we multiplied both side of the equation by the reciprocal of the variable coefficient

$x^2 = \frac{1}{16}$

$x = \pm\frac{\sqrt{1}}{\sqrt{16}}$

$x = \pm\frac{1}{4}$.

Example
Given $f(x) = 9x^2 - 6x$, solve for $f(x) = -1$.

Solution
Given $f(x) = 1$, we have:

$9x^2 - 6x = -1$

$9x^2 - 6x + 1 = 0$

$(3x - 1)^2 = 0$ we applied the "perfect square trinomials

$3x - 1 = 0$

$3x = 1$

$\frac{1}{3} \times 3x = \frac{1}{3} \times 1 = 0$ we multiplied both side of the equation by the reciprocal of the variable coefficient

$x = \frac{1}{3}$.

2.6. Functions, Horizontal Line and Vertical Line Tests

Graph of functions displays a great many input-output pairs in a small space. The visual information they provide makes relationships easier to understand. Graphs are typically constructed with the input values along the horizontal axis, x-axis and the output values along the vertical axis, y- axis.

The conventional graphs name the input value x and the output value y, and it is read y is a function of x, or $y = f(x)$ when the function is named f. The graph of the function is the set of all points (x, y) in the plane that satisfies the equation $y = f(x)$.

The **vertical line test** can be used to determine whether a graph represents a function. To do this, we draw any vertical line that intersects a graph more than once, then we conclude that the graph does not define a function because a function has only one output value for each input value. The visual information they provide makes relationships easier to understand.

Some functions have a given output value that corresponds to more the one input value. However, some functions have only one input value for each output value, as well as having only one output for each input. We call these functions one-to-one functions.

A Function

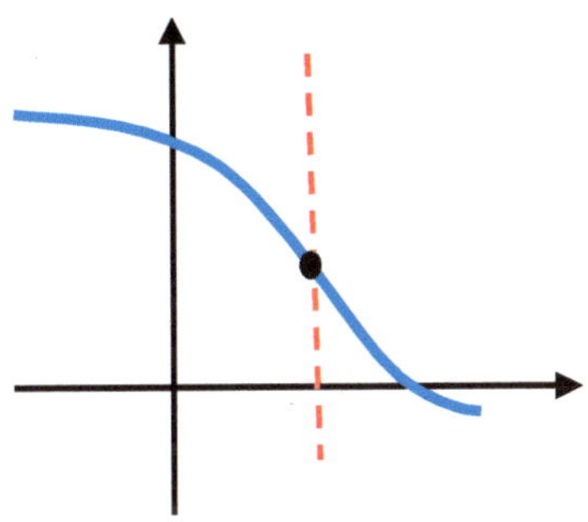

Not a Function

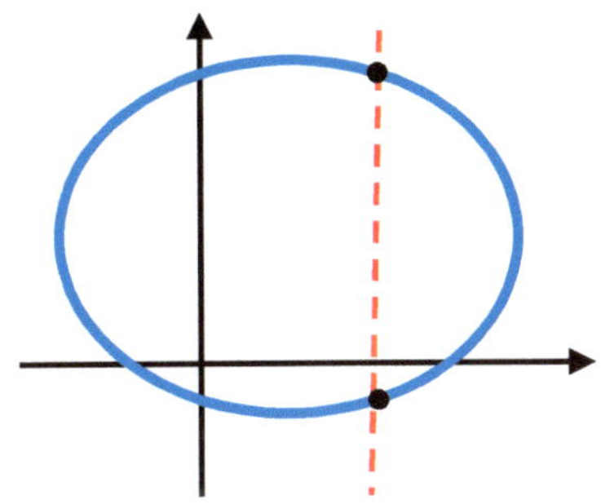

A Function

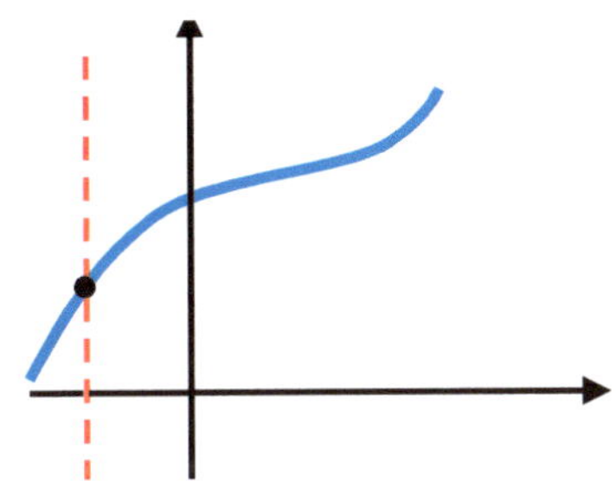

Once it is determined that a graph defines a function, an easy way to determine if it is a one-to-one function is to use the horizontal line test. Draw horizontal lines through the graph. If any horizontal line intersects the graph more than once, then the graph does not represent a one-to-one function.

A One-to-one Function

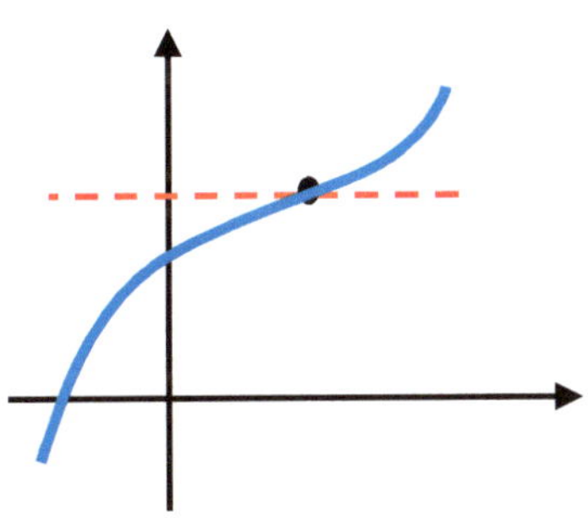

Not a One-to-One Function

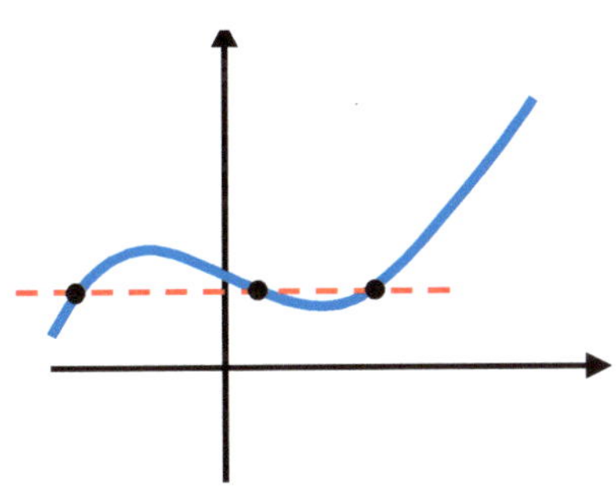

Not a One-to-One Function

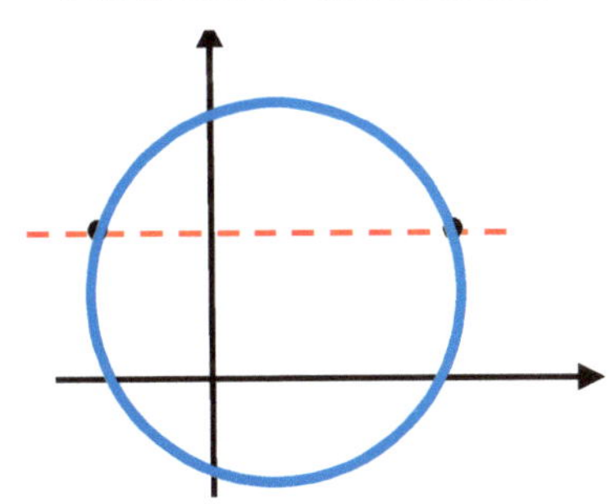

> **One-to-One Function**
> A one-to-one function is a function in which each output value corresponds to exactly one input value.

2.7. Functions, Rate of Change

A **rate of change** describes how an output quantity changes relative to the change in the input quantity. The units on a rate of change are "output units per input units". The average rate of change between two input values is the total change of the function values (output values) divided by the change in the input values. The average rate of change is finding how much something changes over time.

For a linear function, this is very much like finding the slope of a line.

$\frac{\Delta x}{\Delta y} = \frac{f(x_2)-f(x_1)}{x_2-x_1}.$

Finding the Average Rate of Change - Given the value of a function at different points, to calculate the average rate of change of a function for the interval between two values x_1 and x_2, is:

1. Calculate the difference $y_2 - y_1 = \Delta y$.
2. Calculate the difference $x_2 - x_1 = \Delta x$.
3. Find the ratio $\frac{\Delta x}{\Delta y}$.

Example

Given the data in the table below, find the average rate of change of the price of gasoline between,

a) 2007 and 2009.

b) 2009 and 2012.

y = Year	2005	2006	2007	2008	2009	210	2011	2012
$C(y)$ = Price ($)	2.31	2.62	2.87	3.30	2.41	2.84	3.58	3.67

Solution

a) In 2007, the price of gasoline was $2.84. In 2009, the price was reduced and became $2.41. The average rate of change is.

$$\frac{\Delta x}{\Delta y} = \frac{f(x_2) - f(x_1)}{x_2 - x_1}$$

$$\frac{\Delta x}{\Delta y} = \frac{2.41 - 2.87}{2009 - 2007}$$

$$\frac{\Delta x}{\Delta y} = \frac{-\ 0.46}{2 \text{ years}}$$

$$\frac{\Delta x}{\Delta y} = -\ 0.23 \text{ per year.}$$

Note that a decrease is expressed by a negative change or "negative increase." The rate of change is negative when the output decreases as the input increases or when the output increases as the input decreases.

b) Again, in 2009, the price of gasoline was $2.41. In 2012, the price was $3.68. The average rate of change is

$$\frac{\Delta x}{\Delta y} = \frac{f(x_2) - f(x_1)}{x_2 - x_1}$$

$$\frac{\Delta x}{\Delta y} = \frac{3.67 - 2.41}{2012 - 2009}$$

$$\frac{\Delta x}{\Delta y} = \frac{1.26}{4 \text{ years}}$$

$$\frac{\Delta x}{\Delta y} = 0.315 \text{ per year.}$$

2.8. Increasing and Decreasing Function

A function is "**increasing**" when the y-value increases as the x-value increases. In other words, a function $f(x)$ is an **increasing function** on an open interval if $f(x_2) \geq f(x_1)$ for any two input values x_1 and x_2 in the given interval where $x_2 > x_1$. An increasing function is depicted as the figure (a) below.

Also, a function $f(x)$ is said to be a **strictly increasing function** on open interval if $f(x_2) > f(x_1)$ for any two input values x_1 and x_2 in the given interval where $x_2 > x_1$.

A function is "**decreasing**" when the y-value decreases as the x-value increases. In other words, a function f is a **decreasing function** on an open interval if $f(x_2) \leq f(x_1)$ for any two input values x_1 and x_2 in the given interval where $x_2 > x_1$. A decreasing function is depicted as the figure (b) below.

Similarly, a function $f(x)$ is said to be a **strictly decreasing function** on open interval if $f(x_2) < f(x_1)$ for any two input values x_1 and x_2 in the given interval where $x_2 > x_1$.

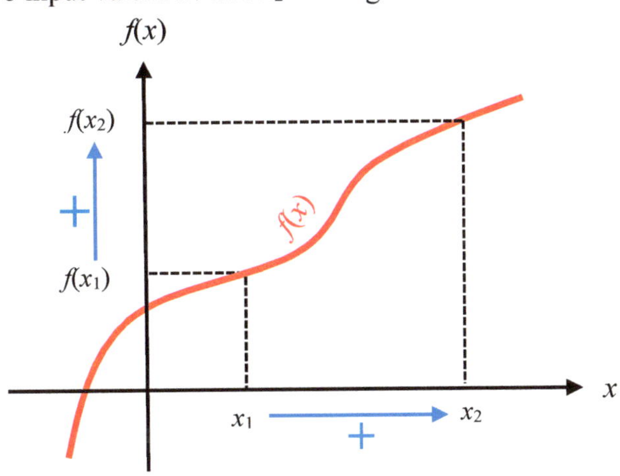

Figure (a)

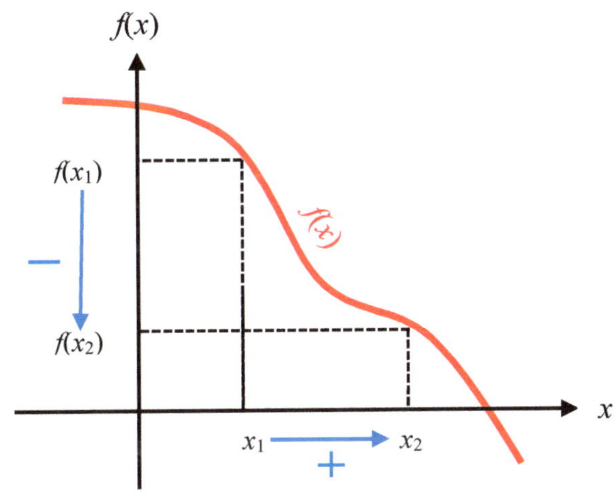

Figure (b)

2.9. Minima and Maxima of a Function

The **minima** and **maxima** (the respective plural of minimum and maximum) of a function which are also known as **extrema** (the plural of extremum) are the smallest and largest value of a function, either within a given interval, that is called the **local or relative extrema**, or on the entire domain which is called the **global or absolute extrema**.

In other words, a **local minimum** point on a function is a point (x, y) on the graph of the function whose y coordinate is smallest than all other y coordinates on the graph at points close to (x, y). More precisely, a function f has a local minimum at $x = t$ if there exists an interval (a, b) with $a < t < b$ such that, for any x in the interval (a, b), $f(x) \geq f(t)$, as seen below.

Similarly, f has a local maximum at $x = k$ if there exists an interval (b, c) with $b < k < c$ such that, for any x in the interval (b, c), $f(x) \leq f(k)$, as seen below.

The absolute or global minimum of the function f at $x = l$ is $f(l)$ where $f(l) \leq f(x)$ for all x in the domain of f, as seen below.

The absolute or global maximum of the function f at $x = m$ is $f(m)$ where $f(m) \geq f(x)$ for all x in the domain of f, as seen below.

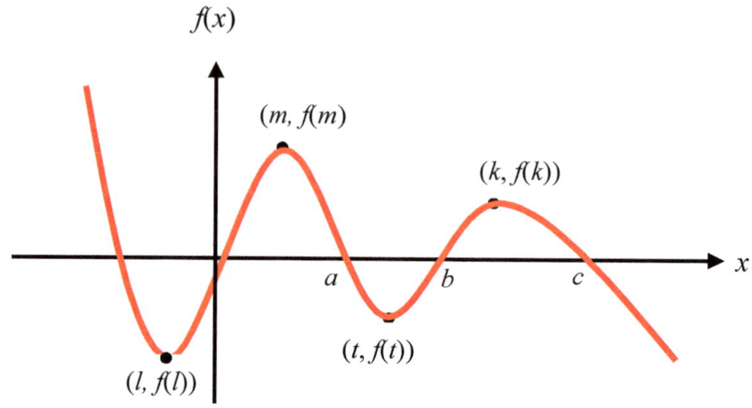

2.10. Exercises

1. Given $f = \{(2, 4), (3, 9), (4, 9), (5, 8), (6, 7)\}$, and $g = \{(4, 13), (8, 12), (9, 11), (7, 5)\}$, find the composite function of g and f, that is $g \circ f$.

2. Given $f(x) = 3x^2 - 7$, $g(x) = 4x + 9$, and $h(x) = x^2$ find
   a) The composite function of $f(g(x))$.
   b) The composite function of $g(f(x))$.
   c) The composite function of $f(h(x))$.
   d) The composite function of $h(f(x))$.
   e) The composite function of $h(g(x))$.
   f) The composite function of $g(h(x))$.

3. Given $f(u) = \frac{5}{4u - 108}$ and $g(u) = 3u^2$ find the domain of $f(g(u))$.

4. Given $f(x) = x^2 - 5$ and $h(x) = \sqrt{2x - 3}$, find each of the following:
   a) $f(h(x))$.
   b) $h(f(x))$.
   c) $f(h(3))$.
   d) $h(f(11))$.

5. Given $f(x) = 6x - 9$ and $h(x) = \frac{2x + 6}{6}$, find each of the following:
   a) $f(h(x))$.
   b) $h(f(x))$.
   c) $f^{-1}(h(x))$.
   d) $h^{-1}(f(x))$.

6. Given the two functions $f(x)$ and $g(x)$ as shown in the figures below, sketch $f(x) - 2$ and $g(x) + 3$ in the same xy-axis.

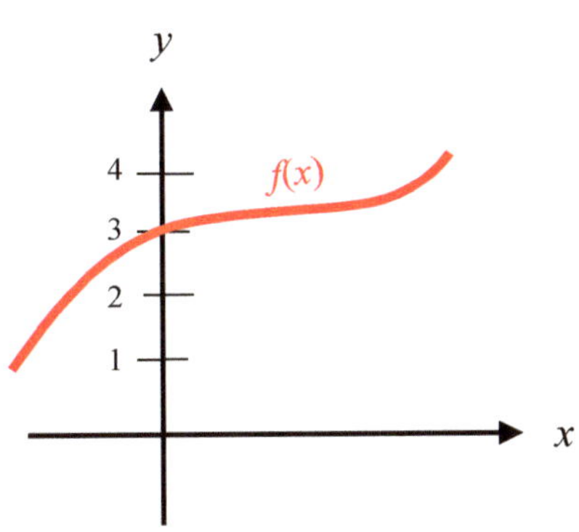

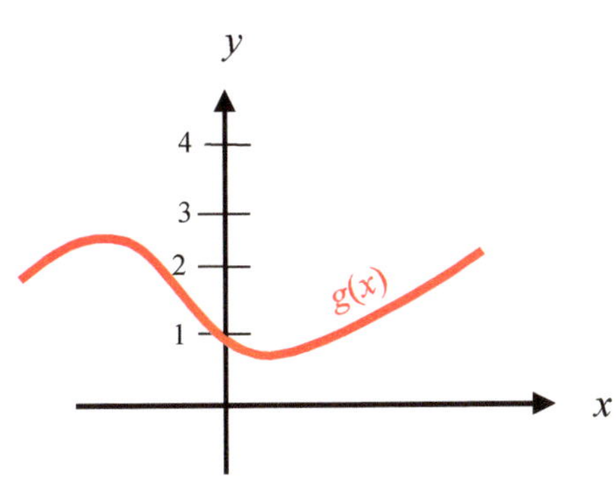

7. Given tow functions $f(x) = x^2 + 1$ and $g(x) = 3x - 12$, sketch $f(x) = x^2 + 1$, $f(2x)$ and $g(x + 3)$ in the same xy -axis.

8. Given the function $f(x) = |x|$, graph $g(x) = |3x - 2|$, and $h(x) = 2|x + 1|$.

9. Given the function $f(x) = |x|$, graph and find the domine and the range of each new function by the following combined transformations:
   a) The function by 2 units horizontal shift left, and then by 1-unit vertical shift up.
   b) The function by 1-unit vertical shift up, and then by 2 units horizontal shift right.
   c) The function by 1-unit vertical shift down, horizontally stretch by $\frac{1}{2}$ and then by 2 units horizontal shift right.

10. Given $f(x) = 5x - 2$ find $f^{-1}(x)$.

11. Given $f(x) = \frac{3x + 1}{2x - 5}$ find $f^{-1}(x)$ and $f^{-1}(2)$.

12. Given $f(x) = \frac{2x + 1}{4 - 5x}$, find $f(x - 2)$.

13. Given $f(x) = 4^{x+1}$, find $f(2x - 4)$ and $f(9)$.

14. Given $f(x) = \frac{3x - 7}{4x - 9}$, find $f(x^2)$ and $f^{-1}(x)$.

15. Given $f(x) = \frac{2x^2 + 1}{1 - 5x^4}$, find $f(\sqrt{x})$.

16. Given $f(x) = \frac{\sqrt{4x^2 - 8}}{12x - 9x^2}$, find $h(x) = f(x^2 - 3)$, and $h(6)$.

17. Determine whether each graph (a), (b) or (c) represents a function.

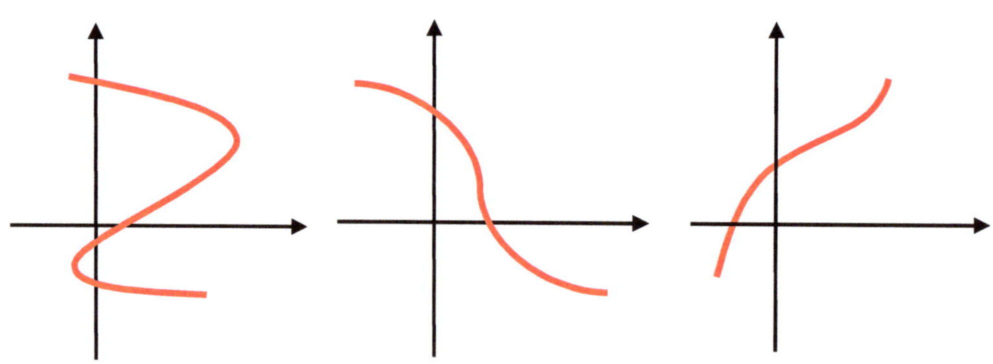

(a) (b) (c)

18. Evaluate $f(x) = 2x^2 + 7x - 11$ at:
   a) $x = 7$.
   b) $x = t + 1$.
   c) $f(9t) - f(3t)$.

19. In 2001, Helen purchased a house for \$238,000. In 2019, the house was worth \$416,000. Find the average annual rate of change in dollars per year in the value of the house.

20. A girl rides away from home in an automobile at the rate of 38 miles an hour and walks back at the rate of 4 miles an hour. The round trip requires 1 hour and 30 minutes. How far did she ride?

21. Given $f(v) = 5v^2 - 6v + 2$, solve for v when $f(v) = 2$.

22. Given functions $f(x)$ and $g(x)$ in figures below, designate which one is "one-to one" function.

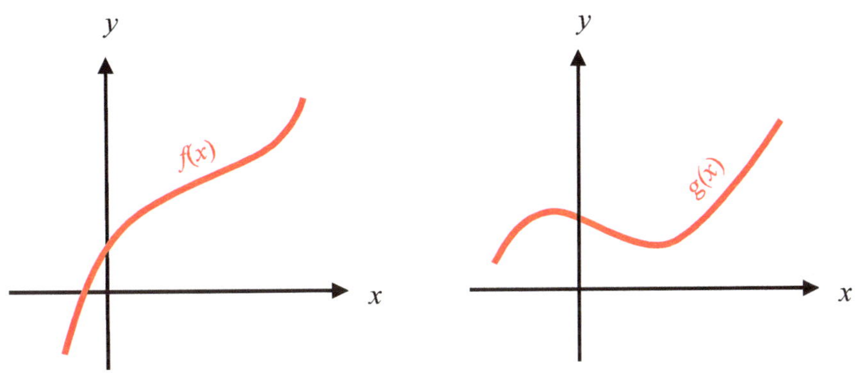

23. Given $f(x)$ and two points, x_1 and x_2, on the $f(x)$, determine which of function below is increasing, decreasing or flat.
    a) If $x_1 = 5$, $x_2 = 8$, $f(5) = 32$, and $f(8) = 45$.
    b) If $x_1 = 6$, $x_2 = 13$, $f(6) = 17$, and $f(13) = 8$.
    c) If $x_1 = -4$, $x_2 = 1$, $f(-4) = 12$, and $f(1) = 17$.
    d) If $x_1 = -1$, $x_2 = 2$, $f(-1) = 26$, and $f(2) = 26$.
    e) If $x_1 = 9$, $x_2 = 18$, $f(9) = 54$, and $f(18) = 33$.
    f) If $x_1 = 11$, $x_2 = 14$, $f(11) = 12$, and $f(14) = 17$.

24. Given $f(x)$ and two points, x_1 and x_2, on the $f(x)$ with the domain of $(-\infty, \infty)$, designate the local minima, local maxima, global minima, and global maxima of the function $f(x)$.
    a) If $x_1 = 1$, $x_2 = 7$, $f(c) > f(x)$ for every $x \epsilon [1, 7]$.
    b) If $x_1 = -1$, $x_2 = 11$, $f(c) < f(x)$, for every $x \epsilon [-1, 1]$.
    c) If $x_1 = 2$, $x_2 = 15$, $f(c) < f(x)$, for every $x \epsilon [2, 15]$ and $f(a) > f(x)$ for every $x \epsilon (-\infty, \infty)$.
    d) If $x_1 = -5$, $x_2 = -2$, $f(c) < f(x)$, for every $x \epsilon [-5, -2]$ and $f(a) < f(x)$ for every $x \epsilon (-\infty, \infty)$.
    e) If $x_1 = 9$, $x_2 = 18$, $f(c) > f(x)$, for every $x \epsilon [9, 18]$ and $f(a) < f(x)$ for every $x \epsilon (-\infty, \infty)$.
    f) If $x_1 = 10$, $x_2 = 16$, $f(c) > f(x)$, for every $x \epsilon [10, 16]$ and $f(a) > f(x)$ for every $x \epsilon (-\infty, \infty)$.

25. Given $f(x)$ and two points, x_1 and x_2, on the $f(x)$ with the domain of $(-9, 22)$, designate the local minima, local maxima, global minima, and global maxima of the function $f(x)$.
    a) If $f(a) > f(x)$ for every $x \epsilon [x_1 = -9, x_2 = 7]$, $f(b) > f(x)$ for every $x \epsilon [x_1 = 7, x_2 = 12]$, $f(c) > f(x)$ for every $x \epsilon [x_1 = 12, x_2 = 22]$ and $f(\mathrm{d}) > \text{maximum}\{f(a), f(b), f(c)\}$.

b) If $f(a) < f(x)$ for every $x \epsilon [x_1 = -9, x_2 = -2]$, $f(b) < f(x)$ for every $x \epsilon [x_1 = -2, x_2 = 14]$, $f(c) < f(x)$ for every $x \epsilon [x_1 = 14, x_2 = 22]$, and $f(\mathrm{d}) >$ minimum$\{f(a), f(b), f(c)\}$.

c) If $f(a) < f(x)$ for every $x \epsilon [x_1 = -9, x_2 = -5]$, $f(b) > f(x)$ for every $x \epsilon [x_1 = -5, x_2 = 1]$, $f(c) < f(x)$ for every $x \epsilon [x_1 = 1, x_2 = 8]$, $f(d) < f(x)$ for every $x \epsilon [x_1 = 8, x_2 = 13]$, $f(e) > f(x)$ for every $x \epsilon [x_1 = 13, x_2 = 19]$, $f(k) > f(x)$ for every $x \epsilon (-9, 22)$, $f(t) <$ minimum$\{f(a), f(c), f(d)\}$, and $f(h) >$ maximum$\{f(b), f(e), f(k)\}$.

3. Derivatives of Functions

In the previous section we defined the rate of change between two output function values with respect to their corresponding input values. In this section we look at the definition, concept, and applications of derivative and its relations to the rate of change of a function. A derivative is defined as the instantaneous rate of change in function based on one of its variables. It is like finding the slope of a tangent to the function at a point.

Definition of Derivative of a Function - The derivative of a function of a real variable determines the sensitivity to change of the function value, output value, with respect to a change in its input value. For example, consider a case in which drivers of a car want to measure how quickly the position of the car changes as time advances. The derivative of how far the car has gone with respect to time is the car's velocity. The derivative measures how quickly the position of the car changes when time advances.

The derivative of a at a chosen input value, when it exists, is the slope of the tangent line to the graph of the function at that point. The tangent line is the best linear approximation of the function near that input value. For this reason, the derivative is often described as the "instantaneous rate of change", the ratio of the instantaneous change in the dependent variable to that of the independent variable.

3.1. Definition of a Derivative, Scant Lines, Tangent Lines and Limit

A scant line, L, is as straight line joining two points on a function. It is also equivalent to the average rate of change, or simply the *slope* between two points, slope (see Figure (a) below). The slope of the tangent line, T (see Figure (b) below).

$$\text{Slope} = m = \frac{f(a+h)-f(a)}{h} = \text{average rate of change}$$

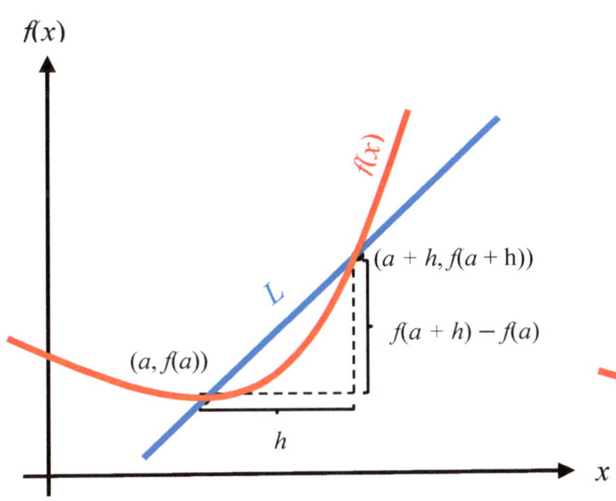

Figure (a)

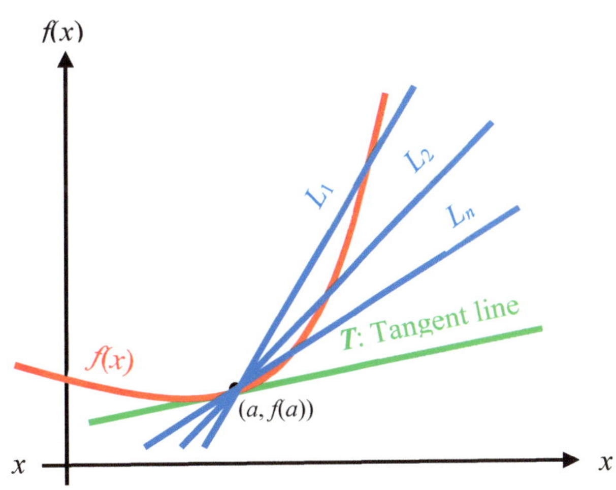

Figure (b)

s Figure (b) illustrate the tangent line at $(a, f(a))$ seems to be the limit of these "scant lines, $L_1, L_2, \cdots, L_n$" as h approaches 0. It is to be noted that we have never looked at the limit of lines, but we can look at the limit of their slope. The slope of the tangent line through the point $(a, f(a))$ is:

$$\lim_{h \to 0} \frac{f(a+h)-f(a)}{h}.$$

Now, we can present a definition for differentiability of a function.

Definition of Function Differentiability

The function $f(x)$ is differentiable at $x = a$, if $\lim_{h \to 0} \frac{f(a+h) - f(a)}{h}$ exists. The limit is denoted by $f'(a)$ and is called the derivative of $f(x)$ at a, shortly, the function $f'(x)$ is called the derivative of $f(x)$.

Definition of the Average Rate of Change of a Function

The average rate of change of f with respect to a is also called the difference quotient. It is given as,

$$\frac{f(a+h) - f(a)}{h}, \text{ where } h \neq 0.$$

The difference quotient is equal to the slope of the secant line from $(a, f(a))$ to $(a + h, f(a + h))$.

Definition of a Function Derivative

For a function $f(x)$ its derivative at x is the function $f'(x)$ defined as,

$$f'(x) = \lim_{h \to 0} \frac{f(x+h) - f(x)}{h},$$

provided that the limit exists. If $f'(x)$ exists, we say that f is differentiable at x.

Example

Rickey PPE, Inc., is a manufacturing firm producing safety goggles. The following graph shows the total production of goggles by Rickey PPE, Inc., during one of the Monday mornings of work.

- What was the number of goggles produced at Rickey PPE, Inc., from 10*A.M* to 11*A.M*?
- What was the average number of goggles produced per hour from 8*A.M* to 11*A.M*?

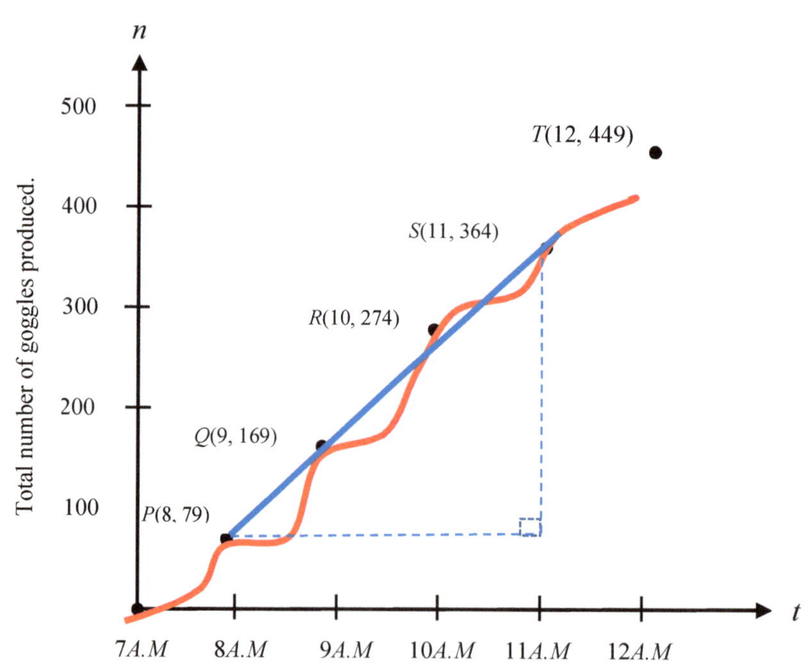

Solution

At 10*A.M*, 274 goggles had been produced. At 11*A.M* 364 goggles had been produced. In the hours from 10*A.M* to 11*A.M*, the number of goggles produced was:

The number of goggles produced from 10*A.M* to 11*A.M* = 364 – 274 = 90 goggles.

Therefore, the number of goggles produced from 10*A.M* is 90 goggles. It is to be noted that 90 is the slope of the line from *R* to *S*.

Again, at 8*A.M*, 79 goggles had been produced. At 11*A.M*, 364 goggles had been produced. In the hours from 8*A.M* to 11*A.M*, the average number of goggles, *AV*, produced per hour was:

The average number of goggles produced per hour from 8*A.M* to 11*A.M*.

$$AV = \frac{364 \text{ goggles} - 79 \text{ goggles}}{11A.M. - 8A.M.}$$

$$AV = \frac{285 \text{ goggles}}{3 \text{ hours}} = 95 \text{ goggles.}$$

Therefore, the average number of goggles produced per hour from 8*A.M* to 11*A.M* is 95 goggles.

Example

Suppose we have the function $f(x) = 3x^2$.

a) Find the derivative at x of the function.

b) Find the derivative $f'(x)$ at $x = 1$ and at $x = -1$.

Solution

a) To find the derivative of the function $f(x)$ at x, we learned that $f'(x) = \lim_{h \to 0} \frac{f(x+h) - f(x)}{h}$. Also we learned for the function $f(x) = 3x^2$, $f(x + h) = 3(x + h)^2$. So,

$$f'(x) = \lim_{h \to 0} \frac{f(x+h) - f(x)}{h}$$

$$f'(x) = \lim_{h \to 0} \frac{3(x+h)^2 - 3x^2}{h}$$ we replaced $f(x + h) = 3(x + h)^2$, and $f(x) = 3x^2$ into the equation

$$f'(x) = \lim_{h \to 0} \frac{3(x^2 + 2h \ \ + h^2) - 3x^2}{h}$$ we applied binomial rule

$$f'(x) = \lim_{h \to 0} \frac{3x^2 + 6hx + 3h^2 - 3x^2}{h}$$

$$f'(x) = \lim_{h \to 0} \frac{6hx + 3h^2}{h}$$

$$f'(x) = \lim_{h \to 0} \frac{h(6x + 3h)}{h}$$

$$f'(x) = \lim_{h \to 0} (6x + 3h)$$

$$f'(x) = 6x.$$

Therefore, the derivative at x of the function $f(x) = 3x^2$ is $f'(x) = 6x$.

b) To find the derivative $f'(x)$ at $x = 1$, we simply need to replace x with the number 1 in the function $f'(x) = 6x$. So,

$f'(x) = 6x$

$f'(1) = 6(1) = 6$ we plugged $x = 1$ into the equation.

Similarly, to find the derivative $f'(x)$ at $x = -1$, we again simply need to replace x with the number -1.2 in the function $f'(x) = 6x$. So,

$f'(x) = 6x$

$f'(-1) = 6(-1) = -6$ we plugged $x = -1$ into the equation.

Therefore, the derivative $f'(x)$ at $x = 1$ and at $x = -1$ are 6 and -6, respectively.

By sketching the graph of the function $f(x) = 3x^2$ and the tangent lines L_1 and L_2 at points $x = 1$ and at $x = -1$ (see the figure below), we can say:

The tangent line, L_1, to the curve at the point (1, 3) has slope 6.

The tangent line, L_2, to the curve at the point $(-1, 3)$ has slope -6.

The instantaneous rate of change at $x = 1$ is 6.

The instantaneous rate of change at $x = -1$ is -6.

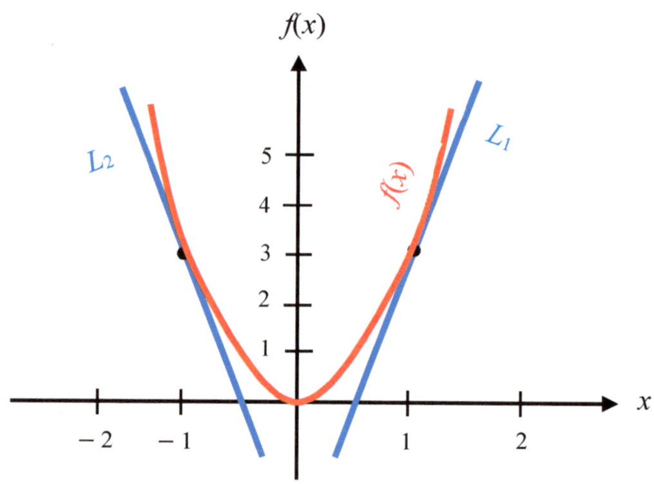

Example

Suppose we have the function $f(x) = \frac{2}{x}$.

a) Find the derivative at x of the function, $f(x)$.

b) Find the derivative, $f'(x)$, at $x = 2$ and at $x = -2$.

Solution

a) To find the derivative of the function, we use the formula $f'(x) = \lim_{h \to 0} \frac{f(x+h) - f(x)}{h}$. The function $f(x + h)$ is equal to $\frac{2}{x+h}$. So,

$$f'(x) = \lim_{h \to 0} \frac{f(x+h) - f(x)}{h}$$

$f'(x) = \lim\limits_{h \to 0} \frac{\frac{2}{x+h} - \frac{2}{x}}{h}$ we replaced $f(x+h) = \frac{2}{x+h}$, and $f(x) = \frac{2}{x}$ into the equation

$$f'(x) = \lim_{h \to 0} \frac{\frac{2x - 2(x+h)}{x(x+h)}}{h}$$

$$f'(x) = \lim_{h \to 0} \frac{\frac{2x - 2x - 2h}{x(x+h)}}{h}$$

$$f'(x) = \lim_{h \to 0} \frac{\frac{-2h}{x(x+h)}}{h}$$

$$f'(x) = \lim_{h \to 0} \frac{-2h}{x(x+h)} \times \frac{1}{h}$$

$$f'(x) = \lim_{h \to 0} \frac{-2}{x(x+h)} = \frac{-2}{x(x+0)}$$

$$f'(x) = \frac{-2}{x^2}.$$

Therefore, the derivative at x of the function $f(x) = \frac{2}{x}$ is $f'(x) = -\frac{2}{x^2}$.

b) To find the derivative $f'(x)$ at $x = 2$, we simply need to replace x with the number 2 in the function $f'(x) = -\frac{2}{x^2}$. So,

$$f'(x) = -\frac{2}{x^2}$$

$f'(2) = -\frac{2}{2^2}$. we plugged $x = 2$ into the equation

$$f'(2) = -\frac{2}{4} = -\frac{1}{2}.$$

Similarly, to find the derivative $f'(x)$ at $x = -2$, we again simply need to replace x with the number -1.2 in the function $f'(x) = -\frac{2}{x^2}$. So,

$$f'(x) = -\frac{2}{x^2}$$

$f'(-2) = -\frac{2}{(-2)^2}$ we plugged $x = -2$ into the equation

$$f'(-2) = -\frac{2}{4} = -\frac{1}{2}.$$

Therefore, the derivative $f'(x)$ at $x = 2$ and at $x = -2$ are both equal to $-\frac{1}{2}$.

To sketch the graph of the function $f(x) = \frac{2}{x}$. and the two tangent lines L_1 and L_2 at points $x = 2$ and at $x = -2$, we first need to find the equations of the tangent lines. We have found that the slops of both the tangent lines L_1 and L_2 at points $x = 2$ and $x = -2$ are both equal to $-\frac{1}{2}$. We also have a point $x = 2$ on the tangent line L_1 and a point $x = -2$ on the tangent line L_2. We hence can use the point-slope formula, $y - y_0 = m(x - x_0)$, for finding the equations of the line L_1 and line L_2. So:

For equation of L_1, with $m = -\frac{1}{2}$, $x_0 = 2$, and $y_0 = f(2) = \frac{2}{2} = 1$, we will have:

$y - y_0 = m(x - x_0)$

$y - 1 = -\frac{1}{2}(x - 2)$ — we plugged $m = -\frac{1}{2}$, $x_0 = 2$, and $y_0 = 1$ into the equation

$y - 1 = -\frac{1}{2}x + \frac{2}{2}$

$y - 1 = -\frac{1}{2}x + 1$

$y - 1 + 1 = -\frac{1}{2}x + 1 + 1$ — to isolate y we added 1 to both sides

$y = -\frac{1}{2}x + 2.$

Therefore, the equation of the line L_1 is $y = -\frac{1}{2}x + 2$. For equation of L_2, with $m = -\frac{1}{2}$, $x_0 = -2$, and $y_0 = f(2) = \frac{2}{-2} = -1$, we will have:

$y - y_0 = m(x - x_0)$

$y - (-1) = -\frac{1}{2}(x - (-2))$ — we plugged $m = -\frac{1}{2}$, $x_0 = -2$, and $y_0 = -1$ into the equation

$y + 1 = -\frac{1}{2}(x + 2)$

$y + 1 = -\frac{1}{2}x - \frac{2}{2}$

$y + 1 = -\frac{1}{2}x - 1$

$y + 1 - 1 = -\frac{1}{2}x - 1 - 1$ — to isolate y we added -1 to both sides

$y = -\frac{1}{2}x - 2.$

Therefore, the equation of the line L_2 is $y = -\frac{1}{2}x - 2$. Now, we can sketch the graph of the function $f(x) = \frac{2}{x}$ and the tangent lines L_1 and L_2 at points $x = 2$ and at $x = -2$ (see the figure below). Note that the scales on the x-axis and the y-axis are different.

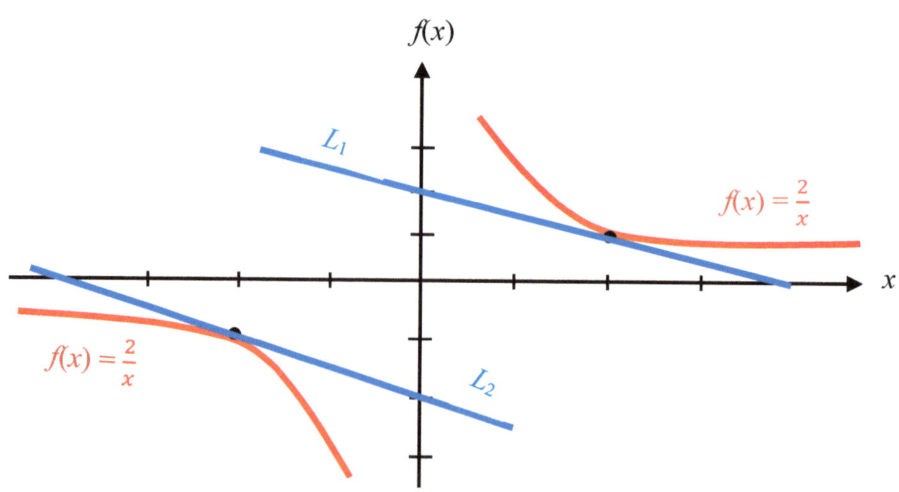

3.2. Leibniz 's Notation

When dealing with mathematical expressions that contains derivatives it is convenient to denote the prime notation of derivative of a function $y = f(x)$, as a functional relationship between dependent variable y and independent variable x, by $\frac{dy}{dx}$. That is,

$$\frac{dy}{dx} = \frac{d}{dx}f(x) = f'(x) = y'.$$

This notation is called Leibniz notation (in honor of the 17th German mathematician G. W. Leibniz). For example, the derivative of the function $y = f(x) = x^2$, we can write $\frac{dy}{dx} = \frac{d}{dx}f(x) = 2x$ or $y' = f'(x) = 2x$.

When we wish to evaluate a derivative at a point $x = 2$, we write,

$$\frac{dy}{dx}\Big|_{x=2} = f'(2),$$

Where the vertical line is interpreted as "evaluated at". Hence the above expression is read as "the derivative of y with respect to x evaluated at $x = 2$ is the value $f'(2)$".

3.3. Elementary Rules of Differentiation

The following are differentiation rules, that is, rules for computing the derivative of all functions of real numbers that return real values.

3.3.1. The Power Rule in Polynomial in Derivative of a Function

For the power rule if $f(x) = x^k$, for any real number $k \neq 0$, then

$$f'(x) = kx^{k-1}.$$

As a special case, when $k = 1$, we will have $f(x) = x$, then $f'(x) = 1$.

For the polynomial rule, the combination of the power rule, the sum rule, and the product rule, enables us to compute the derivative of any polynomial.

Example

Given the two functions $f(x) = 6x^3$ and $g(x) = 8\sqrt[4]{x^3}$, find:

a) The derivative of $f(x)$.

b) The derivative of $g(x)$.

Solution

a) We have learned that the derivative of $f(x) = x^k$ is $f'(x) = kx^{k-1}$. Hence, the derivative of $f(x) =$ 6x3 is:

$$f'(x) = \frac{d}{dx}f(x) = 3(6x^{3-1})$$

$$f'(x) = 3(6x^2)$$

$$f'(x) = 18x^2.$$

Therefore, the derivative of $f(x) = 6x^3$ is $f'(x) = 18x^2$.

b) First, we rewrite $g(x) = 8\sqrt[4]{x^3}$ as $g(x) = 8x^{\frac{3}{4}}$. Then the derivative of $g(x) = 8\sqrt[4]{x^3}$ is: So,

$$g'(x) = \frac{d}{dx}g(x) = \frac{3}{4}(8x^{\frac{3}{4}-1})$$

$$g'(x) = \frac{3}{4}(8x^{\frac{3}{4}-1\times\frac{4}{4}})$$

to make the exponent with the same denominator, we multiplied 1 by $\frac{4}{4}$

$$g'(x) = \frac{3}{4}(8x^{\frac{3}{4}-\frac{4}{4}})$$

$$g'(x) = \frac{3}{4}(8x^{\frac{3-4}{4}})$$

$$g'(x) = \frac{3}{4}(8x^{\frac{-1}{4}})$$

$$g'(x) = \frac{24}{4}x^{-\frac{1}{4}}$$

$$g'(x) = 6 \times \frac{1}{x^{\frac{1}{4}}}$$

$$g'(x) = 6 \times \frac{1}{\sqrt[4]{x}} = \frac{6}{\sqrt[4]{x}}.$$

Therefore, the derivative of $g(x) = 8\sqrt[4]{x^3}$ is $g'(x) = \frac{6}{\sqrt[4]{x}}$.

3.3.2. Linearity of Differentiation Rule in Derivative of a Function

The derivative of any linear combination of functions equals the same linear combination of the derivatives of the functions; this property is known as linearity of differentiation, the rule of linearity, or the superposition rule for differentiation.

For any functions f and g and any real numbers a and b, the derivative of the function $h(x) = af(x) + bg(x)$ with respect to x is:

$$h'(x) = af'(x) + bg'(x).$$

In Leibniz's notation this is written as:

$$\frac{d}{dx}h(x) = \frac{d}{dx}(af(x) + bg(x)) = a\frac{d}{dx}f(x) + b\frac{d}{dx}g(x).$$

Special cases include:

The constant factor rule,

$$(af(x))' = af'(x).$$

The sum factor rule,

$$(f(x) + g(x))' = f'(x) + g'(x).$$

The subtraction rule,

$$(f(x) - g(x))' = f'(x) - g'(x).$$

Example
Given the two functions $f(x)$ and $g(x)$, find the derivative of $h(x) = 7f(x) + 9g(x)$.

Solution
The derivative of $h(x)$ is,

$$h'(x) = af'(x) + bg'(x)$$

$$h'(x) = 7f'(x) + 9g'(x).$$

or

$$\frac{d}{dx}h(x) = \frac{d}{dx}(af(x) + bg(x)) = a\frac{d}{dx}f(x) + b\frac{d}{dx}g(x)$$

$$\frac{d}{dx}h(x) = \frac{d}{dx}(7f(x) + 9g(x)) = 7\frac{d}{dx}f(x) + 9\frac{d}{dx}g(x).$$

Example
Given the two functions $f(x) = 13x^2$ and $g(x) = 9x$, find the derivative of $h(x) = 2f(x) + 5g(x)$.

Solution
The derivative of $h(x)$ is,

$$h'(x) = af'(x) + bg'(x)$$

$$h'(x) = 2f'(x) + 5g'(x)$$

$$h'(x) = 2(13 \times 2x^{2-1}) + 5(9 \times 1\mathbf{x}^{\mathbf{1-1}})$$

$$h'(x) = 2(13 \times 2x^1) + 5(9 \times 1x^0)$$

$$h'(x) = 2(13 \times 2x) + 5(9 \times 1 \times 1)$$

$$h'(x) = 2(26x) + 5(9)$$

$$h'(x) = 52x + 45.$$

Or in Leibniz's notation we will have:

$$\frac{d}{dx}h(x) = \frac{d}{dx}(af(x) + bg(x)) = a\frac{d}{dx}f(x) + b\frac{d}{dx}g(x)$$

$$\frac{d}{dx}h(x) = \frac{d}{dx}(2f(x) + 5g(x)) = 2\frac{d}{dx}f(x) + 5\frac{d}{dx}g(x)$$

$$\frac{d}{dx}h(x) = 2(13 \times 2x^{2-1}) + 5(9 \times 1x^{1-1})$$

$$\frac{d}{dx}h(x) = 52x^1 + 45x^0$$

$$\frac{d}{dx}h(x) = 52x + 45.$$

3.3.3. The Product Rule in Derivative of a Function

For the functions $f(x)$ and $g(x)$, the derivative of the function $h(x) = f(x) \times g(x)$ with respect to x is:

$$h'(x) = (f(x) \times g(x))'(x) = f'(x) \times g(x) + f(x) \times g'(x).$$

In Leibniz's notation this is written as:

$$\frac{d}{dx}h(x) = \frac{d}{dx}(f(x) \times g(x)) = \frac{d}{dx}f(x) \times g(x) + f(x) \times \frac{d}{dx}g(x).$$

Example

Given the two functions $f(x) = 4x^2$ and $g(x) = 6x^3$, find the derivative of $h(x) = f(x) \times g(x)$.

$$h'(x) = (f(x) \times g(x))'(x) = f'(x) \times g(x) + f(x) \times g'(x)$$

$$h'(x) = (4x^2 \times 6x^3)'(x) = (4 \times 2x^{2-1}) \times (6x^3) + (4x^2) \times (6 \times 3x^{3-1})$$

$$h'(x) = (8x^1) \times (6x^3) + (4x^2) \times (18x^2)$$

$$h'(x) = 48x^{1+3} + 72x^{2+2}$$

$$h'(x) = 48x^4 + 72x^4 = 120x^4.$$

In Leibniz's notation we will have:

$$\frac{d}{dx}h(x) = \frac{d}{dx}(f(x) \times g(x)) = \frac{d}{dx}f(x) \times g(x) + f(x) \times \frac{d}{dx}g(x)$$

$$\frac{d}{dx}h(x) = (4 \times 2x^{2-1}) \times (6x^3) + (4x^2) \times (6 \times 3x^{3-1})$$

$$\frac{d}{dx}h(x) = (8x^1) \times (6x^3) + (4x^2) \times (18x^2)$$

$$\frac{d}{dx}h(x) = 48x^{1+3} + 72x^{2+2}$$

$$\frac{d}{dx}h(x) = 48x^4 + 72x^4 = 120x^4.$$

Example

Given the two functions $f(x) = 7x$ and $g(x) = x^2$, find the derivative of $h(x) = 3f(x) \times 2g(x)$.

Solution

The derivative of $h(x)$ is,

$$h'(x) = (3f(x) \times 2g(x))'(x) = 3f'(x) \times 2g(x) + 3f(x) \times 2g'(x)$$

$$h'(x) = (3 \times 7x^{1-1}) \times (2 \times x^2) + (3 \times 7x) \times (2 \times 2x^{2-1})$$

$$h'(x) = (21x^0) \times (2x^2) + (21x) \times (4x^1)$$

$$h'(x) = 21 \times (2x^2) + 21x \times 4x$$

$$h'(x) = 42x^2 + 84x^2 = 126x^2.$$

In Leibniz's notation we will have:

$$\frac{d}{dx}h(x) = \frac{d}{dx}(3f(x) \times 2g(x)) = (3 \times \frac{d}{dx}f(x)) \times (2g(x)) + (3f(x))(2 \times \frac{d}{dx}g(x))$$

$$\frac{d}{dx}h(x) = (3 \times \frac{d}{dx}(7x)) \times (2 \times x^2) + (3 \times 7x) \times (2 \times (\frac{d}{dx}x^2))$$

$$\frac{d}{dx}h(x) = (3 \times 7x^{1-1}) \times (2 \times x^2) + (3 \times 7x) \times (2 \times 2x^{2-1})$$

$$\frac{d}{dx}h(x) = (21x^0) \times (2x^2) + (21x) \times (4x^1)$$

$$\frac{d}{dx}h(x) = (21 \times 1) \times (2x^2) + (21x) \times (4x)$$

$$\frac{d}{dx}h(x) = 42x^2 + 84x^2 = 126x^2.$$

3.3.4. The Chain Rule in Derivative of a Function

The derivative of the function $h(x) = f(g(x))$ is,

$$h'(x) = f'(g(x)) \times g'(x).$$

In Leibniz's notation this is written as:

$$\frac{d}{dx}h(x) = \frac{d}{dz}f(z)|_{z=g(x)} \times \frac{d}{dx}g(x).$$

Example

Given the two functions $f(x) = 7x^2$ and $g(x) = 2x$, find the derivative of $h(x) = f(g(x))$.

Solution

We have learned the derivative of the function $h(x) = f(g(x))$ is, $h'(x) = f'(g(x)) \times g'(x)$. So,

$$f'(x) = 7 \times 2x^{2-1} = 14x^1 = 14x$$

$$f'(g(x)) = f'(2x) = 14(2x) = 28x.$$

$$g'(x) = 2x^{1-1} = 2x^0 = 2 \times 1 = 2.$$

Now, since we have $h'(x) = f'(g(x)) \times g'(x)$, we just need to replace $f'(g(x)) = 28x$ and $g'(x) = 2$ into this equation. So,

$$h'(x) = f'(g(x)) \times g'(x) = 28x \times 2 = 56x.$$

In Leibniz's notation we will have:

$$\frac{d}{dx}h(x) = \frac{d}{dz}f(z)|_{z=g(x)} \times \frac{d}{dx}g(x)$$

$$\frac{d}{dx}h(x) = \frac{d}{dz}f(7z^2)|_{z=2x} \times \frac{d}{dx}2x$$

$$\frac{d}{dx}h(x) = (7 \times 2x^{2-1})|_{z=2x} \times (2x^{1-1})$$

$$\frac{d}{dx}h(x) = (14x^1)|_{z=2x} \times (2x^0)$$

$$\frac{d}{dx}h(x) = 14x^1|_{z=2x} \times (2 \times 1)$$

$$\frac{d}{dx}h(x) = 14\,(2x) \times 2$$

$$\frac{d}{dx}h(x) = 28x \times 2 = 56x.$$

Chain rule of product in derivative Product

$$\frac{d}{dx}[f(x)h(x)] = h(x)\frac{d}{dx}f(x) + f(x)\frac{d}{dx}h(x).$$

Example

Find the derivative of $f(x) = (2x + 2)^4(5x - 3)^5$.

Solution

$$\frac{d}{dx}f(x) = 4(2)(2x + 2)^3(5x - 3)^5 + 5(5)(2x + 2)^4(5x - 3)^4$$

$= 8(2x + 2)^3(5x - 3)^5 + 25(2x + 2)^4(5x - 3)^4$

$= (2x + 2)^3(5x - 3)^4[8(5x - 3) + 25(2x + 2)]$

$= (2x + 2)^3(5x - 3)^4(40x - 24 + 50x + 50)$

$= (2x + 2)^3(5x - 3)^4(90x + 26).$

3.3.5. The Reciprocal Rule in Derivative of a Function

The derivative of $g(x) = \frac{1}{f(x)}$ for any nonvanishing function f is:

$g'(x) = -\frac{f'(x)}{(f(x))^2}$ whenever f is nonzero.

In Leibniz's notation this is written as:

$\frac{d}{dx}(g(x)) = \frac{d}{dx}(\frac{1}{f(x)}) = -\frac{1}{(f(x))^2}\frac{d}{dx}f(x).$

Example

Given the function $f(x) = 8x^3$, find the derivative of $g(x) = \frac{1}{f(x)}$.

Solution

We have learned that the derivative of $g(x) = \frac{1}{f(x)}$ is $g'(x) = -\frac{f'(x)}{(f(x))^2}$ whenever f is nonzero. So,

$g(x) = \frac{1}{f(x)} = \frac{1}{8x^3}$

$g'(x) = -\frac{f'(x)}{(f(x))^2}$

$f'(x) = 8(3x^{3-1})$

$f'(x) = 8(3x^2)$

$f'(x) = 24x^2$

$g'(x) = -\frac{24x^2}{(8x^3)^2}$ we plugged $f'(x) = 24x^2$ and $f(x) = 8x^3$ into the equation

$g'(x) = -\frac{24x^2}{64x^6}$

$g'(x) = -\frac{8 \times 3 \times x^2}{8 \times 8 \times x^4 \times x^2}$

$g'(x) = -\frac{3}{8x^4}.$

In Leibniz's notation we will have:

$\frac{d}{dx}(g(x)) = \frac{d}{dx}(\frac{1}{f(x)}) = -\frac{1}{(f(x))^2}\frac{d}{dx}f(x).$

$\frac{d}{dx}f(x) = \frac{d}{dx}(8x^3) = 8(3x^{3-1})$

$\frac{d}{dx}f(x) = 24x^2$

$\frac{d}{dx}(g(x)) = -\frac{1}{(f(x))^2}\frac{d}{dx}f(x) = (-\frac{1}{(8x^3)^2})(24x^2)$ we plugged $f'(x) = 24x^2$ and $f(x) = 8x^3$ into the equation

$\frac{d}{dx}(g(x)) = (-\frac{1}{64\ \ {}^{6}})(24x^2)$

$\frac{d}{dx}(g(x)) = -\frac{24\ \ {}^{2}}{64x^6}$

$\frac{d}{dx}(g(x)) = -\frac{8 \times 3 \times x^2}{8 \times 8 \times x^4 \times x^2}$

$\frac{d}{dx}(g(x)) = -\frac{3}{8\,x^4}.$

3.3.6. The Quotient Rule in Derivative of a Function

If $f(x)$ and $g(x)$ are functions, and the function $h(x) = \frac{f(x)}{g(x)}$, then the derivative of $h(x)$ is:

$$h'(x) = \frac{f'(x)g(x) - g'(x)f(x)}{(g(x))^2}.$$

In Leibniz's notation this is written as:

$$\frac{d}{dx}(h(x)) = \frac{(\frac{d}{dx}f(x))g(x) - (\frac{d}{dx}g(x))f(x)}{(g(x))^2}.$$

Note that the reciprocal rule is a special case of the quotient rule. It is to be noted that the reciprocal rule can be derived also from the combination of power rule and chain rule.

Example

Given the two functions $f(x) = 3x^2 + 4$ and $g(x) = 5x^2 - 8$, find the derivative of $h(x) = \frac{\boldsymbol{f(x)}}{\boldsymbol{g(x)}}$.

Solution

We are given that if $f(x)$ and $g(x)$ are functions, and $h(x) = \frac{f(x)}{g(x)}$, then, $h'(x) = \frac{f'(x)g(x) - g'(x)f(x)}{(g(x))^2}$. So,

For $f(x) = 3x^2 + 4$:

$f'(x) = 6x.$

For $g(x) = 5x^2 - 8$:

$g'(x) = 10x.$

Now, for $h'(x) = \frac{f'(x)g(x) - g'(x)f(x)}{(g(x))^2}$, we will have:

$$h'(x) = \frac{f'(x)g(x) - g'(x)f(x)}{(g(x))^2}$$

$h'(x) = \frac{6x(5x^2 - 8) - 10\ \ (3x^2 + 4)}{(5x^2 - 8)^2}$ we plugged $f(x) = 3x^2 + 4$, $g(x) = 5x^2 - 8$, $f'(x) = 6x$, and $g'(x) = 10x$ into the equation

$h'(x) = \frac{30x^3 - 48x - 30x^3 - 40}{25x^4 - 80\ \ {}^{2} + 64}$ we applied the binomial rule in the denominator

$$h'(x) = \frac{-48x - 40}{25x^4 - 80x^2 + 64}.$$

In Leibniz's notation we will have:

$$\frac{d}{dx}(h(x)) = \frac{(\frac{d}{dx}f(x))g(x) - (\frac{d}{dx}g(x))f(x)}{(g(x))^2}$$

$$\frac{d}{dx}(h(x)) = \frac{(6x)(5x^2 - 8) - (10x)(3x^2 + 4)}{(5x^2 - 8)^2}$$

we plugged $f(x) = 3x^2 + 4$, $g(x) = 5x^2 - 8$, $f'(x) = 6x$, and $g'(x) = 10x$ into the equation

$$\frac{d}{dx}(h(x)) = \frac{30x^3 - 48x - 30\ ^3 - 40}{25\ ^4 - 80x^2 + 64}$$

we applied the binomial rule in the denominator

$$\frac{d}{dx}(h(x)) = \frac{-48x - 40}{25x^4 - 80\ ^2 + 64}.$$

3.4. Derivatives of Exponential and Logarithmic Functions

The derivative of an exponential function $f(x) = a^{bx}$, for $a > 0$ is.

$$\frac{d}{dx}(f(x)) = \frac{d}{dx}\left(a^{bx}\right) = ba^{bx}\ ln\ a, \quad a > 0.$$

And the derivative of the function $f(x) = a^{h(x)}$, for $a > 0$ is.

$$\frac{d}{dx}(f(x)) = \frac{d}{dx}\left(a^{h(x)}\right) = a^{h(x)}\ ln\ a\ h'(x), \qquad a > 0.$$

Example

Find the derivative of $f(x) = 3^{kx}$.

Solution

We have learned that the derivative of $f(x) = a^{bx}$ is $f'(x) = b(a^{bx})ln\ a$, for $a > 0$. In this case we have $a = 3$ which is greater than zero. So,

$$f'(x) = k3^{kx}ln3.$$

In Leibniz's notation we will have:

$$\frac{d}{dx}(f(x)) = \frac{d}{dx}\left(a^{bx}\right) = ba^{bx}\ ln\ a, \quad a > 0$$

$$\frac{d}{dx}(f(x)) = \frac{d}{dx}\left(3^{kx}\right) = k3^{kx}\ ln3.$$

The derivative of a logarithmic function $f(x) = \log_a x$, for $a > 0$ and $a \neq 1$ is.

$$\frac{d}{dx}(f(x)) = \frac{d}{dx}(\log_a x) = \frac{1}{x\ lna}. \qquad a > 0 \text{ and } a \neq 1.$$

Example

Find the derivative of $f(x) = \log_4 \frac{3}{x} + \frac{3}{x}$.

Solution

Using the chain rule and linearity properties of the derivative we have:

$$f'(x) = (\log_4 \tfrac{3}{x})' + \left(\frac{3}{x}\right)'$$

$$f'(x) = \frac{1}{\frac{3}{x}ln} \times \left(\frac{3}{x}\right)' + 3 \times \left(\frac{1}{x}\right)'$$

$$f'(x) = \frac{x}{3ln4} \times 3 \times (-\frac{1}{x^2}) + 3 \times (-\frac{1}{x^2})$$

$$f'(x) = -\frac{3}{x^2}(\frac{x}{3ln} + 1)$$

$$f'(x) = -\frac{3}{x^2} \times \frac{x + 3ln4}{3ln4}$$

$$f'(x) = -\frac{x + 3ln4}{x^2 ln4}.$$

In Leibniz's notation we will have:

$$\frac{d}{dx}(f(x)) = \frac{d}{dx}\left(\log_4\frac{3}{x} + \frac{3}{x}\right)$$

$$\frac{d}{dx}(f(x)) = \frac{d}{dx}(\log_4\frac{3}{x}) + \frac{d}{dx}(\frac{3}{x})$$

$$\frac{d}{dx}(f(x)) = \frac{1}{\frac{3}{x}ln} \times \frac{d}{dx}\left(\frac{3}{x}\right) + 3 \times \frac{d}{dx}(\frac{1}{x})$$

$$\frac{d}{dx}(f(x)) = \frac{x}{3ln4} \times 3 \times (-\frac{1}{x^2}) + 3 \times (-\frac{1}{x^2})$$

$$\frac{d}{dx}(f(x)) = -\frac{3}{x^2}(\frac{x}{3ln4} + 1)$$

$$\frac{d}{dx}(f(x)) = -\frac{3}{x^2}(\frac{x + ln4}{3ln4})$$

$$\frac{d}{dx}(f(x)) = -\frac{x + ln4}{x^2 ln4}.$$

The derivative of the function $f(x) = lnx$, for $x > 0$ is.

$$\frac{d}{dx}(f(x)) = \frac{d}{dx}(lnx) = \frac{1}{x} \qquad x > 0.$$

Example

Find the derivative of $f(x) = x^2 lnx$ at $x = e$.

Solution

Using the power rule and the product rule, we have:

$$f'(x) = (x^2)' \times ln\, x + x^2 \times (ln\, x)'$$

$$f'(x) = (2x) \times ln\, x + x^2 \times \frac{1}{x}$$

$$f'(x) = 2xln\, x + \frac{x^2}{x}$$

$$f'(x) = 2xln\, x + x$$

$f'(x)|_{x=e} = 2eln\, e + e$ we plugged $x = e$ into the right-hand side

$$f'(x)|_{x=e} = 2e \times 1 + e$$

$$f'(x)|_{x=e} = 2e + e$$

$$f'(x)|_{x=e} = 3e.$$

In Leibniz's notation we will have:

$$\frac{d}{dx}(f(x)) = \frac{d}{dx}(x^2) \times ln\, x + (x^2) \times \frac{d}{dx}(lnx)$$

$$\frac{d}{dx}(f(x)) = (2x)\times ln\, x + x^2 \times \frac{1}{x}$$

$$\frac{d}{dx}(f(x)) = 2xln\, x + \frac{x^2}{x} = 2xln\, x + x$$

$$\frac{d}{dx}(f(x))|_{x=e} = 2e\, ln\, e + e$$

$$\frac{d}{dx}(f(x))|_{x=e} = 2e + e = 3e.$$

The derivative of the function $f(x) = ln|x|$, for $x \neq 0$ is.

$$\frac{d}{dx}(f(x)) = \frac{d}{dx}(ln|x|) = \frac{1}{x} \qquad x \neq 0.$$

And the derivative of the function $f(x) = ln\,|h(x)|$, for $x \neq 0$ is.

$$\frac{d}{dx}(f(x)) = \frac{d}{dx}(ln|x|) = \frac{1}{h(x)}h'(x) \qquad x \neq 0.$$

Example

Find the derivative of $f(x) = ln|3x^2 - x|$.

Solution

We have learned that the derivative of the function $f(x) = ln|h(x)|$, for $x \neq 0$ is $f'(x) = \frac{1}{h(x)}h'(x)$. So,

$$f'(x) = \frac{1}{3x^2 - x}(3x^2 - x)'$$

$$f'(x) = \frac{1}{3x^2 - x}(6x - 1)$$

$$f'(x) = \frac{6x - 1}{3x^2 - x} \qquad x \neq 0.$$

In Leibniz's notation we will have:

$$\frac{d}{dx}(f(x)) = \frac{1}{3x^2 - x}\frac{d}{dx}(3x^2 - x)$$

$$\frac{d}{dx}(f(x)) = \frac{1}{3x^2 - x}(6x - 1)$$

$$\frac{d}{dx}(f(x)) = \frac{6x - 1}{3x^2 - x} \qquad x \neq 0.$$

The derivative of the function $f(x) = e^{ax}$ is.

$$\frac{d}{dx}(e^{ax}) = ae^{ax}.$$

And the derivative of the function $f(x) = e^{h(x)}$ is

$$\frac{d}{dx}\left(e^{h(x)}\right) = e^{h(x)}\frac{d}{dx}h(x).$$

Example

Find the derivative of $f(x) = e^{4x^2 - 2}$.

Solution

We learned that the derivative of the function $f(x) = e^{ax}$ is $f'(x) = e^{h(x)}\, h(x)'$. So,

$$f'(x) = (e^{4x^2 - 2})(4x^2 - 2)'$$

$$f'(x) = (e^{4x^2 - 2})\,(8x)$$

$$f'(x) = 8xe^{4x^2 - 2}.$$

In Leibniz's notation we will have:

$$\frac{d}{dx}\left(e^{4x^2 - 2}\right) = (e^{4x^2 - 2})\frac{d}{dx}(4x^2 - 2)$$

$$\frac{d}{dx}\left(e^{4x^2 - 2}\right) = (e^{4x^2 - 2})\,(8x)$$

$$\frac{d}{dx}\left(e^{4x^2 - 2}\right) = 8xe^{4x^2 - 2}.$$

The derivative of the function $f(x) = x^x$ is.

$$\frac{d}{dx}(f(x)) = \frac{d}{dx}(x^x) = x^x\,(1 + \ln x).$$

And the derivative of the function $f(x) = x^{h(x)}$ is

$$\frac{d}{dx}\left(x^{h(x)}\right) = x^{h(x)}\frac{d}{dx}h(x)(1 + \ln x).$$

Example

Find the derivative of $f(x) = x^{2x}$.

Solution

We learned that the derivative of the function $f(x) = x^{h(x)}$ is $f'(x) = x^{h(x)}\, h'(x)(1 + \ln x)$. So,

$$f'(x) = x^{2x}\, h'(x)(1 + \ln x)$$

$$f'(x) = x^{2x}\, (2x)'(1 + \ln x)$$

$$f'(x) = x^{2x}\, (2)(1 + \ln x)$$

$$f'(x) = 2x^{2x}\, (1 + \ln x).$$

In Leibniz's notation we will have:

$$\frac{d}{dx}(x^{2x}) = x^{2x} \times \frac{d}{dx}(2x) \times (1 + \ln x)$$

$$\frac{d}{dx}(x^{2x}) = x^{2x} \times (2) \times (1 + \ln x)$$

$$\frac{d}{dx}(x^{2x}) = 2x^{2x}(1 + \ln x).$$

The derivative of the function $h(x) = \ln(f(x))$ is.

$$\frac{d}{dx}(h(x)) = \frac{d}{dx}(\ln(f(x)) = \frac{\frac{d}{dx}(f(x))}{f(x)}.$$

Example
Find the derivative of $f(x) = ln(2x^3 - 13x)$.

Solution
We learned that the derivative of the function $f(x) = ln(f(x))$ is $f'(x) = \frac{f'(x)}{f(x)}$. So,

$$f'(x) = \frac{(2x^3 - 13x)'}{2x^3 - 13x}$$

$$f'(x) = \frac{3 \times 2x^{3-1} - 13}{2x^3 - 13x}$$

$$f'(x) = \frac{6x^2 - 13}{2x^3 - 13x}.$$

In Leibniz's notation we will have:

$$\frac{d}{dx}[ln(f(x))] = \frac{\frac{d}{dx}(2x^3 - 13\ )}{2x^3 - 13x}$$

$$\frac{d}{dx}[ln(f(x))] = \frac{3 \times 2x^{3-1} - 13}{2x^3 - 13x}$$

$$\frac{d}{dx}[ln(f(x))] = \frac{6x^2 - 13}{2x^3 - 13x}.$$

Example
Find the derivative of $f(x) = ln(x + \sqrt{x^2 + 4b^3})$.

Solution
We learned that the derivative of the function $f(x) = ln(f(x))$ is $f'(x) = \frac{f'(x)}{f(x)}$. So,

$$f'(x) = \frac{(x + \sqrt{x^2 + 4b^3})'}{x + \sqrt{x^2 + 4b^3}}$$

$$f'(x) = \frac{1 + \frac{2x}{2\sqrt{x^2 + 4b^3}}}{x + \sqrt{x^2 + 4b^3}}$$

$$f'(x) = \frac{1 \times \frac{2\sqrt{x^2 + 4b^3}}{2\sqrt{x^2 + 4b^3}} + \frac{2x}{2\sqrt{x^2 + 4b^3}}}{x + \sqrt{x^2 + 4b^3}}$$

we multiplied the first term in the nominator by $\frac{2\sqrt{x^2 + 4b^3}}{2\sqrt{x^2 + 4b^3}}$

$$f'(x) = \frac{\frac{2\sqrt{x^2 + 4b^3} + 2x}{2\sqrt{x^2 + 4b^3}}}{x + \sqrt{x^2 + 4b^3}}$$

$$f'(x) = \frac{1}{\sqrt{x^2 + 4b^3}}.$$

In Leibniz's notation we will have:

$$\frac{d}{dx}[ln(x + \sqrt{x^2 + 4b^3})] = \frac{\frac{d}{dx}(x + \sqrt{x^2 + 4b^3})}{x + \sqrt{x^2 + 4b^3}}$$

$$\frac{d}{dx}[ln(x+\sqrt{x^2+4b^3})] = \frac{1+\frac{2x}{2\sqrt{x^2+4b^3}}}{x+\sqrt{x^2+4b^3}}$$

$$\frac{d}{dx}[ln(x+\sqrt{x^2+4b^3})] = \frac{1\times\frac{2\sqrt{x^2+4b^3}}{2\sqrt{x^2+4b^3}}+\frac{2x}{2\sqrt{x^2+4b^3}}}{x+\sqrt{x^2+4b^3}}$$

we multiplied the first term in the nominator by $\frac{2\sqrt{x^2+4b^3}}{2\sqrt{x^2+4b^3}}$

$$\frac{d}{dx}[ln(x+\sqrt{x^2+4b^3})] = \frac{\frac{2\sqrt{x^2+4b^3}+2x}{2\sqrt{x^2+4b^3}}}{x+\sqrt{x^2+4b^3}}$$

$$\frac{d}{dx}[ln(x+\sqrt{x^2+4b^3})] = \frac{2\sqrt{x^2+4b^3}+2x}{2\sqrt{x^2+4b^3}}\times\frac{1}{x+\sqrt{x^2+4b^3}}$$

$$\frac{d}{dx}[ln(x+\sqrt{x^2+4b^3})] = \frac{x+\sqrt{x^2+4b^3}}{\sqrt{x^2+4b^3}}\times\frac{1}{x+\sqrt{x^2+4b^3}}$$

$$\frac{d}{dx}[ln(x+\sqrt{x^2+4b^3})] = \frac{1}{\sqrt{x^2+4b^3}}.$$

The table below presents some common derivatives.

Function	Derivative
$f(x) = c$	$\frac{d}{dx}(c) = 0$, c is any constant.
$f(x) = x$	$\frac{d}{dx}(x) = 1$
$f(x) = cx$	$\frac{d}{dx}(cx) = c$, c is any constant
$f(x) = x^n$	$\frac{d}{dx}(x^n) = nx^{n-1}$, n is any number
$f(x) = \frac{1}{x}$	$\frac{d}{dx}\left(\frac{1}{x}\right) = -\frac{1}{x^2}$
$f(x) = a^x$	$\frac{d}{dx}(a^x) = a^x ln\ (a)$
$f(x) = a^{h(x)}$	$\frac{d}{dx}\left(a^{h(x)}\right) = a^{h(x)}\ln\ (a)h'(x)$
$f(x) = e^x$	$\frac{d}{dx}(e^x) = e^x$
$f(x) = e^{h(x)}$	$\frac{d}{dx}\left(e^{h(x)}\right) = h'(x)e^{h(x)}$
$f(x) = \ln x$	$\frac{d}{dx}(\ln x) = \frac{1}{x}, x > 0$
$f(x) = \ln\|x\|$	$\frac{d}{dx}(\ln\|x\|) = \frac{1}{x}, x \neq 0$
$f(x) = \ln(h(x))$	$\frac{d}{dx}(\ln(h(x)) = \frac{h'(x)}{h(x)}.$
$f(x) = \log x$	$\frac{d}{dx}f(x) = \frac{1}{x}$
$f(x) = \log_a x$	$\frac{d}{dx}\log_a x = \frac{1}{x\ \ln a}, x > 0$
$f(x) = \log_a(h(x))$	$\frac{d}{dx}\log_a(h(x)) = \frac{h'(x)}{h(x)\ \ln a}, h(x) > 0$
$f(x) = \sin x$	$\frac{d}{dx}\sin x = \cos x$

$f(x) = \cos x$	$\frac{d}{dx}\cos x = -\sin x$
$f(x) = \tan x$	$\frac{d}{dx}\tan x = \sec^2 x$
$f(x) = \sec x$	$\frac{d}{dx}\sec x = \sec x \tan x$
$f(x) = \csc x$	$\frac{d}{dx}\csc x = -\csc x \cot x$
$f(x) = \cot x$	$\frac{d}{dx}\cot x = -\csc^2 x$
$f(x) = \sin kx$	$\frac{d}{dx}\sin kx = k\cos x$
$f(x) = \cos kx$	$\frac{d}{dx}\cos kx = -k\sin x$
$f(x) = \tan kx$	$\frac{d}{dx}\tan kx = k\sec^2 x$
$f(x) = \sec kx$	$\frac{d}{dx}\sec kx = k \sec x \tan x$
$f(x) = \csc kx$	$\frac{d}{dx}\csc kx = -k\csc x \cot x$
$f(x) = \cot kx$	$\frac{d}{dx}\cot kx = -k\csc^2 x$

3.5.1. Second Derivative

The **second derivative** of a function $f(x)$ is the derivative of the derivative of $f(x)$. Generally, addressing, the second derivative measures how the rate of change of a quantity is itself changing. For example, consider a case in which drivers of a car want to measure how quickly the position of the car changes as time advances. The second derivative of how far the car has gone with respect to time is the instantaneous acceleration of the car or rate at which the velocity of the car is changing with respect to time. In Leibniz notation

$$a = \frac{dv}{dt} = \frac{d^2x}{dt^2}$$

where a is acceleration, v is velocity, t is time, x is position, and d is the instantaneous "delta" or change. The last expression $\frac{d^2x}{dt^2}$ is the second derivative of position (x) with respect to time.

3.5.1. Second Derivative Power Rule

The second derivative power rule is obtained by applying the power rule of for the first derivative twice. In mathematical words, the second derivative power rule is:

$$\frac{d^2}{dx^2}(x^n) = \frac{d}{dx}\frac{d}{dx}(x^n) = \frac{d}{dx}(nx^{n-1}) = n(n-1)x^{n-2}.$$

The second derivative of a function $f(x)$ is usually denoted $f''(x)$. That is:

$$f''(x) = (f')'(x).$$

In Leibniz's notation, the second derivative of a dependent variable y with respect to an independent variable x is written as:

$$\frac{d^2y}{dx^2} = \frac{d}{dx}\left(\frac{dy}{dx}\right), \text{ or}$$

$y'' = \frac{d^2y}{dx^2} = \frac{d}{dx}\left(\frac{dy}{dx}\right).$

Example
Given the function $f(x) = x^4$, find the second derivative.

Solution
The derivative of $f(x) = x^4$ is:

$f'(x) = 4x^3.$

the second derivative of $f(x)$ is the derivative of $f'(x)$. So,

$(f')'(x) = f''(x) = 12x^2.$

In Leibniz's notation we will have:

$$\frac{d^2f(x)}{dx^2} = \frac{d}{dx}\left(\frac{df(x)}{dx}\right)$$

$$\frac{d^2f(x)}{dx^2} = \frac{d}{dx}(4x^3)$$

$$\frac{d^2f(x)}{dx^2} = 12x^2.$$

3.5. Higher-Order Derivatives of a Function

Given a function $y = f(x)$ having a finite derivative, we have already looked at the second order derivative of $f(x)$ as $f''(x) = (\frac{dy}{dx})' = \frac{d}{dx}\left(\frac{dy}{dx}\right) = \frac{d^2y}{dx^2}$. Similarly, if f'' exists and is differentiable, we can calculate the third derivative of the function $f(x)$:

$f''' = \frac{d^3y}{dx^3} = y'''.$

Hence, taking the derivative of n times is called the nth derivative of $y = f(x)$ with respect to x and is denoted as:

$\frac{d^nf(x)}{dx^n} = \frac{d^ny}{dx^n}$ (in Leibniz's notation)

$f^{(n)}(x) = y^{(n)}(x)$ (in Lagrange's notation).

Example
Find the first five derivatives of the function $f(x) = 7x^3 + 5x^2 - 2x^{\frac{1}{2}} + e^x$

Solution
Simply we just need to take the derivatives. So,

$$f'(x) = 21x^2 + 10x - 2(\tfrac{1}{2})x^{\frac{1}{2}-1} + e^x = 21x^2 + 10x - x^{-\frac{1}{2}} + e^x$$

$$f''(x) = 42x + 10 - (-\tfrac{1}{2})x^{-\frac{1}{2}-1} + e^x = 42x + 10 + \tfrac{1}{2}x^{-\frac{3}{2}} + e^x$$

$$f'''(x) = 42 + \tfrac{1}{2}(-\tfrac{3}{2})x^{-\frac{3}{2}-1} + e^x = 42 - \tfrac{3}{4}x^{-\frac{5}{2}} + e^x$$

$$f^{(4)}(x) = (-\frac{3}{4})(-\frac{5}{2})x^{-\frac{5}{2}-1} + e^x = \frac{15}{8}x^{-\frac{7}{2}} + e^x$$

$$f^{(5)}(x) = (\frac{15}{8})(-\frac{7}{2})x^{-\frac{7}{2}-1} + e^x = \frac{105}{16}x^{-\frac{9}{2}} + e^x.$$

In Leibniz's notation we will have:

$$\frac{df(x)}{d} = \frac{d}{dx}\left(7x^3 + 5x^2 - 2x^{\frac{1}{2}} + e^x\right)$$

$$\frac{df(x)}{d} = 3 \times 7x^{3-1} + 2 \times 5x^{2-1} - 2 \times x^{\frac{2}{2}-1} + e^x = 21x^2 + 10x - x^{-\frac{1}{2}} + e^x$$

$$\frac{d^2f(x)}{dx^2} = \frac{d}{dx}\left(21x^2 + 10x - x^{-\frac{1}{2}} + e^x\right)$$

$$\frac{d^2f(x)}{dx^2} = 2 \times 21x^{2-1} + 1 \times 10x^{1-1} - (-\frac{1}{2}) \times x^{-\frac{1}{2}-1} + e^x = 42x + 10 + \frac{1}{2}x^{-\frac{3}{2}} + e^x$$

$$\frac{d^3f(x)}{dx^3} = \frac{d}{dx}\left(42x + 10 + \frac{1}{2}x^{-\frac{3}{2}} + e^x\right) = 42 - \frac{3}{4}x^{-\frac{5}{2}} + e^x$$

$$\frac{d^3f(x)}{dx^3} = 1 \times 42x^{1-1} + \frac{1}{2} \times \left(-\frac{3}{2}\right)x^{-\frac{3}{2}-1} + e^x = 42 - \frac{3}{4}x^{-\frac{5}{2}} + e^x$$

$$\frac{d^4f(x)}{dx^4} = \frac{d}{dx}\left(42 - \frac{3}{4}x^{-\frac{5}{2}} + e^x\right)$$

$$\frac{d^4f(x)}{dx^4} = (-\frac{5}{2}) \times (-\frac{3}{4}x^{-\frac{5}{2}-1}) + e^x = \frac{15}{8}x^{-\frac{7}{2}} + e^x$$

$$\frac{d^5f(x)}{dx^5} = \frac{d}{dx}\left(\frac{15}{8}x^{-\frac{7}{2}} + e^x\right) = \frac{105}{16}x^{-\frac{9}{2}} + e^x.$$

$$\frac{d^5f(x)}{dx^5} = (-\frac{7}{2})(\frac{15}{8}x^{-\frac{7}{2}-1}) + e^x = \frac{105}{16}x^{-\frac{9}{2}} + e^x.$$

Applications of the derivative identifies was that this concept is used in everyday life such as optimization, determining concavity, curve sketching, etc.

3.6. Use of the Second Derivative Test for Concavity

Here we look at the use of the **second derivative test** for determining where a function is **concave upward or downward**. Additionally, the second derivative can also be thought of as whether the function has an increasing or decreasing slope over a period.

Concavity and The Second Derivative

Suppose that $f(x)$ is twice differentiable on the open interval (a, b).

- If $f(x)$ is concave up on (a, b) then $f''(x) > 0$ on (a, b).
- If $f(x)$ is concave down on (a, b) then $f''(x) < 0$ on (a, b).
- If $f''(x) > 0$ on (a, b), then $f(x)$ is concave up on the interval (a, b).
- If $f''(x) < 0$ on (a, b), then $f(x)$ is concave down on the interval (a, b).

The following table provides a summary of the tests, determining whether the function $f(x)$ is increasing or decreasing as well as whether $f(x)$ is concave upward or concave downward, and can be a useful guide in sketching graphs.

Signs of 1st and 2nd derivatives	Conclusion	Shape of the graphs
$f'(x) > 0+$	$f(x)$ is increasing	
$f''(x) > 0+$	$f(x)$ is concave up	
$f'(x) > 0+$	$f(x)$ is increasing	
$f''(x) < 0-$	$f(x)$ is concave down	
$f'(x) < 0-$	$f(x)$ is decreasing	
$f''(x) > 0+$	$f(x)$ is concave up	
$f'(x) < 0-$	$f(x)$ is decreasing	
$f''(x) < 0-$	$f(x)$ is concave down	

A function, $f(x)$, whose second derivative is positive will be concave upward, (also referred to as convex function), meaning that the tangent line will lie below the graph of the function. In this case $f'(x)$ is increasing. Similarly, a function whose second derivative is negative will be concave downward (also simply called concave function), and its tangent lines will lie above the graph of the function. In this case $f'(x)$ is decreasing (see the figure below).

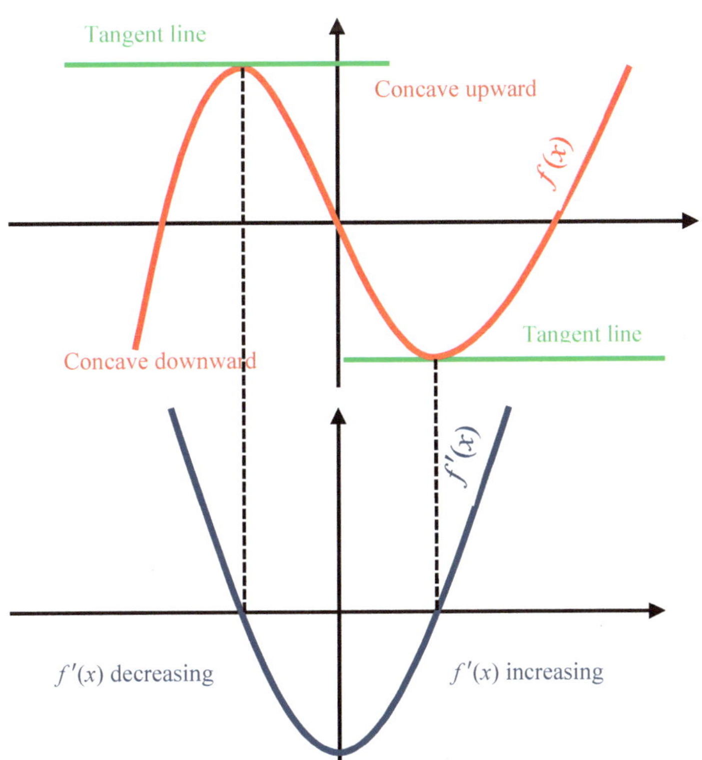

Example
Determine the regions in which the following function is concave upward or downward:

$$f(x) = 4x^3 + 2x^2 - 2x.$$

Solution
Let us take the first two derivatives,

$f'(x) = 12x^2 + 4x - 2$ — we used the power rule of derivatives.

$f''(x) = 24x + 4$ — again, we used the power rule of derivatives.

Now let us set $f''(x) = 0$. So,

$$f''(x) = 24x + 4 = 0$$

$$24x + 4 = 0$$

$24x + 4 - 4 = 0 - 4$ — to isolate the variable, x, we added -4 to both sides

$$24x = -4$$

$$24x = -4$$

$\frac{1}{24} \times 24x = \frac{1}{24} \times (-4)$ — to make the coefficient of x equal to 1, we multiplied both sides by the reciprocal of the variable

$$x = -\frac{4}{24} = -\frac{1}{6}.$$

Now to find the sign of $f''(x) = 24x + 4$, before and after the point $x = -\frac{1}{6}$, we use two check points. Let us choose the two-check points $x = -1$ and $x = 0$. The value of $f''(-1)$ at $x = -1$ is $f''(x) = 24(-1) + 4 = -24 + 4 = -20$. The value of $f''(x)$ at $x = 0$ is $f''(0) = 24(0) + 4 = 4$. It can be realized that the second derivative before the point $x = -\frac{1}{6}$ is negative and after $x = -\frac{1}{6}$ is positive. So,

$f(x)$ is concave downward up to the point $x = -\frac{1}{6}$

$f(x)$ is concave upward from the point $x = -\frac{1}{6}$ on.

Figure below illustrates the concavity of $y = f(x)$.

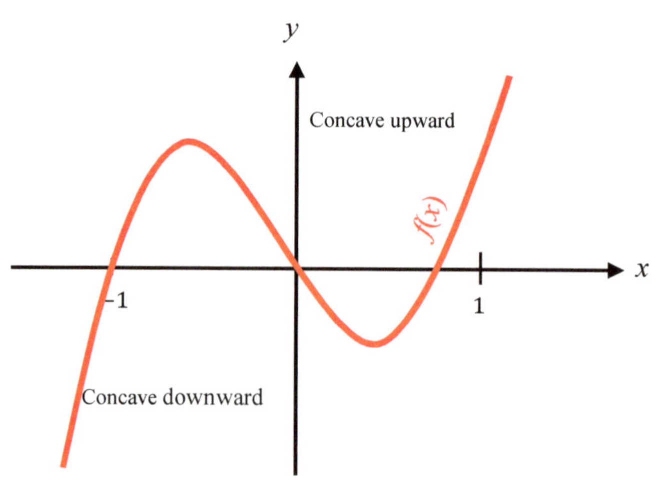

Concavity and The Second Derivative
Suppose that $f(x)$ is twice differentiable on the open interval (a, b).

- If $f(x)$ is concave up on (a, b) then $f''(x) > 0$ on (a, b).
- If $f(x)$ is concave down on (a, b) then $f''(x) < 0$ on (a, b).
- If $f''(x) > 0$ on (a, b), then $f(x)$ is concave up on the interval (a, b).
- If $f''(x) < 0$ on (a, b), then $f(x)$ is concave down on the interval (a, b).

3.7. Sketching a Function and Finding Extrema Using the First and Second Derivative

In optimization problems, our objective is to find either the largest value (maximum) or the smallest value (minimum) that a function can achieve. To achieve this, we utilize a derivative test, which involves analyzing the derivative of a function to pinpoint the extreme or critical points and determine whether each point corresponds to a local maximum, a local minimum, or a saddle point.

Extreme, namely maximum, and minimum points, represent the highest and lowest values of the function, either within specific intervals or across the entire domain.

A relative maximum occurs at a point higher than the surrounding points within each interval, while a relative minimum occurs at a point lower than the surrounding points. On the other hand, global or absolute extrema, referred to as a global (absolute) maximum and a global (absolute) minimum, denote the largest and smallest values of the function over the entire domain.

The behavior of the graph changes at points where the second derivative of a function changes sign, indicating a transition from concave down to concave up, or vice versa.

Relative maxima and minima are pivotal in curve sketching and can be identified using either the first or second derivative test. An inflection point marks the location where this change occurs. It's important to note that assuming the second derivative is continuous, it must be zero at any inflection point, although not every point where the second derivative is zero necessarily qualifies as an inflection point.

In optimization problems, the initial step involves identifying the quantity to be optimized and the associated constraint. The constraint refers to a condition that must be satisfied regardless of the solution obtained.

Example
Given the following function identify the intervals where the function is concave up and concave down and the intervals where the function is increasing and decreasing. Use this information to sketch the graph and designate extreme points.

$f(x)=3x^5 - 5x^3 + 3.$

Solution
First, we need to take the first two derivatives. So,
For the first derivative:

$f'(x) = 15x^4 - 15x^2$

$f'(x) = 15x^2(x^2 - 1)$ we factored out $5x^2$

$f'(x) = 15x^2(x - 1)(x + 1)$ we applied the difference of squared rule.

For the second derivative:

$$f''(x) = (f'(x))' = (15x^4 - 15x^2)'$$

$$f''(x) = 60x^3 - 30x$$

$$f''(x) = 30x(2x^2 - 1).$$

Now, let us determine the intervals where the function is increasing and decreasing. We have learned that the interval in which $f'(x) > 0$ the function is increasing and the interval in which $f'(x) < 0$ the function is decreasing. To find these, we first need to obtain the roots of $f'(x)$.

Hence, we set $f'(x) = 0$. So,

$f'(x) = 15x^2(x - 1)(x + 1) = 0 \quad 15x^2 = 0 \quad$ or $\quad x - 1 = 0 \quad$ or $\quad x + 1 = 0.$

For $15x^2 = 0 \quad x = 0.$

For $x - 1 = 0 \quad x = 1.$

For $x + 1 = 0 \quad x = -1.$

There are three roots for the function $f'(x)$, $x = 0$, $x = 1$, and $x = -1$. To designate the requested intervals, we provide a number line for the increasing/decreasing information. We use four check points, $x = -1.5$, $x = -0.5$, $x = 0.5$, and $x = 1.5$ as shown below:

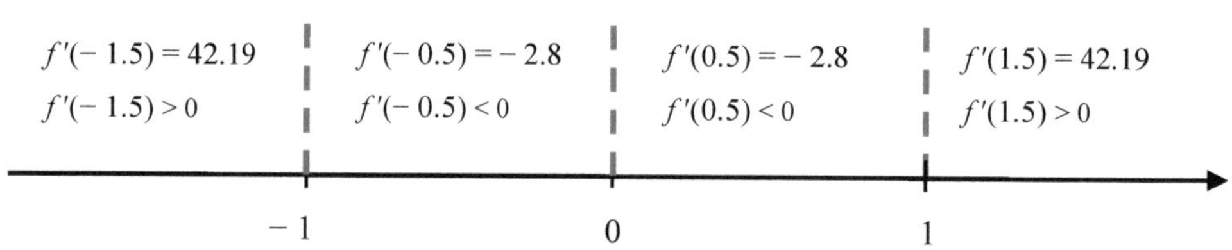

Therefore, we have got the following intervals of increasing and decreasing.

For the x values within the interval $-\infty < x < -1$, the function $f(x) = 3x^5 - 5x^3 + 3$ is increasing.

For the x values within the interval $-1 < x < 1$, the function $f(x) = 3x^5 - 5x^3 + 3$ is decreasing.

For the x values within the interval $1 < x < \infty$, the function $f(x) = 3x^5 - 5x^3 + 3$ is increasing.

Now, let us find the interval for concavity of the function. The procedure is like for identifying the intervals of increasing and decreasing except that in this case we need to use the second derivative instead of the first derivative.

Here we need to set the function $f''(x) = 30x(2x^2 - 1)$ equal to zero. So,

$$f''(x) = 30x(2x^2 - 1) = 0$$

$$30x(2x^2 - 1) = 0$$

$$x(2x^2 - 1) = 0$$

$x = 0 \quad$ or $\quad 2x^2 - 1 = 0.$

For $2x^2 - 1 = 0$:

$2x^2 - 1 + 1 = 0 + 1$	to isolate the variable we added 1 to both sides
$2x^2 = 1$	
$\frac{1}{2} \times 2x^2 = \frac{1}{2} \times 1$	we multiplied both sides by the reciprocal of the variable
$x^2 = \frac{1}{2}$	

$$x = -\sqrt{\frac{1}{2}} \quad \text{or} \quad x = +\sqrt{\frac{1}{2}}$$

$$x = -\sqrt{\frac{1}{2}} \times \frac{\sqrt{2}}{\sqrt{2}} \quad \text{or} \quad x = +\sqrt{\frac{1}{2}} \times \frac{\sqrt{2}}{\sqrt{2}}$$

$$x = -\frac{\sqrt{2}}{2} \quad \text{or} \quad x = \frac{\sqrt{2}}{2}.$$

There are three roots for the function $f''(x)$, $x = 0$, $x = -\frac{\sqrt{2}}{2}$, and $x = \frac{\sqrt{2}}{2}$. To designate the requested intervals, we provide a number line for the concavity information.

We use four check points, $x = -1$, $x = -0.5$, $x = 0.5$, and $x = 1$ as shown below:

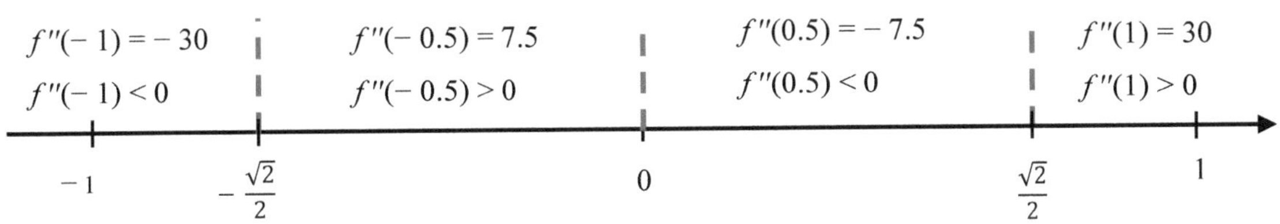

Therefore, we have got the following concavity intervals for the function.

For the x values within the interval $-\infty < x < -\frac{\sqrt{2}}{2}$, the function $f(x)$ is concave downward.

For the x values within the interval $-\frac{\sqrt{2}}{2} < x < 0$, the function $f(x)$ is concave upward.

For the x values within the interval $0 < x < \frac{\sqrt{2}}{2}$, the function $f(x)$ is concave downward.

For the x values within the interval $\frac{\sqrt{2}}{2} < x < \infty$, the function $f(x)$ is concave upward.

Using the increasing/decreasing information and realizing concavity behavior of the function we now can sketch the graph of the $f(x)$. Hence, if we start with $x < -1$ we know that we have an increasing function.

We also know that the function $f(x)$ is concave downward in this range. So, we can start off with sketching an increasing curve that has is also concave down until we reach $x = -1$. From this point the graph starts to decrease and will continue to decrease until it hits $x = 1$.

Note, as the function decreases the concavity needs to switch to concave up at $x = -\frac{\sqrt{2}}{2}$ and then switches back to concave down at $x = 0$ with a final switch to concave up at $x = \frac{\sqrt{2}}{2}$. Once it hits $x = 1$ it will start to increase while it is still concave up and both behaviors continue for the rest of the graph.

Compiling all this information will provide us with the following graph of the function.

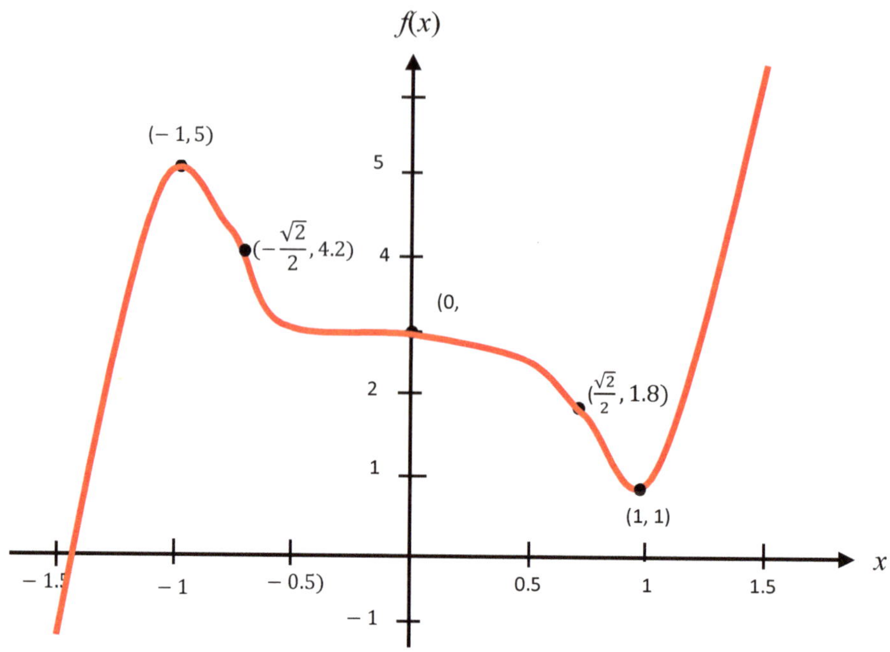

3.8. Finding Extreme Points Using Derivatives

The first derivative test is the process of analyzing functions using their first derivatives to find their extremum point.

Finding the relative extremum points of $f(x)$ requires the following steps:

Step 1: Finding $f'(x)$

Step 2: Finding all extreme points and all points where f is undefined. The extreme points of a function f are the x-values, within the domain of f for which $f'(x)= 0$ or where f' is undefined.

Step 3: Analyzing intervals of increase or decrease. This can be done in many ways, but we like using a sign chart. In a sign chart, we pick a test value at each interval that is bounded by the points we found in *Step* 2 and check the derivative's sign on that value.

Step 4: Finding extremum points. Now that we know the intervals where f increases or decreases, we can find its extremum points. An extremum point would be a point where f is defined and f', prime changes signs.

Another way to find extreme points of a function $f(x)$ is:

If $f'(c) = 0$, then the following three statements are true:

1. If $f''(c) > 0$, then $f(c)$ is a relative minimum.
2. If $f''(c) < 0$, then $f(c)$ is a relative maximum.

3. If $f''(c) = 0$, then we must determine if the point is a relative maximum, relative minimum, or inflection point by using the higher order derivative test.

The higher-order derivative test or general derivative test can determine whether a function's extreme points are maxima, minima, or points of inflection for a wider variety of functions than the second-order derivative test. As shown below, the second-derivative test is mathematically identical to the special case of $n = 1$ in the higher-order derivative test.

Let $f(x)$ be a real-valued, differentiable function on an interval , and c be any real value let all the derivatives of f at c be zero up to and including the nth derivative, but with the $(n + 1)^{\text{th}}$ derivative being non-zero. That is:

$$f'(c) = f''(c) = f^{(3)}(c) = \cdots = f^{(n)}(c) = 0 \text{ and } f^{(n+1)}(c) \neq 0.$$

Then there are four cases in which the first two cases where c is an extremum, the second two where c is an inflection point or a saddle point:

- If ***n* is odd** and $f^{(n+1)}(c) < 0$, then c is a **local maximum**.
- If ***n* is odd** and $f^{(n+1)}(c) > 0$, then c is a **local minimum**.
- If ***n* is even** and $f^{(n+1)}(c) < 0$, then c is a strictly **decreasing point of inflection**.
- If ***n* is even** and $f^{(n+1)}(c) > 0$, then c is a strictly **increasing point of inflection**.

Suppose $f^{(n+1)}(c)$ derivative of $f(x)$ at c. Then, we have the following table which summarizes the higher order derivative test.

Case on parity of n	Case on sign of $f^{(n+1)}(c)$	Conclusion about f at c	Prototypical example (words)	Prototypical example (picture)
odd	negative	strict local maximum.	$f(x) = -x^2$, $c = 0$ (more generally: $f(x) = -x^{2k}$, k is a natural number, and $c = 0$).	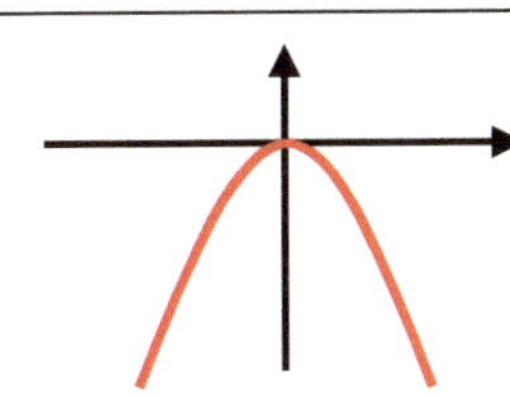
odd	positive	strict local minimum.	$f(x) = x^2$, $c = 0$ (more generally: $f(x) = x^{2k}$, k is a natural number, and c = 0).	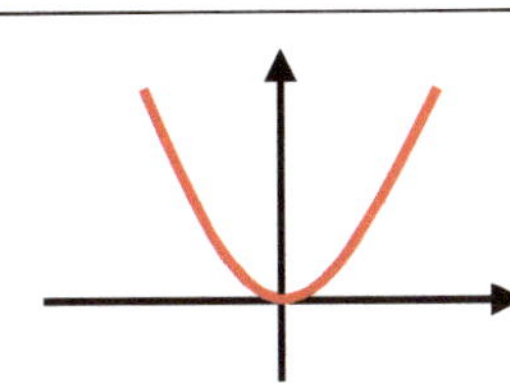
even	negative	neither local maximum nor local minimum. It is a strictly decreasing point of inflection.	$f(x) = -x^3$, $c = 0$ (more generally: $f(x) = -x^{2k+1}$ for k is a natural number, and $c = 0$).	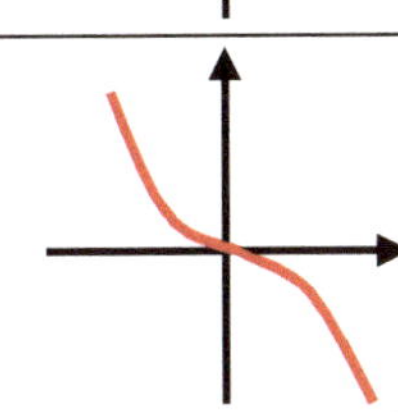

even	positive	neither local maximum nor local minimum. It is a strictly increasing point of inflection.	$f(x) = x^3$, $c = 0$ (more generally: $f(x) = x^{2k+1}$ for k is a natural number, and $c = 0$).	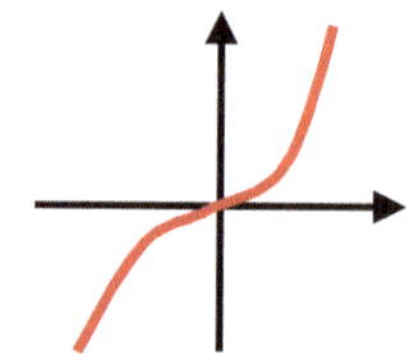

Suppose that $x = c$ is an extreme point of $f(x)$ such that $f'(c) = 0$ and that $f''(c)$ is continuous in a region around $x = c$. Then,

- If $f''(c) < 0$ then $x = c$ is a relative maximum.
- If $f''(c) > 0$ then $x = c$ is a relative minimum.
- If $f''(c) = 0$ then $x = c$ can be a relative maximum, relative minimum or neither. In this case we say the test is inconclusive. There might be a local maximum or minimum, or there might be a point of inflection.

Example

Find the extreme points and determine whether they are a minimum, maximum, or point of inflection for the function $f(x) = \frac{1}{20}x^5 + 18$.

Solution

We need to follow these steps:

1. Find the derivative of the function and set the derivative equal to zero to find critical points.
2. Determine the nature of these critical points using the second derivative test.

Let us start with the first step.

$$f(x) = \frac{1}{20}x^5 + 18$$

$$f'(x) = (5) \times \frac{1}{20}x^{5-1} = 0$$

$$f'(x) = \frac{5}{20}x^4 = \frac{1}{4}x^4 = 0$$

$$x = 0.$$

Therefore, there is only one extreme point at $x = 0$. Next, as the second step, we need to take the higher order derivatives. So,

$$f''(x) = f'(\tfrac{1}{4}x^4) = (4) \times \frac{1}{4}x^{4-1} = \frac{4}{4}x^3 = x^3 \qquad f''(0) = 0$$

$$f^{(3)}(x) = f'(x^3) = 3x^2 \qquad f^{(3)}(0) = 0$$

$$f^{(4)}(x) = f'(3x^2) = 6x \qquad f^{(4)}(0) = 0$$

$$f^{(5)}(x) = f'(6x) = 6 \qquad f^{(5)}(0) = 6 > 0.$$

As shown above, at the point the function $f(x) = \frac{1}{20}x^5 + 18$ has all its derivatives equal to 0, except for the 5th derivative, which is positive. Thus, for the fourth derivative, $n = 4$, is even and by the test there is a strictly increasing point of inflection at .

Example

Given $f(x) = \frac{1}{30}x^6 + 12$, find the extreme points and determine which one is a minimum, a maximum, or a point of inflection.

Solution

First, we need to determine where $f'(x) = 0$. So,

$$f(x) = \frac{1}{30}x^6 + 12$$

$$f'(x) = (6) \times \frac{1}{30}x^{6-1}$$

$$f'(x) = \frac{6}{30}x^5$$

$$f'(x) = \frac{1}{5}x^5$$

$$f'(x) = 0 \Longrightarrow x = 0.$$

Therefore, there is only one extreme point at $x = 0$. Next, we need to take the higher order derivatives. So,

$$f''(x) = f'(\frac{1}{5}x^5) = (5) \times \frac{1}{5}x^{5-1} = \frac{5}{5}x^4 = x^4 \qquad f''(0) = 0$$

$$f^{(3)}(x) = f'(x^4) = 4x^3 \qquad f^{(3)}(0) = 0$$

$$f^{(4)}(x) = f'(4x^3) = 12x^2 \qquad f^{(4)}(0) = 0$$

$$f^{(5)}(x) = f'(12x^2) = 24x \qquad f^{(5)}(0) = 0$$

$$f^{(6)}(x) = f'(24x) = 24 \qquad f^{(6)}(0) = 24 > 0.$$

As it is seen, at the point the function $f(x) = \frac{1}{30}x^6 + 12$ has all its derivatives at equal to 0, except for the 6th derivative, which is positive. Thus $n = 5$, is odd and by the test there is a local minimum at .

Example

Given the function $f(x) = -x^3 + 2x^2 + 7x + 13$. Determine all critical points of f, using the first and the second derivatives and decide whether a relative maximum, relative minimum, or neither occurs at each.

Solution

We need to determine where $f'(x) = 0$. So,

$$f(x) = -x^3 + 2x^2 + 7x + 13$$

$$f'(x) = -3x^2 + 4x + 7 = 0$$

$$x = \frac{-b \pm \sqrt{b^2 - 4ac}}{2a}$$

$$x = \frac{-4 \pm \sqrt{4^2 - 4 \times (-3) \times 7}}{2 \times (-3)} \qquad \text{we plugged } b = 4,\ a = -3, \text{ and } c = 7$$

$$x = \frac{-4 \pm \sqrt{16 - (-84)}}{2 \times (-3)}$$

$$x = \frac{-4 \pm \sqrt{16 + 84}}{-6}$$

$$x = \frac{-4 \pm \sqrt{100}}{-6}$$

$$x = \frac{-4 \pm 10}{-6}$$

$$x = \frac{-14}{-6} = \frac{7}{3} \qquad \text{or} \qquad x = \frac{6}{-6} = -1.$$

Now, we need to take the second derivative and evaluated at $x = \frac{7}{3}$ and $x = -1$. So,

$$f'(x) = -3x^2 + 4x + 7$$

$$f''(x) = f'(-3x^2 + 4x + 7)$$

$$f''(x) = -6x + 4.$$

For $x = \frac{7}{3}$ we will have:

$$f''(x)|_{x=7/3} = (-6)\times\frac{7}{3} + 4 \qquad \text{we plugged } x = \frac{7}{3} \text{ into the equation}$$

$$f''(\tfrac{7}{3}) = -\frac{42}{3} + 4$$

$$f''(\tfrac{7}{3}) = -14 + 4$$

$$f''(\tfrac{7}{3}) = -10 < 0.$$

The second derivative of the $f(x) = -x^3 + 2x^2 + 7x + 13$ evaluated at $x = \frac{7}{3}$ is negative. Hence there is a relative maximum for the function f at $x = \frac{7}{3}$.

For $x = -1$ we will have:

$$f''(x)|_{x=-1} = (-6)\times(-1) + 4 \qquad \text{we plugged } x = -1 \text{ into the equation}$$

$$f''(\tfrac{7}{3}) = 6 + 4 = 10 > 0.$$

The second derivative of the $f(x) = -x^3 + 2x^2 + 7x + 13$ evaluated at $x = -1$ is positive. Hence there is a relative minimum for the function f at $x = -1$.

Example

Given the function $f(x) = \frac{3}{5}x^5 - 3x^3 + 1$. Determine all critical points of f and decide whether a relative maximum, relative minimum, or neither. Then_sketch the graph of f.

Solution

We can use two approaches for this problem. Following we present each one:

Approach I

In this approach we construct both first and second derivative sign charts for f, to determine where f is increasing and decreasing and concave up and concave down, and then to identify all relative extreme values of f. Here is the first derivative of f.

$$f(x) = \frac{3}{5}x^5 - 3x^3 + 1$$

$$f'(x) = 5\times\frac{3}{5}x^4 - 3\times 3x^2$$

$$f'(x) = 3x^4 - 9x^2.$$

we can find the critical points of f by solving $3x^4 - 9x^2 = 0$.

$$3x^4 - 9x^2 = 0$$

$3x^2(x^2 - 3) = 0$ we factored $3x^2$ out

$3x^2 = 0$ or $x^2 - 3 = 0$

$x = 0$ or $x^2 = 3$

$x = 0$ or $x = \pm\sqrt{3}$.

There are three roots for the function $f'(x)$, $x = -\sqrt{3}$, $x = 0$, and $x = \sqrt{3}$. To designate the intervals in which f is increasing or decreasing, we provide a number line for the increasing/decreasing information. We use four check points, $x = -2$, $x = -1$, $x = 1$, and $x = 2$ as shown below:

Therefore, we have got the following intervals of increasing and decreasing.

For the x values within the interval $-\infty < x < -\sqrt{3}$, f is increasing.

For the x values within the interval $-\sqrt{3} < x < \sqrt{3}$, f is decreasing.

For the x values within the interval $\sqrt{3}, < x < \infty$, f is increasing.

So far, we can conclude that at three critical points, f has a maximum at $x = -\sqrt{3}$, a minimum at $x = \sqrt{3}$, and neither it has a maximum nor it has a minimum at $x = 0$.

Now, let us find the interval for concavity of the function. Here we need to set the function $f''(x)$ equal to zero. So,

$f''(x) = f'(3x^4 - 9x^2) = 0$

$f''(x) = 12x^3 - 18x = 0$

$12x^3 - 18x = 0$

$6x(2x^2 - 3) = 0$

$x = 0$ or $2x^2 - 3 = 0$

$x = 0$ or $2x^2 = 3$

$x = 0$ or $x^2 = \frac{3}{2}$

$x = 0$ or $x = \pm\sqrt{\frac{3}{2}} = \pm\sqrt{1.5}$.

There are three roots for the function $f''(x)$, $x = -\sqrt{1.5}$, $x = 0$, and $x = \sqrt{1.5}$.

To designate the concavity intervals, we provide a number line for the concavity information. We use four arbitrary check points, $x = -2$, $x = -1$, $x = 1$, and $x = 2$ as shown below:

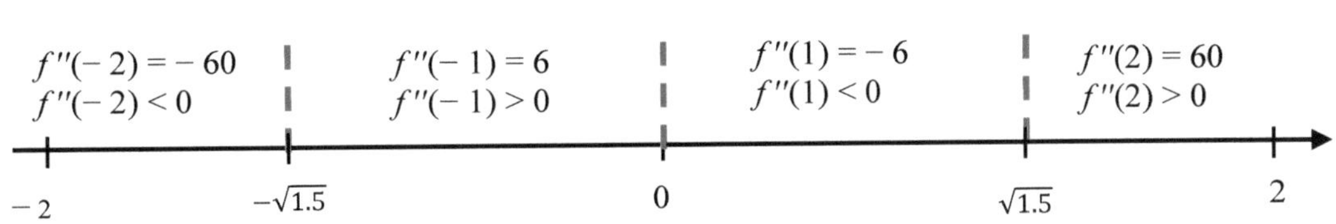

Therefore, we have got the following concavity intervals for the function.

For the x values within the interval $-\infty < x < -\sqrt{1.5}$, the function $f(x)$ is concave downward.

For the x values within the interval $-\sqrt{1.5} < x < 0$, the function $f(x)$ is concave upward.

For the x values within the interval $0 < x < \sqrt{1.5}$, the function $f(x)$ is concave downward.

For the x values within the interval $\sqrt{1.5} < x < \infty$, the function $f(x)$ is concave upward.

Putting all the above information together, we present a graph of f in the figure below, in which:

The point $P = (-\sqrt{3}, 7.2)$ is a local maximum, as f is increasing prior to P and decreasing after.

The point $T = (\sqrt{3}, -5.2)$ is a local minimum, as f is decreasing prior to T and decreasing after.

Also, note that f is concave down at P and concave up at Q, which is consistent both with our second derivative sign chart and the second derivative test. At points Q and S, concavity changes, as we saw in the results of the second derivative sign chart. Finally, at point R, the function f has a critical point with a horizontal tangent line, but neither a maximum nor a minimum occurs there since f is decreasing both before and after R. Hence, f has an inflection (saddle) point at R.

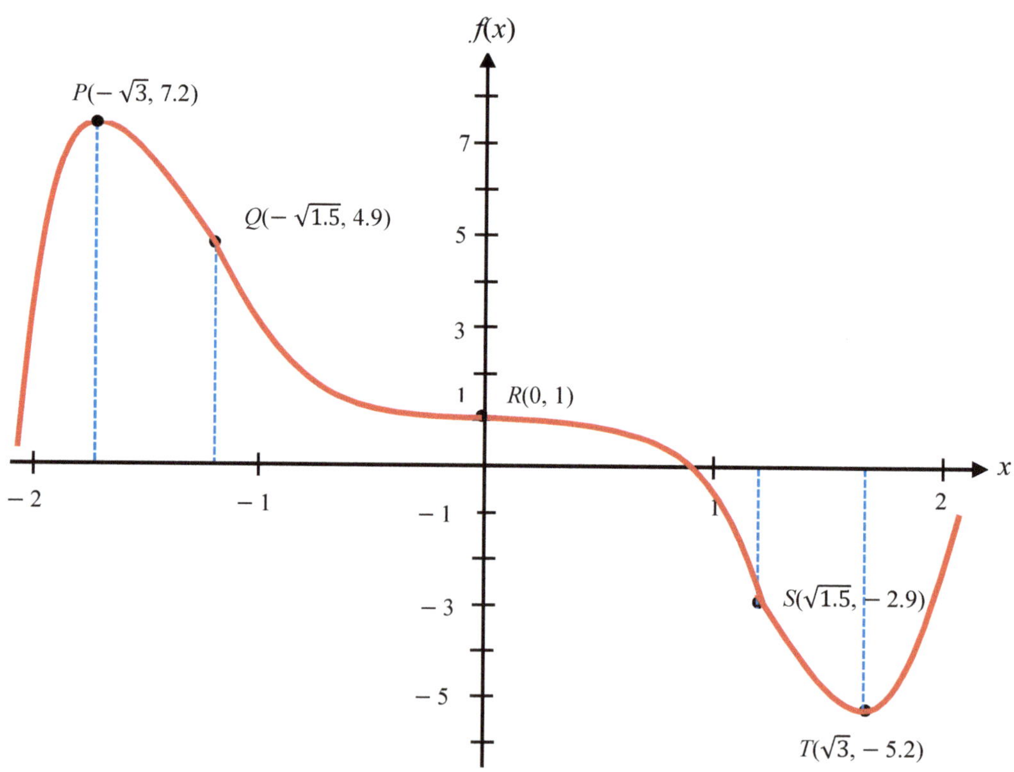

Approach II

In this approach we use the higher order derivatives for finding the critical points of the function $f(x) = \frac{3}{5}x^5 - 3x^3 + 1$. Hence, first we need to set $f'(x)$ equal to zero for obtaining the critical points of f. So,

$$f(x) = \frac{3}{5}x^5 - 3x^3 + 1$$

$$f'(x) = (5) \times \frac{3}{5}x^{5-1} - (3) \times 3x^{3-1} = 0$$

$$f'(x) = 3x^4 - 9x^2 = 0$$

$$3x^4 - 9x^2 = 0$$

$$3x^2(x^2 - 3) = 0 \qquad \text{we factored out } 3x^2$$

$$3x^2 = 0 \quad \text{or} \quad x^2 - 3 = 0$$

$$x = 0 \quad \text{or} \quad x^2 = 3$$

$$x = 0 \quad \text{or} \quad x = \pm\sqrt{3}.$$

Therefore, there are three critical points, $x = -\sqrt{3}$, $x = 0$, and $x = \sqrt{3}$ for f. To determine which is the local maximum, the local minimum, or the point of inflection, we need to take the higher derivatives and evaluate them at the critical points. So,

$$f''(x) = f'(3x^4 - 9x^2)$$

$$f''(x) = 12x^3 - 18x.$$

For $x = -\sqrt{3}$ we have:

$$f''(-\sqrt{3}) = 12 \times (-\sqrt{3})^3 - 18 \times (-\sqrt{3})$$

$$f''(-\sqrt{3}) = -62.35 + 31.18 = -31.17 < 0.$$

Therefore, f has a maximum at $x = -\sqrt{3}$.

For $x = 0$ we have:

$$f''(0) = 12 \times (0)^3 - 18 \times (0) = 0.$$

Since $f''(0) = 0$, f has a critical point with a horizontal tangent line at $x = 0$, but it is neither a maximum nor a minimum. We need to evaluate the higher derivative of f. So,

$$f^{(3)}(x) = f'(12x^3 - 18x)$$

$$f^{(3)}(x) = (3) \times 12x^2 - 18$$

$$f^{(3)}(x) = 36x^2 - 18$$

$$f^{(3)}(0) = 36 \times (0)^2 - 18 = -18 < 0.$$

As it is seen, at the point the function $f(x) = \frac{3}{5}x^5 - 3x^3 + 1$ has all its derivatives equal to 0, except for the 3rd derivative, which is negetive. Thus, for the second derivative, $n = 2$, is even and by the test there is a strictly decreasing point of inflection at .

For $x = \sqrt{3}$ we have:

$$f''(\sqrt{3}) = 12(\sqrt{3})^3 - 18(\sqrt{3})$$

$$f''(\sqrt{3}) = 62.35 - 31.18 = 31.17 > 0.$$

Therefore, f has a minimum at $x = \sqrt{3}$.

It is to be noted that the first approach provided more information regarding the behavior of the graph of the function $f(x) = \frac{3}{5}x^5 - 3x^3 + 1$. It had shown that the function had concavity changes at $x = -\sqrt{1.5}$ and $x = \sqrt{1.5}$. However, by the second approach the critical points of the function had been concluded more rapidly.

3.9. Exercises

1. The following graph shows the weather temperature in Fahrenheit for the past 6 days.
   a) Find the weather temperature changes from Wednesday to Thursday.
   b) Find the average weather temperature from Monday to Friday?

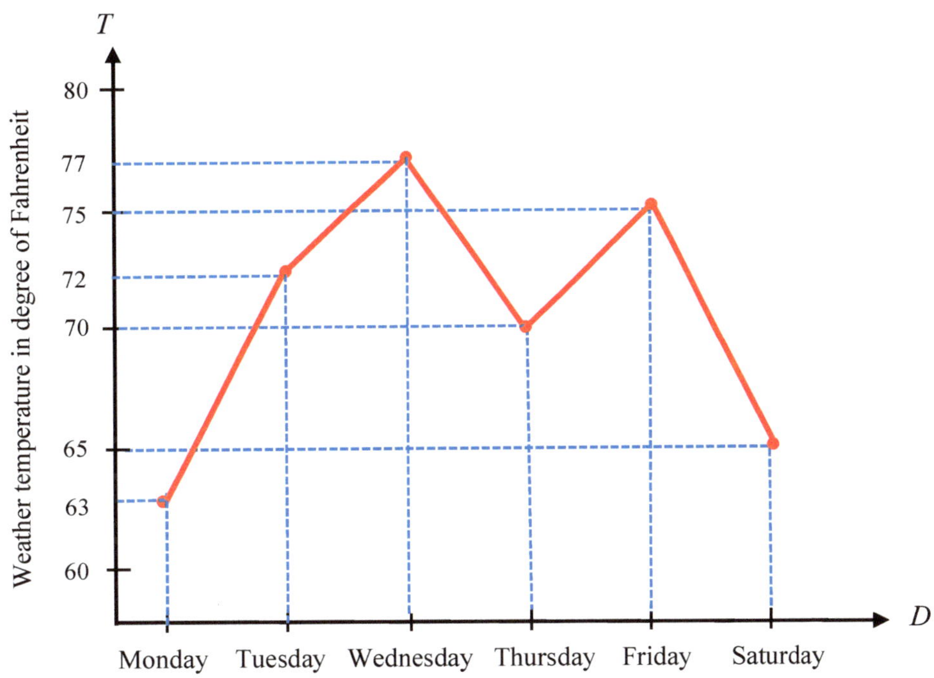

2. Find the average rate of change of the following functions.
   a) $f(x) = 3x^2 + 4$ at $x = 2$, and $h = 5$.
   b) $f(x) = \frac{5}{x} - 3$ at $x = 1$, and $h = 4$.

3. Sketch the graph of function $f(x) = \frac{3}{x}$. and the two tangent lines at points $x = 1$ and $x = -1$.

4. Find the derivatives of the given functions.
   a) $f(x) = 2x^2 - 5x + 12$.
   b) $g(x) = 3x^3 + x^2 - 6x + 7$.
   c) $h(x) = = \frac{1}{2x} - 5x$.
   d) $f(t) = 3at^4 + 4bt^2 - ct$.
   e) $f(h) = \frac{5}{h^2 + 3} - 3h^3 - 4h$.

5. Find the derivatives of the given functions.
   a) $f(h) = \frac{5h^2}{3h + 1} - 6h^2 - 1$.
   b) $f(x) = \frac{3}{x + 1} - 4\sqrt{x} - 17x$.
   c) $f(x) = 2\sqrt{x - 7} - \sqrt[3]{3x^2}$.

6. Given the two functions $f(x) = 5x^3$ and $g(x) = 3x^2 - 2x$, find the derivative of $h(x) = 3f(x) + 4g(x)$.

7. Given the two functions $f(x) = 11x$ and $g(x) = 7x$, find the derivative of $h(x) = f(x) \times g(x)$.

8. Given the two functions $f(x) = x^2$ and $g(x) = 2x - 1$, find the derivative of $h(x) = \frac{2f(x)}{g(x)}$.

9. Find the tangent line to the given curve at the specified point.
   a) $f(x) = 3x + \frac{1}{x}$ at (1, 2).
   b) $g(x) = \frac{1}{9}x^3$ at (3, 6).
   c) $h(x) = \frac{1}{100}x^4$ at (5, 4).

10. Find the derivative of the following functions.
    a) $f(x) = 2^x$.
    b) $f(x) = 5^{3x}$.

11. Find the derivative of the following functions.
    a) $f(x) = \log_3 \frac{1}{x} + 7x^4$.
    b) $g(x) = \log_2(3x - 1) - \log_4 \frac{2}{x}$.
    c) $h(t) = \log_4 \frac{3}{t-2} + \frac{3}{5}\log(5t - 1)$.

12. Find the derivative of given functions.
    a) $f(x) = 2x^3 lnx$ at $x = e$.
    b) $g(x) = ln(4x^2 - 8x)$.
    c) $h(x) = e^{5x^3 + 2x - 2}$.
    d) $f(x) = e^{11x^5 + 2x^3 - 2x^2}$.

13. Find the derivative of the given functions.
    a) $f(x) = 2^{5x}$.
    b) $f(x) = 5^{6x + 7} - 7x^4 + 3$.
    c) $f(x) = 3^{2x^2 - 9} - \frac{1}{x} + 9x^3$.

14. Find the first four derivatives of the following functions.
    a) $f(x) = 7x^6 + 2x^4 - 7x^2$.
    b) $f(x) = 9x^5 + 3x^4 - 4x^{\frac{3}{2}}$.
    c) $f(x) = x^7 + 8x^5 - \frac{1}{x-3}$.

15. Determine the regions in which the following function is concave upward or downward:
$f(x) = 6x^3 + 6x^2 - x + 2.$

16. Given the following function identify the intervals where the function is concave up and concave down and the intervals where the function is increasing and decreasing. Use this information to sketch the graph and designate extreme points.
$f(x) = \frac{2}{3}x^3 - x^2 - 4x.$

17. Find the local maximum and local minimum values associated with the given curves.
   a) $y = 3x^2 - 2x + 5.$
   b) $y = x^3 - 4x^2 + 16x.$
   c) $y = \frac{x^2 - 2x + 4}{2 - x}.$
   d) $y = \frac{1}{5}x^5 - 2x^4 + 4x^3.$

18. Find the global maximum and global minimum value of the given functions over the domain D, given by the interval notation $D = [a, b]$.
$f(x) = \frac{1}{3}x^3 - x^2 - 3x\ ,\ D = [-2, 4].$

$f(x) = \frac{x^2}{1 + x},\ D = [1, 2].$

$f(x) = \frac{x}{x^2 + 1},\ D = [0, 2].$

19. Determine the regions in which the following function is concave upward or downward:
$f(x) = 5x^3 + 3x^2 - x.$

20. Given the following function identify the intervals where the function is concave up and concave down and the intervals where the function is increasing and decreasing. Use this information to sketch the graph and designate extreme points.
$f(x)=2x^3 - 6x^3 + 6x.$

21. Given $f(x) = \frac{1}{42}x7 + 13$, find the extreme points and determine which one is a minimum, a maximum, or a point of inflection.

22. Given $f(x) = \frac{1}{3}x^3 - \frac{5}{2}x^2 + 4x$, find the extreme points and determine which one is a minimum, a maximum, or a point of inflection.

23. Given $f(x) = \frac{2}{3}x^3 - 4x^2 + 6x$, find the extreme points and determine which one is a minimum, a maximum, or a point of inflection.

24. Sketch the given curves and where appropriate describe the domain of the function, x and y–intercepts, relative maximum and minimum points, points of inflection and how the concavity changes.
   a) $y = x^3 - 5x^2 + 4.$
   b) $y = 2x^5 - 3x^3 + 4x.$

c) $y = 4x + \frac{2}{x}$.

d) $y = (x - 4)^2(x - 1)$

e) $y = \frac{x}{x - 3}$.

4. Function Integration

Integration is the process of bringing together smaller pieces into a single structure that functions as one. Integration is a process of finding integral. It is a representation of the differentiation inverse. Along with differentiation, integration is a fundamental operation of mathematics, and serves as a tool to solve problems in physics involving the volume of an arbitrary shape, the area under a curve, center of a mass, and the volume of a solid object, etc.

Differentiation and integration are the two major branches used in mathematics. Although it is difficult to understand the difference between differentiation and integration. Many students and even scholars are not able to understand its difference. The difference between differentiation and integration is that differentiation is used to find out the instant rates of change and the slopes of curves, whereas if you need to calculate the area under curves then make use of integration. As it can be seen, both differentiation and integration are opposite to each other in mathematical significance.

The table below illustrates the comparison between differentiation and integration.

Comparison factor	Differentiation	Integration
Purpose and functions	Differentiation is used to find out the instant rates of change from one point to another.	Integration is used to calculate the area under or between the curves.
To solve the real-world application	• To calculate the profit and loss in business using graphs. • To check the temperature variation. • To determine the speed or distance per hour. • To calculate instant velocity. • To find whether a curve is increasing or decreasing. • To find whether a curve is concave up or down. • ...	• To find an area under a curve or area between two curves. • To find volume of solid or revolution. • To calculate the average value of a function. • To determine force by liquid pressure. • To calculate probabilities, expected values, etc. • To find the work done on an object by variable force. • ...
Directly opposite	Differentiation is the reversed process of integration.	Integration is the reversed process of differentiation.
Method of calculations	The calculation results of differentiation are obtained through division	The calculation results of integration are obtained by addition.

Integration is a method to find **definite and indefinite integrals**. The integrals enumerated here are those termed definite integrals, which can be interpreted formally as the signed area of the region in the plane that is bounded by the graph of a given function between two points in the real line. Integrals may also refer to the concept of an antiderivative, a function whose derivative is the given function. In this case, they are called indefinite integrals.

Suppose we do the reverse of differentiation, given a function, we find another function whose derivative is the given function. This is called antidifferentiation, and it is a part of the larger process of integration. Mathematically, given a function $f(x)$ we determine another function such that $\frac{d}{dx}F(x) = f(x)$.

For example, let $f(x) = 2x$. The function $F(x) = x^2$ is an antiderivative of $f(x)$ since $\frac{d}{dx}x^2 = 2x$. However, other functions also have a derivative of $2x$. For example, $y = x^2 + 1$, $y = x^2 + 2$, $y = x^2 + 100$, and $y = x^2 - 100$ also differentiate to $2x$; the x^2 term differentiates to $2x$, and the constant term differentiates to zero. Therefore, an antiderivative of $f(x) = 2x$ is any function that can be written in the form $F(x) = x^2 + C$, where C is a constant.

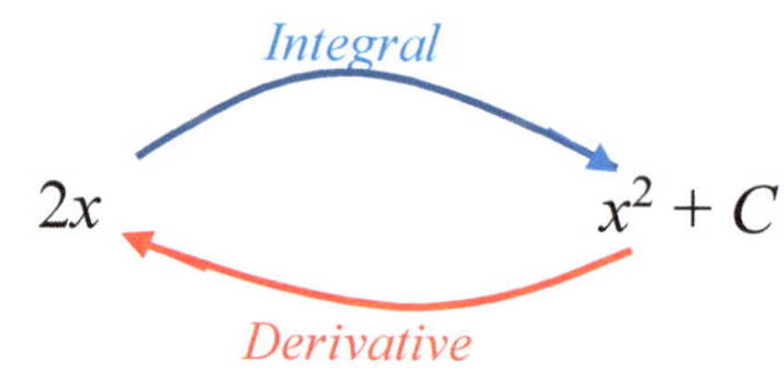

an integral of $2x$ is x^2

Also given $h(x) = 3x^2$ its antiderivative is any function that can be written in the form $H(x) = x^3 + C$, where C is a constant.

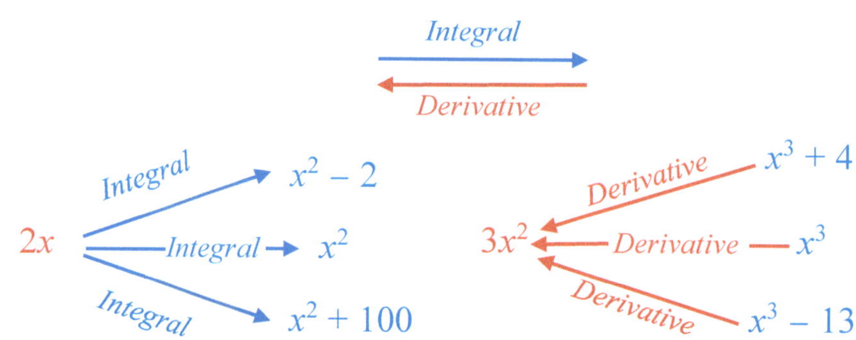

Integral and Derivative are opposite

This leads us to the following theorem.

Theorem
The antiderivative of $f(x)$ is the set of functions $F(x) + C$ such that:

$$\frac{d}{dx}F(x) + C = f(x).$$

The constant C is called constant integration.

the above theorem can be restated as; if two functions $F(x)$ and $H(x)$ have the same derivative $f(x)$ then $F(x)$ and $H(x)$ differ by at most a constant. That is. $F(x) = H(x) + C$.

If $F(x)$ is an antiderivative of a function $f(x)$ we write

$$\int f(x)dx = F(x) + C.$$

This equation is read as "the antiderivative of $f(x)$ with respect to x, is the set of functions $F(x) + C$." The expression on the left side is called an indefinite integral. The symbol $\int$ is the integral sign and is a command for antidifferentiation.

The function $f(x)$ is called the integrand, and the symbol dx, called the differential of the variable x, indicates that the variable of integration is x.

In general, the integral of a function $f(x)$ with respect to a real variable x on an interval $[a, b]$ is written as:

$$\int_a^b f(x)dx.$$

Where a and b are called the limits of integration, and the integral is said to be over the interval $[a, b]$, called the interval of integration. A function is said to be "integrable" if its integral over its domain is finite, and when limits are specified, the integral is called a "definite integral".

Example

Find the following indefinite integrals.

a) $\int 7dx.$

b) $\int 5x^4\,dx.$

c) $\int \frac{7}{x}\,dx.$

d) $\int e^x\,dx.$

e) $\int \frac{1}{x\,lna}dx.$

Solution

We have seen the integrand as derivatives of other functions presented in the previous section (Derivatives of Functions) of this text. It is recommended that you always check your answer by taking its derivative. So,

$\int 7dx = 7x + C$ Check: $\frac{d}{dx}(7x + C) = 7$

$\int 5x^4\,dx = x^5 + C$ Check: $\frac{d}{dx}(x^5 + C) = 5x^4$

$\int \frac{1}{x}\,dx = ln\,x + C$ Check: $\frac{d}{dx}(lnx + C) = \frac{1}{x}$

$\int e^x\,dx = e^x + C$ Check: $\frac{d}{dx}(e^x + C) = e^x$

$\int \frac{1}{x\,lna}dx = \log_a x + C$ Check: $\frac{d}{dx}(\log_a x + C) = \frac{1}{x\,lna}.$

4.1. Integration Rules and Common Formulae

The integral of many functions is well known. However, there are useful rules to work out the integral of more complicated functions. Below we present these rules.

4.1.1. Integration, Multiplication by Constant Rule

$$\int c\, f(x)dx = c \int f(x)dx,\ c \text{ is a constant.}$$

Example
Evaluate the integral below.

$$\int 18 f(x)dx.$$

Solution
We have learned that $\int c\, f(x)dx = c \int f(x)dx$, c is a constant. So,

$$\int 18 f(x)\, dx = 18 \int f(x)\, dx.$$

4.1.2. Integration, Sum Rule

$$\int [f(x) + h(x)]dx = \int f(x)dx + \int h(x)dx.$$

Example
Given the two functions $f(x) = x$ and $g(x) = e^x$. Evaluate the integral of $h(x) = f(x) + g(x) = x + e^x$.

Solution
We have learned that $\int [f(x) + g(x)]dx = \int f(x)dx + \int g(x)dx$. So

$$\int h(x)\, dx = \int [f(x) + g(x)]dx = \int f(x)dx + \int g(x)dx$$

$$\int h(x)\, dx = \int (x + e^x)dx = \int x dx + \int e^x dx.$$

$$\int h(x)\, dx = \frac{1}{2}x^2 + C_1 + e^x + C_2 = \frac{1}{2}x^2 + e^x + C.$$

Check:

$$\frac{d}{dx}\left(\frac{1}{2}x^2 + e^x + C\right) = \frac{d}{dx}\left(\frac{1}{2}x^2\right) + \frac{d}{dx}(e^x) + \frac{d}{dx}C$$

$$= x + e^x + 0 = x + e^x.$$

4.1.3. Integration, Difference Rule

$$\int [f(x) - h(x)]dx = \int f(x)dx - \int h(x)dx.$$

Example
Given the two functions $f(x) = 5x^4$ and $g(x) = 3e^x$. Evaluate the integral of $h(x) = f(x) - g(x) = 5x^4 - 3e^x$.

Solution
We have learned that $\int [f(x) - g(x)]dx = \int f(x)dx - \int g(x)dx$. So,

$$\int h(x)\, dx = \int [f(x) - g(x)]dx = \int f(x)dx - \int g(x)dx$$

$$\int h(x)\, dx = \int (5x^4 - 3e^x)dx = \int 5x^4\, dx - \int 3e^x dx$$

$$\int h(x)\, dx = x^5 + C_1 - (3e^x + C_2) = x^5 - 3e^x + C.$$

Check:

$$\frac{d}{dx}(x^5 - 3e^x + C) = \frac{d}{dx}(x^5) - \frac{d}{dx}(3e^x) + \frac{d}{dx}C$$

$= 5x^4 - 3e^x + 0 = 5x^4 - 3e^x.$

4.1.4. Integration by Part Rule

$$\int udv = uv - \int vdu, \text{ or } \int uv' = uv - \int vu'$$

$$\int [f(x)\, h'(x)]dx = f(x)\, h(x) - \int f'(x)h(x)dx.$$

Example

Evaluate the integral of $\int ln(x)dx$

Solution

We have learned that the rule of integration by part states: $\int uv' = uv - \int vu'$. Let $u = ln(x)$ and $v' = 1$. Having $v' = 1$, we get $v = x$ and having $u = ln(x)$ we get $u' = \frac{1}{x}$. So,

$$\int ln(x)dx = uv - \int vu' = ln(x) \times x - \int x \times \frac{1}{x} dx$$

$$\int ln(x)dx = x\, ln(x) - \int dx$$

$$\int ln(x)dx = x\, ln(x) - x + C.$$

Check:

$$\frac{d}{dx}(x\, ln(x) - x + C) = \frac{d}{dx}[x\, ln(x)] - \frac{d}{dx}x + \frac{d}{dx}C$$

$$= \frac{d}{dx}x \times ln(x) + \frac{d}{dx}ln(x) \times x - \frac{d}{dx}x + \frac{d}{dx}C$$

$$= (1) \times ln(x) + (\frac{1}{x}) \times x - 1 + 0$$

$$= ln(x) + \frac{x}{x} - 1$$

$$= ln(x) + 1 - 1 = ln(x).$$

4.1.5. Integration by Substitution

$$\int [f(u)\frac{du}{dx}]dx = \int f(u)d, \text{ or } \int u\frac{du}{dx} = \int u\, du.$$

Example

Evaluate the integral of $\int ln(2x + 3)dx$.

Solution

We have learned by the rule of substitution $\int u\frac{du}{dx} = \int u\, du$. Let $u = 2x + 3$, and $\frac{du}{dx} = 2$ which leads to $dx = \frac{1}{2}du$. Then we will have:

$$\int ln(2x + 3)dx = \int ln(2x + 3)dx \int (2x + 3) \times 2\frac{du}{dx} = \int (2x + 3)\, du = \int (2x + 3) \times 2dx$$

$\int ln(2x + 3)dx = \int ln(2x + 3) \times \frac{1}{2} du$ — we replace dx with $\frac{1}{2}du$

$ln(2x + 3)dx = \frac{1}{2}\int ln\, u\, du$ — we replace $2x + 3$ with u

$ln(2x + 3)dx = \frac{1}{2}[u \, ln \, u - u] + C$ we had $\int ln \, u \, du = u \, ln \, u - u$

$ln(2x + 3)dx = \frac{1}{2}[(2x + 3) \, ln(2x + 3) - (2x + 3)] + C$ we replaced u with $2x + 3$

$$ln(2x + 3)dx = \frac{1}{2} \times 2x \times ln(2x + 3) + \frac{1}{2} \times 3 \, ln(2x + 3) - \frac{1}{2} 2x + \frac{1}{2} \times 3 + C$$

$$ln(2x + 3)dx = \frac{2}{2} x \, ln(2x + 3) + \frac{3}{2} ln(2x + 3) - \frac{2}{2} x + \frac{3}{2} + C$$

$$ln(2x + 3)dx = x \, ln(2x + 3) + \frac{3}{2} ln(2x + 3) - x + \frac{3}{2} + C.$$

Check:

$$\frac{d}{dx}[x \, ln(2x + 3) + \frac{3}{2} ln(2x + 3) - x + \frac{3}{2} + C]$$

$$= \frac{d}{dx} x \times ln(2x + 3) + x \times \frac{d}{dx} ln(2x + 3) + \frac{3}{2}[\frac{d}{dx} ln(2x + 3)] - \frac{d}{dx} x + \frac{d}{dx}\frac{3}{2} + \frac{d}{dx} C$$

$$= (1) \times ln(2x + 3) + x \times 2 \times \frac{1}{2x + 3} + \frac{3}{2} \times 2 \times \frac{1}{2x + 3} - 1 + 0 + 0$$

$$= ln(2x + 3) + \frac{2x}{2x + 3} + \frac{3}{2x + 3} - 1$$

$$= ln(2x + 3) + \frac{2x + 3}{2x + 3} - 1$$

$$= ln(2x + 3) + 1 - 1$$

$$= ln(2x + 3).$$

4.1.6. Integration, Constant Function Rule

$$\int a \, \mathrm{dx} = ax + C.$$

Example

Evaluate the integral below.

$$\int 12 \, \mathrm{dx}.$$

Solution

We have learned $\int a \, \mathrm{dx} = ax + C$. So,

$$\int 12 \, \mathrm{dx} = 12x + C.$$

Check:

$$\frac{d}{dx}(12x + C) = 12.$$

4.1.7. Integration, Variable to the Power Function Rule

$$\int x^n \, dx = \frac{1}{n+1} x^{n+1} + C, \; n \neq -1.$$

As two special cases, we have "variable function rule, $\int xdx$" and "square function rule, $\int x^2 \, dx$" which respectively we will have:

$$\int xdx = \frac{1}{2} x^2 + C$$

$\int x^2\, dx = \frac{1}{3}x^3 + C.$

Example

Evaluate the integral below.

$\int x^8 dx.$

Solution

We have learned $\int x^n dx = \frac{1}{n+1}x^{n+1} + C$. So,

$\int x^8 dx = \frac{1}{8+1}x^{8+1} + C = \frac{1}{9}x^9 + C.$

Check:

$\frac{d}{dx}(\frac{1}{9}x^9 + C) = x^8.$

4.1.8. Integration, Reciprocal Function Rule

$$\int (\frac{a}{x})\, dx = a\, ln\, |x| + C.$$

Example

Evaluate the integral below.

$\int (\frac{11}{x})\, dx.$

Solution

We have learned $\int (\frac{a}{x})\, dx = a\, ln\, |x| + C$. So,

$\int (\frac{11}{x})\, dx = 11 ln\, |x| + C.$

Check:

$\frac{d}{dx}(11 ln\, |x| + C) = \frac{11}{x}.$

4.1.9. Integration, Exponential Function Rule

$$\int e^{ax} dx = \frac{1}{a}e^x + C.$$

Example

Evaluate the integral below.

$\int e^{7x} dx.$

Solution

We have learned $\int e^{ax} dx = \frac{1}{a}e^x + C$. So,

$\int e^{7x} dx = \frac{1}{7}e^{7x} + C.$

Check:

$\frac{d}{dx}(\frac{1}{7}e^{7x} + C) = e^{7x}.$

4.1.10. Integration, Power Function Rule

$$\int a^{bx}\, dx = \frac{a^{bx}}{b\, ln\, a} + C.$$

Example
Evaluate the integral below.

$$\int 15^{8x} dx.$$

Solution
We have learned $\int a^{bx}\, dx = \frac{a^{bx}}{b\, ln\, a} + C$. So,

$$\int 15^{8x} dx = \frac{1}{8\, ln\, 15} a^{8x} + C.$$

Check:

$$\frac{d}{dx}(\frac{1}{8\, ln\, 15} a^{8x} + C) = 15^{8x}.$$

4.1.11. Integration, Natural Logarithmic Function Rule

$$\int ln(x)\, dx = xln\,(x) - x + C.$$

Example
Evaluate the integral of $\int ln(14x - 1)dx$.

Solution
We have learned by the rule of substitution $\int u\frac{du}{dx} = \int u\, du$. Let $u = 14x - 1$, and $\frac{du}{dx} = 14$ which leads to $dx = \frac{1}{14} du$. Then we will have:

$$\int ln(14x - 1)dx = \int ln(14x - 1)dx \int (14x - 1) \times 14 \frac{du}{dx} = \int (14x - 1)\, du = \int (14x - 1) \times 14dx$$

$\int ln(14x - 1)dx = \int ln(14x - 1) \times \frac{1}{14} du$ we replace dx with $\frac{1}{14} du$

$ln(14x - 1)dx = \frac{1}{14} \int ln\, u\, du$ we replace $14x - 1$ with u

$ln(14x - 1)dx = \frac{1}{14} [u\, ln\, u - u] + C$ we had $\int ln\, u\, du = u\, ln\, u - u$

$ln(14x - 1)dx = \frac{1}{14} [(14x - 1)\, ln(14x - 1) - (14x - 1)] + C$ we replaced u with $14x - 1$

$$ln(14x - 1)dx = \frac{1}{14} \times 14x \times ln(14x - 1) + \frac{1}{14} \times (-1)\, ln(14x - 1) - \frac{1}{14} \times 14x + \frac{1}{14} \times (-1) + C$$

$$ln(14x - 1)dx = \frac{14}{14} x\, ln(14x - 1) - \frac{1}{14} ln(14x - 1) - \frac{14}{14} x - \frac{1}{14} + C$$

$$ln(14x - 1)dx = x\, ln(14x - 1) - \frac{1}{14} ln(14x - 1) - x - \frac{1}{14} + C.$$

Check:

$$\frac{d}{dx}[x\, ln(14x - 1) - \frac{1}{14} ln(14x - 1) - x - \frac{1}{14} + C]$$

$$= \frac{d}{dx} x \times ln(14x - 1) + x \times \frac{d}{dx} ln(14x - 1) - \frac{1}{14} [\frac{d}{dx} ln(14x - 1)] - \frac{d}{dx} x + \frac{d}{dx} (-\frac{1}{14}) + \frac{d}{dx} C$$

$= (1)\times ln(14x - 1) + x \times 14 \times \frac{1}{14x-1} - \frac{1}{14} \times 14 \times \frac{1}{14x-1} - 1 + 0 + 0$

$= ln(14x - 1) + \frac{14x}{14x-1} - \frac{1}{14x-1} - 1$

$= ln(14x - 1) + \frac{14x-1}{14x-1} - 1$

$= ln(14x - 1) + 1 - 1$

$= ln(14x - 1).$

So far, we have looked at some integration rules, each has followed by its associated example. Now, let us look at examples using more than one integration rule.

Example

Evaluate the integral below by substitution rule.

$\int(x^2 - 1)^9\, 2x\, dx.$

Solution

Let $u = x^2 - 1$, $\frac{du}{dx} = 2x$, then $\int(x^2 - 1)^9\, 2x\, dx$ is written as:

$\int(x^2 - 1)^9\, 2x\, dx$

$= \int u^9\, (\frac{du}{dx})dx = \int u^9 du.$

Here, we simply can use the power rule, $\int x^n\, dx = \frac{1}{n+1}x^{n+1} + C$. So,

$\int u^9 du = \frac{1}{9+1}u^{9+1} + C = \frac{1}{10}u^{10} + C.$

Now, since $u = x^2 - 1$ and $\frac{du}{dx} = 2x$ or $du = 2xdx$, we have:

$\int u^9 du = \int(x^2 - 1)^9\, 2xdx = \frac{1}{10}(x^2 - 1)^{10} + C.$

Check:

$\frac{d}{dx}[\frac{1}{10}(x^2 - 1)^{10} + C] = \frac{d}{dx}[\frac{1}{10}(x^2 - 1)^{10}] + \frac{d}{dx}C$

$= \frac{1}{10} \times \frac{d}{dx}(x^2 - 1)^{10} + 0$

$= \frac{1}{10} \times (10) \times (2x)\, (x^2 - 1)^{10-1}$

$= \frac{10}{10}\, (2x)\, (x^2 - 1)^9$

$= 2x(x^2 - 1)^9.$

Example

Evaluate the integral below.

$\int(\frac{21}{\sqrt[4]{x}} + \frac{18}{\sqrt[3]{x}})dx.$

Solution

Using the power rule for integrals, we will have:

$\int(\frac{21}{\sqrt[4]{x}} + \frac{18}{\sqrt[3]{x}})dx = \int \frac{21}{\sqrt[4]{x}}dx + \int \frac{18}{\sqrt[3]{x}}dx$

$$\int(\frac{21}{\sqrt[4]{x}}+\frac{18}{\sqrt[3]{x}})dx = 21\int\frac{1}{\sqrt[4]{x}}dx + 18\int\frac{1}{\sqrt[3]{x}}dx$$

$$\int(\frac{21}{\sqrt[4]{x}}+\frac{18}{\sqrt[3]{x}})dx = 21\int x^{-\frac{1}{4}}dx + 18\int x^{-\frac{1}{3}}dx$$

$$\int(\frac{21}{\sqrt[4]{x}}+\frac{18}{\sqrt[3]{x}})dx = 21\int x^{-\frac{1}{4}}dx + 18\int x^{-\frac{1}{3}}dx$$

$$\int(\frac{21}{\sqrt[4]{x}}+\frac{18}{\sqrt[3]{x}})dx = 21\times\frac{x^{-\frac{1}{4}+1}}{-\frac{1}{4}+1} + 18\times\frac{x^{-\frac{1}{3}+1}}{-\frac{1}{3}+1} + C$$

$$\int(\frac{21}{\sqrt[4]{x}}+\frac{18}{\sqrt[3]{x}})dx = 21\times\frac{x^{-\frac{1}{4}+1\times\frac{4}{4}}}{-\frac{1}{4}+1\times\frac{4}{4}} + 18\times\frac{x^{-\frac{1}{3}+1\times\frac{3}{3}}}{-\frac{1}{3}+1\times\frac{3}{3}} + C$$

$$\int(\frac{21}{\sqrt[4]{x}}+\frac{18}{\sqrt[3]{x}})dx = 21\times\frac{x^{\frac{-1+4}{4}}}{\frac{-1+4}{4}} + 18\times\frac{x^{\frac{-1+3}{3}}}{\frac{-1+3}{3}} + C$$

$$\int(\frac{21}{\sqrt[4]{x}}+\frac{18}{\sqrt[3]{x}})dx = 21\times\frac{x^{\frac{3}{4}}}{\frac{3}{4}} + 18\times\frac{x^{\frac{2}{3}}}{\frac{2}{3}} + C$$

$$\int(\frac{21}{\sqrt[4]{x}}+\frac{18}{\sqrt[3]{x}})dx = 21\times\frac{4}{3}\times x^{\frac{3}{4}} + 18\times\frac{3}{2}\times x^{\frac{2}{3}} + C$$

$$\int(\frac{21}{\sqrt[4]{x}}+\frac{18}{\sqrt[3]{x}})dx = \frac{21\times 4}{3}\times x^{\frac{3}{4}} + \frac{18\times 3}{2}\times x^{\frac{2}{3}} + C$$

$$\int(\frac{21}{\sqrt[4]{x}}+\frac{18}{\sqrt[3]{x}})dx = \frac{3\times 7\times 4}{3}\times x^{\frac{3}{4}} + \frac{2\times 9\times 3}{2}\times x^{\frac{2}{3}} + C$$

$$\int(\frac{21}{\sqrt[4]{x}}+\frac{18}{\sqrt[3]{x}})dx = 28\times x^{\frac{3}{4}} + 27\times x^{\frac{2}{3}} + C$$

$$\int(\frac{21}{\sqrt[4]{x}}+\frac{18}{\sqrt[3]{x}})dx = 28\sqrt[4]{x^3} + 27\sqrt[3]{x^2} + C.$$

Check

$$\frac{d}{dx}(28\sqrt[4]{x^3} + 27\sqrt[3]{x^2} + C)$$

$$= 28\frac{d}{dx}\sqrt[4]{x^3} + 27\frac{d}{dx}\sqrt[3]{x^2} + \frac{d}{dx}C$$

$$= 28\frac{d}{dx}x^{\frac{3}{4}} + 27\frac{d}{dx}x^{\frac{2}{3}} + 0$$

$$= 28\times\frac{3}{4}x^{\frac{3-4}{4}} + 27\times\frac{2}{3}x^{\frac{2-3}{3}}$$

$$= \frac{4\times 7\times 3}{4}x^{\frac{-1}{4}} + 27\times\frac{3\times 9\times 2}{3}x^{\frac{-1}{3}}$$

$$= 21x^{-\frac{1}{4}} + 18x^{-\frac{1}{3}} = \frac{21}{\sqrt[4]{x}} + \frac{18}{\sqrt[3]{x}}.$$

4.2. Definite Integral

An integral with upper and lower limits is called a definite integral. The definite integrals we look at here are those which can be interpreted formally as the signed area of the region in the plane that is bounded by the graph of a given function between two points in the *x*-axis.

As a simple example, consider the velocity of a moving object is given by the function $v(t) = 2.5t$, where t is in hours and v is in kilometer per hour(km/h). If we want to find the distance the object has traveled during the first 2 hours, we need to find the area under the graph of v, where $0 \leq t \leq 2$.

The graph of v is shown below. We see the region corresponding to the time interval is a triangle with base $b = 2$ and height $h = 5$ (since $v(2) = 2.5 \times 2 = 5$). Therefore, the area of this region is $A = \frac{1}{2} \times 5 \times 2 = 5$. The object traveled $5km$. during the first 2 hours.

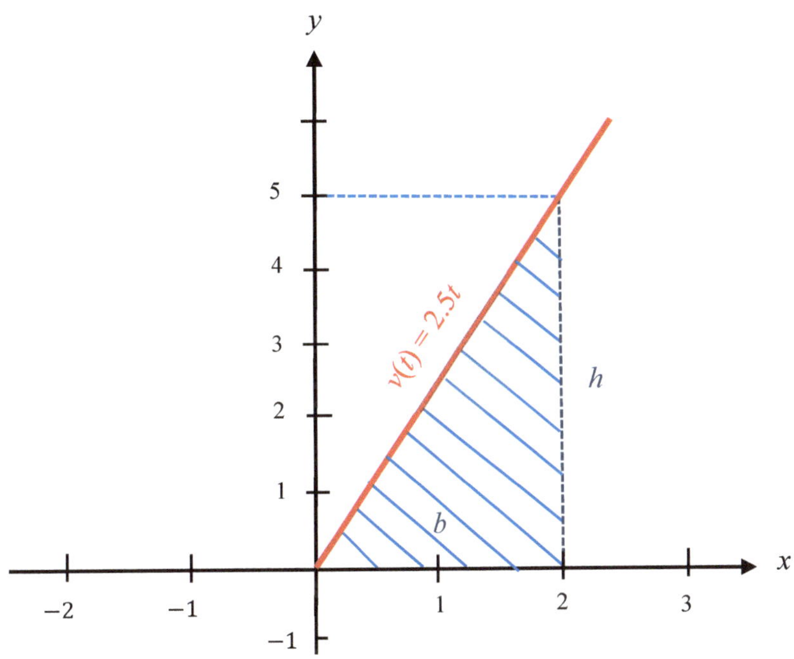

4.2.1. Evaluating Definite Integrals

The standard method of evaluating a definite integral is based on the following theorem.

Theorem
Given the continuous function f on the closed interval $[a, b]$, then:

$$\int_a^b f(x)dx = F(b) - F(a),$$

where F is antiderivative of the function f.

Example
Evaluate the definite integral $\int_2^4 5x^4 dx$.

Solution
Using the above theorem, we first note that x^5 is an antiderivative of $5x^4$. Hence we will have,

$$\int_2^4 5x^4 dx = x^5 \Big|_2^4 = (4^5) - (2^5)$$

$$\int_2^4 5x^4 dx = (1{,}024) - (32)$$

$$\int_2^4 5x^4 dx = 992.$$

Example

Evaluate the definite integral $\int_1^5 (3x^2 + 4x + 3)dx$.

Solution

First, we need to find the antiderivative of $3x^2 + 4x + 3$. So,

$$\int 3x^2 + 4x + 3 = \frac{1}{3} \times 3x^3 + \frac{1}{2} \times 4x^2 + 3x$$

$$\int 3x^2 + 4x + 3 = x^3 + 2x^2 + 3x.$$

Next, we use the above theorem. So,

$$\int_1^5 (3x^2 + 4x + 3)dx = (x^3 + 2x^2 + 3x)|_1^5$$

$$\int_1^5 (3x^2 + 4x + 3)dx = [(5^3) + 2(5^2) + 3(5)] - [(1^3) + 2(1^2) + 3(1)]$$

$$\int_1^5 (3x^2 + 4x + 3)dx = [(125) + 2(25) + 15] - [(1) + 2 + 3]$$

$$\int_1^5 (3x^2 + 4x + 3)dx = (125 + 50 + 15] - 6$$

$$\int_1^5 (3x^2 + 4x + 3)dx = 190 - 6 = 184.$$

4.2.2. Properties of Definite Integral

Replacing Variable: The variable x that appears in definite integrals is called the dummy variable and we can replace this with another to get the same result:

$$\int_a^b f(x)dx = \int_a^b f(t)dt.$$

Sum/Difference of Two Functions: If f and h are continuous functions on the interval $[a, b]$, then we can split up definite integrals with a sum or difference:

$$\int_a^b [f(x)dx \pm h(x)]dx = \int_a^b f(x)dx \pm \int_a^b h(x)dx.$$

Constant Multiplier: If f is a continuous function on $[a, b]$, and k is a real constant, then we have:

$$\int_a^b kf(x)dx = k\int_a^b f(x)dx.$$

Reverse interval: We can interchange the limits on any definite integral, then we have:

$$\int_a^b f(x)dx = -\int_b^a f(x)dx.$$

Zero - Length Interval: If the limits of integration are the same, the integral is just a line and contains no area. The zero length interval stats as:

$$\int_a^a f(x)dx = 0.$$

Adding Intervals: We can split up the integral with the limits $[a, c]$ for some value $b \in \mathbb{R}$ over two adjacent intervals $[a, b]$ and $[b, c]$, then we have:

$$\int_a^c f(x)dx = \int_a^b f(x)dx + \int_b^c f(x)dx.$$

Integral of a Constant: The definite integral of a constant $c \in \mathbb{R}$ is proportional to the width of the interval, then we have:

$$\int_a^b c dx = c(b - a).$$

Even/Odd Function: We have a property for even and odd functions when integrating over the interval $[-a, a]$, then we have:

For even functions $f(-x) = f(x)$, we have:

$$\int_{-a}^a f(x)dx = 2\int_0^a f(x)dx.$$

For odd functions $f(-x) = -f(x)$, we have:

$$\int_{-a}^a f(x)dx = 0.$$

Functions ≥ 0: If $f(x) \geq 0$ in the interval $[a, b]$, then we have:

$$\int_a^b f(x)dx \geq 0, \text{ if } f(x) \geq 0 \text{ in the interval } [a, b].$$

Comparing two functions: If $0 \leq f(x) \leq h(x)$ for all $x \in [a, b]$, then we have:

$$0 \leq \int_a^b f(x)dx \leq \int_a^b h(x)dx.$$

Definite Integral Properties

We assume below that $f(x)$ and $h(x)$ are continuous functions on the closed interval $[a, b]$.

1. $\int_a^b c dx = c(b - a)$.
2. $\int_a^b f(x)dx = -\int_b^a f(x)dx$.
3. $\int_a^a f(x)dx = 0$.
4. $\int_a^b kf(x)dx = k\int_a^b f(x)dx$, where k is any constant number.
5. $\int_a^b [f(x) \pm h(x)]dx = \int_a^b f(x)dx \pm \int_a^b h(x)dx$.
6. $\int_a^c f(x)dx = \int_a^b f(x)dx + \int_b^c f(x)dx$, where $a < b < c$.
7. $\int_a^b f(x)dx \geq 0$, if $f(x) \geq 0$ in the interval $[a, b]$.
8. $0 \leq \int_a^b f(x)dx \leq \int_a^b h(x)dx$, if $0 \leq f(x) \leq h(x)$ for all $x \in [a, b]$.

4.3. Average Value of a Function

The **average or mean** of a function $f(x)$ over interval (a, b) is defined by:

$$\bar{f} = \frac{1}{b-a}\int_a^b f(x)dx.$$

Example

The temperature t in centigrade recorded during working hours of a day followed the function below.

$$f_t(h) = 0.0015h^4 - 0.30h^2 + 26,$$

where h is the number of hours from noon ($8A.M. \leq h \leq 4P.M.$). What was the average temperature during the working day?

Solution

We learned that the average of a function $f(x)$ over interval (a, b) is defined by, $\bar{\boldsymbol{f}} = \frac{\mathbf{1}}{\boldsymbol{b-a}}\int_{\boldsymbol{a}}^{\boldsymbol{b}} \boldsymbol{f(x)dx}$. In this case $f(x)$ is given as $f_t(h) = 0.0015h^4 - 0.3h^2 + 26$, $a = 8$, and b is $4P.M.$ which need to be represented as $12 + 4 = 16$. So,

$$\overline{f_t} = \frac{1}{16-8}\int_4^{16} f_t(h)dh = \frac{1}{16-8}\int_4^{16}(0.0015h^4 - 0.3h^2 + 26)\,dh$$

$$\overline{f_t} = \frac{1}{16-8}\left(\frac{1}{5} \times 0.0015h^5 - \frac{1}{3} \times 0.3h^3 + 26h\right)\Big|_4^{16}$$

$$\overline{f_t} = \frac{1}{16-8}(0.00032h^5 - 0.1h^3 + 26h)\Big|_4^{16}$$

$$\overline{f_t} = \frac{1}{16-8}\left[\left(0.0003(16)^5 - 0.1(16)^3 + 26(16)\right) - \left(0.0003(4)^5 - 0.1(4)^3 + 26(4)\right)\right]$$

$$\overline{f_t} = \frac{1}{16-8}[(314.6 - 409.6 + 416) - (0.3 - 6.4 + 104)]$$

$$\overline{f_t} = \frac{1}{8}(321 - 97.9)$$

$$\overline{f_t} = \frac{223.1}{8} = 27.9.$$

Therefore, the average temperature during the working day is 27.9°C.

Let us look at the additive property of definite integrals as well as the area under a graph. We have seen that the definite integral $\int_a^c f(x)dx$ can be regarded as the area under the graph of f over the closed interval $[a, c]$.

Thus, if we let b to be point within the closed interval of $[a, c]$, that is $a < b < c$ the above integral can be expressed as a sum of two integrals.

The additive property of definite integrals is stated as:

$$\int_a^c f(x)dx = \int_a^b f(x)dx + \int_b^c f(x)dx.$$

As a generalization, imagine a set of points that are in ascending numerical value: $a < b < c < \cdots < k < l$, the additive property of definite integrals is stated as:

$$\int_a^l f(x)dx = \int_a^b f(x)dx + \int_b^c f(x)dx + \cdots + \int_k^l f(x)dx.$$

Graphically, for any number b between a and c, the integral from a to c is the integral from a to b plus the integral from b to c and the area bounded by function $f(x)$ within the closed interval $[a, c]$ is the sum of areas bounded by function $f(x)$ within the closed intervals $[a, b]$ and $[b, c]$.

Here is a graph for additive property of definite integrals.

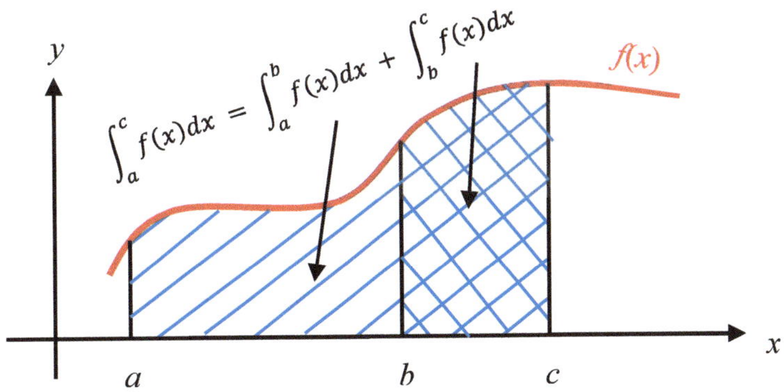

Example
Given the function $f(x)$ below, find the area under its graph from -5 to 5.

$$f(x) = \begin{cases} 4 - 2x & x < -2 \\ 8 & x \geq -2 \end{cases}.$$

Solution
Let us denote the requested area as A. We have learned that the definite integral $\int_a^b f(x)dx$ can be regarded as the area under the graph of over the interval $[a, b]$. So,

$$A = \int_{-5}^{5} f(x)dx$$

$$A = \int_{-5}^{-2} f(x)dx + \int_{-2}^{5} f(x)dx$$

$$A = \int_{-5}^{-2} (4 - 2x)\, dx + \int_{-2}^{5} 8dx \qquad \text{we replaced } f(x) \text{ with its expressions}$$

$$A = (4x - \tfrac{1}{2} \times 2x^2)|_{-5}^{-2} + 8x|_{-2}^{5}$$

$$A = (4x - x^2)|_{-5}^{-2} + 8x|_{-2}^{5}$$

$$A = 4(-2) - 4(-5) - [(-2)^2 - (-5)^2] + 8(5) - 8(-2)$$

$$A = 4(-2) - 4(-5) - (4 - 25) + 8(5) - 8(-2)$$

$$A = 4(-2) - 4(-5) - (-21) + 8(5) - 8(-2)$$

$$A = -8 + 20 + 21 + 40 + 16 = 89.$$

Let us sketch the graph of the given function. Below we present the graph of the function.

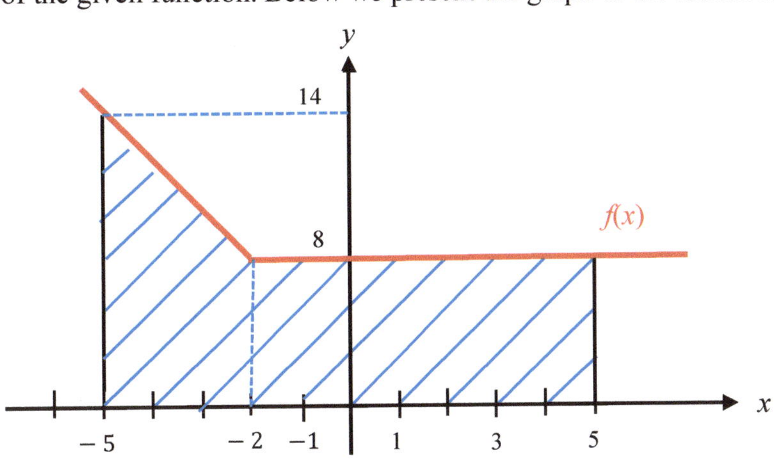

Therefore, the area under the graph of $f(x)$ from -5 to 5 is 89.

Example

Evaluate the definite integral of the function $f(x) = |x|$, from -2 to 3, $\int_{-2}^{3}|x|$.

Solution

First, we rewrite $f(x) = |x|$ as a piecewise function as below:

$$f(x) = |x| = \begin{cases} -x, & x < 0 \\ x, & x \geq 0 \end{cases}.$$

Next, we denote the requested area as A. We have learned that the definite integral $\int_a^b f(x)dx$ can be regarded as the area under the graph of over the interval $[a, b]$. So,

$$A = \int_{-2}^{3} f(x)dx$$

$$A = \int_{-2}^{0} f(x)dx + \int_{0}^{3} f(x)dx$$

$$A = \int_{-2}^{0} -xdx + \int_{0}^{3} xdx \qquad \text{we replaced } f(x) \text{ with its expressions}$$

$$A = (-\tfrac{1}{2} \times x^2)|_{-2}^{0} + (\tfrac{1}{2} \times x^2)|_{0}^{3}$$

$$A = (-\tfrac{x^2}{2})|_{-2}^{0} + \tfrac{x^2}{2}|_{0}^{3}$$

$$A = [-(\tfrac{0^2}{2}) - (-\tfrac{(-2)^2}{2})] + (\tfrac{3^2}{2} - \tfrac{0^2}{2})$$

$$A = [-(\tfrac{0}{2}) - (-\tfrac{4}{2})] + (\tfrac{9}{2} - \tfrac{0}{2})$$

$$A = 0 - (-2) + \tfrac{9}{2} - 0$$

$$A = 2 + 4.5 = 6.5.$$

Now, to visualize the problem, let us sketch the graph of the given function. Below we present the graph of the function.

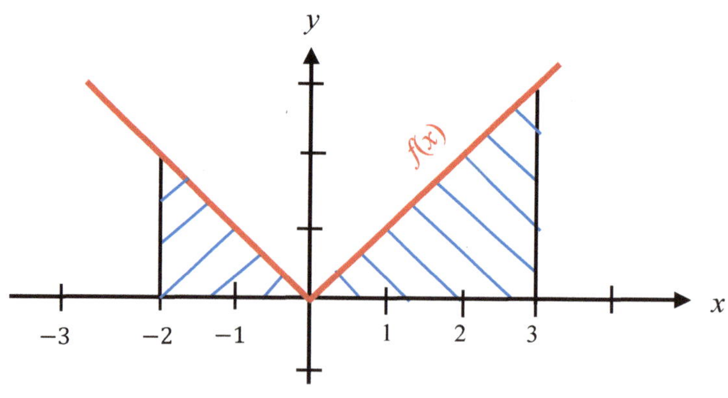

Therefore, the area under the graph of $f(x)$ from -2 to 3 is 6.5.

Let us look at the difference of two definite integrals as well as the area of a region bounded by two graphs. Suppose that we want to find the area of a region bounded by the graphs of two functions, and as shown below. Let us denote the area of the desired region A is the area of A_2 minus that of A_1.

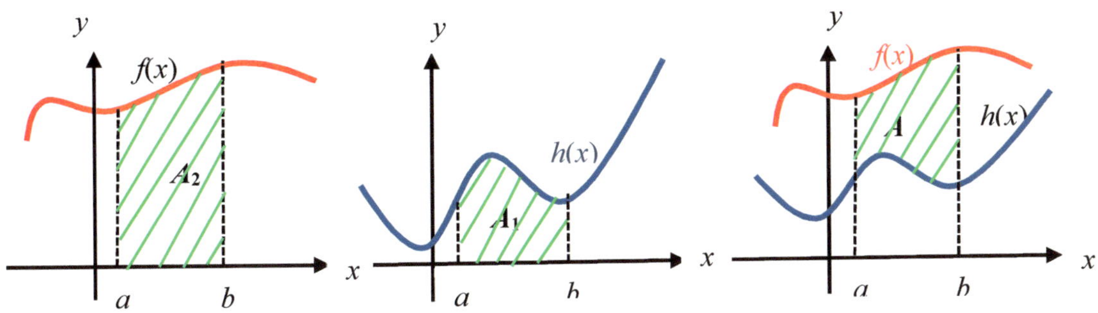

Thus,

$$A = A_2 - A_1$$

$$A = \int_a^b f(x)dx - \int_a^b h(x)\,dx$$

$$A = \int_a^b [f(x) - h(x)]dx.$$

Finding the Area Between Two Curves

Let $f(x)$ and $g(x)$ be two continues functions such that $f(x) \geq g(x)$ over an interval of $[a, b]$. Let R denote the region bounded above the graph of $f(x)$, below by the graph of $g(x)$, and the left and right by the lines $x = a$ and $x = b$, respectively. Then the area, A, of R is given by,

$$A = \int_a^b |f(x) - h(x)|dx.$$

Example

Suppose A is the region bounded above by the graph of the function $f(x) = 1 - \left(\frac{x}{2}\right)^2$ and below by the graph of the $h(x) = -\frac{1}{2}x - 1$. Find the area of region A.

Solution

Region A is depicted in the figure below.

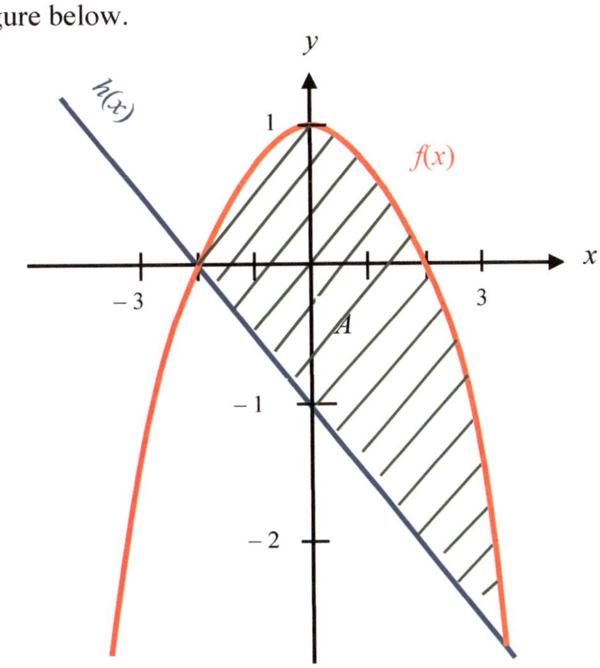

We have learned that the area of a region bounded by the graphs of two function in a given interval, we need to compute the difference of two definite integrals in that interval. In this case the interval is where the graph of the functions intersects. Hence, we first need to determine where the graph of the functions intersect. The intersect is obtained by setting $f(x) = h(x)$. So,

$$f(x) = h(x)$$

$$1 - \left(\frac{x}{2}\right)^2 = -\frac{1}{2}x - 1$$

$$1 - \frac{x^2}{4} = -\frac{1}{2}x - 1$$

$$(-4) \times (1 - \frac{x^2}{4}) = (-4) \times (-\frac{1}{2}x - 1)$$

to convert the x^2 coefficient to a positive integer, we multiplied both sides by 4

$$-4 + x^2 = 2x + 4$$

$$-4 + x^2 - (2x + 4) = 4x + 4 - (2x + 4)$$

to move all terms in one side, we added $-(2x + 4)$ to both sides

$$-4 + x^2 - 2x - 4 = 0$$

$$x^2 - 2x - 8 = 0$$

$$x = \frac{-b \pm \sqrt{b^2 - 4ac}}{2a}$$

$$x = \frac{-(-2) \pm \sqrt{(-2)^2 - 4(1)(-8)}}{2(1)}$$

$$x = \frac{2 \pm \sqrt{4 - (-32)}}{2}$$

$$x = \frac{2 \pm \sqrt{36}}{2}$$

$$x = \frac{2 \pm 6}{2}$$

$$x = \frac{-4}{2} = -2 \qquad \text{or} \qquad x = \frac{8}{2} = 4.$$

So, the graphs of the functions intersect at $x = -2$ or $x = 4$.

Now, we need to integrate the difference of the two functions from -2 to 4. Refereeing to the graphs f and h we realize that $f(x) \geq h(x)$ for $-2 \leq x \leq 4$. Hence, with $a = -2$ and $b = 4$, we will have:

$$A = \int_a^b [f(x) - h(x)]dx.$$

$$A = \int_{-2}^4 \left[1 - \left(\frac{x}{2}\right)^2 - \left(-\frac{1}{2}x - 1\right)\right] dx$$

$$A = \int_{-2}^4 (1 - \frac{x^2}{4} + \frac{1}{2}x + 1)dx$$

$$A = \int_{-2}^4 (2 - \frac{x^2}{4} + \frac{1}{2}x)dx$$

$$A = \left(2x - \frac{1}{3} \times \frac{1}{4}x^3 + \frac{1}{2} \times \frac{1}{2}x^2\right)\Big|_{-2}^{4}$$

$$A = \left(2x - \frac{1}{12}x^3 + \frac{1}{4}x^2\right)\Big|_{-2}^{4}$$

$$A = [2(4) - \frac{1}{12}(4)^3 + \frac{1}{4}(4)^2] - [2(-2) - \frac{1}{12}(-2)^3 + \frac{1}{4}(-2)^2]$$

$$A = (8 - \frac{64}{12} + \frac{16}{4}) - (-4 - \frac{-8}{12} + \frac{4}{4})$$

$$A = (8 - \frac{16}{3} + 4) - (-4 + \frac{2}{3} + 1)$$

$$A = 8 - \frac{16}{3} + 4 + 4 - \frac{2}{3} - 1$$

$$A = 15 - \frac{18}{3} = 15 - 6 = 9.$$

Therefore, the region bounded by the graphs of the two functions is 9.

4.4. Tables of Integration Formulas

From what we have learned you have probably noticed that integration is more challenging than differentiation. Due to this notification, we have listed integral formulas that are reasonable and/or important into a table.

The table below presents some common integral formulae. However, some algebra or a technique such as substitution or integration by parts may be needed as well as a table.

Function	Integral
$f(x) = k$	$\int k\,dx = kx + C$
$f(x) = kx$	$\int kx\,dx = \frac{k}{2}x^2 + C$
$f(x) = x^n$	$\int x^n\,dx = \frac{1}{n+1}x^{n+1} + C$
$f(x) = \frac{1}{x}$	$\int \frac{1}{x}\,dx = \int x^{-1}\,dx = ln\lvert x\rvert + C$
$f(x) = \frac{1}{ax+b}$	$\int \frac{x}{ax+b}dx = \frac{1}{a}ln\lvert ax+b\rvert + C$
$f(x) = \frac{x}{a+bx}$	$\int \frac{1}{a+bx}dx = \frac{a}{b^2} + \frac{x}{b} - \frac{a}{b^2}ln\lvert a+bx\rvert + C$
$f(x) = x^{\frac{r}{s}}$	$\int x^{\frac{r}{s}}\,dx = \frac{1}{\frac{r}{s}+1}x^{\frac{r}{s}+1} + C = \frac{s}{r+s}x^{\frac{r+s}{s}} + C$
$f(x) = e^{ax}$	$\int e^{ax}\,dx = \frac{1}{a}e^{ax} + C$
$f(x) = k^x$	$\int k^x\,dx = \frac{k^x}{ln\,k} + C$
$f(x) = xk^x$	$\int xk^x\,dx = \frac{k^x(x\log(k) - 1)}{\log^2(k)} + C$
$f(x) = \log_a x$	$\int \log_a x dx = x\log_a x - \frac{x}{lna} = \frac{xlnx - x}{lna} + C$
$f(x) = ln(ax)$	$\int ln(ax)\,dx = x\ln(ax) - x + C$
$f(x) = ln(ax+b)$	$\int ln(ax+b)\,dx = \frac{(ax+b)ln(ax+b) - (ax+b)}{a} + C$
$f(x) = (lnx)^2$	$\int (lnx)^2 dx = x(lnx)^2 - 2xlnx + 2x + C$

$f(x) = xe^x$	$\int xe^x\,dx = (x-1)e^x + C$
$f(x) = xe^{ax}$	$\int xe^{ax}\,dx = \frac{1}{a^2}(ax-1)e^{ax} + C$
$f(x) = \frac{1}{x\,lnx}$	$\int \frac{1}{x\,lnx}dx = ln\mid ln\,x\mid + C$
$f(x) = \frac{1}{a^2-x^2}$	$\int \frac{1}{a^2-x^2}dx = \frac{1}{2a}ln\mid\frac{x+a}{x-a}\mid + C$
$f(x) = \frac{1}{x^2-a^2}$	$\int \frac{1}{x^2-a^2}dx = \frac{1}{2a}ln\mid\frac{x-a}{x+a}\mid + C$
$f(x) = \sqrt{x-a}$	$\int \sqrt{x-a}\,dx = \frac{2}{3}(x-a)^{\frac{3}{2}} + C$
$f(x) = \frac{1}{\sqrt{x\pm a}}$	$\int \frac{1}{\sqrt{x\pm a}}dx = 2\sqrt{x\pm a} + C$
$f(x) = \frac{1}{\sqrt{a-x}}$	$\int \frac{1}{\sqrt{a-x}}dx = -2\sqrt{a-x} + C$
$f(x) = \sqrt{a^2-x^2}$	$\int \sqrt{a^2-x^2}\,dx = \frac{a^2}{2}arc\sin\frac{x}{a} + \frac{x}{2}\sqrt{a^2-x^2} + C$
$f(x) = \frac{1}{\sqrt{a^2-x^2}}$	$\int \frac{1}{\sqrt{a^2-x^2}}dx = arc\sin\frac{x}{a} + C$
$f(x) = x^2\sqrt{a^2-x^2}$	$x^2\sqrt{a^2-x^2}dx = \frac{x}{8}(2x^2-a^2)\sqrt{a^2-x^2} + \frac{a^4}{8}\sin^{-1}\frac{x}{a} + C$
$f(x) = \sqrt{x^2\pm a^2}$	$\int \sqrt{x^2\pm a^2}\,dx = \frac{x}{2}\sqrt{x^2\pm a^2} \pm \frac{a^2}{2}ln\mid x+\sqrt{x^2\pm a^2}\mid + C$
$f(x) = \frac{1}{\sqrt{x^2\pm a^2}}$	$\int \frac{1}{\sqrt{x^2\pm a^2}}dx = ln\mid x+\sqrt{x^2\pm a^2}\mid + C$
$f(x) = x^2\sqrt{x^2-a^2}$	$^2\sqrt{x^2-a^2}dx = \frac{x}{8}(2x^2-a^2)\sqrt{x^2-a^2} - \frac{a^4}{8}ln\mid x+\sqrt{x^2-a^2} + C$
$f(x) = \sin x$	$\int \sin x dx = -\cos x + C$
$f(x) = \cos x$	$\int \cos x dx = \sin x + C$
$f(x) = \tan x$	$\int \tan x dx = ln\mid\sec x\mid + C$
$f(x) = \sec x$	$\int \sec x dx = \ln\mid\tan x + \sec x\mid + C$
$f(x) = \text{cosec}\, x$	$\int \text{cosec}\, x dx = \ln\mid\csc x - \cot x\mid + C = \ln\mid\tan(\frac{x}{2})\mid + C$
$f(x) = \cot x$	$\int \cot x dx = \ln\mid\sin x\mid + C$
$f(x) = \sec 2x$	$\int \sec 2x dx = \tan x + C$
$f(x) = \csc 2x$	$\int \text{cosec}\, 2x dx = -\cot x + C$
$f(x) = \sec x\tan x$	$\int \sec x\tan x dx = \sec x + C$
$f(x) = \csc x\cot x$	$\int \text{cosec}\, x\cot x dx = -\csc x + C$
$f(x) = \sin kx$	$\int \sin kx\,dx = -(\frac{1}{k}\cos kx) + C$
$f(x) = \cos kx$	$\int \cos kx\,dx = (\frac{1}{k}\sin kx) + C$

$f(x) = \sin^2 x$	$\int \sin^2 x dx = \frac{x}{2} - \frac{1}{4}\sin 2x + C = \frac{1}{2}(x - \sin x \cos x) + C$
$f(x) = \cos^2 x$	$\int \cos^2 x dx = = \frac{x}{2} + \frac{1}{4}\sin 2x + C = \frac{1}{2}(x + \sin x \cos x) + C$
$f(x) = \tan^2 x$	$\int \tan^2 x dx = \tan x - x + C$
$f(x) = \sec^2 x$	$\int \sec^2 x dx = \tan x + C$
$f(x) = \mathrm{cosec}^2 x$	$\int \mathrm{cosec}^2 x dx = \cot x + C$
$f(x) = \cot^2 x$	$\int \cot^2 x dx = -\cot x - x + C$

Example

Given the function $f(x) = x^2(1 + 2x^3)^5$, use substitution to evaluate the integral of the function $f(x)$ over the interval of [0, 1].

Solution

Let us define $u = 1 + 2x^3$. Then $du = 6x^2dx$, or $x^2dx = \frac{1}{6}\,du$. Also the interval limits are adjusted as:

For $x = 0$ $\quad u = 1 + 2x^3 = 1 + 2(0)^3 = 1$.

For $x = 1$ $\quad u = 1 + 2x^3 = 1 + 2(1)^3 = 3$.

Now substituting $u = 1 + 2x^3$ and $x^2dx = \frac{1}{6}\,du$ we will have:

$$\int_0^1 x^2(1 + 2x^3)^5 dx = \int_1^3 \frac{1}{6}\,(u)^5 du$$

$$\int_0^1 x^2(1 + 2x^3)^5 dx = \frac{1}{6}\int_1^3 (u)^5 du$$

we adjusted the integral limits and substituted $u = 1 + 2x^3$ and $x^2dx = \frac{1}{6}\,du$ into the equation

$$\int_0^1 x^2(1 + 2x^3)^5 dx = \frac{1}{6} \times \frac{1}{6} u^6 \Big|_1^3$$

$$\int_0^1 x^2(1 + 2x^3)^5 dx = \frac{1}{36} u^6 \Big|_1^3$$

$$\int_0^1 x^2(1 + 2x^3)^5 dx = \frac{1}{36}[(3)^6 - (1)^6]$$

$$\int_0^1 x^2(1 + 2x^3)^5 dx = \frac{1}{36}[(729 - 1]$$

$$\int_0^1 x^2(1 + 2x^3)^5 dx = \frac{728}{36} = 20.22.$$

Therefore, the evaluation of $\int_0^1 x^2(1 + 2x^3)^5 dx$ leads to 20.22.

4.5. Exercises

1. Evaluate the following indefinite integrals and check the answer.
   a) $y = \int 18dx.$
   b) $y = \int 9x^2\, dx.$
   c) $y = \int (8t^3 + 7t)dt.$
   d) $y = \int (h^3 - 5)^4 dh.$
   e) $t = \int 12y^6 - 3y + (2y^2 - 1)^2 dy.$

2. Evaluate the following indefinite integrals.
   a) $y = \int \sqrt{x^2 + 9}\, dx.$
   b) $y = \int \sqrt{k^2 - 36}\, dk.$
   c) $y = \int u^2\sqrt{4 + u^2} du.$
   d) $y = \int \frac{1}{\sqrt{x^2 - 49}} dx.$
   e) $y = \int \frac{t^2}{\sqrt{t^2 - 81}} dt.$
   f) $y = \int \frac{x^2}{\sqrt{64 - x^2}} dx.$

3. Evaluate the following indefinite integrals.
   a) $y = \int \frac{1}{2x + 6} dx.$
   b) $y = \int \frac{u}{5u - 1} du.$
   c) $y = \int \frac{v^2}{\frac{3}{5}v - \frac{2}{7}} dv.$
   d) $y = \int \frac{1}{t(4 + 5t)} dt.$
   e) $y = \int \frac{1}{k^2(3 + 7k)} dk.$

4. Evaluate the following indefinite integrals.
   a) $y = \int ln(5u)\, du.$
   b) $y = \int xe^{\frac{7}{2}x}.$
   c) $y = \int t^3e^{7t} dt.$
   d) $y = \int v^5\, ln\, v\, dv.$
   g) $y = \int ln\,(7k + 9)dk.$
   h) $y = \int \log_3 udu.$

5. Evaluate each of the indefinite integral.
   a) $y = \int ln(5x + 4)\, dx.$
   b) $y = \int ln(3x + 1)\, dx.$

6. Evaluate the indefinite integral given as $\int x^2 ln(x)\, dx$ by part.

7. Evaluate the indefinite integral given as $\int \frac{\ln(x)}{x^3}\, dx$ by part.

8. Evaluate the indefinite integral given as $\int (4x^3 - 3x^2)\, ln(x)\, dx$ by part.

9. Evaluate the definite integral given as $\int_0^2 4x^3(x^4 - 3)^5 dx$ using substitution.

10. Evaluate the definite integral given as $\int_1^3 x\sqrt{x+3}\, dx$ using substitution.

11. Evaluate the indefinite integral given as $\int \frac{x}{\sqrt{x-5}}\, dx$ using substitution.
   a) $y = \int_0^2 4x^3(x^4 - 3)^5 dx.$
   b) $y = \int_1^3 ln(5t + 2)dt.$
   c) $t = \int_0^2 \frac{3}{y^2} e^{3y-2} dy.$

12. Evaluate the following indefinite integrals.
   a) $y = \int 7^x\, dx.$
   b) $y = \int h e^{6h^2 - 1} dh.$
   c) $y = \int 3v^2 e^{v^3}\, dv.$

13. Evaluate the following definite integrals.
   a) $y = \int_0^4 (9x^2 - 2x - 13)dx.$
   b) $t = \int_1^3 [\frac{1}{(y-8)^2} + y^3 - ye^{6y}]dy.$
   c) $y = \int_2^3 [\frac{t^2}{(4t+1)^2} + ln(t^2 - 1)]dt.$
   d) $y = \int_0^5 \frac{k^2}{\sqrt{k^2 + 25}} dk.$
   e) $y = \int_2^7 \frac{\sqrt[3]{x^2} - \frac{1}{3}}{\sqrt[3]{x}} dx.$

14. Find the area under curve in the interval of $[a, b]$ for each of the following functions.
   a) $y = 8$ for $a = 2$ and $b = 7$.
   b) $y = \frac{1}{4}x^3$ for $a = -\frac{1}{4}$ and $b = \frac{1}{2}$.
   c) $y = 3x$ for $a = 2$ and $b = 5$.
   d) $y = x^2$ for $a = 1$ and $b = 3$.
   e) $y = 4 - x^2$ for $a = -\frac{1}{2}$ and $b = \frac{3}{2}$.
   f) $y = x^3 - 4x$ for $a = 1$ and $b = 3$.

g) $y = e^{2x}$ for $a = 0$ and $b = 1$.

15. Find the area of the region bounded by the graph $f(x) = x + 2$ and $h(x) = 5 - x$ over the interval $[0, 4]$.

16. Find the area of the region bounded by the graph $f(x) = x^2 + 1$ and $h(x) = 5$.

17. Find the average value of

a) $f(x) = 3 - x$ over the interval $[\frac{1}{2}, \frac{3}{2}]$.

b) $h(x) = 1 + x^2$ over the interval $[-1, 1]$.

18. Find each of the following indefinite integrals.

a) $y = \int (2x + \sin^2 x)dx$.

b) $y = \int (\frac{3}{2}x^2 - 4x + \cos^2 x)\,dx$.

c) $y = \int (\frac{4}{3}x^3 - \frac{1}{x^2(3 + 2x)} + \tan^2 x)\,dx$.

19. An oil well in a Middle east initially produces oil at a rate of 1.8 million barrels of oil per year. The production rate will typically fall 12% per year. How much oil is produced over a 4-year period?

20. Suppose a population of tomato flies increases at a rate of $f(t) = \frac{3}{2}e^{0.005t}$, in flies per day. If the initial population of fruit flies is 80 flies, how many flies are in the population after 9 days?

5. Real-World Applications

Functions can be utilized in various real-life scenarios to relate input values to specific output values. Equational functions are employed to determine precise output values corresponding to given input values. Graphical representations of functions offer a holistic view of how input and output data are related, facilitating an understanding of overall trends and patterns. Functions presented in tabular form are particularly useful when specific input-output relationships need to be analyzed or when individual input values are of particular significance.

5.1. Example: Bacteria Count

The number N of bacteria in a refrigerated food is given by $N(T) = 20T^2 - 90T + 480$, $2 \leq T \leq 12$, in which T is the food temperature in degree of Celsius. When the food is removed from refrigeration, the temperature of the food is given by $T(t) = 4t + 3$, $0 \leq t \leq 3$ where t is the time in hours.

a) Find the composition of the two functions and interpret its meaning in context.

b) Find the time when the bacterial count reaches 1800.

Solution

a) To find the composition of the two functions we need to find $N(T(t))$. We need to replace the variable T with $4t + 3$ in the $N(T)$. So,

$$N(T(t)) = 20(4t + 3)^2 - 90(4t + 3) + 480$$

$$= 20(16t^2 + 24t + 9) - 360t - 270 + 480$$ we expanded the square of binomial in the equation

$$= 320t^2 + 480t + 180 - 360t - 270 + 480$$

$$= 320t^2 + 120t + 390.$$

The composite function $N(T(t))$ represents the number of bacteria in the food as a function of the amount of time the food has been out of refrigeration.

b) To find the time when the bacterial count reaches 1800, we need to set the composite function $N(T(t))$ equal to 1800 and solve for t. So,

$$N(T(t)) = 320t^2 + 120t + 390 = 1800$$

$$320t^2 + 120t + 390 - 1800 = 1800 - 1800$$ to move all the terms in one side we add – 1800 to both sides

$$320t^2 + 120t - 1410 = 0$$

$$t = \frac{-b \pm \sqrt{b^2 - 4ac}}{2a},$$ where $b = 120$, $a = 320$, $c = -1410$

$$t = \frac{-120 \pm \sqrt{120^2 - 4(320)(-1410)}}{2(320)}$$

$$t = \frac{-120 \pm \sqrt{14400 - (-1804800)}}{640}$$

$$t = \frac{-120 \pm \sqrt{1819200}}{640}$$

$$t = \frac{-120 \pm 1348.8}{640}$$

$$t = \frac{-120 - 1348.8}{640} \quad \text{or} \quad t = \frac{-120 + 1348.8}{640}$$

$$t = \frac{-1468.8}{640} = -2.3 \quad \text{or} \quad t = \frac{1228.8}{640} = 1.9.$$

As it is seen, there is a negative answer for solution of the above equation. The negative value is rejected because it is not in the domain of the composite function. Therefore, the bacterial count will reach 1800 when $t = 1.92$ or t equal to 1 hour and 54 minutes.

5.2. Example: Filling Up a Container

A cylindrical container with the height H, and the base inner radius R, contained $860cm^3$ of a liquid and is being filled at the constant rate of $120cm^3$ per second. The inner radius of the container is $24cm$.

a) Write a formula for the quantity Q of the liquid in the container after t seconds.

b) Write a formula for the height H of liquid in the container in terms of Q.

c) Find an expression for the composition $(H_o Q)(t)$ and its meaning.

d) How long does it take for the height H of the liquid in the container to reach 30cm?

Hint: The volume of a cylinder is equal to the product of the area of the circular base and the height of the cylinder. $V = \pi r^2 h$, where r is the cylinder base radius and h is the height of cylinder. Also let $\pi = 3.14$.

Solution

a) The formula for the quantity Q is.

$$Q = 860 + 120t.$$

Where t is the time in second.

b) The formula for the height H of water in the container in terms of Q is.

$$Q = \pi R^2 H.$$

This gives $H = \frac{Q}{\pi R^2}$.

c) The expression for the composition $(H_o Q)(t)$ is.

$$(H_o Q)(t) = H(Q(t)) = \frac{860 + 120t}{\pi R^2} = \frac{860 + 120t}{3.14 \times (24^2)} = \frac{860 + 120t}{1808.6}.$$

This gives the height $H(t)$ of the liquid in the cylinder as a function of time t.

d) To find the time takes the height H of the liquid in the container to reach 30cm, we need to set $H(Q(t))$ equal to 30 and solve for t. So,

$$H(Q(t)) = \frac{860 + 120t}{1808.6} = 30$$

$$(1808.6) \times \frac{860 + 120t}{1808.6} = (1808.6) \times 30$$

$$860 + 12t = 1808.6 \times 30$$

$$860 + 12t = 54258$$

$$-860 + 860 + 12t = -860 + 54258$$

$$12t = 53398$$

$$\frac{1}{12} \times 12t = \frac{1}{12} \times 53398$$

$$t = 4449.8.$$

Therefore, the time the height H of the liquid in the container to reach 30cm is 4449.8 seconds or 1.236 hours (1 hour 14 minutes).

5.3. Example: Population Using Inverse Function

The population of a fast-growing town increases according to the formula $P = 100{,}000e^{0.081t}$, where P is the population and t the number of years, with $t = 0$ corresponding to the year 2020. When will the population be 150,000, 200,000 and 250,000?

Solution

In this case we need to use the invers function operation. We first need to find t by solving the given formula for t. So,

$$P = 100{,}000e^{0.081t}$$

$$\frac{1}{100{,}000} \times P = \frac{1}{100{,}000} \times 100{,}000e^{0.081t}$$

$$\frac{P}{100{,}000} = e^{0.081t}$$

$$ln\left(\frac{P}{100{,}000}\right) = ln\left(e^{0.081t}\right)$$ we took a natural logarithm from both sides

$$ln\left(\frac{P}{100{,}000}\right) = (0.081t)\ ln\ e$$

$$ln\left(\frac{P}{100{,}000}\right) = (0.081t)\ (1)$$ remember that the natural logarithm of e is 1

$$ln\left(\frac{P}{100{,}000}\right) = 0.081t$$

$$\frac{1}{0.081} \times ln\left(\frac{P}{100{,}000}\right) = \frac{1}{0.081} \times 0.081t$$ to make the coefficient of t equal to 1, we multiplied both sides by the reciprocal of its coefficient

$$t = \frac{ln\left(\frac{P}{100{,}000}\right)}{0.081}.$$

next, we now can find the value of t for the given P.

For $P = 150{,}000$ $\quad t = \frac{ln\left(\frac{150{,}00}{100{,}000}\right)}{0.081} = \frac{ln\ (1.5)}{0.081} = \frac{0.405}{0.081} = 5$ years.

For $P = 200{,}000$ $\quad t = \frac{ln\left(\frac{200{,}00}{100{,}000}\right)}{0.081} = \frac{ln\ (2)}{0.081} = \frac{0.693}{0.081} = 8.56$ years.

For P = 250,000 $\quad t = \frac{ln\left(\frac{250{,}00}{100{,}000}\right)}{0.081} = \frac{ln\ (2.5)}{0.081} = \frac{0.916}{0.081} = 11.31$ years.

Therefore, the population of the town will be 1500,000, 200,000, and 250,000 after 5, 8.56, and 11.31 years, respectively.

5.4. Example: Demand, Supply, Surplus, and the Equilibrium Points

Supply and demand, in economics, have the relationship between the quantity of a commodity that producers wish to sell at various prices and the quantity that consumers wish to buy. It has been convenient to think of demand and supply as quantities that are functions of price. Considering the consumer's price-demand function $p = D(x)$ which shows the price per unit that the consumer is willing to pay for x units of a product, then the consumer surplus is defined at quantity $x = Q$, as

$$\int_0^Q D(x)dx - QD(x = Q).$$

Note that a consumer surplus happens when the price that consumers pay for a product or service is less than the price that they are willing to pay.

Similarly, if the producer's price-supply function $p = S(x)$ which shows the price per unit the producer is willing to accept for selling x units, then the producer surplus is defined at quantity $x = Q$, as.

$$QS(x = Q) - \int_0^Q S(x)dx.$$

Given the consumer surplus function and the producer supply function, the equilibrium point is the point at which the supply and demand curves intersect. That is to set the consumer surplus function equal to the producer supply function, $D(x) = S(x)$ and then solve to get x.

Suppose for a commodity the demand function given by $D(x) = (x - 10)^2$, for x-thousand unit, and the supply function is given by $S(x) = x^2 + x + 5$ for x-thousand unit. Find each of the following.

a) The consumer surplus when $x = 6$.

b) The producer surplus when $x = 6$.

c) The equilibrium points.

d) The consumer surplus at the equilibrium point.

e) The producer surplus at the equilibrium point.

Solution

Graphing the demand function curve and supply function curve will help to visualize the problem. Figure below depicts these two curves.

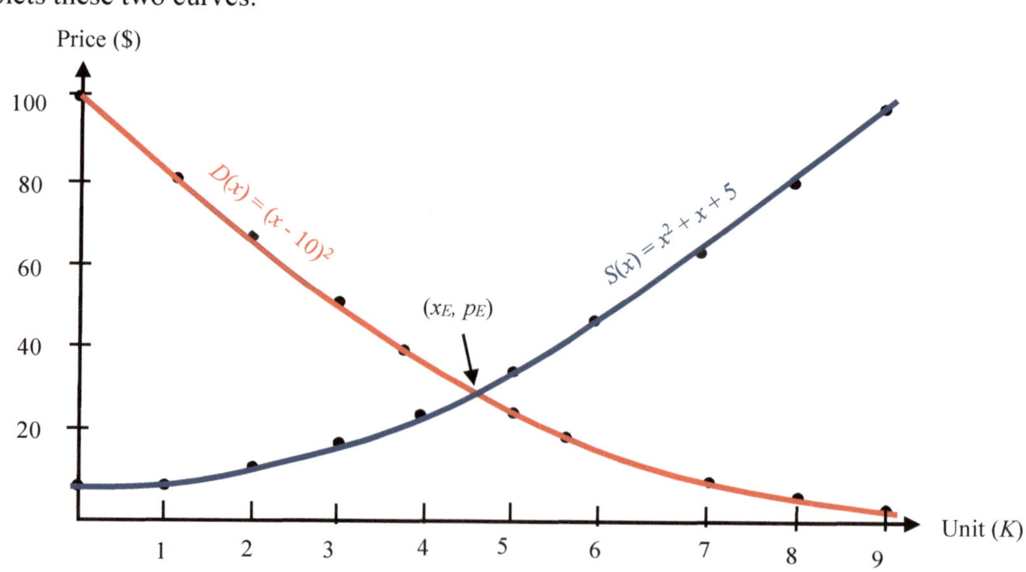

a) We have learned that the consumer surplus at quantity $x = Q$ is $\int_0^Q D(x)dx - QD(x = Q)$. Hence, to find the consumer surplus when $x = 6$, we first need to find the integral of $D(x) = (x - 10)^2$ in the interval of [0, 6]. So,

$$\int_0^6 (x - 10)^2 dx = \int_0^6 (x^2 - 20x + 100)dx$$ we applied the square of binomial rule.

$$= \left(\frac{1}{3}x^3 - \frac{1}{2}20x^2 + 100x\right)\Big|_0^6$$

$$= \left[\frac{1}{3} \times (6)^3 - \frac{1}{2} \times 20 \times (6)^2 + 100 \times (6)\right] - \left[\frac{1}{3} \times (0)^3 - \frac{1}{2} \times 20 \times (0)^2 + 100 \times (0)\right]$$

$$= \left[\frac{1}{3} \times (216) - \frac{1}{2} \times 20 \times (36) + 100 \times (6)\right] - \left[\frac{1}{3} \times (0) - \frac{1}{2} \times 20 \times (0) + 100 \times (0)\right]$$

$$= \left[\frac{216}{3} - \frac{720}{2} + 600\right] - [0]$$

$$= 72 - 160 + 600 = 512.$$

Next, we need to find $QD(x = Q)$. In this case $Q = 6$ and $D(6) = (6 - 10)^2 = 16$. So

$$QD(x = Q) = 6 \times 16 = 96.$$

Now, using the formula $\int_0^Q D(x)dx - QD(x = Q)$ we can get the consumer surplus. So,

consumer surplus $= \int_0^Q D(x)dx - QD(x = Q) = \int_0^6 (x - 10)^2 dx - 6 \times D(6)$

consumer surplus $= 512 - 96 = 416$.

Therefore, the consumer surplus is \$416.

b) To find the producer surplus when $x = 6$, we need to perform similar algebraic operations as we performed in the above. So,

producer surplus $= QS(x = Q) - \int_0^Q S(x)dx$

producer surplus $= 6 \times S(x^2 + x + 5)|_{x=6} - \int_0^Q (x^2 + x + 5)dx$

producer surplus $= 6 \times (6^2 + 6 + 5) - (\frac{1}{3}x^3 + \frac{1}{2}x^2 + 5x)\Big|_0^6$

producer surplus $= 6 \times (36 + 6 + 5) - \{[\frac{1}{3}(6)^3 + \frac{1}{2}(6)^2 + 5(6)] - [\frac{1}{3}(0)^3 + \frac{1}{2}(0)^2 + 5(0)]\}$

producer surplus $= 216 + 36 + 30 - (\frac{216}{3} + \frac{36}{2} + 30 - 0)$

producer surplus $= 216 + 36 + 30 - 72 - 18 - 30$

producer surplus $= 282 - 120 = 162$.

Therefore, the producer surplus is \$162.

c) To find the equilibrium point, we need to set the consumer demand function equal to the producer supply function. So,

$$D(x) = S(x)$$

$$(x - 10)^2 = x^2 + x + 5$$

$x^2 - 20x + 100 = x^2 + x + 5$

$x^2 - 20x + 100 - x^2 = x^2 + x + 5 - x^2$ we added $-x^2$ to both sides

$-20x + 100 = x + 5$

$-20x + 100 - x = x + 5 - x$ to have the variable only in one side, we added $-x$ to both sides

$-21x + 100 = 5$

$-21x + 100 - 100 = 5 - 100$ to isolate the variable, we added -100 to both sides

$-21x = -95$

$-\frac{1}{21} \times (-21x) = -\frac{1}{21} \times (-95)$ to make the coefficient of the variable equal to 1, we multiplied both sides by $-\frac{1}{21}$

$x = \frac{95}{21} = 4.524.$

Therefore, the equilibrium point is $x_E = 4.524$, or 4524 units with the unit price of either $D(4.524) = (4.524 - 10)^2 = (5.476)^2 = 29.99$ or $S(4.524) = 4.524^2 + 4.524 + 5 = 29.99$. Hence the equilibrium point is (4.524, \$29.99) or 4524 units with the unit price of p_E = \$29.99.

d) To find the consumer surplus at the equilibrium point, we need to plug $x_E = 4.524$ and p_E = \$29.99 into the $\int_0^{x_E} D(x)dx - x_E \times p_E$. So,

$\int_0^{x_E} D(x)dx - x_E \times p_E$

$\int_0^{4.524} (x-10)^2 dx - (4.524) \times (29.99)$ we plugged $x_E = 4.524$ and p_E = \$29.99 into the equation

$= \int_0^{4.524} (x^2 - 20x + 100)dx - (135.7)$

$= \left(\frac{1}{3}x^3 - \frac{1}{2}20x^2 + 100x\right)\Big|_0^{4.524} - 135.7$

$= \left[\frac{(4.524)^3}{3} - \frac{20\times(4.524)^2}{2} + 100 \times (4.524)\right] - \left[\frac{(0)^3}{3} - \frac{20\times(0)^2}{2} + 100 \times (0)\right] - 135.7$

$= \left[\frac{92.6}{3} - \frac{409.3}{2} + 452.4\right] - [0] - 135.7$

$= 30.9 - 204.5 + 452.4 - 135.7 = 143.1.$

Therefore, the consumer surplus at the equilibrium point is \$143.1.

e) To find the producer surplus at the equilibrium point, we similarly need to plug $x_E = 4.524$ and p_E = \$29.99 into the $x_E \times p_E - \int_0^{x_E} S(x)dx$. So,

$x_E \times p_E - \int_0^{x_E} S(x)dx$

$(4.524) \times (29.99) - \int_0^{4.524} (x^2 + x + 5)dx$ we plugged $x_E = 4.524$ and p_E = \$29.99 into the equation

$$= 135.7 - \left(\frac{1}{3}x^3 + \frac{1}{2}x^2 + 5x\right)\Big|_0^{4.524}$$

$$= 135.7 - \left[\frac{(4.524)^3}{3} + \frac{(4.524)^2}{2} + 5 \times (4.524)\right] - \left[\frac{(0)^3}{3} + \frac{(0)^2}{2} + 5 \times (0)\right]$$

$$= 135.7 - \left[\frac{92.6}{3} + \frac{20.5}{2} + 22.6\right] - [0]$$

$$=135.7-30.9-10.25-22.6 = 71.95.$$

Therefore, the producer surplus at the equilibrium point is \$71.95.

5.5. Example: Dimension Optimization

Mr. Charleston is the manager of a gardener's supply store in a small town. He wants to build a 480 square foot rectangular enclosure on the store's parking lot to display some equipment. Three sides of the enclosure will be built of decorative whitewood fencing, at a cost of \$15 per running foot. The fourth side will be built of bricks, at a cost of \$11 per running foot. Find the dimensions of the least costly enclosure.

Solution

Let us define w and l as the dimensions of the enclosure, with l being the length of the side made of bricks. Then the area A of enclosure is.

$$A = w \times l = 480.$$

The cost of the decorative white wood side is $\$15(2w + l)$, and the cost of the brick side is $\$11l$. Hence the total costs, C, will be determined as below.

$$C = 15(2w + l) + 11l = 30w + 15l + 11l = 30w + 26l.$$

The object is to minimize the cost function C. we have realized though, it has two variables, so we need to use the area equation. The area equation is the fixed area $A = wl = 480$. We solve A for w to get $w = \frac{480}{l}$, and then substitute it into the cost function C. So,

$$C = 30w + 26l$$

$$C = 30 \times \frac{480}{l} + 26l$$

$$C = \frac{14400}{l} + 26l.$$

Now we have a function of just one variable. We have learned that when dealing with function of one variable, the critical points are the points in the domain of the function where the first derivative is equal to zero. So,

$$C' = \frac{d}{dl}C = \frac{d}{dl}\left(\frac{14400}{l} + 26l\right) = 0$$

$$\frac{l - 14400}{l^2} + 26 = 0$$

$$\frac{l - 14400}{l^2} + 26 = 0$$

$$\frac{l - 14400}{l^2} + \frac{l^2}{l^2} \times 26 = 0$$

for having same denominator, we multiplied the second term by $\frac{l^2}{l^2} = 1$

$$\frac{l - 14400}{l^2} + \frac{26\,l^2}{l^2} = 0$$

$$\frac{l - 14400 + 26l^2}{l^2} = 0$$

$$26l^2 + l - 14400 = 0$$

$$l = \frac{-b \pm \sqrt{b^2 - 4ac}}{2a}, \text{ where } b = 1,\ a = 26,\ c = -14400$$

$$l = \frac{-1 \pm \sqrt{1^2 - 4(26)(-144\ \ )}}{2(26)}$$

$$l = \frac{-1 \pm \sqrt{1 - (-1497600)}}{52}$$

$$l = \frac{-1 \pm \sqrt{1497601}}{52}$$

$$l = \frac{-1 \pm 1223.77}{52}$$

$$l = \frac{-1 - 1223.77}{52} \quad \text{or} \quad l = \frac{-1 + 1223.77}{52}$$

$$l = \frac{-1224.77}{52} \quad \text{or} \quad l = \frac{1222.77}{52}$$

$$l = -23.55 \quad \text{or} \quad l = 23.51.$$

C' is undefined for $l = 0$, and $C' = 0$ when $l = 23.51$ or $l = -23.55$.

Of these three critical points, only $l = 23.51$ is acceptable (is in the domain of the actual function). Note that l is a length, so it cannot be negative.

Now we need to apply the second derivative test to find whether $l = 23.51$ is a minimum, a maximum, or a saddle point. So,

$$C'' = \frac{d^2}{dl^2}C = \frac{d}{dl}(26l^2 + l - 14400) = 2 \times 26l + 1 = 52l + 1$$

$$C''|_{l = 23.51} = 52(23.51) + 1 = 122,52 + 1 = 1223.52 > 0.$$

We have $C'' > 0$ at $l = 23.51$ and hence the length of 23.51 is a local minimum. However, since $l = 23.51$ is the only critical point in the domain, this must be the global minimum. When $l = 23.51$ we can find w by plugging l into the $w = \frac{480}{l}$. Hence $w = \frac{480}{23.51} = 20.42$. The dimensions of the enclosure that minimize the cost are 23.51 feet × 20 .42 feet.

To find value of the minimum total costs of building the enclosure, we need to plug l and w into the cost function. So,

$$C = 30w + 26l$$

$$C = 30(20.42) + 26(23.51)$$

$$C = 612.6 + 611.26$$

$$C = 1223.86.$$

Therefore, the minimum cost for building the enclosure is $1,223.86.

5.6. Example: Profit Maximization

Toys-For-Kids Manufacturing Company sells toys at \$16 each. The production cost of toys consists of two costs, a fixed cost and a variable cost. It is determined that the fixed costs of Toys- For-Kids Company consist of property taxes, a lease, executive salaries, and machines setup costs which add up to \$80,000. The variable cost associated with producing toys is \$4 per unit.

a) Write the cost function for production of x toys.

b) Write the profit function from the sale of x number of toys.

c) Determine the breakeven point of Toys-For-Kids Company.

d) Assuming the demand function for toys is given by $D(x) = 22 - 0.005x$, find the maximize profit from the sale of x number of toys.

Solution

Let us denote $C(x)$, $R(x)$, and $P(x)$ respectively as the total production costs, for manufacturing x units of toys, the revenue from the sale of x number of toys, and the profit from the sale of x number of toys.

a) The cost function for production of x toys can be written below.

$$C(x) = \begin{cases} 0, & x < 0 \\ 8000, & x \geq 0 \end{cases}.$$

b) The profit function is written as the revenue from the sale of x toys minus the cost of manufacturing the x toys. In this case the revenue function is $R(x) = 16x$. So,

$$P(x) = R(x) - C(x)$$

$P(x) = 16x - (8,000 + 4x)$ we replaced $R(x)$ and $C(x)$ into the equation

$$P(x) = 16x - 8,000 - 4x$$

$$P(x) = 12x - 8,000.$$

c) The break-even point, BVP, is determined by the total fixed costs divided by the difference between the unit selling price and variable costs per unit.

In this case, the total fixed costs have been estimated to be \$8,000, the unit sale price of a toy is \$12, and the variable cost of manufacturing toys is \$4 per unit. So,

$$BVP = \frac{\text{total fixed cost}}{\text{unit sale price} - \text{unit variable cost}}$$

$$BVP = \frac{8,000}{12 - 4} = \frac{8,000}{8} = 1,000.$$

Therefore, the break-even point is 1,000 units of toys. Graph of the break-even point is presented after end of this solution.

d) To find the maximum profit from the sale of x number of toys we first need to find the critical points of the profit. We have learned that when dealing with function, a critical point is a point in the domain of the function where the derivative is equal to zero. We then need to apply the second derivative test to find whether the critical point is maximum or minimum or saddle point. Given the demand function, the revenue for selling the toys will be:

$$R(x) = xD(x)$$

$$R(x) = x(22 - 0.005x) = 22x - 0.005x^2.$$

The obtained revenue function leads to the profit function as below.

$$P(x) = R(x) - C(x)$$

$$P(x) = 22x - 0.005x^2 - (8000 + 4x)$$

$$P(x) = 22x - 0.005x^2 - 8000 - 4x$$

$$P(x) = - 0.005x^2 + 18x - 8000.$$

Now the critical points can be obtained by setting the derivative of P equal to zero. So,

$$\frac{d}{dx}P(x) = \frac{d}{dx}(- 0.005x^2 + 18x - 8000) = 0$$

$$- 2\times 0.005x + 18 = 0$$

$$- 0.01x + 18 = 0$$

$$- 0.01x + 18 - 18 = 0 - 18$$ to isolate the variable in one side we added - 18 to both sides

$$- 0.01x = - 18$$

$$(-\frac{1}{0.01})(- 0.01x) = (-\frac{1}{0.01})(- 18)$$ to make the coefficient of the variable 1, we multiplied both sides by the reciprocal of x

$$x = \frac{18}{0.01} = 1800.$$

We have found the critical point of P equal to 1800 units of toys. We now need to apply the second derivative test. So,

$$\frac{d^2}{dx^2}P(x) = \frac{d^2}{dx^2}(- 0.005x^2 + 18x - 8000)$$

$$\frac{d^2}{dx^2}P(x) = \frac{d}{dx}(- 0.01x + 18)$$

$$\frac{d^2}{dx^2}P(x) = - 0.01 < 0.$$

The second derivative of P is negative. Hence, the critical point $x = 1,800$ is the maximum of P. Therefore, the maximize profit when we sale 1,800 toys. To find the maximum profit value, we simply need to plug $x = 1,800$ into the profit function. So,

$$P(x) = - 0.005x^2 + 18x - 8000$$

$$P(1,200) = - 0.005(1800)^2 + 18(1800) - 8000$$

$$P(1,200) = - 16,200 + 32,400 - 8000$$

$$P(1,200) = - 24,200 + 32,400 = 8200.$$

Therefore, the maximum profit is \$8200 and acquired when we sale 1,800 toys.

Below we present the graph for the breakeven point.

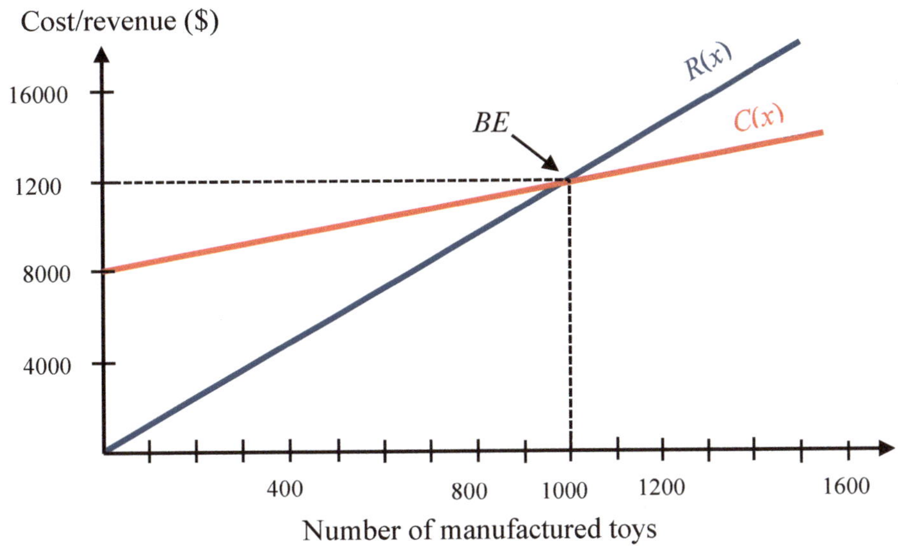

5.7. Example: Car's Density in Traffic Congestion

Suppose that the density of cars in traffic congestion on a freeway change linearly from 20 to 160 cars per km per lane on a 5*km* long stretch. Estimate the total number of cars on the freeway stretch:

a) If the freeway has 3 lanes.

b) If the freeway has 4 lanes.

c) If the freeway has 5 lanes.

Solution

Let $D(x)$ represent the density function in which x is the number of cars. First, we need to drive the equation for the density function. Since the density function is linear, we need to use the two-points formula, $\frac{y - y_1}{y_2 - y_1} = \frac{x - x_1}{x_2 - x_1}$, for finding its equation. The number of cars on a freeway, as affected by traffic congestion, undergoes a linear change, ranging from 30 cars to 160 cars. We have for $x_1 = 0$, $y_1 = D(0) = 20$ and for $x_2 = 5$, $y_2 = D(5) = 160$. So,

$$\frac{y - y_1}{y_2 - y_1} = \frac{x - x_1}{x_2 - x_1}$$

$$\frac{y - 20}{160 - 20} = \frac{x - 0}{5 - 0}$$

$$\frac{y - 20}{140} = \frac{x}{5}$$

$$140 \times \frac{y - 20}{140} = 140 \times \frac{x}{5}$$

for converting fraction to whole number expression, we multiplied both sides by 140

$$y - 20 = \frac{140x}{5}$$

$$y - 20 = 28x$$

$$y - 20 + 20 = 28x + 20$$

to isolate y in one side of the equation, we added 20 to both sides

$$y = 28x + 20.$$

Therefore, the density function is $D(x) = y = 28x + 20$. Now we can estimate the total number of cars, N, on the freeway stretch for different number of lanes. So,
For 3 lanes:

$$N = 3\int_0^5 (28x + 20)\, dx$$

$$N = 3\left(\frac{1}{2} \times 28x^2 + 20x\right)|_0^5$$

$$N = 3(14x^2 + 20x)|_0^5$$

$$N = (42x^2 + 60x)|_0^5$$

$$N = [42(5)^2 + 60(5)] - [42(0)^2 + 60(0)]$$

$$N = [42(25) + 300] - [0]$$

$$N = 1050 + 300$$

$$N = 1350.$$

Therefore, the total number of cars on the freeway is $5km$ stretch when it has 3 lanes is 1350 cars.

For 4 lanes:

$$N = 4\int_0^5 (28x + 20)\, dx$$

$$N = 4\left(\frac{1}{2} \times 28x^2 + 20x\right)|_0^5$$

$$N = 4(14x^2 + 20x)|_0^5$$

$$N = (56x^2 + 80x)|_0^5$$

$$N = [56(5)^2 + 80(5)] - [56(0)^2 + 80(0)]$$

$$N = [56(25) + 400] - [0]$$

$$N = 1400 + 400$$

$$N = 1800.$$

Therefore, the total number of cars on the freeway is $5km$ stretch when it has 4 lanes is 1800 cars.

For 5 lanes:

$$N = 5\int_0^5 (28x + 20)\, dx$$

$$N = 5\left(\frac{1}{2} \times 28x^2 + 20x\right)|_0^5$$

$$N = 5(14x^2 + 20x)|_0^5$$

$$N = (70x^2 + 100x)|_0^5$$

$$N = [70(5)^2 + 100(5)] - [70(0)^2 + 100(0)]$$

$N = [70(25) + 500] - [0]$

$N = 1750 + 500$

$N = 2250.$

Therefore, the total number of cars on the freeway is $5km$ stretch when it has 5 lanes is 2250 cars.

5.8. Example: Distance, Velocity and Acceleration

A particle in the first phase starts moving from rest with a constant acceleration $a_1 = 4$ (m/s^2) for $T_1 = 5$ seconds. In the second phase it decelerates uniformly for $T_2 = 10$ seconds and stops. What is the distance covered by the particle in the two phases?

Solution

Let us first determine the velocity v_1 now, $t = T_1$.

$v_1 = a_1 t = a_1 T_1$

$v_1 = 4 \times 5 = 20$ (m/s^2) we plugged $a_1 = 4$ and $T_1 = 5$ into the equation.

On the second phase, the particle's velocity varies according to the equation below.

$v = v_1 + a_2 t,$

where $0 \leq t \leq T_2$. Because $v(T_2) = 0$, we can simply find the value of acceleration a_2 as follow:

$a_2 = -\frac{v_1}{T_2}$

$a_2 = -\frac{a_1 T_1}{T_2}$

$a_2 = -\frac{4 \times 5}{10}$

$a_2 = -\frac{20}{10} = -2$ (m/s^2) we plugged $a_1 = 4$, $T_1 = 5$, and $T_2 = 10$ into the equation.

The value of a_2 is negative as the particle is decelerating.

Now we can calculate the distance. On the first phase, the distance d_1 traveled by the particle can be obtained by:

$d_1 = \int_0^{T_1} a_1 t dt$

$d_1 = a_1 \int_0^{T_1} t dt$

$d_1 = 4 \int_0^{5} t dt$ we plugged $a_1 = 4$ and $T_1 = 5$ into the equation

$d_1 = \frac{4t^2}{2} \Big|_0^5$

$d_1 = \frac{4 \times 5^2}{2} - \frac{5 \times 0^2}{2}$

$d_1 = \frac{4 \times 25}{2}$

$d_1 = \frac{10}{2} = 50m.$

The distance traveled on the second phase is given by:

$$d_2 = \int_0^{T_2} (v_1 + a_2 t) dt$$

$$d_2 = \int_0^{10} (20 + (-2)t) dt$$

we plugged $T_2 = 10$, $v_1 = 20$ and $a_2 = -2$ into the equation

$$d_2 = \int_0^{10} (20 - 2t) dt$$

$$d_2 = (20t + -\frac{1}{2} \times 2t^2)|_0^{10}$$

$$d_2 = (20 \times 10 - \frac{2 \times 10^2}{2}) - (20 \times 0 - \frac{2 \times 0^2}{2})$$

$$d_2 = (200 - 10^2) - (0)$$

$$d_2 = (200 - 100) - (0)$$

$$d_2 = 200 - 100 = 100m.$$

Now that we have obtained the distance traveled by the particle on the first and the second phase, we can sum them up to find the total distance covered by the particle. So,

$$d = d_1 + d_2 = 50 + 100 = 150m.$$

Therefore, the total distance covered by the particle in the two phases is $150m$.

5.9. Vertical Projection Motion

A ball is thrown vertically upward with an initial velocity of v_0. Calculate maximum height and time taken to reach maximum height, when:

a) Ball is traveled double the distance in the 1st second than in the 2nd second.

b) The initial velocity is $v_0 = 30m/s$.

Solution

Let us first depict the trajectory of the ball as shown below.

If a ball is thrown vertically upwards with an initial velocity v_0, then we will have the following set of formulae:

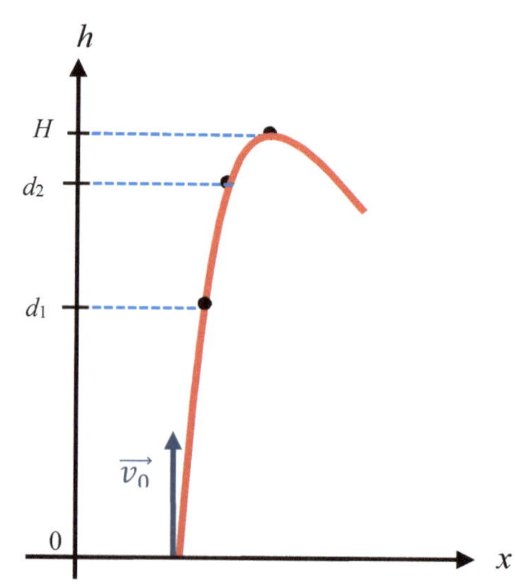

If a ball is thrown vertically upwards with an initial velocity v_0, then we will have the following set of formulae:

1. $H_{max} = \frac{v_0^2}{2g}$ where H_{max} is maximum height and g is acceleration of gravity equal to 9.8 (m/s^2).
2. $v_H = 0$, where v_H is velocity at the highest point.
3. $v = v_0 - gt$ where v is upward velocity at time t, v_0 is initial upward velocity, and g is acceleration of gravity.
4. $T_u = \frac{v_0}{g}$ where T_u is time for upward movement and g is acceleration of gravity.
5. $T_d = \frac{v_0}{g}$ where T_d is time for downward movement and g is acceleration of gravity.
6. $T_t = \frac{2\,v_0}{g}$ where T_t is the total time travel in air and g is acceleration of gravity.

a) The velocity of ball variers by $v(t) = v_0 - gt$. To find the distance ball traveled in the 1st second is:

$$d_1 = \int_0^1 v(t)dt = \int_0^1 (v_0 - gt)dt$$

$$d_1 = (v_0 t - \frac{1}{2}gt^2)|_0^1$$

$$d_1 = [v_0(1) - \frac{1}{2}(9.8)(1)^2] - [v_0(0) - \frac{1}{2}(9.8)(0)^2]$$

$$d_1 = [\,v_0 - 4.9] - [0]$$

$$d_1 = v_0 - 4.9.$$

Similarly, we can find the distance in the 2nd second as:

$$d_2 = \int_1^2 v(t)dt = \int_1^2 (v_0 - gt)dt$$

$$d_2 = (v_0 t - \frac{1}{2}gt^2)|_1^2$$

$$d_2 = [v_0(2) - \frac{1}{2}(9.8)(2)^2] - [v_0(1) - \frac{1}{2}(9.8)(1)^2]$$

$$d_2 = [2v_0 - 4.9(4)] - [v_0 - 4.9]$$

$$d_2 = [2v_0 - 19.6] - [v_0 - 4.9]$$

$$d_2 = 2v_0 - 19.6 - v_0 + 4.9$$

$$d_2 = v_0 - 19.6 - v_0 + 4.9$$

$$d_2 = v_0 - 14.7.$$

Now it is given that $\frac{d_1}{d_2} = 2$. We can find v_0 by substituting d_1 and d_2 into this ratio. So,

$$\frac{d_1}{d_2} = \frac{v_0 - 4.9}{v_0 - 14.7} = 2$$

$$\frac{v_0 - 4.9}{v_0 - 14.7} = 2,$$

$v_0 - 4.9 = 2(v_0 - 14.7)$

$v_0 - 4.9 = 2v_0 - 29.4$

$v_0 - 4.9 - v_0 = 2v_0 - 29.4 - v_0$ — to isolate the variable in one side we added $-v_0$ into the equation

$-4.9 = 2v_0 - 29.4 - v_0$

$-4.9 = v_0 - 29.4$

$-4.9 + 29.4 = v_0 - 29.4 + 29.4$ — again to isolate the variable in one side we added 29.4 into the equation

$v_0 = -4.9 + 29.4$

$v_0 = 24.5.$

It is known that $H_{max} = \frac{v_0^2}{2g}$. So,

$H_{max} = \frac{v_0^2}{2g}$

$H_{max} = \frac{24.5^2}{2(9.8)}$ — we plugged $v_0 = 24.5$ and $g = 9.8$ into the equation

$H_{max} = \frac{600.25}{19.6} = 30.625.$

Therefore, the maximum height is $30.6m$.

To find the time taken the ball reaches to the maximum height, we can use the given formula relating time to velocities, $v = v_0 - gt$. In the case of the maximum height $v = 0$. Also, in this case we found $v_0 = 24.5$. So,

$v = v_0 - gt = 0$

$v_0 - gt = 0$

$24.5 - 9.8t = 0$ — we plugged $v_0 = 24.5$ and $g = 9.8$ into the equation

$24.5 - 9.8t - 24.5 = -24.5$ — to isolate the variable in one side we added -24.5 into the equation

$-9.8t = -24.5$

$(-\frac{1}{9.8})(-9.8t) = (-\frac{1}{9.8})(-24.5)$ — to make the coefficient of variable 1, we multiplied both sides by the reciprocal of the variable

$t = (-\frac{1}{9.8})(-24.5)$

$t = \frac{24.5}{9.8} = 2.4.$

Therefore, the time taken the ball reaches to the maximum height is 2.4 seconds.

b) To find the maximum height and time taken the ball reaches maximum height, when the initial velocity is $v_0 = 30m/s$, we need to plug the value of v_0 into $H_{max} = \frac{v_0^2}{2g}$ and $t = \frac{v_0}{g}$ respectively. So,

For getting the maximum height:

$$H_{max} = \frac{v_0^2}{2g}$$

$$H_{max} = \frac{30^2}{2(9.8)}$$ we plugged $v_0 = 30$ and $g = 9.8$ into the equation

$$H_{max} = \frac{900}{19.6} = 45.9.$$

Therefore, the maximum height is $45.9m$.

$$t = \frac{v_0}{g}$$

$$t = \frac{30}{9.8} = 3.1.$$

Therefore, the time taken the ball reaches to the maximum height is 3.1 seconds.

5.10. Example: Pricing to Maximize Profit

A manufacturing company wants to set a price for its new product to maximize the profit from sale of the product. Market research reveals that the initial price of $7.90 will sale 9000 numbers of the product and sales of the product will increase by another 1500 for every 50 cents of price reduction on the original price of $7.90. Suppose the company production fixed costs (start-up costs) total $8000, and the per item cost of production (marginal cost) is $2.80.

What would be the price which maximizes the profit of the company earning from sale of the product and the amount of the maximum profit?

Solution

We have learned that profit is revenue minus costs. Hence the first step is to write the profit function equation. Let n be the number of products to be sold, $P(x)$ be the profit function, $R(x)$ be the revenue function, and $C(x)$ be the production costs function where x is price per item of the product. So,

$$C(x) = 8000 + 2.80x$$

$$R(x) = nx$$

$$P(x) = R(x) - C(x)$$

$$P(x) = nx - (8000 + 2.80n)$$

$$P(x) = nx - 8000 - 2.80n$$

$$P(x) = nx - 8000 - 2.80n$$

$$P(x) = n(x - 2.80) - 8000$$ we factor out n.

The number of products to be sold, n can itself be written as a function of x. To develop this function:

For the first 50 cents we will have $n = 9000 + (1) \times 1500$

For the second 50 cents we will have $n = 9000 + (2) \times 1500$

For the third 50 cents we will have $n = 9000 + (3) \times 1500$

⋮

For the kth 50 cents we will have $n = 9000 + (k) \times 1500$,

where $k = \frac{\text{initial price} - \text{planned price}}{0.5} = \frac{7.90 - x}{0.5}$, or $n = 9000 + (\frac{7.90 - x}{0.5}) \times 1500 = 9000 + \frac{1500(7.90 - x)}{0.5}$

Now, we need to substitute n in the profit function. So,

$$P(x) = n(x - 2.80) - 8000$$

$$P(x) = (9000 + \frac{1500(7.90 - x)}{0.5})(x - 2.80) - 8000$$

$$P(x) = [9000 + 3000(7.9 - x)](x - 2.8) - 8000$$

$$P(x) = (9000 + 23700 - 3000x)(x - 2.8) - 8000$$

$$P(x) = (32700 - 3000x)(x - 2.8) - 8000$$

$$P(x) = 32700x - 3000x^2 - 91560 + 8400x - 8000$$

$$P(x) = -3000x^2 + 41100x - 99960.$$

To find the maximum profit we need first need to find the critical points of $P(x)$. We learned that critical points of a function are obtained by setting its derivative equal to zero. So,

$$P(x) = -3000x^2 + 41100x - 99960$$

$$\frac{d}{dx}(P(x)) = P'(x) = (2)(-3000x) + 41100 = 0$$

$$-6000x + 41100 = 0$$

$$-6000x + 41100 - 41100 = -41100$$ to isolate the variable we added – 41100 to both sides

$$-6000x = -41100$$

$$-\frac{1}{6000} \times (-6000x) = -\frac{1}{6000} \times (-41100)$$ to make the coefficient of the variable 1, we multiplied both sides by the reciprocal of the variable

$$x = \frac{41100}{6000} = 6.85.$$

The derivative is $P'(x) = -6000x + 41100$, which is zero when $x = 6.85$, and hence it is a critical point of $P(x)$. To find whether it is a maximum or a minimum or a point of inflection, we need to apply the second derivative test. So,

$$\frac{d^2}{dx^2}(P(x)) = P''(x) = \frac{d}{dx}(P'(x)) = \frac{d}{dx}(-6000x + 41100)$$

$$\frac{d^2}{dx^2}(P(x)) = -6000 < 0.$$

The second derivative of $P(x)$ is negative and hence its critical point is a local maximum. Since $x = 6.85$ is the only critical value it must be a global maximum as well.

To find the value of the company maximum profit we simply need to plug $x = 6.85$ into the profit function. So,

$$P(x) = -3000x^2 + 41100x - 99960$$

$$P(6.85) = -3000(6.85)^2 + 41100(6.85) - 99960$$

$P(6.85) = -\ 3000(46.9) + 41100(6.85) - 99960$

$P(6.85) = -\ 140700 + 281535 - 99960$

$P(6.85) = -\ 240660 + 281535$

$P(6.85) = 40875.$

Therefore, the maximum profit is \$40,875, attained when the company set the price at \$6.85 and sells 12,150 number of products.

5.11. Example: Geometry

Find the area of the largest isosceles triangle that fits inside a circle of radius 1, so that the two equal sides of the triangle are along two radiuses of the circle (see figure below).

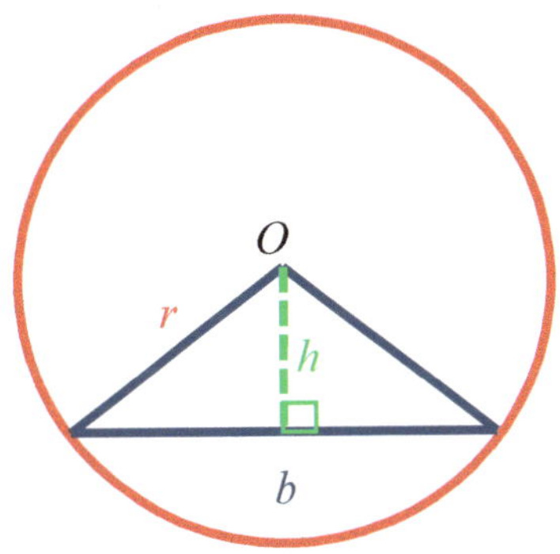

Solution

The area of a triangle, S, is equal to half of the height h times the base, b, that is $S = \frac{1}{2}hb$. Also, by Pythagorean theorem, $r^2 = h^2 + \left(\frac{b}{2}\right)^2$.

Since in the equation of S there are two variables, we use the Pythagorean theorem to find one variable in terms of the other. In this case we have $r = 1$. So,

$h^2 + \left(\frac{b}{2}\right)^2 = r^2 = 1$

$h^2 + \frac{b^2}{4} = 1$

$h^2 + \frac{b^2}{4} - h^2 = 1 - h^2$ — to isolate b^2 in one side we added $-h^2$ to both sides

$\frac{b^2}{4} = 1 - h^2$

$4 \times \frac{b^2}{4} = 4 \times (1 - h^2)$ — to make the coefficient of b^2 equal to 1, we multiplied both sides by 4

$b^2 = 4 \times (1 - h^2)$

$b^2 = 4 - 4h^2$

$b = \pm\sqrt{4 - 4h^2}$

$b = \pm\sqrt{4(1 - h^2)}$ — we factored out 4.

Since b is the length of the triangle base it cannot be negative. Hence $b = \sqrt{4(1-h^2)}$. Now we plug b into the S. So,

$S = \frac{1}{2}hb$

$S = \frac{1}{2}h\sqrt{4(1-h^2)}$ we plugged $b = \sqrt{4(1-h^2)}$ into the S.

$S = \frac{1}{2}h[4(1-h^2)]^{\frac{1}{2}}$

$\frac{d}{dh}S = S' = \frac{d}{dh}\{h[4(1-h^2)]^{\frac{1}{2}}\}$

$S' = 4^{\frac{1}{2}} \times \frac{d}{dh}\{h[(1-h^2)]^{\frac{1}{2}}\}$

$S' = 2 \times \frac{d}{dh}\{h[(1-h^2)]^{\frac{1}{2}}\}$

$S' = 2 \times \{[(1-h^2)]^{\frac{1}{2}} + h \times \frac{1}{2}(-2h)[(1-h^2)]^{\frac{1}{2}-1}\}$

$S' = 2 \times \{[(1-h^2)]^{\frac{1}{2}} - h^2[(1-h^2)]^{-\frac{1}{2}}\}$

$S' = 2 \times [(1-h^2)^{\frac{1}{2}} - \frac{h^2}{(1-h^2)^{\frac{1}{2}}}]$

$S' = 2 \times [(1-h^2)^{\frac{1}{2}} \times \frac{(1-h^2)^{\frac{1}{2}}}{(1-h^2)^{\frac{1}{2}}} - \frac{h^2}{(1-h^2)^{\frac{1}{2}}}]$ to make the same denominator, we multiplied the first term by $\frac{(1-h^2)^{\frac{1}{2}}}{(1-h^2)^{\frac{1}{2}}}$

$S' = 2 \times [\frac{1-h^2}{(1-h^2)^{\frac{1}{2}}} - \frac{h^2}{(1-h^2)^{\frac{1}{2}}}]$

$S' = 2 \times \frac{1-h^2-h^2}{(1-h^2)^{\frac{1}{2}}}$

$S' = 2 \times \frac{1-2h^2}{(1-h^2)^{\frac{1}{2}}}$

$S' = \frac{2-4h^2}{(1-h^2)^{\frac{1}{2}}}$.

We now, set S' equal to zero. So,

$S' = \frac{2-4h^2}{(1-h^2)^{\frac{1}{2}}} = 0$

$2 - 4h^2 = 0$, and $h \neq \pm 1$

$2 - 4h^2 - 2 = 0 - 2$ to isolate $4h^2$ in one side we added -2 to both sides

$-4h^2 = -2$

$\left(-\frac{1}{4}\right) \times (-4h^2) = \left(-\frac{1}{4}\right) \times (-2)$ to make the coefficient of the variable 1, we multiplied both sides by the reciprocal of its coefficient

$h^2 = \left(\frac{2}{4}\right) = 0.5$

$h = \pm\sqrt{0.5}$

$h = \pm 0.7.$

Since h is the length of the height it cannot have a negative value. Hence $h = 0.7$ is a critical point of the S.

Now, we need to apply the second derivative test to find if it is a local maximum or a local minimum or a saddle point. So,

$$\frac{d^2}{dh^2}S = \frac{d}{dh}S' = \frac{d}{dh}\left(\frac{2-4h^2}{(1-h^2)^{\frac{1}{2}}}\right)$$

$$\frac{d^2}{dh^2}S = \frac{(-2\times 4h)(1-h^2)^{\frac{1}{2}} - [\frac{1}{2}(-2h)(1-h^2)^{-\frac{1}{2}}](2-4h^2)}{[(1-h^2)^{\frac{1}{2}}]^2}$$

$$\frac{d^2}{dh^2}S = \frac{(-8h)(1-h^2)^{\frac{1}{2}} - [(-h)(1-h^2)^{-\frac{1}{2}}](2-4h^2)}{1-h^2}$$

$$\frac{d^2}{dh^2}S|_{h=0.7} = \frac{(-8\times 0.7)(1-(0.7)^2)^{\frac{1}{2}} - [(-0.7)(1-(0.7)^2)^{-\frac{1}{2}}](2-4(0.7)^2)}{1-(0.7)^2}$$ we plugged $h = 0.7$ into the equation

$$\frac{d^2}{dh^2}S|_{h=0.7} = \frac{(-5.6)(1-0.5)^{\frac{1}{2}} - [(-0.7)(1-0.5)^{-\frac{1}{2}}](2-4\times 0.5)}{1-0.5}$$

$$\frac{d^2}{dh^2}S|_{h=0.7} = \frac{(-5.6)(0.5)^{\frac{1}{2}} - [(-0.7)(0.5)^{-\frac{1}{2}}](2-2)}{0.5}$$

$$\frac{d^2}{dh^2}S|_{h=0.7} = \frac{(-5.6)(0.5)^{\frac{1}{2}} - [(-0.7)(0.5)^{-\frac{1}{2}}](0)}{0.5}$$

$$\frac{d^2}{dh^2}S|_{h=0.7} = \frac{(-5.6)(0.5)^{\frac{1}{2}}}{0.5}$$

$$\frac{d^2}{dh^2}S|_{h=0.7} = \frac{(-5.6)(0.7)}{0.5}$$

$$\frac{d^2}{dh^2}S|_{h=0.7} = \frac{-3.92}{0.5} = -7.84 < 0.$$

The second derivative is negative, and the test reveals that area S has a local maximum at $h = 0.7$. Since it is the only critical point, therefore, S has a global maximum at $h = 0.7$.

To find the value of the maximum area, we first need to find the length of b. We have $b = \sqrt{4(1-h^2)}$. Hence, when $h = 0.7$ we get $b = \sqrt{4(1-(0.7)^2)} = \sqrt{4(1-0.5)} = \sqrt{4(0.5)} = = 2\sqrt{0.5} = 2 \times 0.7 = 1.4$.

For $h = 0.7$ and $b = 1.4$, we can get the maximum S as:

$$S = \frac{1}{2}hb$$

$$S = \frac{1}{2}(0.7)(1.4)$$

$$S = \frac{0.98}{2} = 0.49.$$

Therefore, the area of the largest isosceles triangle fitted inside a circle of radius 1, when its two equal sides along sided of the two circle radiuses is 0.49.

5.12. Exercises

1. Suppose we are given the graph of function $f(x) = x^2$, the graph of a line $y = 12$ and a rectangle that fits inside the graph of the $f(x)$ below the line $y = 12$ (the top side of the rectangle should be on the horizontal line $y = 12$). Find dimensions of the given rectangle to maximize its area (see figure below).

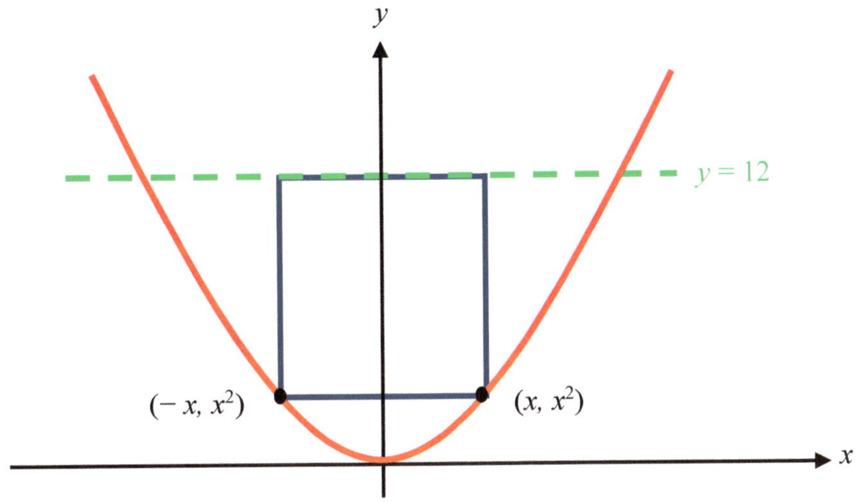

2. A particle starts from rest and moves with an acceleration given by the equation $a(t) = e^t - 2t$ (m/s^2). Find the average speed of the particle over the interval from $t_1 = 1$ second to $t_1 = 3$ seconds.

   Hint: the velocity function is found by integration the acceleration function, $v(t) = \int a(t)dt$.

3. The number N of bacteria in a refrigerated food is given by $N(T) = 19T^2 - 87T + 500$, $3 \le T \le 13$ in which T is the food temperature in degree of Celsius. When the food is removed from refrigeration, the temperature of the food is given by $T(t) = 5t + 1$, $0 \le t \le 180$ where t is the time in minutes.
   a) Find the composition of the two functions and interpret its meaning in context.
   b) Find the time when the bacterial count reaches 2000.

4. Toney plans to start a vegetable garden, for which he decides to fence off in the shape of a rectangle enclosure from the rest of the garden. Toney has only 180m of fencing, so he decides to use the fencing for three sides of enclosure and a brick wall as the fourth border side of the vegetable garden. Calculate the width and length of the garden that corresponds to the largest possible area of enclosure.

5. A car initially moving forward at a velocity of 60m/s approaches a school and slows down to a velocity of 15m/s in 5 seconds. Calculate the car's acceleration.

6. Cans are the most used metal packaging. A company is making cylindrical containers having a volume of 1200$cm3$. How do you find the dimensions that will minimize the cost of metal to manufacture the can, considering the unit cost of the material is applied on the unit area of cylinder surfaces?

7. A square sheet of cardboard with each side a centimeter is to be used to make an open-top box by cutting a small square (x by x) of cardboard from each of the corners and bending up the sides (see figure below) .
   a) What is the side length of the small squares to maximize the volume of the box?
   b) Find the maximum volume.

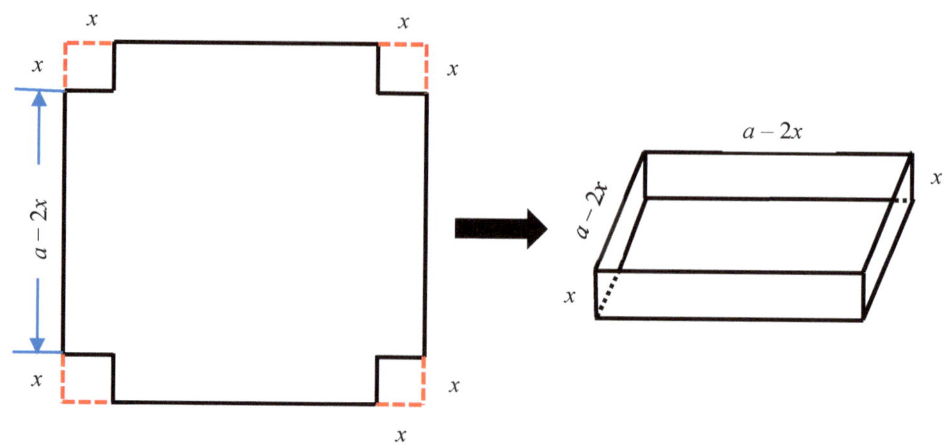

8. Suppose you want to reach a point D that is located across the desert from a nearby road (see figure below). Suppose that the road is straight, and b is the distance from D to the closest point C on the road. Let v_r be your speed on the road, and let v_s, which is less than v_r, be your speed on the desert. Suppose you are at point A, which is a distance a from C. Find point B that you should turn off the road and head across the desert to minimize your travel time from A to D.

   Hint:
   i) Define x to be the distance from B to C.
   ii) By Pythagorean theorem, $BA = \sqrt{x^2 + b^2}$.
   iii) When traveling at constant velocity, time $= \frac{\text{distance}}{\text{velocity}}$.

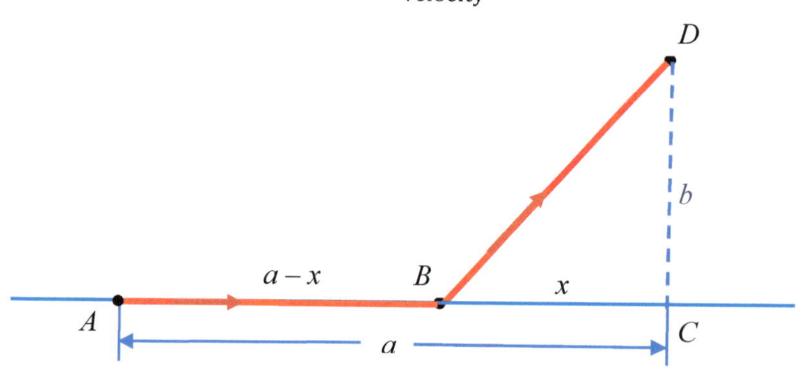

9. A cylindrical can is to be made to hold $1200cm^3$ of oil. How do you find the dimensions that will minimize the cost of metal to manufacture the can, considering the unit cost is applied on the unit area of surfaces

10. Heat treatment is the process of heating and cooling metals, using specific predetermined methods to achieve the desired result such as hardening or softening of a material. Suppose a metallic rod is being heated in an oven where the temperature T varies with the time t as follows: $T = 0.19t + 80$ in which T is the temperature in degree Celsius and t is time in seconds. The length L of the rod varies with temperature and hence with time according to the formula $L = 80 + 10^{-4}T$ in which L is length in centimeter.
    a) Find L in terms of time.
    b) Find L when the time reaches 5 seconds.

11. In the following shape, a rectangle inscribed in the ellipse with equation, $\frac{x^2}{4} + \frac{y^2}{1} = 1$, in which the sides of the rectangle are parallel to the axes. Find the dimensions of the rectangle to maximize its area. Also, find the maximum area.

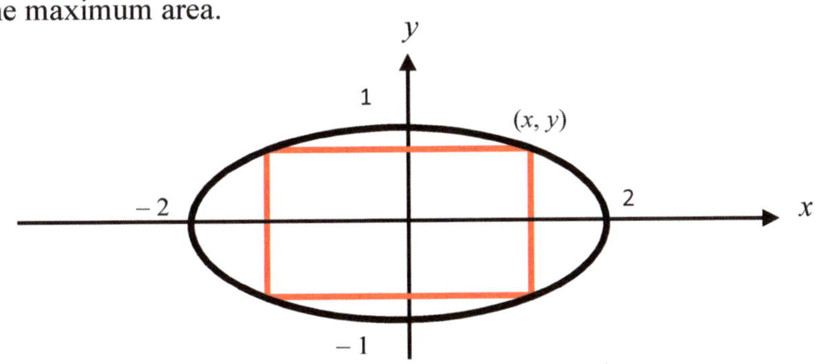

12. A cylindrical container initially has $450cm^3$ of a certain liquid and is being filled at the constant rate of $80(cm^3/second)$. The inner radius of the container is $r = 50cm$.
    a) Write a formula for volume V of the liquid in the container after t seconds.
    b) Write a formula for the height H of the liquid in the container in terms of V.
    c) Find an expression for the composition $(H \circ V)(t)$ and its meaning.
    d) How long does it take for the height H of the liquid in the container to reach $45cm$?

13. In 2020, Alaska gold mine production was 17.9 metric tons, and it was estimated growing exponentially at the rate of 5.9% per year (Source: U.S. Geological Survey, U.S. Dept. of the Interior). If the growth continues at this rate, how many tons of gold will be produced from 2020 to 2024?

14. Katrin has signed an investment contract by which she invests \$8,640 for 10 years at an interest rate of 4.25%, compounded continuously. She is interested to know:
    a) What amount she will get after 3 years (principal plus interest)?
    b) What amount she will get after 6 years?
    c) What amount she will get termination of the investment contract?

    Hint: If P_0 is invested for t years at interest rate r, compounded continuously, then $P(t) = P_0e^{rt}$, where $P = P_0$ at $t = 0$. The value $P(t)$ is called the future value (principal plus interest) of dollars invested at interest rate r, after t years.

15. Given the following shape, find the area of the largest isosceles triangle that fits inside a circle of radius 4, so that the two equal sides of the triangle are along two radiuses of the circle as illustrated in the figure below.

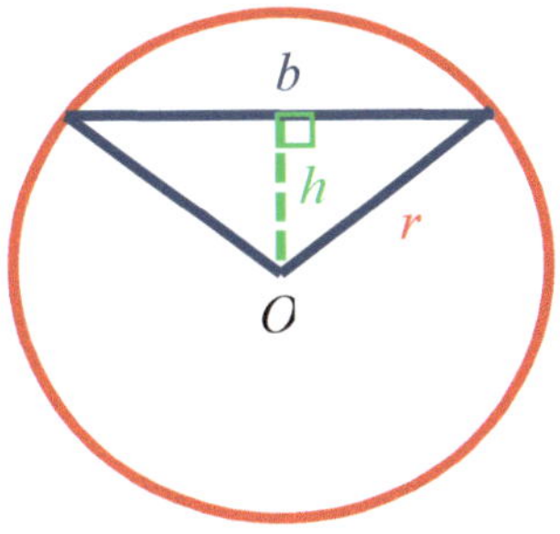

Hint: in an isosceles triangle, the altitude bisects the base.

16. US Appliance Manufacturing Company sells a certain type of toaster, a 4 slice long slot toaster, at \$986 each. The production cost of the toaster consists of two costs, a fixed cost, and a variable cost. It has been determined that the fixed cost of the company is \$600,000. The variable cost associated with producing the toaster is \$386 per unit.
    a) Write the cost function for production of x toasters.
    b) Write the profit function from the sale of x number of toasters.
    c) Determine the breakeven point of the company.

17. Suppose for a product the supply function is given by $S(x) = x^2 + x + 6$ for x-thousand unit and demand function given by $D(x) = (x - 8)^2$, for x-thousand units. Find each of the following.
    a) The consumer surplus when $x = 6$.
    b) The producer surplus when $x = 6$.
    c) The equilibrium points.
    d) The producer surplus at the equilibrium point.
    e) The consumer surplus at the equilibrium point.

18. In the figure below, an arrow is thrown in the X^+ direction with an initial velocity $V_0 = 48m/s$ and landed downrange (X) of $32m$. find the time elapsed during the motion. Also find the height at which the arrow is thrown.

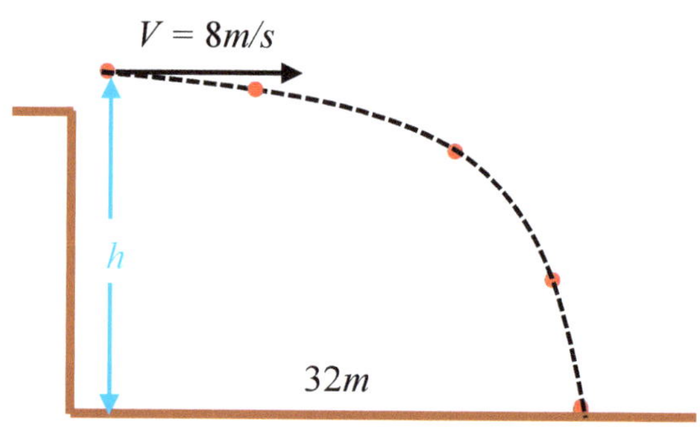

Hint: the formula for the horizontal movement is $x = vt$, and for the vertical movement is $y = \frac{1}{2}gt^2$, where g represents the acceleration due to gravity ($g = 9.8\ m/s^2$).

19. A culture contains initially 12,000 number of bacteria. After half an hour, the bacteria count is 13,200.
    a) Write an exponential model for the number of bacteria as a function of time, t, in hours.
    b) Find the number of bacteria after 3 hours.
    c) Find the number of bacteria between 3 hours and 4 hours.

20. Assum that the number of bacteria growths has an exponential function $N(t) = ae^{kt}$, where a is the initial population, and t is time in minutes. The count in bacteria culture was 1,000 after 30 minutes and 1,700 after 40 minutes.
    a) What was the initial size of the culture?
    b) Find the population after 90 minutes.

c) How many minutes after the start of the experiment will the population reach 33,304?

21. The current, I, in milliamperes and voltage, V, in volts across a circuit element are given by the equations below in which t is the time in milliseconds,

a) $I(t) = 20e^{-100t}$

b) $V(t) = 10 - 5e^{-100t}$

Determine the total energy dissipated by the element between $t = 0$ and $t = 10ms$.

Hint The power delivered to circuit element is given by $P(t) = I(t)V(t)$, where $I(t)$ is current across a circuit element at time t and $V(t)$ is voltage across a circuit element at time t. Also, the energy, E, dissipated in an element during the time interval $[0, T]$ can be found through integration: $E = \int_0^T P(t)dt$.

22. A box having a lid is to be constructed from a square piece of cardboard having sides of length l. The box is to be constructed by cutting squares with sides x from two corners and then cutting rectangles with sides of length x and y from the opposite corners as illustrated in the figure below. The sides are folded up and the lid is folded over with the sides to be taped. If $l = 40$, what would be the dimensions of the box to have the largest volume?

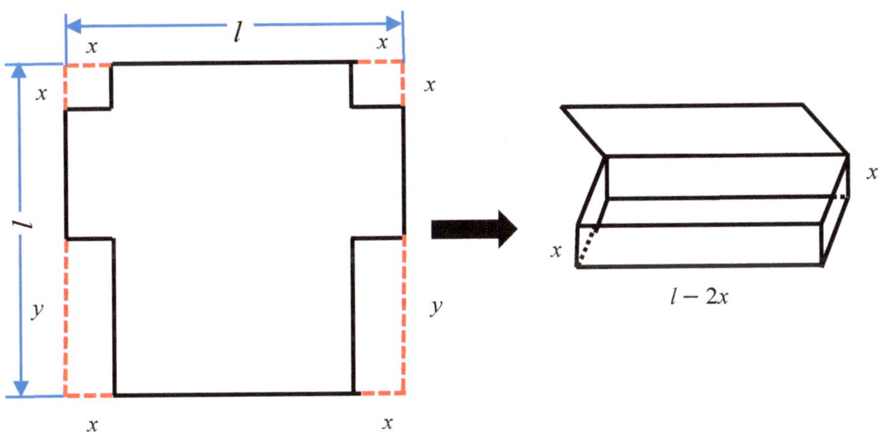

23. The volume of the rectangular prism is $1440cm^3$, and the area of the base of the rectangular prism is $160cm^2$. Find the height of the rectangular prism.

24. The function that describes a continues decay is $A = A_0e^{(\frac{ln(0.5)}{5730})t}$, where A is the amount of carbon-14 remaining and A_0 is the amount of carbon-14 at the start of decaying. Suppose we found a bone fragment that contains 21% of its original carbon-14. To the nearest year, how old is the bone?

25. The standard beverage can hold $12fl.\ oz$, or 354.9 milliliters. Assuming the shape of the can is a right circular cylinder, find:

a) What dimensions yield the minimum surface area?

b) Find the minimum surface area.

6. Multiple - Choice Questions

1. Suppose the function $f(x) = x^2 - 2c$, where c is a constant number. If $f(-2) = 6$, what is the value of c?

   A -5
   B 1
   C -1
   D 5

2. Based on the chart below, which of the following could demonstrate the relationship between x and y?

x	y
-2	7
1	4
8	-3

   A $y = x^2 - 2$
   B $y = 2x - 1$
   C $y = x - 6$
   D $y = -x + 5$

3. Which of the following is NOT representing a function?

A
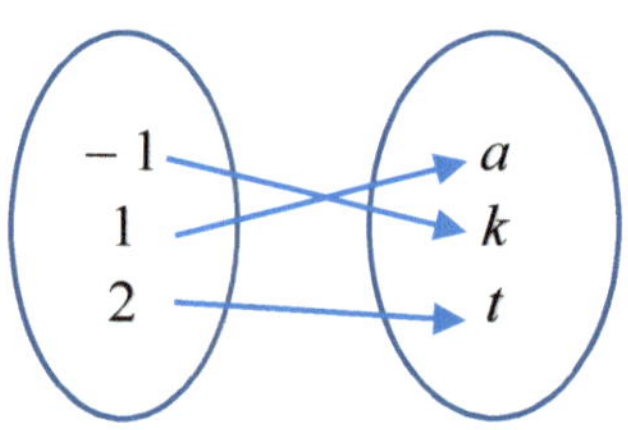

B
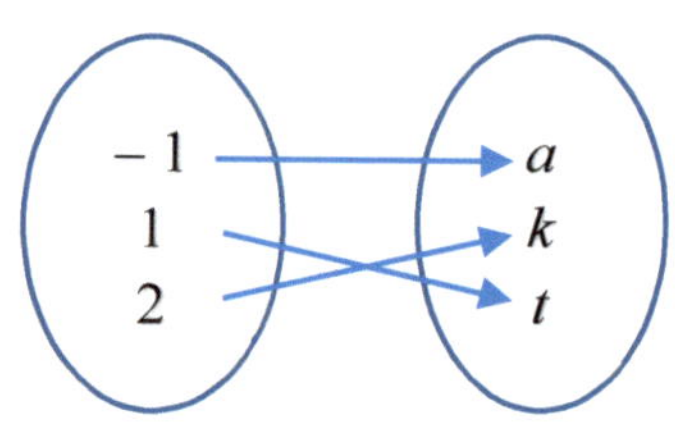

C
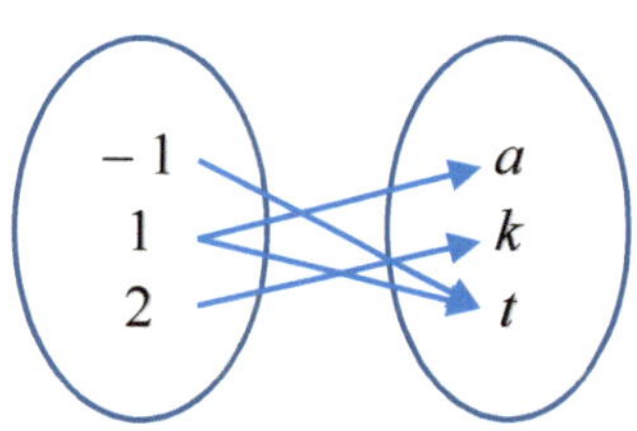

D
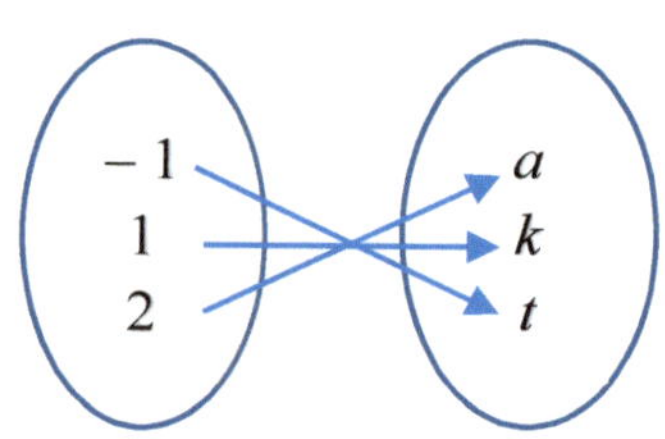

4. Which of the following are domain and range of the function below?
   $(1, \frac{1}{2}), (2, 1), (3, \frac{3}{2}), (4, 2), (5, \frac{5}{2}), (6, 3), (7, \frac{7}{2})$.

   A Domain is $\{1, 2, 3, 4, 5. 6\}$ and range is $\{\frac{1}{2}, 1, \frac{3}{2}, 2, \frac{5}{2}, 3, \frac{7}{2}\}$
   B Domain is $\{1, 2, 3, 4, 5. 6, 7\}$ and range is $\{\frac{1}{2}, 1, \frac{3}{2}, 4, \frac{5}{2}, 3, \frac{7}{2}\}$

C Domain is $\{1, 2, 3, 4, 5. 6, 7\}$ and range is $\{\frac{1}{2}, 1, \frac{3}{2}, 2, \frac{5}{2}, 3, \frac{7}{2}\}$

D Domain is $\{1, 2, 3, 4, 5. 6, 8\}$ and range is $\{\frac{1}{2}, 1, \frac{3}{2}, 2, \frac{5}{2}, 3, \frac{7}{2}\}$

Questions 5 to 7 refer to the figure below.

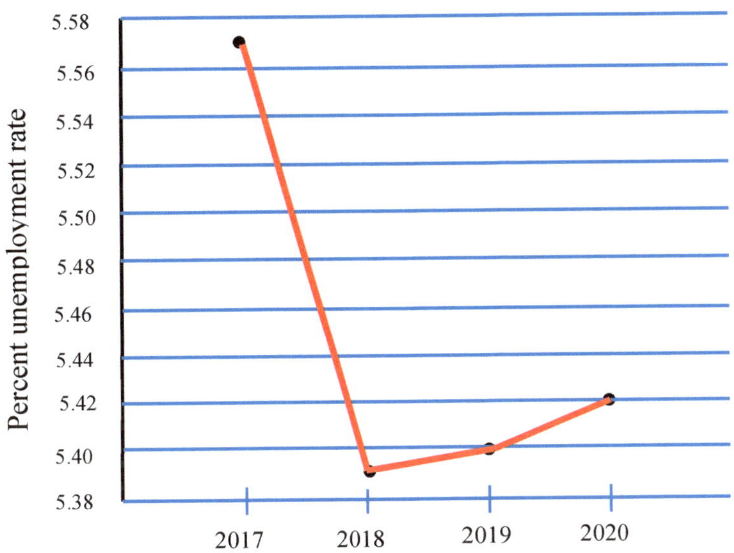

5. Figure above shows worldwide unemployment rates from 2004 to 2013. Which of the following statements is NOT true?

A The unemployment rate of 2017 is 5.57%

B The unemployment rate of 2018 is 5.39%

C The unemployment rate of 2019 is 5.40%

D The unemployment rate of 2020 is 5.41%

6. Figure above shows worldwide unemployment rates from 2004 to 2013. Which of the following statements is true?

A The graph is decreased from 2017 to 2019 by .18%

B The graph is increased from 2018 to 2019 by .02%

C The graph is increased from 2018 to 2020 by .03%

D The graph is decreased from 2017 to 2020 by .16%

7. Figure above shows worldwide unemployment rates from 2004 to 2013. Which of the following statements is true?

A The graph is decreasing everywhere

B The graph is increasing everywhere

C The graph is decreasing from 2017 to 2019 and then increasing

D The graph is increasing from 2018 to 2020

Questions 8 to 10 refer to the figure below.

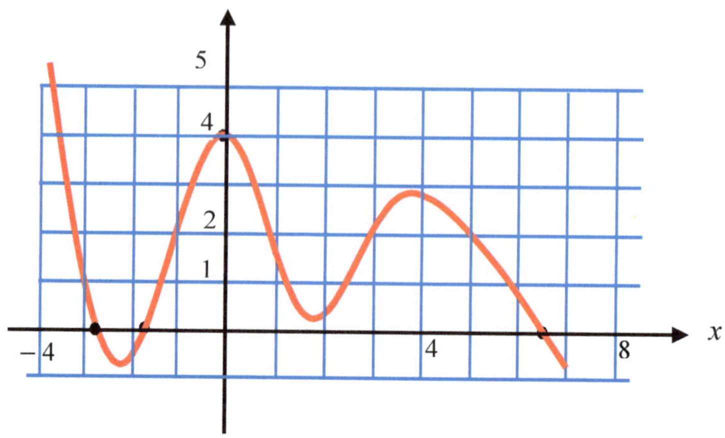

8. Suppose the above figure illustrates the graph of $f(x) - 3$. What is the value of $f(0)$?

A − 3

B 4

C − 3.8

D 6.5

9. Suppose the above figure illustrates the graph of $h(x)$. What is the value of $h(-3)$?

A 1

B − 1

C − 4

D 0

10. Suppose the above figure illustrates the graph of $g(x)$. If $g(x = a) = 0$ then the approximate value of a can be.

A $a = -1.8$, $a = -2.8$ $a = 6.5$

B $a = -1.8$, $a = -2.8$ $a = -6.5$

C $a = 1.8$, $a = -2.8$ $a = -6.5$

D $a = -1.8$, $a = 2.8$ $a = 6.5$

11. Let $f(x) = 7x^2 + 5x - 6$ and $g(x) = -8x - 3$. The function $(f + g)(x)$ is.

A $(f + g)(x) = 7x^2 - 13x - 3$

B $(f + g)(x) = 7x^2 - 3x - 9$

C $(f+g)(x) = 7x^2 - 3x - 3$

D $(f+g)(x) = 7x^2 - 13x - 9$

12. Let $f(x) = 2x^2 - 4x + 6$ and $g(x) = x^2 + 9x - 11$. The function $(f-g)(x)$ is.

A $(f-g)(x) = x^2 - 13x + 17$

B $(f-g)(x) = 3x^2 + 5x - 17$

C $((f-g)(x) = x^2 - 5x - 5$

D $(f-g)(x) = 3x^2 - 13x - 5$

13. Let $f(x) = \frac{3}{10x^2} - \frac{1}{5}x + 1$ and $g(x) = 20x^4$. The function $(f \times g)(x)$ is.

A $(f \times g)(x) = 6x^4 - \frac{4}{5}x^5 + 20x^4$

B $(f \times g)(x) = 6x^2 - \frac{4}{5}x^5 + 20x^4$

C $(f \times g)(x) = \frac{3}{10}x^2 - 4x^5 + 20x^4$

D $(f \times g)(x) = 6x^2 - 4x^5 + 20x^4$

14. Let $f(x) = 6x^2 - 7x$ and $g(x) = 12x^3 - 14x^2$, the function $\frac{f}{g}(x)$ is.

A $\frac{f}{g}(x) = \frac{1}{x}$

B $\frac{f}{g}(x) = \frac{1}{2x^2 - 2x}$

C $\frac{f}{g}(x) = \frac{6x^2 - 7x}{2x^3 - 2x^2}$

D $\frac{f}{g}(x) = \frac{1}{2x}$

15. The vertical asymptotes of $f(x) = \frac{3x - 1}{x - 2}$ is.

A $x = \frac{1}{3}$

B $x = -2$

C $x = 2$

D $x = -\frac{1}{3}$

16. Determine which of the following equation is NOT functions.

A $6x^2 - 9y^2 + 1 = 0$

B $4x^2 - 12y - 13x = 0$

C $3x^4 - 4y + 18 = 0$

D $-\frac{1}{3}x^3 - 8y - 4x^2 = 0$

17. If $f(h(2)) = -5$ and $f(x) = x - 3$, then which of the following could define $h(x)$?

A $h(x) = x - 6$

B $h(x) = x - 5$

C $h(x) = x - 4$

D $h(x) = x - 3$

Questions 18 to 20 refer to the figure below.

18. Which of the following functions best describes the figure?

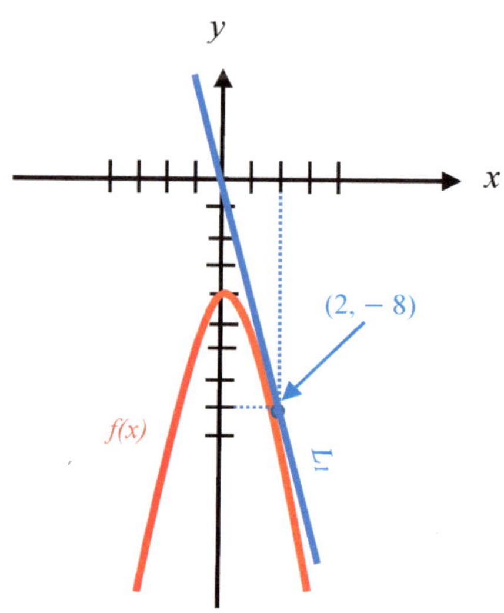

A $f(\mathrm{x}) = -(x^2 + 8)$

B $f(\mathrm{x}) = -(x^2 + 6)$

C $f(\mathrm{x}) = -(x^2 + 4)$

D $f(\mathrm{x}) = -(x^2 + 2)$

19. Which of the following is slope, m, of the tangent line, L_1, to the curve at the point $(2, -8)$?

A $m = 4$

B $m = -6$

C $m = -4$

D $m = 6$

20. Which of the following is equation of the tangent line, L_1, to the curve at the point $(2, -8)$?

A $y = 4x$

B $y = 6x$

C $y = -6x$

D $y = -4x$

21. If f is a function and $f(2) = 6$, which of the following CANNOT be the definition of f?

A $f(x) = x + 4$

B $f(x) = 4x - 2$

C $f(x) = x^2 + 2$

D $f(x) = 5x - 3$

22. Given $f(x) = \frac{6x}{5x^2 - 8}$, which one of the following is $f(2)$.

A $f(2) = 1$

B $f(2) = 2$

C $f(2) = 6$

D $f(2) = \frac{12}{13}$

23. Given $f(u) = \frac{13}{2u - 6}$ and $h(u) = 2u^2$, which one of the following is the function $f(h(x))$.

A $\frac{26u^3}{4u^3 - 6}$

B $\frac{26u^2}{4u^2 - 6}$

C $\frac{13u(u^2)}{2u\,(u^2) - 6}$

D $(\frac{13u}{2u - 6})(2u^2)$

24. Given $f(x) = 3x - 5$, what is $g(y) = f^{-1}(f(x))$.

A $g(y) = \frac{y + 3}{5}$

B $g(y) = \frac{y + 5}{3}$

C $g(y) = \frac{y - 3}{5}$

D $g(y) = \frac{y - 5}{3}$

25. A function is said to be ____ if and only if $f(h) = f(k)$ implies that $h = k$ for all h and k in the domain of f, then a function is said to be:

A One-to-many

B One-to-one

C Many-to-many

D Many-to-one

26. Suppose the monthly pay of salesman is denote as y and includes basic pay of \$380 plus a commission of \$4.50 for every unit he sells. The function for this case can be written as:

A $y = 4.5x + 380$

B $4.5xy = 380$

C $x = 4.5y + 380$

D $y = x + 384.5$

27. Given $f(x) = \sqrt{x+1}$, for $x \geq 0$, which of the following is correct?

A $f(xy) = f(x)\,f(y)$

B $f(xy) \geq f(x)\,f(y)$

C $f(xy) \leq f(x)\,f(y)$

D None of the above

28. Given the function $f(x) = \frac{x+1}{x-1}$, what would be the $f(f(x))$.

A $f(f(x)) = \frac{1}{x-1}$

B $f(f(x)) = 2x$

C $f(f(x)) = \frac{1}{x}$

D $f(f(x)) = x$

29. Given $f(x) = x^2 - 3$ and $h(x) = \sqrt{5x-2}$, evaluate each of the following is $(f \circ h)(1)$.

A $(f \circ h)(1) = \sqrt{3} - 3$

B $(f \circ h)(1) = 6$

C $(f \circ h)(1) = 12$

D $(f \circ h)(1) = 0$

30. Which is the inverse for the function $y = \frac{5x+1}{x-3}$?

A $y = \frac{3x+1}{x-5}$

B $y = \frac{x-5}{3x+1}$

C $y = \frac{5x-3}{x+1}$

D $y = \frac{3x-3}{5x+1}$

31. If $f(x) = \sqrt{6x-2}$, what is the smallest possible value of $f(x)$?

A $\frac{1}{3}$

B $-\frac{1}{3}$

C 0

D -1

32. Given $f(x) = 9x^2$, the solution for $f(x) = 1$ is.

A ± 1

B $\pm \frac{1}{3}$

C ± 3

D $\pm \frac{1}{2}$

33. Given $f(x) = \frac{2x-1}{2}$, what is $f(t^2) - f(2t)$?

A $t^2 - 2t$

B $\frac{t^2 - t - 1}{2}$

C $\frac{t^2 - 2t - 1}{2}$

D $\frac{2t^2 - 2t - 1}{2}$

34. Let $f(x) = (x+1)^2 - (x-1)^2$, then f represent a function which is:

A Cubic

B Quadratic

C Reciprocal

D Linear

35. Given $f(x) = 120$, what is $\frac{d}{dx}f(x)$?

A $-\infty$

B $+\infty$

C 120

D 0

36. Given $f(x) = \frac{x+3}{x-3}$, what is $\lim_{x \to \infty} f(x)$?

A 0

B 1

C 2

D $+\infty$

37. Given $f(x) = \frac{x^2 - 16}{x - 4}$, what is $\lim_{x \to 4} f(x)$?

A -16

B $+\infty$

C 8

D Does not exist

38. Given $f(x) = \sqrt{x^2 + 5}$, what is $\frac{d}{dx} f(x)$?

A $x(x^2 + 5)^{-\frac{1}{2}}$

B $\frac{1}{2}x(x^2 + 5)^{-\frac{1}{2}}$

C $2x(x^2 + 5)^{-\frac{1}{2}}$

D $(x^2 + 5)^{-\frac{1}{2}}$

39. Given $f(x) = (x - 1)^3 - x^3$, what is $\frac{d}{dx} f(x)$?

A $-2x + 1$

B $6x + 3$

C $-3x + 3$

D $-6x + 3$

40. Given $f(x) = \frac{2x}{4x^2 - 1}$, what is $\frac{d}{dx} f(x)$?

A $\frac{8x^2 - 2}{(4x^2 - 1)^2}$

B $-\frac{8x^2 + 2}{(4x^2 - 1)^2}$

C $-\frac{8x^2 - 2}{(4x^2 - 1)^2}$

D $\frac{8x^2 + 2}{4x^2 - 1}$

41. Given $f(x) = x^{\frac{1}{3}}$, what is $\frac{d}{dx}f(x)$?

A $\frac{1}{\sqrt[3]{x^2}}$

B $\frac{1}{3\sqrt[3]{x}}$

C $\frac{\sqrt[2]{x}}{3\sqrt[3]{x^2}}$

D $\frac{\sqrt[3]{x^2}}{3x^2}$

42. Given $f(x) = 3x^5 + 6x^3 - 9x^2 + 11$, what is $\frac{d}{dx}f(x)$?

A $15x^5 + 18x^3 - 18x^2$

B $15x^4 + 18x^2 - 18x$

C $3x^4 + 6x^4 - 9x + 11$

D $15x^4 + 18x^2 - 18x + 11$

43. Given $f(x) = (2x - 4)^6$, what is $\frac{d}{dx}f(x)$?

A $6(2x - 4)^5$

B $6(2x - 4)^6$

C $12(2x - 4)^6$

D $12(2x - 4)^5$

44. Given $f(x) = x^{\frac{2}{x}}$, what is $\frac{d}{dx}f(x)$?

A $\frac{2}{x}\, x^{1-\frac{2}{x}}$

B $\frac{2}{x}\, x^{\frac{2}{x}}$

C $\frac{2}{x}\, x^{\frac{2}{x}-1}$

D $\frac{1}{x}\, x^{1-\frac{2}{x}}$

45. Given $f(x) = (3x - 7)^4$, what is $\int f(x)dx$?

A $\frac{1}{15}(3x - 7)^5 + C$

B $\frac{1}{5}(3x - 7)^4 + C$

C $\frac{1}{5}(3x-7)^5+C$

D $\frac{1}{15}(3x-7)^4+C$

46. Which one is the difference property of definite integrals?

A $\int_a^b[-f(x)-h(x)]dx$

B $\int_a^b[f(-x)-h(x)]dx$

C $\int_a^b[f(x)-h(x)]dx$

D $\int_a^b[f(-x)+h(-x)]dx$

47. Identify the zero-length interval property.

A $\int_a^a f(x)dx=0$

B $\int_a^b f(x)dx-\int_a^d f(x)dx=0$

C $\int_a^b f(x)dx=1$

D $\int_a^b f(x)dx=-1$

48. Identify adding intervals property, where $a<c<b$.

A $\int_a^b f(x)dx=\int_a^c f(x)dx+\int_a^b f(x)dx$

B $\int_a^b f(x)dx=\int_a^c f(x)dx+\int_c^b f(x)dx$

C $\int_a^b f(x)dx=\int_a^c f(x)dx+\int_c^b h(x)dx$

D $\int_a^b f(x)dx=\int_c^a f(x)dx+\int_b^a f(x)dx$

49. What is the name of the property $\int_a^b f(x)dx=-\int_b^a f(x)dx$?

A Zero length interval property

B Adding interval property

C Difference of integrals property

D Reverse interval property

50. What is the area bounded by the function $f(x) = -2x + 3$ and the line $x = 1$?

A 3

B 2

C $\frac{3}{2}$

D $\frac{5}{2}$

51. Find the derivative of e^{x^2}.

A e^{x^2}

B $2x$

C $2e^{x^2}$

D $2xe^{x^2}$

52. Given $f(x) = x^3 - 2$, the coordinate of the saddle point of f is:

A $(-2, 0)$

B $(0, -2)$

C $(2, 0)$

D $(0, 2)$

53. Find the area bounded by the curve $y = x^2$ the line $x = 2$, and $x = -2$?

A 4

B 8

C $\frac{16}{3}$

D $\frac{16}{5}$

54. Given $f(x) = 11x^3 - 7x^2$, find $\int f(x)dx$.

A $\frac{11}{3}x^3 - \frac{7}{2}x^2 + C$

B $\frac{11}{4}x^4 - \frac{7}{5}x^2 + C$

C $\frac{11}{4}x^4 - \frac{7}{3}x^3 + C$

D $7x^4 - \frac{7}{3}x^2 + C$

55. If $f(x) = \sqrt{6x - 2}$, what is the smallest possible value of $f(x)$?

A $\frac{1}{3}$

B 0

C $\frac{2}{3}$

D 1

56. Given $f(x) = x(x - 5)^{12}$, find $\int_5^6 f(x)dx$.

A 0.356

B 0.456

C 0.556

D 0.656

57. Given $f(x) = \frac{e^x - 1}{e^x + 1}$, find $\int_0^2 f(x)dx$.

A 0.57

B 0.67

C 0.77

D 0.87

58. The velocity of an object as a function of time is given as $v(t) = 5e^{-2t} + 6$, where t is in seconds, and v is in miles per seconds.

The acceleration at $t = 0.5s$ is:

A - 3.48

B - 3.48

C - 3.68

D - 3.78

59. What is the area bounded by the curve $y = -\frac{1}{9}x^2$ and the line $y = -1$?

A 2.8 square units

B 3.2 square units

C 4 square units

D 4.6 square units

60. Evaluate $\int 18[5f(x) - g(x)]dx$, if $\int 18f(x)dx = 2$ and $\int 18g(x)dx = 6$.